Section I

The Basics of Cognitive Control

Theoretical Constructs and Behavioural Phenomena

1

Cognitive Control

Core Constructs and Current Considerations

Jonathan D. Cohen

The capacity for cognitive control is perhaps the most distinguishing characteristic of human behaviour. Broadly defined, cognitive control refers to the ability to pursue goal-directed behaviour, in the face of otherwise more habitual or immediately compelling behaviours. This ability is engaged by every faculty that distinguishes human abilities from those of other species, and in virtually every domain of human function from perception to action, decision making to planning, and problem solving to language processing. Understanding the mechanisms that underlie our capacity for cognitive control seems essential to unravelling the mystery of why, on the one hand, we are capable of intelligent, goal-directed behaviour, whereas on the other hand this ability seems so vulnerable to irrational influences and failure. Not surprisingly, the distinction between controlled and automatic processing is one of the most fundamental and long-standing principles of cognitive psychology. However, as fundamental as the construct of cognitive control is, it has been almost equally elusive. Most importantly, the construct on its own says little about the mechanisms involved.

Fortunately, in the half-century since the concept of control was first introduced into psychology (Miller, Galanter, & Pribram, 1960), and afforded a central role in cognitive psychology not long thereafter (Posner & Snyder, 1975; Shiffrin & Schneider, 1977), considerable progress has been made in characterising cognitive control in more precise and mechanistic terms, at both the psychological and neurobiological levels of analysis (e.g., Anderson, 1983; Botvinick & Cohen, 2014; Collins & Frank, 2013; Daw, Niv, & Dayan, 2005; Dayan, 2012; Duncan, 2010; Koechlin & Summerfield, 2007; Miller & Cohen, 2001; O'Reilly, 2006). Much of this progress is reflected in the chapters of this volume. Needless to say, however, considerable progress remains to be made, and formulating a way forward may benefit by revisiting, and carefully reconsidering some of the foundational ideas that originally motivated the construct of cognitive control, and how these have evolved over the past half-century.

In this introduction, I review the original formulations of the distinction between controlled and automatic processing, the issues that this distinction raised, how these have been addressed, and questions that remain. This review is intended to be useful in at least two ways: (a) as a guide to the central constructs and most pressing issues concerning cognitive control for those who are new to this area of research; and (b) as an inventory of challenges that a satisfying account of cognitive control must address for those who are familiar with the area. I have organised the issues into three broad categories: (a) core, defining features of cognitive control; (b) the relationship of cognitive control to other closely related constructs

The Wiley Handbook of Cognitive Control, First Edition. Edited by Tobias Egner.

in psychology; and (c) ways in which the understanding of cognitive control may be informed by theoretical approaches that have proved valuable in other areas such as computer science.

Core Constructs

Definitional Attributes of Controlled Versus Automatic Processing

As a theoretical construct, cognitive control grew out the study of communications and control systems, including the discipline of cybernetics that flourished in the middle of the last century (e.g., Wiener, 1948). In particular, an influential book by Miller et al. (1960) explicitly drew the connection between control theory, the goal-directed nature of human cognition, and its apparently hierarchical structure—topics that have regained attention in modern research (as will be discussed below). However, three articles are generally credited with operationalising the construct of cognitive control, and placing it at the centre of experimental research in cognitive psychology: one by Posner & Snyder (1975), and a pair by Shiffrin and Schneider (Schneider & Shiffrin, 1977; Shiffrin & Schneider, 1977). These articles focused on three attributes that distinguish controlled from automatic processes: (a) controlled processes are slower to execute; (b) are subject to interference by competing automatic processes; (c) and rely on a central, limited-capacity processing mechanism.

The canonical example chosen by Posner & Snyder (1975) to illustrate this was a comparison of colour naming and word reading in the Stroop task (MacLeod, 1991; Stroop, 1935). Adults are almost universally faster to read a word out loud than to name the colour of a stimulus (criterion 1). Critically, when responding to incongruent stimuli (e.g., the word 'RED' displayed in green), the colour of the stimulus has almost no impact on the word reading response, whereas the word invariably interferes with naming the colour. Furthermore, attempts to name the colour while performing another unrelated task (such as mental arithmetic) are likely to be impaired. These properties generally do not apply to word reading. These findings were explained by proposing that colour naming is a controlled process, whereas word reading is automatic. This account of findings in the Stroop task quickly became—and in many areas still remains—a foundational paradigm for studying controlled and automatic processing (for example, the same principles are used to infer the influence of automatic processes using the Implicit Association Task—IAT; Greenwald & Banaji, 1995). However, almost as soon as the construct of controlled processing was introduced, it raised concerns.

Capacity Constraints

Central, limited-capacity mechanism. Perhaps the most important and controversial assertion was that cognitive control relies on a central, limited-capacity processing mechanism that imposes a serial constraint on the execution of controlled processes, as distinct from automatic processes that can be carried out in parallel.[1] The importance of this assumption cannot be overestimated. The idea was paradigmatic in the literal sense. It provided the operational criterion that is used almost universally to identify a process as control demanding: dual-task interference. If performance of a task suffers when another task that is unrelated (i.e., does not involve the same stimuli or responses) must be performed at the same time, then the first task is deemed to be control demanding. However, as practically—and introspectively—appealing as this assertion is, it is equally problematic.

The capacity constraints on control are generally attributed to its reliance on a limited resource; however, neither the nature of the resource, nor the reason for its limitation has yet been identified. Some have argued that the resource may be metabolic (see discussion of effort below). However, there is no reliable evidentiary basis for this (Carter, Kofler, Forster, & McCullough, 2015), and it seems improbable given the importance of the function, the metabolic resources available to the brain, and the scale on which it is able to commit metabolic resources to other processes.

Another suggestion is that the limitation is structural. For example, most models of cognitive control propose that control-demanding processes rely on the activation and maintenance of control representations that are used to guide execution (see discussion below). These representations are considered the 'resource' upon which control relies, and the limited capacity of control is attributed to a limitation in the scope of such representations that can be actively maintained (e.g., a limitation in working memory capacity). However, this begs the question: Why is *that* capacity limited? One possibility is a physical limitation (akin to the limited number of memory registers in a CPU). However, like metabolic constraints, this seems highly improbable. There are 100 billion neurons in the human brain, of which about one third reside in areas thought to be responsible for cognitive control (e.g., the prefrontal and dorsal parietal cortex). With those resources available, evolution would have to be viewed as a poor engineer to be incapable of maintaining more than a single control representation at a time. Another possibility is that there are functional constraints on the system; for example, the number of representations that can be simultaneously maintained in an attractor system, or a tension between their number and resolution (Edin et al., 2009; Ma & Huang, 2009; Usher, Cohen, Haarmann, & Horn, 2001). Such efforts reflect important progress being made in developing quantitative, mechanistically explicit accounts of representation and processing in neural systems, and may well explain constraints within circumscribed domains of processing. However, once again, this begs the question: Why cannot a system as vast as the human neocortex proliferate attractor systems for a function as valuable as cognitive control?

Multiple resources hypothesis. An alternative to the idea that dual-task interference reflects a constraint in the control system *itself* is the idea that, instead, it reflects something about the processes being controlled. This idea has its origins in multiple resource theories of attention (Allport, 1980; Logan, 1985; Navon & Gopher, 1979; also see Allport, Antonis, & Reynolds, 1972; Wickens, 1984). They argued that interference between tasks may reflect cross-talk within *local* resources (e.g., representations or processes) needed to perform different tasks if they must make simultaneous use of those resources for different purposes—a problem that can arise anywhere in the system, and not just within the control system itself. A classic example of such cross-talk (Shaffer, 1975) contrasted two dual-task conditions: repeating an auditory stream ('echoing') while simultaneously typing visually presented text (copy-typing), versus simultaneously reading aloud and taking dictation. The former pair is relatively easy to learn, while the latter is considerably more difficult. The multiple resources explanation suggests that echoing and copy-typing involve non-overlapping local representations and processing pathways (one auditory—phonological—verbal, and the other visual—orthographic—manual). Because they make use of distinct resources there is no risk of cross-talk, and so it is possible to do both at once. In contrast, reading out loud and taking dictation make dual competing use of phonological representations (e.g., the one to be read and the one to be transcribed), and similarly for orthographic representations, and thus are subject to the problem of cross-talk.

This idea has been expressed in a number of models addressing cognitive control (Botvinick, Braver, Carter, Barch, & Cohen, 2001; Cohen, Dunbar, & McClelland, 1990; Meyer & Kieras, 1997; Salvucci & Taatgen, 2008). These models suggest that constraints on the simultaneous execution of multiple tasks can be viewed as the *purpose* of control, rather than a limitation in its ability: A process relies on control whenever it risks coming into conflict with (i.e., is subject to cross-talk from) another process. Such conflict can impair performance (either by slowing processing, or generating overt errors). Thus, a critical function of the control system is to monitor for such indicators (e.g., Botvinick et al., 2001; Holroyd & Coles, 2002), in order to limit processing in such a way as to avoid such conflicts—in just the way that traffic signals are meant to limit collisions among vehicles travelling on intersecting thoroughfares. There is considerable empirical evidence in support of this (e.g., Carter et al., 1998; Egner & Hirsch, 2005; Ridderinkhof, Ullsperger, Crone, & Nieuwenhuis, 2004; Venkatraman, Rosati, Taren, & Huettel, 2009; Yeung, Botvinick, & Cohen, 2004). Recent computational work has suggested that even modest amounts of overlap among processing streams within a system can impose surprisingly strict limitations on the number of processes that can be safely executed at a given time (Feng, Schwemmer, Gershman, & Cohen, 2014; Musslick et al., 2016). Furthermore, these restrictions can be nearly scale invariant, and thus may offer a plausible account for the strikingly strict constraints on control-dependent behaviour in a system as computationally rich as the human brain.

Constraints on the ability to sustain control. Although most discussions of limited capacity focus on a numerical constraint (that is, how *many* control-dependent processes can be executed at once), there is an equally impressive and consistently observed *temporal* constraint (how *long* control can be sustained for a given task). Here again, the constraint has been assumed to reflect a limited resource. One popular version of this account—the 'ego depletion' hypothesis (Baumeister, Bratslavsky, Muraven, & Tice, 1998)—proposes that the resource is literally energetic, and the inability to sustain control reflects metabolic fatigue. Although this concurs with the subjective sense of effort associated with control, recent studies have called into question both the physiological basis (Kurzban, 2010) and empirical support (Carter et al., 2015) for this hypothesis. An alternative is that temporal constraints may reflect motivational factors (Inzlicht & Schmeichel, 2012). For example, as discussed in the section that follows, effort may reflect the signalling of opportunity costs associated with persistent performance of a given task, rather than a metabolic expense.

Effort and Motivation

From the earliest formulations, the construct of cognitive control has been closely associated with effort and motivation, an association that continues to be a focus of modern research (as evidenced by Chapters 9, 10, 23, and 24 in this volume on the topic by Chiew & Braver, Kool et al., Winecoff & Huettel, and Krebs & Woldorff). These terms have frequently elicited concern. An obvious and persistent one is about the qualitative—and potentially irreducibly subjective—nature of the phenomena to which they refer. Another is the awkwardness of the fit, perhaps most notably with regard to the original example: For example, is it really any more *effortful* simply to name the colour of an object, than it is to read a word? Despite these concerns, there are at least two reasons for considering effort and motivation, and their association with control.

Phenomenologically, the experience of effort helps identify and characterise conditions that seem to engage control. For example, while it may not be particularly effortful to name the colour of an apple, it *is* effortful to name the colour of an incongruent stimulus in the Stroop task. What is the difference? In one case, there is no interference, while in the other there is. Thus, the experience of effort—though subjective—offers a clue that control is engaged by conflict between processes (or the potential thereof), rather than by a particular process itself. This aligns with the idea, noted above, that a fundamental function of control is to reduce interference where it can arise.

The association of effort and motivation with control also highlights a theoretical construct that, until recently, has been all but ignored in research on control: the *cost* of control.

Irrespective of the sources of constraint, control mechanisms must manage a limited budget, with the consequence that allocating control to one process incurs an opportunity cost for others. This is something that was recognised in the earliest conceptions of cognitive control (Shiffrin & Schneider, 1977). From this perspective, 'effort' might be viewed as the phenomenological correlate of a signal indexing the opportunity cost associated with the allocation of control, and 'motivation' as the system's 'willingness to pay' the cost of control. These ideas are gaining currency in modern research, as reflected by several chapters in this volume (see also Braver, 2015; Kurzban, Duckworth, Kable, & Myers, 2013), and represent an exciting and important direction for future research. In particular, it offers the promise of bringing research on the mechanisms underlying cognitive control into contact with the study of value-based decision making and disturbances of behaviour specifically related to valuation and motivation, such as savings, drug addiction, and gambling (e.g., Bickel, Jarmolowicz, Mueller, Gatchalian, & McClure, 2012; Westbrook & Braver, 2015; Chapter 23 by Winecoff & Huettel in this volume).

The Continuum of Automaticity and Control

Context dependence. Another major concern with early formulations of controlled versus automatic processing was about the treatment of this distinction as a dichotomy. This quickly met with empirical challenges (Kahneman & Treisman, 1984). In one particularly striking example, each in a set of arbitrary shapes was assigned a colour word as its name, and participants were taught to name the shapes (MacLeod & Dunbar, 1988). When those shapes were presented in colours that conflicted with their newly learned names, shape naming exhibited the attributes of a controlled process (it was slower, and subject to interference from the colour in which the shape was displayed), whereas colour naming appeared to be the automatic process (faster, and unaffected by the shape's name). At the same time, findings were reported suggesting word reading, a process considered to be canonically automatic (LaBerge & Samuels, 1974; Posner & Snyder, 1975) could be shown to rely on attention and/or control (Kahneman & Henik, 1981). Such observations presented a paradox for the assumption that a process was either controlled or automatic; rather, it seemed to depend on the context in which the process was executed.

Learning. A closely related observation was that controlled processes could become automatic with practice. In particular, in one of the most elegant set of studies in cognitive psychology, Schneider and Shiffrin (1977) showed that if a task was practised extensively, it developed all of the signs of automaticity: It became faster, less effortful, and less subject to control. Critically, however, this required that the association between stimuli and responses remained fixed. If these varied, automaticity did not develop even over the course of thousands of trials with the same stimuli and responses. Interestingly, recent evidence suggests that, under some circumstances, processes with the signature of automaticity can develop much more quickly (Meiran, Pereg, Kessler, Cole, & Braver, 2015—see below). The extent—and type—of training required for a process to become automatic (i.e., less reliant on control) remains a critical area of inquiry, and may hold important clues to the mechanisms involved.

Continuum of automaticity and control. Taken together, the observations reviewed above have been interpreted as evidence that controlled and automatic processing define the ends of a continuum, and that the place occupied by a given process along the continuum is a function of both learning (i.e., the number of times that exact process has been executed) and the context in which it occurs (i.e., what other processes are engaged at that same time). This characterisation has been formalised in a variety of models, ranging from symbolic production system models (e.g., Anderson, 1983) to neural network models (e.g., Cohen et al., 1990; Cohen, Servan-Schreiber, & McClelland, 1992). The latter directly address the graded nature of automaticity, attributing this to the strength of the processing pathway required to perform a task (which is directly determined by learning), relative to the strength of pathways supporting any processes with which it must compete. In such models, control augments the sensitivity of a pathway to its inputs, allowing it to compete more effectively with other pathways that carry interfering information.

Flexibility

A hallmark of the human capacity for cognitive control is its flexibility: the remarkable ability to rapidly configure and execute a seemingly limitless variety of behaviours, including ones that have never before been performed (e.g., Meiran et al., 2015). Most current theories of cognitive control assume that this relies on the activation of control representations that serve as internal context, guiding processing in the parts of the system required to implement goal-relevant processes or behaviour. In symbolic architectures (e.g., ones that use production system architectures, such as Anderson, 1983), this is assumed to rely on representations in declarative memory (e.g., goal representations). In neural models, it is generally assumed to rely on the activation of context representations in the prefrontal cortex (e.g., Miller & Cohen, 2001). Symbolic models readily afford flexibility, as they have at their core the ability to bind control representations (e.g., variables) to arbitrary values, and compose these in arbitrary ways to implement new behaviours. However, whether arbitrary variable binding and full compositionality are implemented in the brain is less clear. At the least, humans exhibit limitations in abilities that are trivial to implement in truly symbolic systems, such as multi-digit mental arithmetic or parsing recursively embedded phrases (e.g., the mouse the cat the dog chased scared squealed). Understanding the flexibility of human behaviour in terms of mechanisms that approximate symbol processing, or that use altogether different computational mechanisms, is one of the major challenges for the study of cognitive control (O'Reilly et al., 2013). Efforts to address this challenge have brought into focus four functional requirements that such a system must satisfy: (a) a *representational code* sufficient to span the seemingly limitless range of control-dependent behaviours; (b) the ability to *acquire* (or configure) such a code from experience; (c) the ability to *update* the currently active control representation(s) in a context-appropriate manner; and (d) the ability *select* the appropriate representation to activate. Understanding how these requirements are met has become an important focus of research on cognitive control.

What code do control representations use? This is perhaps the most fundamental question that confronts research on cognitive control: What is the form of the representations used to flexibly guide performance? Some have argued against an explicit code, suggesting that a sufficiently large, randomly connected network may be powerful enough to explain the flexibility of control-dependent behaviour (Rigotti et al., 2013; Susillo, 2014). However, this begs the question of how, out of the vast number of possibilities, the system can be configured immediately (e.g., under instruction) to implement the precise combination of processes needed to implement an arbitrary, novel task. Others have proposed that, as in fully symbolic systems, there must be some more systematic, combinatorial code—a 'vocabulary'—that efficiently spans the space of possibilities (e.g., Eliasmith et al., 2012; Plate, 1995). Whether random or systematic, characterising the nature of the representations on which control relies remains one of the greatest challenges of cognitive neuroscience. Meeting this challenge may rely on progress in addressing another, closely related challenge: understanding how control representations arise through learning and development.

How are control representations acquired? The capacity for cognitive control is clearly one that emerges over the course of development, almost certainly under the regulation of genetic factors, but equally clearly shaped by experience. Understanding this developmental process has been the focus of considerable empirical study (Diamond, 2013; Hanania & Smith, 2010; Munakata, Snyder, & Chatham, 2012; Chapter 26 by Cohen & Casey in this volume). There are at least two lines of theoretical work that address the learning mechanisms underlying cognitive control. One has explored the idea that, with the appropriate neural architecture, simple reinforcement learning algorithms can extract representations from experience that are sufficiently abstract and compositional as to approximate symbol-like processing, even if they are not fully symbolic in the sense of supporting arbitrary variable binding (Kriete, Noelle, Cohen, & O'Reilly, 2013; Rougier, Noelle, Braver, Cohen, & O'Reilly, 2005). Another line of work has explored an extension of simple reinforcement learning—referred to as 'hierarchical reinforcement learning'—that is sensitive to the nested goal–subgoal structure of many tasks, and exploits this to extract representations that can be used to control behaviour over a wide range of tasks (e.g., Botvinick, Niv, & Barto, 2009; Frank & Badre, 2012). These approaches complement one another, and an integration of the insights gained from each promises to advance our understanding of the code used to represent control signals, and how it emerges with experience. This work also points to a third requirement for flexibility of behaviour: the ability to update control representations in a context-appropriate manner.

How and when are control representations updated? All model architectures that address cognitive control implement some mechanism for regulating the activation and maintenance of control representations. In production systems, this is typically managed by the firing of productions that are responsible for updating the contents of working memory. However, several critical features vary across models: How many productions can fire at once; whether they do so synchronously (all updates occur at once) or asynchronously (whenever a relevant production fires); and the 'conflict resolution' mechanisms required to select which of a set of competing productions are allowed to fire at a given time. Similar issues arise in neural network architectures, for example, whether to update representations continuously (as in simple recurrent networks; e.g., Botvinick & Plaut, 2004; Cleermans & McClelland, 1991; Elman, 1990) or at discrete intervals. The latter is usually implemented using a gating mechanism that regulates access to parts of the system responsible for representing and maintaining control signals. Such a mechanism has been implemented both in models used to simulate human performance and brain function (Braver & Cohen, 2000; Chatham & Badre, 2015; Frank, Loughry, & O'Reilly, 2001; Todd, Niv, & Cohen, 2008; Zipser, 1991), as well as in machine learning applications (Hochreiter & Schmidhuber, 1997; LeCun, Bengio, & Hinton, 2015). Such models, applied to human function, make useful contact with the neural mechanisms involved. However, an important challenge for these models is to make contact with the rich set of experimental findings concerning human performance in domains such as task switching that address the dynamics of updating of control (e.g., Collins & Frank, 2013; Gilbert & Shallice, 2002; Reynolds, Braver, Brown, & Van der Stigchel, 2006).

One issue that has come to the fore, concerning the timing of control updating, is whether this happens in anticipation of the need to change the control representation or 'just-in-time' (that is, when the first indication occurs of the need for a new control state). These strategies have been referred to as 'proactive' and 'reactive' control, respectively (Braver, 2012; Chapter 9 by Chiew & Braver in this volume). This issue raises interesting questions about how control can be optimised, as well as its interaction with other memory systems—issues that will be touched on further below.

Another critical question is how the system *learns* when to update control representations. One intriguing answer to this question, suggested by neurophysiological data, is that the gating signal may be tightly coupled with, or may even rely on the same mechanisms used for reinforcement learning. The latter respond to reward prediction errors, which are elicited by a stimulus that itself was not predicted, but that predicts a later reward (Montague, Dayan, & Sejnowski, 1996). This is precisely the timing required for a gating signal: an event that, itself unpredicted, signals the possibility of greater reward if control is redeployed. Thus, coupling the gating signal to the reinforcement learning signal may allow the system to discover when the gating signal should occur (Braver & Cohen, 2000; Frank et al., 2001). Recent work has suggested that gating of the output of the control system may be as important as gating its input (e.g., Chatham, Frank, & Badre, 2014; Kriete et al., 2013). These ideas offer a rich avenue for future research on how the control system develops and how, when it goes awry, it may manifest in clinical conditions.

How are control representations selected for activation? The adaptive value of a control system relies critically on its ability to determine not only *when* control signals should be updated, but specifically *which one(s)* to engage at a particular time. The conflict resolution mechanisms in symbolic models partially address this point, but typically do so on the basis of the structural properties of the opportunities (e.g., which properties are most highly specified) rather than on motivational factors, such as effort and reward. Recent efforts have begun to focus on the latter, suggesting that control signals are selected based on a cost-benefit analysis that takes account of the expected value of the control signals in contention, and selects the one(s) that maximise this value (Shenhav, Botvinick, & Cohen, 2013; see also Chapter 10 by Kool et al. in this volume). This kind of normative approach is a promising avenue for research on control, as discussed in the final section of this introduction.

An intimately related question is how the value of prior experience (i.e., knowledge already accrued about expected outcomes for control signals) is balanced against the value of gaining new experience—a tension that is commonly referred to as the explore–exploit trade-off. Managing this trade-off successfully is a fundamental requirement for any agent that can adapt in non-stationary environments (e.g., Cohen, McClure, & Yu, 2007; Gittins & Jones, 1974; Kaelbling, Littman, & Moore, 1996; Krebs, Kacelnik, & Taylor, 1978; Pratt & Sumpter, 2006; Watkinson et al., 2005). Although there is no fully general solution to the explore–exploit trade-off, recent work has begun to examine how humans manage this problem (Daw, O'Doherty, Dayan, Seymour, & Dolan, 2006; Wilson, Geana, White, Ludvig, & Cohen, in press). These lines of work examine how, in choosing a course of behaviour, agents should weigh the value of acquiring information (i.e., exploration, in the service of learning, and thus as a proxy for future reward) against the value of immediate reward (i.e., exploitation). Here, as with updating mechanisms, there are proposals about the role that neuromodulatory mechanisms may play in regulating this critical function of control (Aston-Jones & Cohen, 2005; Yu & Dayan, 2005) that may help index this function, and provide clues to both its normal operation and the role it plays in clinical conditions. A consideration of the explore–exploit trade-off also suggests an additional point of contact between control and motivational constructs, such as boredom. Recent findings have begun to suggest that, just as effort may reflect a phenomenological correlate of the opportunity cost associated with performance of a given task with respect to *reward*, so boredom may reflect the opportunity cost with respect to *information*—that is, boredom may signal the value of exploration (Geana, Wilson, Daw, & Cohen, under review). These lines of work are beginning to sketch the outline of a more formal, comprehensive, and normative model of control.

Unitary or Multiple Constructs

The earliest theories of cognitive control were explicit in suggesting that it relied on a single, central processing mechanism. This was based in large measure on the observation of capacity constraints (e.g., dual-task interference) and the assumption that this reflects the limited capacity of a central mechanism. As discussed above, this claim has been challenged by the suggestion that controlled processing may lie along a continuum, and that interference effects reflect interactions among the local processes that control is engaged to regulate, rather than an intrinsic limitation of a centralised mechanism. Nevertheless, even if control can be engaged along a continuum and its capacity is not limited, it is meaningful to ask whether control relies on a unitary mechanism, or on disparate domain-specific mechanisms. This question closely parallels ones that have been raised with respect to other constructs, such as intelligence, working memory, and attention. Here, as in those cases, an intermediate answer seems most likely: that control relies on mechanisms of a particular *kind*, implemented in a wide range of domains, with different parameterisations that reflect specific features of those domains, but that share common fundamental attributes. If this is so, then understanding this 'family resemblance' may help guide modelling, hypothesis generation, and empirical inquiry. Thus, it may be most useful to ask: What are the fundamental functions and attributes that define cognitive control, and what are the kinds of mechanisms that can implement these functions and attributes?

The sections above—and many of the chapters of this book—are meant to address these questions. In summary, the following seem to be a core set of functions and attributes of cognitive control: (a) the ability to support control representations as a form of internal context, that serve as signals to guide processing elsewhere in the system; (b) the use of such signals to avert interference that can arise from cross-talk among processes sharing overlapping pathways; (c) constraints on the capacity for control-dependent processing that may reflect more about the propensity for such cross-talk, and the need to limit it, than limitations in the control system *itself*; (d) the ability to update control representations in a manner that is sensitive to the circumstances in which this is needed; and also to (e) the opportunity costs that constraints on control-dependent processing impose, weighed against the opportunities that controlled processes afford for both expected reward (exploitation) and/or to gather useful information (exploration and learning). Progress is being made in understanding the neural architecture and mechanisms underlying these functions and attributes, which provide additional support for the supposition that, at the least, the construct of cognitive control remains a coherent and useful one.

Relationship to Other Psychological Constructs

From its inception, the construct of cognitive control has been inextricably intertwined with several others in the psychological literature, and thus it seems important to consider how these are related. Several of these stand out: executive function, intelligence, volition, attention, working memory, and inhibition. In addition, recent research has begun to focus on the role of cognitive control in two other domains of function: self-control and long-term memory.

Executive Function and Intelligence

Executive function. This term has a long history in the psychological literature (e.g., Bianchi, 1895; Luria, 1966), extending into early work in cognitive psychology (e.g., Baddeley & Hitch, 1974; Shallice, 1982). It seems impossible to distinguish the use of 'executive function' from the construct of cognitive control and, at least within the cognitive psychological literature, 'cognitive control' has largely replaced the use of 'executive function'. This is likely for historical reasons (e.g., the close association of 'executive function' with older neuropsychological constructs and particular batteries of tasks). In any event, it seems reasonable to treat these as synonyms.

Intelligence. Since Spearman introduced the construct of 'general intelligence' (*g*; Spearman, 1904), it has been both captivating and controversial, and inextricably intertwined with the concept of executive function. Like executive function, it has also been associated consistently with frontal lobe function (e.g., Duncan et al., 2000). The question of whether intelligence should be treated as a unitary construct (as proposed by Spearman [1904]), or reflects the operation of a collection of domain-specific faculties parallels, in many ways, a similar question about cognitive control. Answers to these questions may bear on a deeper understanding of the relationship between them, and in particular, whether intelligence reflects the functioning of mechanisms beyond those responsible for cognitive control. What does seem clear, however, is that the capacity for cognitive control represents a necessary, if not sufficient, condition for intelligence (e.g., Kane & Engle, 2002).

Volition

The earliest treatments of cognitive control proposed that the mechanisms involved were responsible for conscious, voluntary action, and intention. This had strong intuitive appeal, and continues to be an interesting and potentially valuable area of inquiry (Dehaene & Naccache, 2001; Graziano & Kastner, 2011). In part, this is because it seeks to satisfy our intrinsic curiosity about the compelling nature of our experience as 'volitional' agents, and because it may lend insights into fundamentally important social and moral questions (such as responsibility for action; e.g., Bratman, 1987; Vargas, 2013). Progress has been made in identifying the correlates of conscious states (e.g., Cleermans, 2007; Dehaene & Changeux, 2011; Schurger, Pereira, Treisman, & Cohen, 2010) and volition (Haggard, 2008; Soon, Brass, Heinze, & Haynes, 2008). However, the extent to which these are co-extensive with those that engage control remains an open and interesting question.

Attention

The construct of cognitive control grew out of the literature on attention, and remains intimately bound to it. One might even argue that these constructs are inseparable, with attention referring to one of the most fundamental functions of control: the selection of some processes for engagement over others (e.g., Ardid, Wang, & Compte, 2007; Cohen et al., 1990; Deco & Rolls, 2005). However, the question of whether 'attention' should simply be considered as a function of control—and nothing more—raises some of the questions discussed above concerning the scope of the construct of control. For example, is it meaningful to assume that the mechanisms responsible for sensory selection (to which the word *attention* is most commonly applied) reflect the operation of cognitive control? Do they share features in common with those responsible for the selection of actions? Similarly, does the 'exogenous' and seemingly automatic engagement of attention (e.g., the 'capture of attention' by salient events, such as a loud sound) engage mechanisms of cognitive control, and are those the same as or in some meaningful way similar to those responsible for 'endogenous' (or 'strategic') forms of attention (e.g., responding to a verbal instruction)? At the least, it has been known for quite some time that these exhibit different dynamics (e.g., Neely, 1977; Petersen & Posner, 2012; Posner & Cohen, 1984). Once again, an answer to these questions, and whether it makes sense to consider attention to be a function of control in all circumstances and at all levels of analysis will only be answered (and may ultimately be rendered irrelevant) with a deeper understanding of the mechanisms involved.

Working Memory

The construct of working memory, first formulated in the context of theories about executive function (Baddeley & Hitch, 1974), referred to a buffer that stored information required for executive processes to operate (e.g., the intermediate products of a computation). Symbolic models proposed that working memory (defined as the activated state of information in long-term memory) served not only to buffer intermediate products of computation, but also goal representations used to guide behaviour (Anderson, 1983). Neural network models of control have emphasised the latter (Cohen et al., 1990; Dehaene & Changeux, 1991; O'Reilly, 2006). In general, there seems to be a consensus that cognitive control relies critically on at least one component of working memory: the activation and maintenance of control representations used to guide processing (Miller & Cohen) and, furthermore, that a critical function of control is to regulate the contents of this component of working memory (O'Reilly, 2006). However, as with attention, it is not yet clear how the boundaries of this construct coincide with those of cognitive control (see also Chapter 3 by Meier & Kane in this volume). This is complicated by overlapping use of the terms 'working memory' and 'short-term memory', both of which refer to the maintenance of information in an activated state. Like the relationship between cognitive control and attention, clarity along these lines is most likely to come from a more detailed understanding of the underlying mechanisms, and their relationship to those involved in attention and control (e.g., Ikkai & Curtis, 2011), rather than from further attempts to refine these definitions in the absence of such an understanding.

Inhibition

Directed versus competitive inhibition. The associations of inhibition with executive function and the frontal lobes dates back to the report about Phineas Gage (Harlow & Martyn, 1868), and neurological studies in the beginning of the last century (Adie & Critchley, 1927; Brain & Curran, 1932; Marinesco & Radovici, 1920) identifying reflexes in infants that disappear in adults, but reappear in patients with damage to the frontal lobes. These 'frontal release signs' continue to be used in clinical practice to identify frontal lobe damage. This phenomenon, together with the observation of primitive (e.g., 'utilisation') behaviours in monkeys and humans with frontal lobe lesions, led to the belief that a cardinal function of the frontal cortex (and the executive function that it supports) is the inhibition of reflexive and/or habitual behaviours (e.g., Bianchi, 1895; Fuster, 1980; Lhermitte, 1983). This belief persists, in many quarters, in the common and steadfast assumption that frontal control mechanisms provide direct and specific inhibition of automatic (e.g., habitual) processes (Buckholtz, 2015). However, there is little direct support for this assumption beyond the domain of neurological reflexes. An alternative to directed inhibition is competitive inhibition: by facilitating selected processes, control allows those processes to compete more effectively with interfering processes (e.g., by way of lateral inhibition, or other 'normalising' mechanisms). It may be difficult, and even impossible, to distinguish between directed and competitive inhibition on theoretical grounds alone. However, competitive inhibition seems both more parsimonious (it should be easier to select and support a single process than direct inhibition towards all potential competitors), and it is consistent with general principles of neural organisation (e.g., the scarcity of long-range inhibitory projections, and the abundance of lateral inhibitory interneurons). Nevertheless, an adjudication of these possibilities, or the identification of an intermediate solution, is likely to require detailed neurophysiological investigation.

Global inhibition and 'stopping'. While the mechanisms of inhibition underlying selection remain to be identified, it has become increasingly clear that the brain implements, and the control system has access to a 'stopping mechanism' that effects a more general or global form of inhibition. This is suggested both by behavioural evidence (e.g., using stop-signal and go-no-go tasks; Logan, 1994; Verbruggen & Logan, 2008), and by neural evidence concerning the basal ganglia, and in particular the subthalamic nucleus (e.g., Aron & Poldrack, 2006; Frank, 2006). One hypothesis is that such systems provide a 'brake' that may reflect an asymmetry in both mechanism and outcome between rewarding and perilous outcomes: In general, the risks associated with erroneous action may be greater or more frequent than those associated with inaction. Such a stopping mechanism may also serve a more refined function in the control of behaviour, by setting response thresholds used to regulate speed–accuracy trade-offs in the service of optimising reward (e.g., Bogacz et al., 2006; Cavanagh et al., 2011; Gold & Shadlen, 2002). Understanding the mechanisms by which control overrides inappropriate responses—whether by directed, competitive, or some more global form of inhibition—remains an important goal not only for basic research, but also for understanding failures of control and, in particular, failures of self-control that are a prevailing social and clinical concern.

Self-Control

The study of self-control has become one of increasing importance, at both the individual and social level (see Chapter 25 by Davisson & Hoyle in this volume). At the individual level, it is obvious that failures of self-control are a fundamental feature of a wide range of clinical disorders, from obsessive-compulsive disorder to drug addiction and gambling (Chapter 32 by Chaarani et al. in this volume). It is also becoming increasingly apparent that problems of self-control are responsible for dysfunctional behaviours in otherwise healthy individuals (such as failures to save adequately for retirement) and at the societal level (e.g., energy use policies). Although, like many of the constructs discussed above, there is no scientifically accepted definition of the term *self-control*, in common use it refers to 'the ability to control oneself, in particular one's emotions and desires'. Once again, it remains to be determined whether the construct of cognitive control, as traditionally defined and studied in the experimental laboratory, extends seamlessly into the domain of self-control in the real world (e.g., Buckholtz, 2015). There are at least two ways in which self-control may involve domain-specific factors: (a) the nature of the process over which control presides; and (b) what appears to be the fundamentally intertemporal nature of decisions involving self-control.

Value-based decision making. If a primary function of control is to choose behaviours that maximise value (see the section above titled 'Effort and Motivation'), then it seems natural to include decisions driven by emotions and desires in its scope (e.g., see Chapter 23 by Winecoff & Huettel in this volume). These can be considered as a class of automatic processes that pose challenges to controlled processing similar to more 'cognitive' ones (e.g., word reading in the Stroop task). Indeed, there is a large literature that exploits this correspondence to infer the influence of emotional and valuation processes on behaviour (e.g., Fazio & Olson, 2003; Greenwald, Poehlman, Uhlmann, & Banaji, 2009; Chapter 24 by Krebs & Woldorff and Chapter 22 by Pessoa in this volume), suggesting that the role of control in regulating cognitive processes may share much in common with mechanisms involved in regulating emotional ones (e.g., Gross & Thompson, 2007). At the same time, affective processes may exhibit dynamics that distinguish them from other types of automatic processes. For example, the 'force' of a desire may increase with time, and the control of such processes often exhibits a stereotyped recovery-and-relapse cycle (e.g., as observed in addiction, dieting, etc.) that is not commonly observed in other domains. These factors pose both theoretical and experimental challenges that may be unique to the domain of self-control.

Intertemporal choice. Decisions about self-control seem invariably to involve choices between one option that is immediately desirable (e.g., a piece of cake on the table) and another that carries greater future value (dieting), a type of decision often referred to as intertemporal choice. Thus, self-control may fundamentally involve not only inhibition (as discussed above), but also intertemporal choice. This is consistent with a growing literature suggesting that intertemporal choice involves a competition between automatic and controlled processes (Figner et al., 2010; Kahneman, 2003; McClure, Laibson, Loewenstein, & Cohen, 2004). Indeed, one intriguing conjecture is that intertemporal choice is fundamental not only to *self*-control, but to control in general. In other words, the immediacy of the reward associated with an outcome may be a critical feature in determining which of two (or more) competing processes demands control. In the Stroop task, the greater ease of reading the word might be viewed as a form of immediate reward, whereas the benefits of overriding this response in order to name the colour come when performance is rewarded, which is usually later. From this perspective, the demands for control might be viewed as intimately bound to intertemporal choice, and control mechanisms as the steward of our future selves. Nevertheless, intertemporal choice involving affective processes and self-control may impose distinct requirements (e.g., due to the greater salience of rewards, and/or greater temporal asymmetries) and may therefore demand distinctive mechanisms of control.

Long-Term Memory, Prospective Memory and Planning

Memory search. Mechanisms that actively represent internal context have come to occupy a central role in theories about the encoding and retrieval of information from long-term memory. Tulving's context-encoding hypothesis proposed that features of the context in which a memory was formed are encoded along with the memory itself, and that retrieval involves a form of 'mental time travel' that reinstates the context in which the memory was formed to facilitate retrieval of the memory (Tulving, 2002). The temporal context model (TCM, Howard & Kahana, 2002; Polyn, Norman, & Kahana, 2009) extends this idea, suggesting that internal, actively maintained representations—such as goals and intentions—are a particularly useful form of context that can identify the time at which a particular memory was encoded. On this view, retrieval involves reinstating the context representation active at the time of memory encoding and, by association, the memory itself. This beautifully ties together the role of context representations in control and long-term memory, and the foundations for a mechanistic understanding of the processes involved.

Prospective memory and planning. Like most studies of cognitive control, TCM focuses on its regulative functions—the selection of a process (in this case, memory retrieval) for current execution. However, an important and growing area of research is on the interaction between control and long-term memory in the service of prospective memory and planning. Imagine the following example: I ask you to perform the colour naming task in 2 s. It is almost certain that you will engage the required task representation(s) immediately. However, if instead I ask you to do it when I return to the room in an hour, it is just as certain that you will *not* engage and maintain those representations while I am gone. Rather, you will do so when I return. Critically, it is likely you will be able to do this without my having to instruct you when I return. This represents a form of prospective memory ('remembering' to do something in the future), and a simple form of planning.

There is increasing evidence that this ability to program a controlled process for future execution relies on an interaction between control mechanisms and episodic memory (e.g., Cohen & O'Reilly, 1996; Einstein & McDaniel, 2005; Gollwitzer, 1996). On this account, when an instruction is presented (or a plan is conceived), an association is formed in episodic memory between the control representation required to execute the necessary behaviour and the future conditions under which it should be executed (e.g., my reappearance in the room). Then, when those conditions occur, the association in episodic memory elicits retrieval of the control representation, which is re-engaged (e.g., gated) in working memory, and the task is executed. Although this idea that control representations can be 'cached' in episodic memory is compelling, it poses many new questions. For example, when is this mechanism used rather than the immediate activation of the control representation (a question closely related to the use of proactive versus reactive forms of control discussed in the previous section titled 'Flexibility'; e.g., Bugg, McDaniel, & Einstein, 2013; Meiran et al., 2015), and what are the factors that influence this decision? When choosing to defer, what features of the future state are chosen to associate with the control representation? To what extent can the association between control representation and episodic memories emerge passively, through experience (cf. Chapter 4 by Egner in this volume)? Answers to these questions promise to unravel one of the greatest mysteries of the human brain: how it supports the ability to plan for the future.

Theoretical Considerations

Normative Theory

The dominant approach to research on cognitive control has been to characterise the properties of control-dependent processing, use these to infer candidate mechanisms, and design experiments to test those mechanisms. Although this has generated considerable progress, there is another, complementary approach that has been conspicuously scarce in research on cognitive control: the construction of normative theory. This is sometimes referred to as rational analysis, or the ideal observer method (e.g., Barlow, 1981; Tanner & Swets, 1954). This seeks to identify the optimal computation for a function of interest, which is then used to generate hypotheses about the mechanisms involved. Although it is rare (though not unprecedented) that natural systems implement fully optimal mechanisms, this approach provides a rational guide for generating hypotheses, and a formally rigorous framework within which to test them. It has driven considerable progress in many domains of science, including psychology (Anderson, 1990; Geisler, 2003; Tenenbaum, Griffiths, & Kemp, 2006). The scarcity of this approach in research on cognitive control is particularly surprising, given that the *optimisation* of behaviour can be viewed as the fundamental purpose of control. This is the definition of control used by systems theory in engineering and, as noted above, inspired the earliest thinking about control in the context of human behaviour (e.g., Wiener, 1948; Miller et al., 1960).

A critical step in normative theory is defining the 'objective function' being optimised; that is, in the context of cognitive control, the goal that the behaviour is intended to achieve. Identifying the objective function for cognitive control poses a serious challenge, given the broad scope of processes, behaviours, and goals it can serve. Once again, this raises questions about the extent to which 'cognitive control' should be treated as a unitary construct, a class of mechanisms sharing a family resemblance, or a disparate set of domain-specific mechanisms. Nevertheless, several lines of work have begun to take on the challenge of this approach.

Task-level optimisation. One approach has been to focus on the role of control in optimising a particular task. An example of this is work on two-alternative forced choice decision-making tasks. Dramatic progress has been made in identifying and characterising the mechanisms involved in such simple decisions, at both the psychological and neural levels of analysis (e.g., Gold & Shadlen, 2007; Ratcliff & McKoon, 2009), and in conducting normative analyses of performance (e.g., Bogacz et al., 2006). This has provided a formally rigorous framework within which to interpret psychological constructs (such as expectations, attention, and the speed–accuracy trade-off), and for quantifying performance associated with optimal control against which human performance can be compared (e.g., Simen et al., 2009). This work serves as an example of how normative theory can be used to study the engagement of cognitive control in a given task domain, of which other examples are now beginning to emerge (e.g., Verguts, Vassena, & Silvetti, 2015; Wiecki & Frank, 2013; Yu, Dayan, & Cohen, 2009).

Meta-level optimisation. A complementary approach has been to consider the problem at the broadest level: How does the control system determine which tasks or goals should be pursued at a given time? One approach is to ascribe this to learning (e.g., Verguts & Notebaert, 2008); another is to consider it as an optimisation problem of its own, nesting the optimisation control for a given task within a higher-level optimisation of the choice among tasks. An example of this is the expected value of control (EVC) theory (Shenhav et al., 2013). This assumes that investment in a control-demanding task improves the probability of reward, but that this carries a cost that scales with the amount of control invested. The EVC theory proposes that control is allocated across tasks based on a cost-benefit analysis of this trade-off, so as to maximise the overall rate of reward. This provides a rigorous framework within which to analyse, and make predictions about the allocation of control in a given task environment.

Optimisation under constraints: Bounded rationality. The cost of control must be taken into account by any normative theory of control. Understanding this cost—its functional form, and whether and how it varies across individuals and domains of behaviour—is an important direction for research. As discussed above, it seems likely that a central factor is the capacity constraint on control, which imposes an opportunity cost: Investing control in one task forgoes the opportunities for reward afforded by others. This reinforces the importance of understanding the nature and source of the constraints on control. Here, computational analyses are proving useful, as discussed further below. Consideration of how control is optimised in the face of a budget represents an instance of an approach to normative theory, broadly referred to as 'bounded rationality' (Simon, 1955, 1992), that has begun to attract growing attention (Gershman, Horvitz, & Tenenbaum, 2015; Griffiths, Lieder, & Goodman, 2015; Howes, Lewis, & Vera, 2009). This assumes that optimisation must take account not only of the system's objectives, but also the constraints under which it must operate—akin to tuning a radio as best as possible to a weak station. One criticism of this approach is that it is possible to explain any pattern of performance post hoc, by conjuring a set of constraints under which the observed performance would be optimal. However, rather than a problem, this can be viewed as a valuable step in the scientific process, the next step of which is to cast those constraints as hypotheses and use them to generate new predictions about performance. This approach has begun to show potential (e.g., Balci et al., 2011; Lieder & Griffiths, 2015), and is a promising avenue of research for the study of control.

The statistics of natural tasks. The properties of the control system itself, the processes over which it presides, and the neural architecture in which these are implemented represent important sources of constraint (Botvinick & Cohen, 2014). However, an equally important one is the *environment* in which the system operates. It is, after all, the environment to which an adaptive system adapts. Therefore, taking account of the structure of the environment should provide important clues about the structure of systems responsive to it. For example, recent progress in understanding the function of the human visual system has been driven in large measure by a characterisation of the statistics of natural scenes that have shaped its architecture (Simoncelli & Olshausen, 2001). The study of cognitive control stands to benefit from a similar approach. The human brain is clearly better at carrying out some kinds of tasks (e.g., crossing a busy road) than others (long division), likely because it has evolved to solve a wide, but not random, set of tasks required to survive in the natural world. Characterising the structure and statistics of tasks should be an important priority for research on cognitive control.

Computational Trade-Offs in Representation and Processing

The constraints on processing considered by the approaches discussed above have largely been physical (e.g., limits in the amount or type of data available to the system).

However, a more fundamental set of constraints may be related to trade-offs inherent to all computational systems. One such trade-off, which has been central to the construct of cognitive control since its inception, is between serial and parallel processing. A second trade-off that has begun to attract attention is between model-free and model-based processing.

Serial versus parallel processing. This distinction was central to the original formulations of controlled and automatic processing (e.g., Shiffrin & Schneider, 1977). Reliance on a central, limited-capacity mechanism was assumed to impose a serial constraint on controlled processing, akin to the sequential execution of a programme by the central processing unit of a standard computer (i.e., one that implements a von Neumann architecture). Conversely, it was assumed that automatic processes can be executed in parallel (i.e., simultaneously without penalty), akin to the 'embarrassing' parallelism common in multi-node computer clusters (that is, running many unrelated jobs on different nodes simultaneously). The serial constraint on controlled processing has also been a central assumption of two of the most fully developed and influential models of cognition (ACT-R and SOAR; Anderson, 1983; Newell, 1990). However, the necessity of this assumption has been challenged by models using similar architectures that weaken or eliminate the serial constraint on controlled processing (e.g., Meyer & Kieras, 1997; Salvucci & Taatgen, 2008). The debate about whether there is a 'central bottleneck' in cognitive control has also been subject to intense empirical inquiry, centred largely around the observation of the psychological refractory period (Pashler, 1984)—a delay in performance associated with the attempt to perform two or more tasks at once, interpreted as evidence that they are being queued for serial execution. The interpretation of this finding in terms of a central bottleneck has been challenged (e.g., Howes et al., 2009; Schumacher et al., 2001); however, several theories continue to assume that control relies on a centralised mechanism (Duncan & Owen, 2000; Roca et al., 2011; Tombu et al., 2011).

The debate about whether controlled processing relies on a central, serially constrained mechanism highlights, and is complicated by, another problem: In practice, it may be very difficult to distinguish between rapid serial and truly (concurrent) parallel processing (e.g., Townsend, 1972). The ability to do so, and the relevance of doing so, depends on the temporal resolution—both of the measurement, and of the outcome of interest. For example, the serial updating of pixels on a computer display could be detected by an oscilloscope with a temporal resolution of greater than 100 Hz, but not the human visual system with a temporal resolution of less than 60 Hz (which perceives the update as synchronous). This suggests that a definitive answer to the question of whether controlled processing is purely sequential, or can support parallel processes, will require, like other questions, finer-grained measurements and possibly neurobiological evidence.

That said, there is another standpoint from which to view the distinction between serial and parallel processing, and its relationship to cognitive control, that may lend clarity and coherence to the array of phenomena associated with controlled processing. This can best be appreciated by considering another kind of parallelism, captured by parallel distributed processing (PDP) or 'connectionist' architectures (Rumelhart, McClelland, & the PDP Research Group, 1986). Here, rather than the number of behavioural tasks that can be performed at once, the appeal of parallelism is the number of constraints that can be taken into account in computing the solution to a single problem—a process sometimes referred to as 'multiple constraint satisfaction'. Each individual unit in a PDP network can be thought of as representing a constraint (or 'micro-process'), and their interaction serves to take account of the mutual influence that these have on each other in parallel, averting the costs of a combinatorial explosion that would be incurred by doing so in serial. Mutual constraint satisfaction is a hallmark of functions at which the human brain excels (such as face perception and natural language processing). PDP models have been used to understand how the human brain computes these functions, and recent advances in machine learning have begun to approximate these abilities using artificial neural networks (LeCun et al., 2015).

The success of PDP networks lies in the extensive interactions among individual processing units—sometimes referred to as fine-grained parallelism. This is in contrast to coarse-grained (or 'embarrassing') parallelism that supports the execution of multiple independent processes at once. Not surprisingly, there is a trade-off between these types of parallelism: The extent to which a network supports fine-grained interactions among its units in performing a task (in the service of mutual constraint satisfaction) and shares the representations involved across multiple tasks (supporting generalisation) is in tension with the extent to which it can support the performance of multiple such tasks at once (Musslick et al., 2016). This suggests there may be a fundamental link between the richness of interactions among the processes involved in performing a task (e.g., recognising a face), and the imposition of a serial constraint on that performance (e.g., finding a face in a crowd). This relationship may also help explain the canonical trajectory during learning from control-dependent to automatic processing (e.g., Schneider & Shiffrin, 1977), in terms of a transition from interactive, generalisable representations that rely on fine-grained parallel processing (and thus demand seriality), to independent, dedicated representations that afford coarse-grained parallel execution (i.e., multitasking). A better understanding of this relationship between representation and the trade-off between serial versus parallel processing may not only offer a new way to frame important, long-standing questions about control, but also new approaches to measurement (e.g., Musslick et al., 2016) and, potentially, intervention.

Model-based versus model-free processing. This distinction, with origins in the work of Tolman (1948), has recently been cast in terms of formal learning algorithms and regained the attention of psychologists and neuroscientists (e.g., Dickinson & Balleine, 2002; Daw et al., 2005; Keramati, Dezfouli, & Piray, 2011; Chapter 11 by de Wit in this volume). In a model-free system, actions are selected based on direct associations from the stimulus to the response, learned through trial and error, and without a representation of potentially mediating factors (Sutton, 1988). In a model-based system, actions can also be evaluated by computing and evaluating *potential* courses of action, using an 'internal model' of the possibilities that can include intervening states (i.e., between the stimulus and response). Although the former is more efficient at processing (i.e., it requires less computation, and thus can respond more quickly in a *given* environment), the latter is more flexible (it can adjust more quickly to *changes* in the environment, by modifying the model rather than relying exclusively on trial-and-error-experience).[2] The construct of model-based processing has played a central role in theories about the role of 'internal replay' in learning and decision making (e.g., Sutton, 1990), in which rehearsal of past sequences of occurrences is used as a proxy for actual experience in learning, and in planning future actions (Gershman, Markman, & Otto, 2014; Redish, 2016; Shohamy & Daw, 2015). Recently, it has been proposed that the distinction between model-based and model-free processing may reflect the differential engagement of controlled and automatic processing—with model-based processing relying on control mechanisms, whereas model-free reflects the operation of more automatic ones—and empirical evidence has begun to accrue in support of this (Deserno et al., 2015; Otto, Skatova,

Madlon-Kay, & Daw, 2015). This offers a rich theoretical framework within which to explore the role of controlled processing in learning and memory, including its role in prospective memory and planning as discussed above.

Bringing the lines of work described above—on the relationship of representation to parallelism, and on model-based versus model-free processing—into contact with one another, and using them as a framework for building models of cognitive control is a particularly promising direction for future theoretical work.

Summary

The construct of cognitive control is a foundational one in cognitive psychology. The phenomenology that initially inspired the construct is compelling—in particular, its association with 'mental effort', the manifest constraints on its capacity (in both number and duration), its apparent sequentiality, and its place in the trajectory of learning—and, for the most part, these are empirically validated. However, defining cognitive control in a more rigorous way and identifying the mechanisms that govern its operation have been a challenge. Recent work at the intersection of cognitive psychology, neuroscience, and computer science has begun to progress in this direction. This introduction was aimed at providing an outline of the theoretical constructs and issues that have emerged from this work, and their relationship to the growing corpus of experimental data—much of which is examined in detail in the remaining chapters of this volume. The most important challenge for the next phase of research will be to integrate the theoretical constructs and empirical findings that have emerged into a coherent, formally rigorous description of the mechanisms involved. The outlines of such a theory are coming into focus: Cognitive control reflects the operation of mechanisms that maintain, and appropriately update internal representations of information needed to guide processes responsible for task execution in a context-relevant, goal-directed manner. Symbolic models have provided a useful high-level description of this system. However, its implementation in the brain imbues it with capabilities (e.g., learning and inference) and constraints (e.g., a tension between generalisation and multitasking) that seem to require a finer grain of analysis and modelling. As in other domains of science, bridging these levels of analysis is a critical step towards the construction of a comprehensive theory. After half a century of research, this synthesis appears to be coming within reach, and doing so promises to provide a scientifically satisfying account of the remarkable and characteristically human capacity for cognitive control.

Acknowledgements

The author would like to thank Todd Braver, Nathaniel Daw, Ida Momennejad, Amitai Shenhav, and Tobias Egner for thoughtful comments on an earlier draft of this chapter, as well as the numerous trainees and colleagues with whom he has had the privilege to work in this rewarding area of research.

Notes

1 This idea was closely related to Broadbent's (1958) highly influential *bottleneck theory*, which asserted that attention should be thought of as a central, limited capacity filter on human information processing (the relationship between attention and control will be discussed further below).

2 This distinction parallels one between compiled versus interpreted processes in computer science.

References

Adie, W. J., & Critchley, M. (1927). Forced grasping and groping. *Brain, 50*, 142–170.

Allport, A., Antonis, B., & Reynolds, P. (1972). On the division of attention: A disproof of the single channel hypothesis. *Quarterly Journal of Experimental Psychology, 24*(2), 225–235.

Allport, D. A. (1980). Attention and performance. In G. I. Claxton (Ed.), *Cognitive psychology: New directions* (pp. 112–153). London: Routledge & Kegan Paul.

Anderson, J. R.(1983). *The architecture of cognition.* Cambridge, MA: Harvard University Press.

Anderson, J. R. (1990). *Rational analysis.* Hillsdale, NJ: Erlbaum.

Ardid, S., Wang, X.-J., & Compte, A. (2007). An integrated microcircuit model of attentional processing in the neocortex. *Journal of Neuroscience, 8*(32): 8486–8495.

Aron, A. R., & Poldrack, R. A. (2006). Cortical and subcortical contributions to stop signal response inhibition: Role of the subthalamic nucleus. *The Journal of Neuroscience, 26*(9), 2424–2433.

Aston-Jones, G., & Cohen, J. D. (2005). An integrative theory of locus coeruleus-norepinephrine function: Adaptive gain and optimal performance. *Annual Review of Neuroscience, 28*, 403–450.

Balci, F., Simen, P., Niyogi, R., Saxe, A., Hughes, J. A., Holmes, P., & Cohen, J. D. (2011). Acquisition of decision making criteria: Reward rate ultimately beats accuracy. *Attention, Perception, & Psychophysics, 73*, 640–657.

Baddeley, A. D., & Hitch, G. (1974). *The psychology of learning and memory.* New York: Academic Press.

Barlow, H. B. (1981). Critical limiting factors in the design of the eye and visual cortex. *Proceedings of the Royal Society of London, Series B, 212*, 1–34.

Baumeister, R. F., Bratslavsky, E., Muraven, M., & Tice, D. M. (1998). Ego depletion: Is the active self a limited resource? *Journal of Personality and Social Psychology, 74*(5), 1252–1265.

Bianchi, L. (1895). *The functions of the frontal lobes.* Oxford: Oxford University Press.

Bickel, W. K., Jarmolowicz, D. P., Mueller, E. T., Gatchalian, K. M., & McClure, S. M. (2012). Are executive function and impulsivity antipodes? A conceptual reconstruction with special reference to addiction. *Psychopharmacology, 221*(3), 361–387.

Bogacz, R., Brown, E. T., Moehlis, J., Hu, P., Holmes, P., & Cohen, J. D. (2006). The physics of optimal decision making: A formal analysis of models of performance in two-alternative forced choice tasks. *Psychological Review, 113*(4), 700–765.

Botvinick, M. M., Braver, T. S., Carter, C. S., Barch, D. M., & Cohen, J. D. (2001). Conflict monitoring and cognitive control. *Psychological Review, 108*(3), 624–652.

Botvinick, M. M., & Cohen, J. D. (2014). The computational and neural basis of cognitive control: Charted territory and new frontiers. *Cognitive Science, 38*, 1249–1285.

Botvinick, M. M., Niv, Y., & Barto, A. C. (2009). Hierarchically organized behavior and its neural foundations: A reinforcement learning perspective. *Cognition, 113*(3), 262–280.

Botvinick, M., & Plaut, D. C. (2004). Doing without schema hierarchies: A recurrent connectionist approach to normal and impaired routine sequential action. *Psychological Review, 111*(2), 395–429.

Buckholtz, J. W. (2015). Social norms, self-control, and the value of antisocial behavior. *Current Opinion in Behavioral Sciences, 3*, 122–129.

Bugg, J. M., McDaniel, M. A., & Einstein, G. O. (2013). Event-based prospective remembering: An integration of prospective memory and cognitive control theories. In Reisberg (Ed.), *The Oxford handbook of cognitive psychology* (pp. 267–282). Oxford: Oxford University Press.

Brain, W. R., & Curran, R. D. (1932). The grasp reflex of the foot. *Brain, 55*, 347–356.

Bratman, M. (1987). Responsibility and planning. *Journal of Ethics, 1*, 27–43.

Braver, T. S., & Cohen, J. D. (2000). On the control of control: The role of dopamine in regulating prefrontal function and working memory. In S. Monsell & J. Driver (Eds.), *Attention and performance XVIII; Control of cognitive processes* (pp. 713–737). Cambridge, MA: MIT Press.

Braver, T. S. (2012). The variable nature of cognitive control: A dual mechanisms framework. *Trends in Cognitive Sciences, 16*(2), 106–113.

Braver, T. S. (2015). *Motivation and cognitive control (Frontiers of Cognitive Psychology.* T. S. Braver (Ed.). Hove: Psychology Press.

Broadbent, D. E. (1958). *Perception and communication.* Elmsford, NY: Pergamon Press.
Carter, C. S., Braver, T. S., Barch, D. M., Botvinick, M. M., Noll, D., & Cohen, J. D. (1998). Anterior cingulate cortex, error detection, and the online monitoring of performance. *Science, 280*(5364), 747–749.
Carter, E. C., Kofler, L. M., Forster, D. E., & McCullough, M. E. (2015). A series of meta-analytic tests of the depletion effect: Self-control does not seem to rely on a limited resource. *Journal of Experimental Psychology: General, 144*(4), 768–815.
Cavanagh, J. F., Wiecki, T. V., Cohen, M. X., Figueroa, C. M., Samanta, J., Sherman, S. J., & Frank, M. J. (2011). Subthalamic nucleus stimulation reverses mediofrontal influence over decision threshold. *Nature Neuroscience, 14*, 1462–1467.
Chatham, C. H., Frank, M. J., & Badre, D. (2014). Corticostriatal output gating during selection from working memory. *Neuron, 81*(4), 930–942.
Cleermans, A., & McClelland, J. L. (1991). Learning the structure of event sequences. *Journal of Experimental Psychology: General, 120*(3), 235–253.
Cleermans, A. (2007). Consciousness: the radical plasticity thesis. *Progress in Brain Research, 168*, 19–33.
Cohen, J. D., Dunbar, K., & McClelland, J. L. (1990). On the control of automatic processes: A parallel distributed processing model of the Stroop effect. *Psychological Review, 97*(3), 332–361.
Cohen, J. D., McClure, S. M., & Yu, A. J. (2007). Should I stay or should I go? How the human brain manages the tradeoff between exploitation and exploration. *Philosophical Transactions of the Royal Society of London Series B (Biological Sciences), 362*(1481), 933–942.
Cohen, J. D., & O'Reilly, R. C. (1996). A preliminary theory of the interactions between prefrontal cortex and hippocampus that contribute to planning and prospective memory. In M. Brandimonte, G. O. Einstein, & M. A. McDaniel (Eds.), *Prospective memory: theory and applications* (pp. 267–295). Hillsdale, NJ: Erlbaum.
Cohen, J. D., Servan-Schreiber, D., & McClelland, J. L. (1992). A parallel distributed processing approach to automaticity. *American Journal of Psychology, 105*, 239–269.
Collins, A. G. E., & Frank, M. J. (2013). Cognitive control over learning: Creating, clustering, and generalizing task-set structure. *Psychological Review, 120*(1), 190–229.
Daw, N. D., Niv, Y., & Dayan, P. (2005). Uncertainty-based competition between prefrontal and dorsolateral striatal systems for behavioral control. *Nature Neuroscience, 8*, 1704–1711.
Daw, N. D., O'Doherty, J. P., Dayan, P., Seymour, B., & Dolan, R. J. (2006). Cortical substrates for exploratory decisions in humans. *Nature, 441*, 876–879.
Dayan, P. (2012). How to set the switches on this thing. *Current Opinion in Neurobiology, 22*(6), 1068–1074.
Deco, G., & Rolls, E. T. (2005). Attention, short-term memory, and action selection: A unifying theory. *Progress in Neurobiology, 76*(4), 236–256.
Dehaene, S., & Changeux, J.-P. (1991). The Wisconsin Card Sorting Test: Theoretical analysis and modeling in a neuronal network. *Cerebral Cortex, 1*(1), 62–79.
Dehaene, S., & Changeux, J.-P. (2011). Experimental and theoretical approaches to conscious processing. *Neuron, 70*(2), 200–227.
Dehaene, S., & Naccache, L. (2001). Towards a cognitive neuroscience of consciousness: Basic evidence and a workspace framework. *Cognition, 79*(1–2), 1–37.
Deserno, L., Huys, Q. J., Boheme, R., Bechert, R., Heinze, H-J., Grace, A. A., … Schlagenhuaf, F. (2015). Ventral striatal dopamine reflects behavioral and neural signatures of model-based control during sequential decision making. *Proceedings of the National Academy of Sciences of the United States of America, 112*(5), 1595–1600.
Diamond, A. (2013). Executive functions. *Annual Review of Psychology, 64*, 135–168.
Dickinson, A., & Balleine, B. (2002). The role of learning in the operation of motivational systems 3. *Stevens' handbook of experimental psychology: Learning, motivation and emotion* (pp. 497–534). New York: John Wiley & Sons.
Duncan, J. (2010). The multiple-demand (MD) system of the primate brain: Mental programs for intelligent behaviour. *Trends in Cognitive Sciences, 14*, 172–179.
Duncan, J., & Owen, A. M. (2000). Common regions of the human frontal lobe recruited by diverse cognitive demands. *Trends in Neurosciences, 23*, 475–483.

Duncan, J., Seitz, R. J., Colony, J., Bor, D., Herzog, H., Ahmed, A., ... Emslie, H. (2000). A neural basis for general intelligence. *Science*, *289*(5478), 457–460.

Edin, F., Klingberg, T., Johansson, P., McNab, F., Tegner, J., & Compte, A. (2009). Mechanism for top-down control of working memory capacity. *Proceedings of the National Academy of Sciences of the United States of America*, *106*(16), 6802–6807.

Egner, T., & Hirsch, J. (2005) Cognitive control mechanisms resolve conflict through cortical amplification of task-relevant information. *Nature Neuroscience*, *8*, 1784–1790.

Einstein, G. O., & McDaniel, M. A. (2005). Prospective memory multiple retrieval processes. *Current Directions in Psychological Science*, *14*(6), 286–290.

Eliasmith, C., Stewart, T. C., Choo, X., Bekolay, T., DeWolf, T., Tang, Y., & Rasmussen, D. (2012). A large-scale model of the functioning brain. *Science*, *338*(6111), 1202–1205.

Elman, J. L. (1990). Finding structure in time. *Cognitive Science*, *14*(2), 179–211.

Fazio, R. H., & Olson, M. A. (2003). Implicit measures in social cognition research: Their meaning and use. *Annual Review of Psychology*, *54*, 297–327.

Feng, S. F., Schwemmer, M., Gershman, S. J., & Cohen, J. D. (2014). Multitasking vs. multiplexing: Toward a normative account of limitations in the simultaneous execution of control-demanding behaviors. *Cognitive, Affective and Behavioral Neuroscience*, *14*(1), 129–146.

Figner, B., Knoch, D., Johnson, E. J., Krosch, A. R., Lisanby, S. H., Fehr, E., & Weber, E. U. (2010). Lateral prefrontal cortex and self-control in intertemporal choice. *Nature Neuroscience*, *13*, 538–539.

Frank, M. J. (2006). Hold your horses: A dynamic computational role for the subthalamic nucleus in decision making. *Neural Networks*, *19*(8), 1120–1136.

Frank, M. J., & Badre, D. (2012). Mechanisms of hierarchical reinforcement learning in corticostriatal circuits 1: Computational Analysis. *Cerebral Cortex*, *22*(3), 509–526.

Frank, M. J., Loughry, B., & O'Reilly, R. C. (2001). Interactions between frontal cortex and basal ganglia in working memory: A computational model. *Cognitive, Affective, & Behavioral Neuroscience*, *1*(2), 137–160.

Fuster, J. M. (1980). *The prefrontal cortex: Anatomy, physiology, and neuropsychology of the frontal lobe.* New York: Raven Press.

Geana, A., Wilson, R. C., Daw, N., & Cohen, J. D. (under review). Boredom, information-seeking and exploration.

Geisler, W. S. (2003). Ideal observer analysis. In L. Chalupa & J. Werner (Eds.), *The visual neurosciences* (pp. 825–837). Boston: MIT Press.

Gershman, S. J., Horvitz, E. J., & Tenenbaum, T. B. (2015). Computational rationality: A converging paradigm for intelligence in brains, minds, and machines. *Science*, *349*(6245), 273–278.

Gershman, S. J., Markman, A. B., & Otto, A.R .(2014). Retrospective revaluation in sequential decision making: A tale of two systems. *Journal of Experimental Psychology: General*, *143*(1),182–194.

Gilbert, S. J., & Shallice, T. (2002). Task switching: A PDP model. *Cognitive Psychology*, *44*(3), 297–337.

Gittins, J. C., & Jones, D. M. (1974). A dynamic allocation index for the sequential design of experiments. In J. Gans (Ed.), *Progress in statistics* (pp. 241–266). Amsterdam, The Netherlands: North-Holland.

Gold, J. I., & Shadlen, M. N. (2002). Banburismus and the brain: Decoding the relationship between sensory stimuli, decisions, and reward. *Neuron*, *89*(5), 881–1120.

Gold, J. I., & Shadlen, M. N. (2007). The neural basis of decision making. *Annual Review of Neuroscience*, *30*, 535–574.

Gollwitzer, P. M. (1996). The volitional benefits of planning. In P. M. Gollwitzer & J. A. Bargh (Eds.), *The psychology of action: Linking cognition and motivation to behavior* (pp. 287–312). New York: Guilford Press.

Graziano, M., & Kastner, S. (2011). Human consciousness and its relationship to social neuroscience: A novel hypothesis. *Cognitive Neuroscience*, *2*(2), 98–113.

Greenwald, A. G., Poehlman, T. A., Uhlmann, E. L., & Banaji, M. R. (2009). Understanding and using the Implicit Association Test: III. Meta-analysis of predictive validity. *Journal of Personality and Social Psychology*, *97*(1), 17–41.

Greenwald, A. G., & Banaji, M. R. (1995). Implicit social cognition: Attitudes, self-esteem, and stereotypes. *Psychological Review*, *102*(1), 4–27.

Griffiths, T. L., Lieder, F., & Goodman, N. D. (2015). Rational use of cognitive resources: Levels of analysis between the computational and the algorithmic. *Topics in Cognitive Science*, *7*(2), 217–229.

Gross, J. J., & Thompson, R. A. (2007). *Emotion regulation: Conceptual foundations. Handbook of emotion regulation* (pp. 3–24). New York: Guilford Press.

Haggard, P. (2008). Human volition: Towards a neuroscience of will. *Nature Reviews Neuroscience*, *9*, 934–946.

Harlow, H., & Martyn, J. (1868). Recovery from the passage of an iron bar through the head. *Publications of the Massachusetts Medical Society*, *2*(3), 327–347.

Hanania, R., & Smith, L. B. (2010). Selective attention and attention switching: towards a unified developmental approach. *Developmental Science*, *13*(4), 622–635.

Hochreiter, S., & Schmidhuber, J. (1997). Long short-term memory. *Neural Computation*, *9*(8), 1735–1780.

Holroyd, C. B., & Coles, M. G. H. (2002). The neural basis of human error processing: Reinforcement learning, dopamine, and the error-related negativity. *Psychological Review*, *109*(4), 679–709.

Howard, M., & Kahana, M. J. (2002). A distributed representation of temporal context. *Journal of Mathematical Psychology*, *46*(3), 269–299.

Howes, A., Lewis, R. L., & Vera, A. (2009). Rational adaptation under task and processing constraints: Implications for testing theories of cognition and action. *Psychological Review*, *116*(4), 717–751.

Ikkai, A., & Curtis, C. E. (2011). Common neural mechanisms supporting spatial working memory, attention and motor intention. *Neuropsychologia*, *49*(6), 1428–1434.

Inzlicht, M., & Schmeichel, B. J. (2012). What Is ego depletion? Toward a mechanistic revision of the resource model of self-control. *Perspectives on Psychological Science*, *7*(5), 450–463.

Kaelbling, L. P., Littman, M. L., & Moore, A. W. (1996). Reinforcement learning: A survey. *Journal of Artificial Intelligence Research*, *4*, 237–285.

Kahneman, D. (2003). A perspective on judgment and choice: Mapping bounded rationality. *American Psychologist*, *58*(9), 697–720.

Kahneman, D., & Henik, A. (1981). Perceptual organization and attention. In M. Kubovy & J. R. Pomerantz (Eds.), *Perceptual organization* (pp. 181–211). Hillsdale, NJ: Erlbaum.

Kahneman, D., & Treisman, A. (1984). Changing views of attention and automaticity. In R. Parasuraman, D. R. Davies, & J. Beatty (Eds.), *Varieties of attention* (pp. 29–61). New York: Academic Press.

Kane, M. J., & Engle, R. W. (2002). The role of prefrontal cortex in working-memory capacity, executive attention, and general fluid intelligence: An individual-differences perspective. *Psychonomic Bulletin & Review*, *9*(4), 637–671.

Keramati, M., Dezfouli, A., & Piray, P. (2011). Speed/accuracy trade-off between the habitual and the goal-directed processes. *PLoS Computational Biology*, *7*(5), e1002055.

Koechlin, E., & Summerfield, C. (2007). An information theoretical approach to prefrontal executive function. *Trends in Cognitive Sciences*, *11*(6), 229–235.

Krebs, J. R., Kacelnik, A., & Taylor, P. (1978) Tests of optimal sampling by foraging great tits. *Nature*, *275*, 27–31.

Kriete, T., Noelle, D. C., Cohen, J. D., & O'Reilly, R. C. (2013). Indirection and symbol-like processing in the prefrontal cortex and basal ganglia. *Proceedings of the National Academy of Sciences of the United States of America*, *110*(41), 16390–16395.

Kurzban, R., Duckworth, A., Kable, J. W., & Myers, J. (2013). An opportunity cost model of subjective effort and task performance. *Behavioral and Brain Sciences*, *36*(6), 661–679.

Kurzban, R. (2010). Does the brain consume additional glucose during self-control tasks? *Evolutionary Psychology*, *8*(2), 244–259.

LaBerge, D., & Samuels, S. J. (1974). Toward a theory of automatic information processing in reading. *Cognitive Psychology*, *6*(2), 293–323.

LeCun, Y., Bengio, Y., & Hinton, G. (2015). Deep learning. *Nature, 521*, 436–444.

Lhermitte, F. (1983). Utilization behavior and its relation to lesions of the frontal lobes. *Brain, 106*, 237–255.

Lieder, F., & Griffiths, T. L. (2015). When to use which heuristic: A rational solution to the strategy selection problem. In D. C. Noelle, R. Dale, A. S. Warlaumont, J. Yoshimi, T. Matlock, C. D. Jennings, & P. P. Maglio (Eds.), *Proceedings of the 37th Annual Conference of the Cognitive Science Society*. Austin, TX: Cognitive Science Society.

Logan, G. D. (1985). Skill and automaticity: Relations, implications, and future directions. *Canadian Journal of Psychology, 39*, 367–386.

Logan, G. D. (1994) On the ability to inhibit thought and action: A users' guide to the stop signal paradigm. In D. Dagenbach & T. H. Carr (Eds.), *Inhibitory processes in attention, memory, and language* (pp. 189–239). San Diego, CA: Academic Press.

Luria, A. (1966). *Higher cortical functions in man*. New York: Basic Books.

Ma, W. J., & Huang, W. (2009). No capacity limit in attentional tracking: Evidence for probabilistic inference under a resource constraint. *Journal of Vision, 9*(3).

MacLeod, C. M. (1991). Half a century of research on the Stroop effect: an integrative review. *Psychological Bulletin, 109*(2), 163–203.

MacLeod, C. M., & Dunbar, K. (1988). Training and Stroop-like interference: Evidence for a continuum of automaticity. *Journal of Experimental Psychology: Learning, Memory, and Cognition, 14*, 126–135.

Marinesco, G., & Radovici, A. (1920). Sur un reflexe cutane nouveau: Reflexe palmo-mentonnier. *Revue neurologique, 27*, 237–240.

McClure, S. M., Laibson, D. I., Loewenstein, G., & Cohen, J. D. (2004). Separate neural systems value immediate and delayed monetary rewards. *Science, 306*(5695), 503–507.

Meiran, N., Pereg, M., Kessler, Y., Cole, M. W., & Braver, T. S. (2015). The power of instructions: Proactive configuration of stimulus–response translation. *Journal of Experimental Psychology: Learning, Memory, and Cognition, 41*(3), 768–786.

Meyer, D. E., & Kieras, D. E. (1997). A computational theory of executive cognitive processes and multiple-task performance: Part 2. Accounts of psychological refractory-period phenomena. *Psychological Review, 104*(4), 749–791.

Miller, E. K., & Cohen, J. D. (2001). An integrative theory of prefrontal cortex function. *Annual Review of Neuroscience, 24*, 167–202.

Miller, G. A., Galanter, E., & Pribram, K. A. (1960). *Plans and the structure of behavior*. New York: Holt, Rhinehart, & Winston.

Montague, P. R., Dayan, P., & Sejnowski, T. J. (1996). A framework for mesencephalic dopamine systems based on predictive Hebbian learning. *Journal of Neuroscience, 16*(5), 1936–1947.

Munakata, Y., Snyder, H. R., & Chatham, C. H. (2012). Developing cognitive control: Three key transitions. *Current Directions in Psychological Science, 21*, 71–77.

Musslick, S., Dey, B., Ozcimder, K., Patwary, M. M. A., Willke, T. L., & Cohen, J. D. (2016). Controlled vs. automatic processing: A graphic-theoretic approach to the analysis of serial vs. parallel processing in neural network architectures. *Proceedings of the 38th Annual Meeting of the Cognitive Science Society*.

Navon, D., & Gopher, D. (1979). On the economy of the human processing system. *Psychological Review, 86*, 214–255.

Neely, J. H. (1977). Semantic priming and retrieval from lexical memory: Roles of inhibitionless spreading activation and limited-capacity attention. *Journal of Experimental Psychology: General, 106*(3), 226–254.

Newell, A. (1990). *Unified theories of cognition*. Cambridge, MA: Harvard University Press.

O'Reilly, R. C. (2006). Biologically based computational models of high-level cognition. *Science, 314*, 91–94.

O'Reilly, R. C., Petrov, A. A., Cohen, J. D., Lebiere, C. J., Herd, S. A., & Kriete, T. (2013). How limited systematicity emerges: A computational cognitive neuroscience approach. In P. Calvo & J. Symons (Eds.), *The architecture of cognition: Rethinking fodor and Pylyshyn's systematicity challenge*. Cambridge: MIT Press.

Otto, A. R., Skatova, A., Madlon-Kay, S., & Daw, N. D. (2015). Cognitive control predicts use of model-based reinforcement learning. *Journal of Cognitive Neuroscience*, *27*(2), 319–333.

Pashler, H. (1984). Processing stages in overlapping tasks: Evidence for a central bottleneck. *Journal of Experimental Psychology: Human Perception and Performance*, *10*, 358–377.

Petersen, S. E., & Posner, M. I. (2012). The attention system of the human brain: 20 years after. *Annual Review of Neuroscience*, *35*, 73–89.

Plate, T. A. (1995). Holographic reduced representation. *IEEE Transactions on Neural Networks*, *6*(3), 623–641.

Polyn, S. M., Norman, K. A., & Kahana, M. J. (2009). A context maintenance and retrieval model of organizational processes in free recall. *Psychological Review*, *116*(1), 129–156.

Posner, M. I., & Cohen, Y. (1984). Components of visual orienting. H. Boom & D. G. Bouwhuis (Eds.), *Attention and performance X*. Hillsdale, NJ: Lawrence Erlbaum.

Posner, M. I., & Snyder, C. R. R. (1975). In R. L. Solso (Ed.), *Information processing and cognition: The Loyola symposium*. New York: Lawrence Erlbaum.

Pratt, S. C., & Sumpter, D. J. T. (2006). A tunable algorithm for collective decision-making. *Proceedings of the National Academy of Sciences of the United States of America*, *103*(43), 15906–15910.

Ratcliff, R., & McKoon, G. (2009). The Diffusion Decision Model: Theory and data for two-choice decision tasks. *Neural Computation*, *20*(4), 873–922.

Redish, A. D. (2016). Vicarious trial and error. *Nature Reviews Neuroscience*, *17*(3), 147–159.

Reynolds, J. R., Braver, T. S., Brown, J. W., & Van der Stigchel, S. (2006). Computational and neural mechanisms of task switching. *Neurocomputing*, *69*(10–12), 1332–1336.

Rigotti, M., Barak, O., Warden, M. R., Wang, X.-J., Daw, N. D., Miller, E. K., & Fusi, S. (2013). The importance of mixed selectivity in complex cognitive tasks. *Nature*, *497*, 585–590.

Roca, M., Torralva, T., Gleichgerrcht, E., Woolgar, A., Thompson, R., Duncan, J., & Manes, F. (2011). The role of area 10 (BA10) in human multitasking and in social cognition: A lesion study. *Neuropsychologia*, *49*, 3525–3531.

Rougier, N. P., Noelle, D. C., Braver, T. S., Cohen, J. D., & O'Reilly, R. C. (2005). Prefrontal cortex and the flexibility of cognitive control: Rules without symbols. *Proceedings of the National Academy of Sciences of the United States of America*, *102*(20), 7338–7343.

Rumelhart, D. E., McClelland, J. L., & the PDP Research Group (1986). *Parallel distributed processing: Explorations in the microstructure of cognition*. Cambridge, MA: MIT Press.

Salvucci, D. D., & Taatgen, N. A. (2008). Threaded cognition: An integrated theory of concurrent multitasking. *Psychological Review*, *115*(1), 101–130.

Schneider, W., & Shiffrin, R. M. (1977). Controlled and automatic human information processing: 1. Detection, search, and attention. *Psychological Review*, *84*(2), 1–66.

Schumacher, E. H., Seymour, T. L., Glass, J. M., Fencsik, D. E., Lauber, E. J., Kieras, D. E., & Meyer, D. E. (2001). Virtually perfect time sharing in dual-task performance: Uncorking the central cognitive bottleneck. *Psychological Science*, *12*, 101–108.

Schurger, A., Pereira, F., Treisman, A., & Cohen, J. D. (2010). Reproducibility of activity characterizes conscious from non-conscious neural representations. *Science*, *327*(5961), 97–99.

Shaffer, L. H. (1975). Multiple attention in continuous verbal tasks. In P. M. A. Rabbitt & S. Dornic (Eds.), *Attention and performance* (pp. 157–167). New York: Academic Press.

Stroop, J. R. (1935). Studies of interference in serial verbal reactions. *Journal of Experimental Psychology*, *18*, 643–662.

Shallice, T. (1982). Specific impairments of planning. *Philosophical Transactions of the Royal Society of London B*, *298*, 199–209.

Shenhav, A., Botvinick, M. M., & Cohen, J. D. (2013). The expected value of control: An integrative theory of anterior cingulate cortex function. *Neuron*, *79*, 217–240.

Shiffrin, R. M., & Schneider, W. (1977). Controlled and automatic human information processing: II. Perceptual learning, automatic attending and a general theory. *Psychological Review*, *84*(2), 127–190.

Shohamy, D., & Daw, N. D. (2015). Integrating memories to guide decisions. *Current Opinion in Behavioral Sciences*, *5*, 85–90.

Simen, P., Contreras-Ros, D., Buck, C., Hu, P., Holmes, P., & Cohen, J. D. (2009). Reward rate optimization in two-alternative decision making: Empirical tests of theoretical predictions. *Journal of Experimental Psychology: Human Perception and Performance, 35*, 1865–1897.

Simon, H. A. (1955). A behavioral model of rational choice. *Quarterly Journal of Economics, 69*, 99–118.

Simon, H. A. (1992). What is an 'explanation' of behavior? *Psychological Science, 3*, 150–161.

Simoncelli, E. P., & Olshausen, B. (2001). Natural image statistics and neural representation. *Annual Review of Neuroscience, 24*, 1193–1216.

Soon, C. S., Brass, M., Heinze, H.-J., & Haynes, J. D. (2008). Unconscious determinants of free decisions in the human brain. *Nature Neuroscience, 11*, 543–545.

Spearman, C. (1904). General intelligence, objectively determined and measured. *The American Journal of Psychology, 15*(2), 201–292.

Susillo, D. (2014). Neural circuits as computational dynamical systems. *Current Opinion in Neurobiology, 25*, 156–163.

Sutton, R. S. (1988). Learning to predict by the methods of temporal differences. *Machine Learning, 3*(1), 9–44.

Sutton, R. S. (1990). Integrated architectures for learning, planning, and reacting based on approximating dynamic programming. *Proceedings of the Seventh International Conference on Machine Learning.*

Tanner, W. P. J., & Swets, J. A. (1954). A decision-making theory of visual detection. *Psychological Review, 61*, 401–409.

Tenenbaum, J. B., Griffiths, T. L., & Kemp, K. (2006). Theory-based Bayesian models of inductive learning and reasoning. *Trends in Cognitive Science, 10*, 309–318.

Todd, M., Niv, Y., & Cohen, J. D. (2008). Learning to use working memory in partially observable environments through dopaminergic reinforcement. *Advances in Neural Information Processing Systems (Vol. 20)*. Cambridge, MA: MIT Press.

Tolman, E. C. (1948). Cognitive maps in rats and men. *Psychological Review, 55*(4), 189–208.

Tombu, M. N., Asplund, C. L., Dux, P. E., Godwin, D., Martin, J. W., & Marois, R. (2011). A unified attentional bottleneck in the human brain. *Proceedings of the National Academy of Sciences of the United States of America, 108*, 13426–13431.

Townsend, J. T. (1972). A note on the identifiability of parallel and serial processes *Perception & Psychophysics, 10*(3), 161–163.

Tulving, E. (2002). Episodic memory: From mind to brain. *Annual Review of Psychology, 53*, 1–25.

Ridderinkhof, K. R., Ullsperger, M., Crone, E. A.,& Nieuwenhuis, S. (2004). The role of the medial frontal cortex in cognitive control. *Science, 306*(5695), 443–447.

Usher, M., Cohen, J. D., Haarmann, H., & Horn, D. (2001). Neural mechanism for the magical number 4: Competitive interactions and nonlinear oscillation. *Behavioral and Brain Sciences, 24*(1), 151–152.

Vargas, M. (2013). *Building better beings.* Oxford: Oxford University Press.

Verbruggen, F., & Logan, G. D. (2008). Automatic and controlled response inhibition: Associative learning in the go/no-go and stop-signal paradigms. *Journal of Experimental Psychology: General, 137*(4), 649–672.

Venkatraman, V., Rosati, A. G., Taren, A. A., & Huettel, S. A. (2009). Resolving response, decision, and strategic control: evidence for a functional topography in dorsomedial prefrontal cortex. *Journal of Neuroscience, 29*, 13158–13164.

Verguts, T., & Notebaert, W. (2008). Hebbian learning of cognitive control: Dealing with specific and nonspecific adaptation. *Psychological Review, 115*(2), 518–525.

Verguts, T., Vassena, E., & Silvetti, M. (2015). Adaptive effort investment in cognitive and physical tasks: a neurocomputational model. *Frontiers in Behavioral Neuroscience, 9.*

Watkinson, S. C., Boddy, L., Burton, K., Darrah, P. R., Eastwood, D., Fricker, M. D., & Tlalka, M. (2005). New approaches to investigating the function of mycelial networks. *Mycologist, 19*, 11–17.

Westbrook, A., & Braver, T. S. (2015). Cognitive effort: A neuroeconomic approach. *Cognitive, Affective, & Behavioral Neuroscience, 15*(2): 395–415.

Wickens, D. D. (1984). Processing resources in attention. In R. Parasuraman, D. R. Davies, & J. Beatty (Eds.), *Varieties of attention* (pp. 63–102). New York: Academic Press.

Wiecki, T. V., & Frank, M. J. (2013). A computational model of inhibitory control in frontal cortex and basal ganglia. *Psychological Review*, *120*(2), 329–355.
Wiener, N. (1948). *Cybernetics, or control and communication in the animal and the machine*. Cambridge: MIT Press.
Wilson, R. C., Geana, A., White, J. M., Ludvig, E. A., & Cohen, J. D. (in press). Humans use directed and random exploration to solve the explore-exploit dilemma. *Journal of Experimental Psychology: General*, *143*(6), 2074–2081.
Yeung, N., Botvinick, M. M., & Cohen, J. D. (2004). The neural basis of error detection: Conflict monitoring and the error-related negativity. *Psychological Review*, *111*(4), 931–959.
Yu, A., & Dayan, P. (2005). Uncertainty, neuromodulation, and attention. *Neuron*, *46*(4), 681–692.
Yu, A. J., Dayan, P., & Cohen, J. D. (2009). Attentional control: Toward a rational Bayesian account. *Journal of Experimental Psychology: Human Perception and Performance*, *35*(3), 700–717.
Zipser, D. (1991). Recurrent network model of the neural mechanism of short-term active memory. *Neural Computation*, *3*(2), 179–193.

3

Attentional Control and Working Memory Capacity

Matt E. Meier and Michael J. Kane

Working memory is a limited-capacity system that operates at the intersection of attention and memory. The capacity of this system (i.e., working-memory capacity [WMC]) is the functional limit on how many memory representations—or how well even one representation—can be kept active in the face of interference and conscious shifts of focus in the service of ongoing cognitive activities. WMC matters: Its variation across people predicts such important skills as reading comprehension (e.g., Daneman & Carpenter, 1980; McVay & Kane, 2012a), following classroom directions (Engle, Carullo, & Collins, 1991), learning computer languages (Kyllonen & Stephens, 1990; Shute, 1991), multitasking (e.g., Bühner, König, Pick, & Krumm, 2006; Hambrick, Oswald, Darowski, Rench, & Brou, 2010), and solving novel problems (Conway, Cowan, Bunting, Therriault, & Minkoff, 2002; Engle, Tuholski, Laughlin, & Conway, 1999; Kane et al., 2004; Martínez & Colom, 2009). The central role WMC plays in vital intellectual activities has made it an attractive target for interventions to increase its capabilities (for a review, see Melby-Lervag & Holme, 2013; Chapter 28 by Spencer-Smith & Klingberg in this volume).

Why does a chapter focused on WMC belong in a handbook of cognitive control? Because WMC impacts task performance, in part, through associated cognitive control processes. The processes covered within this chapter are examined via traditional 'attention' tasks, so we favour the label *attentional control*. Recently, two large-scale studies, using latent-variable techniques to assess the construct-related variance shared across multiple tasks, examined attentional accounts of WMC's contribution to higher-order cognition (Shipstead, Lindsey, Marshall, & Engle, 2014; Unsworth, Fukuda, Awh, & Vogel, 2014). Each study assessed the contributions of attentional control (using tasks such as antisaccade and Stroop), retrieval from long-term memory (using tasks such as source memory and paired-associate recall), and simple short-term memory (STM; using tasks such as change detection and sub-span immediate-recall tasks). In both studies, the best fitting structural model indicated that the relation between WMC and higher-order cognition was fully mediated by these three correlated factors, with all three factors accounting for substantial unique variance. This multifaceted conception of WMC is analogous to the 'unity and diversity' framework of executive functions advanced by Miyake and Friedman (2012), in which executive functions comprise several correlated but separable factors. That is, WMC represents the interactions of multiple

The Wiley Handbook of Cognitive Control, First Edition. Edited by Tobias Egner.

memorial and attentional processes and abilities. Of most importance, here, such studies (see also, Conway et al., 2002; Engle et al., 1999) suggest that attentional control is partly responsible for WMC's prediction of higher-order cognition.

Logic and Function of Individual-Differences Research

Individual-differences research serves an important role in building and testing the nomological net (Cronbach & Meehl, 1955) around the construct of WMC. Instead of viewing between-person variation as unfortunate noise, as in traditional experimental work, individual-difference researchers harness information provided by this variation to reveal critical associations among constructs, to specify boundaries among them, and to test how experimental treatments differentially affect people at different points along some construct (Cronbach, 1957). Underwood (1975) further proposed that individual differences can be used as a 'crucible' for theory construction, whereby individual-differences confirmations of nomothetic-theory predictions provide a 'thumbs up' signal to proceed, but disconfirmations condemn the theory to death. We do not take such a strong stance, as most psychological theories are protected from such falsification via auxiliary hypotheses (Meehl, 1990), but we do regard individual-differences evidence as valuable in testing and sharpening theories, in general, and specifying the relation between WMC and attentional control, in particular.

How Is WMC Measured?

Individual-differences research on WMC is conducted using a variety of tasks: updating (Oberauer, Süß, Schulze, Wilhelm, & Wittmann, 2000), change detection (Luck & Vogel, 1997), and complex span (Conway et al., 2005), most consistently. Complex span tasks have been more intensely studied and validated than other measures (as described below), so this chapter focuses on work operationalising WMC as complex span task performance.

Daneman and Carpenter (1980) developed the first complex span tasks to study the relation between WMC and reading comprehension. The tasks (two visual, one auditory) interleaved a simple memory span task (recall of a short list of items) with a processing task (read aloud or listen to sentences). After each set of 2–6 sentences, subjects recalled the last word from each. Complex span tasks thus required subjects to maintain or retrieve individual words while (or between) comprehending sentences. In two experiments, complex span performance correlated strongly with recalling facts from a story ($rs = .67–.81$), with verbal SAT scores ($rs = .49–.59$), and with passage-ending, pronoun-referent decisions ($rs = .72–.90$). Subsequent work indicated that the relation between complex span tasks and higher-order cognition does not depend on a close match in the stimuli or processing demands between the span task and the higher-order task (e.g., Turner & Engle, 1989). WMC, as measured by complex span tasks, is largely domain general. For example, complex span tasks using varied processing tasks (e.g., arithmetic, spatial rotation, symmetry judgements, counting, judging sensibility of sentences) and memory items (e.g., letters, words, matrix locations, arrow sizes and position) share substantial variance and predict verbal and spatial abilities (Kane et al., 2004).

Complex span scores are reliable, with moderate-to-high internal consistency (0.7–0.9; Conway et al., 2002; Engle et al., 1999; Kane et al., 2004; Unsworth, Heitz, Schrock, & Engle, 2005) and test-retest reliability (0.8–0.9; Klein & Fiss, 1999; Turley-Ames & Whitfield, 2003; Unsworth et al., 2005). Are they also valid? Because individual-differences research is correlational, we must rule out confounding variables that may drive WMC assessments and covariation with other constructs. Two potential confounds that have been investigated deeply are strategy use and effort. Dunlosky and Kane (2007) found that, although subjects performed better on complex span tasks when using normatively effective strategies, strategy usage did *not* statistically mediate the relation between WMC and verbal ability. Strategy use mediated WMC–criterion associations only when the same strategies benefited WMC and criterion tasks (e.g., paired-associate learning but not reading comprehension; Bailey, Dunlosky, & Kane, 2008). Taking a combined differential-experimental approach, Turley-Ames and Whitfield (2003) trained subjects to use an effective strategy and found that training did not close the gap between lower- and higher-WMC groups. Instead, it *increased* it because higher-WMC subjects profited more from training than did lower-WMC subjects. Simply put, strategy use does not drive WMC × complex cognition associations; differential strategy use is, instead, a suppressor variable, obscuring rather than contributing to WMC's covariation with other constructs.

We cannot account for complex span's predictive power by simply appealing to motivation or effort, either. First, Heitz, Schrock, Payne, and Engle (2008) found that when given monetary incentives, previously ascertained lower- and higher-WMC groups both equally improved their complex span performance, thus maintaining the original individual differences; moreover, pupillometry measures indicated equivalent effort expended by the WMC groups in response to both incentives and memory load. Second, Unsworth and McMillan (2013) assessed motivation and interest in a prose-reading task via self-report; they found that WMC was uncorrelated with both and that all three variables made statistically independent contributions to mind wandering while reading. Third, as discussed in detail below, WMC's discriminant validity—its patterns of correlation with some abilities but not others—cannot be explained by mere effort or motivation.

Evidence for the WMC–Attention Association

Evidence for a link between WMC and attentional control comes from two approaches: (1) explorations of the relation between WMC and a particular attentional-control task ('microanalytic'), usually manipulating theoretically relevant variables in the attention task to assess individual-by-treatment interactions; (2) latent-variable studies that use multiple tasks to assess each construct ('macroanalytic'), examining cognitive processes at the construct level by statistically isolating the variance that is shared across the marker tasks.

In an early microanalytic study connecting WMC to attentional control (Kane, Bleckley, Conway, & Engle, 2001), lower- and higher-WMC subjects (defined via complex span) completed an antisaccade task (Hallett, 1978). The antisaccade task required subjects to use an abrupt-onset, flashing cue to direct attention to a backward-masked target (*B*, *P*, or *R*) that subjects identified via key-press. Subjects completed two blocked within-subject conditions: prosaccade, where the target appeared at the cued location, and antisaccade, where the target appeared on the opposite side from the cue. For successful prosaccade performance, subjects could follow the habitual, exogenously driven orienting response. But in the antisaccade condition, subjects had to use endogenous control to either prevent orienting to the cue, or failing that, to quickly disengage from the cue and shift focus to the target. Variation in WMC did *not* predict the more automatic prosaccade performance. But in the antisaccade condition, where endogenous control was necessary to counter prepotency, lower-WMC subjects made more initial saccades towards the cue, and initiated antisaccades more slowly, than did higher-WMC subjects. Of course, the antisaccade task does not require maintaining a large amount of information in memory. WMC's prediction of antisaccade performance thus suggests it reflects more than the sheer amount that can be maintained in memory, but also (or instead) the executive regulation of cognition.

To further explore WMC's association with attention control, Unsworth, Schrock, and Engle (2004, Experiment 2) mixed prosaccade and antisaccade trials in a cued sequence, thus requiring controlled eye movements for each trial type. Lower-WMC subjects were slower and more erroneous on both anti- and prosaccade trials than were higher-WMC subjects. Experiment 3, in a similar vein, manipulated the control requirements of blocked trials by cuing the target location either exogenously (with a peripheral flash) or endogenously (with a central arrow). Lower-WMC subjects performed more poorly than higher-WMC subjects on endogenously cued prosaccade and antisaccade trials, and exogenously cued antisaccade trials, indicating a WMC-related deficit in the voluntary control of attention.

Evidence from latent-variable studies produces a more general type of evidence regarding the processes involved, because the analyses consider the commonality of processes tapped by different measures. Early latent-variable studies of attentional control in WMC examined the relationship between the residual variance from constructs composed of complex span measures after shared variance with STM measures was removed (Conway et al., 2002; Engle et al., 1999). The logic here was that, whereas both STM and WMC tasks require rehearsal and immediate recall of actively maintained stimuli, only WMC tasks do so in the face of the distracting processing task. Both studies found that the residual-WMC variance predicted fluid intelligence (rs = .60 and .49, respectively), supporting the suggestion that *WMC = STM + attentional control + measurement error* (Engle et al., 1999). Although the work discussed above, by Shipstead et al. (2014) and Unsworth et al. (2014), further fractionated WMC and suggests that this conceptualisation was too simple, the evidence is still strong that WMC robustly predicts attentional-control capabilities. Recent research assessing the relation between these constructs (by independently measuring attention control in separate tasks, such as antisaccade, Stroop, and flanker) suggests moderate-to-strong correlations in the range of .5–.7 (e.g., Colom, Abad, Quiroga, Shih, & Flores-Mendoza, 2008; McVay & Kane, 2012a; Schweizer & Moosbrugger, 2004; Unsworth et al., 2014).

Characterising WMC-Related Attentional Control

WMC predicts attentional control, but what can we say (if anything) about the nature of attention-control processes and functions? Here, we highlight evidence that contributes to specifying the amorphous construct of attention (Anderson, 2011). Because prior work had demonstrated WMC-related individual differences in tasks that require attentional control (Kane et al., 2001; Kane & Engle, 2003), and prominent models of visual attention propose a role for endogenous, top-down processes (Treisman & Gelade, 1980; Wolfe, 1994), one might expect that higher-WMC subjects would be faster than lower-WMC subjects at visual search, that is, in locating targets within arrays of perceptually similar distractors. Kane, Poole, Tuholski, and Engle (2006) tested this idea in multiple experiments that included feature-absence, conjunction, and spatial-configuration searches. Regardless of search difficulty, they found no WMC effects. Thus, some effortful, controlled-attention tasks relate to WMC, and others do not.

To explore this boundary condition further, Sobel, Gerrie, Poole, and Kane (2007) manipulated a conjunction search task to examine particular top-down and bottom-up contributions. In displays that permitted bottom-up properties (through perceptual grouping) to drive efficient search, WMC again showed no effect. But in conditions where bottom-up influences had to be overcome by top-down strategies for efficient search, higher-WMC subjects had shallower search slopes than lower-WMC subjects. In three experiments, Poole and

Kane (2009) used a modified search paradigm to test whether WMC would predict search in tasks that cued subjects to possible target locations amid distractor locations. They found that WMC predicted search when—and only when—distractors were presented amid the possible target locations, and subjects had to maintain selective focus on those cued locations over a relatively long interval; WMC had no effect in the absence of distractors or when target arrays appeared almost immediately after cuing.

Taken together, the results from Kane et al. (2006), Sobel et al. (2007), and Poole and Kane (2009) suggest that WMC is not related to prototypical visual search, but search task requirements can be manipulated to produce WMC-related individual differences. When top-down control must override the influence of bottom-up processes, block distractor locations, or maintain focus on probable target locations, higher-WMC subjects outperform lower-WMC subjects. The lack of a relation between WMC and typical visual search demonstrates a critical boundary — that WMC is not related to all types of endogenous attentional control—and provided an early glimpse into the contextual nature of WMC-related individual differences.

A Two-Factor Theory of WMC and Cognitive Control

In five experiments using the Stroop (1935) task, in which subjects were presented with colour words displayed in either matching (congruent) or mismatching (incongruent) colours for colour naming, Kane and Engle (2003) manipulated the congruent trial proportion. For example, in Experiment 1, subjects either completed a task with 0% or 75% congruent trials, and in Experiment 2, all subjects first completed a block of 0% congruent trials and then a block of 75% congruent trials. Kane and Engle hypothesised that with a high proportion of congruent trials there is little environmental support for the goal of colour naming, because on most trials subjects can produce correct responses by reading the word. Support for this goal must be maintained endogenously, and so high-congruency conditions should reveal individual differences in goal-maintenance abilities. In low-congruency conditions, much of the endogenous burden of goal maintenance is removed because the frequent incongruent trials serve as goal reminders, but even when this goal is supported by the task context, it still needs to be executed effectively. (For an extensive review on the effects of proportion congruency on interference effects, see Chapter 5 by Bugg in this volume).

Critically, in high-congruency (low-goal-support) conditions, higher- and lower-WMC subjects differed in both the magnitude of error interference (incongruent minus congruent error rate) and reaction time (RT) facilitation (neutral minus congruent RT), with lower-WMC subjects making more errors and experiencing more facilitation than did higher-WMC subjects. Both findings suggest more frequent goal neglect and word-reading responses by lower-WMC subjects. That is, without support from frequent goal reminders (incongruent trials), the task goal was lost, and lower-WMC subjects resorted to the more habitual word reading (which evokes errors on incongruent trials and fast responses on congruent trials).

In contrast, in conditions that were supportive of the task goal (i.e., low proportion congruency), lower-WMC subjects showed only more RT interference than did higher-WMC subjects. These RT interference differences seem to reflect the ability to resolve response competition; that is, subjects executed the correct response, indicating adequate goal maintenance, but lower-WMC subjects took longer to resolve the actual word–colour conflict presented on those incongruent trials. Taken together, these Stroop results provide evidence for a two-factor theory of WMC's influence on attentional control (Engle & Kane, 2004; Kane, Brown et al., 2007). Higher WMC reflects a superior ability to maintain the novel task goal and also to resolve response competition. Kane and Engle's two-factor theory is similar to Braver and colleagues' proposal for dual mechanisms of control (Braver, Gray, & Burgess, 2007; Chapter 9 by Chiew & Braver in this volume), where WMC is related to performance through *proactive*, or anticipatory, control—akin to goal maintenance—and *reactive* control—similar to response competition).

Although the two-factor theory describes WMC-related control as enacted through two empirically dissociable mechanisms, these may not be wholly independent. Both superior goal maintenance and competition resolution could follow from the degree of goal activation. If the goal is maximally activated, it will affect performance in a top-down, proactive manner; if, however, the goal is less highly activated, it may only facilitate performance reactively, when conflict is actually encountered. That is, not only do goal-maintenance abilities (or activation levels) show themselves in preventing outright lapses of attention, but they may also show themselves in responding to conflict in the moment. Botvinick, Braver, Barch, Carter, and Cohen's (2001) model of cognitive control provides a blueprint of how this may work with a more highly activated goal representation biasing perception towards goal-relevant stimulus features and therefore producing better performance than someone using a less activated goal representation.

The Botvinick et al. (2001) model focuses on goal representations biasing performance on a trial-by-trial level. WMC-related attentional-control processes could affect task performance at different levels, however: at the more global, proactive, task-approach level (e.g., Chapter 5 by Bugg in this volume), or at the more local, reactive, trial-by-trial level (Chapter 4 by Egner in this volume). That is, higher-WMC subjects could show superior performance to lower-WMC subjects because of how they interpret and instantiate the task goal at the global level of the whole task. Or, they could excel because they better adjust cognitive-control settings at the local level in response to immediate demands (as in Botvinick et al.). It is therefore possible that the previously observed global effects of WMC on Stroop task performance (e.g., Kane & Engle, 2003) were the cumulative product of local, trial-level (local) WMC effects. One way to determine the level at which WMC-related attentional control exerts its effects to examine trial-level performance via congruency transitions between trials (e.g., Gratton, Coles, & Donchin, 1992), referred to here as congruency-sequence effects (for a review, see Egner, 2007; Chapter 4 by Egner in this volume). Congruency-sequence effects are often used as markers of in-the-moment cognitive control.

Work that has assessed the dynamics of WMC's association with attentional control has tested for its moderation of congruency-sequence effects. The results have been mixed. Using Stroop tasks, flanker tasks, or both, WMC has not predicted congruency-sequence effects (Keye, Wilhelm, Oberauer, & van Ravenzwaaij, 2009; Meier & Kane, 2013; Unsworth, Redick, Spillers, & Brewer, 2012). In Simon tasks, however, lower-WMC subjects exhibited more reactivity to the congruence of the prior trial when processing the current trial (Keye et al., 2009; Weldon, Mushlin, Kim, & Sohn, 2013). Finally, in a task that contained some trials that presented flanker-like conflict (i.e., conflict between stimulus features) and other trials that presented Simon-like conflict (i.e., conflict between stimulus and response features), Meier and Kane (in press) found no link between WMC and congruency-sequence effects in either trial type. The discrepancies among Simon-task findings from Meier and Kane, Keye et al., and Weldon et al. may be the product of different task approaches cultivated by the mixed trial presentation used by Meier and Kane versus the blocked presentations used by others. That is, subjects may set up different goal representations when performing blocked versus mixed trial tasks, and these may yield downstream performance differences. We suggest, in any case, that trial-to-trial calibrations of control do not drive the relation between WMC and attention-control tasks. The association seems to be caused, instead, by control at the global level of the task.

WMC and Goal Maintenance, Response Time Variability, and Mind Wandering

Off-task thoughts, or mind wandering, may indicate that the task goal has become dislodged or diminished in activation, and so we might expect lower-WMC subjects to mind-wander more frequently. Kane et al. (2007) pretested participants on WMC and later recruited them to carry personal digit assistants (PDAs) for one week. To provide information on the frequency and contextual predictors of mind wandering, the PDA beeped eight times a day prompting subjects to report whether or not their thoughts had wandered from their primary task. This question was followed by Likert-type questions about their context. Kane et al. found that, only as activities required more self-reported concentration, challenge, and effort did higher-WMC subjects better maintain the goal of the task at hand and resist mind wandering. Subsequent laboratory work has found that attentional lapses in the form of mind wandering (i.e., lapses of goal maintenance) partially mediate the relation between WMC and reading comprehension (McVay & Kane, 2012b) and SAT scores (Unsworth, Brewer, & Spillers, 2012; Unsworth, McMillan, Brewer, & Spillers, 2012).

Laboratory studies have also revealed a contextually determined relation between WMC and mind wandering. McVay and Kane (2009) used a variant of a go/no-go task to examine the relation. In this 45 min task, subjects responded via key-press to all stimuli except for infrequent targets (11% of trials). Mind wandering, assessed with task-embedded thought probes, accounted for approximately 50% of the covariation between WMC and task accuracy. A follow-up study (McVay & Kane, 2012a) also found that mind wandering partially mediated the relationship between WMC and go/no-go accuracy, and further investigated whether WMC's relation to performance differed according to task demands. They contrasted WMC relations on the go/no-go task, which required frequent responding to nontargets and rare withholding of responses to targets, to performance on a vigilance version of the task, which required withholding responses to the frequent nontargets and responding to the rare targets. In the go/no-go task, control over habitual responding (repeatedly pressing the space bar) is at a premium on critical no-go trials (see Chapter 6 by Verbruggen & Logan in this volume), whereas in the vigilance task, the level of interference is minimal. WMC variation related positively to accuracy and negatively to mind wandering in the go/no-go task, but was unrelated to both performance and mind wandering in the vigilance task. Here again, it seems that there needs to be a sufficient level of interference, conflict, or challenge in the task for WMC-related individual differences to arise (for similar mind-wandering findings, see Levinson, Smallwood, & Davidson, 2012; Rummel & Boywitt, 2014).

Although goal activation may be continuous, the binary/dichotomy account and continuum accounts have yet to be empirically distinguished. What we do know is that lower-WMC subjects, in addition to reporting more mind wandering and making more overt errors in tasks like the high-congruency Stroop, also respond more variably in RT tasks than do higher-WMC subjects. In both McVay and Kane (2009, 2012a) go/no-go tasks, within-subject RT variability on 'go' trials correlated negatively with WMC and positively with mind wandering. That is, lower-WMC subjects appeared to have trouble maintaining the goal of the task relative to higher-WMC subjects on some, but not all, trials (see also Unsworth, Redick, Lakey, & Young, 2010). Consistent with this idea, formal evidence-accumulation modelling indicated that the parameter reflecting trial-to-trial variation in drift rate correlated with WMC, mind-wandering rate, and RT variability (McVay & Kane, 2012b). Taking a latent-variable approach, Unsworth (2015) analysed data from several studies where subjects completed WMC tasks, mind-wandering probes, attentional-control tasks, and lexical-decision tasks. RT variability from the attentional-control tasks, but not the lexical-decision tasks, correlated significantly with WMC and mind-wandering constructs. WMC measures thus appear to index attentional abilities that predict the consistency with which one can maintain a novel task goal. These findings do not distinguish between dichotomous and continuous accounts of goal maintenance, but they do help visualise goal activation as waxing and waning in strength across attentionally demanding contexts.

WMC and Response Competition

Although considerable work has examined WMC and goal maintenance, much less has studied WMC and response competition. We recently explored whether WMC predicted resolving interference between competing stimulus elements, resolving interference between competing stimulus and response elements, or both (Meier & Kane, 2015). Put differently, we tested whether WMC resolved interference early, late, or throughout the information processing sequence, under the assumption that stimulus-stimulus (S-S) interference affects stimulus selection and/or identification (i.e., early), and stimulus-response (S-R) interference affects response selection and/or motor execution (i.e., late; see Kornblum, 1992; Kornblum, Hasbroucq, & Osman, 1990). To focus on response competition, subjects completed a task with one task goal that was reinforced by many conflict trials, but that elicited different types of interference. In this way, any WMC-related differences that we found between S-S and S-R interference types should primarily reflect competition-resolution processes and not goal-maintenance processes.

Subjects completed a task where different trials presented either S-S or S-R conflict. On each trial, subjects saw an upward- or a downward-pointing arrow that was either below or above, or to the left or right, of fixation (all equidistant from the centre). They reported whether the arrow pointed up or down with a left or right key-press. The irrelevant stimulus dimension of arrow location to the left or right of fixation overlapped with the horizontally oriented key-press responses. Thus, S-R conflict occurred when the arrow appeared on the opposite side of the screen from the appropriate response key. On trials where the arrow was above or below the screen's midpoint, however, the irrelevant (vertical location) and relevant (arrow direction) *stimulus* dimensions overlapped. S-S conflict thus occurred on trials where the arrow direction conflicted with the arrow location (e.g., an upward arrow presented below fixation; a downward arrow above fixation). For both trial types, though, stimulus location was always irrelevant, and the goal was the same.

In two experiments with 50% incongruent trials (half horizontally oriented and half vertically oriented), we found that WMC predicted conflict resolution RTs on S-S trials but not on S-R trials. That is, higher-WMC subjects were better able than lower-WMC subjects to combat the interference created from the S-S conflict of identifying upward arrows located below fixation and downward arrows above fixation. However, higher-WMC subjects were no better than lower-WMC subjects at resolving S-R conflict created from the arrows appearing contralateral to the response key; that is, WMC did not predict resolution of interference when the right key was needed to respond to an arrow on the left, or vice versa (if anything, lower-WMC subjects made fewer errors than higher-WMC subjects under S-R conflict).

Moreover, these S-S versus S-R patterns held on trials that presented combinations of stimulus-stimulus and stimulus-response conflict (Meier & Kane, 2015, Experiment 1). These trials presented an arrow that was slightly above or below the screen's midpoint as well as either slightly to the left or the right of centre (e.g., between 10:00 and 11:00 on a clock face). Here, again, WMC was only related to the resolution of S-S and not S-R interference. WMC's selective association with resolving S-S but not S-R competition indicates that higher-WMC subjects are better able than lower-WMC subjects, in early information processing stages, to identify and select relevant information amid distractions. However, in another experiment in which subjects completed the same task, but with 80% congruent trials to put a premium on goal maintenance for successful performance, a different pattern emerged. Consistent with the two-factor theory of WMC and executive control, higher-WMC subjects exhibited less S-S *and* S-R interference, presumably because they were better able to maintain the task goal in an accessible enough state to influence competition resolution. Task context can clearly tip the balance between the importance of goal-maintenance and response-competition processes, and therefore where WMC has its effects.

In addition to goal-maintenance abilities, the precise competition-resolution mechanisms through which WMC affects performance on S-S but not S-R conflict trials are not well understood. Early speculation has focused on WMC-related interference reduction mechanisms that are focused on removing items from working memory that interfere with stimulus representations but not response representations, and on the relevancy of task features to successful task completion. In Meier and Kane's (2015) arrow task, left-right location codes were necessary for response selection (i.e., action planning) and therefore were not truly task irrelevant, in contrast to location information present on trials where the arrow was presented on the vertical plane. Thus, higher-WMC subjects may be better than lower-WMC subjects at coping with 'pure' task-irrelevant interference.

WMC-Related Constraint of Visual Attention

Although it does not seem to be able to account for the S-S and S-R conflict dissociation, or the findings from antisaccade tasks, WMC also seems related to the precision and speed of selective visual attention. Bleckley, Durso, Crutchfield, Engle, and Khanna (2003) first demonstrated that WMC variation predicts how visual attention is spatially configured. From brief, masked displays, subjects identified a centrally presented target letter and then localised another displaced letter that could appear on one of three concentric rings around the central letter. The ring on which the displaced letter would appear was validly cued on 80% of trials with the words *close*, *medium*, or *distant*. Consider an invalidly cued trial where the subject receives a *distant* cue (i.e., the cue for the farthest ring from the centre). The subject should configure her attention to the screen's centre (for central letter identification) and the distant ring (for displaced target localisation), but not to the close or middle rings. When the target then appears unexpectedly on, say, the middle ring, subjects who were more tightly focused on just the distant ring should perform worse than the subject who was unable to deploy their attention as precisely and instead attended to the whole area from the outer ring to the centre. Bleckley et al. found that, for all subjects, invalidly cued letters occurring outside the cued ring were poorly localised (67% correct). Of most importance, however, lower-WMC subjects localised invalidly cued letters *inside* the cued ring better than did higher-WMC subjects (72% vs. 64%), suggesting that higher-WMC subjects configured their attention discontiguously (encompassing only the cued ring and centre location, but not in between), whereas lower-WMC subjects deployed their visual attention more diffusely, in a spotlight formation extending to the cued ring. Subsequent work by Bleckley, Foster, and Engle (2015) found that under a secondary memory load, higher-WMC subjects were no longer able to deploy their attention discontiguously in the ring-cuing task. The load task thus disrupted the more exquisite control that higher-WMC subjects had been normally able to use to configure their attention discontiguously.

These memory-load findings are consistent with those from Ahmed and de Fockert (2012, Experiment 2), who examined interference effects in a modified flanker task with higher- and lower-WMC groups under different working-memory load conditions. For low load, subjects recalled six digits in ascending order and for high load, six digits in random nonsequential order. In this flanker task, subjects reported whether the target on each trial was an X or a Z, while another (congruent or incongruent) flanking letter appeared above, below, or to the left or right of this target. The flanker was placed at one of four distances from the target on a given trial. Lower-WMC subjects and those under high load showed patterns of interference consistent with a more dispersed field of visual attention. That is, under these conditions, subjects were less able to constrain their attention to just the likely target location than were higher-WMC and low-load subjects, showing greater interference from incongruent flankers that appeared farther from the target.

In a more prototypical flanker task (Eriksen & Eriksen, 1974), subjects must identify a central target stimulus among a fixed horizontal array of flanking nontarget stimuli, and the flankers can match or mismatch the target. In congruent trials, the target and flanking stimuli match (e.g., SSSSS); in incongruent trials, the target and flanking stimuli do not match (e.g., HHSHH). Responding is slowed on incongruent trials compared to congruent trials and, because the target is typically in the centre of the horizontal array, uncertainty of target stimulus location, or visual search, cannot explain this slowing.

Higher-WMC subjects are also faster to constrain their attention to targets amid flanking distractors (Heitz & Engle, 2007). Heitz and Engle's (2007) design was based on previous work by Gratton, Coles, Sirevaag, Eriksen, and Donchin (1988), who examined the time course of information processing. The Gratton et al. results were consistent with a dynamic spotlight (or zoom lens) view of attention, where attention starts out in a diffuse state that permits information from both the target and distractors to enter the system. As time elapses, the focus of attention closes in on the target, limiting the influence of the response-incongruent distractors. Gratton et al. demonstrated this by examining conditional accuracy functions. In both congruent and incongruent trials, the fastest bin of trials was completed with chance accuracy. On incongruent but not congruent trials, the next fastest bin was performed at *below-chance* accuracy. This dip below chance suggests that visual focus has yet to be constrained to the target, so responses were based on the identity of the (incongruent) flanking stimuli. After this below-chance dip, performance gradually increased, and at the slowest RTs, the flanker effect (the difference in accuracy between congruent and incongruent trials) disappeared.

Heitz and Engle's (2007) findings paralleled those of Gratton et al. (1988), but of most importance, higher- and lower-WMC groups differed in performance at the intermediate time points on incongruent trials, while performing similarly on very fast and very slow incongruent trials and congruent trials. This pattern of results—equivalence at the fastest and slowest RTs, but WMC-related differences at the intermediate time points—provides evidence that higher-WMC subjects constrained their focus to the target more quickly than did lower-WMC subjects.

Conclusions

WMC is a strong predictor of higher-order cognition, due in part to the contributions of attention-control capabilities. We propose that these attentional-control mechanisms are arranged hierarchically with goal maintenance at the top and task-specific competition-resolution mechanisms, and the speed and precision of visual attention, below. In many ways, this proposal is similar to the Botvinick et al. (2001) conflict adaptation model, with (at least) one critical difference: WMC-related cognitive control is initiated in a more top-down manner (from goals instantiated from task objectives) rather than in response to local conflict. Variation in WMC predicts who is best able to maintain (or activate) the goal as well as implement some forms of attentional control to combat response competition. A theme that runs through much of the work presented here is that the WMC-attentional-control relationship is moderated by context, with WMC-related individual differences only being revealed in specific contexts, such as those that are rich in interference. More work is needed, however, to refine our understanding of the thresholds between contexts that trigger WMC-related control and those that do not.

References

Ahmed, L., & de Fockert, J. W. (2012). Focusing on attention: The effects of working memory capacity and load on selective attention. *PLoS ONE*, *7*(8), e43101.

Anderson, B. (2011). There is no such thing as attention. *Frontiers in Psychology*, *2*, 246.

Bailey, H., Dunlosky, J., & Kane, M. J. (2008). Why does working memory span predict complex cognition? Testing the strategy affordance hypothesis. *Memory & Cognition*, *36*(8), 1383–1390.

Bleckley, M. K., Durso, F. T., Crutchfield, J. M., Engle, R. W., & Khanna, M. M. (2003). Individual differences in working memory capacity predict visual attention allocation. *Psychonomic Bulletin and Review*, *10*, 884–889.

Bleckley, M. K., Foster, J. L., & Engle, R. W. (2015). Working memory capacity accounts for the ability to switch between object-based and location-based allocation of visual attention. *Memory & Cognition*, *43*(3), 379–388.

Braver, T. S., Gray, J. R., & Burgess, G. C. (2007). Explaining the many varieties of working memory variation: Dual mechanisms of cognitive control. In A. R. A. Conway, C. Jarrold, M. J. Kane, A. Miyake, & J. N. Towse (Eds.), *Variation in Working Memory* (pp. 76–106). New York: Oxford University Press.

Botvinick, M. M., Braver, T. S., Barch, D. M., Carter, C. S., & Cohen, J. D. (2001). Conflict monitoring and cognitive control. *Psychological Review*, *108*(3), 624–652.

Bühner, M., König, C., Pick, M., & Krumm, S. (2006). Working memory dimensions as differential predictors of the speed and error aspect of multitasking performance. *Human Performance*, *19*(3), 253–275.

Colom, R., Abad, F. J., Quiroga, M. Á., Shih, P. C., & Flores-Mendoza, C. (2008). Working memory and intelligence are highly related constructs, but why? *Intelligence*, *36*(6), 584–606.

Conway, A. R. A., Cowan, N., Bunting, M. F., Therriault, D., & Minkoff, S. (2002). A latent variable analysis of working memory capacity, short term memory capacity, processing speed, and general fluid intelligence. *Intelligence*, *30*, 163–183.

Conway, A. R., Kane, M. J., Bunting, M. F., Hambrick, D. Z., Wilhelm, O., & Engle, R. W. (2005). Working memory span tasks: A methodological review and user's guide. *Psychonomic Bulletin & Review*, *12*(5), 769–786.

Cronbach, L. J. (1957). The two disciplines of scientific psychology. *American Psychologist*, *12*, 671–684.

Cronbach, L. J., & Meehl, P. E. (1955). Construct validity in psychological tests. *Psychological Bulletin*, *52*, 281–302.

Daneman, M., & Carpenter, P. A. (1980). Individual differences in working memory and reading. *Journal of Verbal Learning and Verbal Behavior*, *19*, 450–466.

Dunlosky, J., & Kane, M. J. (2007). The contributions of strategy use to working memory span: A comparison of strategy-assessment methods. *Quarterly Journal of Experimental Psychology*, *60*(9), 1227–1245.

Egner, T. (2007). Congruency sequence effects and cognitive control. *Cognitive, Affective, & Behavioral Neuroscience*, *7*(4), 380–390.

Engle, R. W., Carullo, J. J., & Collins, K. W. (1991). Individual differences in working memory for comprehension and following directions. *Journal of Educational Research*, *84*, 253–262.

Engle, R. W., & Kane, M. J. (2004). Executive attention, working memory capacity, and a two-factor theory of cognitive control. *Psychology of Learning and Motivation*, *44*, 145–199.

Engle, R. W., Tuholski, S. W., Laughlin, J. E., & Conway, A. R. A. (1999). Working memory, short-term memory and general fluid intelligence: A latent variable approach. *Journal of Experimental Psychology: General*, *128*, 309–331.

Eriksen, B. A., & Eriksen, C. W. (1974). Effects of noise letters upon the identification of a target in a non-search task. *Perception and Psychophysics*, *16*, 143–149.

Gratton, G., Coles, M. G., & Donchin, E. (1992). Optimizing the use of information: Strategic control of activation of responses. *Journal of Experimental Psychology: General, 121*(4), 480–506.

Gratton, G., Coles, M. G. H., Sirevaag, E. J., Eriksen, C. W., & Donchin, E. (1988). Prestimulus and poststimulus activation of response channels: A psychophyiological analysis. *Journal of Experimental Psychology: Human Perception and Performance, 14*, 331–344.

Hallett, P. E. (1978). Primary and secondary saccades to goals defined by instructions. *Vision Research, 18*(10), 1279–1296.

Hambrick, D. Z., Oswald, F. L., Darowski, E. S., Rench, T. A., & Brou, R. (2010). Predictors of multitasking performance in a synthetic work paradigm. *Applied Cognitive Psychology, 24*(8), 1149–1167.

Heitz, R. P., & Engle, R. W. (2007). Focusing the spotlight: Individual differences in visual attention control. *Journal of Experimental Psychology: General, 136*, 217–240.

Heitz, R. P., Schrock, J. C., Payne, T. W., & Engle, R. W. (2008). Effects of incentive on working memory capacity: Behavioral and pupillometric data. *Psychophysiology, 45*(1), 119–129.

Kane, M. J., Bleckley, M. K., Conway, A. R. A., & Engle, R. W. (2001). A controlled-attention view of working-memory capacity. *Journal of Experimental Psychology: General, 130*, 169–183.

Kane, M. J., Brown, L. E., McVay, J. C., Silvia, P. J., Myin-Germeys, I., & Kwapil, T. R. (2007). For whom the mind wanders, and when: An experience-sampling study of working memory and executive control in daily life. *Psychological Science, 18*, 614–621.

Kane, M. J., Conway, A. R. A., Hambrick, D. Z., & Engle, R. W. (2007). Variation in working memory as variation in executive attention and control. In A. R. A. Conway, C. Jarrold, M. J. Kane, A. Miyake, & J. N. Towse (Eds.), *Variation in working memory* (pp. 21–48). New York: Oxford University Press.

Kane, M. J., & Engle, R. W. (2003). Working-memory capacity and the control of attention: The contributions of goal neglect, response competition, and task set to Stroop interference. *Journal of Experimental Psychology: General, 132*(1), 47–70.

Kane, M. J., Hambrick, D. Z., Tuholski, S. W., Wilhelm, O., Payne, T. W., & Engle, R. W. (2004). The generality of working memory capacity: A latent-variable approach to verbal and visuospatial memory span and reasoning. *Journal of Experimental Psychology-General, 133*(2), 189–217.

Kane, M. J., Poole, B. J., Tuholski, S. W., & Engle, R. W. (2006). Working memory capacity and the top-down control of visual search: Exploring the boundaries of 'executive attention'. *Journal of Experimental Psychology: Learning, Memory, and Cognition, 32*, 749–777.

Keye, D., Wilhelm, O., Oberauer, K., & van Ravenzwaaij, D. (2009). Individual differences in conflict-monitoring: Testing means and covariance hypothesis about the Simon and the Eriksen Flanker task. *Psychological Research, 73*(6), 762–776.

Klein, K., & Fiss, W. H. (1999). The reliability and stability of the Turner and Engle working memory task. *Behavior Research Methods, Instruments and Computers, 31*, 429–432.

Kornblum, S. (1992). Dimensional overlap and dimensional relevance in stimulus-response and stimulus- stimulus compatibility. In G. E. Stelmach & J. Requin (Eds.), *Tutorials in motor behavior II* (pp. 743–777). Amsterdam, North Holland: Elsevier Press.

Kornblum, S., Hasbroucq, T., & Osman, A. (1990). Dimensional overlap: Cognitive basis for stimulus-response compatibility—A model and taxonomy. *Psychological Review, 97*, 253–270.

Kyllonen, P. C., & Stephens, D. L. (1990). Cognitive abilities as determinants of success in acquiring logic skill. *Learning and Individual Differences, 2*, 129–160.

Levinson, D. B., Smallwood, J., & Davidson, R. J. (2012). The persistence of thought evidence for a role of working memory in the maintenance of task-unrelated thinking. *Psychological Science, 23*(4), 375–380.

Luck, S. J., & Vogel, E. K. (1997). The capacity of visual working memory for features and conjunctions. *Nature, 390*(6657), 279–281.

Martínez, K., & Colom, R. (2009). Working memory capacity and processing efficiency predict fluid but not crystallized and spatial intelligence: Evidence supporting the neural noise hypothesis. *Personality and Individual Differences, 46*, 281–286.

McVay, J. C., & Kane, M. J. (2009). Conducting the train of thought: Working memory capacity, goal neglect, and mind wandering in an executive-control task. *Journal of Experimental Psychology: Learning, Memory, and Cognition, 35*(1), 196–204.

McVay, J. C., & Kane, M. J. (2012a). Drifting from slow to 'd'oh!': Working memory capacity and mind wandering predict extreme reaction times and executive control errors. *Journal of Experimental Psychology: Learning, Memory, and Cognition, 38*(3), 525–549.

McVay, J. C., & Kane, M. J. (2012b). Why does working memory capacity predict variation in reading comprehension? On the influence of mind wandering and executive attention. *Journal of Experimental Psychology: General, 141*(2), 302–320.

Meehl, P. E. (1990). Appraising and amending theories: The strategy of Lakatosian defense and two principles that warrant it. *Psychological Inquiry, 1*(2), 108–141.

Meier, M. E., & Kane, M. J. (2013). Working memory capacity and Stroop interference: Global versus local indices of executive control. *Journal of Experimental Psychology: Learning, Memory, and Cognition, 39*(3), 748–759.

Meier, M. E., & Kane, M. J. (2015). Carving executive control at its joints: Working memory capacity predicts stimulus-stimulus, but not stimulus-response, conflict. Journal of Experimental Psychology*: Learning, Memory, and Cognition.*

Melby-Lervåg, M., & Hulme, C. (2013). Is working memory training effective? A meta-analytic review. *Developmental Psychology, 49*(2), 270–291.

Miyake, A., & Friedman, N. P. (2012). The nature and organization of individual differences in executive functions: Four general conclusions. *Current Directions in Psychological Science, 21*, 8–14.

Oberauer, K., Süß, H. M., Schulze, R., Wilhelm, O., & Wittmann, W. W. (2000). Working memory capacity—Facets of a cognitive ability construct. *Personality and Individual Differences, 29*(6), 1017–1045.

Poole, B. J., & Kane, M. J. (2009). Working memory capacity predicts the executive control of visual search among distractors: The influences of sustained and selective attention. *Quarterly Journal of Experimental Psychology, 62*, 1430–1454.

Rummel, J., & Boywitt, C. D. (2014). Controlling the stream of thought: Working memory capacity predicts adjustment of mind-wandering to situational demands. *Psychonomic Bulletin & Review, 21*(5), 1309–1315.

Schweizer, K., & Moosbrugger, H. (2004). Attention and working memory as predictors of intelligence. *Intelligence, 32*(4), 329–347.

Shipstead, Z., Lindsey, D. R., Marshall, R. L., & Engle, R. W. (2014). The mechanisms of working memory capacity: Primary memory, secondary memory, and attention control. *Journal of Memory and Language, 72*, 116–141.

Shute, V. J. (1991). Who is likely to acquire programming skills? *Journal of Educational Computing Research, 7*, 1–24.

Sobel, K. V., Gerrie, M. P., Poole, B. J., & Kane, M. J. (2007). Individual differences in working memory capacity and visual search: The roles of top-down and bottom-up processing. *Psychonomic Bulletin & Review, 14*, 840–845.

Stroop, J. R. (1935). Studies of interference in serial verbal reactions. *Journal of Experimental Psychology, 18*, 643–662.

Treisman, A. M., & Gelade, G. (1980). A feature-integration theory of attention. *Cognitive Psychology, 12*, 97–136.

Turley-Ames, K. J., & Whitfield, M. M. (2003). Strategy training and working memory task performance. *Journal of Memory and Language, 49*, 446–468.

Turner, M. L., & Engle, R. W. (1989). Is working memory capacity task dependent? *Journal of Memory and Language, 28*, 127–154.

Underwood, B. J. (1975). Individual differences as a crucible in theory construction. *American Psychologist, 30*, 128–134.

Unsworth, N. (2015). Consistency of attentional control as an important cognitive trait: A latent variable analysis. *Intelligence, 49*, 110–128.

Unsworth, N., Brewer, G. A., & Spillers, G. J. (2012). Variation in cognitive failures: An individual differences investigation of everyday attention and memory failures. *Journal of Memory and Language, 67*(1), 1–16.

Unsworth, N., Fukuda, K., Awh, E., & Vogel, E. K. (2014). Working memory and fluid intelligence: Capacity, attention control, and secondary memory retrieval. *Cognitive Psychology, 71*, 1–26.

Unsworth, N., Heitz, R. C., Schrock, J. C., & Engle, R. W. (2005). An automated version of the operation span task. *Behavior Research Methods, 37*, 498–505.

Unsworth, N., & McMillan, B. D. (2013). Mind wandering and reading comprehension: Examining the roles of working memory capacity, interest, motivation, and topic experience. *Journal of Experimental Psychology: Learning, Memory, and Cognition, 39*(3), 832–842.

Unsworth, N., McMillan, B. D., Brewer, G. A., & Spillers, G. J. (2012). Everyday attention failures: An individual differences investigation. *Journal of Experimental Psychology: Learning, Memory, and Cognition, 38*(6), 1765–1772.

Unsworth, N., Redick, T. S., Lakey, C. E., & Young, D. L. (2010). Lapses in sustained attention and their relation to executive control and fluid abilities: An individual differences investigation. *Intelligence, 38*, 111–122.

Unsworth, N., Redick, T. S., Spillers, G. J., & Brewer, G. A. (2012). Variation in working memory capacity and cognitive control: Goal maintenance and microadjustments of control. *The Quarterly Journal of Experimental Psychology, 65*(2), 326–355.

Unsworth, N., Schrock, J. C., & Engle, R. W. (2004). Working memory capacity and the antisaccade task: Individual differences in voluntary saccade control. *Journal of Experimental Psychology: Learning, Memory, and Cognition, 30*, 1302–1321.

Weldon, R. B., Mushlin, H., Kim, B., & Sohn, M. H. (2013). The effect of working memory capacity on conflict monitoring. *Acta Psychologica, 142*(1), 6–14.

Wolfe, J. M. (1994). Guided search 2.0: A revised model of visual search. *Psychonomic Bulletin & Review, 1*, 202–238.

4

Conflict Adaptation

Past, Present, and Future of the Congruency Sequence Effect as an Index of Cognitive Control

Tobias Egner

Introduction

An influential conception of cognitive control is that is represents the ability to employ internal goals for the top-down biasing of information processing when there is a potential conflict between competing representations or task demands (e.g., Miller & Cohen, 2001; Chapter 14 by Cohen in this volume). The latter situation is particularly pronounced when goal-relevant processing pertains to a novel or weakly established behaviour that has to compete against a well-learned or habitual response (e.g., Norman & Shallice, 1986). Accordingly, popular tests of cognitive control typically require participants to produce an uncommon, instructed response to a task-relevant stimulus feature (target information) in the presence of task-irrelevant stimulus features (distractor information) that have strong habitual response associations, and which can be either congruent or incongruent (i.e., conflicting) with the former. For instance, in the classic colour-naming Stroop task (MacLeod, 1991; Stroop, 1935), subjects have to indicate the ink colour of written colour words (e.g., RED) while ignoring the words' meaning. The relative success of controlled processing is gauged by contrasting performance on trials where the distractor is congruent (e.g., the word RED written in red ink) with those where it is incongruent (e.g., the word RED written in blue ink) with the target, and may therefore interfere with target processing unless it is effectively ignored. The canonical finding is a marked congruency effect: responses are slower and more error prone to incongruent compared to congruent stimuli. Slowed responses on correct incongruent trials may therefore be taken as an index of the relative effectiveness of cognitive control: A stronger engagement of top-down control should be associated with a smaller congruency effect (e.g., Cohen, Dunbar, & McClelland, 1990).

Importantly, the size of the congruency effect, and by implication, the effectiveness of top-down control, has been shown to be malleable by a variety of contextual factors. One of these consists of the frequency of incongruent stimulus occurrences—for instance, mean congruency effects are reduced in blocks of trials that contain a high proportion of incongruent trials compared to blocks where incongruent trials are rare (Logan & Zbrodoff, 1979). These types of longer-term contextual influences on congruency effects are the topic of Chapter 5 by Bugg in this volume. Another, more short-term factor that modulates the

The Wiley Handbook of Cognitive Control, First Edition. Edited by Tobias Egner.

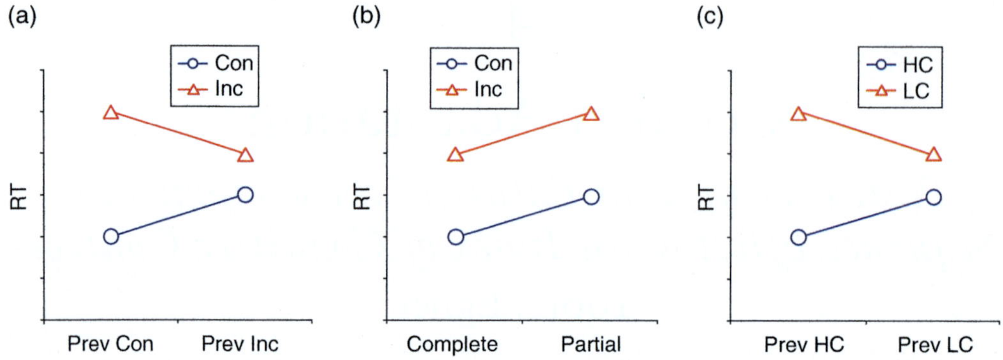

Figure 4.1 Different interpretations of the congruency sequence effect (CSE). (a) Hypothetical data plotted and labelled as a congruency sequence effect (CSE), where response time (RT) varies as a function of an interaction between previous trial congruency and current trial congruency. Con = congruent; Inc = incongruent; Prev = previous trial. (b) The same data as in (a), plotted in line with the feature integration account, whereby the data pattern represents two additive effects of congruency and feature repetitions. Complete = complete feature repetitions or complete feature alternations from previous trial; Partial = partial feature repetitions from previous trial. (**c**) The same data and plotting as in (**a**), but relabelled in line with the interpretation of the contingency learning account, where the data are explained as an interaction between previous trial and current trial contingency effects. LC = low contingency trials; HC = high contingency trials. Source: Egner 2014. Reproduced with permission of Egner.

congruency effect, however, is the congruency of the previous trial (Gratton, Coles, & Donchin, 1992): The influence of distractors on the processing of target information is reduced on trials that follow an incongruent trial compared to those that follow a congruent trial (for reviews, see Duthoo, Abrahamse, Braem, Boehler, & Notebaert, 2014; Egner, 2007), leading to a reduced congruency effect following an incongruent trial (Figure 4.1a). This effect represents the topic of the present chapter; it is often referred to as 'conflict adaptation' (thus, the chapter title), but in the following I will instead employ the theoretically more neutral term 'congruency sequence effect' (CSE). The conflict adaptation moniker derives from the fact that the CSE has been cast as reflecting a key signature of cognitive control regulation by the highly influential *conflict-monitoring model* (Botvinick, Braver, Barch, Carter, & Cohen, 2001). The model's focus on the CSE has steered a huge amount of attention towards this phenomenon over the last 15 years, resulting in a vast literature on the determinants, boundary conditions, and neural correlates of the CSE. An exhaustive review of this literature lies beyond the scope of a brief book chapter, so I will here instead provide a selective summary of key developments in this field, followed by an attempted conceptual integration of some of the varying perspectives on the CSE (see also Egner, 2014).

The central theoretical debate that has driven empirical work on the CSE has revolved around rival explanations for this phenomenon that can be grouped into two major categories, 'top-down control-based' versus 'bottom-up associative' accounts (see, e.g., Blais, Robidoux, Risko, & Besner, 2007; Botvinick et al., 2001; Egner, 2007; Hommel, Proctor, & Vu, 2004; Mayr, Awh, & Laurey, 2003; Schmidt, 2013; Ullsperger, Bylsma, & Botvinick, 2005; Verguts & Notebaert, 2008). I will provide brief summaries of these perspectives, as well as hybrid accounts, followed by a sketch of the current state of play in terms of empirical evidence. I will then briefly discuss findings from the neuroscience literature before trying to place the kind of cognitive control process that might be reflected in the CSE into a broader understanding of the brain's ability to produce adaptive thought and behaviour.

The Past: A Brief History of Rival Theoretical Accounts for the CSE

Control-Based Perspectives

In line with the standard interpretation of congruency effects as reflecting a measure of the effectiveness of cognitive control, the original observation of the CSE was interpreted as the expression of a strategic adjustment in attentional strategy (Gratton et al., 1992). More specifically, Gratton and colleagues (1992) reasoned that encountering a congruent or incongruent trial would engender an expectation for the forthcoming trial to be of the same congruency (cf. Remington, 1969), which in turn would lead subjects to strategically enhance (following incongruent trials) or decrease (following congruent trials) their attentional focus on the target stimulus feature (e.g., ink colour in the Stroop task). This in turn would decrease the influence of distractors (e.g., word meaning) and, *ergo*, the congruency effect, following an incongruent trial, and increase the influence of distractors (and, *ergo*, the congruency effect) following a congruent trial. This strategic, expectation-driven interpretation of the CSE was strengthened by the fact that the same authors observed a similar reduction in the congruency effect when they provided explicit congruency cues to participants prior to stimulus onset (Gratton et al., 1992).

About a decade later, the conflict-monitoring model put forward a related but more formalised (and influential) control-based account for the CSE (Botvinick et al., 2001). Botvinick and colleagues advanced an elegant computational scheme for a 'homunculus-free' regulation of top-down control, positing that the cognitive apparatus detects internal conflict between mutually incompatible stimulus or response representations (e.g., the simultaneous urge to answer 'red' and 'blue' when faced with the incongruent Stroop stimulus described above), and uses the degree of conflict to produce commensurate adjustments in top-down biasing—the more conflict is experienced, the more control will be applied. Thus, conflict caused by incongruent distractors triggers an up-regulation of attentional focus towards the target, which results in more efficient attentional selection (and hence, a smaller congruency effect) on the following trial; the opposite is true when encountering a low-conflict, congruent trial, which results in a relaxation of attention and, thus, less efficient attentional selection (and a larger congruency effect) on the following trial. This represents the 'conflict adaptation' account of the CSE (Figure 4.1a).

Although there are clear conceptual differences between the expectation- and conflict-based accounts (see, e.g., Duthoo, Wuhr, & Notebaert, 2013; Egner, 2007; Egner, Ely, & Grinband, 2010; Jimenez & Mendez, 2013), for the present purpose they can both be considered core members of the 'control-based' model category, along with various proposed refinements and extensions of the basic conflict-monitoring proposal. The latter have included the proposal of multiple, partially independent conflict-control loops that deal with different types of conflicts (e.g., stimulus-based vs. response-based; Egner, 2008; Egner, Delano, & Hirsch, 2007); the idea that conflict is processed as an aversive learning signal (Botvinick, 2007; Dreisbach & Fischer, 2012); the notion that conflict-driven control plays out primarily at the level of task set reinforcement (Hazeltine, Lightman, Schwarb, & Schumacher, 2011); and the inclusion of a flexible, self-adjusting learning rate in the conflict-monitoring model (Jiang, Heller, & Egner, 2014).

Regardless of their differences, however, these views commonly assume that the CSE results from strategic adjustments in top-down attention or task set. Second, and more importantly, these accounts all operate at a level of processing adjustment that is, in principle, independent of what the specific stimulus features or responses are that will comprise the subsequent trial. For instance, conflict- (or expectation-) triggered enhanced attentional filtering in the Stroop task would result in improved ink-colour selection regardless of the exact nature of that colour or the distractor word information in the upcoming trial. As will be discussed in more detail later on, this 'level of abstraction' of the mechanism that is held responsible for the CSE, concerning either a *generalisable cognitive state* (for instance, attentional focus) or *specific stimulus or response characteristics*, represents the key distinguishing feature between control-based and associative accounts of the CSE.

According to this criterion, we can also subsume under the control-based category CSE accounts that focus on control adjustments at the level of response selection rather than perceptual attention. For instance, the 'activation-suppression model' posits a control mechanism that detects and suppresses response activation elicited by distractor stimuli, and assumes this mechanism to work more effectively if it had been recently activated by a prior incongruent trial (e.g., Ridderinkhof, 2002; van den Wildenberg et al., 2010). Crucially, this model nevertheless assumes that the process that is facilitated following an encounter with an incongruent stimulus, namely, the categorical suppression of 'distractor-route' responses, is independent of the specific features of the subsequent stimulus.

Associative Perspectives

Associative accounts have proposed that the CSE may stem from bottom-up memory-driven effects, based on differing frequencies with which specific stimulus and response features repeat over consecutive trials for different congruency sequences (Hommel et al., 2004; Mayr et al., 2003). For instance, Hommel et al.'s (2004) *feature-integration* account is grounded in prior work showing that the specific stimulus and response features that co-occur on a given trial of an alternative forced-choice (AFC) task (say, the word RED in blue ink is responded to with a left button press in the above-mentioned Stroop task) become ***bound together*** in episodic memory as an 'event file' (Hommel, 1998, 2004; cf. Treisman & Gelade, 1980). Moreover, the subsequent re-occurrence of any one of these features (e.g., the word RED) appears to trigger the retrieval of the entire prior event file, presumably to supply a potential shortcut to the correct response associated with a previously seen stimulus (cf. Logan 1988). This feature-binding mechanism leads to a relative facilitation of processing when all of the current trial features match the previous event (***complete repetitions***; see also Mayr et al., 2003; Pashler & Baylis, 1991), or when there is no feature overlap across successive trials (***complete alternations***), relative to cases where some features are repeated but others are not (***partial repetitions***), because in the latter scenario, the retrieved event file has to be either discarded or 'unbound' in order for the currently presented stimulus to be responded to correctly (Hommel, 1998, 2004; see also Neill, 1997).

By way of example, consider once more the Stroop task alluded to earlier, consisting of a stimulus set of the words RED and BLUE, printed in either red or blue ink, thus rendering a total of two congruent and two incongruent stimuli, and two possible responses. Here, congruent-congruent and incongruent-incongruent trial sequences will consist entirely of complete feature repetitions or complete alternations (and thus, result in fast responses), whereas congruent-incongruent and incongruent-congruent trial sequences will consist entirely of partial feature repetitions (thus resulting in slow responses). Hence, the CSE data pattern of reduced congruency effects following an incongruent trial compared to a congruent trial (Figure 4.1a) can be re-interpreted as reflecting a main effect of congruency paired with a relative handicapping of trials where partial feature repetitions impose an 'unbinding cost' on performance (Figure 4.1b; Hommel et al., 2004). Evidently, in contrast to the control-based accounts, this associative perspective requires no trial-by-trial adjustments of selective attention to explain the CSE, and, importantly, the proposed mechanism underlying this effect operates at the level of specific stimulus features and responses.

Given that the feature-integration account highlights potential associative confounds in the CSE that seem to be specifically inherent in small stimulus and response sets, a natural response to these concerns was a movement towards employing conflict tasks with larger sets (typically, moving from 2-AFC to 4-AFC schemes), such that first-order repetitions of stimulus and response features could be either prophylactically prevented from occurring (e.g., Jimenez & Mendez, 2013; Puccioni & Vallesi, 2012), or removed from analysis after the fact (e.g., Akcay & Hazeltine, 2007; Ullsperger et al., 2005). However, as recently highlighted by several authors (Mordkoff, 2012; Schmidt & De Houwer, 2011), this trend may have introduced a new associative confound to the CSE, in the form of *contingency learning*. Specifically, the expansion of the stimulus set (for instance, going from two to four colours in the Stroop task) creates more possible unique incongruent than congruent stimuli. When researchers then present congruent and incongruent trials with the same frequency (i.e., 50%), each congruent stimulus occurs more frequently (and well above chance) than each incongruent stimulus, which creates a contingency linking each distractor to their congruent response (e.g., the word RED is most frequently paired with the colour red, and thus, the response 'red'). Because high-contingency (congruent) trials are responded to faster than low-contingency (incongruent) trials, *and* consecutive trials with the same contingency level appear to facilitate performance (Schmidt & De Houwer, 2011), it is possible that the CSE in typical 4-AFC tasks is a reflection of contingency learning rather than of control-based processing adjustments (Figure 4.1c).

Hybrid Perspectives

In addition to bottom-up associative accounts that see no role for cognitive control in the CSE, some 'hybrid accounts' espouse the notion of conflict-enhanced control, but link this mechanism associatively to specific stimulus features (Blais et al., 2007; Blais & Verguts, 2012; Verguts & Notebaert, 2008). Thus, whereas control-based accounts of the CSE have proposed that the effect derives from an adaptation that occurs at a relatively abstract level (e.g., conflict leading to enhanced top-down control over ink-colour processing), these hybrid models have argued instead that this effect blends control and learning at a more concrete or local level. Namely, these models have shown that the CSE can be successfully simulated based on the assumption of a *stimulus-specific* learning process (Blais et al., 2007; Blais & Verguts, 2012; Verguts & Notebaert, 2008). For instance, the 'adaptation-by-binding' model proposes that the occurrence of conflict triggers an arousal response that enhances the association between top-down task-relevant biasing processes and the currently activated input units, which would correspond to the *specific stimulus features* of the present incongruent stimulus (Verguts & Notebaert, 2008, 2009). The authors propose that this form of 'conflict-modulated Hebbian learning', binding top-down attention to conflict-evoking stimuli, accounts for the CSE as reflecting repeated strengthening of connections between task-relevant top-down biasing and recurring stimulus features (see also Blais et al., 2007). When no stimulus features are repeated over successive trials, the model attributes the occurrence of a CSE to the assumption that input units for task-relevant features (e.g., different colours in the Stroop task) that are not actually part of the current stimulus are also slightly activated, such that even nonpresent stimulus features are held to be subject to conflict-modulated Hebbian learning (Verguts & Notebaert, 2008).

The Present: Empirical Status Quo of Determinants and Boundary Conditions of the CSE

Bottom-Up and Top-Down Contributions to the CSE Can Be Isolated

The empirical evaluation of control-based and associative sources of the CSE has produced a substantial literature over the last decade, the nuances of which are discussed elsewhere in greater detail than I can provide here (e.g., Duthoo et al., 2014; Egner, 2007; Hazeltine et al., 2011; Schmidt, 2013; Weissman, Jiang, & Egner, 2014); instead, the present section summarises what I consider to be up-to-date take-home messages of that literature. First, there is little doubt that the nature of overlap in stimulus and response features over successive trials can profoundly affect performance (Hommel, 1998), and it is *impossible* to circumvent the confounding factor of differential feature overlap between different congruency sequences in the CSE when employing small stimulus sets (e.g., only two or three different target and distractor stimuli). It is therefore possible, or even likely, that CSEs observed in studies with such small stimulus sets are partly, predominantly, or entirely driven by feature-integration effects (e.g., Mayr et al., 2003; Hommel et al., 2004; Nieuwenhuis et al., 2006; Notebaert, Gevers, Verbruggen, & Liefooghe, 2006).

Second, while the movement towards employing larger stimulus sets has resulted in a number of studies reporting CSEs in the absence of feature repetitions (e.g., Ullsperger et al., 2005; Akcay & Hazeltine, 2007, 2011; Hazeltine et al., 2011), almost all of these studies appear to be open to alternative interpretation based on possible contingency-learning confounds because of above-chance occurrence of congruent stimuli (see Schmidt, 2013). Similarly, 2-AFC studies that require subjects to categorise large sets of unique stimuli (e.g., classifying face stimuli according to gender) have produced CSEs in the absence of any stimulus feature repetitions (e.g., Egner et al., 2010; Egner, Etkin, Gale, & Hirsch, 2008; Egner & Hirsch, 2005; Jiang & Egner, 2014; Lee & Cho, 2013), but they have been criticised as being vulnerable to possible feature-integration effects operating at the level of semantic categories (like 'male' and 'female') rather than specific stimulus features (Schmidt, 2013; but see Jiang et al., 2014).

Third, however, a substantial crop of recent papers with designs that specifically control for both feature-integration *and* contingency-learning confounds have in fact reported robust CSEs (Freitas & Clark, 2014; Hengstler, Holland, van Steenbergen, & van Knippenberg, 2014; Kim & Cho, 2014; Kunde & Wuhr, 2006; Schmidt & Weissman, 2014; Weissman, Hawks, & Egner, 2016; Weissman et al., 2014; but see Mayr et al., 2003, Jimenez & Mendez, 2013). A typical design of this recent wave of studies circumvents both stimulus and response feature repetitions, as well as contingency-learning confounds, by splitting a 4-AFC task into two alternating 2-AFC tasks with non-overlapping stimulus and response sets (e.g., presenting alternately Stroop stimuli that are made up either of red/blue or of green/yellow combinations; e.g., Schmidt & Weissman, 2014). This approach has produced robust evidence for the basic presence of a 'memory confound-free' CSE, and current studies are exploring the precise boundary conditions for obtaining the effect under these constraints (e.g., Kim & Cho, 2014; Weissman et al., 2014). For instance, recent studies suggest that obtaining a control-based CSE may require a head-start for the processing of task-irrelevant (distractor) stimulus features, which has been argued to support the idea that the kind of control mechanism mediating the CSE is likely related to the suppression of the responses elicited by distractors (Weissman et al., 2014; Weissman, Egner, Hawks, & Link, 2015).

For the present purpose, the key conclusion from recent developments in this literature is that there is solid evidence that CSEs can be produced by both sources that operate at a *feature-specific* level, driven by (re-)occurrences of particular physical stimulus and response characteristics, as well as by sources that must operate at a more *abstract* level, producing CSEs in a manner that is independent of repetitions of specific stimulus and response characteristics, and therefore presumably reflect a form of conflict-driven control, or conflict adaptation. How these distinct contributors to the CSE may be conceptualised most fruitfully within a single framework is discussed further in later sections of this chapter.

The CSE Is Characterised by Domain Specificity

Apart from the fundamental question of the degree to which the CSE is a reflection of strategic control, probably the most investigated boundary condition of this effect in the literature concerns its generalisability. The basic question many researchers have asked is whether a particular type of conflict—for example, the clashing colour semantics in the Stroop task—would result in subsequent conflict reduction only with respect to that particular conflict (domain specificity) or whether exposure to an incongruent trial would lead to broad performance benefits regardless of the specifics of the task demands on the forthcoming trial. This addresses the question of what *kind* of top-down control process exactly might be involved in the CSE. An exhaustive review of that literature is beyond the scope of this chapter, and the interested reader is referred to detailed reviews and theory papers on this topic (Braem, Abrahamse, Duthoo, & Notebaert, 2014; Egner, 2008). However, the consensus arising from this line of investigation is reasonably clear, in that we can likely reject the notion of a global, or domain-general, adaptation effect. This rejection is based on a large number of studies that have found the CSE to be conflict specific (e.g., Akcay & Hazeltine, 2008, 2011; Egner et al., 2007; Funes, Lupianez, & Humphreys, 2010; Kunde, Augst, & Kleinsorge, 2012; Wendt, Kluwe, & Peters, 2006).

The logic of these studies rests on defining different types of conflict, typically derived from a taxonomy of interference effects developed by Kornblum and colleagues (Kornblum, Hasbroucq, & Osman, 1990), who distinguish between different types of 'dimensional overlap' in the processing of task-relevant versus task-irrelevant stimulus features (Figure 4.2). For example, one can distinguish between conflict generated by an overlap in task-relevant and task-irrelevant stimulus representations (e.g., in the Stroop task) and conflict generated through an overlap between a task-irrelevant stimulus feature and the task-relevant response feature (e.g., in the Simon task; Figure 4.2a, b). Accordingly, if one combines the Stroop and Simon tasks, by presenting colour words to the left or right of a central fixation and requiring subjects to respond to ink colour using left and right response buttons (Figure 4.2c), one obtains additive Simon and Stroop congruency effects (Jiang & Egner, 2014; Kornblum, 1994; Simon & Berbaum, 1990). However, CSEs in this type of task are conflict specific: Stroop congruency effects are reduced following a Stroop-incongruent stimulus, but not following a Simon-incongruent stimulus, and vice versa (Egner et al., 2007; Funes et al., 2010). Note that, given that the basic stimulus features in this type of task are shared between Stroop- and Simon-conflict factors, this domain specificity in the CSE is difficult to explain on the basis of bottom-up associative or hybrid accounts, because according to these perspectives there should be generalisable effects (either due to feature-integration or feature-control binding) across trials involving the same basic stimulus features.

Although there have been a small number of studies reporting generalised CSEs, these seem to either concern very special cases (like 'high personal relevance' of stimuli, (Kleiman, Hassin, & Trope, 2014) or have turned out not to be replicable (Freitas, Bahar, Yang, & Banai, 2007; Runger, Schwager, & Frensch, 2010). On the whole, therefore, much evidence suggests that control-based contributions to the CSE must operate at a relatively 'local', trial-type specific level rather than at a global, highly generalisable level, a conclusion that is also supported by an extensive recent review on this topic (Braem

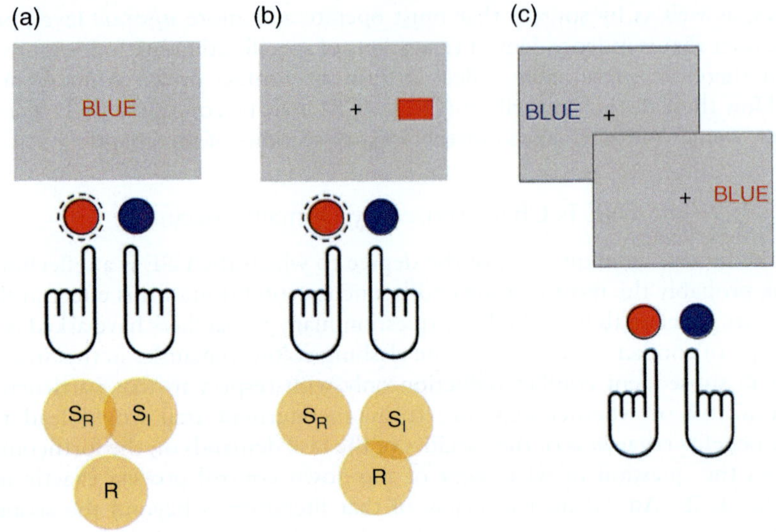

Figure 4.2 Assessing domain specificity of the congruency sequence effect. (a, b) The top panels show example stimuli for a colour word Stroop task (**a**) and a Simon task protocol (b) and associated response options, with the dashed-circled option indicating the correct response to the above stimulus. The bottom panels display potential sources of compatibility (conflict) effects for each task, by showing 'dimensional overlaps' between task-relevant stimulus dimensions (S_R), task-irrelevant stimulus dimensions (S_I), and the response dimension (R). In the colour-naming Stroop task: (a) subjects are required to respond to the ink colour (S_R) of a colourword, while ignoring the word's meaning (S_I). Ink colour and word meaning have a semantic overlap, and can therefore be compatible (e.g., the word RED printed in red) or incompatible (e.g., the word BLUE printed in red) with each other. In the Simon task: (b) subjects are required to respond to the colour of a stimulus (S_R) that can be displayed to the left or right of fixation, while ignoring the stimulus location (S_I). The stimulus location (left/right) has a dimensional overlap with the response dimension (R) (left/right hand button push), and can therefore be compatible (e.g., a red stimulus presented on the left-hand side) or incompatible (e.g., a red stimulus presented on the right-hand side) with the response. (c) A factorial task-crossing design combines Stroop and Simon tasks by presenting colour-word stimuli at peripheral locations. In this example, the first stimulus is compatible with respect to the Stroop stimulus dimension and incompatible with respect to the Simon stimulus dimension, whereas the second stimulus is incompatible with respect to both dimensions. Stroop and Simon compatibility effects in this protocol do not interact, and CSEs are conflict specific: each type of conflict facilitates the resolution of its own kind, but does not affect the resolution of the other kind. Source: Adapted from Egner 2008. Reproduced with permission of Elsevier.

et al., 2014). More specifically, several authors have concluded that the most plausible scenario appears to be that conflict in an ongoing task leads to a reinforcement of settings pertaining to the *task set* that is currently operative in guiding behaviour (Egner, 2008; Hazeltine et al., 2011).

Neural Correlates of the CSE

In addition to being greatly influential in the cognitive psychology literature, the conflict-monitoring model (Botvinick et al., 2001) also entailed specific neuroanatomical claims that have fostered many cognitive neuroscience investigations. Specifically, the model mapped the conflict-monitoring function onto the dorsal anterior cingulate cortex (dACC) and the top-down control function onto the dorsolateral prefrontal cortex (dlPFC). One of the great features of the CSE from the perspective of neuroimaging studies is that it allows for a relatively clean isolation of putative neural substrates of conflict processing and subsequent enhancement of top-down control, on the basis of the assumptions that control is low and conflict is high during incongruent trials that have been preceded by a congruent trial, and conversely that conflict is low owing to high control on incongruent trials that have been preceded by another incongruent trial. Thus, brain regions that are more active during the former than the latter trials are implicated in conflict processing, whereas brain regions that are more active during the latter than the former trials are implicated in conflict-driven top-down control.

Using this logic, a number of influential studies have in fact documented that the dACC is more active during incongruent trials that have been preceded by a congruent trial (e.g., Botvinick, Nystrom, Fissell, Carter, & Cohen, 1999) and that the dlPFC is more active during incongruent trials preceded by an incongruent trial (e.g., Egner & Hirsch, 2005). Moreover, dACC activation during an incongruent trial predicts the magnitude of dlPFC activation on the subsequent trial (Kerns et al., 2004) and that dlPFC activation in turn is associated with a top-down enhancement of task-relevant stimulus feature processing in the sensory cortex (Egner & Hirsch, 2005). An illustration of this putative conflict-control loop is shown in Figure 4.3. An exhaustive evaluation of the literature dedicated to testing these claims lies beyond the scope and purpose of this chapter (but see Chapter 15 by Brown, Chapter 10 by Kool et al., and Chapter 13 by Stokes et al. in this volume), but it is probably fair to say that many neuroimaging studies have supported the basic idea that the dACC plays a central role in performance or outcome evaluation, and the dlPFC in imposing top-down control. Whether the exact computation performed by the dACC corresponds to that of conflict monitoring remains a topic of controversy, however (see Chapter 15 by Brown in this volume). For example, single-neuron studies in the monkey have been equivocal on this

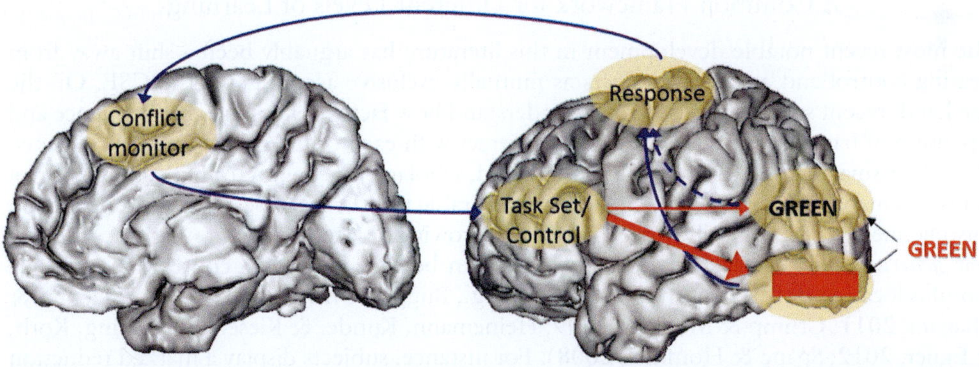

Figure 4.3 Neuroanatomical sketch of a putative conflict-control loop. According to the conflict-monitoring model, and supported by a number of empirical papers described in the main text, the CSE in a Stroop task would occur along the following lines: the instructed task goal ('name ink colour, ignore word meaning') is represented in dlPFC, which implements top-down control over stimulus processing in posterior regions by amplifying the processing of task-relevant (colour) information relative to distractor (word-form) information. As top-down control is imperfect, an incongruent Stroop stimulus (here, GREEN in red ink) will tend to activate both the correct ('red') and incorrect ('green') response, leading to conflict. This conflict is detected by the dACC, which in turn signals to the dlPFC to ramp up top-down biasing, which results in enhanced control (and less conflict) on the subsequent trial.

issue, with some studies finding no evidence of conflict-sensitive neurons in the dACC (e.g., Ito, Stuphorn, Brown, & Schall, 2003) and others locating such neurons in the supplementary eye fields instead (Stuphorn, Taylor, & Schall, 2000). However, more recent studies have produced support for a role of the dACC in conflict processing, with conflict-sensitive neurons observed in this region in both monkeys (Ebitz & Platt, 2015) and humans (Sheth et al., 2012).

As a caveat concerning the above inferences, however, it should be noted that none of these imaging and single-neuron studies have employed the type of memory confound-free experimental protocols discussed above. Another caveat concerns the fact that analysing neural data as a function of first-order sequence effects ignores the likely influence of higher-order sequential effects and does not explicitly tap into the putative learning process that is modelled by the conflict-monitoring theory. An interesting exception is a recent computational model that conceptually extends the conflict-monitoring model by employing a self-adjusting, volatility-modulated learning rate to optimally predict forthcoming conflict (Jiang et al., 2014). By employing the model variables of the learning rate, which determines the degree to which prior trial history impacts model predictions, and conflict prediction itself as trial-based modulators of neural activity in model-based fMRI analyses, these authors found that although conflict-driven adjustments in top-down control did involve the dACC and dlPFC, the learning process that underlies conflict prediction was dependent on the anterior insula and the caudate nucleus—suggesting a subcortical learning machinery underpinning frontal control processes (Jiang, Beck, Heller, & Egner, 2015). This model's focus on the intersection of learning and control processes makes for a coherent transition to what I see as the most likely future focus of work on the concept of conflict adaptation, as described in the last section of this chapter.

The Future? Learning, Sequence Effects, and Adaptive Behaviour

A Common Framework for Different Levels of Learning

The most recent notable development in this literature has arguably been a shift away from treating control and associative effects as mutually exclusive accounts for the CSE. On the one hand, recent studies have sought to understand how factors of congruency sequence and repetition of basic stimulus features might interact with each other (e.g., Spape & Hommel, 2008; Weissman et al., 2016). On the other hand, scholars have begun to think about control states in a manner that renders them more comparable to other features of events encoded in episodic memory. Specifically, a recent but fast-growing research literature has documented that *generalisable* (i.e., abstract) control states can become bound to contextual cues, like stimulus location or specific stimulus features (e.g., Bugg & Chanani, 2011; Bugg, Jacoby, & Chanani, 2011; Crump & Milliken, 2009; Heinemann, Kunde, & Kiesel, 2009; King, Korb, & Egner, 2012; Spape & Hommel, 2008). For instance, subjects display a marked reduction in Stroop congruency effects for stimuli presented at screen locations with a high incidence of incongruent Stroop trials compared to those presented in locations with a low incidence of incongruent trials (Crump & Milliken, 2009; see also Chapter 5 by Bugg in this volume).

One parsimonious way of thinking about these effects is to extend the idea of event files to encompass not just the forming of associations between concrete stimulus and response features, but also of more abstract, categorical stimulus features and, importantly, the linking of these features with co-occurring internal cognitive states, most pertinently, with the *attentional* or *control state* that was being engaged during the processing of said stimulus event. This proposal is, of course, in line with the basic idea expressed in the hybrid CSE models discussed above (Blais et al., 2007; Verguts & Notebaert, 2008). However, as noted in the discussion of conflict- and task-specificity of CSEs, the control states that are being incorporated into the event file appear to reflect *generalisable* control settings pertaining to the current task set (e.g., 'focus on colour') rather than just reinforcing the attentional priority of a specific stimulus feature (e.g., the colour red). A useful reconceptualisation of the traditional dichotomy between control-based and associative CSE accounts would therefore treat stimulus, response, and control features of an event as different in their *level of abstraction* (generalisability), but as being otherwise subject to a common learning process. To wit, the feature-integration and contingency-learning accounts deal with learning that links together *concrete* trial characteristics, namely, particular stimulus features (and perhaps categories) and responses; by contrast, the control-based accounts are concerned with learning aspects of a trial that transcend the physical specifics of the stimuli or responses, dealing instead with more *abstract* properties, like congruency, experienced conflict, and/or the cognitive mechanisms that were recruited for dealing with the latter.

Contextual Cuing of Control States

It is easy to appreciate how this kind of mechanism could account for situations where a specific stimulus or contextual feature, over repeated pairings, becomes associated with a task-appropriate control state. It is less obvious, however, how such context-control binding mechanism would account for a CSE in the absence of any concrete feature overlap (and thus, bottom-up cue) across trials. In order to explain *generalisable* control effects at the trial-by-trial level in the absence of concrete feature repetition, a second conceptual expansion of the event file scheme is required so as to include the concept of a *temporal context*, meaning that the retrieval or priming of particular stimulus-response links and/or control states can be based on a temporally defined frame the organism believes itself to be in (e.g., Braver & Barch, 2002; Koechlin, Ody, & Kouneiher, 2003). This suggestion essentially equates to extending the discrete and instantaneous nature of an 'event file' to a more dynamic and extended form of an 'episode file', which can encode and (upon retrieval) apply temporally extended contingencies and task sets (akin to *schemas*). This broader conception of context would allow for stimuli in the spatial and/or temporal vicinity of the current focus of processing to guide appropriate event/episode file retrieval. Thus, in the context of the CSE, encountering an incongruent trial sets a context under which the subjective likelihood of a forthcoming trial also being incongruent is enhanced, as is inherent in the accounts of Gratton et al. (1992) and Botvinick and colleagues (2001) and Jiang and colleagues (2014).

According to this developing framework, the local temporal context will prime (or maintain) the activation of the control state (as well as the lower-level stimulus and response features) that characterised the context-updating event (i.e., the previous trial). Therefore, even in the absence of any physical feature overlap across consecutive trials in a conflict task (say, moving from 'RED in blue' to 'GREEN in yellow'), the temporal context cue provided by an incongruent trial will facilitate the retrieval (or foster the maintenance) of the control state or task set associated with that trial for a period of time whose extent likely depends on the temporal statistics of the task environment (Egner et al., 2010). Note that the idea that a *temporal* context cue can determine the retrieval and temporary application of suitable processing strategies is not wedded to any particular view of what the exact trigger (e.g., conflict) or nature of the control state might be, and it naturally extends beyond the confines of the CSE. The empirical investigation of this type of integrative episodic learning account of the CSE is at an early stage but has produced some interesting initial findings, including the observation of an over-additive interaction between concrete and abstract feature learning on CSE magnitude, which is anticipated by this model (Weissman et al., 2016). Moreover, a recent fMRI study documented the shared representations of concrete and categorical stimulus features and abstract control state features of a trial event in the anterior hippocampus and dorsal striatum (Jiang, Brashier, & Egner, 2015), thus supplying neural evidence for the existence of event files that integrate external, physical, as well as internal, mental state features of an event.

In sum, the quest for the source of sequential modulations of the congruency effect has stimulated a plethora of accounts and experiments, some of which I have tried to summarise and assess in this chapter. The present consensus seems to be that the CSE *can* index an adaptation of control state, but only under carefully controlled conditions, while it is more typically attributable to a mesh of stimulus and response feature learning in conjunction with control-based adjustments. In the final section, I have laid out a theoretical framework that views these different contributions to the CSE as representing different levels of abstraction in the same learning scheme that serves a single, common goal: forming memory ensembles that will facilitate fast and appropriate responses to recurring stimuli or events in the environment.

References

Akcay, C., & Hazeltine, E. (2007). Conflict monitoring and feature overlap: Two sources of sequential modulations. *Psychonomic Bulletin & Review, 14*(4), 742–748.

Akcay, C., & Hazeltine, E. (2008). Conflict adaptation depends on task structure. *Journal of Experimental Psychology: Human Perception and Performance, 34*(4), 958–973.

Akcay, C., & Hazeltine, E. (2011). Domain-specific conflict adaptation without feature repetitions. *Psychonomic Bulletin & Review, 18*(3), 505–511.

Blais, C., Robidoux, S., Risko, E. F., & Besner, D. (2007). Item-specific adaptation and the conflict-monitoring hypothesis: A computational model. *Psychological Review, 114*(4), 1076–1086.

Blais, C., & Verguts, T. (2012). Increasing set size breaks down sequential congruency: Evidence for an associative locus of cognitive control. *Acta Psychologica (Amst), 141*(2), 133–139.

Botvinick, M. M. (2007). Conflict monitoring and decision making: Reconciling two perspectives on anterior cingulate function. *Cognitive, Affective, & Behavioral Neuroscience, 7*(4), 356–366.

Botvinick, M. M., Braver, T. S., Barch, D. M., Carter, C. S., & Cohen, J. D. (2001). Conflict monitoring and cognitive control. *Psychological Review, 108*(3), 624–652.

Botvinick, M. M., Nystrom, L. E., Fissell, K., Carter, C. S., & Cohen, J. D. (1999). Conflict monitoring versus selection-for-action in anterior cingulate cortex. *Nature, 402*(6758), 179–181.

Braem, S., Abrahamse, E. L., Duthoo, W., & Notebaert, W. (2014). What determines the specificity of conflict adaptation? A review, critical analysis, and proposed synthesis. *Frontiers in Psychology, 5*, 1134.

Braver, T. S., & Barch, D. M. (2002). A theory of cognitive control, aging cognition, and neuromodulation. *Neuroscience & Biobehavioral Reviews, 26*(7), 809–817.

Bugg, J. M., & Chanani, S. (2011). List-wide control is not entirely elusive: Evidence from picture-word Stroop. *Psychonomic Bulletin & Review, 18*(5), 930–936.

Bugg, J. M., Jacoby, L. L., & Chanani, S. (2011). Why it is too early to lose control in accounts of item-specific proportion congruency effects. *Journal of Experimental Psychology: Human Perception and Performance, 37*(3), 844–859.

Cohen, J. D., Dunbar, K., & McClelland, J. L. (1990). On the control of automatic processes: A parallel distributed processing account of the Stroop effect. *Psychological Review, 97*(3), 332–361.

Crump, M. J., & Milliken, B. (2009). The flexibility of context-specific control: Evidence for context-driven generalization of item-specific control settings. *Quarterly Journal of Experimental Psychology (Hove), 62*(8), 1523–1532.

Dreisbach, G., & Fischer, R. (2012). Conflicts as aversive signals. *Brain and Cognition, 78*(2), 94–98.

Duthoo, W., Abrahamse, E. L., Braem, S., Boehler, C. N., & Notebaert, W. (2014). The heterogeneous world of congruency sequence effects: An update. *Frontiers in Psychology, 5*, 1001.

Duthoo, W., Wuhr, P., & Notebaert, W. (2013). The hot-hand fallacy in cognitive control: Repetition expectancy modulates the congruency sequence effect. *Psychonomic Bulletin & Review, 20*(4), 798–805.

Ebitz, R. B., & Platt, M. L. (2015). Neuronal activity in primate dorsal anterior cingulate cortex signals task conflict and predicts adjustments in pupil-linked arousal. *Neuron, 85*(3), 628–640.

Egner, T. (2007). Congruency sequence effects and cognitive control. *Cognitive, Affective, & Behavioral Neuroscience, 7*, 380–390.

Egner, T. (2008). Multiple conflict-driven control mechanisms in the human brain. *Trends in Cognitive Sciences, 12*(10), 374–380.

Egner, T. (2014). Creatures of habit (and control): A multi-level learning perspective on the modulation of congruency effects. *Frontiers in Psychology, 5*, 1247.

Egner, T., Delano, M., & Hirsch, J. (2007). Separate conflict-specific cognitive control mechanisms in the human brain. *Neuroimage, 35*(2), 940–948.

Egner, T., Ely, S., & Grinband, J. (2010). Going, going, gone: Characterizing the time-course of congruency sequence effects. *Frontiers in Psychology, 1*, 154.

Egner, T., Etkin, A., Gale, S., & Hirsch, J. (2008). Dissociable neural systems resolve conflict from emotional versus nonemotional distracters. *Cerebral Cortex, 18*(6), 1475–1484.

Egner, T., & Hirsch, J. (2005). Cognitive control mechanisms resolve conflict through cortical amplification of task-relevant information. *Nature Neuroscience, 8*(12), 1784–1790.

Freitas, A. L., Bahar, M., Yang, S., & Banai, R. (2007). Contextual adjustments in cognitive control across tasks. *Psychological Science, 18*(12), 1040–1043.

Freitas, A. L., & Clark, S. L. (2014). Generality and specificity in cognitive control: Conflict adaptation within and across selective-attention tasks but not across selective-attention and Simon tasks. *Psychological Research, 79*(1), 143–162.

Funes, M. J., Lupianez, J., & Humphreys, G. (2010). Analyzing the generality of conflict adaptation effects. *Journal of Experimental Psychology: Human Perception and Performance, 36*(1), 147–161.

Gratton, G., Coles, M. G., & Donchin, E. (1992). Optimizing the use of information: Strategic control of activation of responses. *Journal of Experimental Psychology General, 121*(4), 480–506.

Hazeltine, E., Lightman, E., Schwarb, H., & Schumacher, E. H. (2011). The boundaries of sequential modulations: Evidence for set-level control. *Journal of Experimental Psychology: Human Perception and Performance, 37*(6), 1898–1914.

Heinemann, A., Kunde, W., & Kiesel, A. (2009). Context-specific prime-congruency effects: On the role of conscious stimulus representations for cognitive control. *Consciousness and Cognition, 18*(4), 966–976.

Hengstler, M., Holland, R. W., van Steenbergen, H., & van Knippenberg, A. (2014). The influence of approach-avoidance motivational orientation on conflict adaptation. *Cognitive, Affective, & Behavioral Neuroscience, 14*(2), 548–560.

Hommel, B. (1998). Event files: Evidence for automatic integration of stimulus-response episodes. *Visual Cognition, 5*, 183–216.

Hommel, B. (2004). Event files: Feature binding in and across perception and action. *Trends in Cognitive Sciences, 8*(11), 494–500.

Hommel, B., Proctor, R. W., & Vu, K. P. (2004). A feature-integration account of sequential effects in the Simon task. *Psychological Research, 68*(1), 1–17.

Ito, S., Stuphorn, V., Brown, J. W., & Schall, J. D. (2003). Performance monitoring by the anterior cingulate cortex during saccade countermanding. *Science, 302*(5642), 120–122.

Jiang, J., Beck, J., Heller, K., & Egner, T. (2015). An insula-frontostriatal network mediates flexible cognitive control by adaptively predicting changing control demands. *Nature Communications, 6*, 8165.

Jiang, J., Brashier, N. M., & Egner, T. (2015). Memory meets control in Hippocampal and striatal binding of stimuli, responses, and attentional control states. *Journal of Neuroscience, 35*(44), 14885–14895.

Jiang, J., & Egner, T. (2014). Using neural pattern classifiers to quantify the modularity of conflict-control mechanisms in the human brain. *Cerebral Cortex, 24*(7), 1793–1805.

Jiang, J., Heller, K., & Egner, T. (2014). Bayesian modeling of flexible cognitive control. *Neuroscience & Biobehavioral Reviews, 46* Pt 1, 30–43.

Jimenez, L., & Mendez, A. (2013). It is not what you expect: Dissociating conflict adaptation from expectancies in a Stroop task. *Journal of Experimental Psychology: Human Perception and Performance, 39*(1), 271–284.

Kerns, J. G., Cohen, J. D., MacDonald, A. W., 3rd, Cho, R. Y., Stenger, V. A., & Carter, C. S. (2004). Anterior cingulate conflict monitoring and adjustments in control. *Science, 303*(5660), 1023–1026.

Kim, S., & Cho, Y. S. (2014). Congruency sequence effect without feature integration and contingency learning. *Acta Psychologica (Amst), 149*, 60–68.

King, J. A., Korb, F. M., & Egner, T. (2012). Priming of control: Implicit contextual cuing of top-down attentional set. *Journal of Neuroscience, 32*(24), 8192–8200.

Kleiman, T., Hassin, R. R., & Trope, Y. (2014). The control-freak mind: Stereotypical biases are eliminated following conflict-activated cognitive control. *Journal of Experimental Psychology General, 143*(2), 498–503.

Koechlin, E., Ody, C., & Kouneiher, F. (2003). The architecture of cognitive control in the human prefrontal cortex. *Science, 302*(5648), 1181–1185.

Kornblum, S. (1994). The way irrelevant dimensions are processed depends on what they overlap with: The case of Stroop- and Simon-like stimuli. *Psychological Research, 56*(3), 130–135.

Kornblum, S., Hasbroucq, T., & Osman, A. (1990). Dimensional overlap: Cognitive basis for stimulus-response compatibility—a model and taxonomy. *Psychological Review, 97*(2), 253–270.

Kunde, W., Augst, S., & Kleinsorge, T. (2012). Adaptation to (non)valent task disturbance. *Cognitive, Affective, & Behavioral Neuroscience, 12*(4), 644–660.

Kunde, W., & Wuhr, P. (2006). Sequential modulations of correspondence effects across spatial dimensions and tasks. *Memory & Cognition, 34*(2), 356–367.

Lee, J., & Cho, Y. S. (2013). Congruency sequence effect in cross-task context: Evidence for dimension-specific modulation. *Acta Psychologica (Amst), 144*(3), 617–627.

Logan, G. D. (1988). Toward an instance theory of automatization. *Psychological Review, 95*, 492–527.

Logan, G. D., & Zbrodoff, N. J. (1979). When it helps to be misled: Facilitative effects of increasing the frequency of conflicting stimuli in a Stroop-like task. *Memory & Cognition, 7*, 166–174.

MacLeod, C. M. (1991). Half a century of research on the Stroop effect: An integrative review. *Psychological Bulletin, 109*(2), 163–203.

Mayr, U., Awh, E., & Laurey, P. (2003). Conflict adaptation effects in the absence of executive control. *Nature Neuroscience, 6*(5), 450–452.

Miller, E. K., & Cohen, J. D. (2001). An integrative theory of prefrontal cortex function. *Annual Review of Neuroscience, 24*, 167–202.

Mordkoff, J. T. (2012). Observation: Three reasons to avoid having half of the trials be congruent in a four-alternative forced-choice experiment on sequential modulation. *Psychonomic Bulletin & Review, 19*(4), 750–757.

Neill, W. T. (1997). Episodic retrieval in negative priming and repetition priming. *Journal of Experimental Psychology: Learning, Memory, and Cognition, 23*, 1291–1305.

Nieuwenhuis, S., Stins, J. F., Posthuma, D., Polderman, T. J. C., Boomsma, D. I., & de Geus, E. J. (2006). Accounting for sequential effects in the flanker task: Conflict adaptation or associative priming? *Memory & Cognition, 34*, 1260–1272.

Norman, D. A., & Shallice, T. (1986). Attention to action: Willed and automatic control of behavior. In G. E. Schwarz & D. Shapiro (Eds.), *Consciousness and self-regulation* (Vol. 4). New York: Plenum Press.

Notebaert, W., Gevers, W., Verbruggen, F., & Liefooghe, B. (2006). Top-down and bottom-up sequential modulations of congruency effects. *Psychonomic Bulletin & Review, 13*(1), 112–117.

Pashler, H., & Baylis, G. (1991). Procedural learning: 2. Intertrial repetition effects in speeded choice tasks. *Journal of Experimental Psychology: Learning, Memory, and Cognition, 17*, 33–48.

Puccioni, O., & Vallesi, A. (2012). Sequential congruency effects: Disentangling priming and conflict adaptation. *Psychological Research, 76*(5), 591–600.

Remington, R. J. (1969). Analysis of sequential effects in choice reaction times. *Journal of Experimental Psychology, 82*, 250–257.

Ridderinkhof, K. R. (2002). Micro- and macro-adjustments of task set: Activation and suppression in conflict tasks. *Psychological Research, 66*(4), 312–323.

Runger, D., Schwager, S., & Frensch, P. A. (2010). Across-task conflict regulation: A replication failure. *Journal of Experimental Psychology: Human Perception and Performance, 36*(1), 136–146.

Schmidt, J. R. (2013). Questioning conflict adaptation: Proportion congruent and Gratton effects reconsidered. *Psychonomic Bulletin & Review, 20*(4), 615–630.
Schmidt, J. R., & De Houwer, J. (2011). Now you see it, now you don't: Controlling for contingencies and stimulus repetitions eliminates the Gratton effect. *Acta Psychologica (Amst), 138*(1), 176–186.
Schmidt, J. R., & Weissman, D. H. (2014). Congruency sequence effects without feature integration or contingency learning confounds. *PLoS ONE, 9*(7), e102337.
Sheth, S. A., Mian, M. K., Patel, S. R., Asaad, W. F., Williams, Z. M., Dougherty, D. D.,...Eskandar, E. N. (2012). Human dorsal anterior cingulate cortex neurons mediate ongoing behavioural adaptation. *Nature, 488*(7410), 218–221.
Simon, J. R., & Berbaum, K. (1990). Effect of conflicting cues on information processing: The 'Stroop effect' vs. the 'Simon effect'. *Acta Psychologica (Amst), 73*(2), 159–170.
Spape, M. M., & Hommel, B. (2008). He said, she said: Episodic retrieval induces conflict adaptation in an auditory Stroop task. *Psychonomic Bulletin & Review, 15*(6), 1117–1121.
Stroop, J. R. (1935). Studies of interference in serial verbal reactions. *Journal of Experimental Psychology, 18*, 643–662.
Stuphorn, V., Taylor, T. L., & Schall, J. D. (2000). Performance monitoring by the supplementary eye field. *Nature, 408*(6814), 857–860.
Treisman, A. M., & Gelade, G. (1980). A feature-integration theory of attention. *Cognitive Psychology, 12*(1), 97–136.
Ullsperger, M., Bylsma, L. M., & Botvinick, M. M. (2005). The conflict adaptation effect: It's not just priming. *Cognitive, Affective, & Behavioral Neuroscience, 5*(4), 467–472.
van den Wildenberg, W. P., Wylie, S. A., Forstmann, B. U., Burle, B., Hasbroucq, T., & Ridderinkhof, K. R. (2010). To head or to heed? Beyond the surface of selective action inhibition: A review. *Frontiers in Human Neuroscience, 4*, 222.
Verguts, T., & Notebaert, W. (2008). Hebbian learning of cognitive control: Dealing with specific and nonspecific adaptation. *Psychological Review, 115*(2), 518–525.
Verguts, T., & Notebaert, W. (2009). Adaptation by binding: A learning account of cognitive control. *Trends in Cognitive Sciences, 13*(6), 252–257.
Weissman, D. H., Egner, T., Hawks, Z., & Link, J. (2015). The congruency sequence effect emerges when the distracter precedes the target. *Acta Psychologica (Amst), 156*, 8–21.
Weissman, D. H., Hawks, Z. W., & Egner, T. (2016). Different levels of learning interact to shape the congruency sequence effect. *Journal of Experimental Psychology: Learning, Memory, and Cognition, 42*(4), 566–583.
Weissman, D. H., Jiang, J., & Egner, T. (2014). Determinants of congruency sequence effects without learning and memory confounds. *Journal of Experimental Psychology: Human Perception and Performance, 40*(5), 2022–2037.
Wendt, M., Kluwe, R. H., & Peters, A. (2006). Sequential modulations of interference evoked by processing task-irrelevant stimulus features. *Journal of Experimental Psychology: Human Perception and Performance, 32*(3), 644–667.

5

Context, Conflict, and Control

Julie M. Bugg

The purpose of this chapter is to provide a critical review of research on contextual influences on cognitive control in conflict paradigms (e.g., Stroop and flanker tests). By no means is the review meant to be exhaustive, as I will focus selectively on one manipulation that has been influential in this line of research, namely, that of proportion congruence. Proportion congruence refers to the proportion of congruent (i.e., compatible) to incongruent (i.e., incompatible) trials and represents the likelihood of conflict for a given context. It critically influences selective attention demands (i.e., the extent to which an irrelevant dimension attracts attention; Melara & Algom, 2003), thereby eliciting conditions that are ripe for investigating control processes that resolve these demands. As I will elucidate in the present chapter, the utility of the proportion congruence manipulation lies in its flexibility, especially the opportunity to isolate what appear to be qualitatively different mechanisms of cognitive control by varying the level (e.g., list wide, context specific, item specific) at which the proportion congruence manipulation is implemented. Borrowing from successful 'user's guides' in the literature (Conway et al., 2005; Jacoby, 1998), this chapter is intended to serve as a user's guide to examining cognitive control via proportion congruence manipulations. At the same time, so as not to be redundant with recent extant reviews on proportion congruence effects (Bugg, 2012; Bugg & Crump, 2012), I will emphasise current theoretical advances in this literature.

The Three Levels at Which Proportion Congruence Is Manipulated

The seminal approach to manipulating proportion congruence targeted entire lists or blocks of 100 trials or so (Logan & Zbrodoff, 1979). As shown in Figure 5.1a, one such block comprised the mostly congruent list (MC list) and had a higher proportion of congruent to incongruent trials, whereas the other block comprised the mostly incongruent list (MI list) and had a higher proportion of incongruent to congruent trials. A smaller congruency effect was observed in the MI list relative to the MC list (i.e., the *list-wide proportion congruence* effect), implicating a contextual influence on cognitive control (see also, e.g., Gratton, Coles, & Donchin, 1992; Hommel, 1994; Kane & Engle, 2003; Lindsay & Jacoby, 1994; Logan, Zbrodoff, & Williamson, 1984;

The Wiley Handbook of Cognitive Control, First Edition. Edited by Tobias Egner.

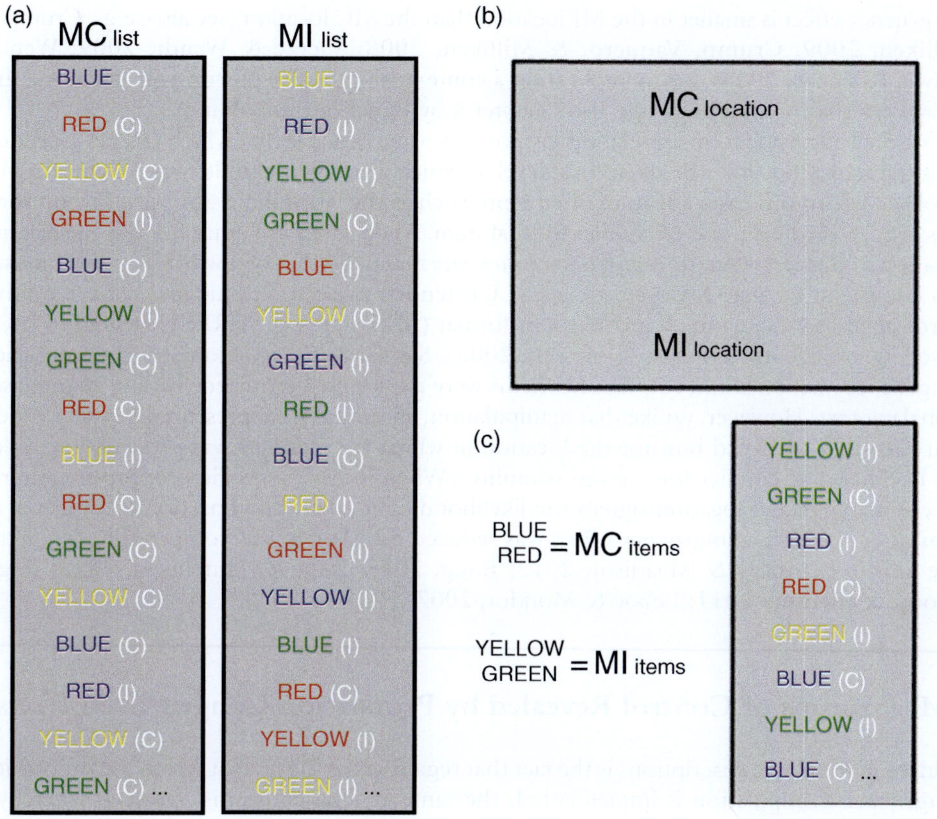

Figure 5.1 Manipulations of proportion congruence targeting the (a) list level, (b) context level, in this case particular locations, and (c) item level. MC = mostly congruent; MI = mostly incongruent; (C) = congruent trial; (I) = incongruent trial. Adapted from Bugg 2012.

Lowe & Mitterer, 1982; Toth et al., 1995; Wendt & Luna-Rodriguez, 2009; West & Baylis, 1998). Presence in a list-wide context in which there was a globally high likelihood of conflict appeared to decrease processing of the irrelevant (conflicting) stimulus dimension relative to a list-wide context in which there was a low likelihood of conflict.

More recently it has been shown that contextual influences on cognitive control also operate at a local level (Corballis & Gratton, 2003), as revealed by context-specific proportion congruence manipulations. For expository reasons, I will emphasise the most common iteration of this manipulation that targets the location in which a stimulus is presented (i.e., the *location-specific proportion congruence* manipulation; Crump, Gong, & Milliken, 2006). Imagine, for example, that the MC list from Figure 5.1a appeared in an upper location on screen and the MI list appeared in a lower location. As depicted in Figure 5.1b, the upper location is then considered an MC location and the lower location an MI location. Importantly, trials from each location are randomly intermixed (rather than one list appearing right after the other as in the list-wide proportion congruence manipulation), resulting in an unbiased global context (i.e., the overall list is 50% congruent). Consequently, the contextual signal (location) that carries information about the likelihood of conflict is available to guide control only upon stimulus onset (i.e., when the location of the stimulus is revealed). Still, the location-specific proportion congruence manipulation yields a pattern whereby the congruency effect is smaller in the MI location than the MC location (see also, e.g., Crump & Milliken, 2009; Crump, Vaquero, & Milliken, 2008; Vietze & Wendt, 2009; Wendt, Kluwe, & Vietze, 2008), again implicating a contextual influence on control, albeit one that varies on a trial-by-trial basis (see also Chapter 4 by Egner in this volume).

The final approach to manipulating proportion congruence is that which targets individual items (stimuli), possibly the most local level at which a contextual influence on control may be observed. In this case, a feature of an item, such as the word dimension of a Stroop stimulus, signals the likelihood of conflict for that item. As depicted in Figure 5.1c, in the original iteration of this *item-specific proportion congruence* manipulation, two words appeared mostly in a congruent format (MC items; e.g., BLUE tended to occur in blue ink), and two other words appeared mostly in an incongruent format (MI items; e.g., GREEN tended to occur in yellow ink; Jacoby, Lindsay, & Hessels, 2003). Similar to the location-specific proportion congruence manipulation, MC and MI items were randomly intermixed yielding an unbiased global context. However, unlike that manipulation, all stimuli were presented in a single location such that the word but not the location in which the stimulus was presented signalled the likelihood of conflict for a given stimulus. (When location-specific proportion congruence is manipulated, location signals the likelihood of conflict, but a particular feature of an item does not.) The congruency effect was reduced for MI items as compared to MC items (see also, e.g., Atalay & Misirlisoy, 2012; Bugg, 2015; Bugg & Hutchison, 2013; Bugg, Jacoby, & Chanani, 2011; Leboe & Mondor, 2007).

Mechanisms of Control Revealed by Proportion Congruence Effects

Evident in the above descriptions is the fact that regardless of the level at which the proportion congruence manipulation is implemented, the same general pattern is observed whereby a smaller congruency effect is found in the MI condition as compared to the MC condition. In spite of this similarity, there is reason to believe that distinct cognitive control mechanisms may be isolated when lists signal proportion congruence (likelihood of conflict) as compared to when proportion congruence is signalled by locations or features of items.

In Table 5.1, I have summarised theoretically important differences among the three approaches to manipulating proportion congruence. Borrowing from the dual mechanisms of control account (Braver, Gray, & Burgess, 2007), the second column refers to two types of mechanisms: proactive and reactive control (see Chapter 9 by Chiew & Braver in this volume). According to Braver et al. (2007), proactive control is the preparatory biasing of attention based on advance information (e.g., such as the likelihood of conflict on upcoming trials).

Table 5.1 Primary Differences Among the Proportion Congruence Manipulations.

Level of PC manipulation	*Type of control mechanism*	*Advance information available?*	*Transfer to novel stimuli?*	*Conflict triggered?*
List-wide	Proactive	Yes	Yes	No
Location-specific	Reactive	No	Yes	???
Item-specific	Reactive	No	No*	Yes

Note. In the literature, the proactive control mechanism supporting the list-wide proportion congruence effect is often referred to as list-wide control, whereas the reactive control mechanisms supporting location-specific and item-specific proportion congruence effects are often referred to as location-specific control and item-specific control, respectively; PC = proportion congruence; * = see text for exception.

This biasing results from active maintenance of task goals and is thought to be associated with tonic (sustained) activity in the lateral prefrontal cortex. By contrast, reactive control is a just-in-time mechanism that acts only after an imperative event based on the reactivation (retrieval) of goal-relevant information. It is thought to be accompanied by transient activation of the anterior cingulate cortex and lateral prefrontal cortex (DePisapia & Braver, 2006).

As can be ascertained from Table 5.1, only the list-wide proportion congruence manipulation may be used to index proactive control. This might seem obvious on logical grounds. Proactive control refers to the *preparatory* biasing of attention, and such biasing depends on advance information regarding the likelihood of conflict for a given context, which is afforded by manipulations that produce globally but not locally biased contexts. That is, after experiencing a number of trials within a given list (e.g., MI), the likelihood of conflict for that context may be learned and can guide proactive control (e.g., attention may be globally oriented away from the word dimension prior to the onset of each subsequent stimulus). In contrast, when proportion congruence is manipulated at the location-specific or item-specific level, the likelihood of conflict for a given context may be learned but is not available to guide control until after a stimulus appears (i.e., prior to stimulus onset, it is unknown as to whether the next stimulus is associated with a low or high likelihood of conflict). Therefore, the biasing of attention in a context-appropriate fashion must occur reactively. A reader might legitimately contend that participants may nevertheless adopt a control setting that corresponds to the 50% congruent list in which the varying locations or items appear, implying use of advance information (i.e., there is a moderate likelihood of conflict within this list). The key point is that neither the location-specific proportion congruence effect nor the item-specific proportion congruence effect can be explained by such an approach. Were a single control setting adopted proactively based on the 50% likelihood of conflict in the list, neither of these effects should be found—equivalent congruency effects should be evidenced across locations or items. Instead, the location- and item-specific proportion congruence effects implicate a flexible (reactive) modulation of control depending on the likelihood of conflict for a given stimulus.

As obvious as a proactive control account of the list-wide proportion congruence effect may seem, it is important to note that the effect is not impervious to interpretations based on reactive mechanisms. This is because the standard list-wide proportion congruence manipulation (see, for example, Figure 5.1a) confounds list-wide and item-specific proportion congruence (see Bugg, 2012, for further explication). Indeed, the findings of two initial forays into the role of reactive influences on the list-wide proportion congruence effect strongly contested the notion that the list-wide manipulation leads to modulations of control based on the global likelihood of conflict within a list (i.e., proactively; Blais & Bunge, 2010; Bugg, Jacoby, & Toth, 2008). Following on the assumption that proactive control entails the preparatory biasing of attention, biasing that should have an effect on *all* items within a list, the strategy these and subsequent studies adopted to evaluate the contribution of proactive control was to examine transfer. Transfer occurs when the list-wide proportion congruence pattern is found for novel, unbiased (50% congruent) stimuli that are presented within a context in which a global (proactive) control mechanism is expected to be operative (see Hutchison, 2011, for an alternative design; see Bugg, McDaniel, Scullin, & Braver, 2011, for an assessment of transfer using neutral stimuli; see also Wuhr, Duthoo, & Notebaert, 2015, for an assessment of transfer from a biased task to an unbiased task).

To illustrate one design used to assess transfer, consider Figure 5.2, which represents Experiment 1a from Bugg's (2014a) study. List-wide proportion congruence was manipulated but with a subtle though important twist: biased items were differentiated from unbiased items. The biased items played the role that stimuli in the traditional list-wide proportion congruence design played, in that the frequency with which they were presented as congruent

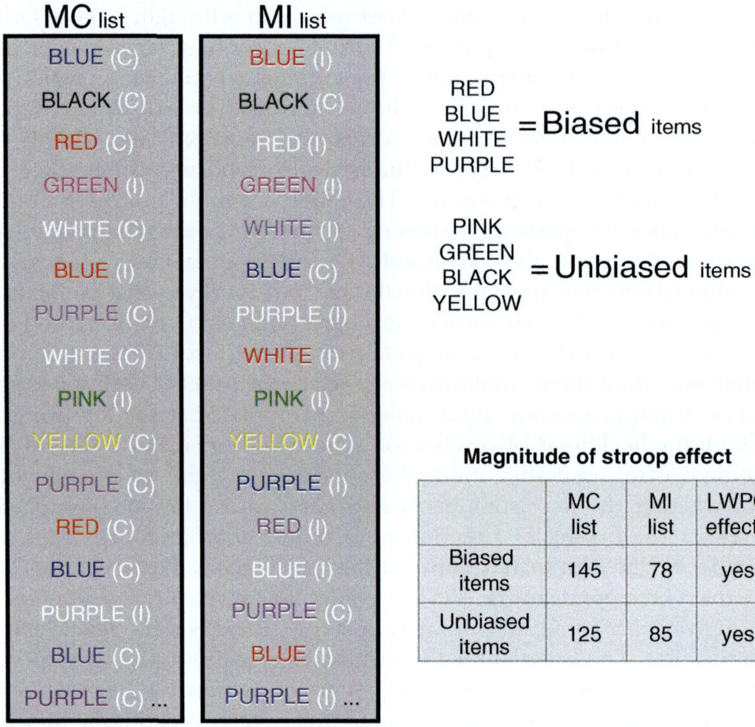

Magnitude of stroop effect

	MC list	MI list	LWPC effect?
Biased items	145	78	yes
Unbiased items	125	85	yes

Figure 5.2 List-wide proportion congruence design used to evaluate the contribution of a globally operating (proactive) control mechanism to the list-wide proportion congruence effect. Biased items were MC in the MC list and MI in the MI list. Unbiased items were 50% congruent in both lists. Biased items were presented disproportionately more frequently than unbiased items. Actual frequencies with which each item occurred may be found in Bugg (2014a, Experiment 1a, Table 1). The list-wide proportion congruence (LWPC) effect was found for the biased and the unbiased items (i.e., there was transfer). Source: Adapted from Bugg 2012.

or incongruent items determined the proportion congruence of a given list (e.g., biased items were presented mostly in the congruent format within the MC list). Consequently, a list-wide proportion congruence effect was expected for these items. More importantly, the unbiased items comprised a unique set of words and colours and were 50% congruent in *both* lists. In other words, they did not share features (words or colours) with the biased items, which might otherwise trigger reactive control, and they were matched across lists on frequency- and item-specific proportion congruence. If the assumption that the LWPC manipulation stimulates engagement of proactive control is valid, then transfer should be found for the unbiased items (i.e., a smaller congruency effect in the MI list compared to the MC list). Bugg (2014a, Experiment 1a) found precisely this transfer pattern as the magnitude of the Stroop effect was reduced from 125 ms in the MC list to 85 ms in the MI list for unbiased items (see Figure 5.2; cf. Bugg, Experiment 2b, in Bugg & Chanani, 2011).

So what might account for the initial evidence indicating a lack of transfer with a similar design (Blais & Bunge, 2010; Bugg et al., 2008)? Bugg (2014a) proposed the Associations as Antagonists to Top-Down Control (AATC) hypothesis. The AATC hypothesis predicts that the availability and use of reliable stimulus–response associations may moderate top-down control engagement (i.e., proactive control). When participants can achieve task goals (e.g., minimise Stroop interference) on most trials via stimulus–response learning, there may not be greater top-down engagement in a high-conflict (MI list) relative to a low-conflict context (MC list); i.e., the Stroop effect may be equivalent for unbiased items in each list. The initial evidence indicating a lack of transfer stemmed from experiments in which this condition was met. When Bugg (2014a) systematically replicated these conditions in Experiment 1b (see Figure 5.3), the transfer effect she observed in Experiment 1a was no longer evident (see also Bugg, 2014a, Experiment 2a). The magnitude of the Stroop effect was instead equivalent for the unbiased items in the MC and MI lists. (The LWPC effect was still found for the biased items, as expected, as such items could be responded to reactively on the basis of item-specific proportion congruence.) The difference between the two experiments was the size of the biased item set. In Experiment 1a, there were four items, whereas in Experiment 1b there were two items (as in Blais & Bunge, 2010; Bugg et al., 2008). In the latter case, participants could quickly predict the associated response on most trials within the lists (e.g., if in MC list, when BLUE, say 'blue'; if in MI list, when BLUE, say 'red'). According to the AATC hypothesis, it is the reliance on such an associative learning process that may preclude engagement of top-down (proactive) control (and therefore transfer). An important point is that the presence of advance

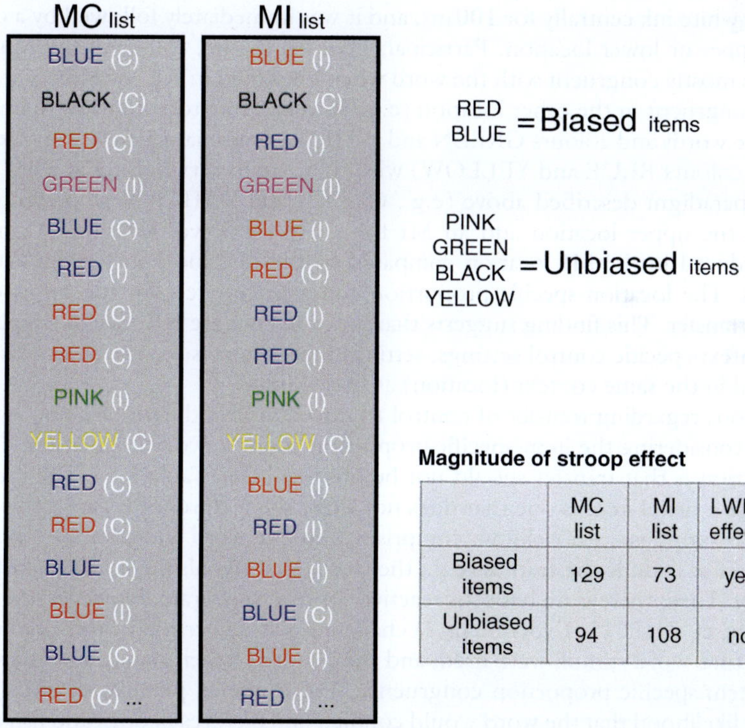

Magnitude of stroop effect

	MC list	MI list	LWPC effect?
Biased items	129	73	yes
Unbiased items	94	108	no

Figure 5.3 List-wide proportion congruence design used to evaluate the contribution of a globally operating (proactive) control mechanism to the list-wide proportion congruence effect. Biased and unbiased items were identical to Bugg (2014a, Experiment 1a) except that the biased item set comprised two rather than four items, therefore promoting reliance on associative stimulus–response learning. Actual frequencies with which each item occurred may be found in Bugg (2014a, Experiment 1b, Table 5). The LWPC effect was found for the biased but not the unbiased items (i.e., there was no transfer). Source: Adapted from Bugg 2012.

information about the global likelihood of conflict is not sufficient to claim the operation of a proactive control mechanism—factors such as associative learning can moderate its influence (for theoretical discussion of possible reasons why this may be the case, see Bugg, 2014a; see Bugg & Smallwood, 2015, for a similar conclusion based on evidence for reactive control in a Stroop paradigm that afforded the engagement of proactive control in response to valid congruency pre-cues).

Turning our attention back to the location-specific and item-specific proportion congruence manipulations, as indicated in Table 5.1, these manipulations are thought to index reactive control. In both cases, modulations of control are triggered by a local contextual signal that conveys information about the likelihood of conflict—a particular location on screen or a particular item feature, respectively. To reiterate, this is not advance information, because it is available only post-stimulus onset (albeit very early as indicated by event-related potential [ERP] evidence that MC and MI items are distinguished 150 ms post-stimulus onset; Shedden, Milliken, Watter, & Monteiro, 2013), and hence the location-specific and item-specific proportion congruence effects cannot reflect proactive control. A reader may initially be surprised by the suggestion in Table 5.1, then, that transfer to novel stimuli may be observed when location-specific proportion congruence is manipulated. A further description of how transfer has been assessed should alleviate this concern. In the single published study that examined transfer of location-specific control, Crump and Milliken (2009) presented a colour word in white ink centrally for 100 ms, and it was immediately followed by a coloured square in an upper or lower location. Participants named the ink colour of the square, and that colour was mostly congruent with the word when presented in one location (e.g., upper) and mostly incongruent in the other location (e.g., lower). Here too, items were designated biased (e.g., the words and colours GREEN and WHITE) or unbiased (50% congruent; e.g., the words and colours BLUE and YELLOW) with these items serving similar roles to those in the LWPC paradigm described above (e.g., GREEN and WHITE were presented in an MC format in the upper location and an MI format in the lower location). Congruency effects were reduced in the MI location compared to the MC location for the biased *and* unbiased items. The location-specific proportion congruence effect for the unbiased items demonstrated transfer. This finding suggests that location cues are effective in triggering the retrieval of context-specific control settings, settings that generalise to novel, unbiased items when presented in the same context (location) as biased items.

The predictions regarding transfer of control to novel, unbiased items are not as straightforward when considering the item-specific proportion congruence manipulation. The most obvious prediction is that transfer should not be observed (see Table 5.1). The rationale is simple—because a novel item is one that does not share the features of biased items (e.g., in a colour-word Stroop task, a novel item comprises a unique word and colour), a novel item necessarily does not include the feature (e.g., the word) that signals the likelihood of conflict for a given item. Thus, there is no basis for reactive control to operate. However, the findings of Bugg, Jacoby, et al. (2011, Experiment 2) challenge a strict version of this prediction. In their study, picture-word stimuli were used, and the picture (rather than the word) served as the signal of item-specific proportion congruence. For example, pictures of birds and cats signalled a low likelihood that the word would conflict, and pictures of dogs and fish signalled a high likelihood of conflict. An item-specific proportion congruence effect was found, with the magnitude of the congruency effect being reduced for dogs and fish (MI items) compared to birds and cats (MC items). Most importantly, for present purposes, in a final block of trials, novel exemplars of each type of animal were presented (e.g., dogs and cats the participant had not seen previously on the biased trials), and these animals were unbiased (50% congruent). A smaller congruency effect was found for the novel pictures of dogs and fish than the novel pictures of birds and cats, suggesting transfer of reactive, item-specific control.

One intriguing interpretation is that participants encoded the biased items categorically such that the animal category served as the signal of the likelihood of conflict for a given item. In item-specific proportion congruence paradigms in which category-level encoding of items is possible, the prediction is therefore that transfer to novel unbiased items (from biased categories) is possible.

What differentiates the transfer that has been observed in location- and item-specific proportion congruence paradigms from transfer in the LWPC paradigm (e.g., Bugg, 2014a, Experiment 1a) is that the modulation of control in these local contexts occurs only after a novel stimulus is presented. Moreover, it is dependent on the availability of a shared signal that conveys that stimulus's likelihood of conflict. In this sense, reactive control may be less optimal than proactive control in contexts in which completely novel stimuli may be encountered that would benefit from the same global attentional bias as usual stimuli.

As Table 5.1 indicates, the three levels at which proportion congruence is manipulated also differ with regard to whether the control mechanism they stimulate is *conflict triggered*. Here, I am referring to the question of whether the control mechanism appears to be selective to trials that entail conflict (incongruent trials). This must be differentiated from the question of whether conflict plays a role in the various proportion congruence effects—the answer to that is yes for each level—the likelihood (history) of conflict (be it global or local) influences adoption of control settings. The evidence to date leans towards the assumption that proactive control in the LWPC paradigm is not evoked selectively on incongruent trials (in the face of conflict). This may relate in part to the muted conflict signals that are present in MI lists when a global attentional bias is present (DePisapia & Braver, 2006; cf. Goldfarb & Henik, 2013). However, reactive control in the item-specific proportion congruence paradigm does appear to be conflict triggered. One line of evidence stems from experimental approaches that have isolated the two types of control within the same conflict task (albeit across different experiments; such evidence should be interpreted cautiously). Consider the two patterns of data from the picture-word Stroop task displayed in Figure 5.4. The pattern on the left represents performance on the unbiased items in a LWPC para-

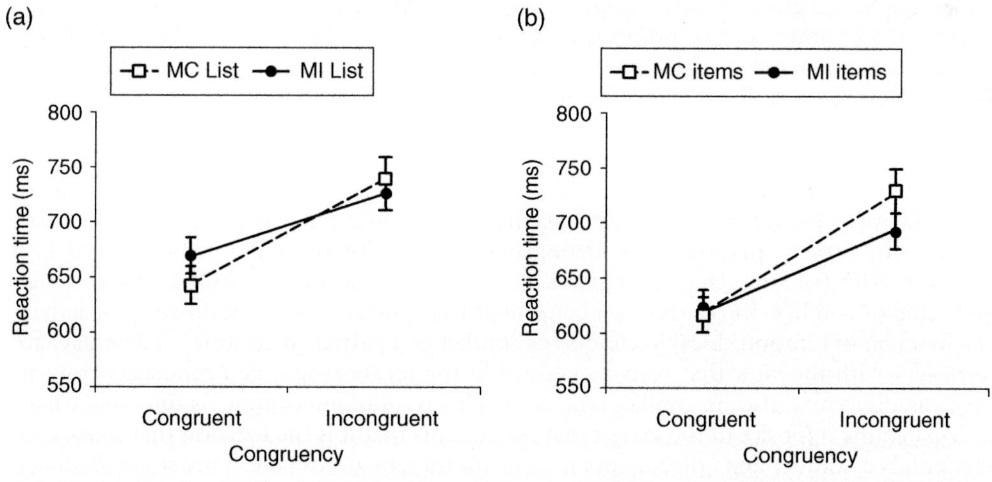

Figure 5.4 Data representing (a) list-wide proportion congruence (LWPC) pattern for unbiased items, indicative of proactive control, adapted from Bugg and Chanani (2011), and (b) item-specific proportion congruence pattern for biased items, indicative of reactive control. Source: Adapted from Bugg 2011.

digm, and thereby the influence of proactive control (Bugg & Chanani, 2011). The pattern on the right represents performance on biased items in an item-specific proportion congruence paradigm that isolated the influence of reactive (item-specific) control (for the details on how this mechanism was isolated, see Bugg, Jacoby, et al., 2011). Notice that proactive control appears to have a more pronounced and potentially selective effect on the congruent trials—that is, there is an apparent cost to performance on congruent trials in the MI compared to the MC list. This may reflect that the preparatory biasing of attention (global shift in attention away from the word dimension) in the MI list reduced the facilitative effect of the word on congruent trials, thereby slowing performance. Why proactive control appears to have a less robust effect on the incongruent trials is less certain. Possibly there are some attentional adjustments that cannot be made until after the target of the control process is known (i.e., what word must be filtered on a Stroop trial; but see Bugg & Smallwood, 2015; Goldfarb & Henik, 2013; for evidence of pre-cue benefits on incongruent trials). Regardless, it stands in contrast to the effect of reactive control, which appears to have a selectively strong influence on incongruent trials and little effect on congruent trials (see Figure 5.4b; see also Bugg, 2015; Bugg & Hutchison, 2013, for replication of this pattern in flanker and colour-word Stroop tasks, respectively). One interpretation is that reactive control is conflict triggered, and hence the effect is apparent on conflict-laden incongruent trials but not on congruent trials. This view accords well with two computational models of item-specific control (Blais, Robidoux, Risko, & Besner, 2007; Verguts & Notebaert, 2008). These models further posit that the history of conflict for a given item dictates the strength of the control signal, which aligns with the pattern of faster responses on incongruent trials from the MI set of items than the MC set of items. These observations paint a picture of a rather optimal control mechanism: Presentation of stimuli that include the feature that signals a high likelihood of conflict does not indiscriminately trigger reactive control. Rather, it appears to be reserved for those stimuli for which its influence is required (those for which conflict must be resolved), else conserving resources. Proactive control, in contrast, appears to operate even in the absence of a conflict signal (and hence the cost on congruent trials), and as such may be considered a less discriminating mode of control that is consequently more resource demanding (e.g., glucose consuming; Braver et al., 2007).

As for the location-specific proportion congruence manipulation, there are no studies in which this manipulation has been employed in a picture-word Stroop task, and therefore data are not available for direct comparison to those shown in Figure 5.4. Nonetheless, King, Korb, and Egner (2012) conducted an fMRI study using a pictorial version of a spatial flanker task involving left- or right-looking faces that yielded relevant data. They found that the medial superior parietal lobule, a region responsible for voluntary shifts of attention, was sensitive to the location-specific proportion congruence manipulation. Most critically for present purposes, on incongruent trials, activity in this region varied in a context-specific fashion such that there was more activity in the location associated with a high likelihood of conflict. In contrast, on congruent trials, there was no modulation of activity as a function of location (local likelihood of conflict in a particular context). These data are consistent with the view that reactive control in the location-specific proportion congruence paradigm may also be conflict triggered. This is quite interesting, as one might have anticipated the opposite to the extent that one accepts that it is the location that signals the likelihood of conflict and adjustments in control, thereby producing transfer (as discussed earlier). Returning to the transfer data of Crump and Milliken (2009), at least in their second experiment, congruent and incongruent transfer stimuli appeared to be *similarly* affected by the context in which these stimuli were presented (though the simple main effects were not reported, and this was not the case in Experiment 1). One possibility is that these patterns may vary depending on the relative speed of two processes, one being retrieval of the location-specific control setting and the other being generation of the conflict signal. Clearly, this is an important issue that merits further investigation (and hence three question marks appear in Table 5.1) as it bears on conclusions regarding the similarity of the reactive control mechanism supporting location-specific and item-specific proportion congruence effects.

To conclude this section, the evidence reviewed above supports that subtle but important differences between the approaches to manipulating proportion congruence are theoretically meaningful in producing patterns of performance that appear to reflect qualitatively different control mechanisms. Corroborating this evidence are single dissociations such as the preservation of reactive control (item-specific and location-specific proportion congruence effects) but impairment of proactive control (no transfer of the LWPC effect to unbiased items) associated with normal ageing (Bugg, 2014a; Bugg, 2014b), and a double dissociation indicating equivalent benefits of proactive and reactive control in reducing the magnitude of congruency effects (on biased trials in list-wide and item-specific proportion congruence paradigms, respectively) but unique costs accompanying these control mechanisms (Gonthier, Braver, & Bugg, 2015). In line with evidence reviewed above, Gonthier et al. (2015) found that proactive control was uniquely associated with a 'congruency cost' (i.e., performance on congruent trials was slowed by its influence; cf. Panel A of Figure 5.4), whereas reactive control was uniquely associated with a 'transfer cost' (i.e., congruency effects on novel, unbiased items were not as benefitted by reactive control as biased items).

Proportion Congruence Effects: The Roles of Expectations and Experience

In the preceding section, I made no mention of assumptions that exist regarding the role of strategies, will, or conscious awareness in the various proportion congruence effects. This is not reflective of the field's interest in this question. I would argue that it is indeed one of the most important and fascinating questions in this domain. However, only recently have data begun to accumulate that truly capture their roles, and the purpose of this section is to review those data.

The term 'expectation-driven' describes modulations in control that reflect explicit and advance knowledge regarding the likelihood of conflict, in other words, strategies participants intentionally employ on the basis of some awareness of the likelihood of conflict within a given context (Bugg, Diede, Cohen-Shikora, & Selmeczy, 2015). Given this definition, it would not be unreasonable to assume that the LWPC effect may be expectation driven. An alternative hypothesis, however, is that the effect is 'experience driven', that is, it is dependent on the accumulation of experience with particular stimuli and reflects any of a number of empirically documented mechanisms that are reliant on such experience (e.g., learned attentional biases, Melara & Algom, 2003; conflict adaptation, Botvinick, Braver, Barch, Carter, & Cohen, 2001; Egner, 2008; contingency learning, Schmidt & Besner, 2008; temporal learning, Schmidt, 2013a; 2013b; for further explication, see Bugg et al., 2015), and may operate implicitly as opposed to explicitly.

A pivotal study that lent support to the experience-driven account of the LWPC effect was that of Blais, Harris, Guerrero, and Bunge (2012). In this study, participants performed 190 hundred-trial lists of a four-choice colour-word Stroop task over several weeks. The lists differed in proportion congruency: Minimally they were 5% congruent, and maximally they were 95% congruent, and proportion congruency varied randomly from list to list. Following each list, they were asked to provide estimates of the proportion of congruent and incongruent trials within the list, and rate their confidence in asserting which trial type was more frequent. Not surprisingly, when they were more confident ('sure'), their estimates were more accurate than when they were 'unsure'. However, and most critically, the magnitude of the LWPC effect (proportion congruency × congruency interaction) did not differ as a function of awareness of the proportion congruency of the lists, challenging the view that the effect reflects an intentional strategy. As Blais et al. concluded, the effect instead appears to reflect implicit adaptations to the environment (i.e., is experience driven).

Using an alternative approach to investigating the role of expectations, Bugg et al. (2015) found a pattern of results that largely supported the view of Blais et al. (2012). In their study, a novel pre-cued lists paradigm was employed in which participants performed a four-choice Stroop task in abbreviated MC and MI lists comprising 10 trials. The key manipulation was that some lists were cued, whereas other lists were uncued. Participants were told about the likelihood of conflict in the upcoming list in the cued lists but not the uncued lists. Comparing performance between the cued and uncued lists offered a means of evaluating the role of expectations while holding experience constant (e.g., an uncued MI list was compared to a cued MI list). To the extent that participants can intentionally engage a particular control setting (strategy) in a given list when the likelihood of conflict is known in advance (rather than having to be learned on the fly), performance was expected to differ between the cued and uncued lists (i.e., indicating a cue-induced shift). A consistent pattern emerged: There was a cue-induced shift in the MC list but no cue-induced shift in the MI list. Participants appeared to use the MC cue to strategically bias attention as indicated by the exacerbation of the Stroop effect in the cued MC list compared to the uncued MC list. Here, expectations led to a relaxation of control. By contrast, the Stroop effect was equivalent (and smaller than in MC lists) when participants knew in advance that a list would be MI as when they encountered an uncued MI list. This pattern suggests that expectations did not lead to a significant heightening of control in the context in which a high likelihood of conflict was anticipated.

Entel, Tzelgov, and Bereby-Meyer (2014, Experiment 1) similarly instructed participants regarding the proportion congruence of an upcoming list in a four-choice colour-word Stroop task. One group of participants was told the list would be MC, and the other group was told it would be MI. The actual list was composed of two halves, and only the second half conformed to the expectations established by the instructions. The first half was 50% congruent for both groups of participants and critically allowed Entel et al. to examine whether expectations alone could guide control. Providing support for a role of expectations in the LWPC effect, the magnitude of the Stroop effect was larger in the first half of the list for participants who expected an MC list relative to participants who expected an MI list. Corroborating this result, Bugg et al. (2015) found a larger Stroop effect in the first half of abbreviated (20 -trial) lists that were 50% congruent and preceded by an invalid MC pre-cue compared to lists that were 50% congruent and preceded by an invalid MI pre-cue. Importantly, inviting the conclusion that the expectation-driven shift was primarily a relaxation of control when an MC list was expected rather than a heightening of control when an MI list was expected, Bugg et al. (2015) found that the Stroop effect was larger in the invalidly cued MC list than in a validly cued 50% congruent list, whereas the Stroop effect was equivalent in the invalidly cued MI list and the validly cued 50% congruent list.

These studies suggest there is a role for experience-driven and expectation-driven adjustments in control in the LWPC effect. (For a discussion of expectation effects on trial-by-trial congruency sequence effects, see Chapter 4 by Egner in this volume). Although additional research is clearly needed, the initial evidence seems to imply a stronger influence of expectations in low-conflict (MC) contexts. An important goal for future research is to understand the conditions under which expectation-driven control plays a role in contexts in which it is presumably most needed (e.g., high-conflict [MI] lists). Bugg et al. (2015) identified two such conditions: Both the use of a speed manipulation and use of a motivational incentive encouraged participants to prepare on the basis of an MI pre-cue. Importantly, however, in both cases the expectation-driven influence on performance was short-lived (i.e., observed on the first trial of the list) and was not sustained across the list. Although the dual mechanisms of control account does not take a stance on whether proactive control is intentionally engaged, claiming only that it is a sustained mechanism that is based on advance information, these patterns cast doubt on the possibility that proactive control, as stimulated by a list-wide MI context, is expectation driven. However, it remains entirely possible that proactive control is based on experience-driven, implicit modulations of control, which may well be sustained.

With respect to the location- and item-specific proportion congruence effects, given that advance knowledge is not available to guide reactive control, it appears certain that neither of these effects reflects expectation-driven control. In line with this view, Entel et al. (2014, Experiment 2) employed a similar instructional manipulation to that described above, but in this case they informed participants about item-specific proportion congruence. Half of the participants were specifically told which words were MC items and which words were MI items before the start of the task. Not surprisingly, regardless of the instructions, they found no evidence of an item-specific proportion congruence effect in the half of the experiment in which all items were actually 50% congruent. Similarly, Crump et al. (2006) found that instructing participants in advance about the proportion congruence of MC and MI contexts (in this case, different shapes served as the critical cue rather than location) and encouraging them to use context-appropriate strategies did not produce a context-specific proportion congruence effect.

The question of whether the (experience-driven) reactive modulations of control that underlie location- and item-specific proportion congruence effects depend on awareness of the proportion congruence manipulations has nonetheless been of great interest. The consensus is that they do not, with the primary approach to addressing this question being the administration of post-experimental questionnaires that gauge awareness. A typical finding is that participants cannot accurately report which location was MC and which was MI or the proportion of congruent trials in a particular location. In other words, the estimates of proportion congruence are very similar for the MC and MI locations, indicating a lack of explicit awareness of the differences in conflict likelihood between contexts (Crump et al., 2006).

Employing a different approach to addressing this question, Schouppe, Ridderinkhof, Verguts, and Notebaert (2014) used a context-specific proportion congruence manipulation in which conflict likelihood was associated with different categories, and then gave participants a choice to perform either the low-conflict (MC) or high-conflict (MI) category. They found that participants were more likely to choose the MC category, indicating an avoidance of the high-conflict context and implying some (implicit) awareness of the proportion congruence manipulation. However, preference was unrelated to the magnitude of the context-specific proportion congruence effect, which calls into question the importance of awareness in the modulations of reactive control. A potential criticism of these approaches, however, is that they fail to consider awareness in an on-line fashion, and rely instead on participant's ability to recollect context-specific experiences from an earlier portion of the experimental session. New approaches are therefore needed to examine this question as it bears on a related issue of theoretical importance, which is the extent to which reactive modulations of control may be considered automatic or relatively effortless, as some have suggested (e.g., Jacoby et al., 2003; see Diede & Bugg, in press, for pupillometry data that challenge this view).

Is Control Necessary for Explaining Proportion Congruent Effects?

One of the most significant points to emerge from attempts to elucidate the role of context in cognitive control is that there are many processes that can masquerade as control in proportion congruence paradigms. Proportion congruence effects are most often evaluated via comparisons of conditions with unequal frequencies of congruent and incongruent trials. Consequently, experience-driven processes tuned to the frequencies with which particular events have occurred, such as instance-based retrieval (Logan, 1988), the creation and retrieval of event files (Hommel, Musseler, Aschersleben, & Prinz, 2001), and the learning of stimulus–response associations (Schmidt & Besner, 2008), may contribute to proportion congruence effects (for an excellent discussion of the need to tease apart stimulus history effects from top-down processes in attention paradigms more generally, see Awh, Belopolsky, & Theeuwes, 2012). Sensitivity to this point is fundamentally important for researchers who utilise proportion congruence manipulations with the goal of advancing theories of cognitive control or studying variations in cognitive control (e.g., individual differences; group differences such as age- or disease-related changes). Realisation of these goals necessitates the use of experimental approaches that reveal the operation of cognitive control mechanisms independent of such processes (see also Chapter 4 by Egner in this volume).

I have already alluded to several approaches that are relevant to this goal (e.g., use of unbiased [50% congruent] items in list-wide paradigms to rule out item-specific effects; use of the pre-cued lists paradigm or instructional manipulations to investigate expectation-driven control processes while holding experience constant). I will conclude this chapter by illustrating the rationale behind a separate set of experimental approaches that has been particularly effective in elucidating the conditions under which the item-specific proportion congruence effect can be comfortably interpreted as indexing reactive control independent of contingency learning. Then I will briefly revisit the modified LWPC design I introduced earlier (see Figure 5.2) to call attention to a present challenge in the field, which is to tease apart the contribution of proactive control from that of other processes that may operate in a global fashion such as temporal learning (Schmidt, 2013a; 2013b).

Isolating Reactive Control in the Item-Specific Proportion Congruence Paradigm

In 2008, Schmidt and Besner presented compelling data that challenged the view that the item-specific proportion congruence effect may be explained by reactive cognitive control. Their view was that participants learned to predict colour naming responses that were highly contingent on the word for select trial types (e.g., congruent items from the MC set and incongruent items from the MI set), yielding the item-specific proportion congruence pattern (see also Jacoby et al., 2003; Musen & Squire, 1993). As depicted in Figure 5.5, this pattern reflected a symmetrical influence of the item-specific proportion congruence manipulation—responses were speeded on high-contingency congruent trials within the MC condition and on high-contingency incongruent trials within the MI condition (relative to low-contingency congruent and incongruent trials, respectively). The contingency account differs from a reactive control perspective that attributes the effect to the learning of stimulus–attention associations as opposed to stimulus–response associations (see Bugg & Crump, 2012; Crump & Milliken, 2009). According to this perspective, participants learn the optimal attentional setting for particular stimuli (items): For an MI item, the setting is one that minimises processing of the (often-conflicting) irrelevant dimension, and for an MC item, the setting is one that more fully processes the (often-facilitative) irrelevant

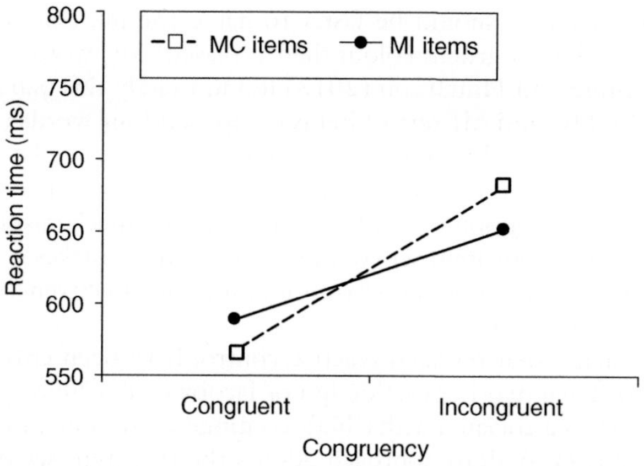

Figure 5.5 The item-specific proportion congruence pattern. Source: Adapted from Jacoby 2003. Note the symmetrical influence of the item-specific proportion congruence manipulation.

dimension. Presentation of a particular item is thought to trigger the associated control settings (see Shedden et al., 2013, for evidence that MI and MC items are distinguished as early as 150 ms post-stimulus onset).

Several experimental approaches have been adopted to isolate conditions under which the item-specific proportion congruence manipulation indexes reactive control (i.e., use of stimulus–attention associations). These approaches have been motivated by the dual item-specific mechanism (dual-ISM) account of item-specific proportion congruence effects (Bugg, 2015; Bugg et al., 2011; Bugg & Hutchison, 2013), which is grounded in tectonic theory (Melara & Algom, 2003), and posits that reactive control and contingency learning both play roles in item-specific proportion congruence effects albeit under different conditions.

One approach is to essentially disrupt the signalling power of the word dimension—that is, to create a design in which the relevant dimension (e.g., colour) strongly signals item-specific proportion congruence but the irrelevant dimension (e.g., word) does not (Bugg et al., 2011; see Bugg & Hutchison, 2013, and Bugg, 2014b, for error analyses that confirm the irrelevant dimension was not used to predict contingent responses in this design). Doing so eliminates the distinction between high- and low-contingency trials (for further explication, see Bugg, 2012, 2015), thereby ruling out a contingency learning interpretation of the item-specific proportion congruence effect. Bugg, Jacoby, et al. (2011, Experiment 2) showed that, under these conditions, an item-specific proportion congruence effect was found, and its pattern (see Figure 5.4b) was qualitatively different from the symmetrical pattern associated with the contingency learning mechanism (see Figure 5.5)—a selective influence of the manipulation was found on the conflict-laden incongruent trials. If contingency learning were operative, an effect on the congruent trials (also) should have been observed (see Bugg, Jacoby, et al., 2011, Experiment 3, for evidence of a selective influence of contingency learning on congruent trials).

A second approach is to allow the irrelevant dimension to signal item-specific proportion congruence such that word–response associations may be learned and then examine performance on transfer trials, defined here as trials on which the irrelevant dimension (word) is not predictive of the correct response, because the colour response is novel. On such trials, a contingency learning mechanism is ineffective. However, a reactive control mechanism is not. Consequently, if reactive control underlies the item-specific proportion congruence effect, participants should be faster to name the ink colour of a (biased) MI word presented in a novel incongruent colour than a (biased) MC word presented in a novel incongruent colour. Bugg and Hutchison (2013) found exactly this pattern selectively in a condition in which the MC and MI sets of items comprised four words (and colours) each and not when the sets comprised two words (and colours) each. The latter, of course, represents the original variant of the item-specific proportion congruence manipulation (see Figure 5.1c), which biases adoption of a contingency learning mechanism (see Bugg, 2015, for further discussion). The four-item set condition, in contrast, makes it impossible to predict responses on incongruent trials, thereby minimising the effectiveness of a contingency learning mechanism in the MI set.

The first two approaches that revealed reactive control have been criticised because they did not preserve the original two-item set design of Jacoby et al. (2003) in which the irrelevant dimension (word) was associated with a high-contingency response for both the MC and MI items (Schmidt, 2014). A third approach adopts the two-item set design and isolates reactive control by manipulating the relative salience of the relevant and irrelevant dimensions. Briefly, the rationale is as follows (see Bugg, 2015, for further explication). In the two-item set design, the relevant *and* irrelevant dimensions strongly signal item-specific proportion congruence. The irrelevant dimension, when attended, allows for learning and use of contingency learning (prediction of response associated with the word) on select trials; the relevant dimension does not. Thus, if attention can be biased towards the relevant dimension, then reactive control and not contingency learning should produce the accompanying item-specific proportion congruence effect. Using various versions of arrow-based and letter-based flanker tasks, Bugg (2015) showed evidence for control-driven item-specific proportion congruence effects when stimuli were used that attracted attention to the relevant dimension (central target), whereas contingency-driven effects were found when stimuli attracted attention to the irrelevant dimension (flankers), consistent with the predictions of the dual-ISM account.

Isolating Proactive Control in the LWPC Paradigm: A New Challenge

As the reader may recall, the modified LWPC design was developed in response to a similar challenge. That challenge was to isolate the influence of the global likelihood of conflict (LWPC) from the likelihood of conflict for a particular item (item-specific proportion congruence including control and contingency learning mechanisms), thereby revealing the apparent contribution of proactive control to the LWPC effect (see, e.g., Figure 5.2). A recent proposal by Schmidt (2013a, 2013b), however, raised the possibility that proactive control may not be the only mechanism that operates globally, thereby affecting performance on the unbiased (50% congruent) items. He proposed that temporal learning may also account for the effect—that is, participants learning the rhythms of responding in MC and MI lists—and provided initial evidence supporting this view by showing that a portion of Hutchison's (2011) list-wide effect could be explained by the speed of responding on the previous trial, a proxy for temporal learning (see Kinoshita, Mozer, & Forster, 2011, for evidence that previous trial history of difficulty may also affect performance under similar conditions). New designs are needed to tease apart a proactive control account from other accounts such as temporal learning that similarly propose a globally operating mechanism. Such designs will be vital for addressing important unresolved questions such as whether all globally operating mechanisms are expected to produce the LWPC pattern for unbiased items shown in Figure 5.4a (e.g., congruency cost), and how, if at all, the conflict-resolving mechanism differs when proactive control is operative relative to temporal learning (or another global mechanism).

Conclusion

The proportion congruence manipulation has played an instrumental role in the progress that has been made in advancing our understanding of contextual influences on cognitive control in conflict tasks. As this chapter illustrates, the manipulation itself is quite simple to employ regardless of the level at which it is implemented. However, responsible use requires an understanding of the theory that informs those conditions under which the manipulation allows for conclusions to be drawn about proactive as opposed to reactive control, expectation-driven as opposed to experience-driven control, as well as non-control mechanisms. It is my hope that this chapter facilitates such use, and accordingly, continued progress in this area of research.

References

Atalay, N. B., & Misirlisoy, M. (2012). Can contingency learning alone account for item-specific control? Evidence from within- and between- language ISPC effects. *Journal of Experimental Psychology: Learning, Memory, and Cognition, 38*, 1578–1590.

Awh, E., Belopolsky, A. V., & Theeuwes, J. (2012). Top-down versus bottom-up attentional control: A failed theoretical dichotomy. *Trends in Cognitive Sciences, 16*(8), 437–443.

Blais, C., & Bunge, S. (2010). Behavioral and neural evidence for item-specific performance monitoring. *Journal of Cognitive Neuroscience, 22*, 2758–2767.

Blais, C., Harris, M. B., Guerrero, J. V., & Bunge, S. A. (2012). Rethinking the role of automaticity in cognitive control. *Quarterly Journal of Experimental Psychology, 65*, 268–276.

Blais, C., Robidoux, S., Risko, E. F., & Besner, D. (2007). Item-specific adaptation and the conflict monitoring hypothesis: A computational model. *Psychological Review, 114*, 1076–1086.

Botvinick, M. M., Braver, T. S., Barch, D. M., Carter, C. S., & Cohen, J. D. (2001). Conflict monitoring and cognitive control. *Psychological Review, 108*(3), 624–652.

Braver, T. S., Gray, J. R., & Burgess, G. C. (2007). Explaining the many varieties of working memory variation: Dual mechanisms of cognitive control. In A. R. A. Conway, C. Jarrold, M. J. Kane, A. Miyake, & J. N. Towse (Eds.), *Variation in working memory* (pp. 76–106). New York: Oxford University Press.

Bugg, J. M. (2012) Dissociating levels of cognitive control: The case of Stroop interference. *Current Directions in Psychological Science, 21*, 302–309.

Bugg, J. M. (2014a). Conflict-triggered top-down control: Default mode, last resort, or no such thing? *Journal of Experimental Psychology: Learning, Memory, and Cognition, 40*, 567–587.

Bugg, J. M. (2014b). Evidence for the sparing of reactive cognitive control with age. *Psychology and Aging, 29*, 115–127.

Bugg, J. M. (2015). The relative attractiveness of distractors and targets affects the coming and going of item-specific control: Evidence from flanker tasks. *Attention, Perception, and Psychophysics, 77*, 373–389.

Bugg, J. M., & Chanani, S. (2011). List-wide control is not entirely elusive: Evidence from picture-word Stroop. *Psychonomic Bulletin & Review, 18*, 930–936.

Bugg, J. M., & Crump, M. J. C. (2012). In support of a distinction between voluntary and stimulus-driven control: A review of the literature on proportion congruent effects. *Frontiers in Psychology: Cognition, 3*.

Bugg, J. M., Diede, N. T., Cohen-Shikora, E. R., & Selmeczy, D. (2015). Expectations and experience: Dissociable bases for cognitive control? *Journal of Experimental Psychology: Learning, Memory, and Cognition*. Advance online publication. 10.1037/xlm0000106

Bugg, J. M., & Hutchison, K. A. (2013). Converging evidence for control of color-word Stroop interference at the item level. *Journal of Experimental Psychology: Human Perception and Performance, 39*(2), 433–449.

Bugg, J. M., Jacoby, L. L., & Chanani, S. (2011). Why it is too early to lose control in accounts of item-specific proportion congruency effects. *Journal of Experimental Psychology: Human Perception and Performance, 37*, 844–859.

Bugg, J. M., Jacoby, L. L., & Toth, J. (2008). Multiple levels of control in the Stroop task. *Memory & Cognition, 36*, 1484–1494.

Bugg, J. M., McDaniel, M. A., Scullin, M. K., & Braver, T. S. (2011). Revealing list-level control in the Stroop task by uncovering its benefits and a cost. *Journal of Experimental Psychology: Human Perception and Performance, 37*, 1595–1606.

Bugg, J. M., & Smallwood, A. (2015). The next trial will be conflicting! Effects of explicit congruency pre-cues on cognitive control. *Psychological Research,* Advance online publication.

Corballis, P. M., & Gratton, G. (2003). Independent control of processing strategies for different locations in the visual field. *Biological Psychology, 64*, 191–209.

Crump, M. J., Gong, Z., & Milliken, B. (2006). The context-specific proportion congruent Stroop effect: Location as a contextual cue. *Psychonomic Bulletin & Review, 13*, 316–321.

Crump, M. J. C., & Milliken, B. (2009). The flexibility of context-specific control: Evidence for context-driven generalization of item-specific control. *The Quarterly Journal of Experimental Psychology, 62*, 1523–1532.

Crump, M. J. C., Vaquero, J. M. M., & Milliken, B. (2008). Context-specific learning and control: The roles of awareness, task relevance, and relative salience. *Consciousness and Cognition, 17*, 22–36.

Conway, A. R. A., Kane, M. J., Bunting, M. F., Hambrick, D. Z., Wilhelm, O., & Engle, R. W. (2005). Working memory span tasks: A review and a user's guide. *Psychonomic Bulletin and Review, 12*, 769–786.

DePisapia, N., & Braver, T. S. (2006). A model of dual control mechanisms through anterior cingulate and prefrontal cortex interactions. *Neurocomputing, 69*, 1322–1326.

Diede, N., & Bugg, J. M. (in press). Cognitive effort is modulated outside of the explicit awareness of conflict frequency: Evidence from pupillometry. *Journal of Experimental Psychology: Learning, Memory, and Cognition.*

Egner, T. (2008). Multiple conflict-driven control mechanisms in the human brain. *Trends in Cognitive Sciences, 12*, 374–380.

Entel, O., Tzelgov, J., & Bereby-Meyer, Y. (2014). Proportion congruency effects: Instructions may be enough. *Frontiers in Psychology, 5.*

Gratton, G., Coles, M. G. H., & Donchin, E. (1992). Optimizing the use of information: Strategic control of activation and responses. *Journal of Experimental Psychology: General, 121*, 480–506.

Goldfarb, L., & Henik, A. (2013). The effect of a preceding cue on the conflict solving mechanism. *Experimental Psychology, 60*, 347–353.

Gonthier, C., Braver, T. S., & Bugg, J. M. (2015). *Dissociating proactive and reactive control the Stroop task.* Manuscript submitted for publication.

Hommel, B. (1994). Spontaneous decay of response-code activation. *Psychological Research, 56*, 261–268.

Hommel, B., Musseler, J., Aschersleben, G., & Prinz, W. (2001). The theory of event coding (TEC): A framework for perception and action planning. *Behavioral and Brain Sciences, 24*, 849–878.

Hutchison, K. A. (2011). The interactive effects of list-based control, item-based control, and working memory capacity on Stroop performance. *Journal of Experimental Psychology: Learning, Memory, and Cognition, 37*, 851–860.

Jacoby, L. L. (1998). Invariance in automatic influences of memory: Toward a user's guide for the process-dissociation procedure. *Journal of Experimental Psychology: Learning, Memory, & Cognition, 24*, 3–26.

Jacoby, L. L., Lindsay, D. S., & Hessels, S. (2003). Item-specific control of automatic processes: Stroop process dissociations. *Psychonomic Bulletin & Review, 10*(3), 638–644.

Kane, M. J., & Engle, R. W. (2003). Working-memory capacity and the control of attention: The contributions of goal neglect, response competition, and task set to Stroop interference. *Journal of Experimental Psychology: General, 132*, 47–70.

King, J. A., Korb, F. M., & Egner, T. (2012). Priming of control: Implicit contextual cueing of top-down attentional set. *The Journal of Neuroscience, 32*, 8192–8200.

Kinoshita, S., Mozer, M. C., & Forster, K. I. (2011). Dynamic adaptation to history of trial difficulty explains the effect of congruency proportion on masked priming. *Journal of Experimental Psychology: General, 140*, 622–636.

Leboe, L. C., & Mondor, T. A. (2007). Item-specific congruency effects in nonverbal auditory Stroop. *Psychological Research, 71*, 568–575.
Lindsay, D. S., & Jacoby, L. L. (1994). Stroop process dissociations: The relationship between facilitation and interference. *Journal of Experimental Psychology: Human Perception and Performance, 20*, 219–234.
Logan, G. D. (1988). Toward an instance theory of automatization. *Psychological Review, 95*, 492–527.
Logan, G. D., & Zbrodoff, N. J. (1979). When it helps to be misled: Facilitative effects of increasing the frequency of conflicting stimuli in a Stroop-like task. *Memory & Cognition, 7*, 166–174.
Logan, G. D., Zbrodoff, N. J., & Williamson, J. (1984). Strategies in the color-word Stroop task. *Bulletin of the Psychonomic Society, 22*, 135–138.
Lowe, D., & Mitterer, J. O. (1982). Selective and divided attention in a Stroop task. *Canadian Journal of Psychology, 36*, 684–700.
Melara, R. D., & Algom, D. (2003). Driven by information: A tectonic theory of Stroop effects. *Psychological Review, 110*, 422–471.
Musen, G., & Squire, L. R. (1993). Implicit learning of color-word associations using a Stroop paradigm. *Journal of Experimental Psychology: Learning, Memory, & Cognition, 34*, 514–523.
Schmidt, J. R. (2013a). The parallel episodic processing (PEP) model: Dissociating contingency and conflict adaptation in the item-specific proportion congruent paradigm. *Acta Psychologica, 142*, 119–126.
Schmidt, J. R. (2013b). Questioning conflict adaptation: Proportion congruent and Gratton effects reconsidered. *Psychonomic Bulletin & Review, 20*, 615–630.
Schmidt, J. R. (2014). Contingencies and attentional capture: The importance of matching stimulus informativeness in the item-specific proportion congruent task. *Frontiers in Psychology, 5*, 540.
Schmidt, J. R., & Besner, D. (2008). The Stroop effect: Why proportion congruence has nothing to do with congruency and everything to do with contingency. *Journal of Experimental Psychology: Learning, Memory, & Cognition, 34*, 514–523.
Schouppe, N., Ridderinkhof, K. R., Verguts, T., & Notebaert, W. (2014). Context-specific control and context selection in conflict tasks. *Acta Psychologica, 146*, 63–66.
Shedden, J. M., Milliken, B., Watter, S., & Monteiro, S. (2013). Event-related potentials as brain correlates of item specific proportion congruent effects. *Consciousness and Cognition, 22*, 1442–1455.
Toth, J. P., Levine, B., Stuss, D. T., Oh, A., Winocur, G., & Meiran, N. (1995). Dissociation of processes underlying spatial S-R compatibility: Evidence for the independent influence of what and where. *Consciousness and Cognition, 4*, 483–501.
Verguts, T., & Notebaert, W. (2008). Hebbian learning of cognitive control: Dealing with specific and nonspecific adaptation. *Psychological Review, 115*, 518–525.
Vietze, I., & Wendt, M. (2009). Context specificity of conflict frequency-dependent control. *The Quarterly Journal of Experimental Psychology, 62*, 1391–1400.
Wendt, M., Kluwe, R. H., & Vietze, I. (2008). Location-specific versus hemisphere-specific adaptation of processing selectivity. *Psychonomic Bulletin & Review, 15*, 135–140.
Wendt, M., & Luna-Rodriguez, A. (2009). Conflict-frequency affects flanker-interference. *Experimental Psychology, 56*, 206–217.
West, R., & Baylis, G. C. (1998). Effect of increased response dominance and contextual disintegration on the Stroop interference effect in older adults. *Psychology and Aging, 13*, 206–217.
Wuhr, P., Duthoo, W., & Notebaert, W. (2015). Generalizing attentional control across dimensions and tasks: Evidence from transfer of proportion-congruence effects. *Quarterly Journal of Experimental Psychology, 68*, 779–801.

6

Control in Response Inhibition

Frederick Verbruggen (University of Exeter) and Gordon D. Logan (Vanderbilt University)

Stopping prepared but no longer relevant responses is a simple act of cognitive control that supports flexible and goal-directed behaviour. In the past two decades, response inhibition has received much attention across research domains. Cognitive psychologists and neuroscientists have explored the cognitive and neural mechanisms of response inhibition, developmental scientists have studied the 'rise and fall' of inhibitory control capacities across the life span, and clinical researchers have examined correlations between individual differences in response inhibition and behaviours such as substance abuse, overeating, and risk taking. In this chapter, we provide a selective review of recent behavioural and computational work on response inhibition (for a review of the neuroscience literature, see Chapter 16 by Forstmann & Alkemade in this volume).

In the response-inhibition literature, individual or group differences are often attributed to variation in the effectiveness of a single inhibitory control function. But many processes contribute to stopping an action. Response inhibition requires an interplay between three basic and computationally well-defined 'reactive' processes (signal detection, action selection, and suppression of motor output), which are regulated and influenced by sets of processes that take place on different timescales: outcome monitoring, advance preparation, rule acquisition and maintenance, associative learning, and development (Verbruggen, McLaren, & Chambers, 2014). Here we will discuss mainly 'reactive', 'proactive', and 'associatively mediated' forms of inhibitory control (for a review of key developmental transitions, see Munakata, Snyder, & Chatham, 2012).

How to Stop a Response

Reactive inhibitory control in response to changes in the environment or internal state is often studied in tasks such as the go/no-go task (Donders, 1868/1969) and the stop-signal task (Lappin & Eriksen, 1966; Logan & Cowan, 1984; Vince, 1948). In the go/no-go task, subjects are instructed to respond when a go stimulus appears (e.g., an 'O'), but to withhold their response when a no-go stimulus appears (e.g., an 'X'). In the stop-signal task, subjects perform a primary go task, such as responding to the identity of a stimulus (e.g., press left

The Wiley Handbook of Cognitive Control, First Edition. Edited by Tobias Egner.

when an 'O' appears, and right when an 'X' appears). On a minority of the trials, an extra visual or auditory signal appears after a variable stop-signal delay (SSD), instructing subjects to withhold the planned go response.

We will focus on the stop-signal paradigm because it allows researchers to estimate the covert latency of response inhibition: the stop-signal reaction time (SSRT). SSRT has become an established marker for reactive inhibitory control (Verbruggen, Chambers, & Logan, 2013), and many studies have used the stop-signal paradigm to develop detailed accounts of the cognitive processes underlying reactive inhibitory control and to examine what factors influence them.

The Race Between Going and Stopping

Since the 1980s, virtually all stop-signal research relied on a mathematical model because SSRT is unobservable without it. The model is the Logan and Cowan (1984) independent race model, which describes response inhibition as a race between a go process, triggered by the presentation of a go stimulus, and a stop process, triggered by the presentation of the stop signal. When the stop process finishes before the go process, response inhibition is successful, and no response is emitted (signal-inhibit trial); when the go process finishes before the stop process, response inhibition is unsuccessful, and the response is incorrectly emitted (signal-respond trial). The independent race model mathematically relates signal-respond reaction time (RT), no-stop-signal RT, and inhibition functions and provides several methods for estimating SSRT. Logan and Cowan wrote it as a race between generic finishing time distributions without specifying the processes that gave rise to the finishing times. This generality is a virtue because it allows the model to be applied to any subject population (including people, monkeys, and rats), effector (hands, eyes, feet), or experimental condition (for a review of models, see Verbruggen & Logan, 2009c). This generality is also a limitation: the model does not specify the underlying processes, so it can only describe and not explain reactive control. Fortunately, several models of the processes underlying the stop-signal task have appeared in the literature to address these issues.

General and Special Race Models Logan, Van Zandt, Verbruggen, and Wagenmakers (2014) extended the independent race model to include choice in the go process and the potential for choice in the stop process. They proposed a general race model that assumed choice was implemented as a race between the alternative responses, assuming one runner for each alternative response in the go task and one runner for the stop process. They proposed special race models in which the runners were characterised as specific stochastic accumulators. The best-fitting model represented each runner as a 'diffusion' to a single bound, characterised by three main parameters: non-decision time, the rate of stochastic accumulation, and the threshold on the accumulator. The first runner to reach its threshold determined the outcome of the trial. If it was a go response, then the response was executed. On no-stop-signal trials, it could be correct or erroneous. If the trial was a stop-signal trial (so there was a stop runner in the race), then the winning go response produced a signal-respond trial. If the trial was a stop-signal trial and the stop runner reached threshold first, the winning stop response produced a signal-inhibit trial. The general and special race models make stronger commitments to underlying processes than the original independent race model. Their assumptions may be controversial and may be falsified (Ratcliff & Smith, 2004; Teodorescu & Usher, 2013), but all specific assumptions are controversial, and falsifiability is a virtue. Importantly, the stronger commitments to underlying processes allow stronger inferences. For example, they allowed Logan et al. (2014) to test hypotheses about capacity sharing between stopping and going.

Research on dual tasks indicates that central processing capacity is limited, resulting in a performance decrement when stimuli for the two tasks appear close together in time (i.e., at the equivalent of short SSDs; Pashler, 1994). Stop and go processes do not seem to share capacity in this way (e.g., Logan, 1981; Logan & Burkell, 1986; Verbruggen, Schneider, & Logan, 2008). Yamaguchi, Logan, and Bissett (2012) measured SSRT in a dual-task experiment in which subjects had to stop one of two go tasks. They found that SSRT was unaffected by the temporal overlap of the two tasks and no different in single- and dual-task conditions. Logan et al. (2014) tested capacity sharing more rigorously by using the diffusion race model. They identified capacity limitations with processing rates in the stochastic accumulators, distinguishing between fixed capacity, in which the rate parameters for all the runners in the race sum to a constant, and unlimited capacity, in which the rate parameters for the different runners are unconstrained and independent (Bundesen, 1990; Townsend & Ashby, 1983). They tested these hypotheses about processing rates in a stop-signal experiment that manipulated the number of choice alternatives (2, 4, or 6) in the go task. They found evidence of strong capacity limitations in the go process: adding more go runners to the race slowed the rate at which each runner accumulated evidence. But they found no evidence of capacity sharing between the go process and the stop process: the rate of accumulation in the stop process was the same no matter how many choice alternatives there were in the go task. This converges on more qualitative analyses in previous research on simple stopping (Logan, 1981; Logan & Burkell, 1986; Yamaguchi et al., 2012) and contrasts with recent findings in selective stopping (reviewed below).

It is important to note that this conclusion could not be reached by applying the Logan and Cowan (1984) race model. The conclusion is based on changes in the parameters of the stochastic accumulators in the special race models. Below we present another application of the special race model. Thus, the additional complexity and stronger commitment to mechanism produces stronger conclusions.

Interactive Race Models and Their Alternatives The turn of the century marked a substantial increase in interest in the neural mechanisms that underlie response inhibition. The Logan and Cowan (1984) race model has been useful in this pursuit, but its failure to specify the underlying computational mechanisms limits its ability to specify the underlying neural mechanisms. Boucher, Palmeri, Logan, and Schall (2007) developed the interactive race model, which explained what happened at the end of the race when the stop process won. They modelled the go process as a single diffusion to a threshold and modelled the stop process as a unit that inhibited the growth of go activation after an afferent delay following a stop signal. They made strong linking propositions that identified the go process with movement-related neurons and the stop process with fixation neurons in frontal eye fields, which control eye movements. The model accounted for behaviour as well as a version of the independent race model but in addition, accounted for firing rates recorded from movement and fixation neurons in monkeys performing a countermanding task (i.e., the saccadic version of the stop-signal task). Movement neuron activity is at baseline at the beginning of a trial and rises to a threshold value 10–20 ms before an eye movement occurs. On trials in which stopping is successful, the activity rises towards a threshold following the same trajectory as on no-stop-signal trials, and then falls abruptly before reaching the threshold. The transition from rising to falling coincides with SSRT measured from the monkeys' behaviour, as if some kind of change happens just before SSRT to drive down go activation and inhibit the response. Boucher et al. (2007) identified the change as inhibition from fixation neurons projected onto the movement neurons.

Logan, Yamaguchi, Schall, and Palmeri (2015) challenged the linking proposition connecting the mechanism of stopping to fixation neurons in frontal eye fields by investigating a blocked input model that inhibits responses by blocking the input that drives them. Like Boucher et al. (2007), they implemented the go process as a single diffusion to a threshold, but unlike Boucher et al., they assumed that the stop process sets the growth rate for the go process to zero, removing the 'drive' that causes the rise to a threshold. They showed that this model accounted for behaviour as well as the interactive race model but fit the neural data better. Specifically, the interactive race model required very strong inhibition from stop (fixation) units to go (movement) units, and that strong inhibition produced much steeper decay in movement unit activation after SSRT than what was observed in the neurons. Logan et al. extended both models to account for steady-state fixation activity, constraining the activity of fixation neurons in the fixation period at the beginning of the trial. This improved the neural predictions of the interactive race model, but worsened behavioural predictions. The blocked input model fit the neural data just as well but fit the behavioural data better. Logan et al. also considered a boosted fixation model, which multiplied the activity in the fixation unit to stop responses and found that it fit the behavioural data as well as the blocked input model, though it predicted steady-state fixation activity following stop trials that was 3 to15 times as high as steady-state fixation activity before stop trials. On balance, the blocked input model did better than interactive race or boosted fixation. Nevertheless, it is possible that responses can be inhibited either by blocking the input or by inhibiting the growth of activation in movement neurons.

Cortical-Basal Ganglia Model Imaging, lesion, and animal studies indicate that response inhibition relies on an extensive cortico-basal-ganglia network (see Chapter 16 by Forstmann & Alkemade in this volume). Wiecki and Frank (2013) proposed a computational model that instantiates this network, extending Frank's (2006) basal ganglia model to include cortical structures involved in response inhibition (see also Chapter 21 by Bhandari, Badre, & Frank in this volume). On go trials, saccadic responses are generated via the frontal eye fields and superior colliculus. On stop-signal trials, responses are inhibited via the 'hyperdirect' pathway: via the right inferior frontal gyrus and the subthalamic nucleus, to the substantia nigra pars reticulata, which then inhibits the superior colliculus. Responses are successfully stopped if the hyperdirect pathway inhibits the superior colliculus before it reaches the threshold at which it releases a response. This model correctly predicts neural dynamics in key structures during stop-signal tasks, correctly predicts behaviour, and correctly predicts the effects of lesions and drugs.

The Wiecki and Frank (2013) model is an important advance because it provides a platform for evaluating hypotheses about interactions between brain structures in response inhibition and other tasks that is strongly constrained by physiology, behaviour, and computation. However, the model is complex (it has to capture the interactions between the many neural structures) and difficult to evaluate by standard model fitting techniques from mathematical psychology. Ratcliff and Frank (2012) fit the diffusion model to the output of the basal ganglia model, showing how diffusion model parameters relate to activation of structures in the basal ganglia model. It would be interesting to fit the Logan et al. (2014) special race model to the output of the Wiecki and Frank (2013) model to relate its parameters to activation of its structures in stop-signal tasks.

Stopping Revisited? The first step in successfully stopping a response is nearly always detecting the stop signal (e.g., a traffic light turning red; Verbruggen, McLaren, et al., 2014). The contribution of non-inhibitory processes is still largely neglected in the 'reactive' inhibitory control literature, even though computational work indicates that most of SSRT is occupied by afferent processes (Boucher et al., 2007; Logan et al., 2014, 2015; Salinas & Stanford,

2013), as indicated in several behavioural studies (e.g., Cavina-Pratesi, Bricolo, Prior, & Marzi, 2001; van den Wildenberg & van der Molen, 2004; Verbruggen, Stevens, & Chambers, 2014). For example, perceptual distractors substantially prolong SSRTs in a visual stop-signal task, especially when the visual stop signals occurred in the periphery (Verbruggen, Stevens, et al., 2014). Other studies have shown that stopping deficits in certain clinical populations could be due to impairments in signal detection rather than in inhibition per se (e.g., Bekker et al., 2005). Thus, a failure to detect the signal quickly can have important consequences and lead to 'stopping' deficits.

More generally, the studies discussed in this and previous sections highlight the importance of focusing on the underlying processes, such as signal detection, rather than general and unitary control functions. They also have important consequences for the interpretation of the SSRT. SSRT describes the chain of processes involved in an act of control that results in a response being withheld (Logan et al., 2014; Verbruggen, McLaren, et al., 2014; Verbruggen & Logan, 2015). It includes the duration of perceptual, decisional, and (inhibitory) motor-related processes. Thus, in simple stop-signal tasks and their many variants, SSRT reflects more than the duration of a single neural inhibitory process, and researchers should consider at which processing stages differences between groups or conditions arise (see Verbruggen, McLaren, et al., 2014).

Are Stopping and Going Really Independent?

Beginning with Logan and Cowan (1984), models of response inhibition have always assumed that stopping and going are independent, at least for most of their durations. The models are based on two assumptions about independence: stochastic independence and context independence. Stochastic independence requires that the finishing times of the stop and go process on a given trial are uncorrelated (i.e., $P(\text{stop} < t \text{ and } \text{go} < t) = P(\text{stop} < t) \times P(\text{go} < t)$). This assumption simplifies the mathematics of the race considerably. Context independence assumes that the stop process is unaffected by the presence of the go process, and vice versa. This assumption allows go RT on no-stop-signal trials to be used as an estimate of go RTs on stop-signal trials, to predict signal-respond RTs and calculate SSRT. If context independence is violated, then estimates of SSRT are invalid. Thus, the assumptions should not be taken lightly. Recently, two lines of evidence have challenged the context independence assumption. One addresses selective versions of the stop signal task (Bissett & Logan, 2014; Verbruggen & Logan, 2015), and one addresses short SSDs in all stop tasks (Bissett, 2014).

Selective Stopping Most stop-signal tasks involve 'simple' stopping: there is only one stop signal and only one response to it, namely, stopping the ongoing response. In selective stopping tasks, subjects perform some kind of discrimination before deciding whether to stop. In stimulus selective stopping tasks, the go task is accompanied by a stop signal or an ignore signal; these signals are usually in the same modality but differ along one or more stimulus dimensions. Subjects are instructed to stop the go response when the stop signal occurs and to continue the go response when the ignore signal occurs (e.g., Bissett & Logan, 2014). In motor selective stopping tasks, there is one stop signal, and subjects are instructed to stop one response (e.g., the left hand) but not the other (e.g., the right hand) when the stop signal occurs (e.g., De Jong, Coles, & Logan, 1995). In conditional motor selective stopping tasks, subjects make two concurrent responses to the go tasks, for example, pressing keys with the index fingers in response to one go stimulus and pressing keys with the middle fingers in response to another go stimulus. The stop signal instructs subjects to stop one of the two responses (e.g., the left hand response) but continue with the other (e.g., the right hand response) when a stop signal occurs (e.g., Aron & Verbruggen, 2008).

Subjects succeed at stopping 'selectively' in all of these tasks, but their success comes with a cost: signal-respond RT is often longer than no-stop-signal RT, and RT for the responses that are not supposed to be inhibited is longer than no-stop-signal RT in most experiments (for a review and a re-analysis of published data, see Bissett & Logan, 2014). The longer signal-respond RTs indicate violations of context independence, which invalidates conclusions based on the application of the independent race model to the data and challenges the common assumption that selective stopping increases the complexity of the stop process without affecting the go process (e.g., van de Laar, van den Wildenberg, van Boxtel, & van der Molen, 2010). The longer RTs on ignore trials suggest that going and stopping share processing capacity (Bissett & Logan, 2014; Verbruggen & Logan, 2015); in some cases, subjects may even stop all responses completely first and then restart them if the trial requires it. This challenges the use of the ignore signal in selective stopping tasks as a control for the requirements to attend to the stop signal (e.g., Sharp et al., 2010).

Verbruggen and Logan (2015) tested the hypothesis that the go and stop processes share capacity in selective stopping tasks by manipulating the consistency of mapping between signals and the requirement to stop or ignore in response to the signal. In consistent mapping conditions, each signal played the same role throughout the experiment; in varied mapping conditions, the role changed repeatedly over the course of the experiment. Following Shiffrin and Schneider (1977) and others, we assumed that varied mapping conditions would demand more capacity than the consistent mapping conditions, and so should produce larger violations of context independence. That is what we found.

These selective stopping results are interesting in contrast with simple stopping, where increasing the capacity demands of the go process has no effect on the stop process (see above). We propose that this is due to the low selection demands in standard stop-signal tasks. This does not imply that capacity sharing can never occur in these tasks. The stop rate parameters depend on the discriminability, intensity, and modality of the stop signal (e.g., van der Schoot, Licht, Horsley, & Sergeant, 2005), which could be interpreted as a capacity limitation (Logan et al., 2014). Furthermore, the influence of visual distractors on stopping (see above) is consistent with the idea that stimuli compete for limited processing capacity (Bundesen, 1990; Duncan, 2006). Finally, SSRT is prolonged on incongruent trials in interference control tasks (e.g., Verbruggen, Liefooghe, & Vandierendonck, 2004). This 'functional dependence' could also be interpreted as a capacity limitation. Thus, it seems that under certain circumstances, capacity sharing may occur in simple stop-signal tasks.

Short SSDs Bissett (2014) looked for violations of context independence in simple stop-signal tasks, and found strong violations at SSDs shorter than 250 ms in almost every experiment he analysed. These violations were unexpected from the stop-signal literature, where the independent race model has reigned supreme since the 1980s, but they would be expected from the dual-task literature (see above). These violations may not have been apparent in published stop-signal studies. Many studies now use a tracking procedure that adjusts SSDs to produce successful inhibition on 50% of the trials. This procedure usually produces a bell-shaped distribution of SSDs with a mode well away from the short SSDs where context independence is violated. Moreover, researchers rarely report performance as a function of SSD. Instead, they often collapse across SSD, masking possible violations at short SSDs. Researchers should not try to avoid violations of context independence by excluding short SSDs from their experiments. Subjects are sensitive to the range of SSDs and routinely slow their go task performance when SSDs are long (Logan, 1981). Instead, we suggest that investigators use a broad range of SSDs (e.g., by using the tracking procedure) and then remove the data from SSDs shorter than 200 ms before calculating SSRT. At present, these recommendations are only provisional. More research needs to be done before we can make stronger recommendations.

Balancing Cognitive Demands

Proactive Control

In response-inhibition tasks, people often adjust their behaviour in anticipation of future acts of inhibitory control (*proactive inhibitory control*). Proactive adjustments are made at the beginning of a block or on a trial-by-trial basis, suggesting a flexible cognitive system that adjusts itself quickly in response to novel contextual information (Verbruggen & Logan, 2009b). Proactive inhibitory control has received much attention in the cognitive neuroscience literature (for reviews, see Aron, 2011; Stuphorn & Emeric, 2012), and researchers have explored to what extent individual differences and developmental trends could reflect differences in the deployment of proactive control (e.g., Chevalier, Kelsey, Wiebe, & Espy, 2014). Here we will primarily focus on studies examining which control adjustments are made.

Proactive Adjustments of Attentional and Response Settings People adjust response settings when they anticipate a stop signal. In a series of experiments, we presented cues indicating whether signals were relevant for the next trial(s). We hypothesised that subjects balance going quickly and stopping by adjusting response thresholds in the go task (see above). Increasing the response threshold increases the finishing time of the go process, but decreases both the probability of an incorrect go response on no-stop-signal trials and the probability of a response on stop-signal trials. Our behavioural results and subsequent diffusion model fits supported the response-threshold account (Logan et al., 2014; Verbruggen & Logan, 2009b).

Further studies suggest that proactive inhibitory control can also directly modulate motor activity (e.g., Jahfari, Stinear, Claffey, Verbruggen, & Aron, 2010; Lo, Boucher, Paré, Schall, & Wang, 2009). For example, Lo et al. (2009) observed high tonic activation of fixation neurons before the onset of a stop signal in a countermanding task, and argued that this was activation modulated by a top-down proactive control signal. Suppression of prestimulus motor activity is also observed in go/no-go and choice RT paradigms, which points to a general proactive control mechanism that prevents premature responding (Frank, 2006).

Most studies have focused on preparatory adjustments of response- or motor-related processes. However, in some tasks, perceptual or attentional settings are also adjusted when subjects anticipate stop signals. The distractor study mentioned above contrasted no-stop-signal blocks with stop-signal blocks in which visual stop signals were presented in the centre of the screen and blocks in which the signals could occur in the periphery (Verbruggen, Stevens, et al., 2014). On half of the no-stop-signal trials in each block type, perceptual distractors were presented throughout the trial. These distractors had the largest effect on go responses in the noncentral signal blocks compared with central-signal blocks and no-stop-signal blocks, suggesting that subjects widened their attentional focus for detecting the signal in the periphery, making them more vulnerable to the distractors on no-stop-signal trials. Another study measured event-related potentials in the go task (Elchlepp, Lavric, Chambers, & Verbruggen, 2016). Both early and late go-ERP components were modulated when subjects expected a stop signal to occur. Thus, proactive inhibitory control involves adjusting the settings of both perceptual and response processes to balance the competing demands of the go and stop tasks (Elchlepp et al., 2016). Importantly, similar modulations of ERP components are observed in non-inhibitory tasks, suggesting similarities between various forms of proactive control (Braver, 2012; Chapter 9 by Chiew & Braver in this volume).

Influences of Proactive Control on Stopping Proactive control does not necessarily influence SSRT (Logan & Burkell, 1986; Verbruggen & Logan, 2009b). Whether or not SSRT is influenced may depend on how proactive control adjustments are implemented (e.g., response slowing vs. enhanced monitoring; Wiecki & Frank, 2013). Proactive inhibitory control can also influence *how* people stop their responses. Brain stimulation studies suggest that both task-relevant and irrelevant muscles are suppressed on stop-signal trials, indicating that stopping has global effects on the motor system (Aron, 2011). However, stopping becomes more selective when it can be prepared in advance (e.g., Aron & Verbruggen, 2008; Smittenaar, Guitart-Masip, Lutti, & Dolan, 2013).

Proactive control could even lead to a 'prepared control reflex' (Verbruggen, Best, Bowditch, Stevens, & McLaren, 2014; Verbruggen, McLaren, et al., 2014). When attention is proactively allocated and responses are prepared, goal-directed behaviour may not require much control anymore (see also Chapter 7 by Hommel in this volume); instead, responses could be activated easily by stimuli in the environment (for a recent review, see Meiran, Cole, & Braver, 2012). Response inhibition can be triggered by task-irrelevant primes (e.g., van Gaal, Ridderinkhof, van den Wildenberg, & Lamme, 2009; Verbruggen & Logan, 2009a), but these priming effects are only observed in contexts in which subjects are instructed to stop occasionally (Chiu & Aron, 2014; Verbruggen & Logan, 2009a). These findings are consistent with the prepared reflex idea: once subjects have proactively adjusted attentional and response settings, the 'stop response' can be activated easily by both task-relevant and task-irrelevant information in the environment. Note that the 'prepared reflex idea' might also explain why going and stopping do not share (much) capacity in standard stop-signal tasks (see above).

Post-Signal Slowing

The proactive control literature indicates that people set detection and response criteria at the beginning of a trial or run of trials. These criteria are further adjusted after each trial. Response latencies are often slower after stop-signal trials than after no-stop-signal trials (Bissett & Logan, 2011, 2012; Emeric et al., 2007; Nelson, Boucher, Logan, Palmeri, & Schall, 2010; Rieger & Gauggel, 1999; Verbruggen & Logan, 2008b; Verbruggen, Logan, Liefooghe, & Vandierendonck, 2008). Bissett and Logan (2011) contrasted several accounts of post-stop-signal slowing, and found most support for a strategic adjustment account that proposes that stop-signal presentation encourages subjects to shift priority from the go task to the stop task. Such a shift produces longer response latencies after a signal trial and can reduce SSRT (Bissett & Logan, 2012). However, the SSRT reduction is only observed when the stop-signal modality remains the same (Bissett & Logan, 2012). This could indicate that the post-signal strategic adjustments influence attentional settings. Alternatively, the modality specificity could indicate stop-signal priming (see below).

Contributions From Memory and Long-Term After-Effects of Stopping

Stimulus-Specific Sequential Effects

The post-signal slowing is usually more pronounced when the stimulus, stimulus category, or stimulus features of the previous trial are repeated (Bissett & Logan, 2011; Rieger & Gauggel, 1999; Verbruggen, Logan, et al., 2008). We have attributed this stimulus-specific slowing to the retrieval of stimulus–stop associations: a go stimulus becomes associated with a 'stop' representation on a stop trial; when it is repeated on the next go trial, the stop representation is activated via associative retrieval, and this will suppress the go response (Verbruggen & Logan, 2008b; Verbruggen, Logan, et al., 2008). Activation of these stop representations has global effects on responding (Giesen & Rothermund, 2014), which suggests that they are general ('stop responding') rather than response-specific (e.g., 'stop the left response').

Associative Learning and Automatic Inhibition

The stimulus-specific stop effects are observed up to 20 trials after the presentation of the stop signal (Verbruggen & Logan, 2008b). Such long-term associations may support the development of automatic response inhibition (Logan, 1988, 1990). In a series of experiments, we trained subjects to stop their response to a specific stimulus, and then reversed the stimulus–stop mappings in a test phase. In this test phase, subjects were slower to respond to stimuli previously associated with stopping compared with stimuli that they had not seen before or stimuli that were inconsistently associated with going and stopping (Verbruggen & Logan, 2008a). On the basis of this finding, we proposed the 'automatic inhibition' hypothesis: inhibitory control in go/no-go and stop-signal tasks can be triggered automatically via the retrieval of stimulus–stop associations from memory.

Subsequent work indicated that different associations can be learned. For example, subjects can learn stimulus-signal associations rather than stimulus–stop associations (Verbruggen, Best, et al., 2014). Such associations between the stop items and the stop signal will prime the representation of the stop-signal detection rather than the stop goal or stop response. As discussed above, signal detection plays a critical role in successful stopping. Thus, by priming the representation of the stop signal, learning could lead to improvements in stopping performance without influencing responding on go trials. The signal learning idea can also account for the modality specificity of post-stop signal improvements in SSRT (see above).

Stopping responses to stimuli can also lead to devaluation of these stimuli and reduce their behavioural incentive (e.g., Kiss, Raymond, Westoby, Nobre, & Eimer, 2008; Wessel, O'Doherty, Berkebile, Linderman, & Aron, 2014). Research on Pavlovian conditioning suggests links between an aversive system and avoidance behaviour (withdrawal, suppression) and between an appetitive system and approach behaviour (Dickinson & Balleine, 2002). Thus, when subjects have to stop their response to a specific stimulus, a link between this stimulus and the aversive/avoidance system may be created, impacting their motivational value. To account for these findings, McLaren and Verbruggen (2016) and Verbruggen, Best, et al. (2014) proposed an associative stop system that combines elements of associative learning theories, the interactive race model, and a Pavlovian model of motivational systems.

In sum, associative learning plays an important role in response-inhibition tasks. It is important to note that the expression of stimulus-specific training effects strongly depends on attentional settings (Best, Lawrence, Logan, McLaren, & Verbruggen, 2016). After learning, subjects also form expectations about the occurrence of a stop signal, which influences subsequent performance. These findings suggest a strong interplay between learning and 'top-down' control processes (Best et al., 2016).

Applications of Automatic Inhibition

Automatic inhibition can have general effects on behaviour. For example, consistent pairing of food-related pictures to stopping in a go/no-go or stop-signal paradigm reduces subsequent food consumption (e.g., Houben & Jansen, 2011; Lawrence, Verbruggen, Morrison, Adams, & Chambers, 2015), and may even lead to weight loss (Lawrence et al., 2015; Veling, van Koningsbruggen, Aarts, & Stroebe, 2014). Recent meta-analyses indicate that no-go/stop training has small but robust effects on food and alcohol consumption (Allom, Mullan, & Hagger, 2015; Jones et al., 2016). Go/no-go training may have larger effects on subsequent appetitive behaviour than stop-signal training. Go/no-go tasks often use a more consistent stimulus–stop pairing, and the probability of responding is usually lower than in stop-signal tasks, which could influence the strength of stimulus–stop associations (Verbruggen & Logan, 2008a, 2008b).

The stimulus-specific stop training opens promising new avenues for treatment. However, more work is required to determine how the training protocols can be optimised to produce large and long-lasting effects. This will also require a careful study of the underlying mechanisms and a better understanding of how task characteristics influence what is learned.

General Learning and Carry-Over Effects

Some studies have also explored whether people can generally learn to stop their responses and whether practice effects can transfer to other domains. The results are mixed. Some studies found that SSRT decreased over practice (e.g., Berkman, Kahn, & Merchant, 2014; Chevalier, Chatham, & Munakata, 2014; Logan & Burkell, 1986), whereas others failed to find such an SSRT decrease (e.g., Enge et al., 2014) General stop-signal task training may induce a shift from reactive to proactive control (Berkman et al., 2014; Chevalier, Chatham, et al., 2014). This idea could help to reconcile the inconsistent training results, as not all forms of proactive control lead to SSRT improvements (see above).

General stop training and encouraging people to be cautious in stop-signal blocks can reduce risk taking in gambling tasks (Stevens et al., 2015; Verbruggen, Adams, & Chambers, 2012) and alcohol-seeking behaviour (Jones, Cole, Goudie, & Field, 2011). However, these effects are small and short-lived (Stevens et al., 2015; Verbruggen et al., 2013), which could potentially explain why some studies failed to observe far-transfer effects (e.g., Enge et al., 2014).

Conclusions

In the past two decades, we have seen an exponential increase in the number of studies examining the neurocognitive processes underlying response inhibition and correlations with behaviours outside the lab (Verbruggen, Chambers, et al., 2013). This work has led to detailed accounts of reactive inhibitory control (Logan et al., 2014). More recently, several studies have also shown how people adjust settings of lower-level perceptual and response systems to find a balance between going quickly and stopping in response to changes in the environment or internal state. Evidence that learning plays a critical role when stopping a response has challenged theory and opened up an exciting new literature on the training and transfer of response inhibition. Above all, the work reviewed in this chapter highlights again that stopping or withholding a response relies on an interplay between many processes that take place on different time scales.

Acknowledgement

FV is a Royal Society Wolfson Research Merit Award holder and currently supported by a starting grant from the European Research Council (ERC) under the European Union's Seventh Framework Programme (Grant Agreement No. 312445). GDL is supported by Grant No. R01 EY021833 from the National Eye Institute.

References

Allom, V., Mullan, B., & Hagger, M. (2015). Does inhibitory control training improve health behaviour? A meta-analysis. *Health Psychol Rev*, 1–38.

Aron, A. R. (2011). From reactive to proactive and selective control: Developing a richer model for stopping inappropriate responses. *Biological Psychiatry*, *69*, e55–68.

Aron, A. R., & Verbruggen, F. (2008). Stop the presses: Dissociating a selective from a global mechanism for stopping. *Psychological Science*, *19*, 1146–1153.

Bekker, E. M., Overtoom, C. C., Kooij, J. J., Buitelaar, J. K., Verbaten, M. N., & Kenemans, J. L. (2005). Disentangling deficits in adults with attention-deficit/hyperactivity disorder. *Archives of General Psychiatry*, *62*, 1129–1136.

Berkman, E. T., Kahn, L. E., & Merchant, J. S. (2014). Training-induced changes in inhibitory control network activity. *The Journal of Neuroscience*, *34*, 149–157.

Best, M., Lawrence, N. S., Logan, G. D., McLaren, I. P. L., & Verbruggen, F. (2016). Should I Stop or Should I Go? The Role of Associations and Expectancies. *Journal of Experimental Psychology: Human Perception and Performance*, *42*(1), 115–137.

Bissett, P. G. (2014). *Evaluating the independent race model for the stop signal paradigm: Context independence is violated at short stop signal delays.* Unpublished dissertation, Vanderbilt University.

Bissett, P. G., & Logan, G. D. (2011). Balancing cognitive demands: Control adjustments in the stop-signal paradigm. *Journal of Experimental Psychology: Learning, Memory, and Cognition*, *37*, 392–404.

Bissett, P. G., & Logan, G. D. (2012). Post-stop-signal adjustments: Inhibition improves subsequent inhibition. *Journal of Experimental Psychology: Learning, Memory, and Cognition*, *38*, 955–966.

Bissett, P. G., & Logan, G. D. (2014). Selective stopping? Maybe not. *Journal of Experimental Psychology: General*, *143*, 455–472.

Boucher, L., Palmeri, T. J., Logan, G. D., & Schall, J. D. (2007). Inhibitory control in mind and brain: An interactive race model of countermanding saccades. *Psychological Review*, *114*, 376–397.

Braver, T. S. (2012). The variable nature of cognitive control: A dual mechanisms framework. *Trends in Cognitive Sciences*, *16*, 106–113.

Bundesen, C. (1990). A theory of visual attention. *Psychological Review*, *97*, 523–547.

Cavina-Pratesi, C., Bricolo, E., Prior, M., & Marzi, C. A. (2001). Redundancy gain in the stop-signal paradigm: Implications for the locus of coactivation in simple reaction time. *Journal of Experimental Psychology: Human Perception and Performance*, *27*, 932–941.

Chevalier, N., Chatham, C. H., & Munakata, Y. (2014). The practice of going helps children to stop: The importance of context monitoring in inhibitory control. *Journal of Experimental Psychology: General*, *143*(3), 959–965.

Chevalier, N., Kelsey, K. M., Wiebe, S. A., & Espy, K. A. (2014). The temporal dynamic of response inhibition in early childhood: An ERP study of partial and successful inhibition. *Developmental Neuropsychology*, *39*, 585–599.

Chiu, Y. C., & Aron, A. R. (2014). Unconsciously triggered response inhibition requires an executive setting. *Journal of Experimental Psychology: General, 143*, 56–61.

De Jong, R., Coles, M. G. H., & Logan, G. D. (1995). Strategies and mechanisms in nonselective and selective inhibitory motor control. *Journal of Experimental Psychology: Human Perception and Performance, 21*, 498–511.

Dickinson, A., & Balleine, B. (2002). The role of learning in the operation of motivational systems. In C. R. Gallistel (Ed.), *Stevens' handbook of experimental psychology* (Vol. *3*, 3rd ed., pp. 497–534). New York, NY: Wiley.

Donders, F. C. (1868/1969). On the speed of mental processes. *Acta Psychologica, 30*, 412–431.

Duncan, J. (2006). EPS mid-career award 2004: Brain mechanisms of attention. *Quarterly Journal of Experimental Psychology, 59*, 2–27.

Elchlepp, H., Lavric, A., Chambers, C. D., & Verbruggen, F. (2016). Proactive inhibitory control: A general biasing account. *Cognitive Psychology, 86*, 27–61.

Emeric, E. E., Brown, J. W., Boucher, L., Carpenter, R. H., Hanes, D. P., Harris, R.,...Schall, J. D. (2007). Influence of history on saccade countermanding performance in humans and macaque monkeys. *Vision Research, 47*, 35–49.

Enge, S., Behnke, A., Fleischhauer, M., Küttler, L., Kliegel, M., & Strobel, A. (2014). No evidence for true training and transfer effects after inhibitory control training in young healthy adults. *Journal of Experimental Psychology: Learning, Memory, and Cognition, 40*, 987–1001.

Frank, M. J. (2006). Hold your horses: A dynamic computational role for the subthalamic nucleus in decision making. *Neural Networks, 19*, 1120–1136.

Giesen, C., & Rothermund, K. (2014). You better stop! Binding 'stop' tags to irrelevant stimulus features. *Quarterly Journal of Experimental Psychology, 67*, 809–832.

Houben, K., & Jansen, A. (2011). Training inhibitory control. A recipe for resisting sweet temptations. *Appetite, 56*, 345–349.

Jahfari, S., Stinear, C. M., Claffey, M., Verbruggen, F., & Aron, A. R. (2010). Responding with restraint: What are the neurocognitive mechanisms? *Journal of Cognitive Neuroscience, 22*(7), 1479–1492.

Jones, A., Cole, J., Goudie, A., & Field, M. (2011). Priming a restrained mental set reduces alcohol-seeking independently of mood. *Psychopharmacology, 218*(3), 557–565.

Jones, A., Di Lemma, L. C. G., Robinson, E., Christiansen, P., Nolan, S., Tudur-Smith, C., & Field, M. (2016). Inhibitory control training for appetitive behaviour change: A meta-analytic investigation of mechanisms of action and moderators of effectiveness. *Appetite, 97*, 16–28.

Kiss, M., Raymond, J. E., Westoby, N., Nobre, A. C., & Eimer, M. (2008). Response inhibition is linked to emotional devaluation: Behavioural and electrophysiological evidence. *Frontiers in Human Neuroscience, 2*, 13.

Lappin, J. S., & Eriksen, C. W. (1966). Use of a delayed signal to stop a visual reaction-time response. *Journal of Experimental Psychology, 72*, 805–811.

Lawrence, N. S., O'Sullivan, J., Parslow, D., Javaid, M., Adams, R. C., Chambers, C. D., ... Verbruggen, F. (2015). Training response inhibition to food is associated with weight loss and reduced energy intake. *Appetite, 95*, 17–28.

Lawrence, N. S., Verbruggen, F., Morrison, S., Adams, R. C., & Chambers, C. D. (2015). Stopping to food can reduce intake. Effects of stimulus-specificity and individual differences in dietary restraint. *Appetite, 85*, 91–103.

Lo, C. C., Boucher, L., Paré, M., Schall, J. D., & Wang, X. J. (2009). Proactive inhibitory control and attractor dynamics in countermanding action: A spiking neural circuit model. *The Journal of Neuroscience, 29*, 9059–9071.

Logan, G. D. (1981). Attention, automaticity, and the ability to stop a speeded choice response. In J. Long & A. D. Baddeley (Eds.), *Attention and performance IX* (pp. 205–222). Hillsdale, NJ: Erlbaum.

Logan, G. D. (1988). Toward an instance theory of automatization. *Psychological Review, 95*, 492–527.

Logan, G. D. (1990). Repetition priming and automaticity: Common underlying mechanisms? *Cognitive Psychology, 22*, 1–35.

Logan, G. D., & Burkell, J. (1986). Dependence and independence in responding to double stimulation—A comparison of stop, change, and dual-task paradigms. *Journal of Experimental Psychology: Human Perception and Performance, 12*, 549–563.

Logan, G. D., & Cowan, W. B. (1984). On the ability to inhibit thought and action: A theory of an act of control. *Psychological Review*, *91*(3), 295–327.

Logan, G. D., Van Zandt, T., Verbruggen, F., & Wagenmakers, E. J. (2014). On the ability to inhibit thought and action: General and special theories of an act of control. *Psychological Review*, *121*, 66–95.

Logan, G. D., Yamaguchi, M., Schall, J. D., & Palmeri, T. J. (2015). Inhibitory control in mind and brain 2.0: Blocked input models of saccadic countermanding. *Psychological Review*, *122*, 115–147.

McLaren, I. P. L., & Verbruggen, F. (2016). Association, inhibition, and action. In R. Murphy & R. Honey (Eds.), *The Wiley Handbook on the Cognitive Neuroscience of Learning* (pp. 489–514). Chichester, England: Wiley-Blackwell.

Meiran, N., Cole, M. W., & Braver, T. S. (2012). When planning results in loss of control: Intention-based reflexivity and working-memory. *Frontiers in Human Neuroscience*, *6*, 104.

Munakata, Y., Snyder, H. R., & Chatham, C. H. (2012). Developing cognitive control: Three key transitions. *Current Directions in Psychological Science*, *21*, 71–77.

Nelson, M. J., Boucher, L., Logan, G. D., Palmeri, T. J., & Schall, J. D. (2010). Nonindependent and nonstationary response times in stopping and stepping saccade tasks. *Attention, Perception & Psychophysics*, *72*, 1913–1929.

Ratcliff, R., & Frank, M. J. (2012). Reinforcement-based decision making in corticostriatal circuits: Mutual constraints by neurocomputational and diffusion models. *Neural Computation*, *24*, 1186–1229.

Ratcliff, R., & Smith, P. L. (2004). A comparison of sequential sampling models for two-choice reaction time. *Psychological Review*, *111*, 333–367.

Rieger, M., & Gauggel, S. (1999). Inhibitory after-effects in the stop signal paradigm. *British Journal of Psychology*, *90*, 509–518.

Salinas, E., & Stanford, T. R. (2013). The countermanding task revisited: Fast stimulus detection is a key determinant of psychophysical performance. *Journal of Neuroscience*, *33*(13), 5668–5685.

Schevernels, H., Bombeke, K., Van der Borght, L., Hopf, J.-M., Krebs, R. M., & Boehler, C. N. (2015). Electrophysiological evidence for the involvement of proactive and reactive control in a rewarded stop-signal task. *NeuroImage*, *121*, 115–125.

Sharp, D. J., Bonnelle, V., De Boissezon, X., Beckmann, C. F., James, S. G., Patel, M. C., & Mehta, M. A. (2010). Distinct frontal systems for response inhibition, attentional capture, and error processing. *Proceedings of the National Academy of Sciences of the United States of America*, *107*, 6106–6111.

Shiffrin, R. M., & Schneider, W. (1977). Controlled and automatic human information processing: II. Perceptual learning, automatic attending, and a general theory. *Psychological Review*, *84*, 127–190.

Smittenaar, P., Guitart-Masip, M., Lutti, A., & Dolan, R. J. (2013). Preparing for selective inhibition within frontostriatal loops. *The Journal of Neuroscience*, *33*, 18087–18097.

Stevens, T., Brevers, D., Chambers, C. D., Lavric, A., McLaren, I. P., Mertens, M.,…Verbruggen, F. (2015). How does response inhibition influence decision making when gambling? *Journal of Experimental Psychology: Applied*, *21*, 15–36.

Stuphorn, V., & Emeric, E. E. (2012). Proactive and reactive control by the medial frontal cortex. *Frontiers in Neuroengineering*, *5*, 9.

Teodorescu, A. R., & Usher, M. (2013). Disentangling decision models: From independence to competition. *Psychological Review*, *120*, 1–38.

Townsend, J. T., & Ashby, F. G. (1983). *Stochastic modeling of elementary psychological processes.* Cambridge: Cambridge University Press.

van de Laar, M. C., van den Wildenberg, W. P. M., van Boxtel, G., & van der Molen, M. W. (2010). Processing of global and selective stop signals: Application of Donders' subtraction method to stop-signal task performance. *Experimental Psychology*, *57*, 149–159.

van den Wildenberg, W. P., & van der Molen, M. W. (2004). Additive factors analysis of inhibitory processing in the stop-signal paradigm. *Brain and Cognition*, *56*, 253–266.

van der Schoot, M., Licht, R., Horsley, T. M., & Sergeant, J. A. (2005). Effects of stop signal modality, stop signal intensity and tracking method on inhibitory performance as determined by use of the stop signal paradigm. *Scandinavian Journal of Psychology*, *46*, 331–341.

van Gaal, S., Ridderinkhof, K. R., van den Wildenberg, W. P., & Lamme, V. A. (2009). Dissociating consciousness from inhibitory control: Evidence for unconsciously triggered response inhibition in the stop-signal task. *Journal of Experimental Psychology: Human Perception and Performance, 35*, 1129–1139.

Veling, H., van Koningsbruggen, G. M., Aarts, H., & Stroebe, W. (2014). Targeting impulsive processes of eating behavior via the internet. Effects on body weight. *Appetite, 78*, 102–109.

Verbruggen, F., & Logan, G. D. (2008a). Automatic and controlled response inhibition: Associative learning in the go/no-go and stop-signal paradigms. *Journal of Experimental Psychology: General, 137*, 649–672.

Verbruggen, F., & Logan, G. D. (2008b). Long-term aftereffects of response inhibition: Memory retrieval, task goals, and cognitive control. *Journal of Experimental Psychology: Human Perception and Performance, 34*, 1229–1235.

Verbruggen, F., & Logan, G. D. (2009a). Automaticity of cognitive control: Goal priming in response-inhibition paradigms. *Journal of Experimental Psychology: Learning, Memory, and Cognition, 35*, 1381–1388.

Verbruggen, F., & Logan, G. D. (2009b). Proactive adjustments of response strategies in the stop-signal paradigm. *Journal of Experimental Psychology: Human Perception and Performance, 35*, 835–854.

Verbruggen, F., & Logan, G. D. (2009c). Models of response inhibition in the stop-signal and stop-change paradigms. *Neuroscience and Biobehavioral Reviews, 33*, 647–661.

Verbruggen, F., & Logan, G. D. (2015). Evidence for capacity sharing when stopping. *Cognition, 142*, 81–95.

Verbruggen, F., Adams, R., & Chambers, C. D. (2012). Proactive motor control reduces monetary risk taking in gambling. *Psychological Science, 23*, 805–815.

Verbruggen, F., Adams, R. C., van't Wout, F., Stevens, T., McLaren, I. P., & Chambers, C. D. (2013). Are the effects of response inhibition on gambling long-lasting? *PLoS ONE, 8*, e70155.

Verbruggen, F., Best, M., Bowditch, W. A., Stevens, T., & McLaren, I. P. L. (2014). The inhibitory control reflex. *Neuropsychologia, 65*, 263–278.

Verbruggen, F., Chambers, C. D., & Logan, G. D. (2013). Fictitious inhibitory differences: How skewness and slowing distort the estimation of stopping latencies. *Psychological Science, 24*(3), 352–362.

Verbruggen, F., Liefooghe, B., & Vandierendonck, A. (2004). The interaction between stop signal inhibition and distractor interference in the flanker and stroop task. *Acta Psychologica, 116*, 21–37.

Verbruggen, F., Logan, G. D., Liefooghe, B., & Vandierendonck, A. (2008). Short-term aftereffects of response inhibition: Repetition priming or between-trial control adjustments? *Journal of Experimental Psychology: Human Perception and Performance, 34*, 413–426.

Verbruggen, F., McLaren, I. P. L., & Chambers, C. D. (2014). Banishing the control homunculi in studies of action control and behavior change. *Perspectives on Psychological Science, 9*, 497–524.

Verbruggen, F., Schneider, D. W., & Logan, G. D. (2008). How to stop and change a response: The role of goal activation in multitasking. *Journal of Experimental Psychology. Human Perception and Performance, 34*(5), 1212–28.

Verbruggen, F., Stevens, T., & Chambers, C. D. (2014). Proactive and reactive stopping when distracted: An attentional account. *Journal of Experimental Psychology: Human Perception and Performance, 40*, 1295–1300.

Vince, M. A. (1948). The intermittency of control movements and the psychological refractory period. *British Journal of Psychology, 38*, 149–157.

Wessel, J. R., O'Doherty, J. P., Berkebile, M. M., Linderman, D., & Aron, A. R. (2014). Stimulus devaluation induced by stopping action. *Journal of Experimental Psychology: General, 143*, 2316–2329.

Wiecki, T. V., & Frank, M. J. (2013). A computational model of inhibitory control in frontal cortex and basal ganglia. *Psychological Review, 120*, 329–355.

Yamaguchi, M., Logan, G. D., & Bissett, P. G. (2012). Stopping while going! Response inhibition does not suffer dual-task interference. *Journal of Experimental Psychology: Human Perception and Performance, 38*, 123–134.

7

Consciousness and Action Control

Bernhard Hommel

One of the key questions that guide and organise research on human consciousness is functional in nature: What is consciousness good for? Many answers to this question are possible, but the one that I shall be dealing with in this chapter is particularly popular. Although many laypeople and researchers would be willing to grant that many perceptual, to some degree even attentional processes occur outside of conscious awareness, the decision how to react to a given stimulus or situation is often assumed to be either necessarily or at least preferentially conscious. This idea is nicely captured by the blurb for the book *Better Than Conscious?* edited by Engel and Singer (2008): 'Conscious control enables human decision makers to override routines, to exercise willpower, to find innovative solutions, to learn by instruction, to decide collectively, and to justify their choices'. If correct, the implications of this characterisation are that decision making and action control are particularly promising targets of human consciousness, as they support and improve the quality of the way we deal with environmental challenges.

One of the origins of this widely held view is rooted in the psychoanalytic approach of Sigmund Freud. He claimed that both conscious and unconscious processes can contribute to the control of human action and considered control to emerge from the interplay between unconscious, automatic action tendencies generated by the pleasure-seeking Id and socially acceptable considerations provided by the rational Ego (Freud, 1923). Whereas processes initiated by the Id were assumed to be inaccessible to consciousness in principle, some, but not all, Ego operations were considered to be conscious. According to Freud, leaving action control to the Id would be problematic in modern societies, which calls for moderation by the Ego, whose task is to seek compromise between the Id's pleasure-driven urges and the strict requirements of the Superego, which represents societal norms, expectations, and principles.

More modern approaches still buy into this psychoanalytic scenario and the idea that socially responsible action requires conscious mediation. Indeed, so-called 'dual-route' or 'dual-process' models can be found in almost all psychological and cognitive-neuroscientific research areas (for an overview, see Evans & Stanovich, 2013), and even in the theoretical foundations and everyday practices of modern law (e.g., Günther, 2003). For instance, in

---

The Wiley Handbook of Cognitive Control, First Edition. Edited by Tobias Egner.

their action-control model, Norman and Shallice (1986) distinguish between habitual, stimulus-driven actions on the one hand and actions that are assumed to be under 'deliberate conscious control'. It is interesting to note that this terminology implies (for no given reason) that unconscious deliberate control is inconceivable. In the same spirit, Libet (1985) has suggested that consciousness might have a 'veto' that prevents unwanted actions from execution. It is true that this veto idea puts consciousness somewhat later into the information-processing chain than other approaches would imply, but it still follows the tradition of (a) identifying responsible decision making with consciousness and (b) locating this process in between perception and action. The folk-psychological idea underlying these choices is that we, as conscious agents, may not be able to control what we perceive but we do control how we react to that.

The possible connection between consciousness, deliberate choice, and action control has many interesting and far-reaching implications, including questions regarding the freedom of choice, social responsibility, and legal accountability. None of these big questions will be addressed in this chapter, which instead will focus on the possible role of consciousness in the planning and selection of rather simple actions, which often consist in just a key-press. The hope is, of course, that insight into this role scales up to more complex and more important actions, but it should be kept in mind that this is just an assumption. Moreover, this chapter will focus on the technical aspects of action control, on the mechanisms underlying the selection and planning of voluntary actions given a particular goal, but not on the processes leading to the formation of a goal or intention. In other words, the chapter will focus on the volitional rather than on the motivational aspects of action control (cf., Heckhausen & Gollwitzer, 1987).

At this point, there is no comprehensive approach that addresses the role of consciousness in action control. Rather, there are a number of research lines that tap into some aspects of this connection, often driven by little more than a working hypothesis or a general question. In the following, I shall try to organise some of the greater and/or more converging lines into what I consider four possible functions of consciousness (goal implementation, conflict resolution, agency, and task sharing) or, perhaps better, of functions that correlate with conscious experience. The reason why I emphasise the correlational aspect is that there is not yet any evidence that speaks to the causal role of consciousness for the functions that I shall address (Hommel, 2013). All that the available findings show is that some functions are often, perhaps always, associated with conscious experience—operationalised as the ability of individuals to report about it. This does not provide watertight evidence that the having of this experience is essential or required for the respective function to be effective. Hence, philosophically speaking, we have no evidence to rule out the zombie argument (Moody, 1994): it may well be possible to imagine a being that is exactly like a human being except for the ability to have conscious experience. Nevertheless, even if a zombie would be conceivable (which would imply that the addition of conscious experience is unnecessary), the observation that some functions are systematically related to conscious experience may tell us something about the characteristics of, and commonalities between, these functions—an implication that I shall discuss in the 'Conclusions' section. Moreover, I shall not follow the widespread custom of speaking of 'conscious processes'—which I consider a misnomer. There is no evidence that people can be conscious of any cognitive process: we have no idea about how we are retrieving a memory or how we attend to a particular stimulus. What people are conscious of are states *resulting* from such processes: We are conscious of the retrieved memory and the attended stimulus. It would be equally misleading to speak of conscious functions for the same reasons, which is why I shall stick with the less elegant but theoretically more transparent reference to cognitive functions that are associated with conscious experience.

Goal Implementation

When searching for the functionality of a phenomenon or process, its absence is often as informative as is its presence. With respect to consciousness, the motor aspects of human actions have been identified as an area of conspicuous absence from early on. Like his contemporaries, William James (1890, p. 499) observed that 'we are only conversant with the outward results of our volition, and not with the hidden inner machinery of nerves and muscles which are what it primarily sets it at work'. Take the example of tying your shoes or riding a bike: If you are to explain how you are doing that, all you can do is describe each single step from an outsider perspective that often does not reveal any 'privileged knowledge'. In other words, you basically describe your own action the same way as any observer might be able to do—except perhaps for occasional references to proprioceptive or affective impressions that are not accessible to an observer.

This kind of 'executive ignorance' (Turvey, 1977) suggests that people have very little insight into how they are performing a particular action. What they focus on instead are the events that their actions are meant to produce. Brunswik (1944) has tried to capture this aspect by distinguishing between the 'distal' goal event and the 'proximal' means to achieve it, which corresponds to the conscious representation of the outcome of an action and the peripheral muscle commands needed to produce that outcome, respectively. A possible reason why conscious representations refer to distal, but not to proximal, aspects of an action has been suggested by Prinz (1992). As he argues, action needs to be informed by perception, and vice versa, which raises the question how perception and action 'communicate' and which internal 'language' (i.e., code, reference frames, etc.) they may use. Proximal representations of perceived and produced events (i.e., early sensory codes and muscle commands, respectively) are not closely related to their central representations and to each other—they are coded in very different ways, which must make direct communication difficult. In contrast, distal representations (i.e., codes that refer to the external attributes of perceived and produced events) can be considered to be of the same format, which should facilitate informational transfer between perception and action. Accordingly, the theory of event coding (Hommel, Müsseler, Aschersleben, & Prinz, 2001) claims that human cognition is based on distal representations that are sensorimotor in nature. The discovery of so-called mirror neurons (Rizzolatti & Craighero, 2004) points to the same direction: If the same neuron responds to both the production of a particular action and the observation of the same action carried out by another person, it must code for the distal, but not the proximal, characteristics of the action. Distal coding, however, leaves the actual motor aspects of actions out of the picture.

It is interesting how we can afford to be so ignorant of the motor aspects. The answer of James (1890) and contemporaries (Lotze, 1852; Harless, 1861) has become known as ideomotor theory (Shin, Proctor, & Capaldi, 2010; Stock & Stock, 2004). The basic idea is that infants and novices first engage in more or less random movements (motor babbling), which generate movement-specific sensory consequences. Contingencies between motor patterns and sensory consequences are automatically picked up and stored. If then the agent wishes to produce one of the consequences voluntarily, he or she only needs to internally reactivate (i.e., imagine) the representation of these consequences, which will spread activation to the associated motor patterns. Hence, motor activities are controlled by operating on codes of their sensory consequences. This means that conscious access to the latter is sufficient to control the former, so that conscious access to motor activities is not necessary—executive ignorance.

According to this ideomotor scenario, action goals are not given at birth, in contrast to what developmental nativists have suggested (e.g., Rochat, 2001), but emerge through sensorimotor experience (Verschoor, Weidema, Biro, & Hommel, 2010). Conscious experience would then be tied to the implementation of a particular goal, which again would be coded in terms of the intended and expected action outcomes, the sensory action effects. Why would that be the case? It has been suggested that consciousness is related, perhaps even restricted, to the integration of knowledge (e.g., Baars, 1988; but see Morsella, 2005, for some limitations of the integration hypothesis). If so, consciousness would be more likely to be associated with processes and functions that consider multiple sources of information, and coding and implementing a goal is arguably meeting this criterion. Action effects are often highly context dependent, as the same movement can lead to many different sensory consequences in different situations—just think of pushing a button, which might create a letter on a monitor, switch on the light, or start a nuclear war. This means that identifying the action effect that is best suited to serve the current goal, or (if the goal is not going beyond producing that action effect) that satisfies current goal constraints, requires the integration of quite a number of informational sources. Once the correct action–effect code has been selected, the associated motor pattern could be more or less automatically primed, and the action could be executed without any further consideration about the context. If so, the hypothesised link between consciousness and information integration would indeed suggest that consciousness is more related to the selection of action–effect codes than to the activation of motor patterns.

The possibility that conscious experience is associated with the implementation of action goals rather than their online control is also consistent with an early observation of Exner (1879). He carried out various informal experiments on himself and noticed that his conscious activity was mainly concerned with setting up a particular task and preparing oneself for carrying it out. Once that was done, an external stimulus would be sufficient to trigger the further performance of the task without much conscious ado. Hence, implementing an action goal could be considered to turn the cognitive system into a 'prepared reflex' (Hommel, 2000; Woodworth, 1938). Later, more systematic experimental research has supported the idea that implementing a goal and a related action plan delegates control to internal and external stimuli, which may operate outside of consciousness (Bargh, 1989).

Among other things, this perspective tends to undermine dual-route theorising. Take, for instance, the notorious Stroop (1935) effect. The effect consists in the demonstration that naming the colour of a word is slower and less accurate if this word refers to an incongruent colour (e.g., the word RED written in green ink) than if it refers to a congruent colour (the word GREEN written in green ink). Dual-route models attribute this effect to a competition between the intentional route, which is responsible for translating word *colour* into the corresponding naming response, and an automatic, habitual route, which in this case represents the previously acquired reading skill (for a review, see MacLeod, 1991). As the automatic route is not under intentional control, the cognitive system needs to counteract the automatic activation of the reading response in incongruent trials, which takes time and effort—as reflected in delayed reaction time and reduced accuracy. While the resulting conflict scenario may be correct, the prepared-reflex view would suggest that the reading response is a result of the task goal to utter colour words in response to colour stimuli. This would fit with the observation that the Stroop effect is very small (less than 10% of its original size) or absent in task versions in which the verbal naming response is replaced by key-pressing responses (McClain, 1983), and even the small remainders disappear entirely if covert naming responses are prevented through concurrent articulation (Chmiel, 1984; Martin, 1978).

A similar scenario holds for the Simon effect (Simon & Rudell, 1967). The effect shows that spatially defined responses, such as pressing a left versus right key, are delayed and less accurate if the stimulus appears at a location that corresponds to an incorrect response. Dual-route models attribute this observation to competition between the intentional route, responsible for translating the relevant stimulus feature (e.g., shape) into the correct response, and an automatic route that unintentionally translates stimulus location into response location (for reviews, see Hommel, 2011; Proctor, 2011). But, again, the prepared-reflex approach suggests another scenario: Preparing for a spatially defined response requires the emphasis on (i.e., the stronger weighting of) location features (Memelink & Hommel, 2013), with the side effect that the nominally irrelevant stimulus location now attracts attention (Wykowska, Schubö, & Hommel, 2009). If so, the Simon effect would be a necessary—and in a sense, artificial—consequence of preparing for a task with spatially defined responses. Indeed, Valle-Inclán and Redondo (1998) were able to show that behavioural and electrophysiological indicators of stimulus-induced response activation in a Simon task are obtained if the relevant S-R mapping appears before the stimulus (so that the goal could be implemented ahead of the stimulus), but disappear if the mapping appears after the stimulus.

Taken together, these and other observations leave the possibility that conscious experience is associated with cognitive functions responsible for the implementation of goals and the preparation of related action plans. These functions are likely to affect both 'intentional' operations foreseen in the planning process and 'automatic' operations that result as a by-product of planning. As preparation can be considered to turn the cognitive system into a kind of prepared reflex, all processes that follow the planning phase (whether they are intended or a by-product) are likely to possess some or all characteristics of 'automaticity'. Nevertheless, as they were all enabled through goal implementation, they can just as well be considered 'intentional'—which basically undermines the common intentional/automatic dichotomy (cf., Hommel, 2000, 2007).

Conflict Resolution

In addition to a possible role related to the implementation of goals, conscious experience has been associated with the resolution of response conflicts (see also Chapter 4 by Egner in this volume). Support for this possibility has been derived from a seminal study of Libet, Wright, and Gleason (1982). In this study, participants carried out simple key-presses whenever they would feel an urge to do so, and made use of a clever self-report procedure to estimate the point in time when the urge would be experienced. The estimated time of experience preceded the actual movement, as one would expect. The more controversial observation was that physiological indicators of action preparation preceded the agent's conscious urge to act by hundreds of milliseconds. This finding has sparked widespread debates and encouraged authors to draw far-reaching conclusions, including the claim that acts of will 'cannot cause voluntary actions' and that the freedom of will 'is an illusion' (Roth, 2003). Although one can argue about the rationality of these and related arguments, the rather 'late' occurrence of conscious experience has been taken to suggest that it might be related to a 'veto' function (Libet, 1985). That is, the function of conscious experience might be related to the monitoring of ongoing action planning and performance, and the ability to prevent the eventual execution of actions that seem inappropriate or in conflict with current goals. Note how closely this view relates to the Freudian scenario of Id–Ego interactions.

Neuroscientific evidence and theorising provide support for the idea that consciousness-related cognitive functions are involved in handling response conflict. Botvinick, Braver, Carter, Barch, and Cohen (2001) have suggested that the anterior cingulate cortex (ACC) serves to monitor the cognitive system for the existence of response conflict (see Chapter 10 by Shenhav et al. in this volume). More specifically, the ACC would register the activation of multiple responses (an indicator of response conflict) and send an alert signal to frontal systems responsible for the maintenance of the action goal. This again would result in the reactivation and strengthening of the goal representation, which would increase the degree of top-down control of information processing—so to prevent further conflict (van Veen & Carter, 2002). Although there is some debate about the details of this information-processing loop and the kind of signals being processed, the basic idea of a performance-monitoring function that informs goal representations in the face of conflict has found ample support from behavioural and neuroscientific studies (Botvinick, 2007). Moreover, the hypothesised neural conflict monitoring system (the ACC) has been implicated in various aspects of conscious experience, including conscious effort (Mulert, Menzinger, Leicht, Pogarell, & Hegerl, 2005) and self-conscious emotional reactivity (Sturm et al., 2013). Along the same lines, Posner and Rothbart (2007) have related conscious experience to the activation of a so-called executive-control network, in which the ACC plays a key role (see Chapter 15 by Brown in this volume). Converging theorising comes from Morsella (2005), who argues that consciousness should be particularly concerned with conflict between what he calls 'skeletal muscle plans'. Although all sorts of conflict may arise and be resolved through integration in the cognitive system, so Morsella argues, it is mainly the competition for the control over effectors (a particularly capacity-limited resource) that consciousness-related functions are concerned with.

Further support for a possible role of consciousness in conflict resolution comes from studies that related the experience of conflict or of conflict-inducing stimuli to the probability of conflict-resolution operations (for a review, see Desender, Van Lierde, & Van den Bussche, 2013). For instance, Kunde (2003) observed that the typical reduction of stimulus-induced response conflict after conflict trials (an effect that has been taken to reflect performance monitoring in the sense of Botvinick, 2007) only occurs if participants were able to consciously perceive the conflict-inducing stimulus. Along the same lines, Desender, van Opstal, and Van den Bussche (in press) found trial-to-trial adjustments only if participants consciously experienced conflict, which fits with the observation that the electrophysiological evidence of error processing (the so-called post-error slowing) only occurs if participants realise that they have made an error (Nieuwenhuis, Ridderinkhof, Blom, Band, & Kok, 2001). It is important to consider that these findings provide only correlational evidence, and thus leave the possibility that it is actually the intensity of conflict that determines the probability of both conscious experience and conflict-resolution operations. But even in this case, the presence of conscious experience would represent a diagnostic indicator for the presence of conflict-induced cognitive-control processes.

To summarise, cognitive functions associated with conscious experience might be particularly concerned with the avoidance of response conflict. As with goal implementation, this function can be considered rather information heavy. Which response is consistent or inconsistent with other responses is not self-evident but depends on the task and the action goal. For instance, moving the left and right index finger simultaneously is considered to constitute a conflict in most psychological experiments but can be perfectly functional in piano playing. This means that successful conflict monitoring cannot rely on simple information, such as the number of concurrently active responses (which would still raise the question what counts as a response). Rather, effective monitoring must integrate information about the current goal and stimulus-response mapping, currently available stimuli and other environmental information, and activated response tendencies.

Agency

Being able to perform a goal-directed action requires knowledge about the events that an action can produce. Without such knowledge being involved, moving one's body cannot be considered to have a particular purpose and to aim at a particular goal. This knowledge does not need to be explicit and consciously represented. Many animals are able to carry out goal-directed actions without necessarily being able to represent their goal consciously, let alone communicating about it. The above-discussed ideomotor mechanism is thought to generate the action–effect knowledge that goal-directed actions need, but this knowledge does not have to be consciously accessible either. In fact, ideomotor theory aims to account for 'executive ignorance', which can be taken to reflect the lack of conscious knowledge about action–effect contingencies. And yet, if they are asked, people can commonly tell whether a particular event was caused by their own action and what their contribution to this event was. This ability and conscious experience has been called 'agency' or 'sense of agency', which, together with the 'sense of body ownership', has been claimed to represent an important constituent of the perceived 'self' (Gallagher, 2012).

Various studies have shown that people are sensitive to action–effect contingencies and can report them rather accurately (e.g., Shanks & Dickinson, 1987; Wasserman, 1990). For instance, if human participants are to judge the causal relationship between pressing a key and the flashing of a visual stimulus, the degree of perceived causality matches the actual contingency quite accurately, at least if the temporal gap between key-press and effect is no longer than 2 s (Shanks, Pearson, & Dickinson, 1989). Such findings may suggest that perceived agency is directly derived from implicit knowledge about action–effect contingencies, but the picture seems to be more complicated. For instance, Elsner and Hommel (2004) exposed participants to novel action–effect relations that varied in temporal continuity and in actual contingency. Participants were then tested for the spontaneous acquisition of action–effect associations (as usual in experiments on ideomotor action–effect acquisition: Elsner & Hommel, 2001), and they were to judge the causality between actions and effect explicitly. Both measures were sensitive to contiguity and contingency, and they were affected by these factors in comparable ways. However, the individual effects for the two measures did not correlate with each other at all, which means that how strongly an individual associated an action with its outcome did not predict his or her experience of causality between action and outcome.

This and other observations are consistent with the rather radical approach to personal agency suggested by Wegner (2003). According to Wegner's 'apparent mental causation model', human actions are always initiated by factors that are not directly accessible to consciousness: the unconscious cause of action, which may or may not be identical with the unconscious cause of thought (about the action). The (theoretically not further defined) unconscious cause of action triggers the overt action directly, without any intervention of processes related to conscious representation. However, initiating an action is commonly accompanied by the triggering of a thought about the action, which means that this thought commonly precedes the overt action in time. As humans often take correlations to imply causal relationships, the personal impression therefore is that it was the thought that produced the action, which, according to Wegner, is the reason why people believe in the causal power of their conscious intentions.

On the one hand, this scenario takes on board, and therefore perfectly fits with the claim of Libet and colleagues (1982), that the physiological causes of voluntary actions precede, and are thus not dependent on the conscious experience of the corresponding action intention. On the other hand, however, the approach does not necessarily exclude that what is being conscious is a relatively valid reflection of the actual cause (even though the approach is often interpreted as implying just that). If and to the degree that conscious thought is accurately informed by, and is thus a valid reflection of the unconscious cause, there would be nothing illusory about this thought's content. What is illusionary would only be the idea that the having of the conscious experience would be the actual causal factor. That this is a real possibility is suggested by findings of Wegner and Wheatley (1999). These authors demonstrated that humans experience more agency for the sudden appearance of an (objectively unpredictable) object on a screen if they are presented with a word prime describing the object about half a second before the appearance. This implies that the conscious experience of agency relies on expectations and predictions.

This latter conclusion is also shared by comparator-model approaches to human agency (e.g., Blakemore, Wolpert, & Frith, 2002; Frith, Blakemore, & Wolpert, 2000). Although such models are not particularly articulate in telling implicit from explicit knowledge, they emphasise the idea that predictions about action outcomes play an important role in assessing the success of an action. There are two ways in which actions can be unsuccessful: the performed movement may produce outcomes that were not predicted, which would provide information about an error or mishap, and the movement can go as expected and yet not lead to a state that matches the desired goal state. It is this latter mismatch that according to comparator models informs about agency, or the lack of it. An interesting implication of comparator models is that they may help gain more insight into delusions of control, such as observed in schizophrenia. If, for instance, the agent fails to predict a particular outcome, producing an action that generates this outcome would lead to a mismatch, which might induce the experience of alien control (Blakemore et al., 2002).

To summarise, the conscious perception of personal agency relies on relating expectations about an action's contribution to an intended goal event to its actual outcome, but it does not seem to be a pure mirror image of this relationship. One reason for the observed discrepancy between implicit and explicit action–effect knowledge might be that the latter is integrating more (and more contextualised) information than the former (Hommel, 2015; Synofzik, Vosgerau, & Newen, 2008). For instance, people may have general expectations about differing degrees of causality in different situations, which may shape, and sometimes perhaps even overrule, registered mismatch between actual and expected effects. Given that conscious representations of agency are unlikely to be functionally involved in actual action control, there is indeed no reason why they should not be richer than the implicit representations that are driving the action machinery. Indeed, there does not seem to be any functional reason why questions of agency should keep an agent consciously busy—as long as his or her goals are achieved. Moreover, even estimates from rather simple tasks suggest that generating a conscious experience takes at least 300–500 ms (e.g., Dehaene, Changeux, Naccache, Sackur, & Sergent, 2006; Libet, 2004), which seems to be much too slow to assume that conscious agency considerations intervene between environmental events and appropriate actions on a daily basis (Hommel, 2013). Explicit representations of agency are therefore likely to fulfil other purposes.

What purposes that might be becomes clear if we consider the cultural relativity of the importance and meaning of the agency concept. It is interesting to consider that most publications on human agency uncritically restrict their analysis to an individualistic 'I-perspective' of a single agent and the way he or she is feeling about his or her causal power. Although that seems to be the obvious perspective for readers with a Western background, members of Eastern cultures tend to have a different, often more extended, perspective that includes family, peers, and colleagues in the perception of agency (Markus & Kitayama, 2003). A possible reason for this discrepancy is that the borders of the conscious 'self' (of which perceived agency is assumed to be an important part: Gallagher, 2012) implied in Western thinking coincide with a person's skin, whereas Eastern self-concepts are often more socially extended (Markus & Kitayama, 1991). For instance, whereas Western sportsmen tend to attribute their extraordinary performances to themselves, Eastern sportsmen tend to ascribe the achievement to entire teams including support staff, family members, and friends (Markus & Kitayama, 2003).

Applied to agency, this means that the Western concept is about individual agency, whereas the Eastern concept is about what Markus and Kitayama (2003, 2010) have called 'conjoint agency'. This again implies that the experience of agency is unlikely to be hardwired and unlikely to be a mere mirror image of implicit action–effect knowledge, as that should be comparable for Eastern and Western agents. Rather, the conscious representation of agency seems to be penetrated by cultural experience and expectations, which again raises the possibility that the purpose of having such representations is to do with communication about agency—a possibility that I will develop further in the next section.

Task Sharing

Up to this point, I have discussed evidence suggesting that a few main functions of action control are systematically related to conscious experience. All of them are characterised by their integrative nature. Even though this is consistent with current theorising about consciousness (see the section titled 'Conclusions'), one may ask what the having of conscious experiences may add to the operation of the associated functions. For instance, there is no reason why an entirely unconscious process should be unable to check for the consistency of an ongoing action with the current action goal and to suppress the action if the discrepancy is above some threshold. In fact, this is what each cybernetic system, such as central heating, is doing (Wiener, 1948).

A philosophical way out would be to grant each cybernetic system consciousness, as suggested by McCarthy (in Searle, 1987). A more pragmatic approach could consider how the presence of conscious experience is typically diagnosed (Hommel, 2013). Arguably, the least controversial method to assess the conscious experience of others is to have them communicate about it—which is why verbal report is so popular in consciousness research. This might be taken as a rather arbitrary methodological convenience but it may also have a deeper meaning. If the conscious representation of some state of affairs allows us to communicate it to others, it may be that it is exactly this purpose that explains why some action control functions are associated with consciousness. Indeed, Baumeister and Masicampo (2010; cf., Dennett, 1991) have claimed that communication might represent a crucial function of human conscious thought, and Baumeister and Bargh (in press) suggest that the ability to communicate may be the main purpose underlying the connection between consciousness and action control.

Agency is a good example in case: Although the having of an agency experience does not seem to be particularly useful for the agent herself, it does allow her to comment on her degree of agenthood for a given action, which among other things relates to the action's social responsibility and acceptability (cf., Freud's Ego function). As these functions are to be interpreted within an existing cultural framework, it makes sense that perceived agency is sensitive to cultural factors—as discussed above. Implementing a goal and preparing for an action is also something useful to communicate to others, be it in an attempt to explain one's own behaviour or in order to instruct others to do the same. This is exactly what happens at the beginning of each psychological experiment: the experimenter explains to the subject what to do, how to do it, and when to do it, and this is the only information that a participant needs to produce the data all psychological studies are based on. How participants translate that information into a functional action goal and task set still remains a mystery (see also Chapter 2 by Monsell in this volume), but it is clear that they do and that their ability to do so requires conscious experience. Verbal communication seems to be particularly suitable, presumably because verbal descriptions provide the easiest way to refer to possible, but currently absent, events, like stimuli and responses.

Although the utility of verbal communication for interpersonal task sharing seems obvious, the verbal coding of task-related information may also have benefits for the individual.

In agreement with earlier considerations of Russian activity theorists (Vygotsky, 1962), researchers have assumed that self-control might be mediated by, or at least benefit from, inner speech and verbal self-instruction. Indeed, there is evidence that people tend to instruct themselves verbally at the beginning of a new task (Miyake, Emerson, Padilla, & Ahn, 2004), which, given the strong connection between inner speech and working memory (Baddeley, 1986), may help to maintain a new, still fragile task set (Goschke, 2003). Instructing another person may thus not only provide the information that this person needs to implement a new and unfamiliar task set, but also the (verbal) means to keep this set active. If so, it would make sense that consciousness is found under conditions that induce uncertainty about how to achieve a particular goal and about which actions are best suited to reach it: goal implementation and conflict resolution.

Conclusion

The evidence discussed in this chapter converges on three points. First, cognitive functions that are systematically accompanied by conscious experiences seem to be more integrative in nature than cognitive functions that are not. This observation does not come as a surprise but corresponds to what Morsella (2005) has called the 'integration consensus'—the widely shared view that conscious experience is related to information integration. Second, and to some degree even because of the higher temporal demands of broader integration, conscious representations seem to take too long to emerge to play a systematic, modulating role between perception and action. Hence, the contribution of consciousness to action control does not seem to be well captured by the folk-psychological idea that we perceive, think, and then act. If anything, we rather think, act, and then perceive or, perhaps even more appropriately, we occasionally think about acting/perceiving. And, third, there are reasons to assume that the main function of consciousness in action control is actually for communication with both others and oneself. Having conscious insight into aspects of action control allows us to verbalise what we are doing, how we are doing, and why we are doing it, which provides the opportunity to share tasks and strategies how to implement and control a task with others, and to use this information for self-control. Hence, although conscious experience does not seem to play a role in ongoing action control, it is likely to provide the basis for interpersonal learning and cultural transmission (Hommel, 2013; Masicampo & Baumeister, 2013).

Acknowledgements

This research was supported by a grant of the Netherlands Research Organization (NWO) to the author (433-09-243).

References

Baars, B. (1988). *A cognitive theory of consciousness.* New York: Cambridge University Press.

Baddeley, A. D. (1986). *Working memory.* Oxford, England: Oxford University Press.

Bargh, J. A. (1989). Conditional automaticity: Varieties of automatic influence in social perception and cognition. In J. S. Uleman & J. A. Bargh (Eds.), *Unintended thought* (pp. 3–51). New York: Guilford Press.

Baumeister, R. F., & Bargh, J. A. (in press). Conscious and unconscious: Toward an integrative understanding of human life and action. In J. Sherman (Ed.), *Dual process theories of the social mind.* New York: Guilford.

Baumeister, R. F., & Masicampo, E. J. (2010). Conscious thought is for facilitating social and cultural interactions: How mental simulations serve the animal–culture interface. *Psychological Review, 117*, 945–971.

Blakemore, S. J., Wolpert, D. M., & Frith, C. D. (2002). Abnormalities in the awareness of action. *Trends in Cognitive Sciences, 6*, 237–242.

Botvinick, M. M. (2007). Conflict monitoring and decision making: Reconciling two perspectives on anterior cingulate function. *Cognitive Affective & Behavioral Neuroscience, 7*, 356–366.

Botvinick, M. M., Braver, T. S., Carter, C. S., Barch, D. M., & Cohen, J. D. (2001). Conflict monitoring and cognitive control. *Psychological Review, 108*, 624–652.

Brunswik, E. (1944). Distal focussing of perception. Size constancy in a representative sample of situations. *Psychological Monographs, 56*(1).

Chmiel, N. (1984). Phonological encoding for reading: The effect of concurrent articulation in a Stroop task. *British Journal of Psychology, 75*, 213–220.

Dehaene, S., Changeux, J.-P., Naccache, L., Sackur, J., & Sergent, C. (2006). Conscious, preconscious, and subliminal processing: A testable taxonomy. *Trends in Cognitive Sciences, 10*, 204–211.

Dennett, D. C. (1991). *Consciousness explained.* Boston: Little, Brown and Company.

Desender, K., Van Lierde, E., & Van den Bussche, E. (2013). Comparing conscious and unconscious conflict adaptation. *PLoS ONE, 8*(2), e55976.

Desender, K., Van Opstal, F., & Van den Bussche, E. (in press). Feeling the conflict. The crucial role of conflict experience in adaptation. *Psychological Science.*

Elsner, B., & Hommel, B. (2001). Effect anticipation and action control. *Journal of Experimental Psychology: Human Perception and Performance, 27*, 229–240.

Elsner, B., & Hommel, B. (2004). Contiguity and contingency in the acquisition of action effects. *Psychological Research, 68*, 138–154.

Engel, C., & Singer, W. (Eds.). (2008). *Strüngmann Forum report: Better than conscious? Decision making, the human mind, and implications for institutions.* Cambridge, MA: MIT Press.

Evans, J. St. B. T., & Stanovich, K. E. (2013). Dual-process theories of higher cognition: Advancing the debate. *Perspectives in Psychological Science, 8*, 223–241.

Exner, S. (1879). Physiologie der Grosshirnrinde. In L. Hermann (Ed.), *Handbuch der Physiologie*, 2. Band, 2. Theil (pp. 189–350). Leipzig: Vogel.

Freud, S. (1923). *Das Ich und das Es.* Leipzig etc.: Internationaler Psycho-analytischer Verlag. English translation: *The ego and the id*, Joan Riviere (trans.), London: Hogarth Press and Institute of Psychoanalysis.

Frith, C. D., Blakemore, S. J., & Wolpert, D. M. (2000). Abnormalities in the awareness and control of action. *Philosophical Transactions of the Royal Society of London B: Biological Sciences, 355*, 1771–1788.

Gallagher, S. (2012). Multiple aspects in the sense of agency. *New Ideas in Psychology, 30*, 15–31.

Goschke, T. (2003). Voluntary action and cognitive control from a cognitive neuroscience perspective. In S. Maasen, W. Prinz, & G. Roth (Hrsg.), *Voluntary action: Brains, minds, and sociality* (pp.49–85). Oxford: Oxford University Press.

Günther, K. (2003). Voluntary action and criminal responsibility. In S. Maasen, W. Prinz, & G. Roth (Eds.), *Voluntary action: Brains, minds, and society* (pp. 263–280). Oxford: Oxford University Press.

Harless, E. (1861). Der Apparat des Willens [The apparatus of will]. *Zeitschrift fuer Philosophie und philosophische Kritik, 38*, 50–73.

Heckhausen, H., & Gollwitzer, P. M. (1987). Thought contents and cognitive functioning in motivational versus volitional states of mind. *Motivation and Emotion, 11*, 101–120.

Hommel, B. (2000). The prepared reflex: Automaticity and control in stimulus-response translation. In S. Monsell & J. Driver (Eds.), *Control of cognitive processes: Attention and performance XVIII* (pp. 247–273). Cambridge, MA: MIT Press.

Hommel, B. (2007). Consciousness and control: Not identical twins. *Journal of Consciousness Studies, 14*, 155–176.

Hommel, B. (2011). The Simon effect as tool and heuristic. *Acta Psychologica, 136*, 189–202.

Hommel, B. (2013). Dancing in the dark: No role for consciousness in action control. *Frontiers in Psychology, 4*, 380.

Hommel, B. (2015). The theory of event coding (TEC) as embodied-cognition framework. *Frontiers in Psychology, 6,* 1318.

Hommel, B., Müsseler, J., Aschersleben, G., & Prinz, W. (2001). The theory of event coding (TEC): A framework for perception and action planning. *Behavioral and Brain Sciences, 24,* 849–878.

James, W. (1890). *The principles of psychology* (Vol. 2). New York: Dover Publications.

Kunde, W. (2003). Sequential modulations of stimulus-response correspondence effects depend on awareness of response conflict. *Psychonomic Bulletin & Review, 10,* 198–205.

Libet, B. (1985). Unconscious cerebral initiative and the role of conscious will in voluntary action. *Behavioral and Brain Sciences, 8,* 529–566.

Libet, B. (2004). *Mind time: The temporal factor in consciousness.* Harvard University Press.

Libet, B., Wright, E. W., & Gleason, C. A. (1982). Readiness potentials preceding unrestricted spontaneous and preplanned voluntary acts. *Electroencephalography and Clinical Neurophysiology, 54,* 322–325.

Lotze, R. H. (1852). *Medicinische Psychologie oder die Physiologie der Seele.* Leipzig: Weidmann'sche Buchhandlung.

MacLeod, C. M. (1991). Half a century of research on the Stroop effect: An integrative review. *Psychological Review, 109,* 163–203.

Markus, H. R., & Kitayama, S. (1991). Culture and the self: Implications for cognition, emotion, and motivation. *Psychological Review, 98,* 224–253.

Markus, H. R., & Kitayama, S. (2003). Models of agency: Sociocultural diversity in the construction of action. In V. M. Berman & J. J. Berman (Eds.), *Nebraska symposium on motivation: Crosscultural differences in perspectives on the self* (Vol. *49*, pp. 1–58). Lincoln: University of Nebraska Press.

Markus, H. R., & Kitayama, S. (2010). Cultures and selves: A cycle of mutual constitution. *Perspectives on Psychological Science, 5,* 420–430.

Martin, M. (1978). Speech recoding in silent reading. *Memory & Cognition, 30,* 187–200.

Masicampo, E. J., & Baumeister, R. F. (2013). Conscious thought does not guide moment-to-moment actions—it serves social and cultural functions. *Frontiers in Psychology, 4,* 478.

McClain, L. (1983). Effects of response type and set size on Stroop color-word performance. *Perceptual & Motor Skills, 56,* 735–743.

Memelink, J., & Hommel, B. (2013). Intentional weighting: A basic principle in cognitive control. *Psychological Research, 77,* 249–259.

Miyake, A., Emerson, M. J., Padilla, F., & Ahn, J. C. (2004). Inner speech as a retrieval aid for task goals: the effects of cue type and articulatory suppression in the random task cuing paradigm. *Acta Psychologica, 115,* 123–142.

Moody, T. C. (1994). Conversations with zombies. *Journal of Consciousness Studies, 1,* 196–200.

Morsella, E. (2005). The function of phenomenal states: Supramodular interaction theory. *Psychological Review, 112,* 1000–1021.

Mulert, C., Menzinger, E., Leicht, G., Pogarell, O., & Hegerl, U. (2005). Evidence for a close relationship between conscious effort and anterior cingulate cortex activity. *International Journal of Psychophysiology, 56,* 65–8010.

Nieuwenhuis, S., Ridderinkhof, K. R., Blom, J., Band, G. P., & Kok, A. (2001). Error-related brain potentials are differentially related to awareness of response errors: Evidence from an antisaccade task. *Psychophysiology, 38,* 752–760.

Norman, D. A., & Shallice, T. (1986). Attention to action: Willed and automatic control of behavior. In R. J. Davidson, G. E. Schwartz, & D. Shapiro (Eds.), *Consciousness and self regulation* (Vol. *4*, pp. 1–18). New York Plenum.

Posner, M. I., & Rothbart, M. K. (2007). Research on attention networks as a model for the integration of psychological science. *Annual Review of Psychology, 58,* 1–23.

Prinz, W. (1992). Why don't we perceive our brain states? *European Journal of Cognitive Psychology, 4,* 1–20.

Proctor, R. W. (2011). Playing the Simon game: Use of the Simon task for investigating human information processing. *Acta Psychologica, 136,* 182–188.

Rizzolatti, G., & Craighero, L. (2004). The mirror-neuron system. *Annual Review of Neuroscience, 27,* 169–192.

Rochat, P. (2001). *The infant's world.* Cambridge: Harvard University Press.
Roth, G. (2003). The interaction of cortex and basal ganglia in the control of voluntary actions. In S. Maasen, W. Prinz, & G. Roth (Hrsg.), *Voluntary action: Brains, minds, and sociality* (pp. 115–132). Oxford: Oxford University Press.
Searle, J. (1987). Minds and brains without programs. In C. Blakemore & S. Greenfield (Eds.), *Mindwaves* (pp. 209–233). Oxford: Blackwell.
Shanks, D. R., & Dickinson, A. (1987). Associative accounts of causality judgment. *Psychology of Learning and Motivation, 21*, 229–261.
Shanks, D. R., Pearson, S. M., & Dickinson, A. (1989). Temporal contiguity and the judgment of causality by human subjects. *Quarterly Journal of Experimental Psychology B, 41*, 139–159.
Shin, Y. K., Proctor, R. W., & Capaldi, E. J. (2010). A review of contemporary ideomotor theory. *Psychological Bulletin, 136*, 943–974.
Simon, J. R., & Rudell, A. P. (1967). Auditory S–R compatibility: Effect of an irrelevant cue on information processing. *Journal of Applied Psychology, 51*, 300–304.
Stock, A., & Stock, C. (2004). A short history of ideo-motor action. *Psychological Research, 68*, 176–188.
Stroop, J. R. (1935). Studies of interference in serial verbal reactions. *Journal of Experimental Psychology, 28*, 643–662.
Sturm, V. E., Sollberger, M., Seeley, W. W., Rankin, K. P., Ascher, E. A., Rosen, H. J., ... Levenson, R. W. (2013). Role of right pregenual anterior cingulate cortex in self-conscious emotional reactivity. *Social Cognitive and Affective Neuroscience, 8*, 468–474.
Synofzik, M., Vosgerau, G., & Newen, A. (2008). Beyond the comparator model: A multifactorial two-step account of agency. *Consciousness and Cognition, 17*, 219–239.
Turvey, M. T. (1977). Preliminaries to a theory of action with reference to vision. In R. Shaw & J. Bransford (Eds.), *Perceiving, acting and knowing: Toward an ecological psychology.* Hillsdale, NJ: Lawrence Erlbaum Associates.
Valle-Inclán, F., & Redondo, M. (1998). On the automaticity of ipsilateral response activation in the Simon effect. *Psychophysiology, 35*, 366–371.
van Veen, V., & Carter, C. S. (2002). Conflict and cognitive control in the brain. *Current Directions in Psychological Science, 15*, 237–240.
Verschoor, S. A., Weidema, M., Biro, S., & Hommel, B. (2010). Where do action goals come from? Evidence for spontaneous action-effect binding in infants. *Frontiers in Psychology, 1*, 201.
Vygotsky, L. S. (1962). *Thought and language.* Cambridge, MA: MIT Press.
Wasserman, E. A. (1990). Detecting response-outcome relations: Toward an understanding of the causal texture of the environment. *Psychology of Learning and Motivation, 26*, 27–82.
Wegner, D. M. (2003). The mind's best trick: How we experience conscious will. *Trends in Cognitive Sciences, 7*, 65–69.
Wegner, D. M., & Wheatley, T. P. (1999). Apparent mental causation: Sources of the experience of will. *American Psychologist, 54*, 480–492.
Wiener, N. (1948). *Cybernetics, or communication and control in the animal and the machine.* Cambridge, MA: MIT Press.
Woodworth, R. S. (1938). *Experimental psychology.* New York: Henry Holt & Co.
Wykowska, A., Schubö, A., & Hommel, B. (2009). How you move is what you see: Action planning biases selection in visual search. *Journal of Experimental Psychology: Human Perception and Performance, 35*, 1755–1769.

Section II

Models of Cognitive Control

Computations, Mechanisms, and Neuroanatomy

8

Computational Models of Cognitive Control

Tom Verguts

Introduction

A tennis player is ready to serve his match point, when lightning flashes across the sky. What should he do? Serve the ball? Run for shelter? Or first pick up his kid from the playground? In a situation like this, cognitive control is required to act adaptively. Broadly, cognitive control refers to organising, selecting, or otherwise operating on more basic cognitive or motor processes (e.g., serving a tennis ball). This is particularly, but not exclusively, needed when something goes wrong (e.g., lightning strikes).

Over the last decades, cognitive control has witnessed a remarkable surge of interest in cognitive neuroscience. In parallel, several computational models of cognitive control have been proposed, which help in formulating precise cognitive mechanisms to both integrate existing data and provide theoretical guidance to steer future empirical work. Especially in a multidisciplinary field like cognitive control, characterised by a large and bewildering number of experimental methods, experimental paradigms, studied organisms, and theories, such guidance from computational modelling is indispensable.

I here review some of these models. To structure the review, I use two core ideas that have figured repeatedly in computational models of cognitive control. The first core idea is that cognitive control is a form of hierarchically higher decision making. The lower level of decision making chooses or optimises actions; the higher level of decision making (cognitive control) is concerned with choosing or optimising the brain or cognitive processes that do so. The second core idea is that the process of learning to make such decisions can be modelled using the same computational machinery as lower-level decision making. One advantage is that the same theoretical frameworks that have been proposed for solving lower-level problems can be used for cognitive control as well. In this respect, I discuss how three influential modelling frameworks (parallel distributed processing, reinforcement learning, and Bayesian inference) have been used to formulate computational models of cognitive control.

Cognitive Control: Solving Higher-Level Problems

The first core idea is that cognitive control is about solving problems, but not all types of problems. As long as the sky is blue, the tennis player can keep on playing and leave control to *primary controllers*. Identifying an object as a house, reading a word, and serving a tennis

The Wiley Handbook of Cognitive Control, First Edition. Edited by Tobias Egner.

ball are not the business of cognitive control, but are handled by visual perception, language, and motor controllers, respectively. A single primary controller corresponds to a relatively well-circumscribed task that can be solved with a single stimulus–action mapping (possibly with hidden representations). An example from actual cognitive control laboratory tasks would be evaluating a number's parity status or identifying which of two numbers is larger. Cognitive control operates to *control* such primary controllers when needed (Braver & Cohen, 2000). I will call such hierarchically higher controllers *secondary controllers.* A class of secondary controllers that are in charge of a similar task (e.g., inhibition) will be called a *cognitive control function.* It is worth stressing that this is not meant to imply that a function (class of secondary controllers) is localised in one brain area dedicated to this task. It simply identifies one computational task (function). Also, my choice of the cognitive control functions is 'literature driven'; the selection was made on the basis of which functions are often discussed (and modelled) in the literature. Furthermore, a full hierarchy may involve many more levels than just those two. For example, there are surely controllers below the 'primary' level of serving a tennis ball (Haruno, Wolpert, & Kawato, 2001). With these caveats addressed, I find the notions of primary and secondary controllers useful for the current purpose, so I will adopt them.

Cognitive control is needed when a problem arises in primary controllers or the responses they generate (see also Chapter 17 by Ullsperger in this volume). In the example above, simply continuing to run the tennis-serving controller may no longer be optimal when lightning strikes, and the tennis player might change his course of action and head for the playground. However, this is not always appropriate. When there is no child in the playground, he can opt to energise the currently operating primary controller (tennis serving) and inhibit the perception of dark clouds across the sky, in the hope of quickly winning the match. Hence, every challenge can be confronted in several ways, requiring different secondary controllers. Here I identify five classes of secondary controllers (i.e., cognitive control functions). The first three roughly correspond to the cognitive control functions as identified by Miyake and colleagues (Miyake et al., 2000); the other two involve energising or optimizing a given primary controller in some way (see below for details).

The first cognitive control function is choosing which of two (or more) primary controllers is allowed to steer behaviour (Rogers & Monsell, 1995; Chapter 2 by Monsell in this volume); the tennis player from the example is about to make such a choice. I will refer to this function as *choosing between controllers.* It has been studied extensively in the task-switching literature, but that term is too narrow for the current purpose since choosing between controllers does not necessarily involve a task switch if the current (primary) controller remains appropriate. The second cognitive control function concerns *inhibition* of the current (primary) controller (Aron, Robbins, & Poldrack, 2004; Chapter 6 by Verbruggen & Logan in this volume) or of the (currently prepared) response. The third cognitive control function is to *provide context* in working memory for primary controllers (Miller & Cohen, 2001; O'Reilly & Frank, 2006).

The fourth cognitive control function is *energising* the current primary controller. Different implementations for this function have been discussed (Botvinick, Braver, Barch, Carter, & Cohen, 2001). A fifth and final cognitive control function is to *choose a learning rate* (of a primary controller; Behrens, Woolrich, Walton, & Rushworth, 2007). Changing the learning rate can involve either an up- or downregulation depending on the context.

When cognitive control is triggered by an actual problem (e.g., a lightning strike), it is called reactive. However, the problems to which cognitive control responds must not be actually experienced; predicting them may suffice. When cognitive control is triggered by a stimulus *predicting* that a problem will occur, cognitive control is said to be proactive (Braver, 2012; Chapter 9 by Chiew & Braver in this volume). The tennis player in the example may either comfort his child when it comes running toward him (reactive); or the lightning strike may lead him to foresee a scared child and fetch the child before it starts crying (proactive).

Solving Higher-Level Problems: Learning, Value, and Inference

I discussed that cognitive control concerns solving higher-order problems; the next issue is how to solve them. Here the basic modelling strategy is to seek inspiration from the literature on primary controllers (e.g., auditory perception, depth processing, motor control), and how these primary controllers know what to do. This question has resulted in a number of theories and algorithms, with varying emphasis on learning, optimisation, or inference. Over the last several decades, three modelling traditions in particular have been very influential.

Modellers in the parallel distributed processing (PDP) tradition emphasise that processing occurs in parallel between several (distributed) controllers. These controllers know what to do from a gradual approach toward an optimal point of a goal function (e.g., the minimum of an error function). This gradual process is called learning (Rumelhart, Williams, & Hinton, 1986; Seidenberg & McClelland, 1989). The parameters that change (i.e., are optimised) during learning are typically weights in a neural network (controller) in this tradition. The second modelling tradition is reinforcement learning (RL): It emphasises the optimisation of value functions to obtain a goal (e.g., finding the maximum of a value function; Montague, Dayan, & Sejnowski, 1996; Sutton & Barto, 1998). These value functions combine the expected reward (reinforcement) and cost attached to available stimuli or actions. The parameters that change during optimisation can again be weights in a neural network, or (more typically) the probabilities of actions the model must take, given the current situation. Finally, Bayesian modellers emphasise that, rather than obtaining a single parameter estimate, one can infer its probability distribution (belief distribution; Anderson, 1991; Tenenbaum, Griffiths, & Kemp, 2006). An example of such a parameter could again be a weight in a neural network or the probability of an action given the current situation. In a normal distribution, obtaining the parameter's belief distribution would entail estimating both the mean and variance of this distribution (Kruschke, 2008). Estimating this mean and variance entails finding the mean and variance that have maximal posterior probability, given the data already experienced.

Various debates have been held about which research framework is most appropriate for understanding cognition and the brain (Griffiths, Chater, Kemp, Perfors, & Tenenbaum, 2010; McClelland et al., 2010). Yet, in actual model implementations (of the PDP, RL, or Bayesian variety), the same key ideas occur repeatedly. In particular, most proposals start from the computational principle that a goal function must be optimised. This goal function may be an error function (to be minimised) in the PDP tradition; a value function (to be maximised) in the RL tradition; or a posterior probability function (to be maximised) in the Bayesian tradition. This goal function and its optimisation are subject to both environmental and cognitive constraints. The controller sets its parameters (e.g., neural network weights) by optimising this goal function. This is how a (primary) controller learns what to do.

Consistent with the second core idea, the same (PDP, RL, or Bayesian) optimisation principles can be applied 'one level up', i.e. to cognitive control. The actions to be optimised for a secondary controller are not (primary) motor acts (like, e.g., in motor control or language processing); instead, the actions consist of modulating the primary controllers.

Combining the two ideas, one might summarise the discussion thus far with the statement that cognitive control involves a form of metalearning (Doya, 2002; Khamassi et al., 2011). Knowing how to exert cognitive control entails learning at the secondary level. I now review how the two ideas have been instantiated in computational models of cognitive control, as applied to the five cognitive control functions that were identified here, in the three modelling frameworks.

Models of Cognitive Control: PDP

General Background

In response to computer-inspired serial processing models of memory and cognition developed in the second part of the 20th century (Atkinson & Shiffrin, 1968), PDP modellers instead emphasised the *parallel and distributed* nature of cognitive processing (Rumelhart, McClelland, et al., 1986). They constructed neural networks that were trained using the delta rule or one of its generalisations. The core of the delta rule (Rescorla & Wagner, 1972; Widrow & Hoff, 1960) is a difference (delta) term, contrasting a desired (target) value injected into a neural unit, with an actual (obtained) value for this unit (i.e., delta = target – obtained). Hence, as long as a neural unit does what it is supposed to do, delta equals zero, and there is no learning. When delta is not zero, then neural network weights attached to that unit will change in proportion to delta.

A crucial breakthrough was the discovery of the backpropagation algorithm (by, among others, Rumelhart, Williams, & Hinton, 1986). This is a generalisation of the delta rule that can train sophisticated (primary) controllers containing hidden layers of units (for which no target values are available). This approach was broadly applied in cognitive psychology (e.g., Seidenberg & McClelland, 1989). Its popularity decreased in the first decade of the 21st century, perhaps because of the stigma of being 'biologically implausible'. However, its aim of letting structure emerge by gradual optimisation of a goal function and recruitment of the hidden units for that purpose has attracted renewed interest in recent 'deep learning' models (Hinton & Salakhutdinov, 2006; Mnih et al., 2015). Because most cognitive control models in this research framework were developed in the era of PDP, I retain that label here.

Choosing Between Controllers

One early conceptualisation for choosing between controllers was the hierarchical mixture-of-experts model. This model considers choosing between primary controllers ('experts') an optimisation problem (higher level), in addition to choosing the parameters of the controllers themselves (lower level). Early applications used backpropagation (in the PDP tradition; Jacobs, Jordan, & Barto, 1991) or a statistical mixture model formulation (Jordan & Jacobs, 1994). The primary and secondary controllers were trained simultaneously and with the same algorithm (cf. second core idea outlined above).

A related model for choosing between primary controllers in the PDP tradition was Cohen, Dunbar, and McClelland's model of the Stroop task (Cohen, Dunbar, & McClelland, 1990). In the model, two primary controllers compete for responding: One controller responds to colour information, and the other controller responds to words. A secondary controller determines whether word or colour reading is allowed to determine verbal output. Key for this secondary controller is learning in which task context which primary (colour or word) controller should determine responding. An important aspect of the Stroop task is that the word reading controller is much more automatised and thus stronger than the colour-reading controller. When colour is the relevant dimension, cognitive control is thus required to inhibit word reading, or alternatively boost the colour-reading controller. In the Stroop model, the secondary controller choice problem was learned simultaneously (using backpropagation) with learning the task-specific mappings by the word- and colour-processing controllers. This again illustrates that the same optimisation principle (here instantiated by backpropagation) can determine the parameters of both primary and secondary controllers.

The choice between controllers is studied more explicitly in the task-switching literature (for a review, see Chapter 2 by Monsell in this volume). A task-switching model that followed up on the Cohen et al. Stroop model was proposed by Gilbert and Shallice (Gilbert & Shallice, 2002). They used a similar architecture, but their model continued processing between trials, with residual activation from the previous trial on the current trial, which allowed modelling task switch costs. In addition, they implemented fast Hebbian learning between stimulus and task demand representations between trials. This led them to account for several aspects of the task switch cost, such as the asymmetry in switching between easy and difficult tasks (Wylie & Allport, 2000).

Inhibition

Sometimes the best you can do is to just stop doing whatever you are doing. A crying child would be a perfect reason for our tennis player to stop playing. In the experimental psychology lab, this is typically studied in stop-signal or go/no-go tasks (Aron, 2007). It is also studied in task switch experiments containing uninformed task switches (Nee, Kastner, & Brown, 2011); here also, an (uninformed) task switch would be a perfect occasion to just stop and ponder your next strategy. To deal with the latter, Brown, Reynolds, and Braver proposed a model that not only monitored conflict between responses but also conflict between tasks (Brown, Reynolds, & Braver, 2007). When there was a high level of conflict within a trial and between responses, the model attended the current task set more strongly (i.e., as in the conflict-monitoring model). In contrast, when there was conflict in the task demand layer, due to a change in task set (or response set) between trials, the model inhibited activation in the response layer. This led to response slowing, but it also allowed exploration of task sets and eventually higher accuracy.

Providing Context

Context determines when and how to operate primary controllers. On a tennis court, it is alright to swing a tennis racket; at home, it is odd; at a party in your friend's house, it is weird. Such contextual information often needs to span long time delays, so working memory is crucial for providing context (Cohen & Servan-Schreiber, 1992). An important challenge in information processing is when to 'load' new information into working memory (input gating); and similarly, when to allow this information to determine processing carried out by primary controllers (output gating). This challenge was identified as an important computational principle by Hochreiter and Schmidhuber (Hochreiter & Schmidhuber, 1997). In the current terminology, input and output gating constitute decisions of a secondary controller: The action of 'keeping information in working memory' or 'updating working memory' are not motor actions, but they support the context for primary controllers to function appropriately. Rougier and colleagues (Rougier, Noelle, Braver, Cohen, & O'Reilly, 2005) developed a model[1] of the Wisconsin card sorting task which included input/output gating for prefrontal cortex neurons. When exposed to a sufficient number of instances of several tasks, it discovered not only abstract rules to solve the tasks (primary controller level), but simultaneously also discovered when to gate information in and out of working memory (secondary controller level).

Energising the Current Controller

Another follow-up to the Cohen et al. model was the conflict-monitoring model (Botvinick et al., 2001). They noted that the Cohen et al. model did not specify *when* the secondary controller must intervene. They made the highly influential proposal that this is determined by tracking response conflict. When there is conflict, for example because two response representations are simultaneously active, the relevant primary controller (e.g., colour reading in the Stroop task) is attended to more strongly. The latter was operationalised in several ways. One was by providing a stronger top-down signal to the colour-reading controller (in the Stroop task). Another was focussing spatial attention more strongly to the centre of the visual field, which is useful in other situations. An example considered by the authors is the flanker task, in which case focussing spatial attention blocks interference from flanker stimuli.

Choosing a Learning Rate

The fact that learning rates may be context dependent was already suggested several decades ago (Pearce & Hall, 1980; Rescorla & Wagner, 1972). In cognitive control, it was suggested as a solution to a theoretical problem in the conflict-monitoring model (Verguts & Notebaert, 2008, 2009; see also Blais, Robidoux, Risko, & Besner, 2007). The authors noted that it needs to be specified how the conflict-monitoring system is supposed to know where exactly to intervene for implementing cognitive control. For example, when extra attention needs to be assigned to the colour dimension, how does the system know how to do that? And how does it know that the colour dimension is the relevant one? The authors demonstrated that this problem can be solved using Hebbian learning. In particular, they proposed that the learning rate of ongoing Hebbian learning processing would be increased by incongruent stimuli (via arousal). This would increase connectivity (perhaps synaptic connections) between the currently active task and stimulus representations. If we consider task and stimulus representations to be part of the same primary controller, this corresponds to choosing an appropriate (Hebbian) learning rate for a primary controller. This is a useful strategy if one keeps on doing the same task. However, it may actually impair performance if a different task needs to be done on the next trial (Notebaert & Verguts, 2008).

Models of Cognitive Control: RL

General Background

The formalism used in most RL models is described in the seminal handbook of Sutton and Barto (Sutton & Barto, 1998). Here, RL is used to seek an optimal policy via the construction and optimisation of value functions. These value functions are both estimated and followed by the acting organism. RL is sometimes called semi-supervised or weakly supervised because the feedback the model (organism) receives from its environment is less detailed (only value-based feedback, e.g., 'right' or 'wrong') than in typical PDP models, where each (output) neuron receives feedback from the environment on what the correct response should have been. RL became an influential framework for understanding cognition in the last several decades. One reason is perhaps the remarkable synergy between neurophysiology (in particular, measurements in dopaminergic nuclei that are proposed to carry the RL signals; Schultz, 1998) and computational models of those dopamine (DA) signals and the (sub)cortical areas they project to (Montague et al., 1996; Sutton & Barto, 1998). Later fMRI work also demonstrated the feasibility of the RL framework in humans (O'Doherty, Dayan, Friston, Critchley, & Dolan, 2003).

The RL framework has also been applied to cognitive control. It has been proposed that anterior cingulate cortex (ACC) calculates the expected value of control, signals to (primary) controllers that an intervention is valuable, and what type of intervention is needed (Shackman et al., 2011; Shenhav, Botvinick, & Cohen, 2013; for a review of ACC models, see Chapter 15 by Brown in this volume). In the remainder of the chapter, we discuss computational models consistent with this idea, and a number of interventions they have considered.

Choosing Between Controllers

An important RL-based computational model of cognitive control was reported by Holroyd and Coles (Holroyd & Coles, 2002). This paper drew ideas from motor control, where a very similar problem appears: Which motor controller must be chosen in which situation? A hierarchical approach (two levels of controllers) was applied in motor control for this purpose (Haruno et al., 2001; Wolpert & Kawato, 1998). Analogously, Holroyd and Coles proposed that for cognitive control the ACC contains a secondary controller that chooses between several primary controllers. To use their example, one primary controller may search for immediate reinforcement, whereas a different primary controller may attempt to avoid pain, and the secondary controller then chooses which one currently steers behaviour. For this purpose, the secondary controller (ACC) was trained by a dopaminergic reward prediction error signal. Consistent with one of the main themes of this chapter, this same signal was used for training the parameters of the primary controllers.

Like Holroyd and Coles, Daw and colleagues (Daw, Niv, & Dayan, 2005) considered how organisms choose between controllers. They started from a rich and multidisciplinary data set suggesting that two qualitatively different systems control behaviour (Balleine & Dickinson, 1998). One system is 'model free' in the sense that it does not construct a model of the world to guide its action choices. Instead, it simply keeps track of how much reward is attached to each individual stimulus or action. The other control system is 'model based' in the sense that it constructs a model of the world to guide its actions. This model of the world is then applied in specific contexts to infer an appropriate action. For example, a rat in a maze may decide to go left because going left was associated with reward in previous trials (model free); or because it predicts (based on its knowledge of the maze) that going left will yield a reward (model based). A perhaps more familiar example would be ordering yet another drink late at night in a bar (model free); versus going home because you predict your cognitive state the next morning if you stay (model based). From the current point of view, the choice between model-free and model-based systems or controllers would be identified as cognitive control. Additionally, one can identify the model-based system as requiring cognitive control (Daw et al., 2005), given that more (cognitive) actions are needed in model-based than in model-free controlling. In particular, model-based acting or reasoning requires iterating through a transition matrix repeatedly until the expected value of an option can be calculated; this must be done for each available action separately. Such operations would require much more input/output gating than simply choosing the action with the highest associated value in a model-free system. The Daw et al. (2005) model was not originally rooted in the cognitive control literature. The current discussion shows that there are nevertheless strong similarities. Consistently, performance on a cognitive control (Stroop) task predicts how often participants make model-based choices (Otto, Skatova, Madlon-kay, & Daw, 2015).

Inhibition

In his model of basal ganglia–cortex interactions, Frank (2005) considered how the corticostriatal direct pathway learns (via RL) to associate specific stimuli and contexts with actions (go learning). In contrast, the corticostriatal indirect pathway learns (also via RL) to inhibit actions for specific stimuli and contexts (no-go learning). In later work, Frank added the hyperdirect pathway (via the subthalamic nucleus rather than the striatum) to this model, which learns to inhibit the direct and indirect pathways when there is response conflict (Frank, 2006). This hyperdirect inhibitory pathway can be considered a secondary controller because it is able to shut down and thus overrule the currently operating controller (e.g., the direct pathway). As noted, this inhibitory pathway is triggered by response conflict; however, an RL model could be developed that is trained using reward and punishment, just as the direct and indirect pathways in Frank's model are.

Providing Context

Early PDP work already considered the role of DA in cognitive control (Cohen & Servan-Schreiber, 1992). However, the link of DA to cognitive control and learning was made explicit and systematic in an RL framework (Braver, Barch, & Cohen, 1999; Braver & Cohen, 2000). These authors proposed that a crucial part of many tasks is deciding to maintain (or not) information in working memory. The key idea of their proposal is that DA has a double role in cognitive control: Both *learning* when to maintain information in working memory and supporting this working memory itself (*gating*). This same idea was developed in a more biologically detailed model (O'Reilly & Frank, 2006). Later work also considered output gating in an RL framework for cognitive control (Hazy, Frank, & O'Reilly, 2007; Kriete & Noelle, 2011).

Energising the Current Controller

Another cognitive control choice that was recently couched in an RL framework is choosing to invest effort in a task. Holroyd and McClure (2015) proposed that the ACC is located in a hierarchy of controllers with the striatum below it and (in the rodent brain) the prelimbic cortex above. They argued that the striatum may oppose effortful (and hence costly) and aversive actions; but in case an effortful action would lead to high value, the ACC overrules striatal tendencies to avoid it. In another model of effort investment (Verguts, Vassena, & Silvetti, 2015), the ACC was considered part of the cortical-basal ganglia limbic loop (Alexander & Crutcher, 1990). Verguts et al. proposed that this limbic loop has its own value function that considers the value of energising primary controllers. The model attempted to optimise the value function by choosing to energize (or not) specific primary controllers depending on the estimated benefits and costs of such energising. Energising was implemented by increasing the gain (or signal-to-noise) parameter of the relevant primary controllers (Aston-Jones & Cohen, 2005; Cohen & Servan-Schreiber, 1992; McClure, Gilzenrat, & Cohen, 2005).

Models of Cognitive Control: Bayesian Inference

General Background

A third influential framework for conceptualising cognitive control is Bayesian inference. As for the PDP and RL models, Bayesian inference for cognitive control is inspired by application of the framework in cognitive science more broadly (e.g., Griffiths, Kemp, &

Tenenbaum, 2008; Körding & Wolpert, 2006). As noted in the introduction, optimisation of a goal function is also a key feature in Bayesian inference (as in the other two frameworks). Yet, two important features characterise Bayesian modelling. The first is the explicit consideration of the statistical distribution of all variables and parameters in the model. I here focus on the parameters. The model of Daw et al. mentioned in the RL section combines RL with Bayesian inference, and can serve as an example. Whereas a standard RL model would compute estimates of the values of various available actions, Daw et al. proposed that an organism may also compute the statistical (belief) distributions of these values (see below).

The second feature typical for Bayesian approaches is model inversion. For the purpose of parameter estimation, a Bayesian perspective envisions the organism as constructing a model of the world (Dayan, 1998). The world generates observations, which must subsequently be 'inverted' (from observations to the model structure) by the organism to know something about the (parameters of the) generating model. Note that the word *model* is used here as in 'model based' above, namely, to denote a structure in the mind of the studied organism, not a structure in the mind of the researcher. The notion of model construction and inversion, or Bayesian inference more generally, has been applied in various branches of cognition, including perception (Dayan, 1998), motor control (Haruno et al., 2001), and categorisation (Kemp & Tenenbaum, 2008).

Choosing Between Controllers

Continuing the discussion of the previous paragraph, computing the belief distribution of action values allows the model (organism) to use this information to arbitrate between choosing a model-based or a model-free controller. In particular, Daw et al. (2005) proposed that the controller with the lowest parameter variance (and thus the lowest uncertainty) would be allowed to choose the current action, because this would eventually lead to the optimal choice. For example, if the model-free system is quite sure that action A is good and action B is bad, the variance around the value estimates of A and B will be low for both A and B. If at the same time the model-based system is not quite sure what to do, then its variances of these value estimates will be high, and the model-free system can decide what the organism will do.

Inhibition

A prominent cognitive control model inspired by Bayesian inference was proposed by Brown and Braver (2005). In their learned error likelihood prediction model, they proposed that the ACC computes the probability that an error is going to occur (given the current stimulus), corresponding to the (Bernoulli) distribution of error occurrence. Controllers attached to the ACC could then use this error information to inhibit responding in case of high error probability.

A more explicitly Bayesian model of inhibition was proposed by Shenoy and Yu (2011). In their 'model of the world', one variable indexed on each experimental trial in a stop task indicated whether that trial was actually a stop trial (or not). The variable indicating the stop status generated a stop signal at each time step in a trial with some probability (second variable), which in itself generated evidence (noticed by the experimental subject) with some probability (third variable). The subject inverted this three-layer model of the world (using the available evidence from current and earlier trials), in order to generate a posterior probability on each time step in a trial that the current trial was a stop trial. This was subsequently used to make a decision on whether to inhibit responding (i.e., stopping) or not. Finally, an RL learner (as described in the previous section) implemented this decision.

Choosing a Learning Rate

Further, choosing an optimal learning rate has been considered in a Bayesian framework. For linear input-output mappings, there is a simple and elegant solution for choosing an optimal learning rate (the Kalman filter; Dayan, Kakade, & Montague, 2000; Kruschke, 2008). A linear mapping may be appropriate for learning (in many environments and in most experiments) about cue–outcome relationships. In this case, the Kalman filter induces an update equation for (beliefs in) cue–outcome associations. Interestingly, its learning rate depends on the uncertainty attached to a given cue. Uncertain cues (i.e., cues for which there is much to learn) receive a higher learning rate. However, environmental contingencies are not always stationary but may instead vary across time (volatility). Behrens and colleagues (Behrens et al., 2007) provided a Bayesian model inversion procedure to estimate volatility. In their model, volatility determined reward probability, which determined actual reward. The computational model estimated reward and volatility distributions on a trial-to-trial basis. Although no convenient prediction-error-type update rule was obtained, fitting a Rescorla–Wagner model to the simulated data demonstrated that the model learned more quickly in uncertain (volatile) situations. Another Bayesian approach for learning rate estimation was proposed by Nassar and colleagues (Nassar, Wilson, Heasly, & Gold, 2010), who considered a task in which subjects must estimate the mean of a (normal) distribution of samples to which they received gradual (trial-by-trial) access. The task was non-trivial because at unknown trials the mean of the distribution could change. They developed an online Bayesian inference model that updated its learning rate such that if the probability of a changed distribution was high, the learning rate increased. A key point in these models is to increase the learning rate when less is known about the quantity of interest. Empirical evidence for such a dependency of learning rate on uncertainty was also recently reported (Nassar et al., 2012; Silvetti, Seurinck, van Bochove, & Verguts, 2013).

Instead of the volatility of reward probability, other authors (Jiang, Heller, & Egner, 2014) considered the volatility of congruency, the congruency distribution, and the actual congruency in cognitive control tasks (e.g., Stroop or flanker tasks). They proposed that subjects track such distributions and use them for the purpose of exerting cognitive control (e.g., focussing more strongly on the middle position in a flanker task).

General Discussion

I discussed how two key principles occur repeatedly under various guises in the cognitive control modelling literature: Cognitive control is a form of higher-level (secondary) control, and its parameters are set by optimising a goal function. In addition, I identified three modelling frameworks that incorporate these principles in five cognitive control functions. For the sake of brevity, I have restricted discussion to the computational modelling literature. However, some of the principles have also been discussed extensively in other theories. A common theme is that cognitive control is about solving higher-level problems (Norman & Shallice, 1986; Oberauer, 2009). Also, the five cognitive control functions have been discussed extensively outside the computational modelling literature. However, the fact that the same (learning, optimisation, inference) principles can be applied at both lower and higher levels is typically emphasised most strongly in the computational modelling literature. Perhaps this is not surprising: When forced to make computational commitments to a cognitive architecture, what better strategy is there than recycle the principles that have been successful before (Dehaene & Cohen, 2007)?

Broadly, and inspired again by the RL literature (Sutton & Barto, 1998), a distinction can be made between *Critic* systems, which monitor the need for cognitive control, and *Actor* systems, which provide the necessary control when such a need is detected. I focused on the Actor (controller) in this review, so I will now just mention briefly that much work in cognitive control has considered the monitoring (Critic) part instead (see e.g., Chapter 17 by Ullsperger in this volume; Chapter 10 by Kool et al. in this volume). Several authors proposed how such a monitoring function could be implemented, for example, by the ACC (Alexander & Brown, 2011; Botvinick et al., 2001; Silvetti, Seurinck, & Verguts, 2011).An emerging idea is that the ACC computes when (and possibly how and how much) control is required (Shackman et al., 2011; Shenhav et al., 2013). This Critic role of the ACC is reviewed more extensively elsewhere (Brown, 2013; Silvetti, Alexander, Verguts, & Brown, 2014; Brown, this volume).

The three modelling frameworks have their own focus and complementary strengths and weaknesses. For example, RL and Bayesian models often adopt a tabular approach in which stimuli and actions are primitives, rather than learned. Stimulus–action transition probabilities (i.e., probabilities of action given the current situation) are stored in a table that can be read out by the appropriate controllers. Unfortunately, estimation of such probabilities can quickly become unstable for combinatorial reasons (LeCun, Bengio, & Hinton, 2015). Instead, in PDP (Rogers et al., 2004) and deep learning models (Stoianov & Zorzi, 2012), stimulus and action representations emerge from training also. However, the construction of such representations (called function approximation) can be combined with RL principles (Sutton & Barto, 1998). In cognitive neuroscience, such cross-fertilisation has only recently begun (Mnih et al., 2015), and it can be fruitful for understanding cognitive control too in future work.

As there are at least five different cognitive control functions, each comprising potentially several (secondary) controllers, an obvious question is: How does the cognitive system know which secondary controller to choose? This is a question at yet a hierarchically higher level, above the primary and secondary levels considered in this chapter. Accordingly, one can consider choosing between secondary controllers as a (tertiary) control problem, presumably governed by the same principles of learning and optimisation. Although the issue of different levels of task control was discussed in earlier work (Koechlin & Summerfield, 2007), the issue of choosing between cognitive controllers in the sense discussed here, has not. This is another important avenue for future computational modelling.

My approach in this chapter has been to consider how research topics from cognitive neuroscience were recycled in cognitive control. Before concluding the chapter, I consider two themes that have been influential in computational cognitive neuroscience, but not yet in computational models of cognitive control.

Binding by Synchrony

All neural models discussed thus far are rate code models, meaning that the (mean) firing rate of their units codes all relevant information. However, besides the firing rate, the timing of firing may constitute a second, temporal code, which can be exploited for information processing (Buzsáki & Wang, 2012). In particular, it has been proposed that a subset of neurons exhibits oscillations whose properties (phase, frequency, and amplitude) may structure neural information flow (Engel, Fries, & Singer, 2001; Fries, 2005; Chapter 14 by M. X. Cohen in this volume). Such models have been applied in a number of domains (Börgers, Epstein, & Kopell, 2008; Li & Hopfield, 1989). One domain related to cognitive control is short-term memory; in these models, slow (theta-frequency) waves structure the precise firing order of other neurons that code the to-be-remembered items (Jensen & Lisman, 2005; Lisman & Idiart, 1995). Because cognitive control is hierarchically placed above primary controllers and is concerned with their communication, structuring information flow is an important issue in cognitive control. For this reason, exploiting oscillatory properties could be very useful in cognitive control (for an in-depth discussion, see Chapter 14 by M. X. Cohen in this volume). Recent empirical data indeed show strong modulations of oscillatory properties (e.g., phase) by cognitive control (Nigbur, Cohen, Ridderinkhof, & Stürmer, 2012; Womelsdorf, Johnston, Vinck, & Everling, 2010). Hence, computational oscillatory models may gain prominence in cognitive control too.

Extending Cognitive Control

Current models of cognitive control typically address rather abstract domains of application with simple stimuli and responses, such as Stroop or flanker tasks. This simplification in both experiment and theory is deliberate and useful for the discovery of basic principles. However, the brain did not evolve to solve the Stroop task; cognitive control is always applied in the context of other tasks and domains like declarative memory, language, mental arithmetic, or motor control. Future models should study how cognitive control operates in such richer contexts. One particularly fruitful context may be the training of robots (Khamassi, Lallée, Enel, Procyk, & Dominey, 2011) to increase their flexibility for interacting with the world. Potentially, such extensions will in the longer term inform us about cognitive control in the real world—like when a tennis player wonders what to do when lightning flashes across the sky.

Acknowledgements

I was supported by Ghent University GOA grant BOF08/GOA/011. I thank Elger Abrahamse for discussions about the ideas presented here, and his critical reading of the text.

Note

1 The model's main learning algorithm was a form of contrastive Hebbian learning, which has many similarities with backpropagation.

References

Alexander, G. E., & Crutcher, M. D. (1990). Functional architecture of basal ganglia circuits: Neural substrates of parallel processing. *Trends in Neurosciences*, *13*(7), 266–271.

Alexander, W. H., & Brown, J. W. (2011). Medial prefrontal cortex as an action-outcome predictor. *Nature Neuroscience*, *14*(10), 1338–1344.

Anderson, J. R. (1991). The adaptive nature of human categorization. *Psychological Review*, *98*(3), 409–429.

Aron, A. R. (2007). The neural basis of inhibition in cognitive control. *The Neuroscientist: A Review Journal Bringing Neurobiology, Neurology and Psychiatry*, *13*(3), 214–228.

Aron, A. R., Robbins, T. W., & Poldrack, R. A. (2004). Inhibition and the right inferior frontal cortex. *Trends in Cognitive Sciences*, *8*(4), 170–177.

Aston-Jones, G., & Cohen, J. D. (2005). An integrative theory of locus coeruleus-norepinephrine function: Adaptive gain and optimal performance. *Annual Review of Neuroscience*, *28*, 403–450.

Atkinson, R. C., & Shiffrin, R. M. (1968). Human memory: A proposed system and its control processes. In K. W. Spence & T. J. Spence (Eds.), *The psychology of learning and motivation* (Vol. 2, pp. 89–195). New York: Academic Press.

Balleine, B. W., & Dickinson, A. (1998). Goal-directed instrumental action: Contingency and incentive learning and their cortical substrates. *Neuropharmacology*, *37*(4–5), 407–419.
Behrens, T. E. J., Woolrich, M. W., Walton, M. E., & Rushworth, M. F. S. (2007). Learning the value of information in an uncertain world. *Nature Neuroscience*, *10*(9), 1214–1221.
Blais, C., Robidoux, S., Risko, E. F., & Besner, D. (2007). Item-specific adaptation and the conflict-monitoring hypothesis: A computational model. *Psychological Review*, *114*(4), 1076–1086.
Börgers, C., Epstein, S., & Kopell, N. J. (2008). Gamma oscillations mediate stimulus competition and attentional selection in a cortical network model. *Proceedings of the National Academy of Sciences of the United States of America*, *105*(46), 18023–18028.
Botvinick, M. M., Braver, T. S., Barch, D. M., Carter, C. S., & Cohen, J. D. (2001). Conflict monitoring and cognitive control. *Psychological Review*, *108*(3), 624–652.
Braver, T. S. (2012). The variable nature of cognitive control: A dual mechanisms framework. *Trends in Cognitive Sciences*, *16*(2), 106–113.
Braver, T. S., Barch, D. M., & Cohen, J. D. (1999). Cognition and control in schizophrenia: A computational model of dopamine and prefrontal function. *Biological Psychiatry*, *46*(3), 312–328.
Braver, T. S., & Cohen, J. D. (2000). On the control of control: The role of dopamine in regulating prefrontal function and working memory. In S. Monsell & J. Driver (Eds.), *Attention & performance XVIII* (pp. 713–737). Cambridge, MA: MIT Press.
Brown, J. W. (2013). Beyond conflict monitoring: Cognitive control and the neural basis of thinking before you act. *Current Directions in Psychological Science*, *22*(3), 179–185.
Brown, J. W., & Braver, T. S. (2005). Learned predictions of error likelihood in the anterior cingulate cortex. *Science (New York, N.Y.)*, *307*(5712), 1118–1121.
Brown, J. W., Reynolds, J. R., & Braver, T. S. (2007). A computational model of fractionated conflict-control mechanisms in task-switching. *Cognitive Psychology*, *55*(1), 37–85.
Buzsáki, G., & Wang, X.-J. (2012). Mechanisms of gamma oscillations. *Annual Review of Neuroscience*, *35*, 203–225.
Cohen, J. D., Dunbar, K., & McClelland, J. L. (1990). On the control of automatic processes: A parallel distributed processing account of the Stroop effect. *Psychological Review*, *97*(3), 332–361.
Cohen, J. D., & Servan-Schreiber, D. (1992). Context, cortex, and dopamine. *Psychological Review*, *99*(1), 45–77.
Daw, N. D., Niv, Y., & Dayan, P. (2005). Uncertainty-based competition between prefrontal and dorsolateral striatal systems for behavioral control. *Nature Neuroscience*, *8*(12), 1704–1711.
Dayan, P. (1998). A hierarchical model of binocular rivalry. *Neural Computation*, *10*(5), 1119–1135.
Dayan, P., Kakade, S., & Montague, P. R. (2000). Learning and selective attention. *Nature Neuroscience*, *3*(Suppl.), 1218–1223.
Dehaene, S., & Cohen, L. (2007). Cultural recycling of cortical maps. *Neuron*, *56*(2), 384–398.
Doya, K. (2002). Metalearning and neuromodulation. *Neural Networks: The Official Journal of the International Neural Network Society*, *15*(4–6), 495–506.
Engel, A. K., Fries, P., & Singer, W. (2001). Dynamic predictions: Oscillations and synchrony in top-down processing. *Nature Reviews Neuroscience*, *2*(10), 704–716.
Frank, M. J. (2005). Dynamic dopamine modulation in the basal ganglia: A neurocomputational account of cognitive deficits in medicated and nonmedicated Parkinsonism. *Journal of Cognitive Neuroscience*, *17*(1), 51–72.
Frank, M. J. (2006). Hold your horses: A dynamic computational role for the subthalamic nucleus in decision making. *Neural Networks: The Official Journal of the International Neural Network Society*, *19*(8), 1120–1136.
Fries, P. (2005). A mechanism for cognitive dynamics: neuronal communication through neuronal coherence. *Trends in Cognitive Sciences*, *9*(10), 474–480.
Gilbert, S. J., & Shallice, T. (2002). Task switching: A PDP model. *Cognitive Psychology*, *44*(3), 297–337.
Griffiths, T. L., Chater, N., Kemp, C., Perfors, A., & Tenenbaum, J. B. (2010). Probabilistic models of cognition: Exploring representations and inductive biases. *Trends in Cognitive Sciences*, *14*(8), 357–364.
Griffiths, T. L., Kemp, C., & Tenenbaum, J. B. (2008). Bayesian models of cognition. In R. Sun (Ed.), *The Cambridge handbook of computational cognitive modeling* (pp. 59–100). Cambridge: Cambridge University Press.

Haruno, M., Wolpert, D. M., & Kawato, M. (2001). Mosaic model for sensorimotor learning and control. *Neural Computation*, *13*(10), 2201–2220.
Hazy, T. E., Frank, M. J., & O'Reilly, R. C. (2007). Towards an executive without a homunculus: Computational models of the prefrontal cortex/basal ganglia system. *Philosophical Transactions of the Royal Society of London. Series B, Biological Sciences*, *362*(1485), 1601–1613.
Hinton, G. E., & Salakhutdinov, R. R. (2006). Reducing the dimensionality of data with neural networks. *Science*, *313*(5786), 504–507.
Hochreiter, S., & Schmidhuber, J. (1997). Long short-term memory. *Neural Computation*, *9*(8), 1735–1780.
Holroyd, C. B., & Coles, M. G. H. (2002). The neural basis of human error processing: Reinforcement learning, dopamine, and the error-related negativity. *Psychological Review*, *109*(4), 679–709.
Holroyd, C. B., & McClure, S. M. (2015). Hierarchical control over effortful behavior by rodent medial frontal cortex: A computational model. *Psychological Review*, *122*(1), 54–83.
Jacobs, R., Jordan, M. I., & Barto, A. G. (1991). Task decomposition through competition in a modular connectionist architecture: The what and where vision tasks. *Cognitive Science*, *15*(2), 219–250.
Jensen, O., & Lisman, J. E. (2005). Hippocampal sequence-encoding driven by a cortical multi-item working memory buffer. *Trends in Neurosciences*, *28*(2), 67–72.
Jiang, J., Heller, K., & Egner, T. (2014). Bayesian modeling of flexible cognitive control. *Neuroscience and Biobehavioral Reviews*, *46*, 30–43.
Jordan, M. I., & Jacobs, R. A. (1994). Hierarchical mixtures of experts and the EM algorithm. *Neural Computation*, *6*(2), 181–214.
Kemp, C., & Tenenbaum, J. B. (2008). The discovery of structural form. *Proceedings of the National Academy of Sciences of the United States of America*, *105*(31), 10687–10692.
Khamassi, M., Lallée, S., Enel, P., Procyk, E., & Dominey, P. F. (2011). Robot cognitive control with a neurophysiologically inspired reinforcement learning model. *Frontiers in Neurorobotics*, *5*(July), 1.
Khamassi, M., Wilson, C., Rothé, R., Quilodran, R., Dominey, P. F., & Procyk, E. (2011). Meta-learning, cognitive control, and physiological interactions between medial and lateral prefrontal cortex. In R. Mars, J. Sallet, M. Rushworth, & N. Yeung (Eds.), *Neural basis of motivational and cognitive control* (pp. 351–370). Cambridge, MA: MIT Press.
Koechlin, E., & Summerfield, C. (2007). An information theoretical approach to prefrontal executive function. *Trends in Cognitive Sciences*, *11*(6), 229–235.
Körding, K. P., & Wolpert, D. M. (2006). Bayesian decision theory in sensorimotor control. *Trends in Cognitive Sciences*, *10*(7), 319–326.
Kriete, T., & Noelle, D. C. (2011). Generalisation benefits of output gating in a model of prefrontal cortex. *Connection Science*, *23*(2), 119–129.
Kruschke, J. K. (2008). Bayesian approaches to associative learning: From passive to active learning. *Learning & Behavior*, *36*(3), 210–226.
LeCun, Y., Bengio, Y., & Hinton, G. (2015). Deep learning. *Nature*, *521*(7553), 436–444.
Li, Z., & Hopfield, J. J. (1989). Modeling the olfactory bulb and its neural oscillatory processings. *Biological Cybernetics*, *61*(5), 379–392.
Lisman, J. E., & Idiart, M. A. (1995). Storage of 7 plus / minus 2 short-term memories in oscillatory subcycles. *Science*, *267*.
McClelland, J. L., Botvinick, M. M., Noelle, D. C., Plaut, D. C., Rogers, T. T., Seidenberg, M. S., & Smith, L. B. (2010). Letting structure emerge: Connectionist and dynamical systems approaches to cognition. *Trends in Cognitive Sciences*, *14*(8), 348–356.
McClure, S. M., Gilzenrat, M. S., & Cohen, J. D. (2005). An exploration-exploitation model based on norepinephrine and dopamine activity. In Y. Weiss, B. Schölkopf, & J. C. Platt (Eds.), *Advances in neural information processing systems* (Vol. *18*, pp. 867–874). Neural Information Processing Systems Foundation.
Miller, E. K., & Cohen, J. D. (2001). An integrative theory of prefrontal cortex function. *Annual Review of Neuroscience*, *24*, 167–202.
Miyake, A., Friedman, N. P., Emerson, M. J., Witzki, A. H., Howerter, A., & Wager, T. D. (2000). The unity and diversity of executive functions and their contributions to complex 'Frontal Lobe' tasks: A latent variable analysis. *Cognitive Psychology*, *41*(1), 49–100.

Mnih, V., Kavukcuoglu, K., Silver, D., Rusu, A. A., Veness, J., Bellemare, M. G.,...Hassabis, D. (2015). Human-level control through deep reinforcement learning. *Nature, 518*(7540), 529–533.

Montague, P. R., Dayan, P., & Sejnowski, T. J. (1996). A framework for mesencephalic predictive Hebbian learning. *Journal of Neuroscience, 16*(5), 1936–1947.

Nassar, M. R., Rumsey, K. M., Wilson, R. C., Parikh, K., Heasly, B., & Gold, J. I. (2012). Rational regulation of learning dynamics by pupil-linked arousal systems. *Nature Neuroscience, 15*(7).

Nassar, M. R., Wilson, R. C., Heasly, B., & Gold, J. I. (2010). An approximately Bayesian delta-rule model explains the dynamics of belief updating in a changing environment. *Journal of Neuroscience, 30*(37), 12366–12378.

Nee, D. E., Kastner, S., & Brown, J. W. (2011). Functional heterogeneity of conflict, error, task-switching, and unexpectedness effects within medial prefrontal cortex. *NeuroImage, 54*(1), 528–540.

Nigbur, R., Cohen, M. X., Ridderinkhof, K. R., & Stürmer, B. (2012). Theta dynamics reveal domain-specific control over stimulus and response conflict. *Journal of Cognitive Neuroscience, 24*(5), 1264–1274.

Norman, D., & Shallice, T. (1986). Attention to action: Willed and automatic control of behaviour. Center for human information processing (Technical Report No. 99). In R. J. Davidson, G. E. Schwartz, & D. Shapiro (Eds.), *Consciouness and self-regulation* (Vol. 4). New York NY: Plenum Press.

Notebaert, W., & Verguts, T. (2008). Cognitive control acts locally. *Cognition, 106*(2), 1071–1080.

O'Doherty, J. P., Dayan, P., Friston, K. J., Critchley, H., & Dolan, R. J. (2003). Temporal difference models and reward-related learning in the human brain. *Neuron, 38*(2), 329–337.

O'Reilly, R. C., & Frank, M. J. (2006). Making working memory work: A computational model of learning in the prefrontal cortex and basal ganglia. *Neural Computation, 18*(2), 283–328.

Oberauer, K. (2009). Design for a working memory. *Psychology of Learning & Motivation, 51*, 45–100.

Otto, A. R., Skatova, A., Madlon-kay, S., & Daw, N. D. (2015). Cognitive control predicts use of model-based reinforcement learning. *Journal of Cognitive Neuroscience, 27*(2), 319–333.

Pearce, J. M., & Hall, G. (1980). A model for Pavlovian learning: Variations in the effectiveness of conditioned but not of unconditioned stimuli. *Psychological Review, 87*(6), 532–552.

Rescorla, R. A., & Wagner, A. (1972). A theory of Pavlovian conditioning: Variations in the effectiveness of reinforcement and nonreinforcement. In A. Black & W. Prokasy (Eds.), *Classical conditioning II: Current research and theory* (pp. 64–99). New York: Appleton Century Crofts.

Rogers, R. D., & Monsell, S. (1995). Costs of a predictable switch between simple cognitive tasks. *Journal of Experimental Psychology: General, 124*(2), 207–231.

Rogers, T. T., Ralph, M. A. L., Garrard, P., Bozeat, S., McClelland, J. L., Hodges, J. R., & Patterson, K. (2004). Structure and deterioration of semantic memory: A neuropsychological and computational investigation. *Psychological Review, 111*(1), 205–235.

Rougier, N. P., Noelle, D. C., Braver, T. S., Cohen, J. D., & O'Reilly, R. C. (2005). Prefrontal cortex and flexible cognitive control: Rules without symbols. *Proceedings of the National Academy of Sciences of the United States of America, 102*(20), 7338–7343.

Rumelhart, D., McClelland, J. L., & Group, T. P. Research. (1986). *Parallel distributed processing: Explorations in the microstructure of cognition. Volume 1: Foundations.* Cambridge, MA: MIT Press.

Rumelhart, D., Williams, R. J., & Hinton, G. E. (1986). Learning internal representations by error propagation. In D. Rumelhart, J. L. Mcclelland, & The PDP research group (Eds.), *Parallel distributed processing; Explorations in the microstructure of cognition.* Cambridge, MA: MIT Press.

Schultz, W. (1998). Predictive reward signal of dopamine neurons. *Journal of Neurophysiology, 80*(1), 1–27.

Seidenberg, M. S., & McClelland, J. L. (1989). A distributed, developmental model of word recognition and naming. *Psychological Review, 96*(4), 523–568.

Shackman, A. J., Salomons, T. V., Slagter, H. A., Fox, A. S., Winter, J. J., & Davidson, R. J. (2011). The integration of negative affect, pain and cognitive control in the cingulate cortex. *Nature Reviews Neuroscience, 12*(3), 154–167.

Shenhav, A., Botvinick, M. M., & Cohen, J. D. (2013). The expected value of control: An integrative theory of anterior cingulate cortex function. *Neuron, 79*(2), 217–240.

Shenoy, P., & Yu, A. J. (2011). Rational decision-making in inhibitory control. *Frontiers in Human Neuroscience*, *5*(May), 48.

Silvetti, M., Alexander, W. H., Verguts, T., & Brown, J. W. (2014). From conflict management to reward-based decision making: Actors and critics in primate medial frontal cortex. *Neuroscience and Biobehavioral Reviews*, *46*, 44–57.

Silvetti, M., Seurinck, R., van Bochove, M. E., & Verguts, T. (2013). The influence of the noradrenergic system on optimal control of neural plasticity. *Frontiers in Behavioral Neuroscience*, *7*, 160.

Silvetti, M., Seurinck, R., & Verguts, T. (2011). Value and prediction error in medial frontal cortex: Integrating the single-unit and systems levels of analysis. *Frontiers in Human Neuroscience*, *5*(August), 75.

Stoianov, I., & Zorzi, M. (2012). Emergence of a 'visual number sense' in hierarchical generative models. *Nature Neuroscience*, *15*(2), 194–196.

Sutton, R. S., & Barto, A. G. (1998). *Reinforcement learning: An introduction*. Cambridge, MA: MIT Press.

Tenenbaum, J. B., Griffiths, T. L., & Kemp, C. (2006). Theory-based Bayesian models of inductive learning and reasoning. *Trends in Cognitive Sciences*, *10*(7), 309–318.

Verguts, T., & Notebaert, W. (2008). Hebbian learning of cognitive control: Dealing with specific and nonspecific adaptation. *Psychological Review*, *115*(2), 518–525.

Verguts, T., & Notebaert, W. (2009). Adaptation by binding: A learning account of cognitive control. *Trends in Cognitive Sciences*, *13*(6), 252–257.

Verguts, T., Vassena, E., & Silvetti, M. (2015). Adaptive effort investment in cognitive and physical tasks: A neurocomputational model. *Frontiers in Behavioral Neuroscience*, *9*, 57.

Widrow, G., & Hoff, M. E. (1960). Adaptive switching circuits. In *IRE Western electric show and convention record*. Hillsdale, NJ: Lawrence Erlbaum Associates.

Wolpert, D. M., & Kawato, M. (1998). Multiple paired forward and inverse models for motor control. *Neural Networks: The Official Journal of the International Neural Network Society*, *11*(7–8), 1317–1329.

Womelsdorf, T., Johnston, K., Vinck, M., & Everling, S. (2010). Theta-activity in anterior cingulate cortex predicts task rules and their adjustments following errors. *Proceedings of the National Academy of Sciences of the United States of America*, *107*(11), 5248–5253.

Wylie, G., & Allport, A. (2000). Task switching and the measurement of 'switch costs'. *Psychological Research*, *63*(3–4), 212–233.

9

Context Processing and Cognitive Control

From Gating Models to Dual Mechanisms

Kimberly S. Chiew and Todd S. Braver

The ability to maintain and flexibly regulate thoughts and action plans in pursuit of internally represented behavioural goals is central to the psychological construct of *cognitive control* and critical to adaptive human behaviour. This ability requires a complex balance between maintaining current goal representations against distracting information, while also flexibly updating these representations as goals and environmental factors change. In the present chapter, we review evidence that the balance between control stability and flexibility is supported by context maintenance in the prefrontal cortex and flexible representation updating via a dopaminergic gating mechanism. We will then discuss how context processing can be understood as supporting dual mechanisms of cognitive control characterised by distinct temporal dynamics, by reviewing evidence for the dual mechanisms of control (DMC) framework (Braver, 2012; Braver, Gray, & Burgess, 2007). We will also outline methodological approaches and tools that have been fruitful in empirical investigation within the DMC framework and discuss future directions for investigation stemming from this perspective. Given past evidence and future directions within this line of research, we propose that the core utility of the DMC framework lies in its ability to account for and generate specific, testable predictions regarding variability in cognitive control dynamics across a broad variety of task paradigms and at multiple levels of analysis.

Context Processing and Gating Models

Context can be broadly described as task-relevant information represented in such a form so as to bias selection of the appropriate task response. Internal, active representations of context in working memory play a critical role in guiding executive, goal-oriented behaviour (Braver, Barch, & Cohen, 2002; J. D. Cohen & Servan-Schreiber, 1992). These representations may bias attention to task-relevant information, promote inhibition of task-irrelevant information, and structure encoding, maintenance, and retrieval of information in working and long-term memory, while generally supporting the planning and execution of adaptive goal-directed actions and behaviour. These cognitive processes correspond closely to the putative functions of the prefrontal cortex (PFC), based on early lesion and neuroimaging studies in humans

The Wiley Handbook of Cognitive Control, First Edition. Edited by Tobias Egner.

(Stuss & Knight, 2013) as well as neurophysiological studies in primates (Goldman-Rakic & Friedman, 1991). Along with identifying the PFC as a neural substrate supporting executive function, studies from this body of work further identified the dopamine (DA) neurotransmitter system as a major modulator of goal-oriented behaviour and PFC activity: that is, studies of disrupted cognitive function and DA systems in schizophrenia (J. D. Cohen & Servan-Schreiber, 1992) as well as studies in primates suggesting that pharmacological manipulations of the DA system alter working-memory-related activity in PFC neurons (Sawaguchi & Goldman-Rakic, 1991). Although this work was critical in identifying functional brain areas supporting context processing and the implementation of goal-directed behaviour, these early studies were largely agnostic on the mechanisms underlying active learning, maintenance, and updating of context representations and their use in the execution of adaptive task behaviour.

Braver and Cohen (Braver & Cohen, 1999, 2000) proposed a gating model aimed at addressing this theoretical gap by identifying and testing a plausible mechanistic means by which DA could modulate active context processing in the PFC, supporting controlled behaviour. This connectionist model posits that selection, updating, and maintenance of context occur through interactions between the PFC and the DA neuromodulatory system. Specifically, this model posits that phasic bursts of DA act as a gating mechanism, regulating information access to active memory mechanisms subserved by the PFC (Braver, Barch, & Cohen, 1999). Further, DA plays a learning function (via phasic firing in response to prediction errors; i.e., situations where reward outcomes are either greater or lesser than anticipated), allowing the system to discover what information is relevant for selection as context (Braver & Cohen, 2000).

These assertions were tested in a simulation of the model using the *AX Continuous Performance Task* (AX-CPT; J. D. Cohen & Servan-Schreiber, 1992). The AX-CPT is a delayed-response task requiring context maintenance and updating for successful performance (see Figure 9.1a). On each experimental trial of the AX-CPT, participants must respond to a cue–probe pair presented sequentially (typically, letter stimuli). One specific combination requires a target response (i.e., the letter 'A' followed by the letter 'X'; AX trial), whereas all other combinations of cue and probe require a non-target response. Target (AX) trials occur at a high frequency (typically 70%), leading to associations between the target cue (the letter A) and target response, and between the target probe (the letter X) and the target response. These associations subsequently lead to interference for two low-frequency cue–probe pairs (typically occurring at 10% each): AY trials (target cue, non-target probe), where contextual cue leads to a bias towards target response that must be overcome; and BX (non-target cue, target probe) trials, where the contextual information must be used to inhibit the probe-related tendency towards target response. BY (non-target cue, non-target probe) trials also occur at a low-frequency (10%) control condition.

Simulations with this model (schematic in Figure 9.1b) suggested that it is biologically plausible for DA to successfully gate information into active memory in the PFC and, in response to reward prediction errors, strengthen stimulus–response associations supporting learning (in turn optimising reward pursuit and goal-oriented behaviour). Phasic DA activity as a gating mechanism was thus posited to provide a means by which context information could be actively maintained in the PFC, remain protected against interference yet flexible to updating, and used to bias action responses in the service of goal-oriented behaviour. The gating model provides an account consistent with behavioural and neurobiological evidence for altered context processing in populations including schizophrenic patients (Braver et al., 2002; J. D. Cohen, Barch, Carter, & Servan-Schreiber, 1999; Chapter 31 by Barch & Sheffield in this volume) and healthy older adults and individuals with Alzheimer's disease (Braver et al., 2002). Interestingly, these empirical studies suggested that impaired cognitive

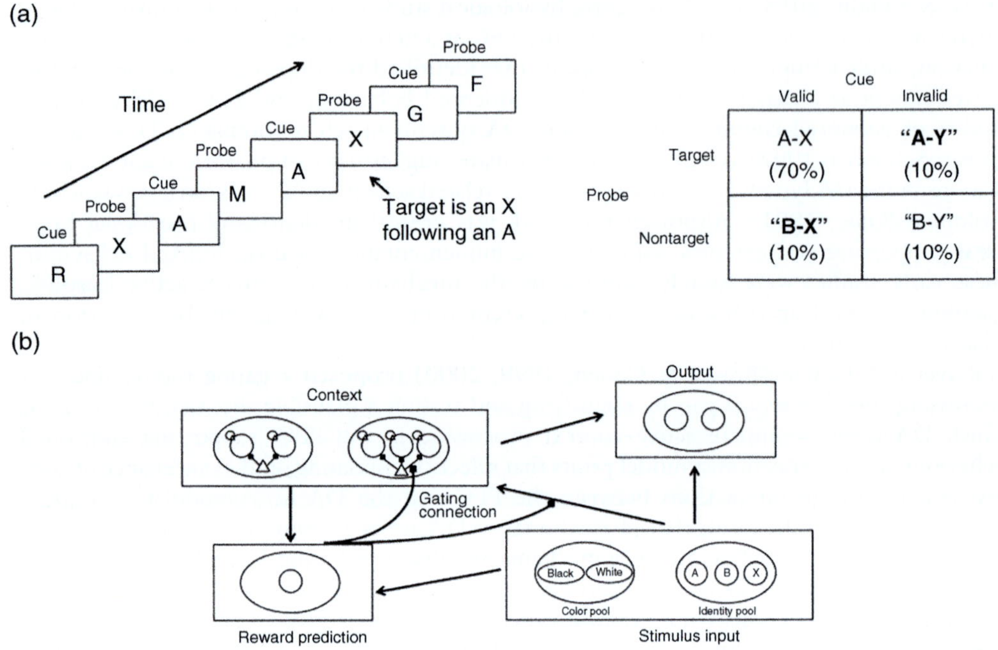

Figure 9.1 (a) Schematic of the AX-CPT paradigm. Single letters are visually displayed as a series of cue–probe pairs. In this example, a target is defined as the occurrence of an 'X' probe immediately following an 'A' cue and occurs at high frequency (70% of trials). Three types of non-target trials occur with equal frequency (10%): AY, BX, and BY (where B refers to any non-A cue, and Y refers to any non-X probe). (b) Schematic of the learning/gating model used in simulation of AX-CPT (following Braver & Cohen, 2000). Excitatory connections exist between layers, as indicated by arrows; lateral inhibitory connections (not pictured) exist within each layer. Context units have self-excitatory connections allowing for active context maintenance. Low levels of baseline activity in the context layer are enforced by local inhibitory bias units (indicated by small triangles). Input and context layers are fully connected to the reward prediction/gating (RPG) unit, which in turn makes a gating connection with both excitatory and inhibitory input to the context layer. The RPG unit also modulates learning in all modifiable network connections. Source: Adapted from Braver and Cohen, 2000.

control performance in such populations, relative to healthy controls, might be best characterised by selective impairments in specific components of context processing, as opposed to a global cognitive deficit. Specifically, both schizophrenic patients and healthy older adults showed worsened BX performance and improved AY performance relative to healthy young adults. Evidence of superior AY performance in these populations relative to young adults is notable, as generally their cognitive performance is comparatively impaired. Successful performance on AY trials in the AX-CPT is specifically attributed to the ability to overcome response bias associated with the contextual cue gated by DA and maintained within the PFC; task performance suggests that in these populations, the cue representation may not be maintained as strongly as in healthy young adults. Early experimental work, using fMRI, provided evidence consistent with this idea, showing cue-related activation in the dorsolateral PFC during performance of the AX-CPT (Braver et al., 2002). Other later work demonstrated reductions in cue-related dorsolateral PFC activity during AX-CPT performance in both patients with schizophrenia (MacDonald & Carter, 2003) and older adults (Paxton, Barch, Racine, & Braver, 2008).

A recent fMRI study provided more direct support for the gating account, both by directly imaging the midbrain DA system during working memory updating, and by using transcranial magnetic stimulation (TMS) to disrupt activity within the PFC (D'Ardenne et al., 2012). Specifically, in this study, activity in the DA midbrain (ventral tegmental area, substantia nigra) was found to phasically increase in response to contextual cues in an AX- CPT variant, and this activation was correlated with cue-related activation observed in the right dorsolateral PFC. Furthermore, phasic TMS applied to this same dorsolateral PFC region, immediately after cue presentation, led to task performance decrements.

Subsequent computational models have built upon the gating model while retaining the core principle that DA may regulate information flow and updating in the PFC. O'Reilly, Frank, and colleagues (Frank, Loughry, & O'Reilly, 2001; O'Reilly & Frank, 2006) developed a computational model of working memory similarly built on the premise that a dopaminergic gating mechanism controls information access to the PFC, but refined this model by positing a specific role for the basal ganglia in releasing inhibition on specific subregions of the frontal cortex (Figure 9.2b). Thus, in this model, known as the prefrontal-cortex basal-ganglia working memory (PBWM) model, phasic DA activity controls the learning of more spatially targeted gating signals within the basal ganglia, which enable selective updating within the PFC (as opposed to the relatively global effect that diffuse DA broadcast would cause; Figure 9.2c). The PBWM gating model has been shown to be especially powerful in understanding hierarchically structured working memory tasks, in which some representations (e.g., subgoals, 'inner loop') need to be updated following presentation of contextual cues, while others must still be maintained during this period (i.e., higher-order goals, 'outer loop'). For example, the PBWM model can simulate a hierarchically structured version of the AX-CPT, known as the 1-2-AX task (Nee & Brown, 2012; O'Reilly & Frank, 2006), in which contextual cues need to be referenced to the higher-order context (1 or 2) present throughout a block of trials (see Figure 9.2a). This type of hierarchically structured context processing task would be difficult to simulate with a global dopamine gating model.

More recently, the gating model has been even further refined, adding not only an input gate to the PFC mediated by corticostriatal interactions, but also a second striatally based output gate that may determine what PFC-maintained active representations are utilised for behaviour (Chatham, Frank, & Badre, 2014) and a third striatally based mechanism that reallocates working memory capacity when representations are no longer relevant (Chatham & Badre, 2015; see also Chapter 21 by Bhandari, Badre, & Frank in this volume).

The DMC Framework

Although clarifying the mechanisms by which context representations are gated into the PFC and selected for action remains an active area of research, models of context maintenance and updating in the PFC have also been expanded through the development of the DMC framework. The DMC framework aims to address and account for variation in cognitive control performance at multiple levels of analysis (Braver, 2012; Braver et al., 2007). This framework proposes that cognitive control can be understood as operating in two primary modes: *proactive* and *reactive*. Proactive control is characterised by the active maintenance of context representations in the PFC; this information enters and is maintained via the phasic DA gating mechanism that the gating model describes. Because the gating mechanism allows for active maintenance of information relatively protected from interference, proactive control is thought to be relatively tonic in nature. In contrast, reactive control is implemented as a

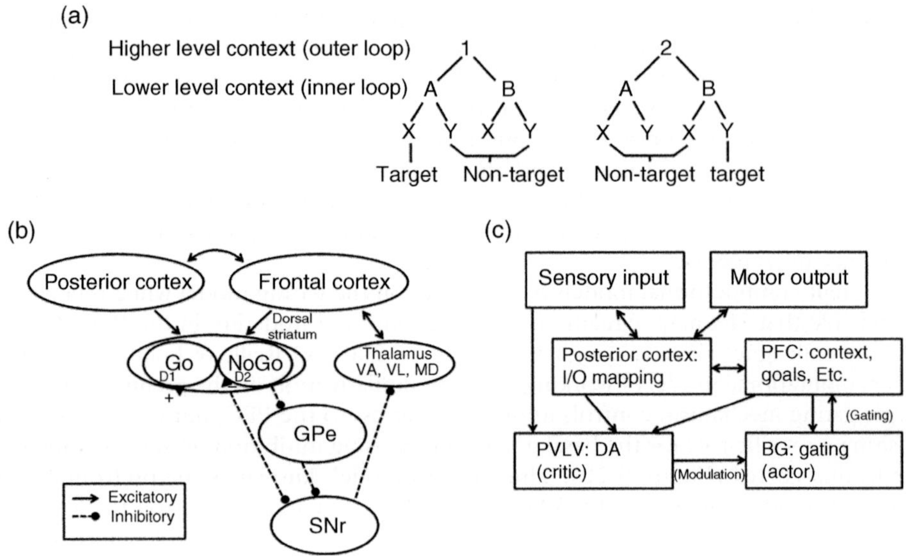

Figure 9.2 (a) Nesting rule structure of the 1-2-AX, hierarchical context updating paradigm (adapted from Nee and Brown, 2012). Subjects respond to 'X' and 'Y' stimuli based upon a nested series of cues. Under the '1' higher-level context (cued prior to trial), subjects make a target response to the letter 'X' if within the 'A' lower-level context (non-target response otherwise). Under the '2' higher-level context, subjects made a target response to the letter 'Y' if within the 'B' lower-level context (non-target response otherwise). (b) Interconnections between the basal ganglia and the frontal cortex through a series of parallel loops (Source: adapted from O'Reilly and Frank, 2006). The thalamus is bidirectionally excitatory with frontal cortex; the SNr (substantia nigra pars reticulata) is tonically active and inhibiting this excitatory circuit. When direct pathway 'Go' neurons in dorsal striatum fire, they inhibit the SNr and thus disinhibit the frontal cortex, producing a gating-like modulation argued to trigger the update of working memory representations in the prefrontal cortex. The indirect pathway 'NoGo' neurons of the dorsal striatum counteract this effect by inhibiting the inhibitory GPe (globus pallidus, external segment). (c) Overall architecture of the PBWM model implemented to capture hierarchical updating tasks such as the 1-2-AX (adapted from O'Reilly and Frank, 2006). Sensory inputs are mapped to motor outputs via posterior cortical ('hidden') layers. The PFC contextualizes this mapping by representing relevant prior information and goals. The basal ganglia (BG) and the primary value learned value (PVLV) system drive DA modulation of BG so it can learn when to update.

transient, phasic form of 'late correction' that occurs in response to changing environmental demands (i.e., in response to performance monitoring) or stimulus-triggered associative retrieval, and is posited to operate in the absence of the gating mechanism. Each mode of control has advantages and drawbacks, leading to an inherent computational trade-off between the two; the emergence of these specialised modes is thought to make possible dynamic shifting of control, towards optimisation of the balance between them. On account of explicit acknowledgement that the two modes of control can dynamically shift as a result of multiple influences, the DMC framework can provide a coherent theoretical account for variability in cognitive control processing from multiple sources, including within-individual, within-group, and between-group variability (Braver, 2012). The DMC framework's ability to account for altered context processing and cognitive performance in different populations and to generate new, testable hypotheses regarding sources of variability in cognitive control has yielded a wealth of empirical research. These studies have utilised different experimental paradigms and methodological approaches, motivational and affective manipulations, and comparisons between population groups and individuals as critical sources of variation in cognitive control. We discuss the highlights of this research below.

Cognitive Paradigms

One of the key contributions of the DMC framework is that it provides a means of understanding the potential sources of cognitive control variability that have been noted and observed across a wide range of experimental paradigms that are typically used in the cognitive control literature. These include context processing, cued task switching, selective attention / stimulus–response compatibility, working memory, prospective memory, and response inhibition. Here we briefly review each of these domains.

As a theoretical expansion of the gating model, the DMC framework has been most extensively employed to interpret variability in context processing tasks, in which contextual cues indicate how to appropriately respond to target items. The most widely investigated variant of context processing is the AX-CPT task. In the DMC interpretation of the AX-CPT, interference on AY trials is attributable to maintenance of contextual target cue ('A') that must be overcome upon presentation of the non-target probe ('Y'), and AY trial interference is thought to reflect a relatively preparatory, or proactive control. In contrast, in BX trials, interference arises via a target response bias to the target probe ('X') that must be inhibited on the basis of a prior contextual cue; BX trial interference is thus thought to reflect relatively reactive control. The utilisation of reactive control (rather than a complete failure of control) would be represented in BX interference coming primarily in the form of reaction time slowing (reflecting the time to engage control following probe presentation), rather than a large increase in BX error rates. Thus, the extent to which interference is experienced in AY and BX trial conditions during task performance may serve as an indicator of relative tendencies towards proactive versus reactive control. Studies of variability in AX-CPT performance and associated neural and physiological activity as a result of experimental manipulations or comparisons between groups have provided evidence that cognitive control during the task can be understood as operating in two primary modes. These findings will be outlined in further detail below (see the sections titled 'Individual Differences' and 'Comparison Between Populations').

A more complex category of the context processing paradigm is cued task switching (for a review, see Chapter 2 by Monsell in this volume). Here, the task varies randomly from trial to trial, and advance cues specify which task is relevant for the upcoming trial. The cognitive control demands of task switching are evidenced by the presence of both switching costs and mixing costs, which reflect performance decrements on, respectively: (a) trials in which the task switches compared to when it repeats; and (b) task-switch blocks compared to single-task blocks (controlling for task-switch trials; Kiesel et al., 2010; Monsell, 2003; Vandierendonck, Liefooghe, & Verbruggen, 2010). The finding that these costs are significantly reduced with increased preparation time following task cues suggests the presence of proactive control (Meiran, Chorev, & Sapir, 2000; Monsell, 2003). Conversely, the fact that the costs are almost never eliminated (i.e., residual), even with long preparation times, and also impacted by target-related factors such as task-rule congruency (whether the target stimulus elicits the same or conflicting responses, on the basis of which task is relevant), suggests the presence of, and demand for, reactive control as well (Kessler & Meiran, 2008; Meiran, Kessler, & Adi-Japha, 2008). Indeed, a primary area of unresolved debate and controversy is the extent to which cued task switching is thought to be accomplished primarily by proactive (e.g., preparatory task set maintenance and biasing) or reactive control mechanisms (e.g., arget-driven priming or conflict-triggered rule retrieval; Arrington, Logan, & Schneider, 2007; Koch & Allport, 2006; Monsell, 2003; Chapter 2 by Monsell in this volume). According to the DMC framework, the weighting or bias towards each control mode is related to a variety of task factors, such as reward motivation or expectation of interference (Bugg & Braver, in press; Bugg, Diede, Cohen-Shikora, & Selmeczy, 2015; Dreisbach, Haider, & Kluwe, 2002; Nieuwenhuis & Monsell, 2002). Indeed, in one computational model, we simulated how shifts between proactive and reactive control might even be implemented on a trial-by-trial basis, in response to random fluctuations of the DA gating mechanism, and how this could account for some of the performance variability observed experimentally (Reynolds, Braver, Brown, & Van der Stigchel, 2006).

Working memory tasks, although typically conceived of as being distinct from context processing and task-switching paradigms, also often have a context-target structure. In this case, the contextual information consists of items that should be stored in short-term memory over a delay interval, but also used to guide responding to the target. Although it is typically assumed that working memory tasks predominantly depend upon proactive control, in order to keep memory items actively maintained over the delay, more recent attention has been given to target-evoked interference effects that also suggest the presence and demand for reactive control (Berman, Jonides, & Lewis, 2009; Irlbacher, Kraft, Kehrer, & Brandt, 2014). Indeed, it is important to note that the DMC framework does not imply an equivalence between short-term memory storage and proactive control. Rather, proactive control refers to processes that select specific items in short-term storage for further attentional processing, such as in the case of the n-back or directed-forgetting-type paradigms, or bias attention towards only task-relevant features of probe items, as in recent negative probe paradigms (Irlbacher et al., 2014). In prior work, we and others have shown that biases towards proactive versus reactive control can be modulated in these paradigms by various manipulations (Irlbacher et al., 2014), such as the expected working memory load (Speer, Jacoby, & Braver, 2003) and the frequency of target-related interference (Burgess & Braver, 2010).

Even in memory tasks that involve much longer delays than standard working memory paradigms, the DMC framework can be used to help understand variability in the cognitive strategies and behavioural markers of cognitive control engaged for task performance. Although there have been very few studies directly examining proactive and reactive control within the episodic memory domain (Dobbins & Han, 2006; Velanova et al., 2003), within the domain of prospective memory, there have been explicit theoretical accounts focusing on variability in the control mechanisms used to support memory for delayed intentions (Einstein et al., 2005). Specifically, it has been recently noted that the influential multi-process account of prospective memory, which postulates a key distinction between sustained attentional monitoring and spontaneous retrieval processes, aligns well to distinctions between proactive and reactive control (Bugg, Scullin, & McDaniel, 2013). A variety of factors are thought to influence whether control is biased towards proactive (attentional monitoring) or reactive (spontaneous retrieval) strategies, but one of the key variables is whether cues indicating the prospective memory target are salient and focally processed as part of the ongoing task or not. In recent work, we have shown that a subtle distinction between the use of focal versus non-focal prospective memory targets had a strong influence on both task performance and the dynamics and location of prefrontal control regions (McDaniel, Lamontagne, Beck, Scullin, & Braver, 2013).

The DMC framework has also been extensively applied in contexts that do not involve cue-target designs or strong memory requirements, but rather primarily tap into selective attention and inhibition or interference control. For example, in the classic Stroop paradigm (Stroop, 1935), cognitive control demands are thought to vary as a function of the proportion of congruent (word name and ink colour match) to incongruent items (word indicates a different colour name than the ink colour [Chapter 5 by Bugg in this volume; Bugg & Crump, 2012; Jacoby, Lindsay, & Hessels, 2003; Logan & Zbrodoff, 1979]). For block-wise manipulations of proportion congruency, it is intuitive to think that proactive control would increase when the tonic, global expectation for interference is high relative to when it is low. Conversely, when proportion congruency is manipulated in an item-specific manner (e.g., certain colours, such as red, are mostly congruent, whereas others, such as blue, are mostly incongruent), but overall block-wise expectations for interference are low, then reactive control might be the dominant mode. Recent work has supported this distinction with reductions of Stroop interference observed following both block-wise and item-specific proportion congruency manipulations (Bugg, Jacoby, & Toth, 2008); however, the two different manipulations have been associated with distinct behavioural and neural signatures (Bugg & Hutchison, 2013; Bugg, Jacoby, & Chanani, 2011; De Pisapia, Slomski, & Braver, 2007; Gonthier, Braver, & Bugg, in revision; Grandjean et al., 2012; Wilk, Ezekiel, & Morton, 2012).

Although less extensively studied, the same types of proportion congruency manipulations can easily be applied and may produce the similar dissociations in related attentional and stimulus–response compatibility paradigms, such as the flanker task (Gratton, Coles, & Donchin, 1992) and the Simon task (Torres-Quesada, Funes, & Lupianez, 2013). These paradigms are also beginning to be explored with precueing manipulations, with the hypothesis that congruency precues might produce preparatory (proactive) control patterns (Bugg & Smallwood, 2014; Chiew & Braver, 2016; Czernochowski, 2015). Finally, a related strategy of precuing control demands has also begun to be explored within the domain of response inhibition, in paradigms such as the go-no go and stop signal (Aron, 2011; King, Korb, von Cramon, & Ullsperger, 2010). In this literature, it has been somewhat surprising that response inhibition, which in some sense represents the most extreme example of reactive control (see Chapter 6 by Verbruggen & Logan in this volume), can show evidence of implementation of proactive control strategies under conditions in which inhibitory demands can be anticipated (Berkman, Kahn, & Merchant, 2014; Verbruggen & Logan, 2009).

Temporal Dynamics and Neural Bases of DMC

The DMC framework makes strong predictions regarding the temporal dynamics of proactive and reactive control. Proactive control is thought to be characterised by sustained and/or anticipatory neural activity, reflecting the active maintenance of context representations, whereas reactive control is thought to be engaged on an as-needed basis, and is thus characterised by rapid engagement of transient neural activity just prior to responding (see Figure 9.3a for a schematic diagram of control dynamics). A range of methods has been used to characterise the anatomical localisation and temporal dynamics of these control mechanisms, with different advantages and disadvantages in terms of spatial and temporal resolution. These techniques have included fMRI, event-related potentials (ERPs), and pupillometry.

Functional neuroimaging using fMRI has played a key role in investigations of cognitive control within the DMC framework. We do not intend to provide an exhaustive review of this line of research here, but instead specifically highlight how fMRI has made neural investigation of separate control modes with distinct temporal dynamics possible. As noted before, the DMC framework predicts that proactive control is associated with relatively sustained and/or anticipatory activity, whereas reactive control is associated with relatively transient activity. fMRI studies of cognitive control have been consistent with the idea that a proactive control mode is associated with sustained activity in the PFC (Braver, 2012), whereas reactive

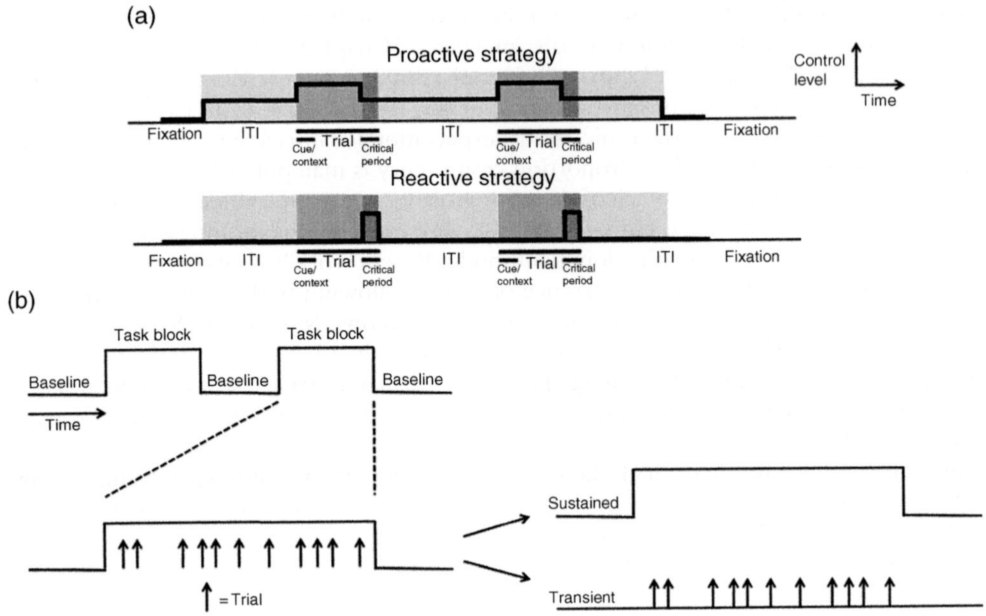

Figure 9.3 (a) Schematic of the temporal dynamics of proactive and reactive control in a context-processing task (such as the AX-CPT), as predicted by the DMC framework. Proactive control is characterised by sustained, preparatory control maintained over time (i.e., increased both tonically throughout a block and/or in an anticipatory fashion, following advance contextual cues), whereas reactive control is characterised by transient activity recruited just prior to responding. (b) The mixed block/event-related experimental design allows for simultaneous modelling of transient, trial-related activity and the sustained task-related activity.

control may rely on transient activity in a more distributed set of brain regions, including the PFC but also implicating an important role for other neural regions including the anterior cingulate, which has been implicated in the detection of conflict, error detection, or performance monitoring signalling the need for changes in control demand (Botvinick, Cohen, & Carter, 2004; Chapter 15 by Brown in this volume; Brown & Braver, 2005; van Veen & Carter, 2002; Chapter 17 by Ullsperger in this volume); all of these control functions can be considered highly reactive in nature.

Block and event-related fMRI designs allow characterisation of sustained and transient neural activity, respectively (Buckner, 1998), but do not allow dissociation of activations at different temporal dynamics. The *mixed block/event design* was developed for use with fMRI to address this issue, permitting disentanglement of relatively block versus event-related effects, originally in the context of visual processing (Visscher et al., 2003), but subsequently with a wealth of cognitive tasks (see Figure 9.3b). Employment of this design in conjunction with fMRI has been very useful in delineating both sustained and transient neural activity associated with cognitive control performance, providing evidence in support of the DMC framework. For example, using a mixed block/event fMRI design in combination with a task-switching paradigm (Braver, Reynolds, & Donaldson, 2003), changes in sustained activity in control-related brain areas were observed to relate to task set (i.e., task switching vs. single-task blocks), but additionally, transient activity in control-related regions was observed on a trial-by-trial basis with changing task demands (i.e., repeat vs. switch trials) and response speed. Likewise, within prospective memory, this design enabled dissociation of a primarily sustained dynamics observed in anterior and dorsolateral PFC under nonfocal (proactive) conditions, but only transient activity in ventral PFC, ACC, and medial parietal regions during focal (reactive) conditions (Braver et al., 2003).

Other event-related designs have enabled dissociation and characterisation of distinct events within a given trial, such as cue/delay versus target periods in cued task switching and context processing paradigms (Ruge, Goschke, & Braver, 2009) and encoding/retention versus probe periods in working memory paradigms (Jimura, Locke, & Braver, 2010). These types of designs have been useful in demonstrating distinct patterns of temporal dynamics in task conditions associated with proactive versus reactive control, as well as shifts in dynamics that can occur even within the same lateral PFC regions, as was observed as a function of both cognitive training and motivational incentives in the AX-CPT task (Braver, Paxton, Locke, & Barch, 2009) (Figure 9.4).

Although fMRI has been a critical tool in investigating the temporal dynamics of cognitive control within the PFC and other neural areas, other methodologies have also

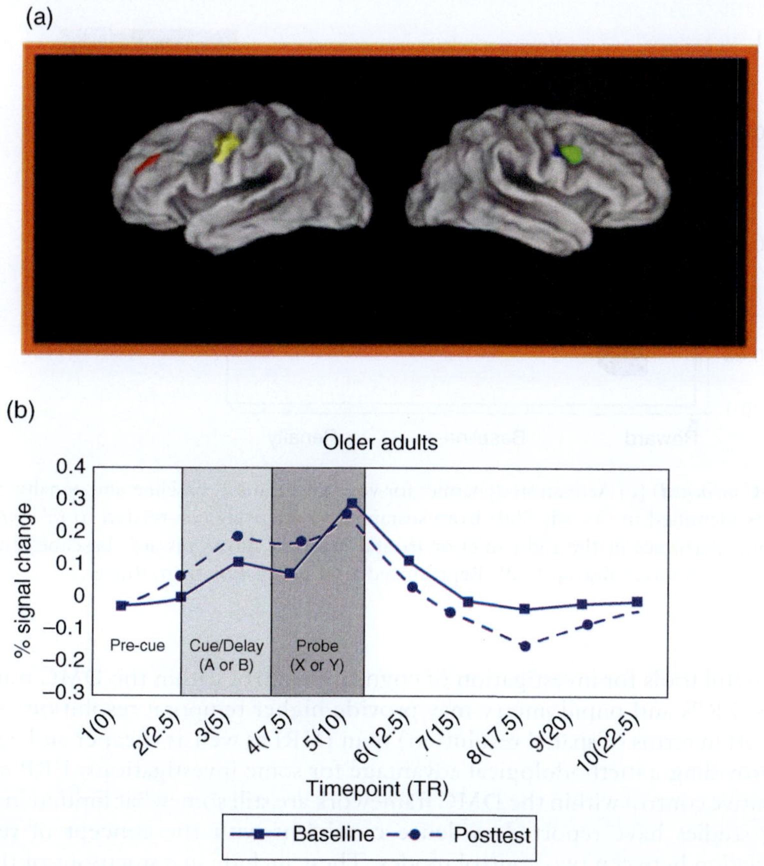

Figure 9.4 Dynamic shifts in cognitive control-related activity in the lateral PFC as a result of training and incentive manipulations. (a) Regions (see Braver et al., 2009, for exact coordinates) demonstrating training-related proactive shift (increased cue-related and decreased probe-related activity) in older adults and penalty-related reactive shift (decreased cue-related and increased probe-related activity) in young adults. (b) Activation dynamics for older adults at baseline and post-test conditions in brain regions identified in (a).

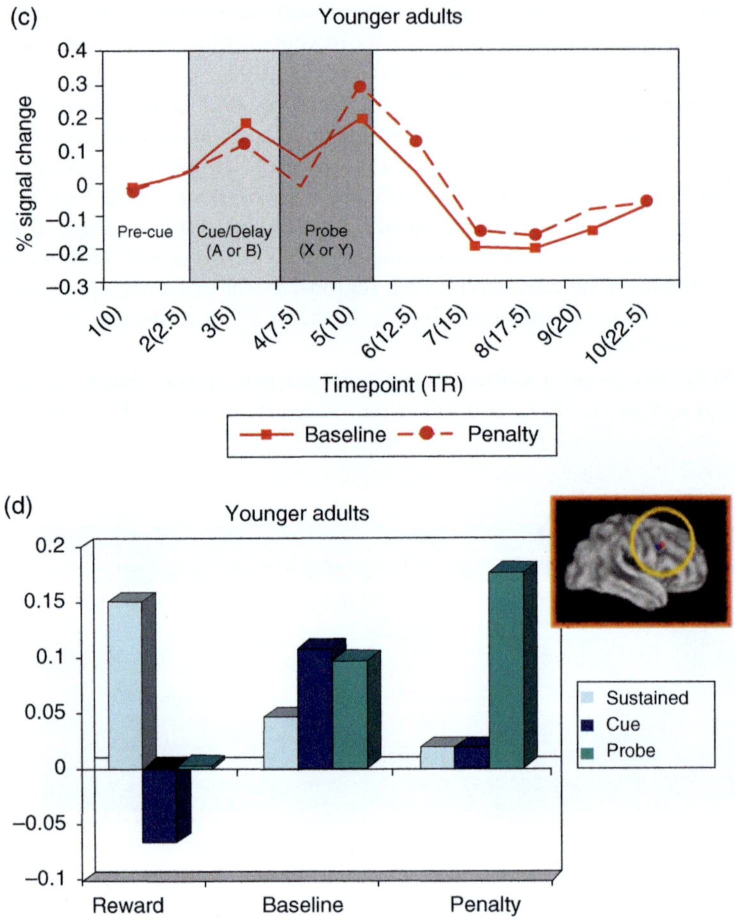

Figure 9.4 (*Continued*) (c) Activation dynamics for younger adults at baseline and penalty conditions in brain regions identified in (a). (d) Shift from sustained to relatively cue-related to relatively probe-related activation dynamics in the right inferior frontal junction across reward, baseline, and penalty AX-CPT conditions. Source: Braver 2009. Reproduced with permission from Braver.

emerged as useful tools for investigation of cognitive control within the DMC framework. Tools such as ERPs and pupillometry may provide higher temporal resolution (although with a trade-off in terms of spatial resolution) than fMRI as well as cheaper and easier data collection, providing a methodological advantage for some investigations. ERP investigations of cognitive control within the DMC framework are still somewhat limited in number, but existing studies have reported evidence consistent with the concept of functional neural dissociation between two control modes. These include investigations of the neural correlates of top-down control versus conflict monitoring (West, 2003) and investigations of conflict and error monitoring and subsequent control adjustment implicating a role for the anterior cingulate (Gehring, Goss, Coles, Meyer, & Donchin, 1993; van Veen & Carter, 2002).

Studies focussed on examining the differential temporal dynamics of cognitive control in relation to the DMC framework have observed changes to ERP component dynamics in response to (a) proportion congruency manipulations in a Stroop task, in line with behavioural evidence of changes in task performance with such manipulations (West & Bailey, 2012); (b) effects of proactive cueing on task-switching performance (Czernochowski, 2015); and (c) age-related changes in the AX-CPT (Hammerer, Li, Muller, & Lindenberger, 2010; Schmitt, Ferdinand, & Kray, 2015). In particular, these studies have revealed double dissociations between distinct negative medial components associated with proactive and reactive control (greater medial frontal negativity associated with proactive control, vs. greater medial posterior negativity associated with reactive control; West & Bailey, 2012) as well as evidence of sustained slow-wave activity while in a proactive control mode, versus the absence of such activity in a reactive control mode (Czernochowski, 2015). Such findings support the utility of ERP measures for identifying neural components of proactive and reactive modes of control operating with distinct temporal dynamics, thus both supporting the DMC framework while clarifying underlying mechanisms of the two proposed control modes. Thus, continued use of this methodology in future work seems especially promising.

In addition to ERP measures, pupillometry has recently emerged as a tool of interest in investigating the temporal dynamics of cognitive control. Based on predictions that proactive control would manifest as preparatory and/or sustained dilation whereas reactive control would manifest as post-stimulus transient dilation, pupil dilation was used to index developmental changes in a modified AX-CPT paradigm, demonstrating a shift from predominantly reactive to proactive control over childhood (Chatham, Frank, & Munakata, 2009). More recently, we demonstrated that pupillometry can be combined with a mixed block/event experimental design to examine changes in cognitive control dynamics (Chiew & Braver, 2013, 2014), specifically by examining the effects of reward incentives on AX-CPT performance and concurrent pupil dilation. We demonstrated that incentives were associated with enhanced performance and pupil dilation (i.e., indicative of increased control), both on a sustained basis (comparing reward vs. baseline blocks) and a transient, trial-evoked basis (comparing incentive trials vs. non-incentive trials within a single block), suggesting a shift towards enhanced proactive control (see Figure 9.5). However, pupillometric investigations of cognitive control are still at a relatively early stage, and underlying neural contributions to the pupil signal are still being elucidated. Nevertheless, a particularly exciting direction for this work is the potential utility of pupillometry (and other oculometric measures) in examining the dynamics of neuromodulatory influences on control, as a potential peripheral marker of both noradrenergic and potentially dopaminergic effects (Gilzenrat, Nieuwenhuis, Jepma, & Cohen, 2010; van Bochove, Van der Haegen, Notebaert, & Verguts, 2013).

Affective and Motivational Influences

A major thrust of the DMC framework has been to advance the understanding of cognitive control in terms of 'non-cognitive' factors, such as affect and motivational influences. Specifically, in the DMC framework, these factors can also contribute to the balance between proactive and reactive control, and thus variability in control performance. Investigations focussed on affective (Chapter 22 by Pessoa in this volume) and motivational influences (Chapter 24 by Krebs & Woldorff in this volume) on cognitive control have become a fast-growing field of study, within which the DMC framework has acted as a useful theoretical tool for interpretation of data and generation of future predictions. There are two theoretical components of the DMC framework that provide relevant conceptual assumptions and hypotheses regarding affect and motivational factors. First, proactive control should maximise harvesting of available rewards, given that a preparatory, planning-based mode should optimise the use of contextual, reward-predictive cues, as well as more tonic indicators of the

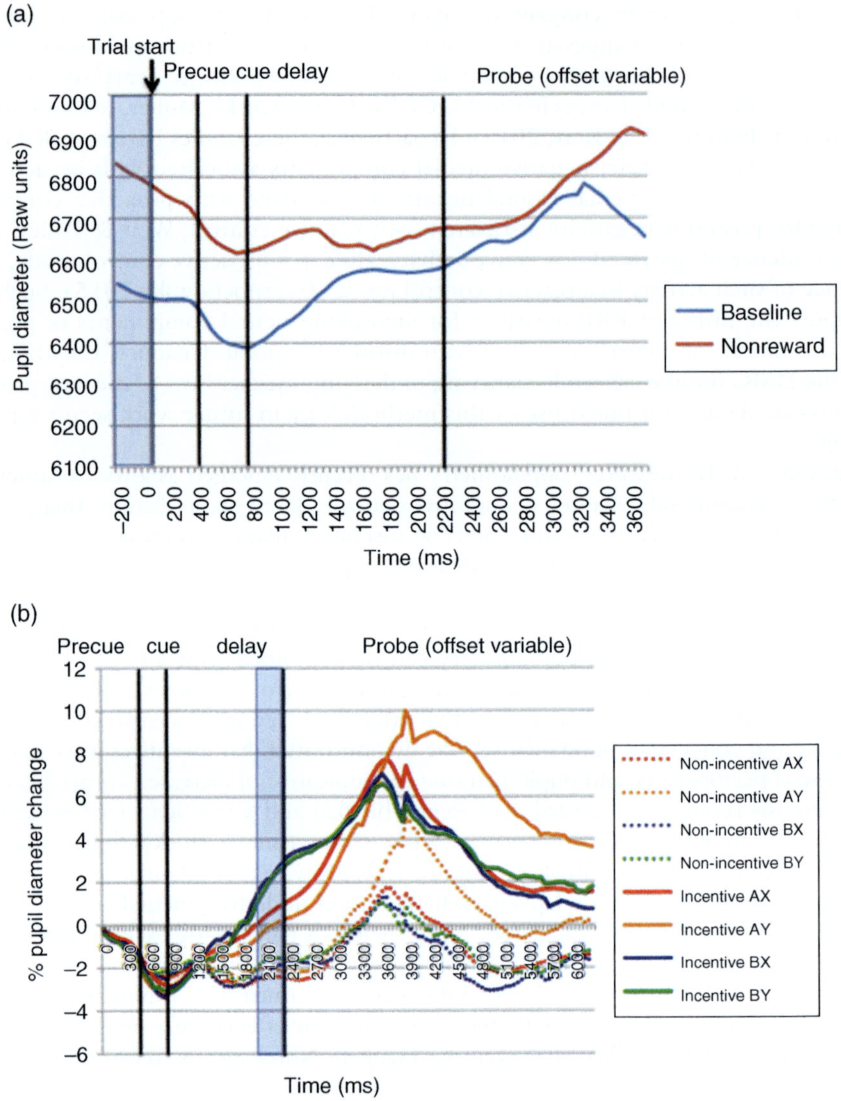

Figure 9.5 Sustained and transient changes in pupil dilation activity as a function of reward incentives in the AX-CPT task. (a) Pupil timecourses as a function of incentive status for the sustained (block) contrast: baseline block trials versus non-incentivized trials within reward block, averaged across trial types. (b) Pupil timecourses as a function of incentive status for the transient contrast: non-incentivized versus incentivized trials within the reward block.

current reward context. Conversely, because such reward signals are known to result in phasic (and potentially tonic; Niv, Daw, Joel, & Dayan, 2007) DA responses (Schultz, Dayan, & Montague, 1997), these should enhance the ability to actively maintain goal representations within the lateral PFC. Thus, according to the DMC framework, enhancement of such reward signals should facilitate implementation of proactive control. Second, the reactive mode should be preferred when control resources need to be deployed towards background monitoring of the external (and internal) environment for the presence of potential threats. This is because proactive control makes the cognitive system less sensitive to goal-incongruent

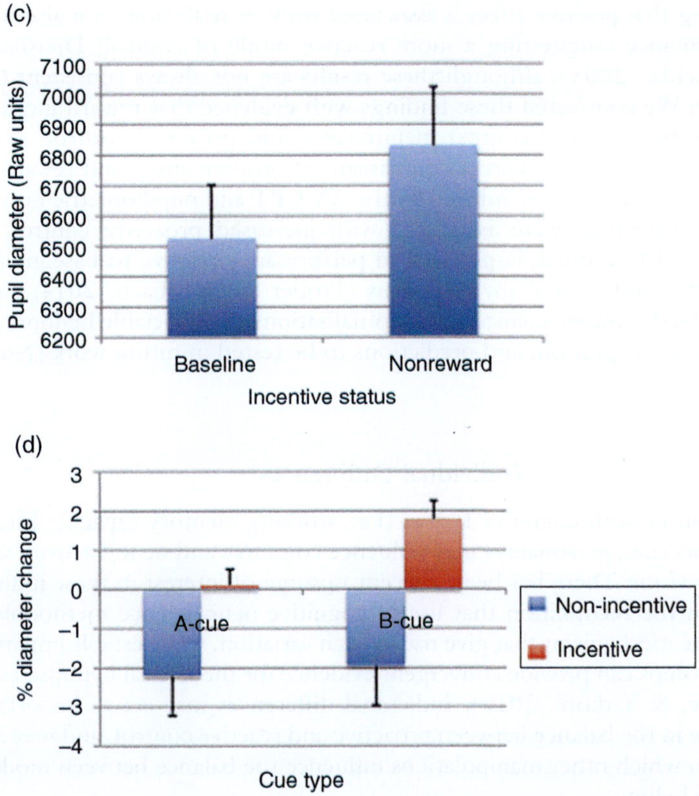

Figure 9.5 (*Continued*) (c) Sustained incentive effects (i.e., an increase in averaged pupil magnitudes) at the pre-trial period (−200 to 0 ms). (d) Transient incentive effects (i.e., an increase averaged pupil magnitudes) during cue maintenance prior to probe onset (1,950 to 2,200 ms). The transient effects also show an additional increase in pupil dilation following the high control-demand B-cues. Source: Adapted from Chiew and Braver 2013.

or incidental features of the environment, which might be threat relevant. Thus, according to the DMC framework, high demands or a pre-existing bias towards background threat monitoring should predispose cognitive control towards the reactive mode.

In recent years, there has been an upsurge in research specifically examining the relationship between reward motivation and cognitive control (Botvinick & Braver, 2015; Braver et al., 2014; Pessoa, 2009; Chapter 22 by Pessoa in this volume; Chapter 24 by Krebs & Woldorff in this volume) in studies utilising manipulation of reward incentives. Observed changes in task performance under reward incentives have been consistent with the hypothesis of enhanced proactive control across several cognitive control tasks, including the AX-CPT (Chiew & Braver, 2013; Locke & Braver, 2008), task-switching (Umemoto & Holroyd, 2014) and working memory tasks (Beck, Locke, Savine, Jimura, & Braver, 2010; Savine, Beck, Edwards, Chiew, & Braver, 2010), and Stroop-type paradigms (Soutschek, Stelzel, Paschke, Walter, & Schubert, 2015).

In addition, our research has utilised the DMC framework to develop and test predictions about the extent to which positive affect (i.e., positively valenced subjective experience) and reward incentive influences on cognitive control should be considered a common construct or dissociable in nature (Chiew & Braver, 2011, 2014). This investigation drew on prior findings suggesting that positive affect is associated with more flexible, but also more easily distracted, performance (suggesting a more reactive mode of control; Dreisbach, 2006; Dreisbach & Goschke, 2004), although these results are not always consistent (Frober & Dreisbach, 2012). We contrasted these findings with evidence that reward incentives have been associated with enhanced context maintenance and proactive control, as described above (Locke & Braver, 2008). Direct comparison of positive affect and reward incentive influences on cognitive control as indexed by the AX-CPT and pupillometric measures suggested that both influences were associated with increased proactive control, although reward incentives led to a much larger shift in performance relative to baseline (Chiew & Braver, 2014). This and related investigations (Frober & Dreisbach, 2014; Goschke & Bolte, 2014) have led to more nuanced conceptualisations of dissociable hedonic and motivational influences on cognition, and predictions to be tested in future work (Notebaert & Braem, 2015).

Individual Differences

Individual variation in both cognitive factors (i.e., working memory capacity; Engle, 2002) and affective factors (i.e., personality) may influence cognitive and neural activity supporting goal-directed behaviour. There has been a recent upsurge of interest in these individual differences, based on the recognition that use of cognitive neuroscience methodologies may help clarify the core mechanisms that give rise to such variation, while establishment of brain–behaviour relationships can provide convergent evidence for theoretical hypotheses of cognition (Braver, Cole, & Yarkoni, 2010). Individual differences may act as important, stable sources of variance in the balance between proactive and reactive control, and may also determine the extent to which other manipulations influence the balance between modes of control, as we outline below.

Working memory capacity (WMC), which can be defined as the executive-attention element of the working memory system allowing information maintenance in the presence of interference, has been associated with PFC function and is of interest as a factor accounting for individual differences in cognitive performance (Kane & Engle, 2002). Individual variation in WMC has been associated with goal maintenance and context processing ability (Redick & Engle, 2011), and thus may act as an important potential determinant of proactive control. Consistent with this prediction, recent studies using adaptations of the AX-CPT paradigm suggest that task performance in high-WMC individuals thus tends towards proactive control (greater interference on AY trials and decreased interference in BX trials, relative to low-WMC individuals; Redick, 2014; Richmond, Redick, & Braver, 2015).

Individual differences in WMC, or related cognitive dimensions such as fluid intelligence (Burgess, Gray, Conway, & Braver, 2011; Kane & Engle, 2002), might also interact with other task factors to influence whether cognitive control shifts between proactive and reactive control modes. In one such study examining the interaction between experimental and individual difference factors (Burgess & Braver, 2010), it was found that increasing interference expectancy in a recent negative probes working memory paradigm led to a general shift towards proactive control, evidenced by reduced interference and a shift of PFC dynamics from being primarily (recent negative) probe driven to present during the cue and delay period. However, this pattern also interacted with individual differences in fluid intelligence (gF), with low-gF individuals showing some evidence of increased reactive control (rather than a shift to proactive control) under conditions of high interference expectancy, whereas in high-gF individuals there was an increased tendency towards the proactive control pattern being used even under low interference expectancy conditions.

Putatively 'non-cognitive' individual differences, such as personality traits, may also play an important role in determining the balance between proactive and reactive control, especially when considered in combination with other influences on control. Such traits may not have a direct influence on cognitive processing, but instead influence the relative costs versus benefits of engaging in one cognitive mode over another. For example, individual variation in reward sensitivity may play a key role in the extent to which reward incentives alter motivation to perform a given task. Experimental evidence suggests that reward-sensitive individuals demonstrate enhanced performance under incentive context on a working memory task, and that this performance change was accompanied by a shift from transient to tonic activity in the right lateral PFC, suggesting that individual variation in reward sensitivity was associated with variation in incentive-related enhancement in proactive control (Jimura et al., 2010).

Individual differences in (non-clinical) trait anxiety have also been found to influence the balance between proactive and reactive control during working memory performance (Fales et al., 2008): A mixed block/event fMRI design revealed a tendency towards reactive control (i.e., decreased sustained and increased transient PFC activity) during task performance as a function of trait anxiety. This is consistent with the idea that individuals with increased anxiety may expend more cognitive resources on worrying or background monitoring for the presence of environmental threats, thus tending towards reactive control as a less efficient but also less effortful task strategy (Eysenck & Calvo, 1992).

Comparisons Between Populations

Initial development of the gating model and subsequent development of the DMC framework drew on observations of impaired context processing and altered DA activity in individuals with schizophrenia (Chapter 31 by Barch & Sheffield in this volume), identifying DA modulation as a candidate gating mechanism regulating information access to the cortex. Subsequently, the DMC framework has provided a powerful means by which to interpret data and generate predictions regarding differences in context processing and cognitive control performance between different populations.

The DMC model has been proved to be especially fruitful for examining age-related changes in cognitive control, both in older adults and during developmental periods (see also Chapter 27 by Zanto & Gazzaley in this volume). Healthy ageing is associated with impairments in cognitive performance and declines in DA neuromodulation due to DA neuron and receptor loss (Backman, Nyberg, Lindenberger, Li, & Farde, 2006; Li, Lindenberger, & Sikstrom, 2001). Systematic investigation of age-related changes in performance in terms of dual control modes has revealed that cognitive performance does not decline globally with ageing, but instead older adults may demonstrate a specific decline in proactive control mechanisms, while reactive control mechanisms remain relatively intact. Behavioural studies of AX-CPT performance in younger and older adults provided have demonstrated improved AY trial performance and worsened BX trial performance with age, consistent with a specific decrement in context maintenance (Braver et al., 2001; Braver, Satpute, Rush, Racine, & Barch, 2005). Following up on these results, a neuroimaging study examined the effects of aging on AX-CPT performance and neural activity using a mixed block/event design to dissociate sustained and transient activations (Paxton et al., 2008). fMRI revealed decreased activity during the cue/delay period and increased activity during the probe period in older relative to younger results, providing neural evidence of decreased proactive and increased reactive control with ageing, complementary to evidence from behavioural performance.

However, it is important to note that the shift from proactive to reactive control might reflect a strategic decision (though not necessarily a conscious or volitional one) made from changes in cost-benefit weightings in the two modes. Support for this idea was found in later studies that examined the degree to which reactive control shifts could be reversed, or at least minimised, following focussed cognitive training. In particular, studies using the AX-CPT task found that training in the explicit utilisation of contextual cues to prepare probe responding led to a shift in both behavioural performance profiles and neural activation dynamics among older adults, to bring them closer to the young adult pattern observed in the task (Braver et al., 2009; Paxton, Barch, Storandt, & Braver, 2006; see Figure 9.4). Work using highly similar training and experimental protocols also observed similar 'normalisation' effects among individuals with schizophrenia in the AX-CPT (Edwards, Barch, & Braver, 2010).

As described earlier, studies using other methodologies, such as ERP, have also been employed to provide evidence for reactive shifts in older adults (Czernochowski, Nessler, & Friedman, 2010; Schmitt et al., 2015). Similar approaches have been used at the other end of the lifespan to demonstrate enhanced reactive control tendencies in younger children and adolescents (Andrews-Hanna et al., 2011). For example, Chatham et al. (2009) used pupillometric and behavioural performance measures to provide evidence for a shift from reactive to proactive control from 3.5- to 8-year-old children in the AX-CPT. These investigations demonstrate the utility and versatility of the DMC framework as a theoretical and experimental tool for understanding how cognitive control processes might change across different population groups. Most critically, the framework provides a more nuanced account of such cognitive control differences, as not simply reflecting intact versus poor cognitive control, but rather a shift in which the type of control strategy tends to be engaged, and how these might be influenced by both task factors and cost-benefit calculations.

Directions for Future Research

As we have attempted to illustrate through this brief review, considering cognitive control in terms of context processing and proactive versus reactive mechanisms has proved to be useful in understanding the adaptive quality of goal-directed behaviour. The DMC framework posits a neurobiologically grounded mechanistic account of cognitive control that can be experimentally tested, across a variety of task paradigms, with a range of cognitive neuroscience methodologies, and in terms of individual and population group differences. These advances have also opened up new questions for future research to address. A number of these outstanding questions, regarding further clarification of gating mechanisms, affective and motivational influences, and the role of individual differences, have been discussed throughout the body of the present chapter. In the present section, we wish to highlight research directions that are somewhat broader than these previously discussed issues, aiming to advance understanding of cognitively controlled behaviour within the DMC framework and within the broader landscape of current cognitive neuroscience research.

An important issue that remains to be resolved is the extent to which proactive and reactive control constitute independent mechanisms. The DMC framework in its present form postulates that the two control modes will be necessarily inversely correlated, in that increased utilisation of proactive control should reduce the demand on reactive control, and vice versa. Likewise, given the distinct computational costs and benefits associated with each mode, the circumstances in which one mode is advantageous are generally circumstances in which the other is not. Nevertheless, it is still an open question whether it is best to think of proactive and reactive control as opposite poles of a single dimension, or rather, two (semi-) independent dimensions of control, that can be modulated in isolation. Thus, an important larger-scale project of interest is to determine whether experimental task manipulations can be designed that influence one control mode, while not affecting the other, across the range of domains and paradigms for which the DMC framework has already proved to be useful (i.e., context processing, cued task switching, working memory, selective attention, prospective memory, and response inhibition).

Initial findings in this endeavour have proved promising. Within the Stroop task, double dissociations have been established, demonstrating that block-wise proportion congruency manipulations show distinct behavioural signatures from item-specific manipulations, with each consistent with selective changes to proactive and reactive control, respectively (Gonthier et al., in revision). In addition, recent work has suggested that positive affect can lead to decreased proactive control without alterations in reactive control (Frober & Dreisbach, 2012). Further work is required, however, to establish whether these control modes can be considered fully independent, and yield not only doubly dissociable behavioural signatures, but also neural ones as well.

Another issue that remains to be addressed is the reconciliation of the DMC framework with other models of cognitive control. One important alternative model posits that cognitive control can be understood in terms of hierarchically organised rule representations and that this organisation is supported neuroanatomically, from low-level to high-level control, by caudal to rostral subregions of the PFC (Badre, 2008; Chapter 12 by Duverne & Koechlin in this volume). Although empirical evidence exists for the hierarchical model framework, initial attempts to reconcile the hierarchical and DMC frameworks have not met with success and, in fact, have shown evidence against a strict hierarchical account in favour of a dynamically flexible one (Reynolds, O'Reilly, Cohen, & Braver, 2012). A recent study by Bahlmann and colleagues (Bahlmann, Aarts, & D'Esposito, 2015) examined the effect of reward incentives on hierarchically organised cognitive control performance and task-related fMRI activity in the PFC and observed that incentive-related benefits in performance were greatest at mid-level (vs. high-level or low-level) control. Given extensive evidence from studies within the DMC framework that reward incentives may enhance proactive control, future studies could potentially explore whether hierarchical and temporal dimensions of control are orthogonal or interactive: That is, it may be possible that control of mid- and high-level task representations (those that are more abstract) involve proactive control in a manner that is not necessary for lower-level ones.

A third issue for future research within the DMC framework to consider is the adoption of a particular cognitive control strategy in terms of cost-benefit decision making. Recent evidence suggests that individuals are inclined to minimise cognitive effort, consistent with the 'law of less work' (Kool, McGuire, Rosen, & Botvinick, 2010; see also Chapter 10 by Kool, Shenhav, & Botvinick in this volume) but may decide to engage cognitive effort based on the anticipated reward value of doing so (Dixon & Christoff, 2012). Moreover, recent findings suggest that the subjective cost of cognitive effort may vary based on both within-individual factors including objective load and anticipated reward, individual difference factors such as trait cognitive motivation, and population differences such as older age (Westbrook, Kester, & Braver, 2013). The sensitivity of subjective cognitive effort costs to all three of these dimensions of variability suggests an important point of potential contact with the DMC framework. In particular, it may be the case that it is specifically the utilisation or demand for proactive control that may underlie the subjective cost of cognitive performance (Westbrook & Braver, 2015). However, direct evidence for this possibility is still lacking. Nevertheless, an important direction for future research is to better understand variability in cognitive control strategies, such as that postulated in the DMC framework, in terms of subjective value and costs, and more generally, from within a framework of cost-benefit decision making (Botvinick & Braver, 2015).

Conclusion

As this review has aimed to demonstrate, the DMC framework has provided an important conceptual account aiming to clarify the computational and neural mechanisms underlying variation in cognitive control, using different cognitive paradigms, methodologies, and comparison groups. Research within this framework has illustrated the importance of considering the temporal dynamics of cognitive and neural processes underlying adaptive, goal-directed behaviour and has illustrated how 'non-cognitive' factors such as affect, motivation, and individual differences may be important determining factors in the engagement of cognitive control, even though they may not influence the efficacy of cognitive function directly. Many questions remain for future research; critical among these is the question of how key principles and findings of the DMC framework can be integrated with evidence from other models and accounts of cognitive control. Addressing this question will be critical in developing an improved and more integrative theoretical framework to further advance our understanding of human higher cognitive function.

References

Andrews-Hanna, J. R., Mackiewicz Seghete, K. L., Claus, E. D., Burgess, G. C., Ruzic, L., & Banich, M. T. (2011). Cognitive control in adolescence: Neural underpinnings and relation to self-report behaviors. *PLoS ONE, 6*(6), e21598.

Aron, A. R. (2011). From reactive to proactive and selective control: Developing a richer model for stopping inappropriate responses. *Biological Psychiatry, 69*(12), e55–68.

Arrington, C. M., Logan, G. D., & Schneider, D. W. (2007). Separating cue encoding from target processing in the explicit task-cuing procedure: Are there 'true' task switch effects? *Journal of Experimental Psychology: Learning, Memory, and Cognition, 33*(3), 484–502.

Backman, L., Nyberg, L., Lindenberger, U., Li, S. C., & Farde, L. (2006). The correlative triad among aging, dopamine, and cognition: Current status and future prospects. *Neuroscience & Biobehavioral Reviews, 30*(6), 791–807.

Badre, D. (2008). Cognitive control, hierarchy, and the rostro-caudal organization of the frontal lobes. *Trends in Cognitive Sciences, 12*(5), 193–200.

Bahlmann, J., Aarts, E., & D'Esposito, M. (2015). Influence of motivation on control hierarchy in the human frontal cortex. *Journal of Neuroscience, 35*(7), 3207–3217.

Beck, S. M., Locke, H. S., Savine, A. C., Jimura, K., & Braver, T. S. (2010). Primary and secondary rewards differentially modulate neural activity dynamics during working memory. *PLoS ONE, 5*(2), e9251.

Berkman, E. T., Kahn, L. E., & Merchant, J. S. (2014). Training-induced changes in inhibitory control network activity. *Journal of Neuroscience, 34*(1), 149–157.

Berman, M. G., Jonides, J., & Lewis, R. L. (2009). In search of decay in verbal short-term memory. *Journal of Experimental Psychology: Learning, Memory, and Cognition, 35*(2), 317–333.

Botvinick, M., & Braver, T. (2015). Motivation and cognitive control: from behavior to neural mechanism. *Annual Review of Psychology, 66*, 83–113.

Botvinick, M. M., Cohen, J. D., & Carter, C. S. (2004). Conflict monitoring and anterior cingulate cortex: An update. *Trends in Cognitive Sciences, 8*(12), 539–546.

Braver, T. S. (2012). The variable nature of cognitive control: A dual mechanisms framework. *Trends in Cognitive Sciences, 16*(2), 106–113.

Braver, T. S., Barch, D. M., & Cohen, J. D. (1999). Cognition and control in schizophrenia: A computational model of dopamine and prefrontal function. *Biological Psychiatry, 46*(3), 312–328.

Braver, T. S., Barch, D. M., & Cohen, J. D. (2002). The role of the prefrontal cortex in normal and disordered cognitive control: A cognitive neuroscience perspective. In D. T. Stuss & R. T. Knight (Eds.), *Principles of frontal lobe function* (pp. 428–447). Oxford: Oxford University Press.

Braver, T. S., Barch, D. M., Keys, B. A., Carter, C. S., Cohen, J. D., Kaye, J. A.,…Reed, B. R. (2001). Context processing in older adults: Evidence for a theory relating cognitive control to neurobiology in healthy aging. *Journal of Experimental Psychology, 130*(4), 746–763.

Braver, T. S., & Cohen, J. D. (1999). Dopamine, cognitive control, and schizophrenia: the gating model. *Progress in Brain Research, 121*, 327–349.

Braver, T. S., & Cohen, J. D. (2000). On the control of control: the role of dopamine in regulating prefrontal function and working memory. In S. Monsell & J. Driver (Eds.), *Attention and performance XVIII* (pp. 713–737). Cambridge, MA: MIT Press.

Braver, T. S., Cole, M. W., & Yarkoni, T. (2010). Vive les differences! Individual variation in neural mechanisms of executive control. *Current Opinion in Neurobiology, 20*(2), 242–250.

Braver, T. S., Gray, J. R., & Burgess, G. C. (2007). Explaining the many varieties of working memory variation: Dual mechanisms of cognitive control. In A. R. A. Conway, C. Jarrold, M. J. Kane, A. Miyake, & J. N. Towse (Eds.), *Variation in working memory* (pp. 76–106). Oxford: Oxford University Press.

Braver, T. S., Krug, M. K., Chiew, K. S., Kool, W., Westbrook, J. A., Clement, N. J.,…Somerville, L. H. (2014). Mechanisms of motivation-cognition interaction: Challenges and opportunities. *Cognitive, Affective, & Behavioral Neuroscience, 14*(2), 443–472.

Braver, T. S., Paxton, J. L., Locke, H. S., & Barch, D. M. (2009). Flexible neural mechanisms of cognitive control within human prefrontal cortex. *Proceedings of the National Academy of Sciences of the United States of America, 106*(18), 7351–7356.

Braver, T. S., Reynolds, J. R., & Donaldson, D. I. (2003). Neural mechanisms of transient and sustained cognitive control during task switching. *Neuron, 39*(4), 713–726.

Braver, T. S., Satpute, A. B., Rush, B. K., Racine, C. A., & Barch, D. M. (2005). Context processing and context maintenance in healthy aging and early stage dementia of the Alzheimer's type. *Psychology and Aging, 20*(1), 33–46.

Brown, J. W., & Braver, T. S. (2005). Learned predictions of error likelihood in the anterior cingulate cortex. *Science, 307*(5712), 1118–1121.

Buckner, R. L. (1998). Event-related fMRI and the hemodynamic response. *Human Brain Mapping, 6*(5–6), 373–377.

Bugg, J. M., & Braver, T. S. (in press). Proactive control of irrelevant task rules during cued task-switching. *Psychological Research.*

Bugg, J. M., & Crump, M. J. (2012). In support of a distinction between voluntary and stimulus-driven control: A review of the literature on proportion congruent effects. *Frontiers in Psychology, 3*, 367.

Bugg, J. M., Diede, N. T., Cohen-Shikora, E. R., & Selmeczy, D. (2015). Expectations and experience: Dissociable bases for cognitive control? *Journal of Experimental Psychology: Learning, Memory, and Cognition, 41*(5), 1349–1373.

Bugg, J. M., & Hutchison, K. A. (2013). Converging evidence for control of color-word Stroop interference at the item level. *Journal of Experimental Psychology: Human Perception and Performance, 39*(2), 433–449.

Bugg, J. M., Jacoby, L. L., & Chanani, S. (2011). Why it is too early to lose control in accounts of item-specific proportion congruency effects. *Journal of Experimental Psychology: Human Perception and Performance, 37*(3), 844–859.

Bugg, J. M., Jacoby, L. L., & Toth, J. P. (2008). Multiple levels of control in the Stroop task. *Memory & Cognition, 36*(8), 1484–1494.

Bugg, J. M., Scullin, M. K., & McDaniel, M. A. (2013). Strengthening encoding via implementation intention formation increases prospective memory commission errors. *Psychonomic Bulletin & Review, 20*(3), 522–527.

Bugg, J. M., & Smallwood, A. (2014). The next trial will be conflicting! Effects of explicit congruency pre-cues on cognitive control. *Psychological Research, 80*(1), 16–33.

Burgess, G. C., & Braver, T. S. (2010). Neural mechanisms of interference control in working memory: Effects of interference expectancy and fluid intelligence. *PLoS ONE, 5*(9), e12861.

Burgess, G. C., Gray, J. R., Conway, A. R., & Braver, T. S. (2011). Neural mechanisms of interference control underlie the relationship between fluid intelligence and working memory span. *Journal of Experimental Psychology, 140*(4), 674–692.

Chatham, C. H., & Badre, D. (2015). Multiple gates on working memory. *Current Opinion in Behavioral Sciences, 1*, 23–31.

Chatham, C. H., Frank, M. J., & Badre, D. (2014). Corticostriatal output gating during selection from working memory. *Neuron, 81*(4), 930–942.

Chatham, C. H., Frank, M. J., & Munakata, Y. (2009). Pupillometric and behavioral markers of a developmental shift in the temporal dynamics of cognitive control. *Proceedings of the National Academy of Sciences of the United States of America, 106*(14), 5529–5533.

Chiew, K. S., & Braver, T. S. (2011). Positive affect versus reward: Emotional versus motivational influences on cognitive control. *Frontiers in Psychology, 2*(279).

Chiew, K. S., & Braver, T. S. (2013). Temporal dynamics of motivation-cognitive control interactions revealed by high-resolution pupillometry. *Frontiers in Psychology, 4*, 15.

Chiew, K. S., & Braver, T. S. (2014). Dissociable influences of reward motivation and positive emotion on cognitive control. *Cognitive, Affective, & Behavioral Neuroscience, 14*(2), 509–529.

Chiew, K. S., & Braver, T. S. (2016). Reward favours the prepared: Incentive and task-informative cues interact to enhance attentional control. *Journal of Experimental Psychology: Human Perception and Performance* , *42*(1), 52–66.

Cohen, J. D., Barch, D. M., Carter, C., & Servan-Schreiber, D. (1999). Context-processing deficits in schizophrenia: Converging evidence from three theoretically motivated cognitive tasks. *Journal of Abnormal Psychology, 108*(1), 120–133.

Cohen, J. D., & Servan-Schreiber, D. (1992). Context, cortex, and dopamine: A connectionist approach to behavior and biology in schizophrenia. *Psychological Review, 99*, 45–77.

Czernochowski, D. (2015). ERPs dissociate proactive and reactive control: Evidence from a task-switching paradigm with informative and uninformative cues. *Cognitive, Affective, & Behavioral Neuroscience, 15*, 117–131.

Czernochowski, D., Nessler, D., & Friedman, D. (2010). On why not to rush older adults—Relying on reactive cognitive control can effectively reduce errors at the expense of slowed responses. *Psychophysiology, 47*(4), 637–646.

D'Ardenne, K., Eshel, N., Luka, J., Lenartowicz, A., Nystrom, L. E., & Cohen, J. D. (2012). Role of prefrontal cortex and the midbrain dopamine system in working memory updating. *Proceedings of the National Academy of Sciences of the United States of America, 109*(49), 19900–19909.

De Pisapia, N., Slomski, J. A., & Braver, T. S. (2007). Functional specializations in lateral prefrontal cortex associated with the integration and segregation of information in working memory. *Cerebral Cortex, 17*(5), 993–1006.

Dixon, M. L., & Christoff, K. (2012). The decision to engage cognitive control is driven by expected reward-value: Neural and behavioral evidence. *PLoS ONE, 7*(12), e51637.

Dobbins, I. G., & Han, S. (2006). Cue- versus probe-dependent prefrontal cortex activity during contextual remembering. *Journal of Cognitive Neuroscience, 18*(9), 1439–1452.

Dreisbach, G. (2006). How positive affect modulates cognitive control: The costs and benefits of reduced maintenance capability. *Brain and Cognition, 60*(1), 11–19.

Dreisbach, G., & Goschke, T. (2004). How positive affect modulates cognitive control: reduced perseveration at the cost of increased distractibility. *Journal of Experimental Psychology: Learning, Memory, and Cognition, 30*(2), 343–353.

Dreisbach, G., Haider, H., & Kluwe, R. H. (2002). Preparatory processes in the task-switching paradigm: Evidence from the use of probability cues. *Journal of Experimental Psychology: Learning, Memory, and Cognition, 28*(3), 468–483.

Edwards, B. G., Barch, D. M., & Braver, T. S. (2010). Improving prefrontal cortex function in schizophrenia through focused training of cognitive control. *Frontiers in Human Neuroscience, 4*, 32.

Einstein, G. O., McDaniel, M. A., Thomas, R., Mayfield, S., Shank, H., Morrisette, N., & Breneiser, J. (2005). Multiple processes in prospective memory retrieval: Factors determining monitoring versus spontaneous retrieval. *Journal of Experimental Psychology, 134*(3), 327–342.

Engle, R. W. (2002). Working memory capacity as executive attention. *Current Directions in Psychological Science, 11*(1), 19–23.

Eysenck, M. W., & Calvo, M. G. (1992). Anxiety and performance: The processing efficiency theory. *Cognition and Emotion, 6*, 409–434.

Fales, C. L., Barch, D. M., Burgess, G. C., Schaefer, A., Mennin, D. S., Gray, J. R., & Braver, T. S. (2008). Anxiety and cognitive efficiency: differential modulation of transient and sustained neural activity during a working memory task. *Cognitive, Affective, & Behavioral Neuroscience, 8*(3), 239–253.

Frank, M. J., Loughry, B., & O'Reilly, R. C. (2001). Interactions between frontal cortex and basal ganglia in working memory: A computational model. *Cognitive, Affective, & Behavioral Neuroscience, 1*(2), 137–160.

Frober, K., & Dreisbach, G. (2012). How positive affect modulates proactive control: Reduced usage of informative cues under positive affect with low arousal. *Frontiers in Psychology, 3*, 265.

Frober, K., & Dreisbach, G. (2014). The differential influences of positive affect, random reward, and performance-contingent reward on cognitive control. *Cognitive, Affective, & Behavioral Neuroscience, 14*(2), 530–547.

Gehring, W. J., Goss, B., Coles, M. G., Meyer, D. E., & Donchin, E. (1993). A neural system for error detection and compensation. *Psychological Science, 4*(6), 385–390.

Gilzenrat, M. S., Nieuwenhuis, S., Jepma, M., & Cohen, J. D. (2010). Pupil diameter tracks changes in control state predicted by the adaptive gain theory of locus coeruleus function. *Cognitive, Affective, & Behavioral Neuroscience, 10*(2), 252–269.

Goldman-Rakic, P. S., & Friedman, H. R. (1991). The circuitry of working memory revealed by anatomy and metabolic imaging. *Frontal lobe function and dysfunction*. Oxford: Oxford University Press.

Gonthier, C., Braver, T. S., & Bugg, J. M. (in revision). Dissociating proactive and reactive control in the Stroop task.

Goschke, T., & Bolte, A. (2014). Emotional modulation of control dilemmas: The role of positive affect, reward, and dopamine in cognitive stability and flexibility. *Neuropsychologia, 62*, 403–423.

Grandjean, J., D'Ostilio, K., Phillips, C., Balteau, E., Degueldre, C., Luxen, A.,...Collette, F. (2012). Modulation of brain activity during a Stroop inhibitory task by the kind of cognitive control required. *PLoS ONE, 7*(7), e41513.

Gratton, G., Coles, M. G., & Donchin, E. (1992). Optimizing the use of information: Strategic control of activation of responses. *Journal of Experimental Psychology, 121*(4), 480–506.

Hammerer, D., Li, S. C., Muller, V., & Lindenberger, U. (2010). An electrophysiological study of response conflict processing across the lifespan: Assessing the roles of conflict monitoring, cue utilization, response anticipation, and response suppression. *Neuropsychologia, 48*(11), 3305–3316.

Irlbacher, K., Kraft, A., Kehrer, S., & Brandt, S. A. (2014). Mechanisms and neuronal networks involved in reactive and proactive cognitive control of interference in working memory. *Neuroscience & Biobehavioral Reviews, 46*(Pt. 1), 58–70.

Jacoby, L. L., Lindsay, D. S., & Hessels, S. (2003). Item-specific control of automatic processes: Stroop process dissociations. *Psychonomic Bulletin & Review, 10*(3), 638–644.

Jimura, K., Locke, H. S., & Braver, T. S. (2010). Prefrontal cortex mediation of cognitive enhancement in rewarding motivational contexts. *Proceedings of the National Academy of Sciences of the United States of America, 107*(19), 8871–8876.

Kane, M. J., & Engle, R. W. (2002). The role of prefrontal cortex in working-memory capacity, executive attention, and general fluid intelligence: An individual-differences perspective. *Psychonomic Bulletin & Review, 9*(4), 637–671.

Kessler, Y., & Meiran, N. (2008). Two dissociable updating processes in working memory. *Journal of Experimental Psychology: Learning, Memory, and Cognition, 34*(6), 1339–1348.

Kiesel, A., Steinhauser, M., Wendt, M., Falkenstein, M., Jost, K., Philipp, A. M., & Koch, I. (2010). Control and interference in task switching—A review. *Psychological Bulletin, 136*(5), 849–874.

King, J. A., Korb, F. M., von Cramon, D. Y., & Ullsperger, M. (2010). Post-error behavioral adjustments are facilitated by activation and suppression of task-relevant and task-irrelevant information processing. *Journal of Neuroscience, 30*(38), 12759–12769.

Koch, I., & Allport, A. (2006). Cue-based preparation and stimulus-based priming of tasks in task switching. *Memory & Cognition, 34*(2), 433–444.

Kool, W., McGuire, J. T., Rosen, Z. B., & Botvinick, M. M. (2010). Decision making and the avoidance of cognitive demand. *Journal of Experimental Psychology, 139*(4), 665–682.

Li, S. C., Lindenberger, U., & Sikstrom, S. (2001). Aging cognition: From neuromodulation to representation. *Trends in Cognitive Sciences, 5*(11), 479–486.

Locke, H. S., & Braver, T. S. (2008). Motivational influences on cognitive control: behavior, brain activation, and individual differences. *Cognitive, Affective, & Behavioral Neuroscience, 8*(1), 99–112.

Logan, G. D., & Zbrodoff, N. J. (1979). When it helps to be misled: Facilitative effects of increasing the frequency of conflicting stimuli in a Stroop-like task. *Memory & Cognition, 7*, 166–174.

MacDonald, A. W., 3rd, & Carter, C. S. (2003). Event-related FMRI study of context processing in dorsolateral prefrontal cortex of patients with schizophrenia. *Journal of Abnormal Psychology, 112*(4), 689–697.

McDaniel, M. A., Lamontagne, P., Beck, S. M., Scullin, M. K., & Braver, T. S. (2013). Dissociable neural routes to successful prospective memory. *Psychological Science, 24*(9), 1791–1800.

Meiran, N., Chorev, Z., & Sapir, A. (2000). Component processes in task switching. *Cognitive Psychology, 41*(3), 211–253.

Meiran, N., Kessler, Y., & Adi-Japha, E. (2008). Control by action representation and input selection (CARIS): A theoretical framework for task switching. *Psychological Research, 72*(5), 473–500.

Monsell, S. (2003). Task switching. *Trends in Cognitive Sciences, 7*(3), 134–140.

Nee, D. E., & Brown, J. W. (2012). Rostral-caudal gradients of abstraction revealed by multi-variate pattern analysis of working memory. *Neuroimage, 63*(3), 1285–1294.

Nieuwenhuis, S., & Monsell, S. (2002). Residual costs in task switching: Testing the failure-to-engage hypothesis. *Psychonomic Bulletin & Review, 9*(1), 86–92.

Niv, Y., Daw, N. D., Joel, D., & Dayan, P. (2007). Tonic dopamine: Opportunity costs and the control of response vigor. *Psychopharmacology (Berl), 191*(3), 507–520.

Notebaert, W., & Braem, S. (2015). Parsing the effects of reward on cognitive control. In T. S. Braver (Ed.), *Motivation and cognitive control*. New York: Psychology Press.

O'Reilly, R. C., & Frank, M. J. (2006). Making working memory work: A computational model of learning in the prefrontal cortex and basal ganglia. *Neural Computation, 18*, 283–328.

Paxton, J. L., Barch, D. M., Racine, C. A., & Braver, T. S. (2008). Cognitive control, goal maintenance, and prefrontal function in healthy aging. *Cerebral Cortex, 18*(5), 1010–1028.

Paxton, J. L., Barch, D. M., Storandt, M., & Braver, T. S. (2006). Effects of environmental support and strategy training on older adults' use of context. *Psychology and Aging, 21*(3), 499–509.

Pessoa, L. (2009). How do emotion and motivation direct executive control? *Trends in Cognitive Sciences, 13*(4), 160–166.

Redick, T. S. (2014). Cognitive control in context: Working memory capacity and proactive control. *Acta Psychologica (Amst), 145*, 1–9.

Redick, T. S., & Engle, R. W. (2011). Integrating working memory capacity and context-processing views of cognitive control. *Quarterly Journal of Experimental Psychology (Colchester), 64*(6), 1048–1055.

Reynolds, J. R., Braver, T. S., Brown, J. W., & Van der Stigchel, S. (2006). Computational and neural mechanisms of task switching. *Neurocomputing, 69*, 1332–1336.

Reynolds, J. R., O'Reilly, R. C., Cohen, J. D., & Braver, T. S. (2012). The function and organization of lateral prefrontal cortex: A test of competing hypotheses. *PLoS ONE, 7*(2), e30284.

Richmond, L. L., Redick, T. S., & Braver, T. S. (2015). Remembering to prepare: The benefits (and costs) of high working memory capacity. *Journal of Experimental Psychology: Learning, Memory, and Cognition, 41*(6), 1764–1777.

Ruge, H., Goschke, T., & Braver, T. S. (2009). Separating event-related BOLD components within trials: The partial-trial design revisited. *Neuroimage, 47*(2), 501–513.

Savine, A. C., Beck, S. M., Edwards, B. G., Chiew, K. S., & Braver, T. S. (2010). Enhancement of cognitive control by approach and avoidance motivational states. *Cognition & Emotion, 24*(2), 338–356.

Sawaguchi, T., & Goldman-Rakic, P. S. (1991). D1 dopamine receptors in prefrontal cortex: Involvement in working memory. *Science, 251*(4996), 947–950.

Schmitt, H., Ferdinand, N. K., & Kray, J. (2015). The influence of monetary incentives on context processing in younger and older adults: an event-related potential study. *Cognitive, Affective, & Behavioral Neuroscience, 15*(2), 416–434.

Schultz, W., Dayan, P., & Montague, P. R. (1997). A neural substrate of prediction and reward. *Science, 275*(5306), 1593–1599.

Soutschek, A., Stelzel, C., Paschke, L., Walter, H., & Schubert, T. (2015). Dissociable effects of motivation and expectancy on conflict processing: An fMRI study. *Journal of Cognitive Neuroscience, 27*(2), 409–423.

Speer, N. K., Jacoby, L. L., & Braver, T. S. (2003). Strategy-dependent changes in memory: Effects on behavior and brain activity. *Cognitive, Affective, & Behavioral Neuroscience, 3*(3), 155–167.

Stroop, J. R. (1935). Studies of interference in serial verbal reactions. *Journal of Experimental Psychology, 18*, 643–662.

Stuss, D. T., & Knight, R. T. (2013). *Principles of frontal lobe function* (2nd ed.). Oxford: Oxford University Press.

Torres-Quesada, M., Funes, M. J., & Lupianez, J. (2013). Dissociating proportion congruent and conflict adaptation effects in a Simon-Stroop procedure. *Acta Psychologica(Amst), 142*(2), 203–210.

Umemoto, A., & Holroyd, C. B. (2014). Task-specific effects of reward on task switching. *Psychological Research.*

van Bochove, M. E., Van der Haegen, L., Notebaert, W., & Verguts, T. (2013). Blinking predicts enhanced cognitive control. *Cognitive, Affective, & Behavioral Neuroscience, 13*(2), 346–354.

van Veen, V., & Carter, C. S. (2002). The anterior cingulate as a conflict monitor: fMRI and ERP studies. *Physiology & Behavior, 77*(4–5), 477–482.

Vandierendonck, A., Liefooghe, B., & Verbruggen, F. (2010). Task switching: Interplay of reconfiguration and interference control. *Psychological Bulletin, 136*(4), 601–626.

Velanova, K., Jacoby, L. L., Wheeler, M. E., McAvoy, M. P., Petersen, S. E., & Buckner, R. L. (2003). Functional-anatomic correlates of sustained and transient processing components engaged during controlled retrieval. *Journal of Neuroscience, 23*(24), 8460–8470.

Verbruggen, F., & Logan, G. D. (2009). Proactive adjustments of response strategies in the stop-signal paradigm. *Journal of Experimental Psychology: Human Perception and Performance, 35*(3), 835–854.

Visscher, K. M., Miezin, F. M., Kelly, J. E., Buckner, R. L., Donaldson, D. I., McAvoy, M. P.,...Petersen, S. E. (2003). Mixed blocked/event-related designs separate transient and sustained activity in fMRI. *Neuroimage, 19*(4), 1694–1708.

West, R. (2003). Neural correlates of cognitive control and conflict detection in the Stroop and digit-location tasks. *Neuropsychologia, 41*(8), 1122–1135.

West, R., & Bailey, K. (2012). ERP correlates of dual mechanisms of control in the counting Stroop task. *Psychophysiology, 49*(10), 1309–1318.

Westbrook, A., & Braver, T. S. (2015). Cognitive effort: A neuroeconomic approach. *Cognitive, Affective, & Behavioral Neuroscience, 15*(2), 395–415.

Westbrook, A., Kester, D., & Braver, T. S. (2013). What is the subjective cost of cognitive effort? Load, trait, and aging effects revealed by economic preference. *PLoS ONE, 8*(7), e68210.

Wilk, H. A., Ezekiel, F., & Morton, J. B. (2012). Brain regions associated with moment-to-moment adjustments in control and stable task-set maintenance. *Neuroimage, 59*(2), 1960–1967.

10

Cognitive Control as Cost-Benefit Decision Making

Wouter Kool,[1] Amitai Shenhav,[1]
and Matthew M. Botvinick

Many everyday situations afford us a set of default behaviours and cognitive processes that could play out automatically in response to stimuli in our environment. Cognitive control enables us to modify our thoughts and actions away from those defaults in a variety of ways, allowing us as a species to perform great intellectual feats such as planning (D. A. Simon & Daw, 2011), reasoning (Christoff et al., 2001), inhibition (Aron, 2011), and working memory maintenance (Goldman-Rakic, 1987). But what is it that determines when we exert control, how much we do so, and what form(s) this control takes? In other words, by what computational and neural mechanisms is the controller itself controlled (Botvinick & Cohen, 2015; Dayan, 2012)?

In this chapter, we address this question by framing it as a reward-based decision-making problem. This approach views the exertion of cognitive effort as being determined by the output of a decision that considers both the costs and benefits of mobilising cognitive control at a given moment. We begin by enumerating a set of factors that weigh in favour of the exertion of control and those that oppose it. We then present a theoretical framework that specifies how these costs and benefits are integrated together to form a decision about whether and how control should be deployed. Finally, we describe the neural underpinnings of this decision process, with a particular focus on the role of the dorsal anterior cingulate cortex (dACC) in determining how best to allocate control.

Why Exert Control? The Demands for (and Benefits of) Control

As mentioned above, some behavioural situations demand control, whereas others permit a more automatic approach to response selection. A key question, central to research on cognitive control, is how the brain detects this difference. How, in other words, is a demand for control detected? Research has pointed to a number of signals, both internal and external, that appear to serve as cues or signals of demand.

[1] The first two authors contributed equally to this work.

The Wiley Handbook of Cognitive Control, First Edition. Edited by Tobias Egner.

Cues to Control Demands

One straightforward way of determining how much control to allocate would be to rely entirely on automatic processes (e.g., habits) until the individual discovers that she has performed the wrong action, for instance because she is told that this is the case or it is obvious as soon as the response is made (e.g., in a situation where time pressure encourages premature responses). After encountering an error, one type of control she can exert is to exercise caution in her subsequent actions (Laming, 1968; Rabbitt, 1966), which can be operationalised as an adjustment in her threshold for responding so as to allow for additional evidence accumulation (and consequently greater accuracy) at the cost of slower responding. This type of sequential adjustment in response to previous-trial errors is commonly observed in the behavioural literature (see Danielmeier & Ullsperger, 2011 for a review, including discussion of inconsistencies in observations of post-error improvements in accuracy). Other kinds of post-error adjustments include reallocating attention away from a stimulus property that is distracting and/or towards a task-relevant property (Danielmeier & Ullsperger, 2011; see Chapter 17 by Ullsperger in this volume).

Rather than relying exclusively on explicit error feedback, which is often unavailable in our daily activities, or expecting errors to otherwise be relatively unambiguous, one can instead simply look out for situations where outcomes are worse than expected (i.e., negative prediction errors) as indicators that greater control is needed. Such a mechanism avoids relying on explicit instruction and provides a continuous rather than binary measure on which to base performance adjustments (i.e., indicating not only that outcomes were worse than expected, but also how much worse). But valuable information can also be garnered from monitoring how different outcomes were from one's predictions, *irrespective* of the direction. Rather than informing specific adjustments in behaviour/control, these *unsigned* prediction errors (or surprise signals) can signal a demand for attending to the elicitor of surprise (Pearce & Hall, 1980), and more generally can act as indicators of volatility/unpredictability in one's environment (Alexander & Brown, 2011; Behrens, Woolrich, Walton, & Rushworth, 2007; Cavanagh & Frank, 2014). Control systems can improve performance in response to these changes in volatility by, for instance, adjusting the rate at which one's world model is adjusted to reflect the changing outcomes, and by extension, the rate at which behaviours are adjusted on the basis of recent feedback (Behrens et al., 2007).

Still, monitoring primarily for prediction errors has the drawback that task feedback can occur too late to adjust control when necessary (i.e., during task performance). One way of avoiding this concern is to monitor for errors relative to richer sets of predictions, for instance, maintaining temporally specific predictions about the occurrence of certain conjunctions of responses and outcomes (Alexander & Brown, 2011). Another alternative to relying solely on outcome-related feedback (whether internal or external) is to take a more direct and online estimate of task difficulty. One such estimate is provided by the current level of processing conflict (Berlyne, 1957), for instance, how much competition exists between different potential responses (Botvinick, Braver, Barch, Carter, & Cohen, 2001). As a proxy for difficulty, levels of conflict on one trial can be used as an indication that greater control should be allocated on a subsequent trial (by, for instance, attending more to the task-relevant stimulus dimension), resulting in lower error rates and faster response times (conflict adaptation; Gratton, Coles, & Donchin, 1992; see Chapter 4 by Egner in this volume). Not only can these conflict signals act as indicators of difficulty, they can also in certain cases obviate error detection altogether by providing an earlier and more sensitive index of potential errors or confidence in one's response in advance of and/or in the absence of feedback (Yeung, Botvinick, & Cohen, 2004).

Cues to Control Incentives

Finally, the amount of overall reward being offered for successful performance of a controlled task can provide a straightforward signal indicating how beneficial it might be to increase control. Accordingly, a number of studies have found that individuals exert greater cognitive effort with higher incentives for performing a cognitive control task (review in Pessoa, 2015; Chapter 22 by Pessoa in this volume; Chapter 24 by Krebs & Woldorff in this volume). For instance, participants performing a Stroop-like task are faster, more accurate, and less influenced by both congruent and incongruent distractors when they are rewarded compared to when they are not rewarded (Etzel, Cole, Zacks, Kay, & Braver, in press; Padmala & Pessoa, 2011). Similarly, participants motivated to either obtain reward or avoid loss are better able to detect a target face appearing against a noisy background (Engelmann, Damaraju, Padmala, & Pessoa, 2009; Engelmann & Pessoa, 2007).

Why Not Exert Control?

There are clearly a number of sources of information that an individual can draw on to determine that there is a demand for greater control. And it seems intuitive that control recruitment *should* be based on such cues to task difficulty and reward. But this intuition hides a thorny problem: Why does control need to be 'recruited' at all? Why must the engagement of control be calibrated to task demands and rewards? Why is control not maximally engaged under all circumstances? Why, in other words, do people not exert maximal mental effort on everything, all the time?

Consider the study by Padmala & Pessoa (2011) cited above. Participants performed a selective attention task in which they had to classify an image while disregarding words that were either unrelated to the image category, or were related and either congruent or incongruent with the classification. Importantly, when a cue indicated that correct performance on the current trial would lead to rewards, performance improved (i.e., reaction times, error rates, and interference effects were reduced). An even more striking example of this effect was provided by Duckworth, Quinn, Lynam, Loeber, and Stouthamer-Loeber (2011), who showed that laboratory participants performing standardised intelligence tests scored more than half a standard deviation higher when performance was incentivised. These findings suggest that performance in the non-incentivised conditions in these studies was marked by less-than-maximal control exertion.

Why did people not exert the maximal amount of cognitive control to improve performance on these tasks? One intuitive answer focuses on the possibility that there is an intrinsic cost attached to the exertion of mental effort. That is, seemingly suboptimal exertion of cognitive control might be the normative result of a rational cost-benefit analysis, weighing the potential rewards of each option against its anticipated costs (Anderson, 1990; H. A. Simon, 1956; Stephens & Krebs, 1986).

Analogous Costs in the Domain of Physical Effort

A similar notion is widely accepted in research on physical-effort-based decision making, where it is assumed that, all else being equal, actions are selected to minimise demands for physical labour. This idea was famously formulated in Hull's (1943) Law of Less Work: all else being equal, agents will prefer lines of action with the smallest demands for physical effort. This basic notion has held currency since at least the 1920s (for a review, see Solomon, 1948) and remains widely influential in modern studies on the neural mechanisms underlying cost-benefit analyses involving physical costs (Salamone, Correa, Farrar, & Mingote, 2007; Walton, Kennerley, Bannerman, Phillips, & Rushworth, 2006). The law of least work is supported by extensive empirical evidence. The most frequent and direct approach has been to place subjects (animals or humans) in situations where they must choose between two courses of action associated with different exertional demands. When rewards are equated, a bias is typically observed towards the less demanding course of action (Solomon, 1948; Thompson, 1944).

Evidence for Cognitive Effort Avoidance

Many psychologists have posited that an analogous law is at work in situations that involve demands for cognitive processing, that is, that mental effort inheres costs just as physical effort does. For example, Allport (1954) famously explained social prejudice as the product of avoidance of effortful controlled processes required for accurate social judgement. Similar assertions have been made in connection with the fields of behavioural economics (Bonner & Sprinkle, 2002; Camerer & Hogarth, 1999), judgement and decision making (Shah & Oppenheimer, 2008; Smith & Walker, 1993), and executive functioning (De Jong, 2000; Engelmann et al., 2009; Westbrook, Kester, & Braver, 2013). In all of these works, the idea that mental effort is costly has been treated as an explanatory principle. However, until very recently, the existence of control costs had not been subjected to a direct empirical test.

Initial progress towards validating the idea of effort costs was made by Kool, McGuire, Rosen, and Botvinick (2010), who introduced the demand-selection task (DST) to test the hypothesis that people avoid the exertion of mental effort. In the DST, participants face a recurring choice between two alternative lines of action, associated with different levels of demands for cognitive control. The most often-used version of the DST manipulates control demands by varying the frequency of shifts between parity and magnitude judgements of numerical digits. Across several versions of this task (using multiple cognitive control tasks), participants are found to consistently favour the option that commits them to fewer demands for cognitive control (Kool et al., 2010; Kool, McGuire, Wang, & Botvinick, 2013; McGuire & Botvinick, 2010; Schouppe, Ridderinkhof, Verguts, & Notebaert, 2014). Follow-up analyses confirmed that this behavioural tendency could not be fully explained by a motivation to avoid errors, or to minimise time on task. The results appear consistent with a law of least *mental* effort, the idea that, all else being equal, actions tend to be selected to minimise cognitive demand.

The idea that demand registers as a cost predicts, additionally, that people should be more willing to exert effort when appropriate incentives are provided. Evidence for this prediction has come from multiple sources. Kool et al. (2010) introduced a new demand-selection paradigm in which participants tended to avoid an effortful task switch, even if this would lead to increased task completion time. However, when monetary incentives were supplied for each completed trial, participants' propensity to avoid task switches disappeared. This finding suggests that the exertional cost of task options is weighed against associated rewards in a cost-benefit analysis. In a series of convergent studies, Westbrook et al. (2013) presented participants with a series of choices between performing a low-demand working memory task for a certain reward and a more demanding task for a higher wage. The latter payment was titrated until an indifference point was reached between the two tasks. Importantly, across participants, at this indifference point, the pay for the high-demand task was consistently higher than the pay for the low-demand task. As discussed before, other data consistent with this idea shows that performance on attention-demanding tasks is higher in the face of appropriate incentives (Aarts et al., 2010; Engelmann et al., 2009; Engelmann & Pessoa, 2007;

see also Chapter 24 by Krebs & Woldorff in this volume). Together, these studies suggest that people discount the value of reward by the amount of cognitive effort that needs to be exerted (see also Dixon & Christoff, 2012; Vassena et al., 2014).

Why Is Control Costly?

So far, we have explored the question of why people do not exert maximal control all of the time. The research that we reviewed provides a straightforward answer to this question, namely, that there is an intrinsic cost associated with the expenditure of mental effort. However, this observation in turn raises a new question: *Why* is the exertion of cognitive control costly? Over the last decades, several researchers have attempted to answer this question from at least two broad perspectives, one focussing on potential limits on control resources, and the other focussing on the opportunity costs associated with the exertion of cognitive control.

Resource Limits

One class of theories that helps address why cognitive effort appears to be underutilised proposes that the exertion of cognitive control depends on a limited resource, and that the amount of available resource determines to what degree control can be implemented (Baumeister, Bratslavsky, Muraven, & Tice, 1998; Bijleveld, Custers, & Aarts, 2009; Kahneman, 1973; Navon & Gopher, 1979). This theory has been supported by a collection of experiments showing that people are less inclined towards controlled behaviour after they have already engaged in a control-demanding activity, a phenomenon referred to as 'ego depletion' (see Hagger, Wood, Stiff, & Chatzisarantis, 2010; but see also Carter, Kofler, Forster, & McCullough, 2015; Blain, Hollard, & Pessiglione, 2016; Hagger & Chatzisarantis, 2016). According to this framework, control can be viewed as costly insofar as it depletes a resource that could otherwise be conserved or husbanded (Muraven, Shmueli, & Burkley, 2006).

However, recent research on self-control has initiated a trend away from this resource-based account of control, focussing instead on motivational explanations (Inzlicht & Schmeichel, 2012; Job, Dweck, & Walton, 2010). This trend has been supported by, among other things, findings suggesting that depletion effects disappear in the face of increased incentives (e.g., Muraven & Slessareva, 2003). These results place doubt on its potential as a mechanistic explanation and cast doubt on the conservation hypothesis more generally (but see also Holroyd, in press).

Opportunity Costs

In an attempt to offer an alternative to resource-based accounts, Kurzban, Duckworth, Kable, and Myers (2013) have recently proposed an alternative account according to which the cost of effort depends directly on the foregone reward from other options. In their opportunity cost framework, the cost of control is directly attached to the presence of task options that are more valuable than the current task. In other words, effort is the perceived output of ongoing cost/benefit computations that monitor for more profitable lines of actions and is not directly linked to the amount of top-down control that is demanded to successfully encode task representations. The opportunity cost framework proposes that the evolutionary purpose of this mechanism is to signal when to engage in adaptive alternative mental activities in order to gain maximal net benefit from the environment (see also Hockey, 1997).

Feng, Schwemmer, Gershman, and Cohen (2014) have used a computational modelling approach to formulate an alternate version of this idea. They argue that the cost of control may arise from cross-talk that occurs when multiple tasks are performed simultaneously. Assuming that these tasks rely on shared rather non-overlapping neural representations (an approach that leads to maximal flexibility and efficiency), they show that even relatively small levels of cross-talk in such a system can lead to large performance decrements. In this context, one can imagine that the cost of control serves as a penalty militating against this level of multitasking to avoid such decrements, while promoting efficient and flexible processing.

Putting It All Together: A Cost-Benefit Trade-Off in Cognitive Control

The work reviewed above suggests that control allocation is jointly determined by the payoffs and the intrinsic costs associated with cognitive control. However, it leaves open the question of how these factors are combined to produce effort-based decisions about control allocation. There clearly needs to be a principled way of integrating these two factors to determine when and how much to adjust control.

The Overall Expected Value of Control

We recently proposed a normative account of control allocation that adapts approaches from computational reinforcement learning for optimal action selection to the goal of determining the value of exerting control (Shenhav, Botvinick, & Cohen, 2013). Specifically, our theory suggests that we select the amount of cognitive control at a given moment that maximises a quantity we refer to as the overall expected value of control (EVC). The EVC represents a combination of the two factors we described earlier: the expected payoff and the cost of control (Figures 10.1 and 10.3). Payoffs consist of both positive and negative outcomes experienced after implementing a given setting of control signals. Positive outcomes that can result from control allocation include intrinsic rewards associated with task performance (Cacioppo & Petty, 1982; Eisenberger, 1992) and extrinsic rewards like money or social approval. Negative outcomes include monetary losses, physical pain, negative peer evaluations, and other states that we may experience as aversive (e.g., conditions of increased uncertainty or conflict; Fritz & Dreisbach, 2013; Inzlicht, Bartholow, & Hirsh, 2015). The salience of both types of outcome could additionally be modulated by the time required to obtain the outcome, as has been found in the sensitivity of humans and other animals to *reward rate* (i.e., rewards normalised by time to reward receipt) rather than simply rewards per se (Niv, Daw, & Dayan, 2006; Simen et al., 2009). In other words, individuals tend to maximise their net reward per unit time.

Deciding How Much Control to Exert

According to the EVC framework, an increase in the intensity of control has two simultaneous effects. The first is to change the probabilities of performance outcomes (success, failure, or more graded intermediate levels of outcome quality). This, in turn, can lead to greater expected payoffs without necessarily influencing the payoffs themselves. For instance, attending more or less to colour naming in the Stroop task influences the likelihood of giving a correct response, but not the rewards associated with giving such a response (versus an incorrect response). Control can also influence how long it takes you to arrive at

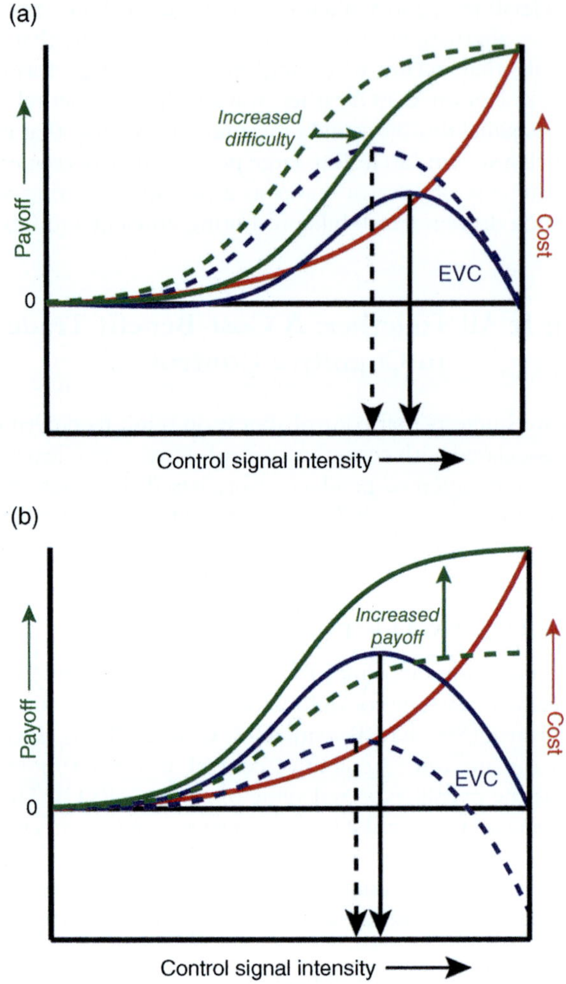

Figure 10.1 Weighing cognitive demands and payoff in the allocation of control. The expected value of control (EVC; blue) reflects the difference between expected payoffs (green) and intrinsic costs (red) associated with increasing control intensity, assuming a task in which the individual makes a binary choice. Control intensity can be selected on the basis of the maximum of the EVC curve. **Panel a** provides an example of increasing task difficulty (e.g., 3-back rather than 2-back), operationalised as a reduction in the efficacy of a given control intensity at achieving a correct response, resulting in a rightward shift of the payoff curve (dashed to solid lines). **Panel b** provides an analogous example of an increase in the payoff for a correct response. These examples show how both manipulations can, under certain conditions, lead to a decision to increase control. Source: Shenhav 2013. Reproduced with permission of Elsevier.

that state (i.e., the denominator in the reward rate calculation), for instance, by modulating response times. Accordingly, it is important to separate the outcomes described above (e.g., amount of reward for correct and loss for incorrect) from the factors that govern the relationship between control investment and the likelihood of obtaining a given outcome. These latter factors are typically what is referred to by the term *cognitive demand*. Signals of task difficulty, like errors or amount of conflict, presage the likelihood of reaching a rewarding state given the current allocation of control, versus a situation in which one increases or decreases their control allocation (e.g., on the next trial).

The second consequence of intensifying control is to produce a subjective 'effort-like' cost for the individual. We assume that this cost increases monotonically with the amount of control being allocated. While the exact function relating these two is still a matter of speculation, there is reason to believe that cost may increase nonlinearly (e.g., exponentially) with control intensity (Fudenberg & Levine, 2006, 2011), as has been observed in research on physical effort (Klein-Flügge, Kennerley, Saraiva, Penny, & Bestmann, 2015).

Because both the payoffs and the costs for control are a function of the intensity of control, one can ascertain the EVC of a given intensity of control by taking the difference between these two quantities (see Figure 10.1 for illustrative examples of this cost-benefit analysis). The optimal set of control signals are those that maximise this difference, that is, that maximise the EVC. Because the EVC integrates both payoffs and costs, those factors that indicate increased demand for control (e.g., error feedback, conflict) can lead to greater control allocation but only up to a point (Figure 10.1a), and likewise for factors that indicate increasing reward for successful task performance (Figure 10.1b).

Deciding How Long to Exert Control

The cost-benefit analysis just described focuses primarily on how intensely to exert control. However, many effort allocation problems incorporate another crucial dimension, namely, how to allocate control over periods of time. Imagine, for example, a graduate student sitting in her office on a weekday. Across time, the student has to repeatedly make a choice between two categories of tasks. She can spend time on activities that will further her research, or she can take a break to daydream or peruse social media networks. The decision here is not just related to intensity of control, but can also be characterised as a graded and time-based choice between cognitive labour and cognitive leisure, or mental work and mental rest. In recent work, we tested a formal framework for understanding such temporal allocation problems. This work draws upon a model from labour economics that addresses the question of how workers choose to allocate time between labour and leisure.

The model, labour supply theory (LST), was originally formulated to account for how workers choose to allocate time between labour and leisure (for an introduction, see Nicholson & Snyder, 2008). Here, the benefits from work (i.e., wage) are weighed against its intrinsic costs in a nonlinear fashion, such that the more hours one is working, the more costly each additional unit of work becomes (Figure 10.2). Through this feature of the model, optimal-effort-based decision making results in time allocations that reflect a balance between the payoffs of work, and the intrinsic value of leisure (i.e., the intrinsic cost of work). LST has been validated in numerous experimental and field studies on labour markets (Charness & Kuhn, 2010; Dickinson, 1999; Fehr & Goette, 2007), and has been extended to decision making based on physical effort in animal learning research (Chen, Lakshminarayanan, & Santos, 2006; Conover & Shizgal, 2005; Kagel, Battalio, & Green, 1995).

We recently provided evidence that LST can be applied to decisions involving allocation of cognitive effort over extended periods (Kool & Botvinick, 2014). In these studies, participants divided their time between a mentally demanding task (labour) and a trivially easy task (leisure) within each of a series of experimental sessions. By varying the incentive structure across sessions, we tested a set of critical predictions from LST and found that these held in the case of cognitive effort. The results mirror findings in real-life labour markets and reveal that the cost of mental effort scales in a nonlinear fashion with time

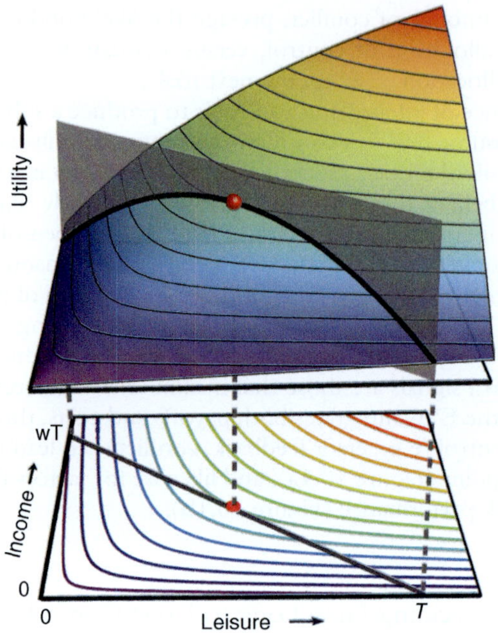

Figure 10.2 The utility function of labour supply theory (LST). As discussed by Kool and Botvinick (2014), LST is a theory of how workers should allocate time between work and leisure. Work is valuable because it yields income via a wage (*w*). Leisure, on the other hand, carries inherent value. Note that this value can also be viewed as the utility of avoiding the effort costs associated with work. Each combination of income and leisure corresponds to a point on the surface at the top, and is associated with one of the iso-utility contours in the bottom of the figure. Allocation decisions are constrained by a time budget, or the total interval of time (*T*) available for the two activities. This time budget is represented by the grey plane that intersects the utility surface in the top part of the figure top, and the diagonal line segment in the lower part. LST assumes that the iso-utility curves of the utility function are convex to the origin, resulting in the inverted-u shaped function on top. The optimal time allocation decision maximises utility over the time budget line. Because the utility is convex to the origin, LST predicts a preference for combinations of income and leisure that balance between the two extreme allocations. Source: Adapted from Kool 2014. Reproduced with permission of APA. Abbreviations: *T*, maximum time available for leisure; *wT*, maximum attainable income.

spent exerting control. This nonlinearity has the effect of motivating individuals to strike a balance between mental labour and leisure.

Deciding What Kinds of Control to Exert

Note that in this last example, the individual is faced with a decision not only over how much control to exert but also to which task or tasks to allocate this control. This highlights a crucial dimension of control that must be selected simultaneously with the amount of control: the type of control that needs adjusting (e.g., attention to different tasks, locations, or stimulus properties; threshold for sampling from episodic memory; affective responses to downregulate). We refer to these as the *identity* and *intensity* of potential control signals, which are the basic unit over which value is maximised within the EVC model described earlier (Figure 10.3). By operationalising cognitive control

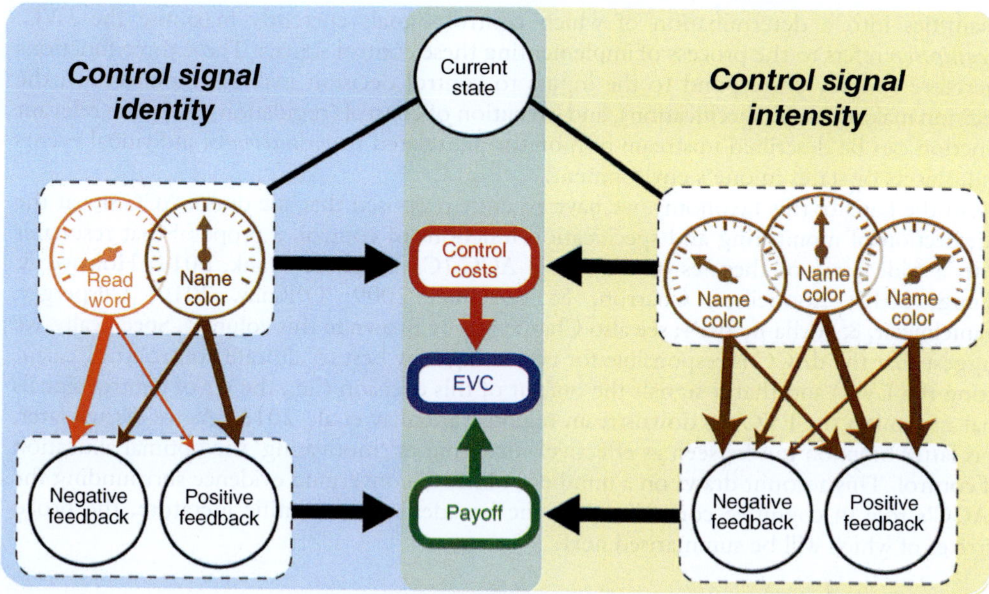

Figure 10.3 Specifying the optimal identity and intensity of control. EVC is simultaneously maximised across possible intensity settings (right) of available control identities (left), for instance, when deciding how much to attend to the colour versus word form when performing the Stroop task. Vertical arrows between gauges and feedback denote likelihoods of obtaining a given outcome under a given control setting. Source: Adapted from Shenhav 2013. Reproduced with permission of Elsevier.

adjustments in this way, the EVC framework is able to extend beyond simple models of reinforcement learning and draw explicit connections to the field of optimal control theory, an area of engineering focused on how to simultaneously optimise the settings of multiple control variables within a dynamical system (e.g., gears and steering-wheel positions on a car) on the basis of feedback from the environment (Kirk, 2012; Wolpert & Landy, 2012).

Because the EVC is simultaneously maximised over the intensities of multiple potential control signal identities, as EVC-relevant variables like difficulty and payoff change, optimal control adaptations can occur within as well as between control signals. For example, increasing difficulty within a certain range may encourage greater engagement of one control signal (e.g., greater attention to task-relevant stimuli), but further increasing task difficulty may lead to engaging a different control signal entirely (e.g., adjustment of decision threshold, or attending to an alternate task).

Under the Hood: Neural Circuitry for Control Evaluation and Specification

According to the EVC framework, describing the mechanisms for control requires delineating (at least) three distinct functions related to control: monitoring, specification, and regulation. *Monitoring* refers to the process of detecting and signalling control-relevant changes in the environment, including changes in the control demands described earlier (e.g., payoff, difficulty). *Specification* refers to the process of integrating these monitored-for quantities into a determination of which control signals currently maximise the EVC. *Regulation* refers to the process of implementing these control signals. These three functions therefore roughly correspond to the inputs to control decision making (monitoring), the decision making itself (specification), and execution of control (regulation). A fourth relevant function can be described upstream of monitoring, related to *valuation* of individual events and objects or states in one's environment.

On the basis of this taxonomy, we have recently proposed that the dorsal ACC sits at the intersection of monitoring and specification functions of control, a proposal that resonates with a wide array of theories of the dorsal ACC (Cavanagh & Frank, 2014; Holroyd & Yeung, 2012; Kouneiher, Charron, & Koechlin, 2009; O'Reilly, 2010; Ullsperger, Danielmeier, & Jocham, 2014; see also Chapter 15 by Brown in this volume). Specifically, we suggest that the dACC is responsible for evaluating how best to allocate control (i.e., calculating the EVC) and that it signals the output of this decision (i.e., the set of control signals that maximises the EVC) to downstream regions (Shenhav et al., 2013). As we discuss later, this latter function can be seen as effectively licensing or 'motivating' the optimal allocation of control. This account draws on a number of lines of converging evidence surrounding the dACC's role in cognitive control and in the broader neuroscientific literature, the broad strokes of which will be summarised next.

dACC and Monitoring

A number of lines of evidence support our proposed role for the dACC. First, dACC activity has been shown to track each of the variables relevant to evaluating potential control signals within one's current state. This region differentiates between available states and control signals, such as rules (Durstewitz, Vittoz, Floresco, & Seamans, 2010; Johnston, Levin, Koval, & Everling, 2007; Matsuzaka, Akiyama, Tanji, & Mushiake, 2012; Womelsdorf, Johnston, Vinck, & Everling, 2010), actions (Cai & Padoa-Schioppa, 2012; Hayden & Platt, 2010; Isomura, Ito, Akazawa, Nambu, & Takada, 2003; Morecraft & Tanji, 2009), and task sets (Forstmann, Brass, Koch, & Von Cramon, 2006; Momennejad & Haynes, 2013; Wisniewski, Reverberi, Tusche, & Haynes, 2015). Moreover, the dACC tracks a variety of sources of cognitive demands (Cavanagh & Frank, 2014; Duncan, 2010; Paus, Koski, Caramanos, & Westbury, 1998; Venkatraman & Huettel, 2012), from errors (Emeric et al., 2008; Holroyd & Coles, 2002; Ito, Stuphorn, Brown, & Schall, 2003) to conflict (Botvinick et al., 2001; Ebitz & Platt, 2015; Sheth et al., 2012) to signals indicating increased surprise or unpredictability in one's environment (Behrens et al., 2007; Bryden, Johnson, Tobia, Kashtelyan, & Roesch, 2011; Cavanagh & Frank, 2014; Wessel, Danielmeier, Morton, & Ullsperger, 2012). dACC activity has also been found to anticipate and/or mark the occurrence of positive outcomes like juice or monetary gain (Bartra, McGuire, & Kable, 2013; Kouneiher et al., 2009; Wallis & Kennerley, 2011) as well as negative outcomes like electric shock or monetary loss (Botvinick, 2007; Shackman et al., 2011), though it should be noted that it is not always easy to distinguish these activations from signals related to salience or prediction error (Bartra et al., 2013; Cavanagh & Frank, 2014; Engelmann et al., 2009; Litt, Plassmann, Shiv, & Rangel, 2011) and therefore may sometimes reflect attentional demands rather than outcomes per se.

Importantly, feedback-related signals in the dACC (e.g., errors, reward) have properties that suggest that they are in some way tied to control requirements rather than simply being part of an all-purpose mechanism for learning about states in the environment. First, the presence and strength of feedback-related dACC signals is influenced by the amount of information that can be gained as well as the effort required of the task. When monkeys (Quilodran, Rothé, & Procyk, 2008) or humans (Amiez, Sallet, Procyk, & Petrides, 2012) perform a task involving phases of exploration (attempting to discover the correct response) versus exploitation (repeating this response until it ceases to provide correct feedback), dACC activity is greatest when the feedback indicates discovery of the correct response or a need to explore once again, and is lowest in response to correct feedback during the exploit phase (Khamassi et al., 2010). These findings are consistent with the observation that reward-related activity in the rodent dACC is also diminished when no effort is required to obtain that reward (Hillman & Bilkey, 2010).

Moreover, although a number of findings tie the dACC to levels of surprise or unpredictability, it has also been shown that these surprise signals depend critically on the relevance of the surprising outcome to future control states. For instance, O'Reilly and colleagues (2013) asked participants to saccade to a target that would tend to occur in a constrained region of the screen, with that region changing over the course of the experiment. They found that the dACC was more active when the target appeared in an unexpected location and the target's colour indicated a general shift in possible future target locations, but not when the colour of that surprising target indicated that this trial had no predictive power. In other words, dACC responses to equally unexpected events are modulated by the degree to which that event signalled the need for an update of an internal model (see also Grundy & Shedden, 2014). In further support of our account of dACC monitoring for the purpose of adjusting relevant control signals, the monkey dACC has been found not only to contain separate and overlapping neuronal populations that track positive and negative outcomes (including errors and now also conflict; see below), but this region further distinguishes between different *sources* of errors. For instance, Shen and colleagues (2015) found that error signals in the dACC differentiated between errors that occurred during different stages of a task trial—related to putative failures in sustaining attention, in inhibiting distractors, or in execution of the incorrect response—consistent with a potential role in assigning credit to the appropriate control signal(s) that need to be adjusted (see also Quilodran et al., 2008).

Although some of these findings might simply argue for a role for the dACC in learning and/or selecting between the values of actions (e.g., Hare, Schultz, Camerer, O'Doherty, & Rangel, 2011), recent findings place limitations on such an account. Despite substantial evidence for action values in the dACC, recent studies suggest that the types and timing of value-related responses in the dACC may make this region better suited for post-choice evaluation and control than for direct comparison between action values. For instance, Cai and Padoa-Schioppa (2012) found that during economic choice, dACC action values tend to reflect post-decision variables (e.g., the value of a chosen option or action) rather than pre-decision variables (e.g., the value of the offered options, which are encoded earlier in the orbitofrontal cortex [OFC]). Similarly, Blanchard and Hayden (2014) found that dACC activity in a sequential foraging-like task signalled one post-decision variable (the value of the unchosen option) and one putatively pre-decision variable (the ratio of the offered reward to the delay to reward receipt), but its encoding of the latter interacted with the choice that was made (accept or reject). In other words, the findings support a role for the dACC in using the outcome of a previous decision to modify future strategies and/or control (see also Cowen, Davis, & Nitz, 2012).

Finally, in addition to studies linking the dACC to the encoding of physical effort demands (Hillman & Bilkey, 2010; Hosokawa, Kennerley, Sloan, & Wallis, 2013; Prévost, Pessiglione, Météreau, Cléry-Melin, & Dreher, 2010; Wallis & Kennerley, 2011) and correlates of cognitive demands discussed above, other findings suggest that the dACC may also track the subjective costs associated with exerting cognitive effort (Botvinick, 2007). For example, in one study, Botvinick, Huffstetler, and McGuire (2009) showed that ventral striatal (vStr) responses to both positive and negative feedback were diminished after completing a more rather than less cognitively demanding block of trials (i.e., one that required more rather than less task switching), and that the vStr response to feedback on high-demand blocks was smaller the greater the dACC responded while performing the task for that block. In a follow-up study, McGuire and Botvinick (2010) found that activity in the dACC and the lateral PFC predicted preferences for avoiding a similarly demanding block of an experimental session (again, involving task switching), though in the dACC this association could be explained in terms of objective performance on the task (i.e., RT, error rate; see also Magno, Foxe, Molholm, Robertson, & Garavan, 2006).

dACC and Control Specification

The aforementioned evidence points to a role for the dACC in evaluating (or at least having access to) quantities relevant to control allocation, and suggests that this region is particularly responsive to those values when they are most relevant. A wide literature further implicates this region in exerting an influence on ongoing control on the basis of these monitored signals (see also Chapter 5 by Brown in this volume).

First, several studies have found overlapping representations in the dACC of specific tasks or rules to be attended (i.e., potential targets of control allocation) and their associated values. This includes the rewards associated with specific saccade directions (Cai & Padoa-Schioppa, 2012; Hayden & Platt, 2010) and with targets of covert attention allocation (Kaping, Vinck, Hutchison, Everling, & Womelsdorf, 2011). Cowen and colleagues (2012) found similar overlaps between the encoding of specific paths along a track and their associated effort demands in the rat dACC (though these responses appeared to be involved in regulating chosen actions rather than planning/choosing). Similarly, while participants performed a task that allowed them to choose which of three task sets to engage on a given trial, and the difficulty of those tasks varied over time (a chosen task grew more difficult over time while the unchosen tasks grew easier), Wisniewski and colleagues (2015) found overlapping regions of the dACC encoded the identities of potential tasks during the choice phase and encoded the difficulty of the chosen task during task performance.

Moreover, changes in dACC activity are associated with and often anticipate changes in task performance. Indirect evidence for this comes from a large body of findings showing that the dACC's responses to salient task events predict subsequent adjustments in control. For instance, the dACC has been implicated in error adaptation, predicting post-error adjustments on the following trial such as slowing (Cavanagh & Frank, 2014; Danielmeier, Eichele, Forstmann, Tittgemeyer, & Ullsperger, 2011; Narayanan, Cavanagh, Frank, & Laubach, 2013; Narayanan & Laubach, 2008) and reallocation of attention between low-level regions representing task-relevant versus task-irrelevant stimulus properties (Danielmeier et al., 2011; King, Korb, von Cramon, & Ullsperger, 2010). The dACC has also been implicated in conflict adaptation, for instance, predicting more efficient resolution of conflict within a high-conflict trial (Sohn, Albert, Jung, Carter, & Anderson, 2007) or on a trial that follows (Forster, Carter, Cohen, & Cho, 2011; Horga et al., 2011; Kerns et al., 2004; Oehrn et al., 2014), often through its connectivity with the lateral PFC. In the first demonstration of conflict-related signals in the monkey dACC, Ebitz and Platt (2015) found that dACC signals reflecting either errors or high-conflict stimuli (distractors that were either incongruent or congruent with a target stimulus direction, evoking what they refer to as *task conflict*) predicted pupillary adjustments on subsequent trials that were separately associated with reduced distractor interference on those trials (Shenhav & Botvinick, 2015; see also Amemori & Graybiel, 2012). Other studies have further linked dACC activity to subsequent choice reversals (Boorman, Rushworth, & Behrens, 2013; Shenhav & Buckner, 2014), changes in learning rate (Behrens et al., 2007; Bryden et al., 2011), and both within- and between-trial adjustments in decision threshold in reaction not only to errors (as described above) but to increased choice conflict as well (Cavanagh & Frank, 2014; Cavanagh et al., 2011).

Tying the findings above to specific representations of control signals in the dACC, some studies have explicitly decoded representations of different task rules from neural activity in this region and examined the relationship between the strength of those representations and performance on a task. For instance, Johnston and colleagues (2007) showed that neuronal activity in the dACC could be classified on the basis of the current task rule (pro- or anti-saccade) and that the accuracy of these classifications predicted accuracy during the task (see also Rothé, Quilodran, Sallet, & Procyk, 2011; Womelsdorf et al., 2010): When dACC activity in preparation of engaging a given task rule was less selective for the correct rule, an error was more likely.

These findings collectively support the idea that the dACC has the representational capacity that would be expected of a region that maps states and potential control signals to their associated value. It also provides correlational evidence that such representations in the dACC are associated with downstream adjustments in control states. In addition to this, there is also mounting causal evidence that the dACC exerts an influence on control. For instance, patients with lesions in (anterior regions) of the dACC are impaired at tasks that require shifting between different task sets but not, for instance, at a value-based decision-making task (Gläscher et al., 2012). dACC lesions have also been found to impair sequential adaptation in response to control-relevant stimuli. Sheth and colleagues (2012) found that patients who had undergone cingulotomy showed diminished conflict adaptation following their surgery (relative to pre-lesion performance), while Newman, Creer, and McGaughy (2015) found analogous impairments in conflict adaptation in ACC-lesioned rats (but see also Mansouri, Tanaka, & Buckley, 2009). Narayanan and colleagues (2013) found that dACC-inactivated rats exhibited analogous impairments in adapting (i.e., slowing) to an error on the previous trial.

dACC lesions have more generally been found to impair behavioural adjustment to recent reversals in reinforcement (Kennerley, Walton, Behrens, Buckley, & Rushworth, 2006) and to produce motivational deficits such as global response slowing (Stuss & Alexander, 2007) and reduced willingness to overcome effortful obstacles (i.e., increased effort discounting; Walton et al., 2006). Far more dramatic consequences of lesioning dACC and surrounding regions of the medial prefrontal cortex (especially the supplementary motor area) include disorders like abulia, anergia, and akinetic mutism, with which patients are technically able to perform effortful behaviours but lack the motivation or drive/energy to engage in such activities (Stuss, 2011; Stuss & Alexander, 2007). Such findings are broadly consistent with our proposed role for this region in specifying the EVC-maximising control (akin to a motivational or 'willingness to pay' signal; see also Holroyd & Yeung, 2012; Kouneiher et al., 2009).

Our account of the dACC as overseeing the decision to engage control, and outputting the quantity of control deemed worthwhile, has also been borne out by a few studies that employ the opposite causal approach, stimulating rather than inhibiting the dACC. Parvizi, Rangarajan, Shirer, Desai, and Greicius (2013) used depth electrodes to electrically stimulate the dACC in two awake human patients as those patients verbalised their subjective experiences. A common feature of their reports was a feeling of an increased 'willingness to persevere' in the face of perceived environmental challenges. Using noninvasive methods (transcranial direct current stimulation [tDCS]) in healthy participants, Reinhart and Woodman (2014) were similarly able to provide evidence consistent with our account. They found that excitatory stimulation (anodal tDCS) of the dACC and surrounding regions on the medial surface (including SMA) increased monitoring-related ERPs (error- and feedback-related negativities) and led to improved overall accuracy, faster learning, and greater post-error slowing on a stop-signal target discrimination task, relative to sham controls. The opposite pattern was observed when this region was instead inhibited (cathodal tDCS; see also Reinhart & Woodman, 2015; Rollnik et al., 2004).

By focussing so far on the dACC's role in evaluating and specifying control, we have failed to emphasise a key point, which is that the dACC is central but by no means singular in its involvement in control decision making. A number of structures are presumed to play critical roles in supporting the valuation of control-relevant processes (acting as inputs to the dACC) and in executing control itself (performing the output functions specified by the dACC). At the intersection of valuation and monitoring functions lie regions that include the insula, ventral striatum, orbitofrontal cortex, and ventromedial PFC (Craig, 2009; Haber & Knutson, 2010). On the output end, the regions that implement the necessary control signals vary according to the type and specificity of control required, for instance, whether it involves attention to particular rules or task sets (e.g., by LPFC; Miller & Cohen, 2001; Oehrn et al., 2014; Rothé et al., 2011; Shen et al., 2015); adjustments to decision thresholds to behave more cautiously (e.g., by the subthalamic nucleus; Aron & Poldrack, 2006; Cavanagh et al., 2011; Jahfari et al., 2011); or a change in exploratory versus exploitative modes (e.g., by the locus coeruleus; Aston-Jones & Cohen, 2005).

Conclusions

Throughout this chapter, we have reviewed a number of important domains that encompass the intersection between decision making and cognitive control. At the core of this overlap lies the observation that there are several motivational factors, such as performance-contingent rewards, that have been shown to modulate the degree to which mental effort is deployed (Botvinick & Braver, 2015; Padmala & Pessoa, 2011; Shenhav et al., 2013). This observation suggests that people are reluctant to exert cognitive control, a claim supported by data we reviewed above (Kool et al., 2010; Schouppe et al., 2014). We have also reviewed a decision-making framework that proposes a set of computational and neural mechanisms by which the costs and benefits of control are integrated in order to decide whether and how cognitive control should be deployed.

These findings have obvious ties with other areas of research in which reward-based decision making and executive functioning are implicated. Research on self-control, the important though fallible ability to resist immediate pleasures in favour of longer-term goals (Mischel, Shoda, & Rodriguez, 1989), seems to fit this description. In fact, recent findings suggest that the ability for self-control is related to the cost for cognitive activity (Kool et al., 2013; Westbrook et al., 2013), and recruits the same neural circuitry that is typically associated with cognitive control (Hare, Camerer, & Rangel, 2009; Heatherton, 2011; McClure, Laibson, Loewenstein, & Cohen, 2004). Other fields of research that could benefit from the current approach include the study of voluntary task selection (Arrington & Logan, 2004; Orr & Weissman, 2011) and perseverance towards goals over longer time scales (i.e., 'grit'; Duckworth, Peterson, Matthews, & Kelly, 2007; Duckworth & Seligman, 2005).

In order to gain traction in these related fields, and to facilitate communication across them, there needs to be a common language for describing the computational mechanisms that underlie the evaluation of tasks in terms of their demands for cognitive control. Throughout this chapter, we have provided a broad framework and particular implementations for such an analysis. Understanding cognitive control requires understanding the particular costs and benefits associated with this process. One way to do this is by employing the kinds of tasks reviewed in this chapter that can be used to measure and test the size of cost for cognitive control (DSTs; Kool & Botvinick, 2014; Kool et al., 2010; Westbrook et al., 2013).

Another way is to establish a comprehensive and empirical comparison of the different models that have been offered on how effort-based cost-benefit analyses are implemented (Holroyd & McClure, 2015; Kurzban et al., 2013; Shenhav et al., 2013; Silvetti, Alexander, Verguts, & Brown, 2014). With these and related opportunities at hand, we foresee a period of rapid progress towards understanding the connections between motivation and cognitive control.

References

Aarts, E., Roelofs, A., Franke, B., Rijpkema, M., Fernandez, G., Helmich, R. C., & Cools, R. (2010). Striatal dopamine mediates the interface between motivational and cognitive control in humans: Evidence from genetic imaging. *Neuropsychopharmacology, 35*, 1943–1951.

Alexander, W. H., & Brown, J. W. (2011). Medial prefrontal cortex as an action-outcome predictor. *Nature Neuroscience, 14*, 1338–1344.

Allport, G. W. (1954). *The nature of prejudice*. New York, NY: Addison Wesley.

Amemori, K.-I., & Graybiel, A. M. (2012). Localized microstimulation of primate pregenual cingulate cortex induces negative decision-making. *Nature Neuroscience, 15*, 776–785.

Amiez, C., Sallet, J., Procyk, E., & Petrides, M. (2012). Modulation of feedback related activity in the rostral anterior cingulate cortex during trial and error exploration. *NeuroImage, 63*, 1078–1090.

Anderson, J. R. (1990). *The adaptive character of thought*. Hillsdale, NJ: Erlbaum.

Aron, A. R. (2011). From reactive to proactive and selective control: Developing a richer model for stopping inappropriate responses. *Biological Psychiatry, 69*, e55–e68.

Aron, A. R., & Poldrack, R. A. (2006). Cortical and subcortical contributions to Stop signal response inhibition: Role of the subthalamic nucleus. *Journal of Neuroscience, 26*, 2424–2433.

Arrington, C., & Logan, G. (2004). The cost of a voluntary task switch. *Psychological Science, 15*, 610–615.

Aston-Jones, G., & Cohen, J. D. (2005). An integrative theory of locus coeruleus-norepinephrine function: Adaptive gain and optimal performance. *Annual Review of Neuroscience, 28*, 403–450.

Bartra, O., McGuire, J. T., & Kable, J. W. (2013). The valuation system: A coordinate-based meta-analysis of BOLD fMRI experiments examining neural correlates of subjective value. *NeuroImage, 76*, 412–427.

Baumeister, R. F., Bratslavsky, E., Muraven, M., & Tice, D. M. (1998). Ego depletion: Is the active self a limited resource? *Journal of Personality and Social Psychology, 74*, 1252–1265.

Behrens, T. E. J., Woolrich, M. W., Walton, M. E., & Rushworth, M. F. S. (2007). Learning the value of information in an uncertain world. *Nature Neuroscience, 10*, 1214–1221.

Berlyne, D. E. (1957). Uncertainty and conflict: A point of contact between information-theory and behavior-theory concepts. *Psychological Review, 64*, 329–339.

Bijleveld, E., Custers, R., & Aarts, H. (2009). The unconscious eye opener: Pupil dilation reveals strategic recruitment of resources upon presentation of subliminal reward cues. *Psychological Science, 20*, 1313–1315.

Blain, B., Hollard, G., & Pessiglione, M. (2016). Neural mechanisms underlying the impact of daylong cognitive work on economic decisions. *Proceedings of the National Academy of Sciences of the United States of America, 113*, 6967–6972.

Blanchard, T. C., & Hayden, B. Y. (2014). Neurons in dorsal anterior cingulate cortex signal postdecisional variables in a foraging task. *Journal of Neuroscience, 34*, 646–655.

Bonner, S. E., & Sprinkle, G. B. (2002). The effects of monetary incentives on effort and task performance: Theories, evidence, and a framework for research. *Accounting, Organizations and Society, 27*, 303–345.

Boorman, E. D., Rushworth, M. F., & Behrens, T. E. (2013). Ventromedial prefrontal and anterior cingulate cortex adopt choice and default reference frames during sequential multi-alternative choice. *Journal of Neuroscience, 33*, 2242–2253.

Botvinick, M. M. (2007). Conflict monitoring and decision making: Reconciling two perspectives on anterior cingulate function. *Cognitive, Affective, & Behavioral Neuroscience, 7*, 356–366.

Botvinick, M. M., & Braver, T. (2015). Motivation and cognitive control: From behavior to neural mechanism. *Annual Review of Psychology, 66*, 83–113.

Botvinick, M. M., Braver, T. S., Barch, D. M., Carter, C. S., & Cohen, J. D. (2001). Conflict monitoring and cognitive control. *Psychological Review, 108*, 624–652.

Botvinick, M. M., & Cohen, J. D. (2015). The computational and neural basis of cognitive control: Charted territory and new frontiers. *Cognitive Science, 38*, 1249–1285.

Botvinick, M. M., Huffstetler, S., & McGuire, J. (2009). Effort discounting in human nucleus accumbens. *Cognitive, Affective, & Behavioral Neuroscience, 9*, 16–27.

Bryden, D. W., Johnson, E. E., Tobia, S. C., Kashtelyan, V., & Roesch, M. R. (2011). Attention for learning signals in anterior cingulate cortex. *Journal of Neuroscience, 31*, 18266–18274.

Cacioppo, J. T., & Petty, R. E. (1982). The need for cognition. *Journal of Personality and Social Psychology, 42*, 116–131.

Cai, X., & Padoa-Schioppa, C. (2012). Neuronal encoding of subjective value in dorsal and ventral anterior cingulate cortex. *Journal of Neuroscience, 32*, 3791–3808.

Camerer, C. F., & Hogarth, R. M. (1999). The effects of financial incentives in experiments: A review and capital-labor-production framework. *The Journal of Risk and Uncertainty, 19*, 7–42.

Carter, E. C., Kofler, L. M., Forster, D. E., & McCullough, M. E. (2015). A series of meta-analytic tests of the depletion effect: Self-control does not seem to rely on a limited resource. *Journal of Experimental Psychology: General, 144*(4), 796–815.

Cavanagh, J. F., & Frank, M. J. (2014). Frontal theta as a mechanism for cognitive control. *Trends in Cognitive Sciences, 18*, 414–421.

Cavanagh, J. F., Wiecki, T. V., Cohen, M. X., Figueroa, C. M., Samanta, J., Sherman, S. J., & Frank, M. J. (2011). Subthalamic nucleus stimulation reverses mediofrontal influence over decision threshold. *Nature Neuroscience, 14*, 1462–1467.

Charness, G., & Kuhn, P. (2010). What can labor economists learn from the lab? *IZA Discussion Paper no. 4941.*

Chen, M. K., Lakshminarayanan, V., & Santos, L. R. (2006). How basic are behavioral biases? Evidence from capuchin monkey trading behavior. *Journal of Political Economy, 114*, 517–537.

Christoff, K., Prabhakaran, V., Dorfman, J., Zhao, Z., Kroger, J. K., Holyoak, K. J., & Gabrieli, J. D. E. (2001). Rostrolateral prefrontal cortex involvement in relational integration during reasoning. *NeuroImage, 14*, 1136–1149.

Conover, K. L., & Shizgal, P. (2005). Employing labor-supply theory to measure the reward value of electrical brain stimulation. *Games and Economic Behavior, 52*, 283–304.

Cowen, S. L., Davis, G. A., & Nitz, D. A. (2012). Anterior cingulate neurons in the rat map anticipated effort and reward to their associated action sequences. *Journal of Neurophysiology, 107*, 2393–2407.

Craig, A. D. B. (2009). How do you feel—now? The anterior insula and human awareness. *Nature Reviews Neuroscience, 10*, 59–70.

Danielmeier, C., Eichele, T., Forstmann, B. U., Tittgemeyer, M., & Ullsperger, M. (2011). Posterior medial frontal cortex activity predicts post-error adaptations in task-related visual and motor areas. *Journal of Neuroscience, 31*, 1780–1789.

Danielmeier, C., & Ullsperger, M. (2011). Post-error adjustments. *Frontiers in Psychology, 2*, 233.

Dayan, P. (2012). How to set the switches on this thing. *Current Opinion in Neurobiology, 22*, 1068–1074.

De Jong, R. (2000). An intention-activation account of residual switch costs. In S. Monsell & J. Driver (Eds.), *Control of cognitive processes: Attention and performance* (pp. 357–376). Cambridge, MA: MIT Press.

Dickinson, D. L. (1999). An experimental examination of labor supply and work intensities. *Journal of Labor Economics, 17*, 638–670.

Dixon, M. L., & Christoff, K. (2012). The decision to engage cognitive control is driven by expected reward-value: Neural and behavioral evidence. *PLoS ONE, 7*, e51637.

Duckworth, A. L., Peterson, C., Matthews, M. D., & Kelly, D. R. (2007). Grit: Perseverance and passion for long-term goals. *Journal of Personality and Social Psychology, 92*, 1087–1101.

Duckworth, A. L., Quinn, P. D., Lynam, D. R., Loeber, R., & Stouthamer-Loeber, M. (2011). Role of test motivation in intelligence testing. *Proceedings of the National Academy of Sciences of the United States of America, 108*, 7716–7720.

Duckworth, A. L., & Seligman, M. E. P. (2005). Self-discipline outdoes IQ in predicting academic performance of adolescents. *Psychological Science, 16*, 939–944.
Duncan, J. (2010). The multiple-demand (MD) system of the primate brain: Mental programs for intelligent behaviour. *Trends in Cognitive Sciences, 14*, 172–179.
Durstewitz, D., Vittoz, N. M., Floresco, S. B., & Seamans, J. K. (2010). Abrupt transitions between prefrontal neural ensemble states accompany behavioral transitions during rule learning. *Neuron, 66*, 438–448.
Ebitz, R. B., & Platt, M. L. (2015). Neuronal activity in primate dorsal anterior cingulate cortex signals task conflict and predicts adjustments in pupil-linked arousal. *Neuron, 85*, 628–640.
Eisenberger, R. (1992). Learned industriousness. *Psychological Review, 99*, 248–267.
Emeric, E. E., Brown, J. W., Leslie, M., Pouget, P., Stuphorn, V., & Schall, J. D. (2008). Performance monitoring local field potentials in the medial frontal cortex of primates: Anterior cingulate cortex. *Journal of Neurophysiology, 99*, 759–772.
Engelmann, J. B., Damaraju, E., Padmala, S., & Pessoa, L. (2009). Combined effects of attention and motivation on visual task performance: Transient and sustained motivational effects. *Frontiers in Human Neuroscience, 3*, 4.
Engelmann, J. B., & Pessoa, L. (2007). Motivation sharpens exogenous spatial attention. *Emotion, 7*, 668–674.
Etzel, J. A., Cole, M. W., Zacks, J. M., Kay, K. N., & Braver, T. S. (in press). Reward motivation enhances task coding in frontoparietal cortex. *Cerebral Cortex*.
Fehr, E., & Goette, L. (2007). Do workers work more if wages are high? Evidence from a randomized field experiment. *American Economic Review, 97*, 298–317.
Feng, S. F., Schwemmer, M., Gershman, S. J., & Cohen, J. D. (2014). Multitasking versus multiplexing: Toward a normative account of limitations in the simultaneous execution of control-demanding behaviors. *Cognitive, Affective, & Behavioral Neuroscience, 14*, 129–146.
Forster, S. E., Carter, C. S., Cohen, J. D., & Cho, R. Y. (2011). Parametric manipulation of the conflict signal and control-state adaptation. *Journal of Cognitive Neuroscience, 23*, 923–935.
Forstmann, B. U., Brass, M., Koch, I., & Von Cramon, D. (2006). Voluntary selection of task sets revealed by functional magnetic resonance imaging. *Journal of Cognitive Neuroscience, 18*, 388–398.
Fritz, J., & Dreisbach, G. (2013). Conflicts as aversive signals: Conflict priming increases negative judgments for neutral stimuli. *Cognitive Affective, & Behavioral Neuroscience, 13*, 311–317.
Fudenberg, D., & Levine, D. K. (2006). A dual-self model of impulse control. *American Economic Review, 96*, 1449–1476.
Fudenberg, D., & Levine, D. K. (2011). Risk, delay, and convex self-control costs. *American Economic Journal: Microeconomics, 3*, 34–68.
Gläscher, J., Adolphs, R., Damasio, H., Bechara, A., Rudrauf, D., Calamia, M.,...Tranel, D. (2012). Lesion mapping of cognitive control and value-based decision making in the prefrontal cortex. *Proceedings of the National Academy of Sciences of the United States of America, 109*, 14681–14686.
Goldman-Rakic, P. S. (1987). Circuitry of primate prefrontal cortex and regulation of behavior by representational memory. In F. Plum (Ed.), *Handbook of physiology* (pp. 373–417). Bethesda, MD: American Physiological Society.
Gratton, G., Coles, M. G., & Donchin, E. (1992). Optimizing the use of information: Strategic control of activation of responses. *Journal of Experimental Psychology: General, 121*, 480–506.
Grundy, J. G., & Shedden, J. M. (2014). Support for a history-dependent predictive model of dACC activity in producing the bivalency effect: An event-related potential study. *Neuropsychologia, 57*, 166–178.
Haber, S. N., & Knutson, B. (2010). The reward circuit: Linking primate anatomy and human imaging. *Neuropsychopharmacology, 35*, 4–26.
Hagger, M. S., & Chatzisarantis, N. L. (2016). *Perspectives on Psychological Science, 11*(4), 546–573.
Hagger, M. S., Wood, C., Stiff, C., & Chatzisarantis, N. L. D. (2010). Ego depletion and the strength model of self-control: A meta-analysis. *Psychological Bulletin, 136*, 495–525.
Hare, T. A., Camerer, C. F., & Rangel, A. (2009). Self-control in decision-making involves modulation of the vmPFC valuation system. *Science, 324*, 646–648.

Hare, T. A., Schultz, W., Camerer, C. F., O'Doherty, J. P., & Rangel, A. (2011). Transformation of stimulus value signals into motor commands during simple choice. *Proceedings of the National Academy of Sciences of the United States of America, 108*, 18120–18125.

Hayden, B. Y., & Platt, M. L. (2010). Neurons in anterior cingulate cortex multiplex information about reward and action. *The Journal of Neuroscience, 30*, 3339–3346.

Heatherton, T. F. (2011). Neuroscience of self and self-regulation. *Annual Review of Psychology, 62*, 363–390.

Hillman, K. L., & Bilkey, D. K. (2010). Neurons in the rat anterior cingulate cortex dynamically encode cost-benefit in a spatial decision-making task. *Journal of Neuroscience, 30*, 7705–7713.

Hockey, G. R. J. (1997). Compensatory control in the regulation of human performance under stress and high workload: A cognitive-energetical framework. *Biological Psychology, 45*, 73–93.

Holroyd, C. B. (in press). The waste disposal problem of effortful control. In T. Braver (Ed.), *Motivation and cognitive control.* New York, NY: Psychology Press.

Holroyd, C. B., & Coles, M. G. H. (2002). The neural basis of human error processing: Reinforcement learning, dopamine, and the error-related negativity. *Psychological Review, 109*, 679–709.

Holroyd, C. B., & McClure, S. M. (2015). Hierarchical control over effortful behavior by rodent medial frontal cortex: A computational model. *Psychological Review, 122*, 54–83.

Holroyd, C. B., & Yeung, N. (2012). Motivation of extended behaviors by anterior cingulate cortex. *Trends in Cognitive Sciences, 16*, 121–127.

Horga, G., Maia, T. V., Wang, P., Wang, Z., Marsh, R., & Peterson, B. S. (2011). Adaptation to conflict via context-driven anticipatory signals in the dorsomedial prefrontal cortex. *The Journal of Neuroscience, 31*, 16208–16216.

Hosokawa, T., Kennerley, S. W., Sloan, J., & Wallis, J. D. (2013). Single-neuron mechanisms underlying cost-benefit analysis in frontal cortex. *Journal of Neuroscience, 33*, 17385–17397.

Hull, C. L. (1943). *Principles of behavior.* New York: Appleton Century.

Inzlicht, M., Bartholow, B. D., & Hirsh, J. B. (2015). Emotional foundations of cognitive control. *Trends in Cognitive Sciences, 19*, 126–132.

Inzlicht, M., & Schmeichel, B. (2012). What is ego depletion? Toward a mechanistic revision of the resource model of self-control. *Perspectives on Psychological Science, 7*, 450–463.

Isomura, Y., Ito, Y., Akazawa, T., Nambu, A., & Takada, M. (2003). Neural coding of 'attention for action' and 'response selection' in primate anterior cingulate cortex. *Journal of Neuroscience, 23*, 8002–8012.

Ito, S., Stuphorn, V., Brown, J. W., & Schall, J. D. (2003). Performance monitoring by the anterior cingulate cortex during saccade countermanding. *Science, 302*, 120–122.

Jahfari, S., Waldorp, L., Van Den Wildenberg, W. P. M., Scholte, H. S., Ridderinkhof, K. R., & Forstmann, B. U. (2011). Effective connectivity reveals important roles for both the hyperdirect (fronto-subthalamic) and the indirect (fronto-striatal-pallidal) fronto-basal ganglia pathways during response inhibition. *Journal of Neuroscience, 31*, 6891–6899.

Job, V., Dweck, C., & Walton, G. (2010). Ego depletion—Is it all in your head? *Psychological Science, 21*, 1686–1693.

Johnston, K., Levin, H., Koval, M., & Everling, S. (2007). Top-down control-signal dynamics in anterior cingulate and prefrontal cortex neurons following task switching. *Neuron, 53*, 453–462.

Kagel, J. H., Battalio, R. C., & Green, L. (1995). *Economic choice theory: An experimental analysis of animal behavior.* Cambridge: Cambridge University Press.

Kahneman, D. (1973). *Attention and effort.* Englewood Cliffs, NJ: Prentice-Hall.

Kaping, D., Vinck, M., Hutchison, R. M., Everling, S., & Womelsdorf, T. (2011). Specific contributions of ventromedial, anterior cingulate, and lateral prefrontal cortex for attentional selection and stimulus valuation. *PLoS Biology, 9*, e1001224.

Kennerley, S. W., Walton, M. E., Behrens, T. E. J., Buckley, M. J., & Rushworth, M. F. S. (2006). Optimal decision making and the anterior cingulate cortex. *Nature Neuroscience, 9*, 940–947.

Kerns, J. G., Cohen, J. D., MacDonald, A. W., Cho, R. Y., Stenger, V. A., & Carter, C. S. (2004). Anterior cingulate conflict monitoring and adjustments in control. *Science, 303*, 1023–1026.

Khamassi, M., Wilson, C. R. E., Rothé, M., Quilodran, R., Dominey, P. F., & Procyk, E. (2010). Meta-learning, cognitive control, and physiological interactions between medial and lateral prefrontal cortex. In R. Mars, J. Sallet, M. F. S. Rushworth & N. Yeung (Eds.), *Neural bases of motivational and cognitive control* (pp. 351–370). Cambridge, MA: MIT Press.

King, J. A., Korb, F. M., von Cramon, D. Y., & Ullsperger, M. (2010). Post-error behavioral adjustments are facilitated by activation and suppression of task-relevant and task-irrelevant information processing. *Journal of Neuroscience, 30,* 12759–12769.

Kirk, D. E. (2012). *Optimal control theory: An introduction.* Englewood Cliffs, NJ: Prentice-Hall.

Klein-Flügge, M. C., Kennerley, S. W., Saraiva, A. C., Penny, W. D., & Bestmann, S. (2015). Behavioral modeling of human choices reveals dissociable effects of physical effort and temporal delay on reward devaluation. *PLoS Computational Biology, 11,* e1004116.

Kool, W., & Botvinick, M. M. (2014). A labor/leisure trade-off in cognitive control. *Journal of Experimental Psychology: General, 143,* 131–141.

Kool, W., McGuire, J. T., Rosen, Z. B., & Botvinick, M. M. (2010). Decision making and the avoidance of cognitive demand. *Journal of Experimental Psychology: General, 139,* 665–682.

Kool, W., McGuire, J. T., Wang, G. J., & Botvinick, M. M. (2013). Neural and behavioral evidence for an intrinsic cost of self-control. *PLoS ONE, 8,* e72626.

Kouneiher, F., Charron, S., & Koechlin, E. (2009). Motivation and cognitive control in the human prefrontal cortex. *Nature Neuroscience, 12,* 939–945.

Kurzban, R., Duckworth, A. L., Kable, J. W., & Myers, J. (2013). An opportunity cost model of subjective effort and task performance. *Behavioral and Brain Sciences, 36,* 661–726.

Laming, D. (1968). *Information theory of choice-reaction times.* New York: Academic Press.

Litt, A., Plassmann, H., Shiv, B., & Rangel, A. (2011). Dissociating valuation and saliency signals during decision-making. *Cerebral Cortex, 21,* 95–102.

Magno, E., Foxe, J. J., Molholm, S., Robertson, I. H., & Garavan, H. (2006). The anterior cingulate and error avoidance. *Journal of Neuroscience, 26,* 4769–4773.

Mansouri, F. A., Tanaka, K., & Buckley, M. J. (2009). Conflict-induced behavioural adjustment: A clue to the executive functions of the prefrontal cortex. *Nature Reviews Neuroscience, 10,* 141–152.

Matsuzaka, Y., Akiyama, T., Tanji, J., & Mushiake, H. (2012). Neuronal activity in the primate dorsomedial prefrontal cortex contributes to strategic selection of response tactics. *Proceedings of the National Academy of Sciences of the United States of America, 109,* 4633–4638.

McClure, S. M., Laibson, D., Loewenstein, G., & Cohen, J. D. (2004). Separate neural systems value immediate and delayed monetary rewards. *Science, 306,* 503–507.

McGuire, J. T., & Botvinick, M. M. (2010). Prefrontal cortex, cognitive control, and the registration of decision costs. *Proceedings of the National Academy of Sciences of the United States of America, 107,* 7922–7926.

Miller, E. K., & Cohen, J. D. (2001). An integrative theory of prefrontal cortex function. *Annual Reviews in Neuroscience, 24,* 167–202.

Mischel, W., Shoda, Y., & Rodriguez, M. L. (1989). Delay of gratification in children. *Science, 244,* 933–938.

Momennejad, I., & Haynes, J.-D. (2013). Encoding of prospective tasks in the human prefrontal cortex under varying task loads. *Journal of Neuroscience, 33,* 17342–17349.

Morecraft, R. J., & Tanji, J. (2009). Cingulofrontal interactions and the cingulate motor areas. In B. A. Vogt (Ed.), *Cingulate neurobiology and disease* (pp. 113–144). Oxford: Oxford University Press.

Muraven, M., Shmueli, D., & Burkley, E. (2006). Conserving self-control strength. *Journal of Personality and Social Psychology, 91,* 524–537.

Muraven, M., & Slessareva, E. (2003). Mechanisms of self-control failure: Motivation and limited resources. *Personality and Social Psychology Bulletin, 29,* 894–906.

Narayanan, N. S., Cavanagh, J. F., Frank, M. J., & Laubach, M. (2013). Common medial frontal mechanisms of adaptive control in humans and rodents. *Nature Neuroscience, 16,* 1888–1895.

Narayanan, N. S., & Laubach, M. (2008). Neuronal correlates of post-error slowing in the rat dorsomedial prefrontal cortex. *Journal of Neurophysiology, 100,* 520–525.

Navon, D., & Gopher, D. (1979). On the economy of the human-processing system. *Psychological Review, 86,* 214–255.

Newman, L. A., Creer, D. J., & McGaughy, J. A. (2015). Cognitive control and the anterior cingulate cortex: How conflicting stimuli affect attentional control in the rat. *Journal of Physiology-Paris, 109,* 95–103.

Nicholson, W., & Snyder, C. M. (2008). *Microeconomic theory: Basic principles and extensions.* Boston, MA: Cengage Learning.

Niv, Y., Daw, N., & Dayan, P. (2006). How fast to work: Response vigor, motivation and tonic dopamine. *Advances in Neural Information Processing Systems, 18,* 1019.

O'Reilly, J. X., Schüffelgen, U., Cuell, S. F., Behrens, T. E. J., Mars, R. B., & Rushworth, M. F. S. (2013). Dissociable effects of surprise and model update in parietal and anterior cingulate cortex. *Proceedings of the National Academy of Sciences of the United States of America, 110,* E3660–3669.

O'Reilly, R. C. (2010). The What and How of prefrontal cortical organization. *Trends In Neurosciences, 33,* 355–361.

Oehrn, C. R., Hanslmayr, S., Fell, J., Deuker, L., Kremers, N. A., Do Lam, A. T.,...Axmacher, N. (2014). Neural communication patterns underlying conflict detection, resolution, and adaptation. *Journal of Neuroscience, 34,* 10438–10452.

Orr, J. M., & Weissman, D. H. (2011). Succumbing to bottom-up biases on task choice predicts increased switch costs in the voluntary task switching paradigm. *Frontiers in Psychology, 2,* 31.

Padmala, S., & Pessoa, L. (2011). Reward reduces conflict by enhancing attentional control and biasing visual cortical processing. *Journal of Cognitive Neuroscience, 23,* 3419–3432.

Parvizi, J., Rangarajan, V., Shirer, W. R., Desai, N., & Greicius, M. D. (2013). The will to Persevere induced by electrical stimulation of the human cingulate gyrus. *Neuron, 80,* 1359–1367.

Paus, T., Koski, L., Caramanos, Z., & Westbury, C. (1998). Regional differences in the effects of task difficulty and motor output on blood flow response in the human anterior cingulate cortex: A review of 107 PET activation studies. *Neuroreport, 9,* R37–R47.

Pearce, J., & Hall, H. (1980). A model for Pavlovian learning: Variations in the effectiveness of conditioned but not of unconditioned stimuli. *Psychological Review, 87,* 532–552.

Pessoa, L. (2015). Multiple influences of reward on perception and attention. *Visual Cognition, 23,* 272–290.

Prévost, C., Pessiglione, M., Météreau, E., Cléry-Melin, M., & Dreher, J. (2010). Separate valuation subsystems for delay and effort decision costs. *Journal of Neuroscience, 30,* 14080–14090.

Quilodran, R., Rothé, M., & Procyk, E. (2008). Behavioral shifts and action valuation in the anterior cingulate cortex. *Neuron, 57,* 314–325.

Rabbitt, P. (1966). Errors and error correction in choice-response tasks. *Journal of Experimental Psychology, 71,* 264–272.

Reinhart, R. M. G., & Woodman, G. F. (2014). Causal control of medial-frontal cortex governs electrophysiological and behavioral indices of performance monitoring and learning. *Journal of Neuroscience, 34,* 4214–4227.

Reinhart, R. M. G., & Woodman, G. F. (2015). Enhancing long-term memory with stimulation tunes visual attention in one trial. *Proceedings of the National Academy of Sciences of the United States of America, 112,* 625–630.

Rollnik, J. D., Schröder, C., Rodríguez-Fornells, A., Kurzbuch, A. R., Däuper, J., Möller, J., & Münte, T. F. (2004). Functional lesions and human action monitoring: Combining repetitive transcranial magnetic stimulation and event-related brain potentials. *Clinical Neurophysiology, 115,* 145–153.

Rothé, M., Quilodran, R., Sallet, J., & Procyk, E. (2011). Coordination of high gamma activity in anterior cingulate and lateral prefrontal cortical areas during adaptation. *Journal of Neuroscience, 31,* 11110–11117.

Salamone, J. D., Correa, M., Farrar, A. M., & Mingote, S. M. (2007). Effort-related functions of nucleus accumbens dopamine and associated forebrain circuits. *Psychopharmacology, 191,* 461–482.

Schouppe, N., Ridderinkhof, K. R., Verguts, T., & Notebaert, W. (2014). Context-specific control and context selection in conflict tasks. *Acta Psychologica, 146,* 63–66.

Shackman, A. J., Salomons, T. V., Slagter, H. A., Fox, A. S., Winter, J. J., & Davidson, R. J. (2011). The integration of negative affect, pain and cognitive control in the cingulate cortex. *Nature Reviews Neuroscience, 12,* 154–167.

Shah, A., & Oppenheimer, D. (2008). Heuristics made easy: An effort-reduction framework. *Psychological Bulletin, 134*, 207.

Shen, C., Ardid, S., Kaping, D., Westendorff, S., Everling, S., & Womelsdorf, T. (2015). Anterior cingulate cortex cells identify process-specific errors of attentional control prior to transient prefrontal-cingulate inhibition. *Cerebral Cortex, 25*(8), 2213–2228.

Shenhav, A., & Botvinick, M. (2015). Uncovering a missing link in anterior cingulate research. *Neuron, 85*, 455–457.

Shenhav, A., Botvinick, M. M., & Cohen, J. D. (2013). The expected value of control: An integrative theory of anterior cingulate cortex function. *Neuron, 79*, 217–240.

Shenhav, A., & Buckner, R. L. (2014). Neural correlates of dueling affective reactions to win-win choices. *Proceedings of the National Academy of Sciences of the United States of America, 111*(30), 10978–10983.

Sheth, S. A., Mian, M. K., Patel, S. R., Asaad, W. F., Williams, Z. M., Dougherty, D. D.,...Eskandar, E. N. (2012). Human dorsal anterior cingulate cortex neurons mediate ongoing behavioural adaptation. *Nature, 488*, 218–221.

Silvetti, M., Alexander, W., Verguts, T., & Brown, J. (2014). From conflict management to reward-based decision making: Actors and critics in primate medial frontal cortex. *Neuroscience and Biobehavioral Reviews, 46*(Pt. 1), 44–57.

Simen, P., Contreras, D., Cara, B., Hu, P., Holmes, P., & Cohen, J. D. (2009). Reward rate optimization in two-alternative decision making: Empirical tests of theoretical predictions. *Journal of Experimental Psychology: Human Perception and Performance, 35*, 1865–1897.

Simon, D. A., & Daw, N. D. (2011). Neural correlates of forward planning in a spatial decision task in humans. *Journal of Neuroscience, 31*, 5526–5539.

Simon, H. A. (1956). A comparison of game theory and learning theory. *Psychometrika, 21*, 267–272.

Smith, V. L., & Walker, J. M. (1993). Monetary rewards and decision cost in experimental economics. *Economic Inquiry, 31*, 245–261.

Sohn, M.-H., Albert, M. V., Jung, K., Carter, C. S., & Anderson, J. R. (2007). Anticipation of conflict monitoring in the anterior cingulate cortex and the prefrontal cortex. *Proceedings of the National Academy of Sciences of the United States of America, 104*, 10330–10334.

Solomon, R. L. (1948). The influence of work on behavior. *Psychological Bulletin, 45*, 1–40.

Stephens, D. W., & Krebs, J. R. (1986). *Foraging theory*. Princeton, NJ: Princeton University Press.

Stuss, D. T. (2011). Functions of the frontal lobes: Relation to executive functions. *Journal of the International Neuropsychological Society, 17*, 759–765.

Stuss, D. T., & Alexander, M. P. (2007). Is there a dysexecutive syndrome? *Philosophical Transactions of the Royal Society of London. Series B, Biological Sciences, 362*, 901–915.

Thompson, M. E. (1944). Learning as a function of the absolute and relative amounts of work. *Journal of Experimental Psychology, 34*, 506–515.

Ullsperger, M., Danielmeier, C., & Jocham, G. (2014). Neurophysiology of performance monitoring and adaptive behavior. *Physiological Reviews, 94*, 35–79.

Vassena, E., Silvetti, M., Boehler, C. N., Achten, E., Fias, W., & Verguts, T. (2014). Overlapping neural systems represent cognitive effort and reward anticipation. *PLoS ONE, 9*, e91008.

Venkatraman, V., & Huettel, S. A. (2012). Strategic control in decision-making under uncertainty. *European Journal of Neuroscience, 35*, 1075–1082.

Wallis, J. D., & Kennerley, S. W. (2011). Contrasting reward signals in the orbitofrontal cortex and anterior cingulate cortex. *Annals of the New York Academy of Sciences, 1239*, 33–42.

Walton, M. E., Kennerley, S. W., Bannerman, D. M., Phillips, P. E. M., & Rushworth, M. F. S. (2006). Weighing up the benefits of work: Behavioral and neural analyses of effort-related decision making. *Neural Networks, 19*, 1302–1314.

Wessel, J. R., Danielmeier, C., Morton, J. B., & Ullsperger, M. (2012). Surprise and error: Common neuronal architecture for the processing of errors and novelty. *Journal of Neuroscience, 32*, 7528–7537.

Westbrook, A., Kester, D., & Braver, T. S. (2013). What is the subjective cost of cognitive effort? Load, trait, and aging effects revealed by economic preference. *PLoS ONE, 22*, e68210.

Wisniewski, D., Reverberi, C., Tusche, A., & Haynes, J.-D. (2015). The neural representation of voluntary task-set selection in dynamic environments. *Cerebral Cortex*, *25*(12), 4715–4726.
Wolpert, D. M., & Landy, M. S. (2012). Motor control is decision making. *Current Opinion in Neurobiology*, *22*, 996–1003.
Womelsdorf, T., Johnston, K., Vinck, M., & Everling, S. (2010). Theta-activity in anterior cingulate cortex predicts task rules and their adjustments following errors. *Proceedings of the National Academy of Sciences of the United States of America*, *107*, 5248–5253.
Yeung, N., Botvinick, M., & Cohen, J. (2004). The neural basis of error detection: Conflict monitoring and the error-related negativity. *Psychological Review*, *111*, 931–959.

11

Control of Behaviour by Competing Learning Systems

Sanne de Wit

Introduction

Are you completely satisfied with your day-to-day behaviours? Or can you think of an undesirable behaviour that you would like to change? Probably most readers can think of at least one example. In 2014, for instance, 422 first-year psychology students at the University of Amsterdam took part in an anonymous survey, in which they were asked to list their three worst 'bad habits' (unpublished data of the author). Commonly reported behaviours concerned an unhealthy diet, lack of exercise, studying routines and procrastination, and (anti-) social behaviour. However, what stood out was that one student reported that he had no bad habits whatsoever. His complete satisfaction with his behaviour illustrates the exception that confirms the more general rule, namely, that most of us struggle sometimes to control our behaviour. Habit learning may play a role in this 'intention-behaviour gap'. In the words of Triandis: '*as behaviour repeatedly takes place, habit increases and this becomes a better predictor of behavior than behavioral intentions*' (1977). Therefore, the present chapter will review experimental research into habits. Most of this experimental work was originally conducted in animals, but more recently the field has witnessed an explosion of studies in human behavioural neuroscience. A Pubmed search reveals that the number of published papers with 'habit' in their title has rapidly increased even during the last 10 years (with 639 in the period 2005–2014 compared to 272 in 1995–2004). These studies offer growing evidence that we need to consider at least two learning processes or systems: one goal directed and the other habitual (Balleine & O'Doherty, 2010; Dickinson, 1985). Therefore, a dual-system theory of action control will be discussed, according to which these learning systems sometimes compete—but also cooperate—to control action. Next, evidence will be presented for a role of habits in persistent maladaptive behaviour in clinical conditions that involve impulsive-compulsive tendencies, for example, obsessive-compulsive disorder, drug abuse, and obesity. Finally, several challenges for future research will be identified.

Defining Goal-Directed Action and Habits

Before we turn to the scientific investigation of action control, we first need to clearly specify what we mean by 'habit' and 'goal-directed action'. First, in the behavioural neuroscience literature, the term 'goal-directed' has been used very broadly to refer to behaviours that involve some form of planning, inhibition, or other higher-order cognitive function, or in a reverse inference, prefrontal

---

The Wiley Handbook of Cognitive Control, First Edition. Edited by Tobias Egner.

cortex involvement. However, in the present chapter, we will employ a more narrow definition that has been derived from folk psychology. According to this definition, goal-directed actions are instrumental behaviours that are only performed when one has a certain goal and believes that this behaviour will increase the likelihood of reaching that goal (or conversely avoid a certain dreaded event). First of all, what is meant by 'instrumental behaviour'? Instrumental behaviours are learnt as a consequence of a causal relationship between the action or response (R) and its outcome (O), as opposed to being controlled purely by predictive Pavlovian relationships between environmental stimuli (S) and the outcome (Pavlov, 1927). Once the instrumental status of behavior has been established, we still need to assess whether it meets the belief and desire criteria of goal-directed action (Heyes & Dickinson, 1990). The Belief criterion stipulates that goal-directed actions are mediated by knowledge of the causal action–outcome relationship. According to the desire criterion, goal-directed actions are only executed when the outcome is currently desirable. In other words, the outcome needs to constitute a goal. For example, buying popcorn is a goal-directed action when one currently desires the taste of popcorn and believes that the act of buying it is a necessary step to achieve this. In associative learning terms, this means that it is a prerequisite for goal-directed action to encode the causal R→O relationship in an R-O association.

When a goal-directed action is repeatedly performed, it can gradually turn into a habit, which is triggered by contextual stimuli even when the consequences are no longer valuable. As a consequence, one may absent-mindedly buy popcorn at the cinema even when already fully sated or when craving an alternative snack more. The law of effect was postulated by Thorndike over 100 years ago, to account for this incremental development of habits (1911). He proposed that the experience of reward following an instrumental response leads to the strengthening of a mental association between contextual stimuli (S) and the response (R), so that on future occasions, the context will directly activate the response through the S-R association. Similarly, the omission of a dreaded aversive event (and accompanying positive experience of relief) may also act to strengthen the S-R link. In contrast, the occurrence of an aversive outcome (or cancellation of an anticipated reward) should weaken the S-R association. The defining difference between habits and goal-directed actions is that only in the latter case is performance mediated by knowledge of the R→O relationship and an evaluation of the anticipated outcome in light of one's current motivation. Because habits are mediated by S-R links, they are 'behaviorally autonomous' of the current desirability of the outcome.

In spite of their relative inflexibility, habits are considered to have adaptive value. As Maudsley formulated it succinctly: '*If an act became no easier after being done several times (…), it is evident that the whole activity of a lifetime might be confined to one or two deeds*' (1867). The advantage of habits is thought to be that they can be executed fast and in an efficient manner. When you consider that almost half of our daily routines are estimated to be performed in the same manner and in the same context (Wood, Quinn, & Kashy, 2002), it makes adaptive sense that we have evolved the ability to act faster and more efficiently as a certain behaviour repeatedly leads to reward in a stable environment. By freeing up cognitive resources, habit formation (or skill learning) thus allows us to attend to other important matters. However, the efficiency of overtrained skills (Jueptner & Weiller, 1998; Lehéricy et al., 2005) has typically been studied separately from the research that the present chapter will focus on, namely, the investigation of behavioural autonomy.

Investigating the Goal-Directed Versus Habitual Status of Behaviour

The force of habit has been studied in many diary investigations of real-life behaviours. A meta-analysis supports the notion that frequent repetition leads to a shift from goal-directed control towards habit, by showing that behavioural intentions are a weaker predictor of actions that are performed frequently (e.g., seat belt use) than those that are performed rarely (e.g., course enrolment; Quellette & Wood, 1998). A related meta-analysis focused on self-reported habit strength in terms of the experience of automaticity and frequency of past performance (as assessed with the Self-Report Habit Index; Bas Verplanken & Orbell, 2003). In line with the idea that habits play a role in the 'intention-behaviour gap', the authors found that this measure moderated the strength of the relation between behavioural intentions and actual behaviour (Gardner, de Bruijn, & Lally, 2011).

In line with the law of effect, social psychologists have also found evidence that habits are contextually dependent. For instance, Verplanken and colleagues found that people with environmental concerns were more likely to act on their good intentions by restricting car use if they had just recently moved (Verplanken, Walker, Davis, & Jurasek, 2008). This finding was interpreted in the framework of the 'habit discontinuity hypothesis', according to which old S-R habits can be disrupted by a change in context, thereby providing a window of opportunity for adapting behaviour in light of one's current goals.

Although diary studies generally underscore the impact that behavioural repetition has on behavioural autonomy, it is important to note that there are certain interpretative issues inherently associated with this approach to the study of habit. One issue is that goal pursuit and behavioural repetition are heavily intertwined. For example, imagine that person A has a longer history of unhealthy snacking than person B, and despite an equally strong intention to curb this 'bad habit' continues to succumb to the temptation. This could be due to a causal interpretation of the relationship between behavioural repetition and a loss of goal-directed control, but there are many other alternative explanations. For example, person A may simply have a stronger preference and craving for unhealthy snacks than person B. This motivational difference could explain both that person A has frequently enjoyed snacking in the past, and that (s)he finds it hard to suppress or change this behaviour. Furthermore, there may be differences in the daily environments of these two individuals, such that person A is relatively frequently confronted with opportunities to snack, or may even feel pressured to snack in certain social surroundings. Again, these factors could independently affect the frequency of repetition and the difficulty of changing the behaviour. A final issue regarding diary investigations concerns the reliance on self-report. The subjective nature of this report is compounded by the fact that people may not always be sufficiently aware of the habitual behaviour to accurately report it. Therefore, to further our understanding of the underlying mechanisms, we need to turn to strictly experimentally controlled investigations of habit. This section will focus on the canonical experimental assay of goal-directed versus habitual action control, namely, the outcome-revaluation test.

Outcome-Revaluation Studies in Animals

Goal-Directed Action in Animals The most frequently used behavioural model of instrumental behaviour is that of lever pressing for food by rats in a Skinner box. For example, Adams and Dickinson (1981) trained hungry rats to lever press for one type of food outcome, either sugar or mixed diet food pellets, while delivering the same number of the other type of pellet noncontingently, and hence independently of lever pressing. They found that their rats would readily learn to press the lever, and at the surface this behaviour looked goal directed. However, the rats' ability to rapidly learn to respond for food when hungry could equally well be supported by a simple S-R reinforcement mechanism, as envisaged in the law of effect. Therefore, Dickinson and colleagues developed an elegant way to empirically test whether instrumental behaviour meets the belief and desire criteria of goal-directed action, namely, through an outcome-revaluation procedure. Following the initial learning phase, they removed the rats from the Skinner boxes, and conditioned an aversion to one of the two food pellets by pairing it with lithium chloride (LiCl)-induced nausea. For one group, they devalued the contingent food pellets, and for the other the noncontingent pellets. Subsequently, they returned the rats to the Skinner boxes, where they were once again given the opportunity to press the lever. During this test, they observed that the rats pressed less if the contingent rather than noncontingent outcome had been devalued. Importantly, this test was conducted in extinction, which means that the outcomes were no longer presented. Because performance was assessed in the absence of outcome presentations, the devaluation effect must have been mediated by knowledge of the R→O contingency as well as evaluation of the anticipated pellet outcome. If the devalued outcomes had instead still been presented contingent upon responding during the test phase, then any decline in responding could have been due to a gradual weakening or inhibition of direct S-R associations between the context and the instrumental response through a punishment mechanism envisaged by Thorndike in his law of effect (1911). Since this first demonstration, it has been firmly established that rats are indeed capable of goal-directed action (Balleine & O'Doherty, 2010).

Habit in Animals However, in a subsequent study, Adams (1982) found that extensive 'overtraining' diminished the effect of devaluation on performance. Whereas after moderate training (100 lever presses), devaluation of a pellet outcome through LiCl pairing decreased responding, rats that were extensively trained (500 lever presses) continued to press for the devalued food. Importantly, this insensitivity did not reflect ineffectiveness of the aversion procedure. When animals were once again offered the pellets in a final reacquisition test, the overtrained animals also reduced their response for the devalued pellets. Therefore, this experiment demonstrated unequivocally that behavioural repetition renders behaviour habitual in the sense that it is no longer dependent on the current desirability of the outcome. The development of behavioural autonomy has since then been replicated in several studies (Balleine & O'Doherty, 2010). Furthermore, demonstrations that behavioural autonomy as a consequence of behavioural repetition is constrained to the learning context directly support the law of effect (Killcross & Coutureau, 2003; Thrailkill & Bouton, 2015).

Determinants of Habit Using outcome-revaluation tests, several determinants of relative goal-directed/habitual control have been identified: training schedule, choice, and Pavlovian influences. First of all, interval schedules—on which rewards are made available only after a certain amount of time has elapsed—lead to relative insensitivity to outcome devaluation even after moderate training when compared to ratio schedules (Dickinson, Nicholas, & Adams, 1983). Another important factor to consider is choice. In contrast to the studies discussed above, some of the early outcome-devaluation studies failed to find evidence for habit formation after overtraining (e.g., Colwill & Rescorla, 1985). Interestingly, the rats in the failure-to-replicate studies received concurrent training of two rewarded responses, instead of just a single response. More recently, Kosaki and Dickinson (2010) showed that it is choice *during training* (as opposed to during test) that leads to enduring goal-directed action control even after extensive training. Therefore, behavioural repetition does not always lead to inflexible habits. Clearly, the circumstances during acquisition play a pivotal role. Finally, one caveat of many outcome-revaluation studies is that it is challenging to rule out that the effect of outcome revaluation on instrumental performance is to some extent driven by Pavlovian influences. Adams and Dickinson's design (1981) controlled for this by equating the number of context–outcome pairings of the instrumental outcome with that between the context and a noncontingent (and therefore purely Pavlovian) reward. This procedure ensured that the context was always associated with one devalued and one still-valuable outcome during the test phase. The devaluation effect after limited training had to be driven, therefore, by R-O associations. Similarly, Colwill and Rescorla achieved this by employing a concurrent training design (1985). However, studies that have demonstrated behavioural autonomy due to overtraining have failed to incorporate this important control (in the form of a noncontingent or concurrent reward) in their designs. In fact, these studies all involved training on a single lever for a single outcome, followed by devaluation and an extinction test. More recently, Jonkman, Kosaki, Everitt, and Dickinson (2010) showed that with such a design, extinction of the context–outcome relation prior to the test phase reduced the outcome-devaluation effect. It remains possible then that Pavlovian influences play an important role in the reduced sensitivity to outcome devaluation after overtraining.

Outcome-Revaluation Studies in Humans

Goal-Directed Action in Humans It is generally assumed that humans are capable of goal-directed action. Perhaps that explains in part why experimental investigation of this issue in human research has lagged behind the field of animal learning. However, more recently, the experimental outcome-revaluation paradigm has been translated to the field of human psychology. In order to mimic the full control over learning history in the Skinner box experiments, these studies have used computerised tasks to train people to perform actions within a carefully controlled setup. In one of the earliest studies (Valentin, Dickinson, & O'Doherty, 2007), participants were asked to choose between two icons on each trial. One icon was always paired with a low probability of orange juice, and the other either with a high probability of chocolate milk or tomato juice. Subsequent satiation on one of the high-probability liquids reduced responding selectively for that outcome in a short extinction test. For example, drinking chocolate milk to satiety reduced responding for chocolate milk relative to tomato juice. The outcome-devaluation effect through satiation on food rewards has subsequently been replicated in several studies (e.g., Watson, Wiers, Hommel, & de Wit, 2014).

An alternative approach is that of 'instructed devaluation', by which participants are explicitly informed that a certain outcome is devalued because it no longer earns the participant points and/or financial credits (de Wit, Niry, Wariyar, Aitken, & Dickinson, 2007). A possible concern with this approach is that symbolic outcomes (and instructed devaluation) fail to engage the basic learning mechanisms that we have evolved to optimise the pursuit of biologically relevant events like food. This concern is to some extent mitigated by neuroscientific investigations that have shown that the same neural substrates underlie responses to natural and symbolic reinforcers (see the section titled 'Neurobiology of Action and Habit' later in the chapter).

The instructed devaluation approach is taken in the 'Fabulous Fruit Game', in which participants learn to press keys in order to gain fruit icons—which are worth points—on a computer screen. In other to establish multiple R-O associations, this task makes use of an instrumental discrimination design, in which discriminative stimuli signal which R→O relation is currently effective. This is achieved by having fruit icons not only serve as outcomes, but also as discriminative stimuli. For example, an icon of an orange signals that right key-presses are rewarded with an icon of cherries and points, whereas an icon of an apple signals that left key-presses are rewarded with an icon of a pear and points. Subsequently, some of these fruit icons are devalued—for example, the cherries—by instructing participants that these icons are no longer worth points, whereas others remain valuable. During the ensuing choice test, subjects are once again offered a choice to press for different fruit icons—but this time in the absence of the discriminative stimuli. For example, when shown the icons of cherries and pear together on the screen, goal-directed control would be evident in preferential pressing on the left key. The test is conducted in nominal extinction, which means that participants no longer receive feedback, but are informed that they are still earning points, and that their total score will be revealed at the end of the experiment. Unsurprisingly, healthy participants are very well able to perform this task in a goal-directed manner (de Wit et al., 2007).

However, this simple choice test does not allow S-R associations a role in response selection, as the original discriminative stimuli from the learning phase were omitted from the test phase. To allow S-R associations to compete with R-O goal-directed control, we developed the *slips-of-action test*. This test was inspired by the 'slips of action' (sometimes also referred to as habit intrusions) that reveal habit learning in everyday life. For example, imagine cycling into town to do some shopping during the weekend. If distracted, you may just take that right turn that takes you to the office instead. This may have happened because the contextual stimuli at the crossroads triggered this habitual response, even though the ultimate outcome is not in line with your current intentions. To model this, following the initial discrimination training described above, participants are instructed that most of the fruit outcomes are still valuable, but that some have been devalued (usually two out of six available outcomes). Continued pressing for those devalued outcomes will result in subtraction of points from the participants' total score (and in some experiments financial credits). Subsequently, during the (nominal) extinction test phase, participants are shown the discriminative stimuli in rapid succession. If the discriminative stimulus signals a still-valuable outcome, they should respond on the corresponding key. On the other hand, if the signalled outcome is devalued, they should withhold the learnt response. In order to perform well during this test, participants need to have knowledge of the outcomes of their actions (belief criterion) and base their decision to respond or not (go or no-go) on the current outcome value (desire criterion). Furthermore, this time, S-R associations may interfere with goal-directed performance, as the discriminative stimuli should trigger learnt responses independently of sudden changes in outcome value. However, overall, healthy participants are able to perform well on this test (de Wit et al., 2012). As will be discussed later in this chapter though, this test has proved to be a useful paradigm for investigating individual differences in action control.

Habit in Humans Overall, the studies discussed so far indicate that, unsurprisingly, humans are capable of goal-directed action. Perhaps the more interesting question is whether—like rats—their behaviour loses its immediate sensitivity to outcome value after extensive repetition. An elegant field experiment by Neal, Wood, Wu, and Kurlander (2011) provided supporting evidence. In this study, participants were offered either fresh or stale popcorn upon entering a cinema to watch a movie. Whereas participants with a weak self-reported habit to eat popcorn in the cinema ate less of the stale popcorn relative to fresh, no significant difference was observed in the group of participants with a strong self-reported habit. This finding could simply be explained by a strong subjective reward value of popcorn in the frequent popcorn eaters. However, when the experimenters tested their participants in a novel context (a meeting room) instead of the cinema, even the group with strong habits consumed much less stale than fresh popcorn. This finding supports the idea that habits become linked to contextual stimuli, such that behaviour is triggered independently of current motivation.

Even stronger evidence for behavioural autonomy as a consequence of extensive repetition comes from a fully experimentally controlled study in which (healthy) participants performed a computerised task in order to gain snack rewards (Tricomi, Balleine, & O'Doherty, 2009). In this study, different fractal stimuli on the computer screen signalled that key pressing would be rewarded with two different food rewards (M&M's and Fritos). One group of participants received moderate training during a single session, while another group was trained extensively during six sessions distributed over three consecutive days. Subsequently, outcome devaluation was achieved by asking them to eat one of the two food rewards until it was no longer pleasant to them. In the final critical test, participants were once again offered the opportunity to press the keys during exposure to the training stimuli. Whereas the moderately trained group was sensitive to the current outcome value, the extensively trained group seemed impervious to the current desirability of its outcome. For example, participants sated on M&M's still tended to persist in responding for this food reward.

Neurobiology of Action and Habit

The development of well-controlled experimental paradigms to assess the goal-directed/habitual status of instrumental action has made it possible to study their neural basis. In animal research, this has been accomplished by studying the effect of lesions and pharmacological manipulations on the outcome sensitivity of instrumental behaviour, thereby allowing for causal inferences concerning the involvement of different brain regions in habit and goal-directed action. In contrast, by necessity, the human line of research is less invasive and therefore mostly correlational. By relating behavioural performance measures to neuroimaging—mostly functional magnetic resonance imaging (fMRI)—cognitive neuroscientists have been able to identify the corresponding neural correlates of action control in humans.

For clarity's sake, this section will focus strictly on outcome-revaluation studies. The aim is to clarify the contributions of neuroscientific research to our understanding of dissociable corticostriatal pathways that are responsible for goal-directed action versus habit (for more complete reviews of the neural basis of instrumental action and neurotransmitter involvement, see, for example, Balleine & O'Doherty, 2010; Hart, Leung, & Balleine, 2014).

Animal Lesioning Research

Animal studies have implicated several regions in goal-directed control, which means that lesioning or inactivating these regions reduces sensitivity to outcome devaluation. First, pre-training lesions of the prelimbic cortex lead to behavioural autonomy—even after only limited training (Balleine & Dickinson, 1998; Corbit & Balleine, 2003; Killcross & Coutureau, 2003). Therefore, the prelimbic cortex appears to be crucially involved in the R-O learning process that underlies goal-directed action. On the other hand, both pre- and post-training lesions of the dorsomedial striatum disrupt goal-directed performance, implicating this subcortical area in both the acquisition and expression of goal-directed action (Yin, Ostlund, Knowlton, & Balleine, 2005). It has been suggested, therefore, that this region is responsible for long-term storage of R-O associations. Importantly, these cortical and striatal areas are in fact heavily anatomically interconnected, suggesting that they are part of a corticostriatal pathway for goal-directed action (Balleine & O'Doherty, 2010). Another component of this goal-directed pathway is the mediodorsal thalamus (Corbit, Muir, & Balleine, 2003).

In contrast, lesions of the dorsolateral striatum disrupt the formation of habits. Yin, Knowlton, and Balleine (2004) were the first to show that animals with lesions to the dorsolateral striatum remained sensitive to devaluation even after extensive training. An even more impressive finding was that habitual responding following extensive training became goal-directed again when the dorsolateral striatum was temporarily inactivated (through direct local infusion of a GABA agonist; Yin, Knowlton, & Balleine, 2006). Therefore, it appears that R-O associations remain to some extent intact following extensive training, but may well be by-passed by the more direct S-R habits. In other words, the ability to act in line with current goals can be compromised by concurrently formed, competing habits. Another important area for habitual control appears to be the infralimbic cortex (Coutureau & Killcross, 2003; Killcross & Coutureau, 2003). Therefore, infralimbic cortex and dorsolateral striatum may be part of a corticostriatal pathway for habitual action control.

Human Neuroimaging Research

In line with animal lesioning research, human neuroimaging research has provided evidence for dissociable corticostriatal pathways that underlie goal-directed versus habitual action control. As described previously, Valentin and colleagues conducted an outcome-devaluation paradigm with liquid food rewards. When they scanned participants during the test phase using fMRI, they found evidence that preferential responding for the still-valuable outcome was associated with activity in the ventromedial prefrontal cortex/medial orbitofrontal cortex, suggesting that this may be the functional homologue of the rodent prelimbic cortex. Accordingly, another fMRI study using instructed outcome devaluation (in the Fabulous Fruit Game), revealed that goal-directed performance during the outcome-devaluation test was predicted by engagement of the ventromedial prefrontal cortex during the preceding learning phase (de Wit, Corlett, Aitken, Dickinson, & Fletcher, 2009). On the other hand, the overtraining (fMRI) study by Tricomi and colleagues (2009) suggests that the posterior putamen is the functional homologue of the rodent dorsolateral striatum, as activity in this area tracked the development of habits.

The question arises whether in the human brain—as in animals—specific lesions to the different corticostriatal pathways would render behaviour either generally habitual (even after only minimal training) or generally goal directed (even after extensive training). Even minor individual differences in the integrity of these pathways could theoretically bias one to be more 'a creature of habit' or generally more goal directed. This is by no means a novel idea. For example, Triandis (1977) already speculated that some people may be more likely to habitually repeat past actions, whereas others may be more likely to carefully consider the consequences of each future action. To investigate this idea, we related behavioural sensitivity to instructed devaluation to inter-individual variability in white-matter connectivity within corticostriatal pathways. Using the technique of diffusion tensor imaging (and probabilistic seeded tractography analyses), we provided evidence that individuals with strong estimated white-matter connectivity between the ventromedial prefrontal cortex and caudate tended to act better—and therefore more goal directed—in the slips-of-action test. On the other hand, strong estimated premotor cortex–posterior putamen connectivity was negatively related with performance, as reflected in persistent stimulus-cued responding for devalued outcomes (de Wit, Watson et al., 2012). Finally, we used the technique of voxel-based morphometry to show that high grey-matter volume of the posterior putamen was also a predictor of habitual control. Therefore, these findings replicate and extend the findings from previous fMRI studies. Taken together, neuroscientific investigations have, therefore, provided evidence for a neural dual-system architecture.

Dual Systems

It is important to note that over the years, there have been many different dual-system/process theories of decision making and learning that vary in their scope and assumptions (for an overview, see Evans & Stanovich, 2013). To illustrate this with a well-known example, consider Stanovich's Systems 1 and 2 (1999). System 1 is described as intuitive, subconscious/implicit, and associative, with the advantage of being fast and requiring relatively little cognitive effort. System 2, on the other hand, is regarded as deliberative, conscious/explicit, and rule based, but slow and having limited capacity. At first glance, it may appear that habits are best placed within System 1, whereas goal-directed action is better captured by the description of System 2. However, several aspects of these systems do not align well with the research described so far in this chapter. For instance, System 1 has been described as similar to animal cognition, whereas System 2 is described as being reserved uniquely for humans. As we have seen in the previous sections, however, the fundamental abilities of goal-directed action and habit learning are shared across laboratory rats and humans. Another problem is that some of the terms used to describe the systems are challenging to clearly operationalise for rigorous experimental investigation. Finally, it is not immediately clear that the fundamental ability of goal-directed action (as defined in this chapter) requires rule-based processes, or whether it can instead be captured by a relatively simple R-O associative learning mechanism. Therefore, this dual-system framework may not be optimally suited to capture the distinction between goal-directed versus habitual control as defined in this chapter.

Another well-known variant of dual-system theory is the hot/cool system analysis of delay of gratification of Metcalfe and Mischel (1999). To account for individual differences in the ability to forego an immediate temptation in order to obtain a greater longer-term reward, Metcalfe and Mischel proposed that a flexible, cognitive cool system competes with a stimulus-controlled emotional hot system for action control. Again, this framework would not be suitable to apply to the dissociation between the fundamental capacities of goal-directed action and habit, as goal-directed actions are driven by desire for the consequences in an even more direct way than habits. Also, both goal-directed actions and habits may to some extent be triggered by stimuli in the environment, as will be discussed below.

The Associative-Cybernetic (AC) Model

The AC model is an integrated associative dual-system account of instrumental action control that was derived from Thorndike's 'representative' or 'ideational' theory of human instrumental behaviour (1931), and further developed by Dickinson and Balleine (1994; see also, de Wit & Dickinson, 2009). According to the AC model, the degree to which behaviour is habitual is determined by the relative strengths of associations in S-R and R-O systems (de Wit & Dickinson, 2009). Both of these associative systems can generate behavioural output by activating the motor system. However, whereas the S-R system does this by activating specific motor units, the R-O system can only generally enhance motor activity to the extent that the anticipated outcome is positively evaluated by the incentive system. As a consequence, the S-R and R-O systems do not simply compete for action control, but they also have to work together—at least after moderate training. After only moderate repetition, the S-R associations should still be relatively weak, and therefore the habit system on its own should be unable to activate specific motor units above a certain threshold that is required for motor output. An additional general 'boost' of the motor system is required through the R-O system. Importantly, this boost is only produced when the outcome is indeed evaluated as positive by the incentive system. A general positive feedback signal will thus be generated, and if it coincides with the specific response activation by the S-R system, then the two signals will summate to activate individual motor units sufficiently to generate a motor response. Therefore, after limited repetition, behaviour is goal directed thanks to cooperation between the S-R and R-O systems. This feature of the AC model also solves the problem of response sampling. If an agent would need to consider all possible actions and their respective outcomes at any given time, this would impose an impossible cognitive burden. However, the integration of the S-R habit mechanism with R-O driven action solves this problem, by letting external stimuli preferentially prime responses that have been reinforced as a consequence of past rewarding experiences. Therefore, this model assumes that, after limited repetition, goal-directed actions are selected through an indirect S-R-O associative chain.

To provide an illustration of this brief description of the AC model (see de Wit & Dickinson for a more elaborate explanation), consider the behaviour of drinking coffee during work breaks. For the 'caffeine addicts' among us, this may be driven by deeply ingrained habits, but let us go back in time together to the moment that it all started: that first cup of coffee during a work break. That first rewarding experience ensured that the incentive system fired and reinforced an association between your work surroundings and the behaviour of drinking coffee. After just a few repetitions, that S-R association was relatively weak, but strong enough to bring to mind this behaviour and make you consider its outcome via the R-O association, and evaluate it in light of your current needs and desires through its links with the incentive system. In the moments that you craved the coffee flavour or needed to stay alert, the incentive system would have led to that general boost to the motor system to produce a goal-directed action. In contrast, after having just drunk three cups of coffee, temporary satiation may have reduced that positive feedback signal, making you refrain from drinking coffee.

However, as the behaviour of drinking coffee at work was performed repeatedly, this led to the gradual reinforcement of stimulus-response (S-R) associations in the habit system. At some point, those associations may have become sufficiently strong to activate the motor units by themselves—without the need for R-O involvement. And in fact, because the S-R pathway is more direct and therefore faster, it may have started to completely bypass the indirect, goal-directed route. And there you have it: The caffeine addict finds himself/herself reaching for the fourth cup of coffee that day, perhaps even despite explicit intentions to curb this unhealthy habit. Therefore, the development of behavioural autonomy of the current incentive value of the goal can be readily accounted for by the AC model.

Related Associative and Computational Accounts

Correlational Account This account is an extension of the AC model. In short, Dickinson proposed that the strength of R-O associations is determined by the perceived correlation between the rate of performance and the rate of reinforcement of individual responses (1985). According to this view, extensive training indirectly leads to weaker R-O associations as a consequence of less variability in response rates and therefore a reduced experience of the R→O contingency. This account can also explain why habit formation is only observed when rats are trained on a single response. Concurrent training schedules should lead to continued experience of the individual R-O correlations. Another observation that can be explained in these terms is that interval schedules—on which there is a relatively low correlation between response rate and reinforcement rate—reduce sensitivity to outcome devaluation (Dickinson, Nicholas, & Adams, 1983).

O-R Accounts According to O-R theories, the thought of an outcome precedes the decision to perform the action that leads to it (O-R instead of R-O; e.g., Asratyan, 1974; Hommel, 2003). This O-R mechanism could, in fact, be readily integrated within the AC model (de Wit & Dickinson 2009). It is an important pathway to consider, as there is mounting evidence that (Pavlovian) stimuli that remind one of a specific reward will bias performance towards that reward. For example, an abstract picture that has been associated with popcorn through Pavlovian conditioning will subsequently enhance responding on a key that leads to popcorn relative to one that leads to chocolate (Watson et al., 2014). Importantly, however, this outcome-specific 'Pavlovian-instrumental transfer' effect is unaffected by outcome devaluation. For example, even when one has just been sated on popcorn, the popcorn signal will still trigger the associated response (Watson et al., 2014; for related animal research, see, e.g., Holland, 2004). Therefore, it appears that this S-O-R associative chain may mediate habit-like responding that is behaviorally autonomous of outcome value—even after only moderate training. The different ways in which Pavlovian stimuli can influence instrumental behaviour are definitely important to consider in future research into action control. Clearly, there are more than two pathways (or systems) to response selection.

Hierarchical Action Sequence Account According to this alternative account of habit, the repeated execution of action sequences leads to the formation of a single response unit or 'action chunk' (Graybiel, 2008). Although the agent may select the chunk in a goal-directed manner, the subsequent execution thereof follows automatically (Dezfouli, Lingawi, & Balleine, 2014). This theory may also offer an explanation for the protective effects of concurrent schedules on habit formation—as under those circumstances, the animals do not perform according to a fixed sequence.

Model-Free Versus Model-Based Decision Making According to recent computational models, goal-directed versus habitual action control can be captured by 'model-based' versus 'model-free' control mechanisms as assessed with the sequential two-step task (Daw, Niv, & Dayan, 2005; Daw, Gershman, Seymour, Dayan, & Dolan, 2011). The model-free mechanism selects whatever action was rewarded in the past (akin to S-R habits), whereas the model-based mechanism makes decisions on the basis of an internal model of the transition structure between states of the world. This model allows the agent to compute which responses are most likely to yield desirable outcomes. This computational modelling approach has been adopted by many recent studies, which have generated interesting results in both the field of neuroimaging and clinical psychology. The advantage of this approach may be that it can disentangle the relative contributions of model-based/goal-directed versus model-free/habitual control to behaviour. A disadvantage (at least for some) is that a full understanding of the results of these studies requires insight into computational modelling. Furthermore, it remains to be assessed directly to what extent performance on the relatively complex two-step task reflects the same mechanisms as the outcome-devaluation extinction test (but for preliminary supportive evidence, see Gillan, Otto, Phelps, & Daw, 2015). A further interesting question that can be addressed in this line of research is the computations that may underlie arbitration between the two systems (Lee, Shimojo, & O'Doherty, 2014).

Clinical Relevance

Dual-system theory provides a theoretical framework for understanding the role of habit in psychopathologies, especially those that involve compulsive behavior that persists despite one's awareness of far-reaching negative consequences. On the one hand, the gradual formation of S-R habits may contribute to treatment resistance of frequently performed maladaptive behaviours. On the other hand, a tendency towards aberrantly strong habit formation and/or weak goal-directed control—may render certain individuals particularly vulnerable to the development of maladaptive habits.

Addiction. Drug abuse was the earliest clinical condition to be characterised in terms of dominant S-R habitual control (Tiffany, 1990). According to Everitt and Robbins, recreational drug use starts out as voluntary, but the highly rewarding effect of drugs can lead to the formation of strong S-R associations, such that goal-directed drug seeking rapidly transits into stimulus-driven compulsive drug-seeking habits. Their research in animals revealed that the gradual shift towards behavioural autonomy of the drug-seeking response is reflected in the progression from ventral to dorsal striatum. Furthermore, weak prefrontal functioning as a vulnerability factor as well as a consequence of prolonged drug use has been proposed to underlie impaired goal-directed and inhibitory control (Everitt, Dickinson, & Robbins, 2001; for a recent review, see Everitt & Robbins, 2016). Relatively little research has directly addressed this idea in humans, but recently Sjoerds et al. (2013) showed that alcohol dependents were impaired at adjusting their performance when instrumental outcomes were devalued through instruction (in the Fabulous Fruit Game). Furthermore, fMRI analyses revealed that during learning they showed less activity in the ventromedial prefrontal cortex and more activity in the posterior putamen.

Obsessive-compulsive disorder. The first direct evidence for a general tendency towards habit propensity in OCD came from an instructed outcome-devaluation study (with the Fabulous Fruit Game). In this study, we found that patients with OCD were impaired at directing their responses towards still-valuable outcomes and away from devalued ones. In order to investigate habit learning in an aversive context, Gillan et al. (2014) next developed a shock avoidance paradigm in which mild shocks to the left and right wrist could be avoided by pressing food pedals on the corresponding side. At the end of the learning phase, the experimenter disconnected one of the electrodes, thereby removing the threat of shock. Following extensive (as opposed to minimal) training, OCD patients, but not healthy controls, continued to press to avoid the 'devalued' shock. In a subsequent fMRI study, it was found that impaired performance of OCD patients on this paradigm was related to hyperactivity in the ventromedial prefrontal-orbitofrontal cortex and caudate (Gillan, Apergis-Schoute, et al., 2015).

Obesity and eating disorders. Excessive consumption of energy-dense food and a lack of physical activity are considered to be important contributors to the recent rise in overweight and obesity. Force of habit may help explain why some people fail to break unhealthy habits despite full knowledge of the adverse consequences of overweight and obesity. Recently, Horstmann et al. (2015) conducted the first outcome-devaluation study with obese participants (without a history of eating disorders). They trained obese and lean men to press keys for two snack rewards and subsequently devalued one of these through specific satiety. The satiation manipulation reduced self-reported desire for the satiated snack, as well as response to this snack during a subsequent extinction test. However, this reduction in response was negatively related to BMI. These results are in line with the notion that obese individuals may show accelerated habit formation (at least in the context of food reinforcers). However, these results should be treated with caution, and certainly an important extension of this study would be to determine whether the observed inflexibility increases with more extensive training.

In an animal model of binge eating, during five consecutive weeks, rats received restricted access to sweetened condensed milk. This was followed by instrumental training with a different food reward. When that reward was subsequently devalued, the rats on the binge-like diet failed to adjust their performance. In contrast, control rats and a group of rats that had received continuous access to the milk did show an outcome-devaluation effect (Furlong, Jayaweera, Balleine, & Corbit, 2014). Another eating disorder that has been suggested to involve habitual compulsive behaviours is anorexia nervosa (Godier & Park, 2014; Steinglass & Walsh, 2006). This hypothesis remains, however, to be directly investigated.

Habit as a Transdiagnostic Compulsive Trait

As just reviewed, there is evidence to suggest that habit learning and habit propensity play a role in drug abuse, OCD, and eating disorders. These three conditions all involve an element of compulsivity, and it could be argued, therefore, that habit propensity is a transdiagnostic trait for compulsive conditions (Robbins, Gillan, Smith, de Wit, & Ersche, 2012). Relatedly, we recently also provided evidence for habit propensity in Tourette's syndrome, which was associated with white-matter connectivity within the habit pathway (Delorme et al., 2016). However, an overreliance on habit learning has also been observed in healthy ageing (de Wit, van de Vijver, & Ridderinkhof, 2014), Parkinson's disease (de Wit, Barker, Dickinson, & Cools, 2011), and has also been related to treatment response in social anxiety disorder

(Alvares, Balleine, & Guastella, 2014) and to schizophrenia (Morris, Quail, Griffiths, Green, & Balleine, 2015), whereas no evidence for habit propensity was found in children with autism spectrum disorder by Geurts and de Wit (2014). The question arises as to what extent habit propensity is characteristic of compulsive conditions specifically—or instead is a common factor across psychopathologies and in individuals with impaired prefrontal functioning. A related question that remains to be clarified is to what extent impaired sensitivity to outcome revaluation is related to a more general impairment in cognitive control functions—most notably working memory and response inhibition. Furthermore, stress may play an important mediating role in the observed deficits in patients. Chronic as well as acute stress levels during testing may often be higher in the patients than in healthy controls, and previous research suggests that stress may lead to dominant habitual control (Schwabe & Wolf, 2009).

Challenges for the Field

Habits play a role in almost any domain of human behaviour: eating, shopping, television watching, phone use, social interactions, public transport use, and perhaps also in skills like driving a car or playing a musical instrument. Its role in both adaptive and maladaptive decision making is pervasive. It is of crucial importance, therefore, to further improve our insight into the conditions that foster habit.

An Integrative Approach

To date, investigations of habit have been conducted separately but in parallel in the fields of social psychology, experimental psychology, and behavioural neuroscience (both in animals and humans). These different approaches have yielded convergent evidence for the force of habit, but have each also offered unique insights. In future research, an integrative approach across these disciplines will allow for a fuller understanding of the role of habit in decision making. To give a concrete example, animal research into the effects of training schedule and choice could be translated to study these factors as possible determinants of habit in humans. Another example concerns the issue of ecological validity of the highly controlled and constrained experiments in the laboratory. To determine whether the mechanisms identified there actually play a significant role in complex everyday (adaptive and maladaptive) behaviour, experimental research needs to be directly related to the diary investigations and structured observations that are being used in social psychology. Another challenge for future research is to integrate research into inflexibility of habits as assessed with outcome-revaluation procedures with that into the efficiency of extensively trained behaviour, in order to better understand how these two important aspects are interrelated.

Habit Interventions

Let us speculate about different ways in which our current thinking about the mechanisms underlying habits might inform attempts to change habits. First, as was stated earlier, habits may require less cognitive effort and attention than goal-directed action, and one should therefore try to turn a desired behaviour into a habit as fast as possible (Rothman, Sheeran, & Wood, 2009). This necessarily requires repetition in a stable context. One way to further accelerate the formation of context-dependent habits may be to formulate a clear 'if-then' plan, sometimes referred to as an 'implementation intention' (see Chapter 25 by Davisson & Hoyle in this volume). For example, 'if I have lunch, then I will eat an apple' is likely to accelerate the process of habit formation compared to a vague goal intention like 'I will eat more fruit'. Another advice would be to create a window of opportunity by removing contextual cues that are associated with an old, undesirable habit (Verplanken et al., 2008). Also, if possible, one should capitalise on existing habits, which can act as (R-R) triggers. For example, one could start a new flossing habit after the already firmly established habit of tooth brushing (instead of the other way around). A different lesson that we can learn from animal research is that habits are slow to form in a situation that offers a choice between multiple response options Judah et al., 2013). In everyday life, there are always multiple choice options available, but salient cues may mask some of these or focus your attention on a single alternative at the expense of others. Therefore, if one wishes to turn a new behaviour into a habit fast, it is important to associate the intended behaviour with a salient cue, for instance, via the formation of an implementation intention. In conclusion, it should be an important aim of future research to translate theoretical insights into habit to practical interventions to help people to break undesirable habits and establish 'good habits'.

References

Adams, C. D. (1982). Variations in the sensitivity of instrumental responding to reinforcer devaluation. *Quarterly Journal of Experimental Psychology*, *34*(2), 77–98.

Adams, C. D., & Dickinson, A. (1981). Instrumental responding following reinforcer devaluation. *Quarterly Journal of Experimental Psychology*, *33*, 109–122.

Alvares, G. A., Balleine, B. W., & Guastella, A. J. (2014). Impairments in goal-directed actions predict treatment response to cognitive-behavioral therapy in social anxiety disorder. *PLoS ONE*, *9*(4), e94778.

Asratyan, E. A. (1974). Conditional reflex theory and motivated behavior. *Acta Neurobiologiae Experimentalis*, *34*, 15–31.

Balleine, B. W., & Dickinson, A. (1998). Goal-directed instrumental action: Contingency and incentive learning and their cortical substrates. *Neuropharmacology*, *37*(4–5), 407–419.

Balleine, B. W., & O'Doherty, J. P. (2010). Human and rodent homologies in action control: Corticostriatal determinants of goal-directed and habitual action. *Neuropsychopharmacology*, *35*(1), 48–69.

Colwill, R. M., & Rescorla, R. A. (1985). Instrumental responding remains sensitive to reinforcer devaluation after extensive training. *Journal of Experimental Psychology: Animal Behavior Processes*, *11*(4), 520–536.

Corbit, L. H., & Balleine, B. W. (2003). The role of prelimbic cortex in instrumental conditioning. *Behavioural Brain Research*, *146*(1–2), 145–157.

Corbit, L. H., Muir, J. L., & Balleine, B. W. (2003). Lesions of mediodorsal thalamus and anterior thalamic nuclei produce dissociable effects on instrumental conditioning in rats. *The European Journal of Neuroscience*, *18*(5), 1286–1294.

Coutureau, E., & Killcross, S. (2003). Inactivation of the infralimbic prefrontal cortex reinstates goal-directed responding in overtrained rats. *Behavioural Brain Research*, *146*(1–2), 167–174.

Daw, N. D., Gershman, S. J., Seymour, B., Dayan, P., & Dolan, R. J. (2011). Model-based influences on humans' choices and striatal prediction errors. *Neuron*, *69*(6), 1204–1215.

Daw, N. D., Niv, Y., & Dayan, P. (2005). Uncertainty-based competition between prefrontal and dorsolateral striatal systems for behavioral control. *Nature Neuroscience*, *8*(12), 1704–1711.

De Wit, S., Barker, R. A., Dickinson, A. D., & Cools, R. (2011). Habitual versus goal-directed action control in Parkinson disease. *Journal of Cognitive Neuroscience*, *23*(5), 1218–1229.

De Wit, S., Corlett, P. R., Aitken, M. R., Dickinson, A., & Fletcher, P. C. (2009). Differential engagement of the ventromedial prefrontal cortex by goal-directed and habitual behavior toward food pictures in humans. *The Journal of Neuroscience*, *29*(36), 11330–11338.

De Wit, S., & Dickinson, A. (2009). Associative theories of goal-directed behaviour: A case for animal-human translational models. *Psychological Research*, *73*(4), 463–476.

De Wit, S., Niry, D., Wariyar, R., Aitken, M. R., & Dickinson, A. (2007). Stimulus-outcome interactions during instrumental discrimination learning by rats and humans. *Journal of Experimental Psychology. Animal Behavior Processes*, *33*(1), 1–11.

De Wit, S., van de Vijver, I., & Ridderinkhof, K. R. (2014). Impaired acquisition of goal-directed action in healthy aging. *Cognitive, Affective & Behavioral Neuroscience, 14*(2), 647–658.

De Wit, S., Watson, P., Harsay, H. A., Cohen, M. X., van de Vijver, I., & Ridderinkhof, K. R. (2012). Corticostriatal connectivity underlies individual differences in the balance between habitual and goal-directed action control. *The Journal of Neuroscience, 32*(35), 12066–12075.

Delorme, C., Salvador, A., Valabrègue, R., Roze, E., Palminteri, S., Vidailhet, M., de Wit, S., Robbins, T. W., Hartmann, A., & Worbe, Y. (2016). Enhanced habits formation in Gilles de la Tourette syndrome. *Brain, 139*(pt2), 605–615.

Dezfouli, A., Lingawi, N. W., & Balleine, B. W. (2014). Habits as action sequences: Hierarchical action control and changes in outcome value. *Philosophical Transactions of the Royal Society of London. Series B, Biological Sciences, 369*(1655), 10.1098/rstb.2013.0482.

Dickinson, A. (1985). Actions and habits: The development of behavioural autonomy. *Philosophical Transactions of the Royal Society B: Biological Sciences, 308*(1135), 67–78.

Dickinson, A., & Balleine, B. (1994). Motivational control of goal-directed action. *Animal Learning & Behavior, 22*(1), 1–18.

Dickinson, A., Nicholas, D. J., & Adams, C. D. (1983). The effect of instrumental contingency on susceptibility to reinforcer devaluation. *Quarterly Journal of Experimental Psychology, 35*(B), 249–263.

Evans, J. S. B. T., & Stanovich, K. E. (2013). Dual-process theories of higher cognition: Advancing the debate. *Perspectives on Psychological Science, 8*(3), 223–241.

Everitt, B. J., Dickinson, A., & Robbins, T. W. (2001). The neuropsychological basis of addictive behaviour. *Brain Research. Brain Research Reviews, 36*(2–3), 129–138.

Everitt, B. J., & Robbins, T. W. (2016). Drug addiction: Updating actions to habits to compulsions ten years on. *Annual Review of Psychology, 67*, 23–50.

Furlong, T. M., Jayaweera, H. K., Balleine, B. W., & Corbit, L. H. (2014). Binge-like consumption of a palatable food accelerates habitual control of behavior and is dependent on activation of the dorsolateral striatum. *The Journal of Neuroscience, 34*(14), 5012–5022.

Gardner, B., de Bruijn, G.-J., & Lally, P. (2011). A systematic review and meta-analysis of applications of the Self-Report Habit Index to nutrition and physical activity behaviours. *Annals of Behavioral Medicine, 42*(2), 174–187.

Geurts, H. M., & de Wit, S. (2014). Goal-directed action control in children with autism spectrum disorders. *Autism: The International Journal of Research and Practice, 18*(4), 409–418.

Gillan, C. M., Apergis-Schoute, A. M., Morein-Zamir, S., Urcelay, G. P., Sule, A., Fineberg, N. A., ... Robbins, T. W. (2015). Functional neuroimaging of avoidance habits in obsessive-compulsive disorder. *The American Journal of Psychiatry, 172*(3), 284–293.

Gillan, C. M., Morein-Zamir, S., Urcelay, G. P., Sule, A., Voon, V., Apergis-Schoute, A. M., ... Robbins, T. W. (2014). Enhanced avoidance habits in obsessive-compulsive disorder. *Biological Psychiatry, 75*(8), 631–638.

Gillan, C. M., Otto, A. R., Phelps, E. A., & Daw, N. D. (2015). Model-based learning protects against forming habits. *Cognitive, Affective & Behavioral Neuroscience, 15*(3), 523–36.

Godier, L. R., & Park, R. J. (2014). Compulsivity in anorexia nervosa: A transdiagnostic concept. *Frontiers in Psychology, 5*, 778.

Graybiel, A. M. (2008). Habits, rituals, and the evaluative brain. *Annual Review of Neuroscience, 31*, 359–387.

Hart, G., Leung, B. K., & Balleine, B. W. (2014). Dorsal and ventral streams: The distinct role of striatal subregions in the acquisition and performance of goal-directed actions. *Neurobiology of Learning and Memory, 108*, 104–118.

Heyes, C., & Dickinson, A. (1990). The intentionality of animal action. *Mind and Language, 5*(1), 87–103.

Holland, P. C. (2004). Relations between Pavlovian-instrumental transfer and reinforcer devaluation. *Journal of Experimental Psychology. Animal Behavior Processes, 30*(2), 104–117.

Hommel, B. (2003). Planning and representing intentional action. *Scientific World Journal, 3*, 593–608.

Horstmann, A., Dietrich, A., Mathar, D., Pössel, M., Villringer, A., & Neumann, J. (2015). Slave to habit? Obesity is associated with decreased behavioural sensitivity to reward devaluation. *Appetite, 87*, 175–183.

Jonkman, S., Kosaki, Y., Everitt, B. J., & Dickinson, A. (2010). The role of contextual conditioning in the effect of reinforcer devaluation on instrumental performance by rats. *Behavioural Processes, 83*(3), 276–281.

Judah, G., Gardner, B., & Aunger, R. (2013). Forming a flossing habit: An exploratory study of the psychological determinants of habit formation. *Br J Health Psychol.*, 2013 May; *18*(2), 338–353.

Jueptner, M., & Weiller, C. (1998). A review of differences between basal ganglia and cerebellar control of movements as revealed by functional imaging studies. *Brain: A Journal of Neurology*, *121*(Pt. 8), 1437–1449.

Killcross, S., & Coutureau, E. (2003). Coordination of actions and habits in the medial prefrontal cortex of rats. *Cerebral Cortex (New York, N.Y.: 1991)*, *13*(4), 400–408.

Kosaki, Y., & Dickinson, A. (2010). Choice and contingency in the development of behavioral autonomy during instrumental conditioning. *Journal of Experimental Psychology. Animal Behavior Processes*, *36*(3), 334–342.

Lee, S. W., Shimojo, S., & O'Doherty, J. P. (2014). Neural computations underlying arbitration between model-based and model-free learning. *Neuron*, *81*(3), 687–699.

Lehéricy, S., Benali, H., Van de Moortele, P.-F., Pélégrini-Issac, M., Waechter, T., Ugurbil, K., & Doyon, J. (2005). Distinct basal ganglia territories are engaged in early and advanced motor sequence learning. *Proceedings of the National Academy of Sciences of the United States of America*, *102*(35), 12566–12571.

Maudsley, H. (1867). *The physiology and pathology of the mind*. New York: D. Appleton & Company.

Metcalfe, J., & Mischel, W. (1999). A hot/cool-system analysis of delay of gratification: Dynamics of willpower. *Psychological Review*, *106*(1), 3–19.

Morris, R. W., Quail, S., Griffiths, K. R., Green, M. J., & Balleine, B. W. (2015). Corticostriatal control of goal-directed action is impaired in schizophrenia. *Biological Psychiatry*, *77*(2), 187–195.

Neal, D. T., Wood, W., Wu, M., & Kurlander, D. (2011). The pull of the past: When do habits persist despite conflict with motives? *Personality & Social Psychology Bulletin*, *37*(11), 1428–1437.

Pavlov, I. P. (1927). *Conditioned reflexes: An investigation of the physiological activity of the cerebral cortex*. Translated and Edited by G. V. Anrep. London: Oxford University Press.

Quellette, J. A., & Wood, W. (1998). Habit and intention in everyday life: The multiple processes by which past behavior predicts future behavior. *Psychological Bulletin*, *124*(1), 54–74.

Robbins, T. W., Gillan, C. M., Smith, D. G., de Wit, S., & Ersche, K. D. (2012). Neurocognitive endophenotypes of impulsivity and compulsivity: Towards dimensional psychiatry. *Trends in Cognitive Sciences*, *16*(1), 81–91.

Rothman, A. J., Sheeran, P., & Wood, W. (2009). Reflective and automatic processes in the initiation and maintenance of dietary change. *Annals of Behavioral Medicine: A Publication of the Society of Behavioral Medicine*, *38*(Suppl. 1), S4–17.

Schwabe, L., & Wolf, O. T. (2009). Stress prompts habit behavior in humans. *Journal of Neuroscience*, *29*(22), 7191–7198.

Sjoerds, Z., de Wit, S., van den Brink, W., Robbins, T. W., Beekman, A. T., Penninx, B. W., & Veltman, D. J. (2013). Behavioral and neuroimaging evidence for overreliance on habit learning in alcohol-dependent patients. *Translational Psychiatry*, *3*, e337.

Stanovich, K. E. (1999). Who is rational? In *Studies of individual differences in reasoning*. Mahwah, NJ: Erlbaum.

Steinglass, J., & Walsh, B. T. (2006). Habit learning and anorexia nervosa: A cognitive neuroscience hypothesis. *International Journal of Eating Disorders*, *39*(4), 267–275.

Thorndike, E. L. (1911). *Animal intelligence*. New York: The Macmillan Company.

Thorndike, E. L. (1931). *Human learning*. New York: Century.

Thrailkill, E. A., & Bouton, M. E. (2015). Contextual control of instrumental actions and habits. *Journal of Experimental Psychology. Animal Learning and Cognition*, *41*(1), 69–80.

Tiffany, S. T. (1990). A cognitive model of drug urges and drug-use behavior: Role of automatic and nonautomatic processes. *Psychological Review*, *97*(2), 147–68.

Triandis, H. C. (1977). *Interpersonal behavior*. Monterey, CA: Brook/Cole.

Tricomi, E., Balleine, B. W., & O'Doherty, J. P. (2009). A specific role for posterior dorsolateral striatum in human habit learning. *The European Journal of Neuroscience*, *29*(11), 2225–2232.

Valentin, V. V., Dickinson, A., & O'Doherty, J. P. (2007). Determining the neural substrates of goal-directed learning in the human brain. *The Journal of Neuroscience*, *27*(15), 4019–4026.

Verplanken, B., & Orbell, S. (2003). Reflections on past behavior: A self-report index of habit strength. *Journal of Applied Social Psychology*, *33*(6), 1313–1330.

Verplanken, B., Walker, I., Davis, A., & Jurasek, M. (2008). Context change and travel mode choice: Combining the habit discontinuity and self-activation hypotheses. *Journal of Environmental Psychology*, *28*, 121–127.

Watson, P., Wiers, R. W., Hommel, B., & de Wit, S. (2014). Working for food you don't desire: Cues interfere with goal-directed food-seeking. *Appetite*, *79*, 139–148.

Wood, W., Quinn, J. M., & Kashy, D. A. (2002). Habits in everyday life: Thought, emotion, and action. *Journal of Personality and Social Psychology*, *83*(6), 1281–1297.

Yin, H. H., Knowlton, B. J., & Balleine, B. W. (2004). Lesions of dorsolateral striatum preserve outcome expectancy but disrupt habit formation in instrumental learning. *The European Journal of Neuroscience*, *19*(1), 181–189.

Yin, H. H., Knowlton, B. J., & Balleine, B. W. (2006). Inactivation of dorsolateral striatum enhances sensitivity to changes in the action-outcome contingency in instrumental conditioning. *Behavioural Brain Research*, *166*(2), 189–196.

Yin, H. H., Ostlund, S. B., Knowlton, B. J., & Balleine, B. W. (2005). The role of the dorsomedial striatum in instrumental conditioning. *The European Journal of Neuroscience*, *22*(2), 513–523.

12

Hierarchical Control of Behaviour in Human Prefrontal Cortex

Sandrine Duverne and Etienne Koechlin

The prefrontal cortex (PFC) subserves cognitive control, that is, the ability to organise thoughts and actions in relation to current mental states, which are in turn the product of cognitive control processes. Cognitive control guides adaptive behaviour in everyday environments featuring uncertain, changing, and open-ended situations, and forms the cornerstone of human higher cognition. From an evolutionary viewpoint, cognitive control in the PFC has likely evolved to overcome critical limitations of basic adaptive behaviour mechanisms usually referred to as reinforcement learning (RL) involving the premotor cortex and basal ganglia (Koechlin, 2014). RL and, more specifically, temporal-difference RL processes are basic on-line adaptive processes that adjust a behavioural strategy mapping stimuli onto actions according to the discrepancy between actual and expected rewards (Daw & Frank, 2009; Sutton & Barto, 1998; see also Chapter 8 by Verguts and Chapter 11 by de Wit in this volume). RL is both a very simple and robust adaptive process that can learn a variety of complex tasks even in uncertain environments. In particular, when rewards depend only upon current external states and actions, and each state is explored sufficiently often, RL converges towards the behavioural strategy maximising rewards (Sutton & Barto, 1998). However, RL-based adaptive behaviour has severe limitations. A crucial limitation is that adapting to new situations gradually erases previously learned behavioural strategies, which prevents the agent from exploiting previously learned strategies in environments exhibiting recurrent situations occurring alongside new ones. Overcoming this critical limitation is likely to form the core function of cognitive control. We thus view the evolution of cognitive control as endowing the agent with the ability to arbitrate between staying with the ongoing behavioural strategy that adjusts to current external contingencies through RL, switching to previously learned strategies when facing previously encountered situations, and even creating new strategies when experiencing new situations (Koechlin, 2014). This evolutionary viewpoint of cognitive control highlights three important intrinsic dimensions of cognitive control: (1) its *inferential* nature, as the arbitration requires resolving the uncertainty bearing upon actual changes in the current state of the environment that possibly reflect the occurrence of previously encountered situations or new ones; (2) its *hierarchical* nature, as the arbitration operates at the superordinate level of behavioural strategies, forming abstract, discrete mental sets comprising subordinate sensorimotor representations that adjust

The Wiley Handbook of Cognitive Control, First Edition. Edited by Tobias Egner.

especially through RL; and (3) its *capacity-limited* nature, given that the optimal arbitration leads to an intractable computational problem with little biological plausibility: namely, monitoring on-line the whole repertoire of learned strategies that constantly grows, and revising off-line this repertoire whenever new information is acquired. On the basis of this outline, we propose here a comprehensive view describing the functional architecture of cognitive control in the human PFC mainly based on computational modelling and empirical evidence from human neuroimaging studies.

Overall Functional Anatomy of Human Frontal Lobes

We first provide a brief overview of the anatomical organisation of the human PFC subserving cognitive control (Figure 12.1). The PFC corresponds to the cortical sector in front of the premotor cortex. The latter extends along the precentral gyri, adjacent to the primary motor cortex, and covers a substantial part of Brodmann area 6 (BA 6). The premotor cortex is reciprocally connected to primary motor areas and caudal PFC regions (Dum & Strick, 1991; Tomassini et al., 2007). The premotor cortex forms cortico-subcortical loop circuits with basal ganglia, and especially the dorsal striatum, implementing RL mechanisms for learning stimulus–action associations (Atallah, Lopez-Paniagua, Rudy, & O'Reilly, 2007; Daw & Doya, 2006; Doya, 2007; Horvitz, 2009; Thorsten Kahnt et al., 2009; O'Doherty et al., 2004; Samejima, Ueda, Doya, & Kimura, 2005). The premotor cortex is thus involved in representing stimulus–action associations guiding action selection in response to stimuli. Adjacent to the lateral premotor regions, the lateral PFC (referred to as the LPFC) is composed of the inferior, middle, and superior frontal gyri. The LPFC is reciprocally connected with the posterior association cortex, including parietal and temporal regions (Barbas & Pandya, 1989; Pandya & Yeterian, 1996; Petrides & Pandya, 2009). The caudal LPFC

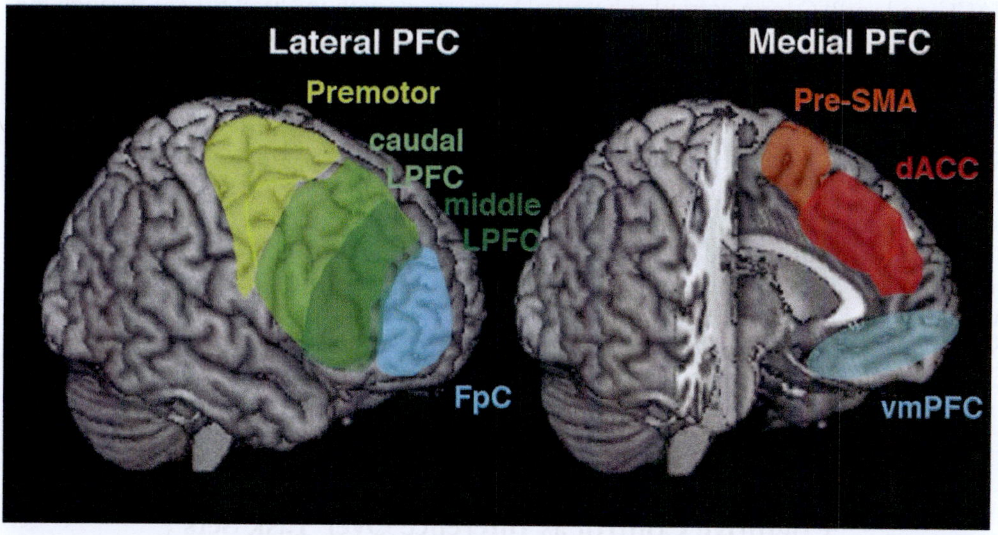

Figure 12.1 Human prefrontal regions involved in cognitive control. Three-dimensional rendering of the main prefrontal areas subserving cognitive control. PFC: prefrontal cortex. Premotor: lateral premotor cortex. LPFC: lateral prefrontal cortex. FpC: frontopolar cortex. Pre-SMA: pre-supplementary motor area. dACC: dorsal anterior cingulate cortex. vmPFC: ventromedial prefrontal cortex.

corresponds to BA 8, 44, and 45, which in the left hemisphere comprise Broca's area, which is involved in human speech and language (Amunts et al., 1999; Broca, 1861). The middle LPFC corresponds to BA 9 and 46. The caudal and middle LPFC are highly interconnected (Pandya & Yeterian, 1996), and as detailed below, reflect a posterior-anterior functional gradient in cognitive control. Finally, the most anterior part of the lateral PFC is usually referred to as the frontopolar cortex (FpC) corresponding to BA 10, which is considerably more developed in human compared to non-human primates (Semendeferi, Armstrong, Schleicher, Zilles, & Van Hoesen, 2001; Teffer & Semendeferi, 2012). The human FpC, which is mainly connected to the adjacent PFC regions (middle and anterior-medial PFC regions), is likely to have no functional homologues in monkey PFC (Koechlin, 2011; Neubert, Mars, Thomas, Sallet, & Rushworth, 2014). Similar to the premotor cortex, all these lateral PFC regions form cortical-subcortical loop circuits cantered on the basal ganglia through the dorsal striatum (Alexander, DeLong, & Strick, 1986).

The medial PFC sector can be subdivided into three main regions along the posterior-anterior axis with distinct roles in cognitive control. In front of and adjacent to the medial premotor cortex is the pre-supplementary motor area (pre-SMA), which is densely and reciprocally connected to the caudal LPFC (Bates & Goldman-Rakic, 1993; Johansen-Berg et al., 2004; Kim et al., 2010). More anteriorly, the large dorsomedial sector extending from the pre-SMA to the head of the corpus callosum is commonly referred to as the dorsal anterior cingulate cortex (dACC). The dACC is densely and reciprocally connected to the adjacent medial PFC regions and especially to the middle LPFC regions (Beckmann, Johansen-Berg, & Rushworth, 2009; Medalla & Barbas, 2009, 2010). Finally, the large medial sector extending anteriorly from the dACC around the head of the corpus callosum to the orbitofrontal cortex is often referred to as the ventromedial PFC (vmPFC). The vmPFC has widespread connections to posterior sensory and associative regions, to LPFC and FpC, and plays an important role in processing action outcomes (Cavada, Company, Tejedor, Cruz-Rizzolo, & Reinoso-Suarez, 2000; Kahnt, Chang, Park, Heinzle, & Haynes, 2012; Ongur & Price, 2000). The pre-SMA, dACC, and vmPFC form cortical-subcortical loop circuits with the basal ganglia through the striatum dorsally (pre-SMA) and ventrally (dACC, vmPFC; Alexander et al., 1986; Lehericy et al., 2004).

It is worth noting that the PFC, present in all mammals (Fuster, 1989; Uylings, Groenewegen, & Kolb, 2003), has evolved from rodents to primates and humans. In rodents, the PFC mainly comprises the paralimbic regions including the anterior cingulate cortex and the orbitofrontal cortex (Uylings et al., 2003). In primates, the PFC has evolved with the development of the lateral PFC regions (Fuster, 1989). In humans, the lateral PFC has further developed with the emergence of left–right asymmetry, yielding to the notion of Broca's area (Schenker et al., 2010; Uylings, Jacobsen, Zilles, & Amunts, 2006) and the considerable development of the FpC (Semendeferi et al., 2001; Teffer & Semendeferi, 2012). This PFC development reflects the evolution and sophistication of cognitive control, with the medial, lateral PFC, and FpC corresponding to the emergence of inferential, hierarchical, and counterfactual dimensions of cognitive control, respectively (Koechlin, 2014).

Cognitive Control as Inference Over Task Sets

According to the view outlined above, the core function of cognitive control is to arbitrate between staying with the current behavioural strategy—and adjusting it through RL—versus retrieving previously learned strategies or exploring new ones (Koechlin, 2014). The arbitration primary requires inferring *when* the current state of the environment changes, thereby reflecting the occurrence of previously encountered situations or new ones. External states especially determine the action outcomes the agent experiences, so that inferring changes in external states primary relies on learning to predict action outcomes and detecting when outcome predictions suddenly fail. Accordingly, learning a *predictive* model mapping stimulus–action pairs onto behavioural outcomes along with the *selective* model mapping stimuli onto actions is a basic requirement for cognitive control (Collins & Koechlin, 2012). Whereas the selective model adjusts through RL and enables selection of actions in response to stimuli, the predictive model learns the likelihood of action outcomes given stimuli and may further alter action selection according to outcome values derived from the immediate needs of the agent (e.g., hunger etc.). Behavioural strategies comprise these two internal models associated together through an abstract superordinate representation usually referred to as *task sets* (Monsell, 2003; Chapter 2 by Monsell in this volume; Sakai, 2008). Task sets are formed and stored as mentally instantiating external states for retrieving the corresponding behavioural strategy when these states reoccur (Collins & Koechlin, 2012). Critically, the predictive model serves to infer on-line the *absolute reliability* of the task set driving ongoing behaviour: that is, the probability that the current external state remains unchanged or, equivalently, still matches the state associated with the ongoing task set according to the match between predicted and actual action outcomes (Collins & Koechlin, 2012). The notion of absolute reliability thus serves to control the arbitration between staying versus switching away from the ongoing task set: While the task set remains reliable, that is, the match with the ongoing task set remains more likely than the no-match, this task set continues to drive behaviour. Otherwise, the task set is deemed unreliable and consequently is inhibited so that another task set is used to drive behaviour (see the section titled 'Cognitive Control as Counterfactual Reasoning' later in this chapter).

There is converging evidence from electrophysiological studies in animals and neuroimaging studies in humans that the vmPFC is involved in predicting action outcomes (Boorman, Behrens, & Rushworth, 2011; Boorman, Behrens, Woolrich, & Rushworth, 2009; Hampton, Bossaerts, & O'Doherty, 2006; Jones et al., 2012; Stalnaker, Cooch, & Schoenbaum, 2015; Zhu, Mathewson, & Hsu, 2012), suggesting that predictive models are encoded in the vmPFC. By contrast, empirical evidence is that the premotor cortex is involved in selecting actions in response to stimuli, suggesting that selective models are encoded in the premotor cortex (Azuar et al., 2014; Badre, Kayser, & D'Esposito, 2010; Ding & Gold, 2012; Koechlin, Ody, & Kouneiher, 2003). Accordingly, task sets consist of large-scale, superordinate neuronal associative representations linking selective and predictive models encoded in the premotor cortex and vmPFC. Recent neuroimaging results further show that the vmPFC monitors the absolute reliability of the task set driving ongoing behaviour according to actual action outcomes (Donoso, Collins, & Koechlin, 2014).

The dorsomedial PFC, by contrast, and more specifically the dACC, transiently responds when the task set driving ongoing behaviour becomes unreliable, and subjects switch away from this task set (Donoso et al., 2014). This neuroimaging result is consistent with other neuroimaging studies showing that in foraging tasks, the dACC is involved in monitoring the opportunity to switch away from an ongoing task (Kolling, Behrens, Mars, & Rushworth, 2012) and in starting new tasks (Dosenbach et al., 2006). Electrophysiological recordings also reveal that neuronal ensembles in the dACC exhibit abrupt activity resetting, when subjects switch away from ongoing behavioural routines for exploring (Durstewitz, Vittoz, Floresco, & Seamans, 2010; Hayden, Pearson, & Platt, 2011; Karlsson, Tervo, & Karpova, 2012). Altogether, these data suggest that the dACC is involved in detecting when the ongoing task set becomes unreliable for inhibiting it and in relation to the middle LPFC, switching to another task set driving subsequent behaviour

(Koechlin, 2014; Kouneiher, Charron, & Koechlin, 2009). (For an in-depth review of dACC models, see Chapter 15 by Brown in this volume.)

There is evidence from human neuroimaging and patient studies that the middle LPFC is involved in maintaining over time the task set driving ongoing behaviour in relation with the occurrence of external cues rather than action outcomes (Azuar et al., 2014; Badre, 2008; Badre & D'Esposito, 2007; Koechlin et al., 2003; Koechlin & Summerfield, 2007b; Kouneiher et al., 2009; Sakai & Passingham, 2003; see also Chapter 13 by Stokes, Buschman, & Miller in this volume). These findings are conceptualised as showing that task sets comprise a third basic internal model, referred to as a *contextual* model, that maps contextual cues directly onto task sets (Collins & Koechlin, 2012). Contextual models learn the likelihood of contextual cues given the ongoing task set and enable its absolute reliability to be updated according to contextual cues (Duverne & Koechlin, submitted; Koechlin, 2014). This function is also termed *episodic* control, in the sense that it controls the maintenance of ongoing task sets over time with respect to the occurrence of external cues (Koechlin et al., 2003; Koechlin & Summerfield, 2007b). Accordingly, the middle LPFC implements episodic control by encoding contextual models indexing task sets according to external cues, possibly through connections with posterior associative cortices. Connected with both the premotor cortex and vmPFC, the middle LPFC thus appears to encode abstract representations of ongoing task sets connecting selective models in premotor regions and predictive models in the vmPFC in relation with contextual models controlling task set reliability and maintenance over time according to external cues (Figure 12.2).

Cognitive Control as Hierarchical Organisation of Behaviour

In the preceding section, selective models composing task sets are described as internal representations mapping stimuli onto motor responses. There is empirical evidence, however, that in the presence of stimuli mixing multiple attributes (e.g., colours and shapes), human subjects spontaneously form hierarchical rather than flat mappings (Badre et al., 2010; Collins, Cavanagh, & Frank, 2014; Collins & Frank, 2013): One attribute (e.g., shapes) is mapped onto responses and forms sensorimotor associations, whereas the other attribute (e.g., colours) is preferentially mapped onto this set of sensorimotor associations. Such hierarchical structures in selective models are built even when there are no immediate behavioural advantages in forming these representations (Collins et al., 2014; Collins & Frank, 2013). Computational modelling indicates that these hierarchical structures favour the generalisation of subordinate sensorimotor mappings to new combinations of perceptual attributes (Collins & Frank, 2013). These structures are formed through inferential processes conditionally upon the assumption that external contingencies remain stable over time, that is, that the task set driving ongoing behaviour is deemed reliable and remains unchanged (Collins & Frank, 2013). Such hierarchical selective models are thus embedded within task sets and enable the selection of consistent sets of sensorimotor associations according to immediate contextual cues, while the same task set drives ongoing adaptive behaviour. This hierarchical form of cognitive control has been termed *contextual* control as an intermediate control level operating across *attribute spaces* and embedded in the episodic control of task sets operating along the *temporal dimension* (Koechlin, 2007; Koechlin et al., 2003; Koechlin & Summerfield, 2007a; Figure 12.2).

Human neuroimaging studies provide evidence that the caudal LPFC in the front of the premotor cortex forms hierarchical selective models within task sets, that is, when external contingencies are inferred as being stable. While the premotor cortex forms sensorimotor

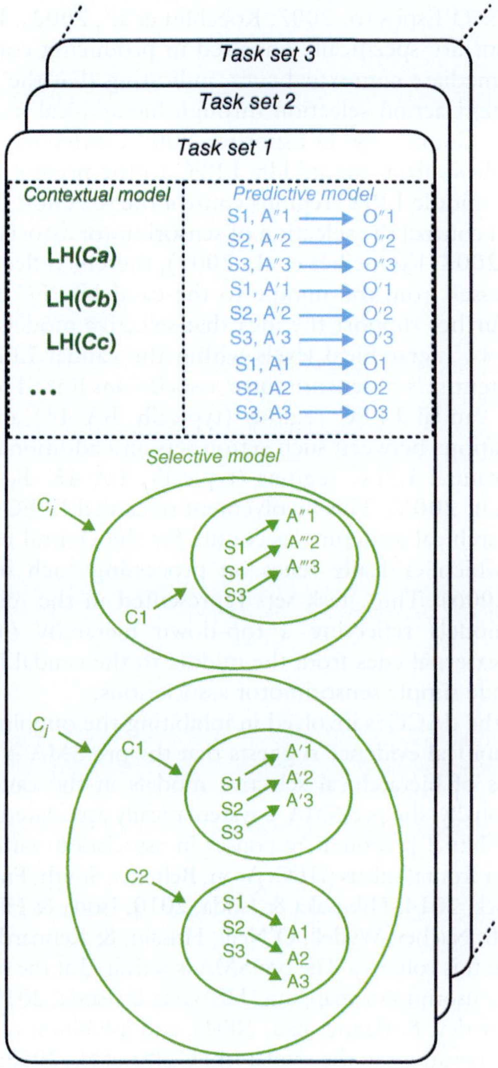

Figure 12.2 The notion of task sets. Task sets are basic executive units subserving cognitive control. Task sets are formalised as instantiating (hidden) states of the environments corresponding to stable external contingencies over time. Task sets comprise a predictive model predicting action outcomes (O) from stimuli (S), a selective model mapping stimuli onto actions (A), possibly organised hierarchically according to additional cues (C) and, finally, a contextual model encoding the likelihood of contextual cues (LH) given external states associated with task sets. Colours refer to brain regions shown in Figure 12.1, indicating critical brain regions representing predictive, selective, and contextual models.

associations (see above), the caudal LPFC is engaged in forming consistent sets of sensorimotor associations corresponding to either action sequences or stimulus–action mappings associated with contextual cues, which for clarity we refer to as *action sets* (Badre et al., 2010; Koechlin, Danek, Burnod, & Grafman, 2002). Moreover, the caudal LPFC is engaged when subjects select responses to stimuli based on such hierarchical selective models or equivalently, when subjects' responses to stimuli are contingent upon immediate contextual cues (Badre & D'Esposito, 2007; Koechlin et al., 2003). Furthermore, patients with caudal LPFC lesions are specifically impaired in producing contingent responses to stimuli according to immediate contextual cues, indicating that the caudal LPFC regions are necessary to implement action selection through hierarchical structures (Azuar et al., 2014; Badre, Hoffman, Cooney, & D'Esposito, 2009). Effective connectivity analyses measuring information flows from the middle LPFC to the premotor cortex further confirm that task sets in the middle LPFC regions control the selection of caudal LPFC representations, which in turn control the selection of sensorimotor associations in the premotor cortex (Koechlin et al., 2003; Kouneiher et al., 2009), thereby reflecting a top-down hierarchy of selection processes from the middle to the caudal LPFC and premotor cortex. Neuroimaging studies further support the idea that selective models driving action selection comprise at least two hierarchical levels within the caudal LPFC: (1) a lower level corresponding to consistent sets of sensorimotor associations linked to contextual cues and involving the posterior caudal LPFC regions (typically BA 44) and (2) a higher level corresponding to associations between such action sets and additional contextual cues and involving the anterior caudal LPFC regions (typically, BA 45; Badre, 2008; Koechlin, 2007; Koechlin & Jubault, 2006). This involvement of caudal LPFC regions in organising behaviour through hierarchical structures accounts for the central role of Broca's area in language production, which critically relies on processing such hierarchical structures (Koechlin & Jubault, 2006). Thus, task sets represented in the middle LPFC comprise hierarchical selective models reflecting a top-down hierarchy of selection processes operating according to external cues from the middle to the caudal LPFC and lateral premotor regions that encode simple sensorimotor associations.

In the same way that the dACC is involved in inhibiting the ongoing task set represented in the middle LPFC, empirical evidence suggests that the pre-SMA is engaged in inhibiting subordinate components of hierarchical selective models in the caudal LPFC within the ongoing task set. Functionally, the pre-SMA is systematically associated with the inhibition of irrelevant responses or habitual prepotent responses in association with the caudal LPFC, in particular, the right inferior frontal sulcus (rIFG; Aron, Behrens, Smith, Frank, & Poldrack, 2007; Aron, Robbins, & Poldrack, 2014; Hikosaka & Isoda, 2010; Isoda & Hikosaka, 2007; Nachev, Kennard, & Husain, 2008; Nachev, Wydell, O'Neill, Husain, & Kennard, 2007; Chapter 16 by Forstmann & Alkemade in this volume). The pre-SMA is activated at the onset of contextual cues priming switches among sensorimotor mapping (Hikosaka & Isoda, 2010; Nachev et al., 2008; Rushworth, Walton, Kennerley, & Bannerman, 2004), and inhibition-related activations in the pre-SMA precede related activations in the caudal LPFC (Jha et al., 2015; Neubert, Mars, Buch, Olivier, & Rushworth, 2010; Rae, Hughes, Anderson, & Rowe, 2015; Swann et al., 2012). Thus, the pre-SMA appears to implement inhibitory mechanisms that facilitate cue-based switches between action sets represented in the caudal LPFC and composing the ongoing task set driving behaviour.

Overall, empirical data support the idea that a functional gradient of inhibitory processes extending from the dACC to the pre-SMA parallels the hierarchical organisation of selective models implemented from the middle to the caudal LPFC. These inhibitory processes from the dACC to the pre-SMA enable the *disengagement* from previously selected action sets and sensorimotor mappings. In parallel, the LPFC implements top-down processes from the middle to the caudal LPFC that enable *selection* of the action sets and sensorimotor mappings according to immediate contextual cues. Importantly, this switching process operates within the same task set represented in the middle LPFC, that is, when external contingencies (or equivalently, the current external situation) are inferred as being unchanged and consequently, hierarchically structured through ongoing learning processes.

Cognitive Control as Counterfactual Reasoning

In the previous section, we described how the ongoing task set is implemented in the PFC and controls behaviour through hierarchical structures. These structures are learned for flexibly switching between sensorimotor mappings and action sets according to immediate contextual cues, provided that external contingencies are inferred as being stable or, equivalently, this task set is inferred as remaining reliable. Switching away from this task set as driving behaviour occurs when the task set is deemed unreliable. A critical remaining issue is how the PFC selects an alternative task set for driving subsequent behaviour (Figure 12.3).

Computational modelling of human performance reveals that humans are able to monitor the absolute reliability of 3–4 task sets concurrently (Collins & Koechlin, 2012; Donoso et al., 2014): The task set driving ongoing behaviour is referred to as the actor, and 2–3 alternative task sets are referred to as counterfactual sets as these task sets neither contribute to ongoing behaviour nor adjust to current external contingencies. This inferential buffer including 3–4

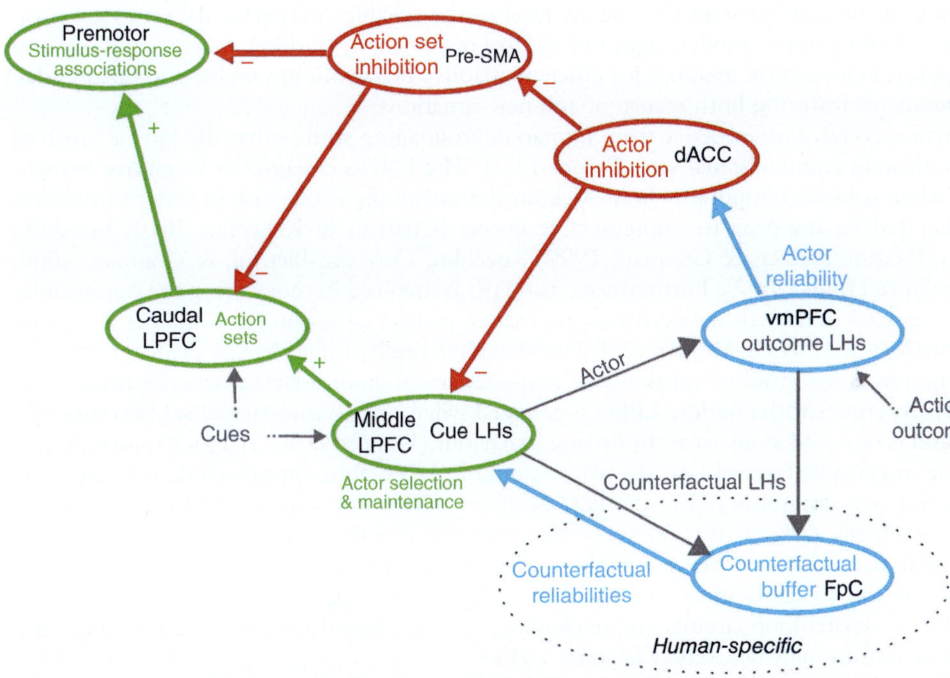

Figure 12.3 Functional architecture of cognitive control in the prefrontal cortex. Schematic diagram showing the main operations and interactions in the prefrontal cortex involved in cognitive control. Actor: the task set driving ongoing behaviour. Counterfactual sets are concurrent task sets monitored along the current actor and possibly used as an actor subsequently. Blue: prefrontal regions monitoring the actor reliability (vmPFC) and the reliability of each counterfactual set (FpC) for arbitrating between staying with the actor, switching to a counterfactual set, or creating a new task set to serve as an actor. Red: prefrontal regions inhibiting the current actor and its subordinate elements when it becomes unreliable. Green: prefrontal regions selecting a counterfactual set and its subordinate elements to serve as an actor when it becomes reliable. Outcome LHs: action outcome likelihoods (predictive models associated with task sets) encoded in vmPFC. Cue LHs: cue likelihoods (contextual models associated with task sets) encoded in the middle LPFC. See Figure 12.1 for other abbreviations and text for explanations.

task sets corresponds to the psychological notion of capacity-limited working memory (Cowan, 2005; Oberauer, 2010; Risse & Oberauer, 2010). When the actor becomes unreliable, one of the counterfactual set may become reliable. In that event, the latter replaces the former as actor. This transition corresponds to the notion of task set switching (Chapter 2 by Monsell in this volume). Conversely, when no counterfactual sets become reliable, a new task set is created from the collection of task sets stored in long-term memory and serves as an actor.

This task set is formed by mixing internal models (selective, predictive, and contextual models) from stored task sets weighted according to contextual models and current contextual cues. This actor is initially deemed unreliable and may subsequently be confirmed or rejected: Confirmation occurs when, thanks to actor learning, the new actor becomes reliable, while the counterfactual sets remain unreliable; rejection occurs when one counterfactual set becomes reliable, while, despite learning, the new actor remains unreliable. Confirmation leads to the consolidation of the new task set in long-term memory—thereby increasing the repertoire of stored task sets—and in case the inferential buffer reaches its capacity limit, confirmation also results in the least recently used task set being discarded from the inferential buffer (which still remains stored in long-term memory). By contrast, rejection leads to disbanding of the new actor and to the retrieval of the reliable counterfactual set as an actor. Thus, according to this model, cognitive control regulates the learning, storage, and retrieval of task sets in long-term memory for efficient adaptive behaviour in variable and open-ended environments featuring both recurrent and new situations.

There is converging evidence from human neuroimaging studies that the FpC is involved in monitoring counterfactual sets (Figure 12.3). The FpC is engaged in 'cognitive branching', when subjects temporarily hold back on the execution of one task in order to perform another task in response to unpredictable events (Charron & Koechlin, 2010; Koechlin, Basso, Pietrini, Panzer, & Grafman, 1999; Koechlin, Corrado, Pietrini, & Grafman, 2000; Koechlin & Hyafil, 2007). Furthermore, the FpC is involved in monitoring the opportunity to switch back and forth between two alternative courses of action (Boorman et al., 2009; Boorman et al., 2011). More recent neuroimaging results even provide evidence that the FpC monitors the absolute reliability of two concurrent counterfactual sets (Donoso et al., 2014). By contrast, the middle LPFC is engaged when one counterfactual set becomes reliable and is retrieved as an actor for driving behaviour (Donoso et al., 2014). Consistent with its role in episodic control (see above), the middle LPFC thus appears to detect when one counterfactual set monitored in the FpC becomes reliable for selection and maintenance as an actor. Finally, task set creation occurring when the ongoing actor becomes unreliable and the counterfactual sets remain unreliable was found to involve the LPFC and premotor regions along with the dorsal striatum (Donoso et al., 2014). This suggests that the premotor-striatal and LPFC-striatal loop circuits are involved in mixing internal models stored in long-term memory to form new task sets (Koechlin, 2014).

It is worth noting that according to the model proposed here, cognitive control in the PFC may potentially guide switching back and forth across a few learned action sets on the basis of contextual cues in two main ways: (1) Action sets are built as stemming from distinct (hidden) states of the environment, which are inferred from contextual cues. In that case, action sets correspond to selective models associated with distinct task sets. As a result, switching between action sets is equivalent to switching between task sets and should involve the middle LPFC. (2) Action sets are built as stemming from the same hidden state of the environment: Action sets are learned as distinct subordinate elements linked to contextual cues forming the selective model of a unique task set. In that case, switching between action sets occurs in response to contextual cues within the same task set and should involve the caudal LPFC. Cases (1) and (2) strongly depend upon how action sets are learned and whether subjects perceive the environment as being stable or variable. In other words, cases (1) and (2) occur according to the framing of the task, which might explain some discrepancies in LPFC activations observed across empirical studies investigating cue-based task switching (Badre, 2008). Although the two cases result in apparently the same cognitive operation, they actually correspond to two distinct control processes based on radically different representations of the environment. And one may speculate that through extensive learning, case (1) may reduce to case (2); that is, distinct task sets may gradually merge into a unique task set driving behaviour. This is consistent with the fact that lateral prefrontal regions gradually disengage from most anterior to posterior areas during learning of action sets (Koechlin et al., 2002; Sakai et al., 1998).

Conclusion

The notion of task sets is central to the present theory of cognitive control: Task sets are internal instantiations of (hidden) states of the environment, whereby external contingencies are deemed stable. Task sets are stored in long-term memory, and cognitive control regulates the creation, learning, storage, and retrieval of task sets from long-term memory for driving behaviour. Cognitive control is organised in the PFC according to three main anatomical dissociations: First, regions anterior to the middle LPFC and dACC, namely, the vmPFC and FpC, make probabilistic inferences about the reliability of a few concurrent task sets measuring to what extent each task set matches the current external state. By contrast, the middle LPFC and dACC along with the more posterior prefrontal regions are involved in switching between task sets. Second, the medial PFC regions control the actor set, that is, the task set driving behaviour and learning external contingencies, while the lateral PFC control counterfactual sets, that is, alternative task sets that make no contributions to ongoing behaviour but might be selected subsequently. Thus, the vmPFC infers actor reliability, whereas the FpC infers the reliability of counterfactual sets. More posteriorly, the dACC inhibits the actor set inferred as becoming unreliable through the vmPFC, whereas the middle LPFC retrieves one counterfactual set inferred as becoming reliable through the FpC to serve as an actor. Third, the posterior-anterior axis from the premotor cortex to the caudal and middle prefrontal regions corresponds to a hierarchy of top-down processes reflecting the hierarchical structure of task sets based on increasingly superordinate associations between external cues and sets of actions. In the medial PFC, this functional gradient extends from the pre-SMA and dACC and is involved in inhibiting subordinate components of the actor task set, when it is inferred as becoming unreliable. In the lateral PFC, the functional gradient extends from the caudal to the middle LPFC and is involved in selecting subordinate components of the counterfactual set that are becoming reliable. Overall, this prefrontal architecture of inferential, inhibition, and selection processes forms a unified control system combining Bayesian inferences and hypothesis testing. The system arbitrates between adjusting the ongoing behavioural strategy, retrieving previously used strategies and, finally, creating new behavioural strategies built from long-term memory for efficient adaptive behaviour in everyday environments featuring uncertain, variable, and open-ended situations.

References

Alexander, G. E., DeLong, M. R., & Strick, P. L. (1986). Parallel organization of functionally segregated circuits linking basal ganglia and cortex. *Annual Review of Neuroscience*, *9*, 357–381.

Amunts, K., Schleicher, A., Burgel, U., Mohlberg, H., Uylings, H. B., & Zilles, K. (1999). Broca's region revisited: Cytoarchitecture and intersubject variability. *Journal of Comparative Neurology*, *412*(2), 319–341.

Aron, A. R., Behrens, T. E., Smith, S., Frank, M. J., & Poldrack, R. A. (2007). Triangulating a cognitive control network using diffusion-weighted magnetic resonance imaging (MRI) and functional MRI. *Journal of Neuroscience*, *27*(14), 3743–3752.

Aron, A. R., Robbins, T. W., & Poldrack, R. A. (2014). Inhibition and the right inferior frontal cortex: one decade on. *Trends in Cognitive Sciences*, *18*(4), 177–185.

Atallah, H. E., Lopez-Paniagua, D., Rudy, J. W., & O'reilly, R. C. (2007). Separate neural substrates for skill learning and performance in the ventral and dorsal striatum. *Nature Neuroscience*, *10*(1), 126–131.

Azuar, C., Reyes, P., Slachevsky, A., Volle, E., Kinkingnehun, S., Kouneiher, F.,...Levy, R. (2014). Testing the model of caudo-rostral organization of cognitive control in the human with frontal lesions. [Research Support, Non-U.S. Gov't]. *Neuroimage*, *84*, 1053–1060.

Badre, D. (2008). Cognitive control, hierarchy and the rostrocaudal organization of the frontal lobes. *Trends in Cognitive Sciences*, *12*(5), 193–200.

Badre, D., & D'Esposito, M. (2007). Functional magnetic resonance imaging evidence for a hierarchical organization of the prefrontal cortex. *Journal of Cognitive Neuroscience*, *19*(12), 2082–2099.

Badre, D., Hoffman, J., Cooney, J. W., & D'Esposito, M. (2009). Hierarchical cognitive control deficits following damage to the human frontal lobe. *Nature Neuroscience*, *12*(4), 515–522.

Badre, D., Kayser, A. S., & D'Esposito, M. (2010). Frontal cortex and the discovery of abstract action rules. *Neuron*, *66*(2), 315–326.

Barbas, H., & Pandya, D. N. (1989). Architecture and intrinsic connections of the prefrontal cortex in the rhesus monkey. *Journal of Comparative Neurology*, *286*(3), 353–375.

Bates, J. F., & Goldman-Rakic, P. S. (1993). Prefrontal connections of medial motor areas in the rhesus monkey. *Journal of Comparative Neurology*, *336*(2), 211–228.

Beckmann, M., Johansen-Berg, H., & Rushworth, M. F. (2009). Connectivity-based parcellation of human cingulate cortex and its relation to functional specialization. *Journal of Neuroscience*, *29*(4), 1175–1190.

Boorman, E. D., Behrens, T. E., & Rushworth, M. F. (2011). Counterfactual choice and learning in a neural network centered on human lateral frontopolar cortex. *PLoS Biology*, *9*(6), e1001093.

Boorman, E. D., Behrens, T. E., Woolrich, M. W., & Rushworth, M. F. (2009). How green is the grass on the other side? Frontopolar cortex and the evidence in favor of alternative courses of action. *Neuron*, *62*(5), 733–743.

Broca, P. (1861). Remarques sur le siège de la faculté du langage articulé suivie d'une observation d'aphémie. *Bulletins de la Société Anatomique (Paris)*, *6*, 330.

Cavada, C., Company, T., Tejedor, J., Cruz-Rizzolo, R. J., & Reinoso-Suarez, F. (2000). The anatomical connections of the macaque monkey orbitofrontal cortex: A review. *Cerebral Cortex*, *10*(3), 220–242.

Charron, S., & Koechlin, E. (2010). Divided representation of concurrent goals in the human frontal lobes. *Science*, *328*(5976), 360–363.

Collins, A., & Koechlin, E. (2012). Reasoning, learning, and creativity: frontal lobe function and human decision-making. *PLoS Biology*, *10*(3), e1001293.

Collins, A. G., Cavanagh, J. F., & Frank, M. J. (2014). Human EEG uncovers latent generalizable rule structure during learning. [Research Support, N.I.H., Extramural Research Support, U.S. Gov't, Non-P.H.S.]. *The Journal of Neuroscience : The Official Journal of the Society for Neuroscience*, *34*(13), 4677–4685.

Collins, A. G., & Frank, M. J. (2013). Cognitive control over learning: creating, clustering, and generalizing task-set structure. *Psychological Review*, *120*(1), 190–229.

Cowan, N. (2005). Working-memory capacity limits in a theoretical context. In C. Izawa & N. Ohta (Eds.), *Human learning and memory: Advances in theory and applications* (pp. 155–175). Erlbaum.

Daw, N. D., & Doya, K. (2006). The computational neurobiology of learning and reward. *Current Opinion in Neurobiology*, *16*(2), 199–204.

Daw, N. D., & Frank, M. J. (2009). Reinforcement learning and higher level cognition: introduction to special issue. [Editorial Introductory]. *Cognition*, *113*(3), 259–261.

Ding, L., & Gold, J. I. (2012). Neural correlates of perceptual decision making before, during, and after decision commitment in monkey frontal eye field. *Cerebral Cortex*, *22*(5), 1052–1067.

Donoso, M., Collins, A. G., & Koechlin, E. (2014). Human cognition. Foundations of human reasoning in the prefrontal cortex. *Science*, *344*(6191), 1481–1486.

Dosenbach, N. U., Visscher, K. M., Palmer, E. D., Miezin, F. M., Wenger, K. K., Kang, H. C., & Petersen, S. E. (2006). A core system for the implementation of task sets. *Neuron, 50*(5), 799–812.

Doya, K. (2007). Reinforcement learning: Computational theory and biological mechanisms. *HFSP Journal, 1*(1), 30–40.

Dum, R. P., & Strick, P. L. (1991). The origin of corticospinal projections from the premotor areas in the frontal lobe. *Journal of Neuroscience, 11*(3), 667–689.

Durstewitz, D., Vittoz, N. M., Floresco, S. B., & Seamans, J. K. (2010). Abrupt transitions between prefrontal neural ensemble states accompany behavioral transitions during rule learning. *Neuron, 66*(3), 438–448.

Fuster, J. M. (1989). *The prefrontal cortex* (2nd ed.). New York: Raven Press.

Hampton, A. N., Bossaerts, P., & O'Doherty, J. P. (2006). The role of the ventromedial prefrontal cortex in abstract state-based inference during decision making in humans. *Journal of Neuroscience, 26*(32), 8360–8367.

Hayden, B. Y., Pearson, J. M., & Platt, M. L. (2011). Neuronal basis of sequential foraging decisions in a patchy environment. *Nature Neuroscience, 14*(7), 933–939.

Hikosaka, O., & Isoda, M. (2010). Switching from automatic to controlled behavior: Cortico-basal ganglia mechanisms. *Trends in Cognitive Sciences, 14*(4), 154–161.

Horvitz, J. C. (2009). Stimulus-response and response-outcome learning mechanisms in the striatum. *Behavioural Brain Research, 199*(1), 129–140.

Isoda, M., & Hikosaka, O. (2007). Switching from automatic to controlled action by monkey medial frontal cortex. *Nature Neuroscience, 10*(2), 240–248.

Jha, A., Nachev, P., Barnes, G., Husain, M., Brown, P., & Litvak, V. (2015). The Frontal Control of Stopping. *Cerebral Cortex, 25*(11), 4392–4406.

Johansen-Berg, H., Behrens, T. E., Robson, M. D., Drobnjak, I., Rushworth, M. F., Brady, J. M., Matthews, P. M. (2004). Changes in connectivity profiles define functionally distinct regions in human medial frontal cortex. *Proceedings of the National Academy of Sciences of the United States of America, 101*(36), 13335–13340.

Jones, J. L., Esber, G. R., McDannald, M. A., Gruber, A. J., Hernandez, A., Mirenzi, A., & Schoenbaum, G. (2012). Orbitofrontal cortex supports behavior and learning using inferred but not cached values. *Science, 338*(6109), 953–956.

Kahnt, T., Chang, L. J., Park, S. Q., Heinzle, J., & Haynes, J. D. (2012). Connectivity-based parcellation of the human orbitofrontal cortex. *Journal of Neuroscience, 32*(18), 6240–6250.

Kahnt, T., Park, S. Q., Cohen, M. X., Beck, A., Heinz, A., & Wrase, J. (2009). Dorsal striatal-midbrain connectivity in humans predicts how reinforcements are used to guide decisions. *Journal of Cognitive Neuroscience, 21*(7), 1332–1345.

Karlsson, M. P., Tervo, D. G., & Karpova, A. Y. (2012). Network resets in medial prefrontal cortex mark the onset of behavioral uncertainty. *Science, 338*(6103), 135–139.

Kim, J. H., Lee, J. M., Jo, H. J., Kim, S. H., Lee, J. H., Kim, S. T.,...Saad, Z. S. (2010). Defining functional SMA and pre-SMA subregions in human MFC using resting state fMRI: functional connectivity-based parcellation method. *NeuroImage, 49*(3), 2375–2386.

Koechlin, E. (2007). The cognitive architecture of human lateral prefrontal cortex. In P. Haggard, Y. Rossetti, & M. Kawato (Eds.), *Sensorimotor foundations of higher cognition: Attention & performance XXII.* Oxford, UK: Oxford University Press.

Koechlin, E. (2011). Frontal pole function: What is specifically human?. *Trends in Cognitive Sciences, 15*(6), 241; author reply 243.

Koechlin, E. (2014). An evolutionary computational theory of prefrontal executive function in decision-making. *Philosophical Transactions of the Royal Society B: Biological Sciences, 369*(1655).

Koechlin, E., Basso, G., Pietrini, P., Panzer, S., & Grafman, J. (1999). The role of the anterior prefrontal cortex in human cognition. *Nature, 399*(6732), 148–151.

Koechlin, E., Corrado, G., Pietrini, P., & Grafman, J. (2000). Dissociating the role of the medial and lateral anterior prefrontal cortex in human planning. *Proceedings of the National Academy of Sciences of the United States of America, 97*(13), 7651–7656.

Koechlin, E., Danek, A., Burnod, Y., & Grafman, J. (2002). Medial prefrontal and subcortical mechanisms underlying the acquisition of motor and cognitive action sequences in humans. *Neuron, 35*, 371–381.

Koechlin, E., & Hyafil, A. (2007). Anterior prefrontal function and the limits of human decision-making. *Science, 318*(5850), 594–598.

Koechlin, E., & Jubault, T. (2006). Broca's area and the hierarchical organization of human behavior. *Neuron, 50*(6), 963–974.

Koechlin, E., Ody, C., & Kouneiher, F. (2003). The architecture of cognitive control in the human prefrontal cortex. *Science, 302*(5648), 1181–1185.

Koechlin, E., & Summerfield, C. (2007a). An information theoretical approach to prefrontal executive function. *Trends in Cognitive Sciences, 11*(6), 229–235.

Koechlin, E., & Summerfield, C. (2007b). An information theoretical approach to prefrontal executive function. *Trends in Cognitive Sciences, 11*(6), 229–235.

Kolling, N., Behrens, T. E., Mars, R. B., & Rushworth, M. F. (2012). Neural mechanisms of foraging. *Science, 336*(6077), 95–98.

Kouneiher, F., Charron, S., & Koechlin, E. (2009). Motivation and cognitive control in the human prefrontal cortex. *Nature Neuroscience, 12*(7), 939–945.

Lehericy, S., Ducros, M., Krainik, A., Francois, C., Van de Moortele, P. F., Ugurbil, K., & Kim, D. S. (2004). 3-D diffusion tensor axonal tracking shows distinct SMA and pre-SMA projections to the human striatum. *Cerebral Cortex, 14*(12), 1302–1309.

Medalla, M., & Barbas, H. (2009). Synapses with inhibitory neurons differentiate anterior cingulate from dorsolateral prefrontal pathways associated with cognitive control. *Neuron, 61*(4), 609–620.

Medalla, M., & Barbas, H. (2010). Anterior cingulate synapses in prefrontal areas 10 and 46 suggest differential influence in cognitive control. *Journal of Neuroscience, 30*(48), 16068–16081.

Monsell, S. (2003). Task-set reconfiguration processes do not imply a control homunuculus: Reply to Altmann. *Trends in Cognitive Sciences, 7*(8), 341–342.

Nachev, P., Kennard, C., & Husain, M. (2008). Functional role of the supplementary and pre-supplementary motor areas. *Nature Reviews Neuroscience, 9*(11), 856–869.

Nachev, P., Wydell, H., O'Neill, K., Husain, M., & Kennard, C. (2007). The role of the pre-supplementary motor area in the control of action. *Neuroimage, 36*(Suppl. 2), T155–T163.

Neubert, F. X., Mars, R. B., Buch, E. R., Olivier, E., & Rushworth, M. F. (2010). Cortical and subcortical interactions during action reprogramming and their related white matter pathways. *Proceedings of the National Academy of Sciences of the United States of America, 107*(30), 13240–13245.

Neubert, F. X., Mars, R. B., Thomas, A. G., Sallet, J., & Rushworth, M. (2014). Comparison of human ventral cortex areas for cognitive control and language with areas in monkey frontal cortex. *Neuron, 81*(3), 700–713.

O'Doherty, J., Dayan, P., Schultz, J., Deichmann, R., Friston, K., & Dolan, R. J. (2004). Dissociable roles of ventral and dorsal striatum in instrumental conditioning. *Science, 304*(5669), 452–454.

Oberauer, K. (2010). Declarative and procedural working memory: Common principles, common capacity limits? *Psychologica Belgica, 50*(3&4), 277–308.

Ongur, D., & Price, J. L. (2000). The organization of networks within the orbital and medial prefrontal cortex of rats, monkeys and humans. *Cerebral Cortex, 10*(3), 206–219.

Pandya, D. N., & Yeterian, E. H. (1996). Morphological correlations of the human and monkey frontal lobe. In A. R. Damasio, H. Damasio, & Y. Christen (Eds.), *Neurobiology of decision-making* (pp. 13–46). Berlin Heidelberg: Springer-Verlag.

Petrides, M., & Pandya, D. N. (2009). Distinct parietal and temporal pathways to the homologues of Broca's area in the monkey. *PLoS Biology, 7*(8), e1000170.

Rae, C. L., Hughes, L. E., Anderson, M. C., & Rowe, J. B. (2015). The prefrontal cortex achieves inhibitory control by facilitating subcortical motor pathway connectivity. *The Journal of Neuroscience: The Official Journal of the Society for Neuroscience, 35*(2), 786–794.

Risse, S., & Oberauer, K. (2010). Selection of objects and tasks in working memory. *Quarterly Journal of Experimental Psychology, 63*(4), 784–804.

Rushworth, M. F., Walton, M. E., Kennerley, S. W., & Bannerman, D. M. (2004). Action sets and decisions in the medial frontal cortex. *Trends in Cognitive Sciences, 8*(9), 410–417.

Sakai, K. (2008). Task set and prefrontal cortex. *Annual Review of Neuroscience, 31*, 219–245.

Sakai, K., Hikosaka, O., Miyauchi, S., Takino, R., Sasaki, Y., & Putz, B. (1998). Transition of brain activation from frontal to parietal areas in visuomotor sequence learning. *Journal of Neuroscience, 18*(5), 1827–1840.

Sakai, K., & Passingham, R. E. (2003). Prefrontal interactions reflect future task operations. *Nature Neuroscience, 6*(1), 75–81.

Samejima, K., Ueda, Y., Doya, K., & Kimura, M. (2005). Representation of action-specific reward values in the striatum. *Science, 310*(5752), 1337–1340.

Schenker, N. M., Hopkins, W. D., Spocter, M. A., Garrison, A. R., Stimpson, C. D., Erwin, J. M., & Sherwood, C. C. (2010). Broca's area homologue in chimpanzees (Pan troglodytes): Probabilistic mapping, asymmetry, and comparison to humans. [Comparative Study. *Cerebral Cortex, 20*(3), 730–742.

Semendeferi, K., Armstrong, E., Schleicher, A., Zilles, K., & Van Hoesen, G. W. (2001). Prefrontal cortex in humans and apes: A comparative study of area 10. *American Journal of Physical Anthropology, 114*(3), 224–241.

Stalnaker, T. A., Cooch, N. K., & Schoenbaum, G. (2015). What the orbitofrontal cortex does not do. [Review]. *Nature Neuroscience, 18*(5), 620–627.

Sutton, R. S., & Barto, A. G. (1998). *Reinforcement learning*. Cambridge, MA: The MIT Press.

Swann, N. C., Cai, W., Conner, C. R., Pieters, T. A., Claffey, M. P., George, J. S.,...Tandon, N. (2012). Roles for the pre-supplementary motor area and the right inferior frontal gyrus in stopping action: Electrophysiological responses and functional and structural connectivity. *NeuroImage, 59*(3), 2860–2870.

Teffer, K., & Semendeferi, K. (2012). Human prefrontal cortex: Evolution, development, and pathology. *Progress in Brain Research, 195*, 191–218.

Tomassini, V., Jbabdi, S., Klein, J. C., Behrens, T. E., Pozzilli, C., Matthews, P. M.,...Johansen-Berg, H. (2007). Diffusion-weighted imaging tractography-based parcellation of the human lateral premotor cortex identifies dorsal and ventral subregions with anatomical and functional specializations. *Journal of Neuroscience, 27*(38), 10259–10269.

Uylings, H. B., Groenewegen, H. J., & Kolb, B. (2003). Do rats have a prefrontal cortex? [Review]. *Behavioural Brain Research, 146*(1–2), 3–17.

Uylings, H. B., Jacobsen, A. M., Zilles, K., & Amunts, K. (2006). Left-right asymmetry in volume and number of neurons in adult Broca's area. *Cortex, 42*(4), 652–658.

Zhu, L., Mathewson, K. E., & Hsu, M. (2012). Dissociable neural representations of reinforcement and belief prediction errors underlie strategic learning. *Proceedings of the National Academy of Sciences of the United States of America, 109*(5), 1419–1424.

13

Dynamic Coding for Flexible Cognitive Control

Mark G. Stokes, Timothy J. Buschman, and Earl K. Miller

Introduction—The Importance of Cognitive Control

A hallmark of cognitive control is its flexibility. Every situation we experience is essentially unique, and yet we are able to act appropriately in different contexts, integrating the current environmental contingencies, our internal goals, and the goals of those around us. Imagine sitting down to play a new card game—you are able to integrate the rules; form an initial, possibly rudimentary, strategy; and then play your hand appropriately.

On top of this incredible ability to coordinate complex rule-guided goal-directed behaviour, we can quickly change our behaviour in response to changes in the world around us or our own desires. Behavioural studies suggest this can happen extremely quickly—within a couple of hundred milliseconds (for a review, see Sakai, 2008). For example, you will likely adapt your strategy as you play the game, learn its nuances, and respond to the strategies of others—perhaps even before the next hand is played.

What are the neural mechanisms that support this flexibility? In this chapter, we suggest that cognitive flexibility depends on the capacity of the prefrontal cortex (PFC) to dynamically encode the task-relevant information. We begin by reviewing the evidence that the lateral PFC (lPFC) plays a particularly important role in flexible cognitive control. Next, we review the evidence for highly dynamic representations in lPFC. Finally, we consider how these representations can be used to establish task-relevant networks throughout the brain.

Cognitive Flexibility

Cognitive flexibility is a defining feature of animal intelligence. This ability has developed throughout evolution, reaching its pinnacle in humans. The size of the PFC has grown in parallel along the phylogenetic tree, from non-primates (e.g., <4% of the total cortical volume in cats) to primates (e.g., ~10% in macaques) to apes (e.g., ~17% in chimpanzees) to its apex in humans (~30%; for review, see Fuster, 2002). This evolutionary trajectory already suggests that the PFC plays a central role in cognitive flexibility, a hypothesis that is supported by the impact of damage to the lPFC on behaviour (see Chapter 29 by Fellows in this volume).

The Wiley Handbook of Cognitive Control, First Edition. Edited by Tobias Egner.

Systematic analysis of lesion extents and behavioural impairments demonstrate that the PFC is critically important for planning and executing sequences of actions, particularly when individuals are faced with abstract or open-ended tasks (Shallice & Burgess, 1991). A particularly powerful paradigm for investigating cognitive flexibility has been the Wisconsin Card-Sorting Task (WCST). The WCST requires individuals to apply a sorting rule to a multidimensional stimulus and then, after an unpredictable amount of time, flexibly switch to another rule. Patients with lesions involving the frontal lobe often display profound difficulties in performing the WCST, particularly at the point where they should switch between different sorting rules on the basis of negative feedback (Barcelo & Knight, 2002; causing perseverative errors, Milner, 1963). Subsequently, lesion studies in non-human primates has confirmed the importance of the PFC in performing a WCST analogue and further parcellated cognitive functions to different subregions in the PFC (Buckley et al., 2009). Such experiments are consistent with the view that the lPFC is particularly important for representing the current task set (Miller & Cohen, 2001).

Neurophysiological studies of the lPFC demonstrate that neurons have a set of properties that may make them uniquely well suited to acting as a cognitive controller. First, individual neurons in the lPFC are known to represent the contents of working memory (WM; Funahashi, Bruce, & Goldman-Rakic, 1989; Fuster & Alexander, 1971; Miller, Erickson, & Desimone, 1996). Although WM is distributed across many different brain regions, coding in the lPFC is more resistant to distraction than visual brain areas (Miller et al., 1996). More recent work also suggests that regions in the PFC have the longest time constants (Murray et al., 2014). In other words, lPFC neurons integrate information over long periods of time (Fuster, 2001). This is a prerequisite for intelligent behaviour—one must respond based on the situation and goals, even if these constraints are not immediately obvious. Second, neurons in the lPFC do not represent just any information from the environment; instead, it prioritises those stimuli that are relevant to the current task (Miller et al., 1996; Rainer, Asaad, & Miller, 1998). Such filtering ensures that only task-relevant information is used to guide behaviour. Third, neurons in the lPFC are able to represent a great diversity of such task-relevant material (Duncan & Miller, 2002). Importantly, this includes information about the current task (Wallis, Anderson, & Miller, 2001; White & Wise, 1999), recent events held in WM, and predictions of future events (Rainer, Rao, & Miller, 1999). In other words, no matter what the task, lPFC neurons are able to represent the relevant information (including what the task is), hold it 'in mind', and then use this task-relevant information to plan future action and guide current behaviour.

Flexible Representations in the PFC

To gain more precise insights into the computational role of the PFC, we need to look more closely at the neurophysiological principles of coding in frontal networks. As noted above, single-cell recordings in the monkey PFC provide extensive evidence that activity encodes the behavioural parameters that are currently relevant. However, it is implausible that the PFC is hardwired to perform all of these functions (Duncan, 2010). Indeed, in a typical experiment, about 30% of neurons represent any given task variable (Duncan & Miller, 2002), implying that the PFC is sufficiently flexible to allow task-relevant information to effectively dominate the available coding space in the PFC.

Duncan and Miller (2002) propose that PFC neurons represent information in a highly dynamic manner, adapting rapidly to the current task. According to the principle of *adaptive coding*, the PFC constitutes a flexible pool of neural resources that can be recruited 'on the fly' to represent whatever information is currently most relevant for achieving behavioural goals.

Real-Time Flexibility

Multivariate analysis tools allow us to characterise population-level neural dynamics that underpin cognitive flexibility. Applied to the PFC, these analytical approaches clearly show that neural coding is not fixed, but depends on the exact context of the current task (Meyers, Freedman, Kreiman, Miller, & Poggio, 2008; Sigala, Kusunoki, Nimmo-Smith, Gaffan, & Duncan, 2008; Stokes et al., 2013). For example, we previously observed that population coding adapts according to a flexibly defined behavioural context. In Stokes et al., monkeys were trained to perform a simple stimulus–stimulus association task. At the beginning of each trial, the animal was given a 'cue' stimulus that would indicate the current context. This context determined which stimulus in an upcoming stream of stimuli would be the current 'target', to which they must respond to receive a reward. Notably, recording neurons in the lPFC, we found the population response to a choice stimulus (target vs. non-target) critically depends on the behavioural context (Figure 13.1a). In context A, the response to stimulus 'A' followed a trajectory towards the 'go' activity state, whereas the response to stimuli 'B' or 'C' followed a different path to the 'no-go' activity state (Figure 13.1b, upper panel). Correspondingly, in context B/C, stimulus 'B'/'C' followed the trajectory towards the 'go' state, whereas the response to stimuli 'A' or 'C'/'B' followed respective path to 'no-go' states. This dynamic mapping can be rephrased as a temporary shift in the decision circuit for flexible accumulation of task-relevant information (see Figure 13.1b, lower panel). How is this dynamic shift in coding established and maintained in the lPFC?

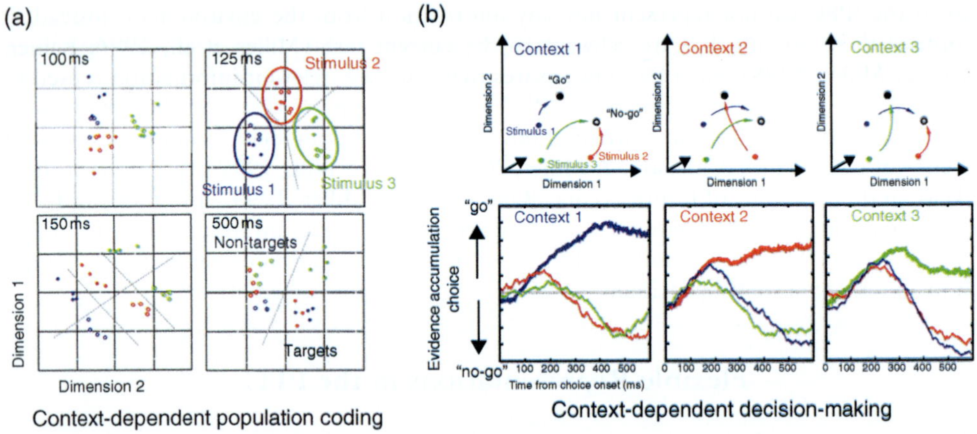

Figure 13.1 Context-dependent population coding in the prefrontal cortex. a. During WM-guided behaviour, population-level activity states quickly evolve from representing the physical properties of the presented stimuli (at 125 ms; stimuli are coloured coded) to the decision-relevant coding from ~150 ms (i.e., target vs. non-target; from Stokes et al. [2013], Open Access). b. This process can be schematised as a context-dependent path through the activity-state space. A functional shift in the response profile of the network effectively maps stimulus-specific activity states to the context-relevant position in the state space (upper panel; from Stokes et al., 2013). This flexible mapping could be framed as a context-dependent decision process. In the lower panel, the results from panel A are replotted as the accumulation of evidence for the 'go' or 'no-go' response as a function of the current rules (i.e., if context 1, then the evidence for stimulus 1 supports a 'go' response, but stimuli 2 and 3 provide counterevidence; upper panel). Source: Adapted from Stokes 2013. Reproduced with permission of Elsevier.

WM States for Guiding Behaviour

WM provides a temporary but stable platform for flexible, context-dependent processing. Early neurophysiological recordings in the lPFC suggested that task-relevant information is maintained via persistent delay activity (e.g., Fuster & Alexander, 1971). Tonic activity states representing task-relevant parameters could, in principle, mix with new input for context-dependent processing (Mante, Sussillo, Shenoy, & Newsome, 2013). However, accumulating evidence suggests that the relationship between delay activity and WM maintenance is more complex. For example, content-specific delay activity has been observed prior to learning a specific WM task, and surprisingly, even reduced as the animals mastered the WM task (Qi, Meyer, Stanford, & Constantinidis, 2011). Even within a single trial, WM-specific activity seems to wax and wane during the delay period, depending on the current focus of attention (reviewed in Stokes, 2015). For example, when the duration of the maintenance period is fixed from trial to trial, robust delay activity may only emerge late during the delay period (Barak, Tsodyks, & Romo, 2010; Watanabe & Funahashi, 2007; Figure 13.2a). So-called 'ramp-up' activity implies that delay activity could optimise behaviour by preparing for the processing demands of the memory probe (e.g., temporal orienting of attention [Nobre, Correa, & Coull, 2007]). However, the relatively silent moments between encoding and response preparation also suggests that the continuity of vigorous delay activity is not always necessary for the continuity of the mental representation (Barak et al., 2010).

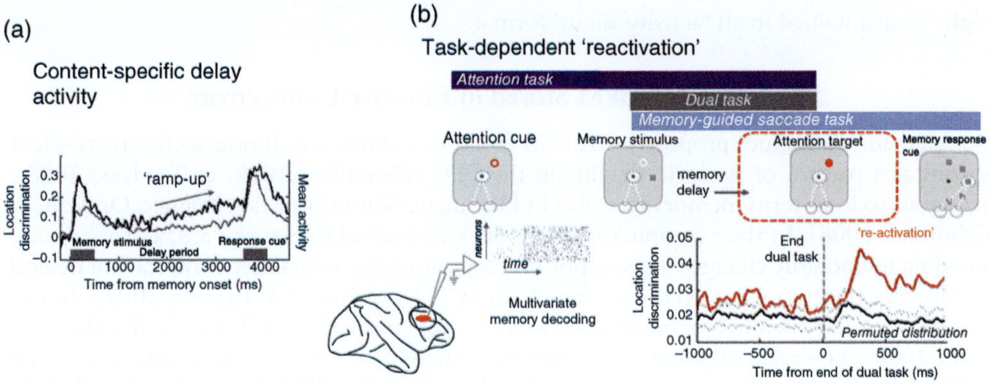

Figure 13.2 Delay activity is modulated by current task relevance. a. Neurophysiological recordings from the PFC during the delay period of a memory-guided saccade task reveal content-specific activity (location-specific activity is shown in black, and population mean activity is shown in grey for reference). However, content-specific delay activity also varies with the current task relevance of the memory item, resulting in a 'ramp up' in anticipation of the response (Source: Adapted from Watanabe 2007. Reproduced with permission of Oxford University Press). b. In a more recent dual-task experiment, an attention task partially overlaid the standard MGS design. Content-specific activity was effectively abolished by the dual task during the memory delay (red trace: from –1000 to onset of the attention target; black trace: permuted distribution). Critically, location-specific information was 'reactivated' at the end of the dual task (red trace, after the onset of the attention target), presumably reflecting a shift in task focus to the MGS (Source: Adapted from Watanabe 2014. Reproduced with permission of Nature Publishing Group). Such evidence suggests that delay activity reflects the task relevance of memoranda. Gaps in content-specific activity further suggest that WM could be maintained in an 'activity-silent' neural state.

Recent electrophysiological studies provide specific evidence that suggests persistent delay activity is not strictly necessary for accurate WM-guided behaviour. Warden and Miller (2007) trained monkeys to perform a two-item WM task: The animals were presented two stimuli in sequence, and then, after a memory delay, they were asked to report the identity and order of the remembered stimuli. Recordings in the lPFC demonstrated that neurons encoding the identity of the first stimulus in the sequence were suppressed during the presentation of the second stimulus. After the second stimulus, memory-specific delay activity for the first stimulus was 'reawakened', presumably in time for WM-guided behaviour. This was true, even when accounting for the high dimensionality of the representations (Rigotti et al., 2013). However, we should note that information was phase-locked to an ~30 Hz population oscillation (Siegel, Warden, & Miller, 2009), which might also be important for information coding (more on this later).

Watanabe and Funahashi recently reported similar results in a dual-task paradigm (Watanabe & Funahashi, 2014). As in the standard variant of a delayed saccade task (Funahashi et al., 1989), monkeys were required to encode and maintain the location of a saccade target for execution at the end of the trial. However, under a dual-task condition, they were also required to attend to a specific spatial location until part way through the WM delay period (Figure 13.2b). This cognitive manipulation almost abolished WM-specific delay activity during the dual-task period, even on trials in which WM performance was preserved (WM-correct trials). Moreover, when the competing task demands were completed, robust delay activity was again reinstated (see also Lebedev, Messinger, Kralik, & Wise, 2004). Together, these findings suggest that mnemonic delay activity in the PFC is not always critical for maintaining the continuity of WM, but can be dynamically re-established when attention is refocused to the task-relevant content. WM does not seem to depend on delay activity, but might be maintained in an 'activity-silent' format.

'Activity-Silent' WM Stored in Effective Connectivity

Mongillo and colleagues proposed that WM could be maintained during such activity-silent periods as a pattern of synaptic weights in the PFC (Mongillo, Barak, & Tsodyks, 2008), analogous to long-term memory (see also in IT; Sugase-Miyamoto, Liu, Wiener, Optican, & Richmond, 2008). In their computational model schematised in Figure 13.3, activity during encoding temporarily changes the synaptic efficacy within the neural network, leaving behind a temporary trace via short-term synaptic plasticity (STSP; Zucker & Regehr, 2002). In this particular model, calcium kinetics provide a window of approximately 2 s for STSP (Mongillo et al., 2008); however, different time constants inherent to other synaptic processes could underlie the diversity of mnemonic time scales observed in the PFC (Bernacchia, Seo, Lee, & Wang, 2011). The essential point is that task-relevant input changes the *effective connectivity* of the network to construct a temporary task-dependent circuit for WM-guided behaviour.

Mongillo et al. (2008) originally proposed that an 'activity-silent' coding scheme is more efficient than persistent firing models. A coding format that does not depend on an unbroken chain for persistent firing would clearly have a cost advantage (i.e., *sparsification* in time). However, here we focus on another possible application of such 'silent' memories: They can provide a neurobiologically plausible mechanistic account for WM-guided behaviour as generalised state-dependent processing (see below for further discussion). Memories are not stored as an active representation, but rather change the functional architecture of the neural network for future processing. WM is therefore expressed by the way the network responds to new input. State-dependent readout would avoid the need for explicit mechanisms for comparing the internal representation with a separate representation reflecting the new input (Mongillo et al., 2008; Sugase-Miyamoto et al., 2008). Moreover, because memories are

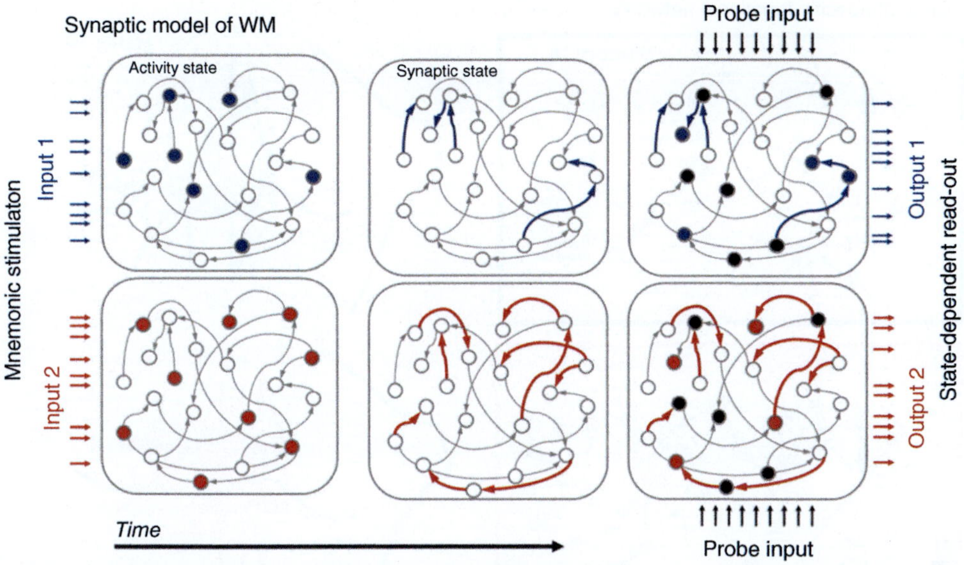

Figure 13.3 Maintaining 'activity-silent' working memory in effective connectivity. Schematic of the synaptic model of working memory described in Mongillo et al. (2008). Task-relevant input (left-side horizontal arrows, blue for 'Memory A' and red for 'Memory B') drives a stimulus-specific activity state (filled circles), which in turn triggers a specific pattern of short-term synaptic plasticity between cells (bold arrows). Memory is read out from this synaptic trace via the context-dependent response at retrieval (solid black circles, same for 'Memory A' and 'Memory B'). The probe-driven response will be patterned by the hidden state of synaptic efficacy, resulting in a discriminable output pattern (right-side horizontal arrows). Source: Adapted from Stokes 2015.

stored in a format that is qualitatively different from more transient representations (e.g., ongoing perception and action), their informational content could be less confusable with other activity-based representations (Olivers, Peters, Houtkamp, & Roelfsema, 2011).

Despite considerable theoretical appeal, a synaptic model of WM is relatively difficult to test empirically. Extensive evidence confirms the prevalence of STSP (Zucker & Regehr, 2002), but its functional role has not yet been fully established. Extracellular recordings can only indirectly infer connectivity via interactions between simultaneous spike trains. The gold-standard evidence for a monosynaptic connection is correlated spiking between cell pairs. In practice, however, the probability of sampling any two neurons with a measurable monosynaptic connection is extremely low (~1–2% of all recorded pairs in Fujisawa, Amarasingham, Harrison, & Buzsaki, 2008). Such a poor yield has been effectively prohibitive for testing synaptic plasticity within a standard primate single neurophysiological study. Fujisawa and colleagues were able to collect sufficient samples of simultaneous recordings in the rat frontal cortex to show that the pattern of effective synaptic connections was dynamically modulated during a WM-dependent maze task (Fujisawa et al., 2008; Figure 13.4 left panel). Although this observation is compatible with a synaptic WM hypothesis, the rodent model is not ideal. The rat frontal cortex lacks key features of primate PFC cytoarchitecture (Wang et al., 2013). Moreover, the rodent WM analogue is difficult to compare with standard behavioural tasks used in primate WM experiments. Therefore, further research in the primate PFC is required to obtain more specific evidence for a synaptic theory of WM.

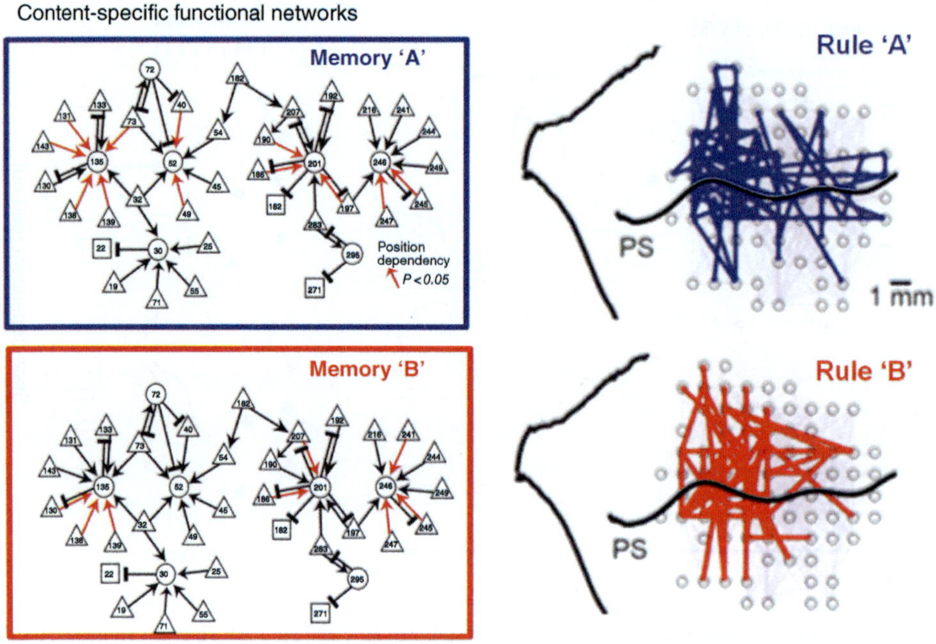

Figure 13.4 Evidence for content-specific functional networks. Simultaneous recordings in the rat frontal cortex revealed direction-specific patterns of synaptic efficacy (red arrows) between cells (putative pyramidal [triangles]; putative interneuron [circle]; unclassified [square]) coding for direction during a WM-based maze task (left panel, Source: Adapted from Fujisawa 2008. Reproduced with permission of Nature Publishing Group). This is consistent with a role for short-term synaptic plasticity in WM. In the monkey PFC (PS = principal sulcus), different task rules are associated with specific functional networks (synchronised electrodes for rule 'A' in blue; rule 'B' in red) coupled by synchrony at ~30 Hz (right panel, Source: Adapted from Buschman 2012. Reproduced with permission of Elsevier). This is consistent with the idea that coherence could also play a role in constructing functional networks for flexible behaviour.

Synaptic plasticity is not the only candidate mechanism for an effective connectivity-coding scheme for WM. For example, synchronous oscillations could provide an alternative, complementary, or even supportive, mechanism for rapid and temporary shifts in the effective connectivity between neurons. The idea is simple: If two neurons are rhythmically synchronised in phase with each other, then they will share the same periods of relative excitability and relative inhibition. This in turn will ensure that any action potentials from one neuron will have a maximal impact on the other neuron (Fries, 2005). Conversely, if two neurons are out of phase, the opportunity for successful information transfer between nodes will be reduced. In this way, functional pathways can be rapidly constructed (and dissolved) simply by modulating synchrony along pre-existing structural paths of the network (see also Chapter 14 by M. X Cohen in this volume).

Buschman et al. (2012) recently provided evidence for this kind of model. They trained monkeys to follow two different rules: Attend and respond to either the colour or orientation of an imperative stimulus. If synchrony acts to shape the effective connectivity in the PFC in support of the task rule, then one would expect the synchrony between populations of neurons to change depending on the current rule. To test this prediction, Buschman et al. used about a dozen electrodes to simultaneously record the activity of a population of neurons in the lPFC. As predicted, they found two distinct patterns of synchrony—one for the colour rule and one for the orientation rule—within the lPFC at 'beta' frequencies (~25 Hz; Figure 13.4, right). Importantly, they also found that neurons carrying task-relevant information, such as the identity of the stimulus, were synchronised to the currently relevant task ensemble. When the animal switched from one rule to the next, these neurons changed their synchronous association, suggesting that synchronous ensembles pull together all of the relevant neurons to execute a task. A similar mechanism for the rapid configuration of content-specific network architectures could be used to keep other forms of task-relevant material in WM (Salazar, Dotson, Bressler, & Gray, 2012). As described above for STSP, such a rapid shift in the functional architecture of the PFC network could directly modulate the response dynamics, allowing for context-appropriate input-output behaviour (Fries, 2005).

Dynamic Coding

Activity-dependent dynamics in connectivity patterns (e.g., STSP) predict a cascading interaction between past and present input, resulting in a complex, but reproducible trajectory through the state space (Figure 13.5a). Consequently, the response profile of the network is in constant flux during active processing. While input drives a specific response to a network according to the current connectivity state, the resultant activity state in turn shifts the connectivity state. Therefore, the response to subsequent stimulation will trigger a unique response pattern according to the new connectivity state. Moreover, this new pattern will further modulate the new connectivity state of the system, thus determining the response to the next input, and so on (Buonomano & Maass, 2009). The reciprocal interaction between the activation state and the underlying connectivity states of the network should result in a complex spatiotemporal trajectory through the state space observed throughout different experimental contexts (Crowe, Averbeck, & Chafee, 2010; Meyers et al., 2008; Stokes et al., 2013; see Figure 13.5b). The full trajectory is reproducible across trials to the extent that the temporary connectivity state returns to the baseline connective state after some time period. The baseline state is determined by more stable connections established via long-term plasticity (i.e., long-term potentiation/depression).

By allowing activity and connectivity states to interact across time, dynamic coding dramatically expands the dimensionality of population coding. This can be thought of as a temporal extension of the mixed selectivity view outlined above (Rigotti et al., 2013). Neurons in the lPFC do not map neatly to distinct behavioural conditions, but tend to show some form of *mixed selectivity* (Rigotti et al., 2013). The usefulness of such mixed selectivity approaches is well established and has been used extensively in computer science (for example, it is directly analogous to the kernel trick used in support vector machines (Vapnik, 1998). For example, the flexibility of mapping 'A' or 'B' to a 'Go' response is facilitated by having a high-dimensional representation that allows for space to be easily divided in different ways. By the same token, the information potential of a dynamic network representation is directly proportional to the statistical independence between time points (Stokes, 2011).

Perhaps more importantly, dynamic coding could condition the network to exhibit context-dependent response properties before the connectivity state relaxes back to the default configuration. This idea underpins probe-driven readout in synaptic models of WM (Mongillo et al., 2008; Sugase-Miyamoto et al., 2008), but could provide a mechanistic account for dynamic tuning profiles that could mediate adaptive coding in the lPFC (Duncan & Miller, 2002). Task-relevant input temporarily shifts the response profile of the lPFC (Stokes, 2015).

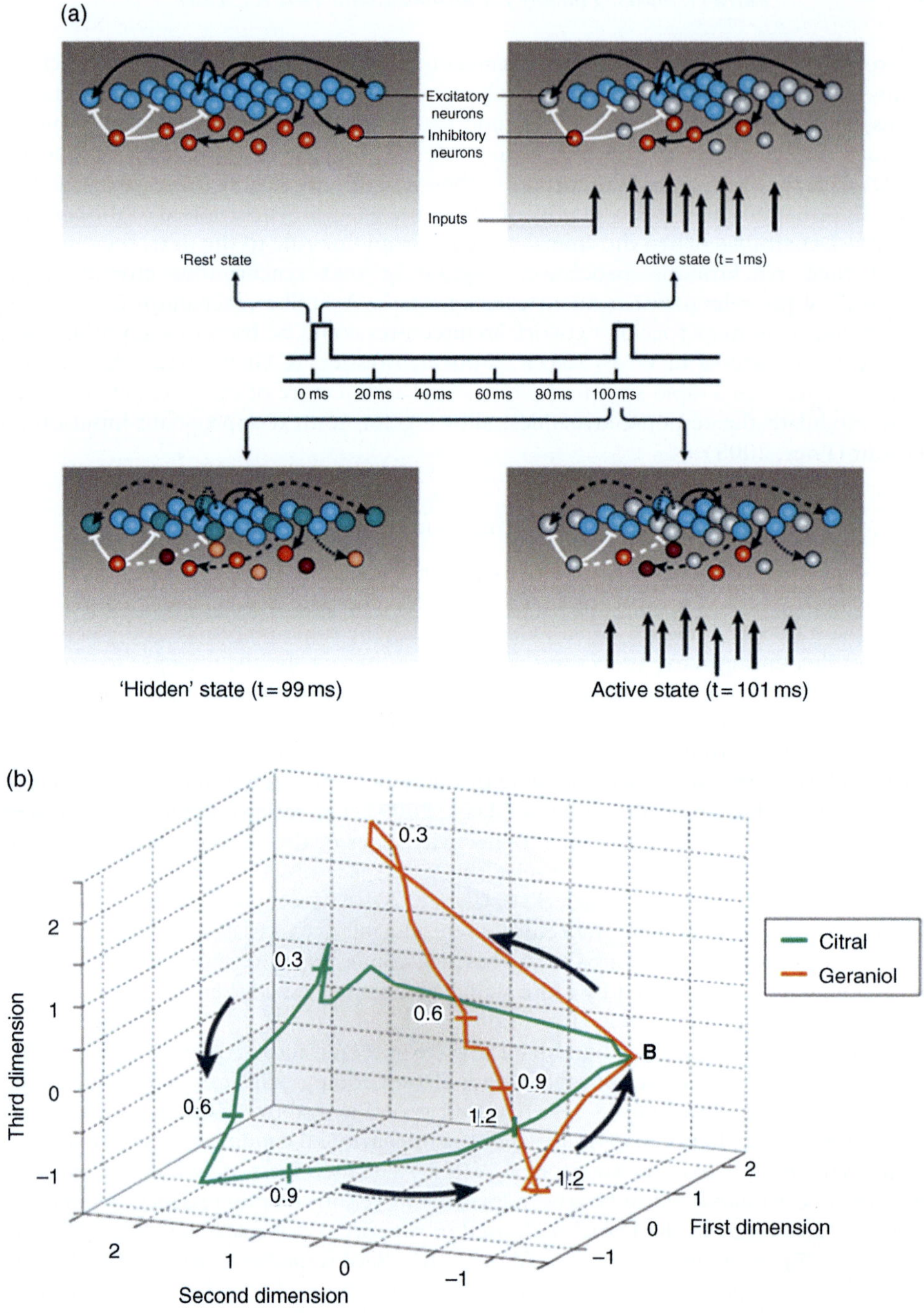

Figure 13.5 Dynamic coding predicts non-stationary activity states. a. Activity-dependent changes in effective connectivity (e.g., STSP) will result in a cascading interaction between past and present activity states. The activity state of the system depends on the input drive and the current connectivity of the network. At each moment, activity changes the connectivity profile, which in turn alters the subsequent response of the network, which in turn results in a different response to the same input drive, and so on (with permission, from Buonomano & Maass, 2009). b. This mutual interaction between activity states and the underlying connectivity states means that even a fixed input to the system will drive a dynamic trajectory through activity-state space, as illustrated by the complex trajectories observed during odour processing in the locust olfactory system for two specific smells: citral and geraniol. Source: Buonomano 2009. Reproduced with permission of Nature Publishing Group.

Although it is difficult to measure connectivity states with standard extracellular recordings, it is possible to infer changes in network architecture from systematic changes in network behaviour. As already mentioned, the complex spatiotemporal trajectory observed during cue processing is a characteristic feature of dynamic changes in network behaviour associated with activity-dependent effective connectivity. However, it is difficult to rule out other kinds of dynamics, including different cell latencies (Harvey, Coen, & Tank, 2012) or nonlinear interaction between activity states (Mante et al., 2013). A specific hidden state should also emit a signature hum as a result of ambient spontaneous activity (Sugase-Miyamoto et al., 2008); however it would be difficult to differentiate such patterned activity from the classic persistent activity states previously ascribed to WM (Shafi et al., 2007).

One possible way to explore otherwise hidden states is to 'ping' the network with a neutral input and see how it responds. As an illustration, consider echolocation (e.g., active sonar), where a simple impulse (e.g., 'ping') is used to probe the hidden contours of an unseen structure (Figure 13.6a). Analogously, the impulse response to neural perturbation should co-depend on the pattern of input activity and the current neural state (activity and connectivity state). If the input pattern is held constant, then any response-related difference could be attributed to underlying changes in network behaviour. Exploiting this logic in Stokes et al. (2013), we found that a neutral task-irrelevant stimulus presented during a WM delay period generated WM-specific patterns of activity in the lPFC. We suggested that this context-dependent response pattern could reflect differences in the hidden neural state, such as effective connectivity. We have recently extended this impulse–response approach for non-invasive EEG (Figure 13.6b), providing new opportunities to study hidden states in the human brain (Wolff, Ding, Myers, & Stokes, 2015).

To summarise, a dynamic coding framework proposes that context-defining input configures a temporary functional architecture in the lPFC, perhaps via effective connectivity (schematic Figure 13.7). Although this kind of state could be 'activity silent', such hidden neural states can influence subsequent processing, including dynamic trajectories observed during encoding (see Figure 13.7). Such a shift in the underlying hidden state would also emit a signature hum from ambient spontaneous activity, and condition the response to an impulse perturbation, and/or establish specific activity states when attended. Most importantly, however, the temporary shift in network behaviour could provide a mechanistic account for WM-guided behaviour specifically (Stokes, 2015), and perhaps context-dependent processing more generally (Stokes et al., 2013). The flow of information is effectively re-routed through the prefrontal network. This can be reformulated as a 'dynamic decision circuit', where evidence accumulation is flexibly determined by the current task context (Deco, Rolls, & Romo, 2010).

Flexibility in Networks

So far we have considered the physiological properties of neurons in the lPFC and how their dynamic nature can support cognitive flexibility. However, cognitive flexibility relies on using these representations to guide neural activity throughout the brain. Indeed, what may give the lPFC its special role in cognitive flexibility may be its connections throughout the brain. Here, we will discuss anatomical and physiological evidence that the lPFC plays a key role in biasing representations throughout the brain.

The PFC is especially well situated anatomically to play a controlling role over cognition. First, subregions within the PFC are densely interconnected, both within and across

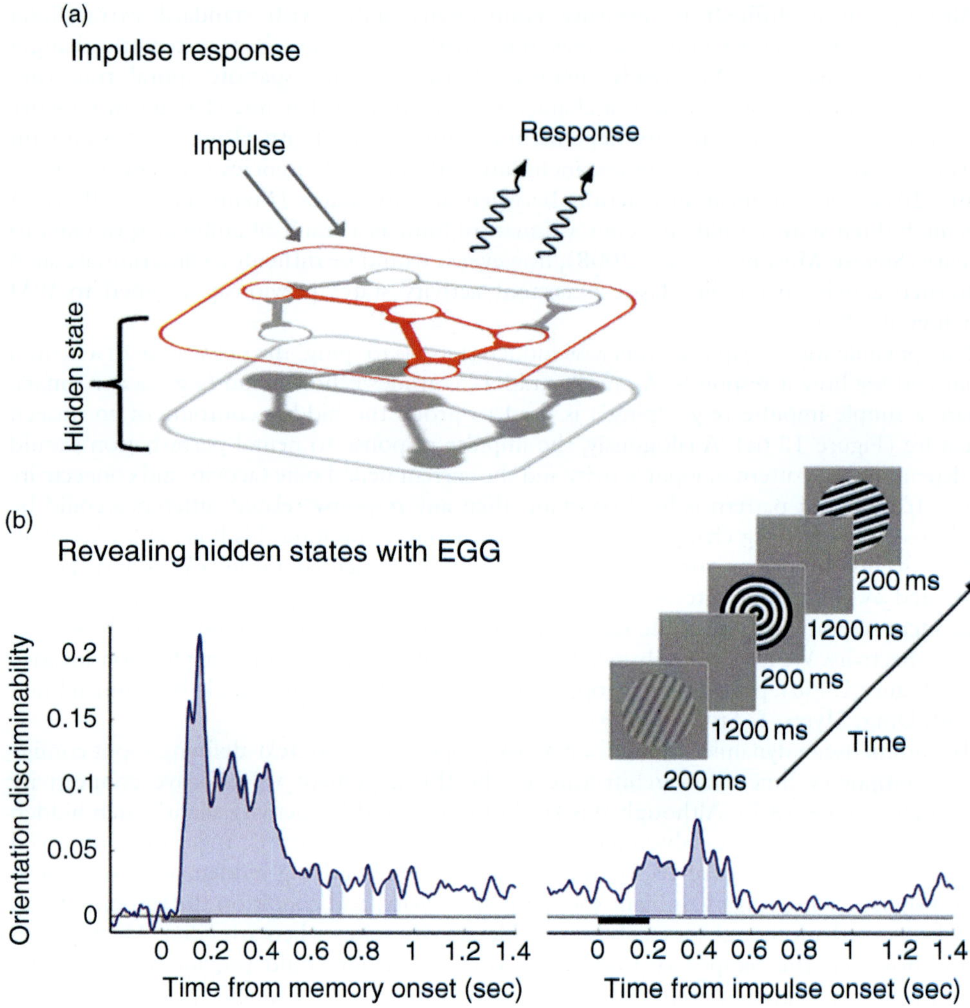

Figure 13.6 Revealing hidden brain states. a. Synaptic working memory is effectively 'activity silent'; however, the information content can be inferred from a shift in the behaviour of the network. This can be probed using an impulse of activity to drive the network. Source: Stokes 2015, Reproduced with permission of Elsevier. b. Source: Wolff et al. (2015) Frontiers in Neuroscience recently developed an approach for measuring impulse-driven reactivation of WM using non-invasive EEG in human participants. Using multivariate statistics, we could show that a neutral driving stimulus triggers a distinct pattern of activity that reflects the content of WM (blue trace), even though the driving stimulus was exactly the same for all WM conditions.

hemispheres (Barbas, Hilgetag, Saha, Dermon, & Suski, 2005; Barbas & Pandya, 1989). This means that even if different subregions within the PFC show an initial preference for different cognitive variables, this information will be rapidly shared throughout the circuit, ensuring that all regions have the necessary task-relevant information.

Second, the PFC is widely interconnected with many other cortical regions, with especially prominent connections to the temporal, parietal, and cingulate cortex (Kunzle & Akert,

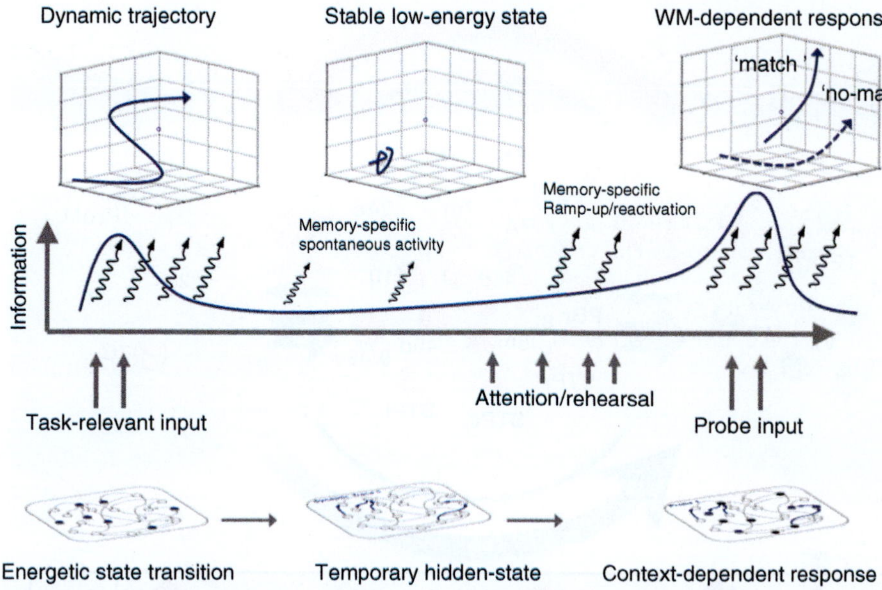

Figure 13.7 Summary schematic of dynamic coding for WM-dependent behaviour. The initial input triggers a specific pattern in the activity state, which in turn alters the underlying hidden state of the network via a temporary shift in effective connectivity (e.g., short-term synaptic plasticity or coherence). Activity-dependent changes in the hidden state drive a dynamic trajectory during the initial high-energy phase via the reciprocal interaction between hidden states and the activity states that modulate them (Meyers et al., 2008; Stokes et al., 2013). After activity has relaxed to baseline levels, the hidden state remains patterned according to the WM item. Although, in principle this temporary hidden state could be 'activity silent', spontaneous activity in the network will be patterned according to the WM context, resulting in a WM-specific activity state during spontaneous firing (Stokes et al., 2013). This kind of 'baseline emission' could help explain content-specific delay activity observed under some circumstances (Sugase-Miyamoto et al., 2008). Increasing the level of network activity, via attention/rehearsal mechanisms, could increase the discriminability of the activity state, resulting in 'ramp-up' delay activity (Barak et al., 2010; Watanabe & Funahashi, 2007) or task-dependent 'reactivations' (Watanabe & Funahashi, 2014). Finally, when the critical memory probe is presented, the context-dependent response maps activity states for WM-guided behaviour (e.g., match/non-match decision; see Stokes et al., 2013). Source: Adapted from Stokes 2015.

1977; Petrides & Pandya, 1999). Graph analysis of a large database of connections between brain regions shows how the PFC is centrally located in the large neural network that is the brain (Markov et al., 2013; Figure 13.8).

Finally, the PFC is closely interconnected with several subcortical regions. In particular, the PFC is the largest source of projections into the basal ganglia via the striatum (Haber, Kunishio, Mizobuchi, & Lynd-Balta, 1995; Jarbo & Verstynen, 2015; McFarland & Haber, 2002), forming cortico-ganglio-thalamocortical loops (Haber, 2003). We have previously suggested that these loops may allow the brain to 'bootstrap' from simple representations to more complex ones, allowing for the complexity of cognition to be built through learning (Buschman & Miller, 2014). The basal ganglia is a 'fast' learner, for example, quickly capturing associations between a stimulus and response (e.g., a bullmastiff barks). In contrast, the 'slower' learning of the PFC allows it to integrate over many experiences. Importantly,

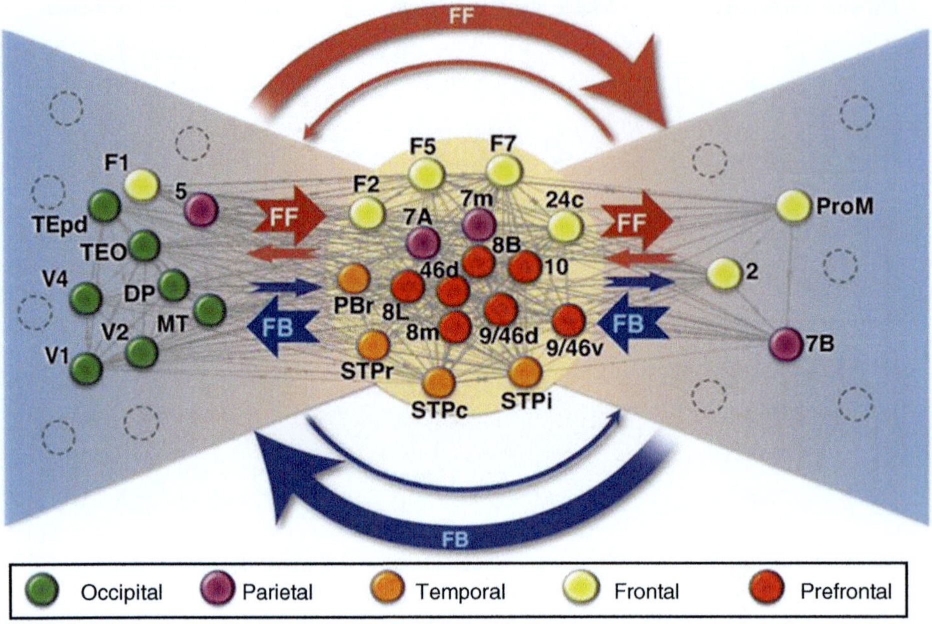

Figure 13.8 PFC is a 'hub' in the brain. Detailed anatomical work has shown that the brain is a highly interconnected network with both feedforward (FF) and feedback (FB) connections between regions. Detailed graph analysis of these connections suggests a core 'hub' of regions is more densely interconnected with other brain regions. Prefrontal regions (shown in red) are a prominent member of this hub. Source: Adapted from Markov 2013; region labels follows. (Source: Markov 2013, reproduced with permission of AAAS.)

the representations learned by the basal ganglia are available to the PFC during learning (given the strong connections between the regions). This allows the PFC to integrate these associations when forming a more generalised representation (e.g., a bullmastiff is a type of dog; and all dogs bark). This generalised representation then becomes available to further fast learning in the basal ganglia, allowing the brain to continually bootstrap increasingly complex and abstract associations.

Evidence for Top-Down Control Over Other Brain Areas

Given the anatomical position of the PFC within the broader network of brain structures, it seems natural that it plays an important role in cognitive control and, thus, cognitive flexibility. However, structural anatomy is necessary but not sufficient to determine what role the PFC plays in cognition. Critically, we must also consider how these network properties are modulated for different task contexts. Again, we need to consider the role of effective connectivity for cognitive flexibility.

Detailed studies of visual attention provide rich insights into the role of the PFC in cognitive control. Attention is the ability to selectively enhance task-relevant representations. Typically, this is studied in the domain of visual perception, where decades of research have demonstrated that attention can improve the perception of a selected visual stimulus while simultaneously suppressing distracting stimuli (for a recent review, see Carrasco, 2011). In this way, visual attention is a case study for cognitive control: Contextual/task variables are used to guide what information is represented in the brain. Just as in cognitive control more broadly, lesions of the PFC cause deficits in controlling attention (Rossi, Bichot, Desimone, & Ungerleider, 2007). In particular, deficits can be characterised as causing a reduction in the ability to shift attention—much like in the WCST, patients with PFC lesions tend to get stuck on attending to a single location or object (Knight, 1997; Knight, Grabowecky, & Scabini, 1995; see also Chapter 29 by Fellows in this volume).

However, it is important to note that the PFC does not act alone in controlling attention. In particular, lesions in the parietal cortex will also lead to problems in controlling attention that are almost identical to the impact of a PFC lesion (Eglin, Robertson, & Knight, 1991). Electrophysiological recordings in the PFC and the parietal cortex find that neurons in both regions reflect shifts in attention (for a detailed review, see Miller and Buschman, 2013). However, by using multiple electrodes to record simultaneously from both regions, Buschman and Miller (2007) found that the PFC plays a leading role when attention is wilfully directed, using internal knowledge. Once these signals are observed in the lPFC, they supply feedback in a 'top-down' manner to other prefrontal regions (the frontal eye fields, FEF) and the parietal cortex (the lateral intraparietal region, LIP). Recent work has extended these findings to humans (Li, Gratton, Yao, & Knight, 2010). These results provide evidence that the lPFC is the source of top-down control, using internal knowledge to control other brain regions.

Causal evidence that the PFC acts to bias posterior representations comes from a highly influential series of experiments using electrical microstimulation to directly stimulate neurons in the FEF (Moore, 2006). Microstimulation of FEF cells tuned to visuospatial coordinates will boost the animal's behavioural discriminability at the corresponding spatial location (Moore & Fallah, 2001, 2004). In other words, microstimulation of the FEF is functionally equivalent to shifting attention to a location. FEF microstimulation was also found to have attention-like effects on V4 neurons (Moore & Armstrong, 2003), providing direct evidence that the FEF modulates activity in the visual cortex to bias sensory processing. Similar results have been observed in humans using a combination of transcranial magnetic stimulation (TMS) and electrophysiology (Taylor, Nobre, & Rushworth, 2007).

More recently, Siegel et al. (2015) found electrophysiological evidence that the PFC plays a leading role in cognitive control during a decision-making task. Two monkeys were trained to perform a decision-making task that required the task-dependent integration of sensory information to make a choice. To trace sensory, task, and choice information across the brain, Siegel and colleagues used multiple electrodes to record simultaneously from six brain regions: the sensory cortex (V4, MT, and IT), associative cortex (lPFC and LIP), and premotor cortex (FEF) cortex. As expected, sensory information flowed in a bottom-up manner, starting in the sensory regions (V4, MT, and IT) and then propagating into the parietal and frontal regions (Figure 13.9a). In contrast, sustained task information and choice information flowed in a top-down direction, starting in the frontal and parietal cortex and then feeding back into the sensory regions (Figures 13.9b and 13.9c). The timing of the signals suggests that task information represented in fronto-parietal networks is used to selectively integrate the currently relevant sensory information and transform this signal into an action. These results complement the results from the prefrontal and parietal cortex in the control of attention, demonstrating the PFC represents context information in order to support task-relevant computations throughout the brain.

PFC as the Central Executive

Taken together, all of these results suggest that the PFC is ideally suited to play the role of a 'central executive'. As reviewed above, lPFC neurons have the ability to adaptively and dynamically encode the currently task-relevant information. Furthermore, because of its

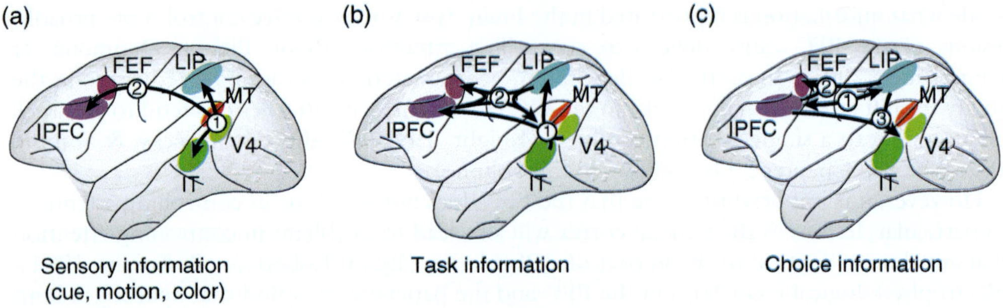

Figure 13.9 Dynamics of task information across six different brain regions. (a) Sensory information flows from the sensory cortex into the frontal-parietal associative areas. In contrast, (b) sustained task information and (c) decision information are seen first in the frontal-parietal network. Source: Siegel 2015. Reproduced with permission of AAAS.

anatomical connections, it is well positioned to use this activity to influence much of the brain. In this way, the PFC can act as a 'central executive', guiding activity throughout the brain in a way that accomplishes the current task.

But what are the neural mechanisms that would allow the PFC to act as a central executive? When considering possible mechanisms, it is important to remember that they must meet two competing criteria. First, any mechanism of cognitive control must be able to selectively boost those representations that are relevant to accomplishing the current task (while also suppressing undesired representations). However, this selection cannot be absolute—behaviour in the real world is highly dynamic, changing from moment to moment. Therefore, the selection mechanism must be flexible enough to allow for rapid changes between behaviours.

The first requirement can be met via experience-dependent changes in network connections. Learning could establish new anatomical pathways that are optimised for repeated scenarios and behavioural routines. However, the second requirement means that we need to consider more rapid changes in network behaviour. The brain simply cannot rewire on a moment-to-moment basis. Instead, flexibility demands that some changes must be rapid and temporary. In other words, anatomy acts as the substrate that defines the set of all possible behaviours, and temporary changes in effective connectivity ride on top to select the currently relevant behaviour.

Miller and Cohen (2001) outline one biophysically plausible mechanism through which this might occur. They propose that, at any given moment in time, we are faced with a plethora of stimulus inputs, all of which can lead to a large set of possible responses. As we reviewed above, there is strong evidence that the PFC is the source of internally directed attention and that attention selects a specific desired stimulus from competing stimuli. Miller and Cohen propose that this same selection mechanism can be generalised from stimulus inputs to all possible stimulus–response mappings. In other words, the model suggests that neural activity in the PFC is distributed across the brain in a manner that boosts the representations of the contextually appropriate behaviour. Competition between representations leads to suppression of inappropriate behaviours. Finally, because control depends only on the pattern of activity in the PFC, it can be highly flexible, changing as rapidly as the patterns of activity in the PFC.

However, the biasing mechanism underlying the Miller and Cohen model may not be the only mechanism supporting cognitive control. We suggest that cognitive control (and its flexibility) is the result of changes in effective connectivity. In particular, we suggest that the same changes in effective connectivity that act to support dynamics within the PFC can be generalised to the whole-brain network in order to support cognitive flexibility.

Effective Connectivity as a General Mechanism for Dynamic Coding Within and Across Brain Areas

As reviewed above, changes in effective connectivity between neurons can alter the dynamics of a population of neurons within a cortical region. This could occur through either STSP and/or through changes in the synchrony between populations of neurons. These same mechanisms could also apply at the scale of whole-brain networks, with changes in the effective connectivity between regions acting to bias connections in a way that supports the current behaviour.

This is exemplified in many of our everyday experiences. For example, when walking home from work, you follow a typical path that has been learned over many experiences: You know to go straight when approaching a familiar intersection (Figure 13.10a). Once well learned, such behaviours are reliable and operate effortlessly, reflected in the fact that one's mind often wanders as one traverses the route. However, if you suddenly remember you have to get milk on the way home, you need to override this default path, deviating towards the corner store. Now the same stimulus (the familiar intersection) must lead to a different response (turning left; Figure 13.10b). In this case, the PFC becomes engaged and represents the current rule in a pattern of activity across a population of neurons. Synchrony within the PFC couples the task-relevant neurons into an ensemble, whereas synchrony between regions may act to bias currently relevant associations while suppressing alternatives.

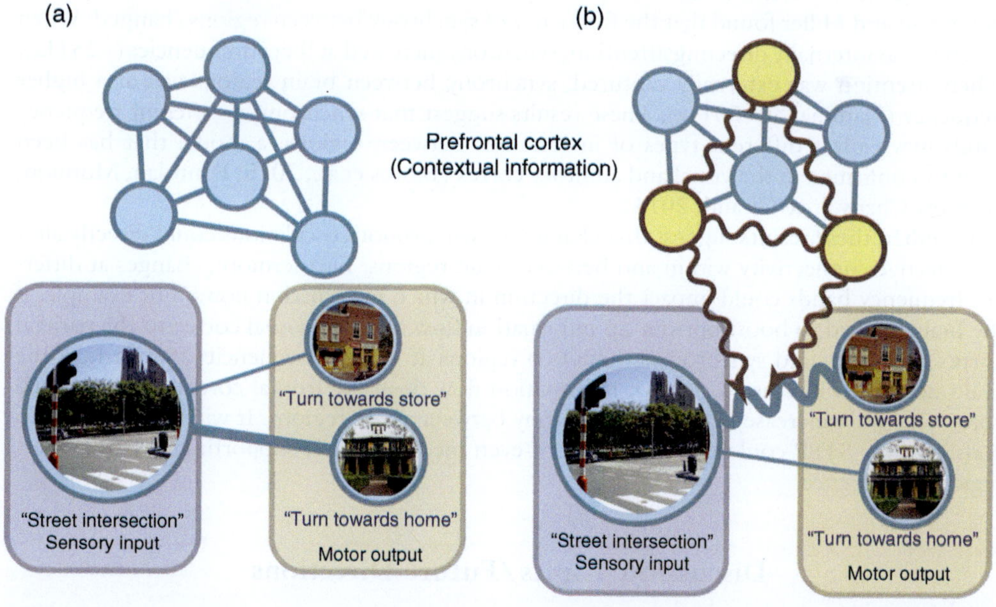

Figure 13.10 Model of cognitive control. (a) At baseline, stimuli lead to habitual responses. (b) If one is in a situation where the context requires alternative responses, then the PFC represents the current task and uses this to bias representations throughout the brain.

Recent neuroimaging experiments provide general support for such a model. For example, Rowe et al. (2005) found that the effective connectivity between the PFC and the posterior cortex changed as subjects switched between performing different tasks. Subjects were asked to perform two 'free-choice' tasks: one where they selected a colour of their choosing and one where they selected an action of their choosing. Using structural equation modelling, they measured the effective connectivity between the PFC, motor cortex, and visual cortex during both tasks. Interestingly, they found PFC–motor connectivity during the free-choice action task and greater PFC–visual connectivity during the free-choice colour task. When subjects were instructed as to which colour/action to choose, this was no longer the case, suggesting that effective connectivity is only seen when cognitive control is needed to select among competing representations. Similar results have also been seen during visual discrimination tasks (Chadick & Gazzaley, 2011).

In addition, there has been growing evidence that changes in synchrony within the lPFC may support flexible cognitive control. As reviewed above, Buschman et al. (2012) found that synchrony within the PFC acts to carve task-relevant ensembles of neurons from the PFC population. This could also act to manage the highly dynamic representations within the PFC—if neurons are multidimensional, then understanding their current meaning requires integrating their activity with a larger population (in order to disambiguate between multiple preferred cases). Synchrony could act to bring such a population together, allowing downstream neurons to accurately decode the current state (see also 'synapsembles'; Buzsaki, 2010)

Such changes in synchrony have also been observed across different brain regions. For example, Pesaran et al. (2008) found that synchrony between the PFC and the parietal cortex was increased when animals were making a free choice regarding what response to make in comparison to when they were being instructed as to their choice. Similarly, Buschman and Miller (2007) showed that synchrony between the PFC and the parietal cortex changes depending on whether attention is internally directed or externally captured. Furthermore, Buschman and Miller found that the frequency of synchrony between regions changed: When the PFC was internally directing attention, synchrony increased at 'beta' frequencies (~25 Hz). When attention was externally captured, synchrony between brain regions was at a higher frequency ('gamma' at ~50 Hz). These results suggest that synchrony at different frequency bands may reflect different types of interactions between regions, a model that has been recently confirmed in the visual and auditory cortex (Bastos et al., 2015; Fontolan, Morillon, Liegeois-Chauvel, & Giraud, 2014).

Together, these results suggest that changes in synchronous oscillations could directly alter the effective connectivity within and between brain regions. Furthermore, changes at different frequency bands could impact the direction in which information flows. For example, if the brain needed to boost bottom-up information flow from the visual cortex to the parietal cortex, then it would synchronise these two regions at gamma frequencies. Similarly, if the brain needed to increase top-down information flow from the frontal cortex to the parietal cortex, it would increase beta-band synchrony between these regions. It will be important to consider how STSP could play a parallel, or even mechanistically supporting, role for interregional coherence.

Discussion Topics/Future Directions

In this chapter, we have reviewed evidence that the PFC plays a central role in flexible cognitive control. We have highlighted the dynamic nature of PFC representations and how such representations may support the flexibility needed in high-level cognition. Finally, we reviewed evidence that the PFC can use these dynamic representations to bias activity throughout the brain in order to achieve the task at hand. In particular, we highlighted the importance of flexible networks, suggesting that either short-term changes in synaptic plasticity or changes in synchrony could be effective mechanisms for temporarily altering the connectivity between and within populations of neurons.

However, there are several remaining questions. First, it is not clear how the dynamics observed within and between brain regions relate to the specific underlying anatomy. In this chapter, we have assumed that changes in effective connectivity act within the framework of the available anatomical connections. However, it remains to be seen to what degree this is true—what proportion of connections are malleable in this way? It is also unclear what neural mechanisms drive changes in effective connectivity. For example, if changes in synchrony are important for guiding the flow of information within and between networks, what are the neural mechanisms that alter this synchrony? How are specific patterns of effective connectivity established and maintained?

Cognitive neuroscience has come a long way since the first PFC recordings in awake, behaving monkeys (e.g., Fuster & Alexander, 1971). In this chapter, we have attempted to highlight the dynamic nature of PFC representations, and consider potential mechanisms that may allow for prefrontal neurons to alter responses throughout the brain to support the current behaviour. Together, these ideas could help shed some light on the most important aspects of intelligent behaviour: cognitive flexibility.

References

Barak, O., Tsodyks, M., & Romo, R. (2010). Neuronal population coding of parametric working memory. *Journal of Neuroscience*, *30*(28), 9424–9430.

Barbas, H., Hilgetag, C. C., Saha, S., Dermon, C. R., & Suski, J. L. (2005). Parallel organization of contralateral and ipsilateral prefrontal cortical projections in the rhesus monkey. *BMC Neuroscience*, *6*, 32.

Barbas, H., & Pandya, D. N. (1989). Architecture and intrinsic connections of the prefrontal cortex in the rhesus monkey. *Journal of Comparative Neurology*, *286*(3), 353–375.

Barcelo, F., & Knight, R. T. (2002). Both random and perseverative errors underlie WCST deficits in prefrontal patients. *Neuropsychologia*, *40*(3), 349–356.

Bastos, A. M., Vezoli, J., Bosman, C. A., Schoffelen, J. M., Oostenveld, R., Dowdall, J. R.,...Fries, P. (2015). Visual areas exert feedforward and feedback influences through distinct frequency channels. *Neuron*, *85*(2), 390–401.

Bernacchia, A., Seo, H., Lee, D., & Wang, X. J. (2011). A reservoir of time constants for memory traces in cortical neurons. *Nature Neuroscience*, *14*(3), 366–372.

Buckley, M. J., Mansouri, F. A., Hoda, H., Mahboubi, M., Browning, P. G., Kwok, S. C.,...Tanaka, K. (2009). Dissociable components of rule-guided behavior depend on distinct medial and prefrontal regions. *Science*, *325*(5936), 52–58.

Buonomano, D. V., & Maass, W. (2009). State-dependent computations: Spatiotemporal processing in cortical networks. *Nature Reviews Neuroscience*, *10*(2), 113–125.

Buschman, T. J., Denovellis, E. L., Diogo, C., Bullock, D., & Miller, E. K. (2012). Synchronous oscillatory neural ensembles for rules in the prefrontal cortex. *Neuron*, *76*(4), 838–846.

Buschman, T. J., & Miller, E. K. (2007). Top-down versus bottom-up control of attention in the prefrontal and posterior parietal cortices. *Science*, *315*(5820), 1860–1862.

Buschman, T. J., & Miller, E. K. (2014). Goal-direction and top-down control. *Philosophical Transactions of the Royal Society B: Biological Sciences*, *369*(1655).

Buzsaki, G. (2010). Neural syntax: Cell assemblies, synapsembles, and readers. *Neuron*, *68*(3), 362–385.

Carrasco, M. (2011). Visual attention: The past 25 years. *Vision Research*, *51*(13), 1484–1525.

Chadick, J. Z., & Gazzaley, A. (2011). Differential coupling of visual cortex with default or frontal-parietal network based on goals. *Nature Neuroscience, 14*(7), 830–832.

Crowe, D. A., Averbeck, B. B., & Chafee, M. V. (2010). Rapid sequences of population activity patterns dynamically encode task-critical spatial information in parietal cortex. *Journal of Neuroscience, 30*(35), 11640–11653.

Deco, G., Rolls, E. T., & Romo, R. (2010). Synaptic dynamics and decision making. *Proceedings of the National Academy of Sciences of the United States of America, 107*(16), 7545–7549.

Duncan, J. (2010). The multiple-demand (MD) system of the primate brain: Mental programs for intelligent behaviour. *Trends in Cognitive Sciences, 14*(4), 172–179.

Duncan, J., & Miller, E. K. (2002). Cognitive focusing through adaptive neural coding in the primate prefrontal cortex. In D. Stuss & R. T. Knight (Eds.), *Principles of frontal lobe function* (pp. 278–291). New York: Oxford University Press.

Eglin, M., Robertson, L. C., & Knight, R. T. (1991). Cortical substrates supporting visual search in humans. *Cerebral Cortex, 1*(3), 262–272.

Fontolan, L., Morillon, B., Liegeois-Chauvel, C., & Giraud, A. L. (2014). The contribution of frequency-specific activity to hierarchical information processing in the human auditory cortex. *Nature Communications, 5*, 4694.

Fries, P. (2005). A mechanism for cognitive dynamics: Neuronal communication through neuronal coherence. *Trends in Cognitive Sciences, 9*(10), 474–480.

Fujisawa, S., Amarasingham, A., Harrison, M. T., & Buzsaki, G. (2008). Behavior-dependent short-term assembly dynamics in the medial prefrontal cortex. *Nature Neuroscience, 11*(7), 823–833.

Funahashi, S., Bruce, C. J., & Goldman-Rakic, P. S. (1989). Mnemonic coding of visual space in the monkey's dorsolateral prefrontal cortex. *Journal of Neurophysiology, 61*(2), 331–349.

Fuster, J. M. (2001). The prefrontal cortex—An update: Time is of the essence. *Neuron, 30*(2), 319–333.

Fuster, J. M. (2002). Frontal lobe and cognitive development. *Journal of Neurocytology, 31*(3–5), 373–385.

Fuster, J. M., & Alexander, G. E. (1971). Neuron activity related to short-term memory. *Science, 173*(3997), 652–654.

Haber, S. N. (2003). The primate basal ganglia: Parallel and integrative networks. *Journal of Chemical Neuroanatomy, 26*(4), 317–330.

Haber, S. N., Kunishio, K., Mizobuchi, M., & Lynd-Balta, E. (1995). The orbital and medial prefrontal circuit through the primate basal ganglia. *Journal of Neuroscience, 15*(7 Pt 1), 4851–4867.

Harvey, C. D., Coen, P., & Tank, D. W. (2012). Choice-specific sequences in parietal cortex during a virtual-navigation decision task. *Nature, 484*(7392), 62–68.

Jarbo, K., & Verstynen, T. D. (2015). Converging structural and functional connectivity of orbitofrontal, dorsolateral prefrontal, and posterior parietal cortex in the human striatum. *Journal of Neuroscience, 35*(9), 3865–3878.

Knight, R. T. (1997). Distributed cortical network for visual attention. *Journal of Cognitive Neuroscience, 9*(1), 75–91.

Knight, R. T., Grabowecky, M. F., & Scabini, D. (1995). Role of human prefrontal cortex in attention control. *Advances in Neurology, 66*, 21–34; discussion 34-26.

Kunzle, H., & Akert, K. (1977). Efferent connections of cortical, area 8 (frontal eye field) in Macaca fascicularis: A reinvestigation using the autoradiographic technique. *Journal of Comparative Neurology, 173*(1), 147–164.

Lebedev, M. A., Messinger, A., Kralik, J. D., & Wise, S. P. (2004). Representation of attended versus remembered locations in prefrontal cortex. *PLoS Biology, 2*(11), e365.

Li, L., Gratton, C., Yao, D., & Knight, R. T. (2010). Role of frontal and parietal cortices in the control of bottom-up and top-down attention in humans. *Brain Research, 1344*, 173–184.

Mante, V., Sussillo, D., Shenoy, K. V., & Newsome, W. T. (2013). Context-dependent computation by recurrent dynamics in prefrontal cortex. *Nature, 503*(7474), 78–84.

Markov, N. T., Ercsey-Ravasz, M., Van Essen, D. C., Knoblauch, K., Toroczkai, Z., & Kennedy, H. (2013). Cortical high-density counterstream architectures. *Science, 342*(6158), 1238406.

McFarland, N. R., & Haber, S. N. (2002). Thalamic relay nuclei of the basal ganglia form both reciprocal and nonreciprocal cortical connections, linking multiple frontal cortical areas. *Journal of Neuroscience, 22*(18), 8117–8132.

Meyers, E. M., Freedman, D. J., Kreiman, G., Miller, E. K., & Poggio, T. (2008). Dynamic population coding of category information in inferior temporal and prefrontal cortex. *Journal of Neurophysiology, 100*(3), 1407–1419.

Miller, E. K., & Cohen, J. D. (2001). An integrative theory of prefrontal cortex function. *Annual Review of Neuroscience, 24*, 167–202.

Miller, E. K., Erickson, C. A., & Desimone, R. (1996). Neural mechanisms of visual working memory in prefrontal cortex of the macaque. *Journal of Neuroscience, 16*(16), 5154–5167.

Mongillo, G., Barak, O., & Tsodyks, M. (2008). Synaptic theory of working memory. *Science, 319*(5869), 1543–1546.

Moore, T. (2006). The neurobiology of visual attention: Finding sources. *Current Opinion in Neurobiology, 16*(2), 159–165.

Moore, T., & Armstrong, K. M. (2003). Selective gating of visual signals by microstimulation of frontal cortex. *Nature, 421*(6921), 370–373.

Moore, T., & Fallah, M. (2001). Control of eye movements and spatial attention. *Proceedings of the National Academy of Sciences of the United States of America, 98*(3), 1273–1276.

Moore, T., & Fallah, M. (2004). Microstimulation of the frontal eye field and its effects on covert spatial attention. *Journal of Neurophysiology, 91*(1), 152–162.

Murray, J. D., Bernacchia, A., Freedman, D. J., Romo, R., Wallis, J. D., Cai, X.,...Wang, X. J. (2014). A hierarchy of intrinsic timescales across primate cortex. *Nature Neuroscience, 17*(12), 1661–1663.

Nobre, A. C., Correa, A., & Coull, J. (2007). The hazards of time. *Current Opinion in Neurobiology, 17*(4), 465–470.

Olivers, C. N., Peters, J., Houtkamp, R., & Roelfsema, P. R. (2011). Different states in visual working memory: When it guides attention and when it does not. *Trends in Cognitive Sciences, 15*(7), 327–334.

Pesaran, B., Nelson, M. J., & Andersen, R. A. (2008). Free choice activates a decision circuit between frontal and parietal cortex. *Nature, 453*(7193), 406–409.

Petrides, M., & Pandya, D. N. (1999). Dorsolateral prefrontal cortex: Comparative cytoarchitectonic analysis in the human and the macaque brain and corticocortical connection patterns. *European Journal of Neuroscience, 11*(3), 1011–1036.

Qi, X. L., Meyer, T., Stanford, T. R., & Constantinidis, C. (2011). Changes in prefrontal neuronal activity after learning to perform a spatial working memory task. *Cerebral Cortex, 21*(12), 2722–2732.

Rainer, G., Asaad, W. F., & Miller, E. K. (1998). Selective representation of relevant information by neurons in the primate prefrontal cortex. *Nature, 393*(6685), 577–579.

Rainer, G., Rao, S. C., & Miller, E. K. (1999). Prospective coding for objects in primate prefrontal cortex. *Journal of Neuroscience, 19*(13), 5493–5505.

Rigotti, M., Barak, O., Warden, M. R., Wang, X. J., Daw, N. D., Miller, E. K., & Fusi, S. (2013). The importance of mixed selectivity in complex cognitive tasks. *Nature, 497*(7451), 585–590.

Rossi, A. F., Bichot, N. P., Desimone, R., & Ungerleider, L. G. (2007). Top-down attentional deficits in Macaques with lesions of lateral prefrontal cortex. *Journal of Neuroscience, 27*(42), 11306–11314.

Rowe, J. B., Stephan, K. E., Friston, K., Frackowiak, R. S., & Passingham, R. E. (2005). The prefrontal cortex shows context-specific changes in effective connectivity to motor or visual cortex during the selection of action or colour. *Cerebral Cortex, 15*(1), 85–95.

Sakai, K. (2008). Task set and prefrontal cortex. *Annual Review of Neuroscience, 31*, 219–245.

Salazar, R. F., Dotson, N. M., Bressler, S. L., & Gray, C. M. (2012). Content-specific fronto-parietal synchronization during visual working memory. *Science, 338*(6110), 1097–1100.

Shafi, M., Zhou, Y., Quintana, J., Chow, C., Fuster, J., & Bodner, M. (2007). Variability in neuronal activity in primate cortex during working memory tasks. *Neuroscience, 146*(3), 1082–1108.

Shallice, T., & Burgess, P. W. (1991). Deficits in strategy application following frontal lobe damage in man. *Brain, 114*(Pt. 2), 727–741.

Siegel, M., Buschman, T. J., & Miller, E. K. (2015). Cortical information flow during flexible sensorimotor decisions. *Science, 348*(6241), 1352–1355.

Siegel, M., Warden, M. R., & Miller, E. K. (2009). Phase-dependent neuronal coding of objects in short-term memory. *Proceedings of the National Academy of Sciences of the United States of America, 106*(50), 21341–21346.

Sigala, N., Kusunoki, M., Nimmo-Smith, I., Gaffan, D., & Duncan, J. (2008). Hierarchical coding for sequential task events in the monkey prefrontal cortex. *Proceedings of the National Academy of Sciences of the United States of America, 105*(33), 11969–11974.

Stokes, M. G. (2011). The spatiotemporal structure of population coding in monkey parietal cortex. *Journal of Neuroscience, 31*(4), 1167–1169.

Stokes, M. G. (2015). 'Activity-silent' working memory in prefrontal cortex: A dynamic coding framework. *Trends in Cognitive Sciences, 19*(7), 394–405.

Stokes, M. G., Kusunoki, M., Sigala, N., Nili, H., Gaffan, D., & Duncan, J. (2013). Dynamic coding for cognitive control in prefrontal cortex. *Neuron, 78*(2), 364–375.

Sugase-Miyamoto, Y., Liu, Z., Wiener, M. C., Optican, L. M., & Richmond, B. J. (2008). Short-term memory trace in rapidly adapting synapses of inferior temporal cortex. *PLoS Computational Biology, 4*(5), e1000073.

Taylor, P. C., Nobre, A. C., & Rushworth, M. F. (2007). FEF TMS affects visual cortical activity. *Cerebral Cortex, 17*(2), 391–399.

Vapnik, V. N. (1998). *Statistical learning theory*. New York: Wiley.

Wallis, J. D., Anderson, K. C., & Miller, E. K. (2001). Single neurons in prefrontal cortex encode abstract rules. *Nature, 411*(6840), 953–956.

Wang, M., Yang, Y., Wang, C. J., Gamo, N. J., Jin, L. E., Mazer, J. A.,...Arnsten, A. F. (2013). NMDA receptors subserve persistent neuronal firing during working memory in dorsolateral prefrontal cortex. *Neuron, 77*(4), 736–749.

Watanabe, K., & Funahashi, S. (2007). Prefrontal delay-period activity reflects the decision process of a saccade direction during a free-choice ODR task. *Cerebral Cortex, 17*(Suppl. 1), i88–100.

Watanabe, K., & Funahashi, S. (2014). Neural mechanisms of dual-task interference and cognitive capacity limitation in the prefrontal cortex. *Nature Neuroscience, 17*(4), 601–611.

White, I. M., & Wise, S. P. (1999). Rule-dependent neuronal activity in the prefrontal cortex. *Experimental Brain Research, 126*(3), 315–335.

Wolff, M., Ding, J., Myers, N., & Stokes, M. G. (2015). Revealing hidden states in working memory using electroencephalography. *Frontiers in Neuroscience*.

Zucker, R. S., & Regehr, W. G. (2002). Short-term synaptic plasticity. *Annual Review of Physiology, 64*, 355–405.

14

Neurophysiological Oscillations and Action Monitoring

Michael X. Cohen

To Err Is Neural

Mistakes can be expensive, and in some cases, difficult to clean up. It is in the best interests of every organism to minimise mistakes and to avoid situations in which mistakes are likely to be made, or have been made in the past. On the other hand, biological organisms have been evolving here on planet Earth for approximately 3.5 billion years, and humans are generally believed to be the most intellectually advanced species on the planet. It is therefore puzzling that we make mistakes so frequently. Indeed, acknowledging mistakes has become a natural part of language ('sorry', 'stuff happens', 'mistakes were made', and so on), and people make near-mistakes much more often than they are aware of, perhaps even dozens of times each day (Cohen & van Gaal, 2014; Rochet, Spieser, Casini, Hasbroucq, & Burle, 2014).

Fortunately, the brain has the remarkable ability to very quickly detect that a mistake is about to be committed, and then engage rapid but flexible strategies in order to avoid making that mistake. And in the unfortunate event of actually making a mistake, the brain has the ability to learn from that experience to avoid making the same mistake in the future. We have made so many mistakes over the past hundreds of millions of years that evolution has endowed us with brains containing dedicated circuits that are specialised for the sole purpose of detecting when we are about to make a mistake, adapting behaviour when mistakes are made, and learning from those mistakes.

Action monitoring and adjustments have captured the attention of psychologists for a long time, and, more recently, that of neuroscientists. The purpose of this chapter is to review the literature on electrophysiology and action monitoring, and to discuss insights into the neural computations that might underlie action monitoring and how those computations might be implemented by neurons and neural circuits. More specifically, this chapter will focus almost exclusively on findings revealed by analysing EEG, MEG, and LFP signals using analysis methods that allow for interpretations of neural oscillations. As explained below, this focus reflects the assumption that neural oscillations are the strongest available link between non-invasive neuroimaging and neurophysiological processes. For a review of error-related processing assessed by other means (e.g., fMRI), see Chapter 17 by Ullsperger in this volume.

The Wiley Handbook of Cognitive Control, First Edition. Edited by Tobias Egner.

Studying Action Monitoring and Action Adjustments in the Laboratory

Studying human action monitoring/adjustments in the laboratory involves having volunteer human subjects perform computerised tasks while their electrical brain activity is monitored using electroencephalography (EEG).

Action monitoring processes are most often elicited by using computerised tasks that produce response conflict. Generally speaking, response conflict occurs when one action (e.g., a right-hand button press) is automatically activated by a task-irrelevant feature, such as the physical location of the stimulus, while a different action (e.g., a left-hand button press) is activated by the task-relevant feature that is given by instruction. The task-relevant feature often entails arbitrary stimulus–response mappings such as pressing the left-hand button for green stimuli and the right-hand button for purple stimuli. The competition between the fast automatic response and the slow task-relevant response generates conflict (Ridderinkhof, 2002), particularly when subjects are encouraged to respond quickly.

Over the past several decades, psychologists have developed a variety of tasks that allow careful manipulation of the strength of response conflict. Three tasks have emerged as the most typical. In the 'Simon task' (Leuthold, 2011), coloured stimuli are briefly presented on the right or the left side of the screen. Subjects respond as quickly as possible to the colour (task-relevant feature) while ignoring the location (task-irrelevant feature) of the coloured stimulus. In 'congruent' trials, the stimulus is presented on the same side as the required response; in 'incongruent' trials (those with response conflict), the stimulus is presented on the opposite side as the required response. The incongruent trials are associated with longer reaction times, increased error rates, and post-trial adjustments in control (Ullsperger, Danielmeier, & Jocham, 2014), all behavioural effects attributable to response conflict (see also Chapter 4 by Egner in this volume). In the 'flankers task' (Eriksen & Eriksen, 1974), an array of letters appears on the screen, and subjects respond according to the central letter while ignoring the surrounding flanking letters. Conflict is induced when the target and flanking letters are associated with different responses, for example, 'X X Y X X' where the letter 'Y' is associated with a right-hand response, whereas the letter 'X' is associated with a left-hand response. The third commonly used task is the 'Stroop task' (Stroop, 1935), in which subjects report the colour of a font rather than the meaning of the word. For example, when the word BLUE is printed in red font, one must inhibit the automatic urge to pronounce 'blue' in favour of responding 'red'.

Some of these tasks have been adapted for use in non-human animals. For example, in a rodent model of the Simon task, the rat can poke its nose into a left or right hole in the hope of obtaining a food reward. Prior to the nose poke, a high- or low-pitched auditory cue informs the animal whether it will be rewarded for a left or a right nose poke. The auditory cue is played either from the left or from the right side of the animal. Response conflict is induced, for example, when the tone is played on the left side of the animal but instructs a rightward nosepoke (Marx et al., 2012). Rats are slower to respond on these trials, suggesting that they may experience some kind of 'conflict' that is analogous to what humans experience during the Simon task. Advantages and limitations of using non-human animals to understand human cognitive control processes will be discussed later.

One limitation of standard cognitive control tasks such as Stroop, Simon, and flankers tasks is that they are poor models of real-world and naturalistic behaviours. They are characterised by discrete trials with minimal action requirements that are separated by seconds of inactivity. These kinds of tasks are useful for EEG data analyses because they simplify data analysis protocols. However, they may not reflect cognitive control processes exactly as they are utilised when we are outside the sterile and highly controlled confines of the research laboratory. Significant advances in this field may arise from the development and use of more real-world-relevant tasks.

Neural Oscillations to Study Neural Mechanisms of Cognition

Neural oscillations are rhythmic fluctuations in the activity of populations of neurons. Individual neurons will often exhibit oscillatory behaviours such as subthreshold voltage fluctuations or rhythmic spiking, although typically when neuroscientists speak of 'neural oscillations' they refer to circuit- or network-level oscillations: the aggregated activity of hundreds or thousands, or—in the case of EEG or MEG, hundreds of thousands—of neurons.

The study of the roles of oscillations in the brain is a large field of research, and there are many compelling findings and several dominant theories that suggest important contributions of neural oscillations in behaviourally relevant neural computations. Readers interested in general reviews of neural oscillations in brain function may find the following references to be useful starting points (Buzsáki & Draguhn, 2004; Wang, 2010). In the interest of brevity, this section will discuss only those aspects of neural oscillations that are most relevant for understanding the possible neural mechanisms underlying action monitoring.

There are two themes that permeate the literature on neural oscillations that seem most relevant for this discussion. One theme is that neural oscillations facilitate information transfer across spatially disparate neural networks (e.g., different brain regions). In other words, oscillations can be used to route or gate the flow of information in the brain (Fries, 2005; Jensen & Mazaheri, 2010). In part, this occurs because strong oscillations constrain the timing of action potentials (Lisman & Jensen, 2013; Reimann et al., 2013; Vinck, Oostenveld, van Wingerden, Battaglia, & Pennartz, 2011), and when a downstream neuron receives convergent and synchronised input from many afferent neurons, the resulting burst provides a nonlinear boost in input strength (Eyherabide, Rokem, Herz, & Samengo, 2009; Kepecs, Wang, & Lisman, 2002). This rapid and flexible gating mechanism is both faster and less permanent than, for example, synaptic plasticity, which means that oscillations can regulate neural information flow in the time frame of cognitive processes (tens to hundreds of milliseconds). This is relevant for action monitoring/adaptation because in order for the brain to act flexibly on conflict signals, those signals must be rapidly transmitted to task-relevant regions such as sensory or motor regions.

A second theme is that neural oscillations provide an internal 'clock' for timing and coordinating neural computations. Neural information processing is highly temporally precise (Cohen, 2011b), and oscillations—particularly their phases—provide a temporally precise framework in which the upcoming state of a neural network or circuit configuration can be predicted. This clock-like process is particularly important for computations that must be repeated regularly in a cyclic fashion. As an analogy, think of the day–night cycle of the Earth spinning on its axis as an oscillation, and a watch as marking the phase of that oscillation; many important events such as eating and sleeping recur at specific phases of the oscillation, and their timing can be predicted on the basis of phase. If the neuromechanical implementations of some aspects of action monitoring are fast-repeated processes, oscillations might be used to coordinate the timing of those processes.

Aside from their contributions to brain function and information processing, neural oscillations are also an attractive feature of the brain for neuroscientists to study. Neural oscillations have been observed at and are implicated in brain function at a large range of spatial-temporal scales, from the 'microscopic' scale of synapses and individual neuron action potentials, to the 'mesoscopic' scale of neural circuits, columns, and networks (generally speaking, the coordinated activity of hundreds to thousands of neurons), to the 'macroscopic' scale of patches of brain tissue that are measurable with non-invasive imaging such as M/ EEG or fMRI. Furthermore, neural oscillation speeds range from very slow (<1 Hz oscillations during sleep) to very fast (>500 Hz oscillations in the somatosensory cortex). Neural oscillations are therefore the best candidate functional feature of the brain for understanding how multiple spatial-temporal scales are interconnected (Cohen & Gulbinaite, 2013; Le Van Quyen, 2011; Palva & Palva, 2012).

Relatedly, neural oscillations are also a unique feature of brain function that can be used to compare findings across different species. This is because over the course of evolution, and despite the huge differences in the size of the brain, the speeds of neural oscillations have remained remarkably constant over species (Buzsáki, Logothetis, & Singer, 2013). Even zebrafish and insects have neural oscillations in the same frequency bands as those in humans (Kay, 2014). This indicates that oscillations have a very fundamental role in brain function that has been conserved across many species.

Investigating neural oscillations in EEG data involves testing for temporal patterns in the data that are sinusoidal, and are both temporally brief (hundreds of milliseconds to a few seconds) and frequency band limited. Readers interested in in-depth overviews and instructions on analysing neural time series data may consult Cohen (Cohen, 2014b).

Midfrontal Theta Oscillations and Action Monitoring

The EEG literature has identified several electrophysiological signatures of action monitoring, including time-domain event-related potential components such as the N200 (or N2), error-related negativity (ERN), and feedback-related negativity (FRN) (Ullsperger, Fischer, Nigbur, & Endrass, 2014; Chapter 17 by Ullsperger in this volume). Time-domain trial averaging became the standard cognitive electrophysiology analysis approach several decades ago when computing power was very limited and when the neurobiological understanding of neurons, neural circuits, and local field potentials was relatively impoverished. In addition to obscuring a considerable amount of the dynamics present in EEG data, time-domain averaging also has a poorly understood link to neurophysiological processes. In contrast, studies utilising time-frequency-based analyses reveal additional dynamics, often have higher statistical power and signal-to-noise characteristics, and allow interpretations of neural oscillations. Time-frequency-based analyses therefore provide stronger grounding on which to develop neurophysiologically inspired and neurobiologically plausible theories concerning how action monitoring/adjustment processes are implemented by neurons and neural circuits in the brain. For these reasons, this chapter will focus exclusively on reviewing findings obtained from time-frequency analyses.

There have been several dozens of publications over the past 15 years that have reported time-frequency analyses of EEG activity during action monitoring/adjustment tasks. Taking a bird's eye view of this literature suggests that these studies can be fairly accurately summarised in four key findings that are consistently reported.

The first key finding is that situations requiring action monitoring/adjustments elicit theta-band (4–8 Hz) oscillations over the midfrontal cortex (Cohen & Cavanagh, 2011; Cohen & van Gaal, 2014; Hanslmayr et al., 2008; Pastötter, Hanslmayr, & Bäuml, 2010). The conflict-related increase in midfrontal theta is a robust phenomenon and is seen using a variety of experimental tasks (e.g., Stroop, flankers, Simon) that were described earlier in this chapter.

The frequency peak of this response is around 6.5 Hz and is maximal at EEG electrode FCz (sometimes Cz, but rarely lateral or more anterior). The response-conflict-modulated theta-band response is frequency band limited and typically does not extend below ~3–4 Hz or above ~8–9 Hz. Some of the apparent frequency dispersion seen in publication figures results from a combination of individual differences and frequency smoothing inherent in time-frequency analyses. Thus, individual subjects often have narrower frequency peaks than what is observed in the group-averaged plots. Our preliminary correlational analyses (data not published) have not identified any consistent relationships between individual peak frequency and, for example, task performance.

Conflict-modulated theta tends to be fairly brief, lasting a few hundred milliseconds between stimulus onset and button press. On the basis of studies that have measured finger muscle movements, button presses, and EEG, it seems that the theta response is most strongly time locked to the initial muscle twitches that immediately precede button presses (Cohen & van Gaal, 2014). The brevity of conflict-related midfrontal theta can be interpreted as indicating either that the underlying cognitive control signals are brief, or that the tasks that are used in the laboratory only require brief instants of action monitoring/adjustments. As mentioned earlier, a general limitation of this field is that the computerised laboratory tasks tend to require only brief pulses of action monitoring, whereas many real-world situations require more extended periods of action monitoring and adjustments. Recent studies using a pre-trial warning stimulus that predicts the likelihood of upcoming conflict (van Driel, Swart, Egner, Ridderinkhof, & Cohen, 2015) show extended periods of midfrontal theta that scale with the duration with which action monitoring can be anticipated. Better and more real-world relevant experiments will be required to understand the significance of the temporally brief nature of midfrontal theta. Indeed, it is possible that the reason why the conflict modulation of theta power is so brief is that the tasks used to study conflict require only one response with few microactions.

Conflict-related midfrontal theta is a robust feature of the data. It can easily be observed in individual subjects, and is obtained using a variety of data analysis methods (wavelet convolution, filter-Hilbert, short-time Fourier transform), and a variety of spatial filters (surface Laplacian, beamforming, single-dipole fitting, independent components analysis, and various electrode montage referencing schemes). The robustness to analysis method and data processing strategy is important because it shows that the conflict modulation of midfrontal theta is not a subtle or quirky result of one specific data processing protocol.

Conflict-related midfrontal theta manifests as amplitude modulations of ongoing theta waves. In other words, conflict-related theta is non-phase-locked to stimulus onset or to button press. Therefore, it does not reflect band-pass filtering an event-related potential or other non-oscillatory transient (Cohen & Donner, 2013; Nigbur, Cohen, Ridderinkhof, & Stürmer, 2012; Pastötter, Dreisbach, & Bäuml, 2013). Indeed, correlation and partial correlation analyses reveal that time-domain signatures of cognitive control such as the N2 are not significantly correlated with midfrontal theta across subjects (Cohen & Donner, 2013), suggesting that they reflect different, though perhaps related, EEG signatures. Conflict-related midfrontal theta also has high statistical power (>0.9 with 18 subjects) compared to the N2 (~0.2 with 18 subjects; Cohen & Donner, 2013).

The second key finding is that midfrontal theta correlates both with single-trial task performance (as measured through reaction times) and individual differences in the ability to adapt behaviour after errors or learning from negative feedback (Cohen & Cavanagh, 2011; Cohen & Donner, 2013; Cohen & Van Gaal, 2013; Cohen & van Gaal, 2014). Consistent brain–behaviour correlations provide more compelling evidence of a strong link between midfrontal theta and action monitoring, compared to data analyses that involve averaging over hundreds of trials and 15–30 subjects (Kosslyn et al., 2002; Pernet, Sajda, & Rousselet, 2011).

Single-trial reaction times are often assumed to be a behavioural manifestation of the amount of response conflict or cognitive control that was experienced on that trial, with longer reaction times indicating stronger conflict (enhanced need for control). Trial variations in midfrontal theta predict trial variations in reaction time starting a few hundred milliseconds before the button press (Cohen & van Gaal, 2014; Cohen, 2015). This brain–behaviour correlation does not simply reflect response preparation or 'time on task', because these correlations are significantly stronger during trial with conflict compared to trials that do not contain response conflict (Cohen & Nigbur, 2013).

The topography, time course, and frequency characteristics of the correlations between midfrontal theta and reaction time in high-conflict trials are similar to the characteristics observed in the trial-averaged conflict condition comparisons (which ignore reaction times), suggesting that the same neurocognitive process mediates the main effect of conflict detection as well as the variability over trials.

The third key finding is that convergent lines of evidence suggest an anatomical source of the conflict modulation of theta in the supplementary motor area (SMA) or pre-SMA. Researchers occasionally refer to the localisation of these activations as the anterior cingulate cortex, but a careful analysis of available evidence suggests that the pre-SMA is a more likely source (Nachev, Kennard, & Husain, 2008). This localisation comes from non-invasive source localisation techniques in humans, including single-dipole localisation studies (Luu, Tucker, & Makeig, 2004; Nigbur, Ivanova, & Stürmer, 2011), and distributed source localisation such as minimum norm imaging, LORETA, and beamforming (Cohen & Ridderinkhof, 2013; Huang, Rossi, Hämäläinen, & Ahveninen, 2014). These findings are supported by the results of intracranial EEG studies in epilepsy patients (Cohen, Ridderinkhof, Haupt, Elger, & Fell, 2008; Oehrn et al., 2014). One study using laminar electrodes in the pre-SMA found that theta oscillations are most pronounced in the superficial cortical layers (C. Wang, Ulbert, Schomer, Marinkovic, & Halgren, 2005).

Research in non-human primates has shown cognitive-control-related theta activity in the pre-SMA and neighbouring anterior cingulate cortex (Tsujimoto, Shimazu, Isomura, & Sasaki, 2010; Womelsdorf, Johnston, Vinck, & Everling, 2010), although these studies often do not use the same tasks that are used in humans, impeding direct comparisons. There is some evidence suggesting that action monitoring in the rat also relies on structures in the medial prefrontal cortex (Marx et al., 2012).

In humans, conflict-modulated theta is most often studied using EEG. This likely reflects the fact that EEG is less expensive and more accessible than MEG. We recently conducted a simultaneous MEG–EEG study to compare conflict-modulated theta between the two modalities. Strikingly, conflict-related theta is easily observable with EEG but is much smaller with MEG (Cohen, Gulbinaite, Hinkley, & Nagarajan, unpublished observations). This suggests that the anatomical source producing the conflict modulation of theta is spatially restricted with a dipole oriented radially with respect to the skull.

The fourth key finding is that EEG signatures of action monitoring/adjustments have been linked to individual differences in brain structural anatomy. For example, morphological properties of the midcingulate cortex predict behavioural and event-related potential markers of response conflict during the Stroop task (Huster et al., 2009; Huster, Enriquez-Geppert, Pantev, & Bruchmann, 2012). EEG signatures of action monitoring have also been linked to white matter properties of the brain, as measured through diffusion-weighted MRI. Individuals with stronger error-related midfrontal theta have stronger white matter connections linking the medial prefrontal cortex to lateral prefrontal cortex, motor cortex, and ventral striatum (Cohen, 2011a).

Linking brain structure to behaviour and neurophysiology is important because it provides insights into the networks that support cognitive control functions, and helps account for individual differences in genetic and environmental developmental factors (Kanai & Rees, 2011).

Midfrontal Theta and Other Cognitive Control Processes

Several cognitive control processes have been associated with oscillations in the theta band, including working memory and long-term memory retrieval, valenced feedback and feedback-driven learning, decision making, and so on. Theta oscillations in the prefrontal cortex are observed even during quiet resting. It is likely that the morphological and biophysical properties of neurons in the prefrontal cortex support theta-band rhythmogenesis, and thus different prefrontal-mediated cognitive processes may be associated with theta-band EEG fluctuations. Nonetheless, the spatial-temporal-spectral patterns of EEG signatures of different cognitive control processes appear to be distinguishable, suggesting that there may be unique EEG signatures of subcomponents of cognitive control, such as action monitoring and conflict detection. Unique EEG signatures of cognitive processes imply that there are dedicated neural circuit configurations underlying those cognitive processes, and also provide the possibility of circuit-specific manipulations in causality-based experimental research or clinical treatments.

Response errors and feedback learning are associated with frontal theta (Cavanagh, Zambrano-Vazquez, & Allen, 2012; Gehring, Liu, Orr, & Carp, 2012; Marco-Pallares et al., 2008; van de Vijver, Ridderinkhof, & Cohen, 2011). Errors and conflict can be difficult to disambiguate because many errors occur when conflict is experienced. However, their EEG signatures appear to be differentiable. For example, the error-related theta comprises both phase-locked and non-phase-locked power (Trujillo & Allen, 2007), whereas conflict-related theta contains only non-phase-locked oscillations (Cohen & Donner, 2013). Furthermore, within-subject brain–behaviour correlations show that conflict-related theta power is positively correlated with reaction times, whereas error-related delta/theta power is negatively correlated with reaction times (Cohen & van Gaal, 2014; Cohen, 2015). The theta associated with negative feedback during stimulus–response learning has a more anterior and somewhat right-lateralised topographical distribution (Christie & Tata, 2009) compared to response conflict-related theta. Theta oscillations during memory encoding and retrieval are observed in topographically diverse regions and are generally temporally sustained, unlike conflict-related theta (Raghavachari et al., 2006; Staudigl & Hanslmayr, 2013). Thus, it seems that the EEG manifestation of action monitoring/adjustments, particularly as it relates to response conflict, differs from that of other cognitive control processes associated with prefrontal theta.

Why Is So Little Known About Midfrontal Theta?

Where do EEG midfrontal theta oscillations come from, and what types of neurons are critical to their generation? Is midfrontal theta an epiphenomenal index of neuronal spiking, and if so, which types of neurons in which cortical layers does midfrontal theta predominantly reflect? Or is midfrontal theta an integral aspect of the spatiotemporal organisation of neural information processing, and if so, which neural computations are supported by theta? Does the association between midfrontal theta and action monitoring have implications for the timing of sequenced actions (many of which occur with theta-like rhythmicity, as will be explained later)? There are no answers to these questions, and yet we cannot understand how the brain implements action monitoring without understanding the neural microcircuit dynamics that midfrontal theta reflects.

The majority of the research on midfrontal theta and cognitive control has been done in humans using EEG, but EEG provides only correlative evidence at a fairly large spatial scale. Research at smaller spatial scales (animal and computational models) is necessary to understand the origin and significance of midfrontal theta, but this type of research is rarely pursued in the context of action monitoring. One reason for the lack of knowledge about the microcircuits that underlie midfrontal theta is that most psychological theories of cognitive control use box-and-arrow models to account for behaviour or fMRI activations (Page, 2006); these models were an important first step towards understanding the neurophysiological implementations of cognition, but lack sufficient neurobiological plausibility to account for MFC theta or to make predictions that can be confirmed or rejected using neurophysiology data (Cohen & Gulbinaite, 2013).

The Promise and Pitfalls of Animal Models for Understanding Human Cognitive Control

In the broad disciplines of neuroscience and cognitive neuroscience, scientists and students are trained and generally expected to pronounce that animal models can be highly useful for gaining powerful insights into the neural mechanisms underlying human cognitive functions. This argument has become more prominent over the past decade, as neuroscience technologies for measuring and manipulating the brains of small animals (mostly mice) has expanded enormously, and previously science-fiction-sounding techniques like optogenetics are becoming commonplace tools. However, in terms of understanding the neurophysiological mechanisms underlying action monitoring processing, it is worth approaching this dogma with some measure of critical scepticism.

The question is not whether other animal species have the abilities to monitor their behaviour and make real-time adjustments when mistakes could be made; the question is whether other species utilise the same neural mechanisms as we humans use to achieve this goal. If not, then using animal models is not only uninformative, it is detrimental because it gives a misleading impression of putative mechanisms in humans.

Spiking activity in monkeys might provide limited insights, because action potentials of individual neurons are but one dimension of brain activity (Singh, 2012) and may be largely redundant with fMRI activity in humans. LFPs or electrocorticography might not provide additional insights beyond what can be obtained with high-density EEG or intracranial EEG in human epilepsy patients. The cognitive tasks used in monkey studies are often different from those used in humans, precluding direct comparison. Furthermore, at the cognitive level, it has been argued that monkeys may solve cognitive control tasks using different cognitive strategies or using different neural circuitry compared to humans (Cole, Yeung, Freiwald, & Botvinick, 2009). One specific example: The monkey medial prefrontal cortex shows increased theta to rewards compared to punishments, whereas human medial prefrontal cortex shows increased theta to punishments compared to rewards (i.e., the exact opposite pattern of results; Tsujimoto, Shimazu, & Isomura, 2006; van de Vijver et al., 2011). It is not clear how to interpret this species difference, but it clearly highlights the difficulty of deceptively simple cross-species comparisons.

Rodents might provide better models for piecing apart the neural circuitry, because it is easier to perform large-scale electrophysiological recordings in rodents compared to monkeys, and because there are more possibilities for highly specific manipulations (e.g., activating or suppressing specific classes of cells with high spatial and temporal precision via optogenetics). However, the architecture of the rodent medial prefrontal cortex is quite different from that of humans (though there are also important similarities; Vogt & Paxinos, 2014), and it is not clear whether even analogous behavioural tasks elicit statistically similar patterns of electrophysiological activity in rats and in humans (Narayanan, Cavanagh, Frank, & Laubach, 2013). Furthermore, it is becoming clear that rodent neurons function by a different set of rules than human neurons, particularly with respect to trans-neuronal information flow and how that is regulated by long-term potentiation (Testa-Silva et al., 2014; Verhoog et al., 2013). If rodent neurons are a poor model for human neurons, it seems reasonable to question whether rodent action monitoring is a good experimental model for human action monitoring.

Despite these arguments, research exploring the neural machinery underlying action monitoring should be pursued in non-human animals. Research in humans will necessarily always be limited in terms of the precision of neuroscientific measurement and manipulation. The point of this section is to argue that studies in non-human animals should be approached with a critical perspective, and that findings obtained in non-human animals should, whenever possible, be anchored to a comparable finding that can be obtained in humans.

Laminar and Microcircuit Organisation of the Medial Prefrontal Cortex

This section will review key points about the functional/anatomical organisation of the medial prefrontal cortex that will facilitate thinking about how neural circuits in the cortex might produce theta oscillations and the computations underlying response conflict processes.

The medial prefrontal cortex, similar to the rest of cortex, is organised in a laminar fashion, with specific types of cells and patterns of connectivity within and across layers. The anatomical and neural subtype specialisations shape the functions of each layer: Superficial layers (L2/3) are thought to sample and integrate a range of input patterns, whereas deeper layers (L5/6) are thought to select specific patterns and send output messages to neighbouring and long-range circuits. Layer 4 is the primary thalamic input layer; it is relatively thick in primary sensory regions (V1, S1, A1) and becomes thinner or is absent in cortical association areas. Different cortical layers can produce different oscillations, and it is not known which layers contribute to what extent to EEG. Medial prefrontal regions implicated in conflict processing, unlike sensory processing regions, have relatively low cell density and few parvalbumin-positive interneurons (these fast-spiking cells are implicated in generating high-frequency gamma oscillations), and contain no layer 4.

In addition to its vertical (laminar) organisation, the cortex is also characterised by its horizontal organisation. At the 'macroscopic' level, horizontal organisation is defined as different lobes or different Brodmann (or other) areas. At the 'mesoscopic' scale of hundreds to thousands of neurons, the cortex is organised into 'microcircuits'. A microcircuit, similar to a column, is a small vertically oriented ensemble of neurons that have strong intra-circuit connectivity and work together towards a common computation or set of related computations. Some neuroscientists have proposed that microcircuits reflect a fundamental principle of cortical organisation and that microcircuits subserve a basic unit of neural computation (Bastos et al., 2012; Mountcastle, 1997), although the universality of such microcircuits is debated, at least with regards to the significance of their columnar organisation (Horton & Adams, 2005). Microcircuits exhibit oscillations and other complex spatiotemporal dynamics, perform computations that could be the basic building blocks of cognition, and are studied using computational, *in vitro*, and *in vivo* techniques. They are thus at an optimal spatial scale to bridge neurobiology, systems neuroscience, and cognitive neuroscience because they reflect a scale of functional organisation that is larger than a neuron but smaller than a Brodmann area.

There are fundamental principles of canonical microcircuits, but microcircuit variability within and across brain regions highlights that there are specialised microcircuits for specific computations. For this reason, it is necessary to study the microcircuit functional organisation of the medial prefrontal cortex rather than assume that the microcircuit organisation from, for example, the primary visual cortex applies to the prefrontal cortex. For example, a recent study of neural laminar responses to flashes of light in the primate supplementary eye field (Godlove, Maier, Woodman, & Schall, 2014) found that predictions from the canonical circuit model (based on sensory cortex) were only partly confirmed, with notable differences in the timing and spatial spread of activity.

Studying the laminar and microcircuit dynamics in humans is a challenge. M/EEG lack the spatial accuracy to resolve laminar dynamics, intracranial laminar probes in epilepsy patients are rare, and laminar fMRI is in its infancy. Furthermore, the relationship between fMRI and neural oscillations is unclear and complex, and might depend on cortical state and brain region. Thus, some hypotheses about microcircuits can be tested only in non-human animals, despite the caveats listed earlier.

A Possible Neural Microcircuit Underlying Action Monitoring and Adjustments

This section will provide an overview of a recent model of how action monitoring might be implemented in the medial prefrontal cortex (see Cohen, 2014a, for more details). Like all good models, this one is probably wrong. The value of a model in neuroscience lies in its ability to make specific and testable predictions that will push the field forward, even if (when) the predictions turn out to be incorrect. Unfortunately, ambiguous, non-specific, and non-falsifiable theories can linger for decades, and such fuzzy models arguably plague the psychological and cognitive neuroscientific literature.

In brief, this model (Cohen, 2014a) proposes that small pyramidal cells with distributed and convergent synaptic input in layer 2/3 collect and integrate information regarding possible conflicts of actions and produce the theta oscillations measurable with EEG; the detection of conflicting actions is implemented by large neurons in layer 5, and the signalling of conflict to other brain regions for implementing specific and strategic adjustments is implemented by neurons in layer 6 with long-range axonal projections. As will be explained below, this model makes several novel predictions, some of which can be tested in humans and others which can be tested only in non-human animals or biophysically plausible computational models.

The theta oscillations measured by EEG are hypothesised to be driven by oscillations in layer 2/3. Layer 2/3 contains pyramidal cells that are generally smaller than those in deeper layers, with branching apical dendrites ideally suited to integrate, filter, and amplify incoming information. Subcortical inputs to layer 2/3 dendrites include the basal ganglia, amygdala, and hippocampus, and cortical inputs include lateral prefrontal areas 9 and 46, the insula, anterior cingulate, and several regions along the medial and lateral walls of the parietal cortex. The medial prefrontal layer 2/3 can generate theta oscillations independently of other cortical layers. Furthermore, the rodent layer 2/3 contains inhibitory interneurons that maintain theta rhythmicity under a variety of conditions and chemical environments (Fanselow, Richardson, & Connors, 2008). It is possible that the primate brain contains similar theta-band 'pacing' interneurons in superficial layers.

The purpose of these superficial theta oscillations may be to integrate incoming information from motor cortical areas, modulated by inputs from prefrontal and subcortical areas providing information regarding motivation, goals, and contexts. Synchronous inputs from multiple motor areas would then boost theta oscillations and trigger the apical dendrites of large layer 5 pyramidal cells.

Layer 5 is notable for its large pyramidal cells with apical dendrites that terminate in superficial layers, and switch from slow bursting to regular theta spiking when driven by input currents. These large layer 5 pyramidal cells are proposed to implement conflict detection as coincidence detection. Detecting temporally coincident synaptic events is one of the fundamental dendritic computations (London & Häusser, 2005). Coincidence detection relies on precise timing of synaptic inputs; recent findings in humans show that the timing of competing motor activations, rather than the amplitude of their corresponding muscle responses, is the crucial determinant of conflict-related theta (Cohen & van Gaal, 2014).

The strong interconnectivity of MFC layer 5 pyramidal cells ensures that activation of a small number of cells will quickly ignite an avalanche of synchronous theta-band spiking in a larger population. However, layer 5 pyramidal cells can also inhibit their neighbours indirectly via inhibitory interneurons (Murayama et al., 2009). This dynamic would effectively shut down the deep-layer theta rhythmogenesis a few cycles after it begins, and thus may partly cause EEG conflict-related theta power to occur in brief, large-amplitude bursts.

Medial prefrontal layer 6 contains a variety of output pyramidal cells with long-range projections to the superficial layers of distant cortical regions, such as the lateral prefrontal cortex and premotor cortex, as well as to subcortical regions including the striatum, subthalamic nucleus, and midbrain dopamine regions. These neurons can be activated by layer 5 cells, causing theta-band phase locking between the medial prefrontal and other cortical and subcortical networks. These output signals are likely to be activity dependent, such that the 'conflict' signal is transmitted to all distal axonal targets, but whether that signal impacts local processing depends on the activity levels at each target site. Such activity-dependent responses are a hallmark of top-down signals.

Thus, the predictions from this model for scalp EEG are that EEG midfrontal theta power reflects input integration in layer 2/3, while EEG inter-regional theta phase synchronisation reflects the transmission of the conflict signal from layer 6 to other brain circuits to be used to modify task-relevant processing. To test this prediction in humans, novel experiment paradigms may be necessary to dissociate the integration of information that leads to a conflict signal (midfrontal theta oscillations stemming from layer 2/3) from the signalling of that conflict information to other brain systems (phase synchronisation with layer 6).

Why Is Action Monitoring Associated With the Theta Band?

Although it is indisputable that midfrontal theta oscillations are a signature of action monitoring/adjustments, it is unclear what functional role—if any—theta might have in the neural implementations of conflict computations. That is, from the available data, midfrontal theta could be epiphenomenal—a useful signature for identifying the timing and characteristics of the underlying neural circuitry but otherwise irrelevant for the biophysical implementations. If theta oscillations are not a necessary component of conflict processing, they likely reflect the constraints governed by morphological and biophysical properties of medial prefrontal neurons and microcircuitry, including dendritic lengths and time constants of various currents that result in band-pass filtering of synaptic inputs. Although this is a real possibility, progress in this field will benefit from maintaining the assumption that oscillations are part of the 'neural code', and thus the discussion below provides some speculation as to what the role of midfrontal theta might be during action monitoring processes.

As mentioned earlier, one of the prominent themes that permeates modern theories on the role of brain oscillations in neural computation is that oscillations form a cyclic temporal reference frame for organising information processing. In this light, midfrontal theta may serve to provide a temporal reference frame for monitoring and adjusting temporally sequenced actions. This raises the question: Why would the brain use the theta band instead of, for example, the alpha or gamma bands? Indeed, neither human EEG nor intracranial studies in monkeys show cognitive control modulations of midfrontal activity in non-theta frequency bands. Of course, absence of evidence is not evidence of absence, but the available data suggest that the non-theta frequencies have little or no significant role in the primate midfrontal neurophysiology underlying conflict and related cognitive control operations.

A speculative answer to why theta may be the preferred frequency band for conflict computations comes from the curious observation that many everyday and naturalistic behaviours that must be monitored and adjusted comprise temporally sequenced microactions in the 'theta band' (scare quotes are used here to differentiate neural theta-band oscillations from rhythmic muscle activity in a similar frequency band). Examples include keyboard typing, saccades, speech (most languages are uttered at a rate of 5–8 syllables per second), and visual–motor coordination (Graziano, Polosecki, Shalom, & Sigman, 2011; Pellegrino, Coupe, & Marsico, 2011; Schillings, van Wezel, Mulder, & Duysens, 2000; Vallbo & Wessberg, 1993; Yamaguchi, Crump, & Logan, 2013; not all 'theta-band' actions require monitoring and adjusting: Giggling and shivering also occur with 'theta' rhythmicity). It is tempting to speculate that the medial prefrontal cortex utilises theta as a dominant speed for action monitoring because human muscle movements and microactions are dominated by 'theta-band' rhythmicity.

Because the medial prefrontal cortex is involved not only in the passive detection of multiple competing responses but also in the signalling of that conflict, theta oscillations may additionally serve to separate distinct computations in time. For example, some theta phases might be preferentially associated with the integration of multiple inputs and detection of coincidence, whereas other phases might be preferentially associated with the generation and release of the output 'conflict' signal.

Important Future Directions

Over the past decade, midfrontal theta has emerged as a specific and sensitive marker of the activation of a cognitive system dedicated to detecting and resolving conflict. It has high statistical and single-trial-behavioural predictability, and can readily be linked to putative neurophysiological mechanisms. There will always be room for additional insights into the characterisation of midfrontal theta from scalp EEG-computerised-task studies, but it seems unlikely that major scientific gains will be made from minor variations on existing experiments (e.g., is midfrontal theta the same for squares vs. triangles, or faces vs. coloured circles, or words vs. rectangles).

Instead, important scientific gains will likely come from a deeper understanding of how midfrontal theta is related to the biophysical implementations of underlying neural computations, and from how midfrontal theta is related to real-world and naturalistic human behaviours. Below is a non-exhaustive list of research topics that may lead to advances in the scientific understanding of the functional significance of midfrontal theta for computations underlying action monitoring/adjustments.

One of the deepest mysteries in this field is what the neurocomputational significance midfrontal theta might be. The basic theoretical framework outlined above (expanded in Cohen,

2014a) provides some initial falsifiable predictions that can be evaluated in small animal models and in biophysically detailed computational models. Unfortunately, several key predictions concerning cortical layer and cell type cannot be verified or falsified in humans using current neuroimaging technologies.

That said, whether midfrontal theta is a meaningful aspect of the computations underlying action monitoring can be tested in humans using rhythmic exogenous stimulation, such as transcranial alternating current stimulation (TACS). Behavioural performance or EEG dynamics during cognitive control tasks can be tested using various stimulation protocols. Some initial evidence provides support for theta-band stimulation boosting cognitive task performance (Polanía, Nitsche, Korman, Batsikadze, & Paulus, 2012). A suitable control condition should be stimulation at a different frequency band, to control for general rhythmic stimulation.

Another avenue for future research is to test for a relationship between midfrontal theta and conflict during real-world behaviours that comprise 'theta-band' microactions. As mentioned earlier, decades-old computerised cognitive psychology tasks that were introduced earlier are poorly suited to testing this hypothesis owing to their discrete trials with minimal action requirements that are separated by seconds of inactivity.

The idea that subprocesses of action monitoring/resolution may be computed in different phases of the theta cycle can be tested in humans with combined EEG-TMS. Single TMS pulses can be applied during conflict tasks, and the data can be sorted offline according to the phase of midfrontal theta at which the TMS pulse was given. This categorisation can then be used to determine whether the TMS pulse modulated different behavioural or EEG aspects of the task.

Aside from scientific research, midfrontal theta might be relevant for clinical environments. Cognitive control impairments—and alterations in accompanying EEG patterns—are observed in a variety of disorders, ranging from obsessive-compulsive disorder to schizophrenia to Parkinson's disease to attention deficit/hyperactivity disorder (ADHD; Nieuwenhuis, Nielen, Mol, Hajcak, & Veltman, 2005; Schmiedt-Fehr, Schwendemann, Herrmann, & Basar-Eroglu, 2007; van Veen & Carter, 2006). It is possible that cognitive-control-related changes in midfrontal theta will help improve clinical diagnoses, predict optimal treatment options, or track the progress/decline of these disorders.

References

Bastos, A. M., Usrey, W. M., Adams, R. A., Mangun, G. R., Fries, P., & Friston, K. J. (2012). Canonical microcircuits for predictive coding. *Neuron*, *76*(4), 695–711.

Buzsáki, G., & Draguhn, A. (2004). Neuronal oscillations in cortical networks. *Science* (New York, N.Y.), *304*(5679), 1926–1929.

Buzsáki, G., Logothetis, N., & Singer, W. (2013). Scaling brain size, keeping timing: Evolutionary preservation of brain rhythms. *Neuron*, *80*(3), 751–764.

Cavanagh, J. F., Zambrano-Vazquez, L., & Allen, J. J. B. (2012). Theta lingua franca: A common midfrontal substrate for action monitoring processes. *Psychophysiology*, *49*(2), 220–238.

Christie, G. J., & Tata, M. S. (2009). Right frontal cortex generates reward-related theta-band oscillatory activity. *NeuroImage*, *48*(2), 415–422.

Cohen, M. X. (2011a). Error-related medial frontal theta activity predicts cingulate-related structural connectivity. *NeuroImage*, *55*(3), 1373–1383. Retrieved from

Cohen, M. X. (2011b). It's about Time. *Frontiers in Human Neuroscience*, *5*(January), 2.

Cohen, M. X. (2014a). A neural microcircuit for cognitive conflict. *Trends in Neurosciences*, *37*(9), 480–490.

Cohen, M. X. (2014b). *Analyzing neural time series data: Theory and practice*. Cambridge: MIT Press.

Cohen, M. X. (2015). Comparison of different spatial transformations applied to EEG data: A case study of error processing. *International Journal of Psychophysiology, 97*(3), 245–257.

Cohen, M. X., & Cavanagh, J. F. (2011). Single-trial regression elucidates the role of prefrontal theta oscillations in response conflict. *Frontiers in Psychology, 2*, 30.

Cohen, M. X., & Donner, T. H. (2013). Midfrontal conflict-related theta-band power reflects neural oscillations that predict behavior. *Journal of Neurophysiology, 110*(12), 2752–2763.

Cohen, M. X., & Gulbinaite, R. (2013). Five methodological challenges in cognitive electrophysiology. *NeuroImage, 85*, 702–710.

Cohen, M. X., & Nigbur, R. (2013). Reply to 'Higher response time increases theta energy, conflict increases response time'. *Clinical Neurophysiology: Official Journal of the International Federation of Clinical Neurophysiology, 124*(7), 1479–1481.

Cohen, M. X., & Ridderinkhof, K. R. (2013). EEG source reconstruction reveals frontal-parietal dynamics of spatial conflict processing. *PLoS ONE, 8*(2), e57293.

Cohen, M. X., Ridderinkhof, K. R., Haupt, S., Elger, C. E., & Fell, J. (2008). Medial frontal cortex and response conflict: Evidence from human intracranial EEG and medial frontal cortex lesion. *Brain Research, 1238*, 127–142.

Cohen, M. X., & Van Gaal, S. (2013). Dynamic interactions between large-scale brain networks predict behavioral adaptation after perceptual errors. *Cerebral Cortex, 23*(5), 1061–1072.

Cohen, M. X., & van Gaal, S. (2014). Subthreshold muscle twitches dissociate oscillatory neural signatures of conflicts from errors. *NeuroImage, 86*, 503–513.

Cole, M. W., Yeung, N., Freiwald, W. A., & Botvinick, M. (2009). Cingulate cortex: Diverging data from humans and monkeys. *Trends in Neurosciences, 32*(11), 566–574.

Eriksen, B. A., & Eriksen, C. W. (1974). Effects of noise letters upon the identification of a target letter in a nonsearch task. *Perception & Psychophysics, 16*(1), 143–149.

Eyherabide, H. G., Rokem, A., Herz, A. V. M., & Samengo, I. (2009). Bursts generate a non-reducible spike-pattern code. *Frontiers in Neuroscience, 3*(1), 8–14.

Fanselow, E. E., Richardson, K. A., & Connors, B. W. (2008). Selective, state-dependent activation of somatostatin-expressing inhibitory interneurons in mouse neocortex. *Journal of Neurophysiology, 100*(5), 2640–2652.

Fries, P. (2005). A mechanism for cognitive dynamics: Neuronal communication through neuronal coherence. *Trends in Cognitive Sciences, 9*(10), 474–480.

Gehring, W. J., Liu, Y., Orr, J. M., & Carp, J. (2012). The error-related negativity (ERN/Ne). In S. J. Luck & E. S. Kappenman (Eds.), *The Oxford handbook of event-related potentials* (pp. 231–291). New York: Oxford University Press.

Godlove, D. C., Maier, A., Woodman, G. F., & Schall, J. D. (2014). Microcircuitry of agranular frontal cortex: Testing the generality of the canonical cortical microcircuit. *The Journal of Neuroscience: The Official Journal of the Society for Neuroscience, 34*(15), 5355–5369.

Graziano, M., Polosecki, P., Shalom, D. E., & Sigman, M. (2011). Parsing a perceptual decision into a sequence of moments of thought. *Frontiers in Integrative Neuroscience, 5*, 45.

Hanslmayr, S., Pastötter, B., Bäuml, K.-H., Gruber, S., Wimber, M., & Klimesch, W. (2008). The electrophysiological dynamics of interference during the Stroop task. *Journal of Cognitive Neuroscience, 20*(2), 215–225.

Horton, J. C., & Adams, D. L. (2005). The cortical column: A structure without a function. *Philosophical Transactions of the Royal Society of London. Series B, Biological Sciences, 360*(1456), 837–862.

Huang, S., Rossi, S., Hämäläinen, M., & Ahveninen, J. (2014). Auditory conflict resolution correlates with medial-lateral frontal theta/alpha phase synchrony. *PLoS ONE, 9*(10), e110989.

Huster, R. J., Enriquez-Geppert, S., Pantev, C., & Bruchmann, M. (2012). Variations in midcingulate morphology are related to ERP indices of cognitive control. *Brain Structure & Function.*

Huster, R. J., Wolters, C., Wollbrink, A., Schweiger, E., Wittling, W., Pantev, C., & Junghofer, M. (2009). Effects of anterior cingulate fissurization on cognitive control during Stroop interference. *Human Brain Mapping, 30*(4), 1279–1289.

Jensen, O., & Mazaheri, A. (2010). Shaping functional architecture by oscillatory alpha activity: Gating by inhibition. *Frontiers in Human Neuroscience, 4*, 186.

Kanai, R., & Rees, G. (2011). The structural basis of inter-individual differences in human behaviour and cognition. *Nature Reviews Neuroscience, 12*(4), 231–242.

Kay, L. M. (2014). Olfactory system oscillations across phyla. *Current Opinion in Neurobiology*, *31C*, 141–147.

Kepecs, A., Wang, X.-J., & Lisman, J. (2002). Bursting neurons signal input slope. *The Journal of Neuroscience : The Official Journal of the Society for Neuroscience*, *22*(20), 9053–9062. Retrieved from http://www.ncbi.nlm.nih.gov/pubmed/12388612

Kosslyn, S. M., Cacioppo, J. T., Davidson, R. J., Hugdahl, K., Lovallo, W. R., Spiegel, D., & Rose, R. (2002). Bridging psychology and biology. The analysis of individuals in groups. *The American Psychologist*, *57*(5), 341–351. Retrieved from http://www.ncbi.nlm.nih.gov/pubmed/12025764

Le Van Quyen, M. (2011). The brainweb of cross-scale interactions. *New Ideas in Psychology*, *29*, 57–63.

Leuthold, H. (2011). The Simon effect in cognitive electrophysiology: A short review. *Acta Psychologica*, *136*(2), 203–211.

Lisman, J. E., & Jensen, O. (2013). The θ-γ neural code. *Neuron*, *77*(6), 1002–1016.

London, M., & Häusser, M. (2005). Dendritic computation. *Annual Review of Neuroscience*, *28*, 503–532.

Luu, P., Tucker, D. M., & Makeig, S. (2004). Frontal midline theta and the error-related negativity: Neurophysiological mechanisms of action regulation. *Clinical Neurophysiology*, *115*, 1821–1835.

Marco-Pallares, J., Cucurell, D., Cunillera, T., García, R., Andrés-Pueyo, A., Münte, T. F., & Rodríguez-Fornells, A. (2008). Human oscillatory activity associated to reward processing in a gambling task. *Neuropsychologia*, *46(*1), 241–248.

Marx, C., Lex, B., Calaminus, C., Hauber, W., Backes, H., Neumaier, B.,...Endepols, H. (2012). Conflict processing in the rat brain: Behavioral analysis and functional μPET imaging using [F]fluorodeoxyglucose. *Frontiers in Behavioral Neuroscience*, *6*, 4.

Mountcastle, V. B. (1997). The columnar organization of the neocortex. *Brain: A Journal of Neurology*, *120*(Pt. 4), 701–722. Retrieved from http://www.ncbi.nlm.nih.gov/pubmed/9153131

Murayama, M., Pérez-Garci, E., Nevian, T., Bock, T., Senn, W., & Larkum, M. E. (2009). Dendritic encoding of sensory stimuli controlled by deep cortical interneurons. *Nature*, *457*(7233), 1137–1141.

Nachev, P., Kennard, C., & Husain, M. (2008). Functional role of the supplementary and pre-supplementary motor areas. *Nature Reviews. Neuroscience*, *9*(11), 856–869.

Narayanan, N. S., Cavanagh, J. F., Frank, M. J., & Laubach, M. (2013). Common medial frontal mechanisms of adaptive control in humans and rodents. *Nature Neuroscience*, *16*(12), 1888–1895.

Nieuwenhuis, S., Nielen, M. M., Mol, N., Hajcak, G., & Veltman, D. J. (2005). Performance monitoring in obsessive-compulsive disorder. *Psychiatry Research*, *134*(2), 111–122.

Nigbur, R., Cohen, M. X., Ridderinkhof, K. R., & Stürmer, B. (2012). Theta dynamics reveal domain-specific control over stimulus and response conflict. *Journal of Cognitive Neuroscience*, *24*(5), 1264–1274.

Nigbur, R., Ivanova, G., & Stürmer, B. (2011). Theta power as a marker for cognitive interference. *Clinical Neurophysiology: Official Journal of the International Federation of Clinical Neurophysiology*, *122*(11), 2185–2194.

Oehrn, C. R., Hanslmayr, S., Fell, J., Deuker, L., Kremers, N. A., Do Lam, A. T.,...Axmacher, N. (2014). Neural communication patterns underlying conflict detection, resolution, and adaptation. *The Journal of Neuroscience : The Official Journal of the Society for Neuroscience*, *34*(31), 10438–10452.

Page, M. P. A. (2006). What can't functional neuroimaging tell the cognitive psychologist? *Cortex; A Journal Devoted to the Study of the Nervous System and Behavior*, *42*(3), 428–43. Retrieved from http://www.ncbi.nlm.nih.gov/pubmed/16771054

Palva, S., & Palva, J. M. (2012). Discovering oscillatory interaction networks with M/EEG: Challenges and breakthroughs. *Trends in Cognitive Sciences*, *16*(4), 219–230.

Pastötter, B., Dreisbach, G., & Bäuml, K.-H. T. (2013). Dynamic adjustments of cognitive control: Oscillatory correlates of the conflict adaptation effect. *Journal of Cognitive Neuroscience*, *25*(12), 2167–2178.

Pastötter, B., Hanslmayr, S., & Bäuml, K.-H. T. (2010). Conflict processing in the anterior cingulate cortex constrains response priming. *NeuroImage*, *50*(4), 1599–1605.

Pellegrino, F., Coupe, C., & Marsico, E. (2011). Across-language perspective on speech information rate. *Language*, *87*(3), 539–558.

Pernet, C. R., Sajda, P., & Rousselet, G. A. (2011). Single-trial analyses: Why bother? *Frontiers in Psychology*, *2*, 322.

Polanía, R., Nitsche, M. A., Korman, C., Batsikadze, G., & Paulus, W. (2012). The importance of timing in segregated theta phase-coupling for cognitive performance. *Current Biology : CB*, *22*(14), 1314–1318.

Raghavachari, S., Lisman, J. E., Tully, M., Madsen, J. R., Bromfield, E. B., & Kahana, M. J. (2006). Theta oscillations in human cortex during a working-memory task: Evidence for local generators. *Journal of Neurophysiology*, *95*(3), 1630–1638.

Reimann, M. W., Anastassiou, C. A., Perin, R., Hill, S. L., Markram, H., & Koch, C. (2013). A biophysically detailed model of neocortical local field potentials predicts the critical role of active membrane currents. *Neuron*, *79*(2), 375–390.

Ridderinkhof, K. R. (2002). Micro- and macro-adjustments of task set: Activation and suppression in conflict tasks. *Psychological Research*, *66*(4), 312–323.

Rochet, N., Spieser, L., Casini, L., Hasbroucq, T., & Burle, B. (2014). Detecting and correcting partial errors: Evidence for efficient control without conscious access. *Cognitive, Affective & Behavioral Neuroscience*, *14*(3), 970–982.

Schillings, A. M., van Wezel, B. M., Mulder, T., & Duysens, J. (2000). Muscular responses and movement strategies during stumbling over obstacles. *Journal of Neurophysiology*, *83*(4), 2093–2102. Retrieved from http://www.ncbi.nlm.nih.gov/pubmed/10758119

Schmiedt-Fehr, C., Schwendemann, G., Herrmann, M., & Basar-Eroglu, C. (2007). Parkinson's disease and age-related alterations in brain oscillations during a Simon task. *Neuroreport*, *18*(3), 277–281.

Singh, K. D. (2012). Which 'neural activity' do you mean? fMRI, MEG, oscillations and neurotransmitters. *NeuroImage*, *62*(2), 1121–1130.

Staudigl, T., & Hanslmayr, S. (2013). Theta oscillations at encoding mediate the context-dependent nature of human episodic memory. *Current Biology : CB*, *23*(12), 1101–6.

Stroop, J. R. (1935). Studies of interference in serial verbal reactions. *Journal of Experimental Psychology*, *18*, 643–662.

Testa-Silva, G., Verhoog, M. B., Linaro, D., de Kock, C. P. J., Baayen, J. C., Meredith, R. M., ... Mansvelder, H. D. (2014). High bandwidth synaptic communication and frequency tracking in human neocortex. *PLoS Biology*, *12*(11), e1002007.

Trujillo, L. T., & Allen, J. J. B. (2007). Theta EEG dynamics of the error-related negativity. *Clinical Neurophysiology : Official Journal of the International Federation of Clinical Neurophysiology*, *118*(3), 645–668.

Tsujimoto, T., Shimazu, H., & Isomura, Y. (2006). Direct recording of theta oscillations in primate prefrontal and anterior cingulate cortices. *Journal of Neurophysiology*, *95*, 2987–3000.

Tsujimoto, T., Shimazu, H., Isomura, Y., & Sasaki, K. (2010). Theta oscillations in primate prefrontal and anterior cingulate cortices in forewarned reaction time tasks. *Journal of Neurophysiology*, *103*, 827–843.

Ullsperger, M., Danielmeier, C., & Jocham, G. (2014). Neurophysiology of performance monitoring and adaptive behavior. *Physiological Reviews*, *94*(1), 35–79.

Ullsperger, M., Fischer, A. G., Nigbur, R., & Endrass, T. (2014). Neural mechanisms and temporal dynamics of performance monitoring. *Trends in Cognitive Sciences*, *18*(5), 259–267.

Vallbo, A. B., & Wessberg, J. (1993). Organization of motor output in slow finger movements in man. *The Journal of Physiology*, *469*, 673–691. Retrieved from http://www.pubmedcentral.nih.gov/articlerender.fcgi?artid=1143894&tool=pmcentrez&rendertype=abstract

Van de Vijver, I., Ridderinkhof, K. R., & Cohen, M. X. (2011). Frontal oscillatory dynamics predict feedback learning and action adjustment. *Journal of Cognitive Neuroscience*, *23*(12), 4106–4121.

Van Driel, J., Swart, J. C., Egner, T., Ridderinkhof, K. R., & Cohen, M. X. (2015). (No) Time for control: Frontal theta dynamics reveal the cost of temporally guided conflict anticipation. *Cognitive, Affective & Behavioral Neuroscience, 15*(4), 787–807.

Van Veen, V., & Carter, C. S. (2006). Error detection, correction, and prevention in the brain: A brief review of data and theories. *Clinical EEG and Neuroscience, 37*(4), 330–335. Retrieved from http://www.ncbi.nlm.nih.gov/pubmed/17073172

Verhoog, M. B., Goriounova, N. A., Obermayer, J., Stroeder, J., Hjorth, J. J. J., Testa-Silva, G.,... Mansvelder, H. D. (2013). Mechanisms underlying the rules for associative plasticity at adult human neocortical synapses. *The Journal of Neuroscience: The Official Journal of the Society for Neuroscience, 33*(43), 17197–17208.

Vinck, M., Oostenveld, R., van Wingerden, M., Battaglia, F., & Pennartz, C. M. A. (2011). An improved index of phase-synchronization for electrophysiological data in the presence of volume-conduction, noise and sample-size bias. *NeuroImage, 55*(4), 1548–1565.

Vogt, B. A., & Paxinos, G. (2014). Cytoarchitecture of mouse and rat cingulate cortex with human homologies. *Brain Structure & Function, 219*(1), 185–192.

Wang, C., Ulbert, I., Schomer, D. L., Marinkovic, K., & Halgren, E. (2005). Responses of human anterior cingulate cortex microdomains to error detection, conflict monitoring, stimulus-response mapping, familiarity, and orienting. *The Journal of Neuroscience: The Official Journal of the Society for Neuroscience, 25*(3), 604–613.

Wang, X.-J. (2010). Neurophysiological and computational principles of cortical rhythms in cognition. *Physiological Reviews, 90*(3), 1195–1268.

Womelsdorf, T., Johnston, K., Vinck, M., & Everling, S. (2010). Theta-activity in anterior cingulate cortex predicts task rules and their adjustments following errors. *Proceedings of the National Academy of Sciences of the United States of America, 107*(11), 5248–5253.

Yamaguchi, M., Crump, M. J. C., & Logan, G. D. (2013). Speed-accuracy trade-off in skilled typewriting: Decomposing the contributions of hierarchical control loops. *Journal of Experimental Psychology. Human Perception and Performance, 39*(3), 678–699.

15

Models of Anterior Cingulate Cortex Function in Cognitive Control

Joshua W. Brown

The anterior cingulate cortex (ACC) shows a variety of effects and has been found to be active in over 20% of all fMRI studies (Poldrack, 2011). Among the earliest neuropsychological findings was a pronounced akinetic mutism that resulted from focal lesions of the ACC (Barris & Schumann, 1953; Nemeth, Hegedus, & Molnar, 1988). Neuroscience research on the ACC began in the late 1970s, when it was shown that individual monkey cingulate neurons are more active when errors are committed (Niki & Watanabe, 1979). This finding was supported by subsequent human ERP studies in the early 1990s showing that human medial frontal activity is greater when errors are committed (Falkenstein, Hohnsbein, Hoorman, & Blanke, 1991; Gehring, Coles, Meyer, & Donchin, 1990; Gehring, Goss, Coles, Meyer, & Donchin, 1993). In turn, this led to some speculation about what mechanisms could account for the newly discovered error effects. Some argued that the error signal in the ACC reflected a comparison of the actual versus intended response (Bernstein, Scheffers, & Coles, 1995), whereas others suggested that the error may reflect a comparison of predicted versus actual consequences (Ito, Stuphorn, Brown, & Schall, 2003).

While some were envisioning the error signal as a kind of comparison to detect discrepancies, another series of studies argued that the ACC signalled a co-activation of mutually incompatible (conflicting) signals. An initial study showed that events and responses with a low prior probability led to greater ACC activation, presumably as a result of the more likely but incorrect response representation competing with the less likely but correct response representation (Carter et al., 1998). This key finding showed that the ACC could be activated even when an error did not occur. A follow-up study showed that the ACC activation reflected a transient monitoring function, but it did not remain active as a driver of greater cognitive control during subsequent task performance. This indicated that the ACC might serve more as a performance monitor than a controller (Botvinick, Nystrom, Fissel, Carter, & Cohen, 1999). Later studies revealed a dissociation in which the ACC provided a monitoring signal, while the lateral prefrontal cortex (lPFC) provided a control signal over ongoing behaviour (MacDonald, Cohen, Stenger, & Carter, 2000). The following year, a formal model of conflict detection was proposed. Table 15.1 summarises a number of models of error detection in the ACC (see also Chapter 17 by Ullsperger in this volume).

The Wiley Handbook of Cognitive Control, First Edition. Edited by Tobias Egner.

Table 15.1 Summary of ACC Error Monitoring Theories.

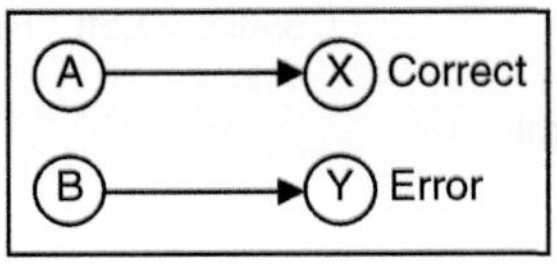

Model	*Method*	*Error detection pseudocode*	*References*
Discrepancy	Subtractive (reactive)	Y-X+1	Scheffers & Coles (2000)
Response selection	Learned conjunction (reactive, proactive)	$V_r(A, Y) - V_{r-1}(A)$	Holroyd et al. (2005)
Correction detection	Sequence detection (reactive)	X after Y?	Steinhauser et al. (2008)
Conflict	Multiplicative (reactive, proactive)	A * B	Botvinick et al. (2001)
Error likelihood	Learned probability (proactive)	P(Y | A,B)	Brown & Braver (2005)
PRO	Learned prediction, comparison (reactive, proactive)	abs(Y-P(Y|A))	Alexander & Brown (2011)
Double integrator	Integration of task cue integrator	$\dot{x}_1 = A + B + X + Y - const$ $\dot{x}_2 = \beta x_1$	Bekolay et al. (2014)

Note. Consider two representations of cued responses with continuous activities A and B in [0, 1] that lead to binary 0 or 1 responses X and/or Y, representing correct and error responses, respectively. For concreteness, A may correspond to a valid cue in the Stroop task and drive correct response X, whereas B may represent an incongruent distractor that drives incorrect response Y. Source: Adapted from Alexander 2010. Reproduced with permission of Wiley.

The Conflict Model

According to the conflict computational model (Botvinick, Braver, Barch, Carter, & Cohen, 2001), the ACC activity reflects a simple multiplication of the activity representing two mutually incompatible response processes (Figure 15.1). If 'A' represents the activity of response A, and 'B' represents the activity of a competing response B, then the ACC activation will simply be the product A×B. The simplicity of this computational model, along with its ability to account for a considerable array of effects found in the ACC, led to it rapidly becoming a leading computational account of ACC function, to the point that ACC activation was nearly synonymous with conflict. The conflict model was applied to various problems and subdivided into various kinds of conflict detection, for example, conflict between task sets and conflict across sequences of changing responses (Brown, Reynolds, & Braver, 2007).

The first question to be addressed by the conflict model is that if the ACC computes conflict, what is the source of the error signals? According to a subsequent analysis of the conflict model, error signals may result simply from 'post-error conflict'. That is, once an error response is committed, there is still motor-related activity representing the planned correct response. The co-activation of the motor activities representing the planned error and the planned (but not executed) correct response leads to a state of conflict. This was proposed as the origin of the error signal (Yeung, Botvinick, & Cohen, 2004).

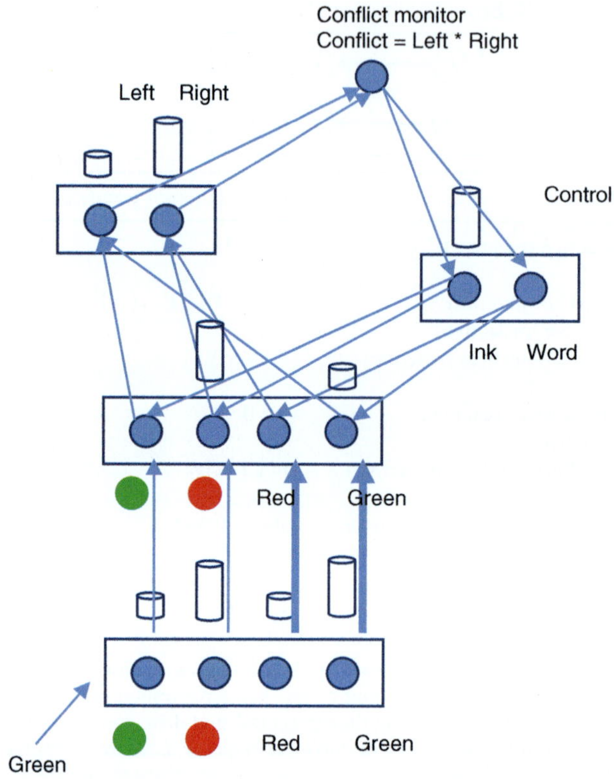

Figure 15.1 Conflict model of the Stroop task. The Stroop task requires subjects to name the ink color of a word and ignore the semantic meaning. When an incongruent Stroop stimulus is presented, both the left and right response representations are simultaneously activated. The conflict model suggests that the ACC detects the simultaneous co-activation of the mutually incompatible responses, that is, Conflict = LEFT * RIGHT response activation levels (Botvinick et al., 2001). The conflict detector in turn increases the gain of the control signals, which bias the network to produce the correct response according to the task rules.

Challenges to the Conflict Model

In the wake of the conflict model proposal, a number of challenges arose. One key challenge to the conflict model was a finding that the ACC activation (as indexed by the error-related negativity, or ERN) could be elicited by feedback that an error has occurred, and not simply by the commission of an error per se. This is an essential distinction. Once a response has been made, the response-related activity presumably decays rapidly. Nevertheless, Holroyd and Coles found that the ERN could be elicited by feedback that an error occurred, even when feedback was given well after the responses were made (Holroyd & Coles, 2002). This was termed the 'feedback ERN', or fERN. It is unclear how the conflict model could account for this effect, because the response-related activations would have decayed well before the outcome was revealed (Holroyd, Yeung, Coles, & Cohen, 2005).

Aside from the conflict model's difficulty in accounting for the fERN, a direct test for conflict cells in the macaque monkey ACC failed to turn up any cells that signalled response conflict (Ito et al., 2003), although conflict-related activity was apparently found more dorsally in the supplementary eye fields (Stuphorn, Taylor, & Schall, 2000). By 2009, some had suggested that perhaps the monkey and human ACC regions were simply different in this regard, and that the monkey ACC did not compute conflict, whereas the human ACC did (Cole, Yeung, Freiwald, & Botvinick, 2009). The lack of findings of conflict in monkey ACC cells has persisted until very recently (Ebitz & Platt, 2015). Unfortunately, the recent claim of monkey conflict cells actually found what turned out to be 'task conflict' cells, and no actual response conflict cells were found (Ebitz & Platt, 2015). The absence of single units that apparently compute response conflict in the monkey ACC remains.

Likewise, several papers challenged the conflict model on the basis of ERP or fMRI results. One paper found that in a Stroop task, ACC activity was greater for neutral than for congruent stimuli, in which neither condition entailed response conflict (Roelofs, van Turennout, & Coles, 2006). An ERP study showed that as conflicting response representations overlapped more in time, the ERN amplitude actually decreased rather than increased, contrary to the predictions of the conflict model (Burle, Roger, Allain, Vidal, & Hasbroucq, 2008). Another set of papers proposed that the ACC activity simply correlates with response time (Carp, Kim, Taylor, Fitzgerald, & Weissman, 2010; Grinband et al., 2011), and that apparent response conflict effects are an epiphenomenal reflection of this effect, because incongruent trials generally entail longer response times.

Reinforcement Theories of the ERN

Shortly after the conflict model was put forward, an alternative model of ACC function was proposed, based on reinforcement learning. This new model suggested how the ACC might compute a signal that indicates an error has occurred on the basis of the fERN, rather than on the basis of internal motor signals. According to the RL-ERN reinforcement model (Holroyd & Coles, 2002), the ACC is activated by pauses in dopamine cell activity (Figure 15.2a), which occur when the expected reward is not received (Schultz, 1998). Thus, an error results in reward omission, which pauses dopamine cell firing, which disinhibits the ACC. The RL-ERN model further posits that the role of the ACC is to select which other frontal areas will be permitted to drive action generation, although the same function has elsewhere been ascribed to the basal ganglia (Brown, Bullock, & Grossberg, 2004). In a follow-up ACC model (Figure 15.2b), the ACC is cast as learning to respond to reduced value, which again may originate as reduced dopamine signalling from the basal ganglia (Holroyd et al., 2005). The later model simulates how the ACC may learn to respond to conjunctions of events that collectively indicate that an error has occurred. In this way, the fERN can be accounted for as being activated when the events indicate an error.

The Error Likelihood Model

Around the same time, an alternative computational model of the ACC was proposed, in which the ACC was cast as learning to predict the likelihood of an error (Brown & Braver, 2005). According to this model, the ACC learns to represent conjunctions of events for which the probability of an error was high (Figure 15.2c). Building on the RL-ERN model (Holroyd & Coles, 2002), the model proposed that dopamine pauses not only activated the ACC, but also trained it to respond in anticipation of errors in situations where they were likely to occur, on the basis of the modulatory influence of dopamine on cortical plasticity (Bao, Chan, & Merzenich, 2001). The main evidence for this model came from an fMRI study in which task conditions associated with a higher likelihood of errors led to greater ACC activity. Critically, this contrast held even when both error commission and

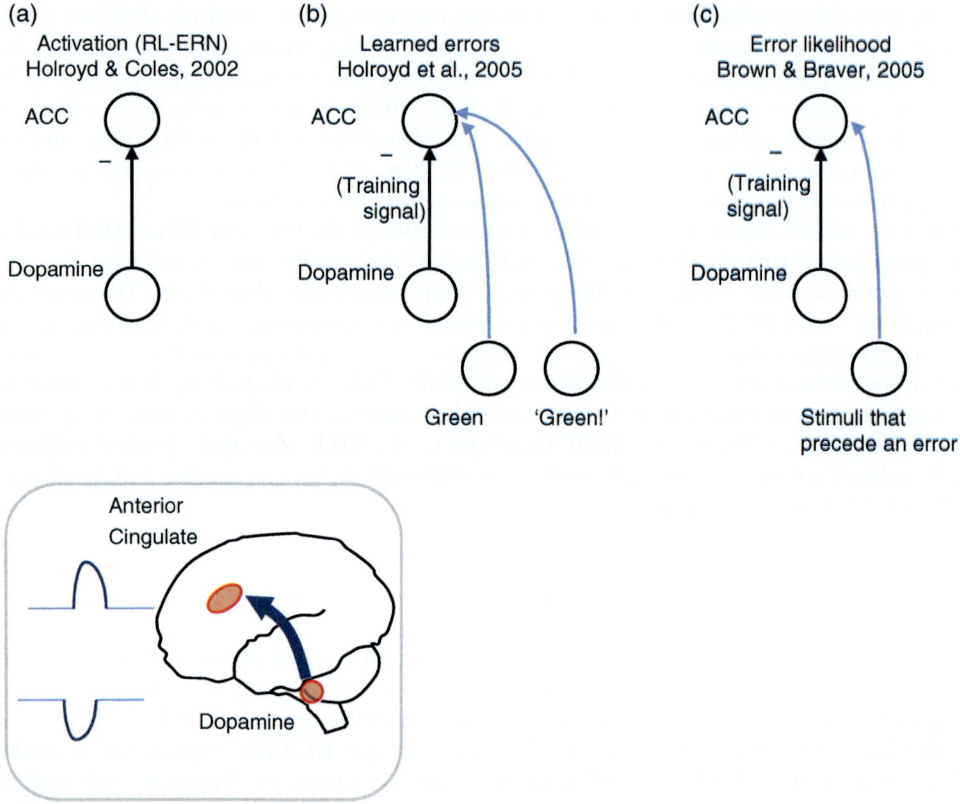

Figure 15.2 Reinforcement learning models of ACC. Left: Dopaminergic (DA) signals from the midbrain disinhibit the anterior cingulate cortex. (a) The RL-ERN model (Holroyd & Coles, 2002) simulates the ACC error activity as resulting from a pause of DA activity, which disinhibits the ACC. (b) The Holroyd et al. (2005) model, in which the DA pause trains the ACC to recognize conjunctions of events that indicate an error has *already* occurred. (c) The error likelihood model of Brown & Braver (2005), in which a DA pause trains the ACC to be activated by cues that reliably *precede* an error.

conflict were controlled for (Brown & Braver, 2005). According to the error likelihood model, error signals could be accounted for in that when an error occurred, the probability of an error was essentially P(error) = 1, so the ACC would signal that an error was maximally likely (as it had, in fact, occurred).

The error likelihood model was quickly challenged, and the results highlighted the importance of individual differences. First, a nearly direct replication study of the ACC error likelihood effect failed (Yeung & Nieuwenhuis, 2009). Subsequent studies revealed that the ACC error likelihood effect was strongest in those who are most risk averse, but virtually absent in those who are risk seeking (Brown & Braver, 2007). Thus, a population sample that is more risk tolerant overall may fail to show an error likelihood effect in the ACC. Subsequent computational neural models formalised this idea—as the ability of the model ACC to learn from dopamine cell activity pauses decreased, the model ACC cells showed progressively weaker error likelihood effects (Brown & Braver, 2008). Overall this suggested that greater ACC activity led to greater risk or loss avoidance, a finding that has been replicated in several fMRI studies (Fukunaga, Bogg, Finn, & Brown, 2013; Fukunaga, Brown, & Bogg, 2012; Krawitz, Fukunaga, & Brown, 2010).

More Constraints on ACC Models

While the conflict, RL-ERN, and error likelihood models gained traction, new studies of the ACC reported additional effects that challenged the existing models. One study showed that simply manipulating whether subjects were required to detect or correct their own errors, by a certain deadline, led to measurable effects on responding; this in turn motivated a model of ACC error detection in which the ACC was proposed to detect not the errors, nor the likelihood of errors, but rather a correction response that follows an error (Steinhauser, Maier, & Hübner, 2008). In particular, the model posited ACC activity when a correction response followed an initial error response.

Other new effects followed. Several studies showed that ACC activity has a strong positive correlation with response time (RT), so that when this correlation is accounted for, no response conflict effects remain. This was construed to mean that perhaps the ACC does not compute response conflict but rather signals a measure of the RT (Carp et al., 2010; Grinband et al., 2011). Other earlier studies found similar effects, although they were construed as either simply inconsistent with the conflict model (Burle et al., 2008) or not relevant to the conflict model (Yeung & Nieuwenhuis, 2009).

Still another influential study showed that ACC activity correlated with the degree of volatility (non-stationarity) in the environment, which in turn correlated with the effective learning rates seen in task performance (Behrens, Woolrich, Walton, & Rushworth, 2007). It is unclear how any of the existing models could have accounted for this. Relatedly, a number of studies began to show effects of prediction errors in the medial prefrontal cortex and ACC in particular. One study in monkeys showed that some ACC cells signal unsigned prediction errors (Hayden, Heilbronner, et al., 2011). A human study showed that error effects in the ACC could actually be inverted, so that there was more activation for correct responses than errors when errors were very likely (Jessup, Busemeyer, & Brown, 2010). This challenged the prevailing view that the ACC responded primarily to negatively valenced events such as errors, and instead, it suggested that the ACC responds to surprising events, whether affectively positive or negative, a finding that has been confirmed by other fMRI and ERP studies (Ferdinand, Mecklinger, Kray, & Gehring, 2012; Garofalo, Maier, & di Pellegrino, 2014; Nee, Kastner, & Brown, 2011; Oliveira, McDonald, & Goodman, 2007).

The Prediction of Responses and Outcomes (PRO) Model

With the mounting empirical evidence, it became clear that the ACC yielded signals consistent with both predictions of outcomes and surprise. Two groups (including the author) began working independently to develop computational neural models of the ACC as predicting outcomes and then comparing the outcomes against the predictions (Alexander & Brown, 2011; Silvetti, Alexander, Verguts, & Brown, 2014; Silvetti, Seurinck, & Verguts, 2011). The two models are remarkably similar (Silvetti et al., 2014). Each builds on the temporal difference model. The Reward Value and Prediction Model (RVPM; Silvetti et al., 2011) casts the ACC in the role of learning to predict reward, and then separately detecting better versus worse than expected reward. The PRO model (Alexander & Brown, 2011) casts the ACC as learning to predict the variety of possible outcomes (good, bad, or neutral), each with some probability. Both models predict the timing of when the reward (RVPM) or various possible outcomes (PRO) might occur.

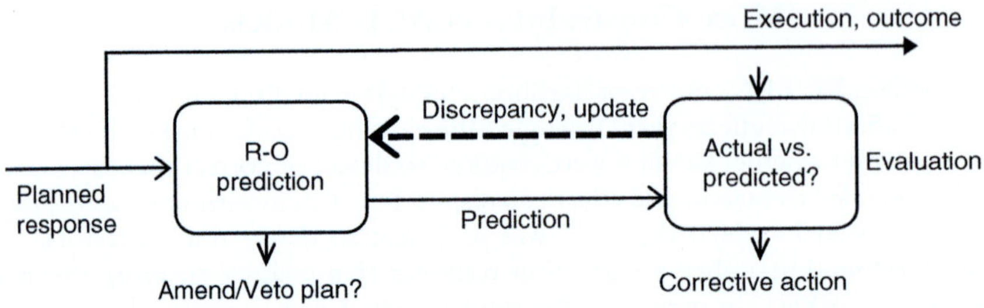

Figure 15.3 PRO model. Planned responses activate learned response–outcome predictions. These predicted outcome signals can feed back to amend or veto a planned action. Once an action is generated, the actual outcome (the movement itself or the feedback from the environment) is compared against the expected outcome, and any discrepancy leads to an update of the learned response–outcome predictions. Source: Adapted from Alexander 2010. Reproduced with permission of Wiley.

Similarly, both models compare the actual outcomes against the prior predictions. The RVPM model has cells that specifically signal when an outcome was more rewarding than expected, versus other cells that signal specifically when a reward was less rewarding than expected. These are simply the positively and negatively rectified signals from the temporal difference signal in the model.

The PRO model is shown in more detail in Figure 15.3. Specifically, the R-O prediction layer has cells that represent specific possible conjunctions of responses and outcomes, whether good or bad. Greater activity signals a greater likelihood of the outcome. When the timing of the expected outcome approaches, the cells increase their activities and peak around the time of the expected outcome. These prediction signals in turn activate an evaluation layer, which compares the actual versus expected outcomes. A large number of empirical effects can be simulated when the evaluation units are excited by the prediction, and inhibited by the actual outcome (Alexander & Brown, 2011). This is referred to as *negative surprise*, that is, the omission of an expected outcome whether subjectively good or bad. In this case, the evaluation cells are most active when an outcome is predicted, but it does not occur. This is how the PRO model generates greater activity when errors occur: It signals not an error per se, but rather an expected correct outcome that failed to occur. Conversely, if errors are more common than correct trials, then correct trials will elicit greater activity when the expected error fails to occur. Such effects have been found empirically (Ferdinand et al., 2012; Garofalo et al., 2014; Jessup et al., 2010; Nee et al., 2011; Oliveira et al., 2007).

The PRO model has been able to simulate and account for a large body of data. The original PRO model publication (Alexander & Brown, 2011) showed that it could simulate human ERP and fMRI effects of error, conflict, greater error likelihood, greater error unexpectedness (Brown & Braver, 2005), volatility (Behrens et al., 2007), positive correlations with RT (Carp et al., 2010; Grinband et al., 2011), and individual difference effects (Brown & Braver, 2008, 2007), and also monkey single-unit effects of reward, unexpected reward, and prediction error (Amador et al., 2000; Hayden, Heilbronner, et al., 2011; Ito et al., 2003).

A subsequent set of simulations (Alexander & Brown, 2014) showed that the PRO model could account for general predictions of subsequent events, including stimuli as 'outcomes'. This generalisation of the model allowed it to simulate ACC effects of trial type frequency, item-specific versus global control (Blais & Bunge, 2010), the mismatch negativity (Crottaz-Herbette & Menon, 2006), informative versus uninformative cues

(Aarts, Roelofs, & van Turennout, 2008), Bayesian surprise (Ide, Shenoy, Yu, & Li, 2013), and additional monkey neurophysiology studies showing surprising gain and loss effects (Sallet et al., 2007).

One of the distinctive features of the PRO model is that it generates timed predictions of expected outcomes. In other words, the prediction units signal not only *what* outcomes are likely to occur and with what probability, but also *when* those outcomes are likely to occur. The R-O prediction unit activities peak at the time when the expected outcome is most likely to occur. This is consistent with empirical results, which show that an expected outcome can still generate ACC activity if it is received at an unexpected time (Forster & Brown, 2011). The PRO model uses a tapped delay line, which is a mechanism in which an array of cells are activated at specific delay intervals following a stimulus. Thus, the timing of an expected outcome (relative to an earlier stimulus) can be learned by training a specific delay interval cell to activate a prediction unit. The PRO model is agnostic about where these delay intervals may be represented, but one possibility is the cerebellum via the thalamus, as the cerebellum is known to both represent adaptive timing (Fiala, Grossberg, & Bullock, 1996) and play an essential role in cognitive as well as motor functions (Middleton & Strick, 1994). In this scenario, the adaptive timing functions of the ACC would be derived from an *external* basis in the cerebellum. An alternative approach has recently been suggested, in which the prediction timing is computed by mechanisms *intrinsic* to the ACC. In this double integrator model (Bekolay, Laubach, & Eliasmith, 2014), the 'ramping up' signals within the ACC are generated by two integrators. The first integrates a signal that the trial is under way, and thus the magnitude of the integrated signal correlates with the elapsed trial time. Likewise, the second signal integrates the first integrator, thus providing an approximate measure of how much time has elapsed since the first integrator changed. The timing of events in correct versus error trials is different, leading to divergent states of the two integrators. The authors showed that the two integrators yielded signals that appeared qualitatively similar to the principal components found in a population of units from the rat prelimbic cortex, which may be the rat equivalent of the ACC (Bekolay et al., 2014).

Value Models

So far, the models discussed have mainly focused on cognitive constructs such as conflict detection, error detection, and error likelihood. Another class of models constructs the ACC as representing different kinds of affective and motivational values in a way that influences decision making, providing a kind of motivational signal to exert cognitive control over decision making. Cognitive control essentially biases decisions away from the immediate demands of strong stimuli and towards the longer-term, higher-level goals of the agent. There is some evidence that ACC activity in humans provides a kind of motivational value to exert effort towards a goal—humans receiving direct electrical stimulation of the dorsal ACC report an impetus to 'push harder to try and get through this' (Parvizi, Rangarajan, Shirer, Desai, & Greicius, 2013).

In some ways, the models above reflect aspects of such values. The error likelihood model could be thought of as representing an aversive signal of impending errors, which in turn inhibits responding (Brown & Braver, 2005, 2007, 2008). Similarly, other models can be viewed as learning states associated with a negative reward value (Holroyd et al., 2005).

A recent review argues that the ACC computes the expected value of control, which is defined as the expected rewards of exerting control over decision making, less the 'costs' of exerting such control. This has been called the 'expected value of control' (EVC) model (Shenhav, Botvinick, & Cohen, 2013; see also Chapter 10 by Kool, Shenhav, & Botvinick in this volume). The costs in this case are internal, subjective, and therefore somewhat difficult to quantify, although there is some qualitative evidence to support the general notion (Kool, McGuire, Wang, & Botvinick, 2013), and the model has not been simulated or used to generate predictions thus far.

A recent paper has proposed that the ACC represents the value of foraging (Kolling, Behrens, Mars, & Rushworth, 2012). According to this view, animals must generally choose whether to engage (exploit) a known but diminishing resource or else forage (explore) for a better resource, with no guarantee that such a better resource will be found. Humans performing this kind of foraging task generally have a bias to engage, even when the expected value is lower relative to foraging. To overcome this bias, the ACC was proposed to generate a control signal whose activity correlates positively with the relative value of foraging instead of engaging. When the activity is higher, it will bias the animal to forage for a better resource. This idea is consistent with monkey single-unit findings, showing that animals forage when ACC cell activity exceeds a specific threshold (Hayden, Pearson, et al., 2011). A recent paper has shown that ACC activity correlates positively with the relative value of foraging (Kolling et al., 2012). A subsequent paper has challenged this interpretation (Shenhav, Straccia, Cohen, & Botvinick, 2014). In it, the authors show that as the relative value of foraging increases, a region in the dorsal ACC shows activity that peaks at a certain relative foraging value and then decreases as the relative value of foraging increases further. The authors argued that the ACC therefore may not represent relative foraging value but instead may represent a more cognitive signal of choice difficulty. Specifically, ACC activity peaks when the choice probabilities of foraging versus engaging are near the indifference point, that is, each 50%, which is the point at which the decision to forage or engage is most difficult. The authors did not specify a computational model of how choice difficulty may be computed, but choice difficulty is not the only possible explanation of this phenomenon. In the PRO model, activity in the evaluation layer (Figure 15.3) might be expected to peak at the indifference point. This is due to the fact that at the indifference point, whichever action is not chosen nevertheless was 50% likely, and therefore the fact that it was not chosen is somewhat surprising. In contrast, when the relative value of foraging is either very high or very low, unlikely actions are rarely chosen (by definition, because they are unlikely), so there is on average less prediction error and therefore less overall activity.

Oscillatory Models

EEG studies of the ACC show oscillatory activity especially in the theta band (approximately 4–8 Hz). The prominent theta signals have been found in EEG signals with a likely ACC source (see also Chapter 14 by M.X. Cohen in this volume). These include the N2 effects of novelty and conflict, and the ERN and feedback ERN effects (Cavanagh & Frank, 2014). A number of models of the ACC simulate neural activity as a rate code rather than by simulating individual spiking neurons. This simplifies the computation, but it also renders such models unable to simulate the moment-by-moment spectrum of oscillatory power. A recent theoretical model (Cohen, 2014) proposes that conflict detection is implemented as coincidence detection in cortical layer 5 of the ACC, whereas the EEG conflict-related theta oscillations may be generated in layers 2 and 3. The idea is that layer 5 cells detect the coincident activation that represents conflict, and these in turn activate layers 2 and 3 cells. The input from layer 5 has the effect of amplifying the processing in layers 2 and 3. The layer 2 and 3 signals may be the source of the observed theta rhythms, but it is unclear whether the theta rhythm plays an important functional role or is merely an epiphenomenal reflection of afferent signals.

Control Signals: How Does the ACC Influence Other Brain Regions?

So far this chapter has considered mainly what signals are generated within the ACC and how they are generated. The next question is exactly what effects those signals have on other parts of the brain and behaviour. Here again, a variety of effects have been proposed. Before considering those, it must be noted that animals and humans generally show surprisingly little effects of ACC lesions. In the most severe cases, massive lesions result in akinetic mutism (Barris & Schumann, 1953; Nemeth et al., 1988). More recent studies show that human ACC lesions do not impair cognitive control in general (Fellows & Farah, 2005) and only moderately slow error correction (Modirrousta & Fellows, 2008). Animal studies show that cingulate lesions impair the ability to take the recent reward history into account (Kennerley, Walton, Behrens, Buckley, & Rushworth, 2006) and an unwillingness to exert extra effort to obtain a larger reward (Walton, Bannerman, & Rushworth, 2002). Any proposed model of ACC effects on other brain regions and behaviour must take these findings into account.

The proposed influences of the ACC on behaviour can generally be categorised into proactive and reactive control signals (Braver, Gray, & Burgess, 2007; see also Chapter 9 by Chiew & Braver in this volume), and this chapter will view such effects through this lens of proactive versus reactive control. Proactive control signals refer to those that are activated prior to a cognitively demanding event, typically in response to cues that precede and correlate with demanding events. Reactive control signals, on the other hand, are activated in response to errors or unexpected events. The ACC may generate both kinds of signals. Error likelihood signals would be an example of a signal that is sufficient to provide proactive control (Brown & Braver, 2005), especially as a means of inducing risk avoidance (Brown & Braver, 2007). Error signals (Gehring et al., 1990; Hohnsbein, Falkenstein, & Hoorman, 1989) would be an example of signals sufficient to drive reactive control.

One proposed way of understanding the ACC is that it does not so much learn specific behaviours as motivate or energise those behaviours, especially to achieve larger task goals (Holroyd & Yeung, 2012). This idea of proactive control was recently fleshed out in a computational model (Holroyd & Mcclure, 2015). According to the model, the ACC represents the values and costs of various higher-level options and then exerts a control signal specifically over the striatum. This proposal is similar to an earlier proposal, which, however, did not focus on a role of the striatum and did not include computational model simulations (Shenhav et al., 2013). The model of Holroyd and Yeung (2012) casts the ACC in the role of an actor rather than a performance monitor, and the effects of such ACC activity would amount to a proactive control signal. This control signal in the model ensures that higher-level goals appropriately guide lower-level action selection, so that behaviour remains directed towards higher-level goals. Simulations of rat prelimbic cortex lesions show that, as with rat empirical data, lesions of the computational model led to a higher number of perseverations and a longer time to criterion in a maze task with regularly switching rules.

Another recent model argues that the ACC reacts to strategy failures by forcing the brain to discard current behavioural strategies in favour of newly synthesised strategies (Donoso, Collins, & Koechlin, 2014). This bears a resemblance to the proposed role of the ACC in driving foraging (Kolling et al., 2012), except that here the foraging would be done at the cognitive level in terms of strategies, rather than at the behavioural level in terms of afforded actions. In this model, the brain creates approximately three possible strategies for the current behavioural task at any given time. One strategy is used to guide behaviour, while the other two strategies are held in reserve and evaluated to see if they would have provided a more successful match to the current task environment (see also Chapter 12 by Duverne and Koechlin in this volume). The frontopolar cortex activity correlated with the reliabilities of the alternative strategies, whereas the left lateral PFC was activated when the current strategy was rejected in favour of an alternative strategy. It is noteworthy that this model does not cast the ACC as driving a strategy switch in every case. Instead, the ACC is active only when both the current *and the alternative* strategies are all found wanting, so that a completely new set of strategies must be implemented.

The PRO model as discussed above (Alexander & Brown, 2011) focuses much on how the ACC generates the empirically observed activity patterns, but it is relatively silent on what functional role those activity patterns have on other brain regions. The only proposed control effect is on learning. Specifically, when the environmental contingencies are relatively stable, the environment can be described as statistically more stationary or less volatile. When the environment is more stationary, the effective learning rate that can be recovered from observed behaviour is relatively lower, and this correlates with reduced ACC activity (Behrens et al., 2007). Conversely, in more non-stationary environments, the effective learning rates are larger. The PRO model captures these phenomena by modelling surprise signals (i.e., prediction errors) as reducing the overall learning rate. Although this may seem counterintuitive, it turns out that in the PRO model, non-stationary environments cause the predicted outcomes of actions to be less certain, and therefore any particular outcome is less surprising. Thus, in non-stationary environments, surprise has a weaker effect on reducing the learning rate, which allows the learning rate to be effectively higher.

The ACC may also provide a proactive control signal that drives risk avoidance. Several studies have shown that ACC activity is greater when subjects choose to engage in a riskier option (Fukui, Murai, Fukuyama, Hayashi, & Hanakawa, 2005; Fukunaga et al., 2012; Krawitz et al., 2010). The error likelihood model simulates how the ACC may learn to represent conditions when errors are more likely, including incongruent stimuli that may entail response conflict (Brown & Braver, 2005). Still, the initial error likelihood proposal did not investigate the potential role of the ACC in risk avoidance. To study this, subjects were asked to self-report their risk-taking preferences. Those who were more risk averse showed stronger error likelihood effects, whereas the more risk-seeking subjects showed essentially no error likelihood effect (Brown & Braver, 2007). This individual difference effect was modelled with the error likelihood model simply by manipulating the learning rate from errors. When the learning rate was higher, the model learned more quickly from errors and showed stronger error likelihood effects in a way that matched the more risk-averse subjects (Brown & Braver, 2008). All of this is consistent with a role for the ACC in driving risk avoidance.

Conclusion

Table 15.1 summarises the essence of a number of computational models that have been proposed over the years (see also Chapter 8 by Verguts in this volume). The ACC shows a number of effects that have attracted significant interest from modellers. The effects variously capture aspects of performance monitoring and cognitive control, including errors, conflict, and more general predictions and prediction errors among many other effects. There is currently substantial controversy around the question of whether the ACC reflects more cognitive signals (such as conflict or choice difficulty) or affective signals (such as action value and reward; Shenhav et al., 2014). Other open questions concern whether predictive timing signals are extrinsic or intrinsic to the ACC, and whether ACC activity correlations with the RT are sufficient to account for other effects such as response conflict. For many years now, the response conflict model in particular has dominated the literature on models of the ACC, to the point that ACC activity has been treated as nearly synonymous with response conflict.

As the above discussion suggests, the modelling literature is much richer now than when the conflict model was introduced, and there is currently a healthy interaction between models and animal and human studies. Future progress will be most efficient if experimental work can be aimed at discriminating among the multiple existing models of the ACC (Alexander & Brown, 2015).

References

Aarts, E., Roelofs, A., & van Turennout, M. (2008). Anticipatory activity in anterior cingulate cortex can be independent of conflict and error likelihood. *Journal of Neuroscience, 28*, 4671–4678.

Alexander, W. H., & Brown, J. W. (2010). Computational models of performance monitoring and cognitive control. *Topics in Cognitive Science, 2*, 658–677.

Alexander, W. H., & Brown, J. W. (2011). Medial prefrontal cortex as an action-outcome predictor. *Nature Neuroscience, 14*, 1338–1344.

Alexander, W. H., & Brown, J. W. (2014). A general role for medial prefrontal cortex in event prediction. *Frontiers in Computational Neuroscience, 8.*

Alexander, W. H., & Brown, J. W. (2015). Reciprocal interactions of computational modeling and empirical investigation. In B. U. Forstmann & E.-J. Wagenmakers (Eds.), *Model-based cognitive neuroscience* (pp. 321–338). New York: Springer.

Amador, N., Schlag-Rey, M., & Schlag, J. (2000). Reward-predicting and reward-detecting neuronal activity in the primate supplementary eye field. *Journal of Neurophysiology, 84*, 2166–2170.

Bao, S., Chan, V. T., & Merzenich, M. M. (2001). Cortical remodelling induced by activity of ventral tegmental dopamine neurons. *Nature, 412*, 79–83.

Barris, W., & Schumann, H. (1953). Bilateral anterior cingulate gyrus lesion: Syndrome of the anterior cingulate gyri. *Neurology, 3*, 44–80.

Behrens, T. E., Woolrich, M. W., Walton, M. E., & Rushworth, M. F. (2007). Learning the value of information in an uncertain world. *Nature Neuroscience, 10*, 1214–1221.

Bekolay, T., Laubach, M., & Eliasmith, C. (2014). A spiking neural integrator model of the adaptive control of action by the medial prefrontal cortex. *Journal of Neuroscience, 34*, 1892–902.

Bernstein, P. S., Scheffers, M. K., & Coles, M. G. (1995). 'Where did I go wrong?' A psychophysiological analysis of error detection. *Journal of Experimental Psychology: Human Perception and Performance, 21*, 1312–1322.

Blais, C., & Bunge, S. (2010). Behavioral and neural evidence for item-specific performance monitoring. *Journal of Cognitive Neuroscience, 22*, 2758–2767.

Botvinick, M. M., Braver, T. S., Barch, D. M., Carter, C. S., & Cohen, J. C. (2001). Conflict monitoring and cognitive control. *Psychological Review, 108*, 624–652.

Botvinick, M. M., Nystrom, L., Fissel, K., Carter, C. S., & Cohen, J. D. (1999). Conflict monitoring versus selection-for-action in anterior cingulate cortex. *Nature, 402*, 179–181.

Braver, T. S., Gray, J. R., & Burgess, G. C. (2007). Explaining the many varieties of working memory variation: Dual mechanisms of cognitive control. In A. Conway, M. Kane, A. Miyake, & J. Towse, C. J. (Eds.), *Variation of working memory.* Oxford: Oxford University Press.

Brown, J. W., & Braver, T. S. (2005). Learned predictions of error likelihood in the anterior cingulate cortex. *Science, 307(5712)*, 1118–1121.

Brown, J. W., & Braver, T. S. (2007). Risk prediction and aversion by anterior cingulate cortex. *Cognitive, Affective, & Behavioral Neuroscience, 7*, 266–277.

Brown, J. W., & Braver, T. S. (2008). A computational model of risk, conflict, and individual difference effects in the anterior cingulate cortex. *Brain Research, 1202*, 99–108.

Brown, J. W., Bullock, D., & Grossberg, S. (2004). How laminar frontal cortex and basal ganglia circuits interact to control planned and reactive saccades. *Neural Networks, 17*, 471–510.

Brown, J. W., Reynolds, J. R., & Braver, T. S. (2007). A computational model of fractionated conflict-control mechanisms in task switching. *Cognitive Psychology, 55*, 37–85.

Burle, B., Roger, C., Allain, S., Vidal, F., & Hasbroucq, T. (2008). Error negativity does not reflect conflict: A reappraisal of conflict monitoring and anterior cingulate cortex activity. *Journal of Cognitive Neuroscience*, *20*, 1637–1655.

Carp, J., Kim, K., Taylor, S. F., Fitzgerald, K. D., & Weissman, D. H. (2010). Conditional differences in mean reaction time explain effects of response congruency, but not accuracy, on posterior medial frontal cortex activity. *Frontiers in Human Neuroscience*, *4*, 231.

Carter, C. S., Braver, T. S., Barch, D. M., Botvinick, M. M., Noll, D. C., & Cohen, J. D. (1998). Anterior cingulate cortex, error detection, and the online monitoring of performance. *Science*, *280*, 747–749.

Cavanagh, J. F., & Frank, M. J. (2014). Frontal theta as a mechanism for cognitive control. *Trends in Cognitive Sciences*, *18*(8), 414–421.

Cohen, M. X. (2014). A neural microcircuit for cognitive conflict detection and signaling. *Trends in Neurosciences*. *37*(9), 480–490.

Cole, M. W., Yeung, N., Freiwald, W. A., & Botvinick, M. (2009). Cingulate cortex: Diverging data from humans and monkeys. *Trends in Neurosciences*, *32*, 566–574.

Crottaz-Herbette, S., & Menon, V. (2006). Where and when the anterior cingulate cortex modulates attentional response: Combined fMRI and ERP evidence. *Journal of Cognitive Neuroscience*, *18*, 766–780.

Donoso, M., Collins, A. G. E., & Koechlin, E. (2014). Human cognition. Foundations of human reasoning in the prefrontal cortex. *Science*, *344*, 1481–1486.

Ebitz, R. B., & Platt, M. L. (2015). Neuronal activity in primate dorsal anterior cingulate cortex signals task conflict and predicts adjustments in pupil-linked arousal. *Neuron*, *85*, 628–640.

Falkenstein, M., Hohnsbein, J., Hoorman, J., & Blanke, L. (1991). Effects of crossmodal divided attention on late ERP components: II. Error processing in choice reaction tasks. *Electroencephalography and Clinical Neurophysiology*, *78*, 447–455.

Fellows, L. K., & Farah, M. J. (2005). Is anterior cingulate cortex necessary for cognitive control? *Brain*, *128*, 788–796.

Ferdinand, N. K., Mecklinger, A., Kray, J., & Gehring, W. J. (2012). The processing of unexpected positive response outcomes in the mediofrontal cortex. *Journal of Neuroscience*, *32*, 12087–12092.

Fiala, J. C., Grossberg, S., & Bullock, D. (1996). Metabotropic glutamate receptor activation in cerebellar Purkinje cells as substrate for adaptive timing of the classically conditioned eye-blink response. *Journal of Neuroscience*, *16*, 3760–3774.

Forster, S. E., & Brown, J. W. (2011). Medial prefrontal cortex predicts and evaluates the timing of action outcomes. *Neuroimage*, *55*, 253–265.

Fukui, H., Murai, T., Fukuyama, H., Hayashi, T., & Hanakawa, T. (2005). Functional activity related to risk anticipation during performance of the Iowa Gambling Task. *Neuroimage*, *24*, 253–259.

Fukunaga, R., Bogg, T., Finn, P. R., & Brown, J. W. (2013). Decisions during negatively-framed messages yield smaller risk-aversion-related brain activation in substance-dependent individuals. *Psychology of Addictive Behaviors*, *27*, 1141–1152.

Fukunaga, R., Brown, J. W., & Bogg, T. (2012). Decision making in the Balloon Analogue Risk Task (BART): Anterior cingulate cortex signals loss aversion but not the infrequency of risky choices. *Cognitive, Affective, & Behavioral Neuroscience*. *12*(3), 479–490.

Garofalo, S., Maier, M. E., & di Pellegrino, G. (2014). Mediofrontal negativity signals unexpected omission of aversive events. *Scientific Reports*, *4*, 4816.

Gehring, W. J., Coles, M. G. H., Meyer, D. E., & Donchin, E. (1990). The error-related negativity: An event-related potential accompanying errors. *Psychophysiology*, *27*, S34.

Gehring, W. J., Goss, B., Coles, M. G. H., Meyer, D. E., & Donchin, E. (1993). A neural system for error detection and compensation. *Psychological Science*, *4*, 385–390.

Grinband, J., Savitskaya, J., Wager, T. D., Teichert, T., Ferrera, V. P., & Hirsch, J. (2011). The dorsal medial frontal cortex is sensitive to time on task, not response conflict or error likelihood. *Neuroimage*, *57*, 303–311.

Hayden, B. Y., Heilbronner, S. R., Pearson, J. M., & Platt, M. L. (2011). Surprise signals in anterior cingulate cortex: Neuronal encoding of unsigned reward prediction errors driving adjustment in behavior. *Journal of Neuroscience, 31*, 4178–4187.

Hayden, B. Y., Pearson, J. M., & Platt, M. L. (2011). Neuronal basis of sequential foraging decisions in a patchy environment. *Nature Neuroscience, 14*, 933–939.

Hohnsbein, J., Falkenstein, M., & Hoorman, J. (1989). Error processing in visual and auditory choice reaction tasks. *Journal of Psychophysiology, 3*, 32.

Holroyd, C. B., & Coles, M. G. (2002). The neural basis of human error processing: Reinforcement learning, dopamine, and the error-related negativity. *Psychological Review, 109*, 679–709.

Holroyd, C. B., & Mcclure, S. M. (2015). Hierarchical control over effortful behavior by rodent medial frontal cortex. *Psychological Review, 122*, 54–83.

Holroyd, C. B., & Yeung, N. (2012). Motivation of extended behaviors by anterior cingulate cortex. *Trends in Cognitive Sciences, 16*, 122–128.

Holroyd, C. B., Yeung, N., Coles, M. G., & Cohen, J. D. (2005). A mechanism for error detection in speeded response time tasks. *Journal of Experimental Psychology: General, 134*, 163–191.

Ide, J. S., Shenoy, P., Yu, A. J., & Li, C. R. (2013). Bayesian prediction and evaluation in the anterior cingulate cortex. *Journal of Neuroscience, 33*, 2039–2047.

Ito, S., Stuphorn, V., Brown, J. W., & Schall, J. D. (2003). Performance monitoring by anterior cingulate cortex during saccade countermanding. *Science, 302*(5642), 120–122.

Jessup, R. K., Busemeyer, J. R., & Brown, J. W. (2010). Error effects in anterior cingulate cortex reverse when error likelihood is high. *Journal of Neuroscience, 30*, 3467–3472.

Kennerley, S. W., Walton, M. E., Behrens, T. E., Buckley, M. J., & Rushworth, M. F. (2006). Optimal decision making and the anterior cingulate cortex. *Nature Neuroscience, 9*, 940–947.

Kolling, N., Behrens, T. E. J., Mars, R. B., & Rushworth, M. F. S. (2012). Neural mechanisms of foraging. *Science, 336*, 95–98.

Kool, W., McGuire, J. T., Wang, G. J., & Botvinick, M. M. (2013). Neural and behavioral evidence for an intrinsic cost of self-control. *PLoS ONE, 8*.

Krawitz, A., Fukunaga, R., & Brown, J. W. (2010). Anterior insula activity predicts the influence of positively framed messages on decision making. *Cognitive, Affective, & Behavioral Neuroscience, 10*, 392–405.

MacDonald, A. W., Cohen, J. D., Stenger, V. A., & Carter, C. S. (2000). Dissociating the role of the dorsolateral prefrontal cortex and anterior cingulate cortex in cognitive control. *Science, 288*(5472), 1835–1838.

Middleton, F. A., & Strick, P. L. (1994). Anatomical evidence for cerebellar and basal ganglia involvement in higher cognitive function. *Science, 266*(5184), 458–461.

Modirrousta, M., & Fellows, L. K. (2008). Dorsal medial prefrontal cortex plays a necessary role in rapid error prediction in humans. *Journal of Neuroscience, 28*, 14000–14005.

Nee, D. E., Kastner, S., & Brown, J. W. (2011). Functional heterogeneity of conflict, error, task-switching, and unexpectedness effects within medial prefrontal cortex. *Neuroimage, 54*, 528–540.

Nemeth, G., Hegedus, K., & Molnar, L. (1988). Akinetic mutism associated with bicingular lesions: Clinicopathological and functional anatomical correlates. *European Archives of Psychiatry and Neurological Sciences, 237*, 218–222.

Niki, H., & Watanabe, M. (1979). Prefrontal and cingulate unit activity during timing behavior in the monkey. *Brain Research, 171*, 213–224.

Oliveira, F. T., McDonald, J. J., & Goodman, D. (2007). Performance monitoring in the anterior cingulate is not all error related: Expectancy deviation and the representation of action-outcome associations. *Journal of Cognitive Neuroscience, 19*(12), 1994–2004..

Parvizi, J., Rangarajan, V., Shirer, W. R., Desai, N., & Greicius, M. D. (2013). The will to persevere induced by electrical stimulation of the human cingulate gyrus. *Neuron, 80*, 1359–1367.

Poldrack, R. A. (2011). Inferring mental states from neuroimaging data: From reverse inference to large-scale decoding. *Neuron, 72*, 692–697.

Roelofs, A., van Turennout, M., & Coles, M. G. (2006). Anterior cingulate cortex activity can be independent of response conflict in Stroop-like tasks. *Proceedings of the National Academy of Sciences of the United States of America, 103*, 13884–13889.

Sallet, J., Quilodran, R., Rothe, M., Vezoli, J., Joseph, J. P., & Procyk, E. (2007). Expectations, gains, and losses in the anterior cingulate cortex. *Cognitive, Affective, & Behavioral Neuroscience*, *7*, 327–336.

Scheffers, M. K., & Coles, M. G. H. (2000). Performance monitoring in a confusing world: Error-related brain activity, judgments of response accuracy, and types of errors. *Journal of Experimental Psychology*, *26*, 141–151.

Schultz, W. (1998). Predictive reward signal of dopamine neurons. *Journal of Neurophysiology*, *80*, 1–27.

Shenhav, A., Botvinick, M. M., & Cohen, J. D. (2013). The expected value of control: An integrative theory of anterior cingulate cortex function. *Neuron*, *79*, 217–240.

Shenhav, A., Straccia, M. A., Cohen, J. D., & Botvinick, M. M. (2014). Anterior cingulate engagement in a foraging context reflects choice difficulty, not foraging value. *Nature Neuroscience*, *17*(9), 1249–1254.

Silvetti, M., Alexander, W., Verguts, T., & Brown, J. W. (2014). From conflict management to reward-based decision making: Actors and critics in primate medial frontal cortex. *Neuroscience & Biobehavioral Reviews*, *46*(Pt. 1), 44–57.

Silvetti, M., Seurinck, R., & Verguts, T. (2011). Value and prediction error in medial frontal cortex: Integrating the single-unit and systems levels of analysis. *Frontiers in Human Neuroscience*, *5*, 75.

Steinhauser, M., Maier, M., & Hübner, R. (2008). Modeling behavioral measures of error detection in choice tasks: Response monitoring versus conflict monitoring. *Journal of Experimental Psychology: Human Perception and Performance*, *34*, 158–176.

Stuphorn, V., Taylor, T. L., & Schall, J. D. (2000). Performance monitoring by the supplementary eye field. *Nature*, *408*, 857–860.

Walton, M. E., Bannerman, D. M., & Rushworth, M. F. (2002). The role of rat medial frontal cortex in effort-based decision making. *Journal of Neuroscience*, *22*, 10996–11003.

Yeung, N., Botvinick, M. M., & Cohen, J. D. (2004). The neural basis of error detection: Conflict monitoring and the error-related negativity. *Psychological Review*, *111*, 931–959.

Yeung, N., & Nieuwenhuis, S. (2009). Dissociating response conflict and error likelihood in anterior cingulate cortex. *Journal of Neuroscience*, *29*, 14506–14510.

16

The Neuroanatomy and Neurochemistry of Response Inhibition

Birte U. Forstmann and Anneke Alkemade

Introduction

In everyday life, we are continuously faced with situations in which we need to inhibit a planned response. Stopping for an unexpected car turning the corner just as you are about to cross the street, or when someone exits the door at the exact same moment that you were planning to enter. In our daily life, we do not give this stopping behaviour much thought. But what happens when the required brain systems in control of stopping behaviour are damaged? A case study of a 71-year-old man provides some answers to these questions (Trillet, Vighetto, Croisile, Charles, & Aimard, 1995). The man presented with a hematoma (accumulation of blood) in the subthalamic nucleus (STN), a vitally important nucleus in the basal ganglia (BG) forming part of the stopping network (Trillet et al., 1995). As a result of the hematoma, he showed disinhibition of motoricity as well as a severe neurobehavioral disinhibition. The clinical presentation was associated with signs of euphoria as well as an excessive, uncontrollable flow of words. Within two months, the hematoma dissolved and symptoms regressed. No frontal signs or mental confusion was present, and the man had full insight into his mental change.

It is intriguing that this patient was fully aware of his disinhibited behaviour, yet was unable to control his responses. This case report also illustrates the central role of the STN in the complex network of response inhibition. Evidently, the disinhibition reported in this case study illustrates an extreme situation of daily life behaviour that is otherwise variable in healthy human individuals. To understand this behaviour, it is important to understand what underlying neuroanatomical brain structures and networks are involved. In more general terms, in order to understand [brain] function, it is important to study [brain] structure (Crick, 1988).

In this chapter, we will first give a brief overview of commonly used inhibition and interference tasks used in the cognitive neurosciences. This will be followed by an in-depth discussion of brain structures and networks subserving inhibitory functioning in inhibition tasks. In particular, we will highlight the anatomo-functional make-up of several cortical and subcortical brain regions and networks in humans and non-human primates as well as rodents. For a review of behavioural work on response inhibition, see Chapter 6 by Verbruggen and Logan in this volume, and for a survey of the possible role of impaired response inhibition in substance abuse, see Chapter 32 by Chaarani et al. in this volume.

The Wiley Handbook of Cognitive Control, First Edition. Edited by Tobias Egner.

Measuring Inhibition in the Cognitive Neurosciences

Several paradigms have been successfully introduced in the cognitive neurosciences capturing inhibition in the so-called stop-signal task, go/no-go task, and interference tasks such as, for example, the Stroop task (for a review see, e.g., Nee, Wager, & Jonides, 2007; see also Chapter 4 by Egner and Chapter 5 by Bugg in this volume). Six of the most prominent paradigms are briefly outlined below.

Go/no-go task: The go/no-go paradigm requires execution or inhibition of a motor response. Two separate stimuli require either a motor response (go-stimulus) or to withhold a motor response (no-go stimulus; Donders, 1969).

Stop-signal task: In the stop-signal task, participants have to respond to a stimulus feature, for example, an arrow pointing leftward or rightward requiring a left or right response, respectively. In a minority of the trials, the go-stimulus is followed by a cue (e.g., a tone) instructing participants to withhold the ongoing motor response (Logan, 1994; see also Chapter 6 by Verbruggen and Logan in this volume).

Stroop task: In the Stroop task, participants are presented with colour names (e.g., 'red') in various font colours (e.g., blue, green, or red). The participants are instructed to name the colour of the font while ignoring the word's meaning. In incongruent trials, colour names and font colours differ (e.g., 'red' is presented in blue)—that is, interfere with each other—and the highly automatic reading of the word has to be suppressed (Stroop, 1935).

Simon task: In the Simon task, participants are typically presented with coloured stimuli in the left or right visual field. Participants respond via a button press on a horizontally aligned keyboard or button box, indicating the colour of the stimulus (e.g., pressing a right-hand button for red stimuli and a left-hand button for blue stimuli). At the same time, they are trying to ignore the task-irrelevant stimulus location, which may either be spatially matched (e.g., a red stimulus shown in the right visual field) or mismatched (e.g., a red stimulus shown in the left visual field) with the correct response. As a sudden-onset stimulus appears to 'activate' a spatially corresponding response, congruent trials are the result, where both relevant (colour) and irrelevant (location) stimulus dimensions activate the same, correct action; in incongruent trials, the irrelevant dimension activates an incorrect response tendency, which interferes with the selection of the correct response (Simon, 1989).

Flanker task: In the flanker task, participants are presented with a centrally presented stimulus (e.g., an arrow pointing to the left or right), and surrounding flanker stimuli taken from the same stimulus set. Participants respond via a left or right button press indicating the direction of the stimulus. The flankers can be congruent (arrows pointing in the same direction as the stimulus), incongruent (arrows pointing in the opposite direction), or neutral (no direction; Eriksen & Eriksen, 1974).

Stimulus–response (S-R) compatibility task: In S-R compatibility tasks, participants are required to switch between two stimulus–response mappings. In a compatible mapping, the response is directly suggested by the stimulus, for example, an arrow pointing to the right requiring a right button press. An incompatible mapping requires a left button press with a right-pointing stimulus (Alluisi & Warm, 1989).

Brain Regions and Networks Involved in Inhibition

The brain regions involved in the execution or inhibition of a planned response in the go/no-go task, stop-signal task, Stroop task, Simon task, flanker task, and stimulus–response compatibility tasks are a number of cortical and subcortical brain areas including brain circuitries that display complex interactions which may inhibit each other (Figure 16.1).

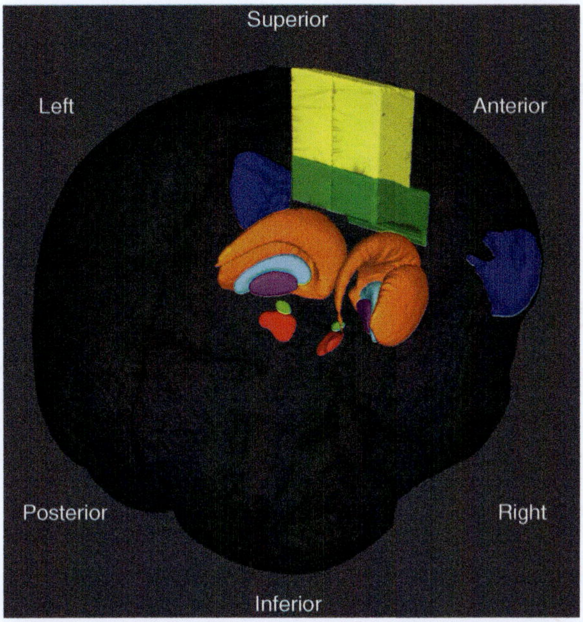

Figure 16.1 3D rendering of the 0.4 mm MNI CBS template including several cortical and subcortical brain regions consistently found in paradigms tapping into response inhibition (see, e.g., Nee et al., 2007; Swick et al., 2011; Rae, Hughes, Weaver, Anderson, & Rowe, 2014; Swick, Ashley, & Turken, 2011). Yellow: Pre-SMA; green: anterior cingulate cortex (ACC); dark blue: inferior frontal cortex (IFC); orange: striatum; light blue: GPI; purple: GPe; green: STN; red: SN. Source: Reproduced with permission of Max Keuken.

At the centre of the anatomical stopping network are the basal ganglia (BG) (Aron, 2007). The BG present a number of closely connected subcortical nuclei located at the base of the forebrain, and include the striatum (caudate nucleus and the putamen); the globus pallidus pars externa (GPe) and interna (GPi); the substantia nigra, pars reticulate (SNr), and compacta (SNc); and the subthalamic nucleus (STN). These structures have many connections with the cerebral cortex, the thalamus, and the brain stem, as well as a number of other brain areas. Together these structures and their connections give rise to the direct, indirect, and hyperdirect pathway (Figure 16.2), transferring information from the cortex to the thalamus (Nambu, Tokuno, & Takada, 2002).

In the direct pathway, glutamatergic, excitatory signals from the cortex to the striatum result in inhibition of the SNr-GPi complex, which releases the inhibition of the thalamus. Subsequently the thalamus provides an excitatory output resulting in a response. In the indirect pathway, glutamatergic signals from the cortex enter the BG at the level of the striatum, which releases the inhibition of the GPe. In turn, the STN is released from inhibition resulting in SNr-GPi inhibition, which then subsequently inhibits the thalamus, resulting in a smaller excitatory output. In addition to being part of the indirect pathway, the STN is part of a hyperdirect pathway in which the STN receives direct input from the cortex, inducing short-latency excitation followed by inhibition and late excitation of the pallidal neurons (Nambu et al., 2002). The schematic description of the brain networks involved in stopping behaviour provides insight in the major brain circuitries involved in response inhibition.

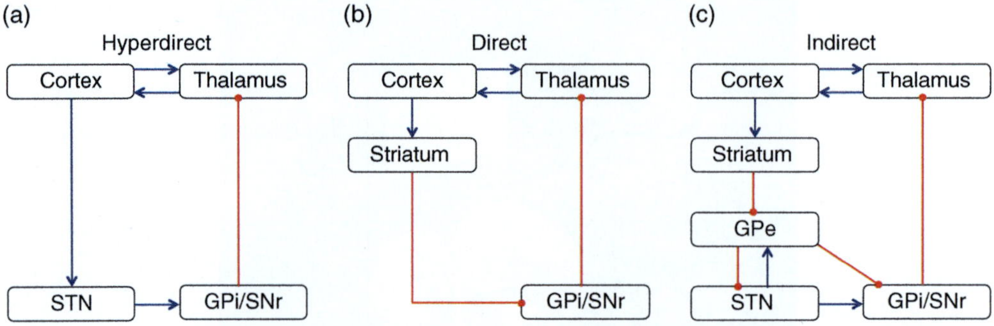

Figure 16.2 Schematic diagram of the (a) cortico-STN-GPi/SNr 'hyperdirect' pathway, (b) cortico-striato-GPi/SNr 'direct' pathway, and (c) cortico-striato-GPe-STN-GPi/SNr 'indirect' pathway. Blue lines represent excitatory glutamatergic (glu), and red lines inhibitory Gamma-aminobutyric acid (GABA)-ergic projections. GPe: external segment of the globus pallidus; GPi: internal segment of the globus pallidus; SNr: substantia nigra pars reticulate; STN: subthalamic nucleus. Source: Adapted from Alexander 1990; Nambu 2002.

Anatomo-Functional Role of Brain Regions and Networks Involved in Inhibition

Functional Anatomy

Human neurocognitive studies using functional magnetic resonance imaging (fMRI) have revealed a common number of activated foci in the brain using a number of response inhibition tasks (Rubia et al., 2010). It is clear that the prefrontal cortex (PFC) is crucial for planning of complex cognitive behaviour, including decision making and inhibition. The attribution of cognitive functions to discrete regions within the PFC is controversial (Aron, Robbins, & Poldrack, 2004, 2014; Hampshire, 2015), but there are a number of consistent findings. Response inhibition is associated with increased activation in dorsolateral PFC, inferior frontal cortex (IFC), and orbitofrontal cortex (OFC). The PFC has been postulated to exert top-down modulatory influences on subcortical and posterior cortical regions. The PFC can be divided into three major areas: (a) The OFC and the limbic pathways are implicated in reward- and affect-based decisions (Rosenbloom, Schmahmann, & Price, 2012); (b) the anterior cingulate cortex (ACC), which is important for handling conflicting options, as well as outcome processing; and (c) the dorsolateral PFC, which can represent and integrate multiple sources of information. These cortical structures are closely connected to each other as well as to subcortical areas including the BG, thalamus, and cerebellum.

The function of the PFC is largely related to its involvement in executive control. It is not clear whether subregions of the PFC show functional differentiation. One view is that PFC performs a single operation (working memory), but differs in sensory input (Goldman-Rakic, 1987). Variants of this hypothesis include the PFC as a having a bias towards posterior cortical or subcortical areas (O'Reilly, Noelle, Braver, & Cohen, 2002) in which signals get directed to either structure. Bilateral activation of the dorsolateral PFC, IFC, and ACC, but not other frontal regions, appears to be consistently found across response inhibition tasks (Duncan & Owen, 2000).

Standard models of the go/no-go tasks include the involvement of the frontal lobes in the inhibition of responses to inappropriate stimuli, signalling an override to the motor system inhibiting the planned response. A number of neuroimaging studies have implicated right hemisphere PFC regions, and more particularly the dorsolateral PFC and (right) inferior frontal gyrus to be predominant in response inhibition (Aron & Poldrack, 2006; Congdon et al., 2010; Garavan, Ross, & Stein, 1999; Kawashima et al., 1996; Konishi et al., 1999; Konishi, Nakajima, Uchida, Sekihara, & Miyashita, 1998; Whelan et al., 2012). In addition, activations were observed in the bilateral and ventrolateral PFC, as well as the medial PFC (Casey et al., 1997; Leung & Cai, 2007; Liddle, Kiehl, & Smith, 2001; Menon, Adleman, White, Glover, & Reiss, 2001; Rubia et al., 2001; Wager et al., 2005; Watanabe et al., 2002). It is clear that frontal lobe activation is consistently indicated, but the exact location of the cortical activation may differ substantially between studies (Simmonds, Pekar, & Mostofsky, 2008). The right dorsolateral PFC and inferior parietal cortex become active when increased working memory is required. The go/no-go task generally elicits activation in the pre-supplementary motor area (pre-SMA) and fusiform gyrus. This finding supports the notion that the pre-SMA is critical for the selection of appropriate behaviour, which may be either response execution or inhibition (Simmonds et al., 2008). The Stroop task has been shown to activate the dorsolateral PFC, ACC, inferior frontal gyrus (IFG), inferior parietal gyrus (IPG), and inferior temporal cortices, whereas the Simon task involves dorsal premotor, posterior, and superior parietal cortices (Liu, Banich, Jacobson, & Tanabe, 2004).

Structural Anatomy

Before the development of modern non-invasive neuroimaging techniques, our understanding of the relation between brain structure and function was based largely on case studies on brain lesions induced by accidents, and on developmental defects and neurological disease. The development of brain imaging techniques has given cognitive neuroscientists the tools required for localisation of function within the healthy brain. In addition, these techniques provide information on functional connectivity between distinct brain areas involved in the execution of cognitive processes. The anatomy underlying functional connectivity is not always known, but could potentially provide additional information that helps to understand cognitive function and to interpret fMRI data. Information on the underlying functional neuroanatomy is available from rodents, non-human primates, as well as post-mortem human tissues. These studies have proved valuable in the understanding of human brain processes; however, one should keep in mind that caution should be used when interpreting data obtained from animal studies. Although data may appear largely in agreement, rodent data cannot always be translated directly to the human situation in view of interspecies differences (e.g., Afsharpour, 1985a, 1985b; Rafols & Fox, 1976).

In the next sections, we will highlight the anatomo-functional make-up of several cortical and subcortical brain regions and networks in humans and non-human primates as well as rodents.

Cortex The human neocortex shows an extensive folding pattern, whereas the rodent brain has a smooth appearance without sulci and gyri, which is in line with the differences between rodents and humans in their higher cognitive functioning. Whether rodents can provide models of prefrontal cortical function is debatable (Brown & Bowman, 2002). Interspecies differences also remain in animals that are closer to humans from an evolutionary point of view. Brodmann's human brain map is in general agreement with that of monkeys, although differences are also present, in particularly in prefrontal and cingulate regions (Brodmann, 1909; Garey, 1994; Neubert, Mars, Sallet, & Rushworth, 2015; Patel et al., 2015). The cortex contains excitatory and inhibitory neurons. Excitation occurs through the release of glutamate, opening α-amino-3-hydroxy-5-methyl-4-isoxazolepropionic acid (AMPA), kainate, and N-Methyl-D-aspartate (NMDA) types of receptors, resulting in neuronal firing. Gamma-aminobutyric acid (GABA), the inhibitory neurotransmitter, opens GABA receptors, causing hyperpolarisation and thereby increasing the threshold for neuronal firing (Petersen, 2014). The ratio of inhibitory to excitatory neurons varies between cortical areas and layers.

Anterior Cingulate Cortex The connectivity between the ACC and the PFC and motor cortices is of crucial importance for the interaction between cognitive and motor systems (see also Chapter 15 by Brown in this volume). There is a strong functional connectivity between the supracallosal ACC and the lateral PFC, which was demonstrated using transcranial magnetic stimulation (TMS) during positron emission tomography (PET; Paus, Castro-Alamancos, & Petrides, 2001. The ACC is a target area for the mesocortical dopamine system, which signals from the ventral tegmental area. Dopaminergic fibres are most prominent in the ACC (Gaspar, Berger, Febvret, Vigny, & Henry, 1989; Lewis, 1992). In addition, the ACC receives serotonergic input, although these innervations have a much lower density (Berger, Trottier, Verney, Gaspar, & Alvarez, 1988). Noradrenergic input originating from the locus coeruleus (LC) mainly targets the deeper layers of the ACC, thereby modulating the input at an additional level (Aston-Jones & Cohen, 2005). Interestingly, in addition to the more well-known neurotransmitters, the ACC also receives (modulatory) input from corticotropin-releasing hormone (CRH; Lewis, Foote, & Cha, 1989), neurotensin (Gaspar, Berger, & Febvret, 1990), and prosomatostatin-derived peptides (Campbell, Lewis, Benoit, & Morrison, 1987; Satoh & Matsumura, 1990), and substance P (Iritani, Fujii, & Satoh, 1989), indicating additional levels of modulation (Paus et al., 2001).

Subthalamic Nucleus The subthalamic nucleus is a small biconvex nucleus located on top of the cerebral peduncle. This nucleus provides a strong glutamatergic (excitatory) signal to the SNr and the GPi, the output nuclei of BG, as well as the GPe (Albin et al., 1992; Alexander & Crutcher, 1990). The STN has motor, cognitive, and limbic functions. These functions have been proposed to show a tripartite subdivision within the STN (Haynes & Haber, 2013; Parent & Hazrati, 1995b; Temel, Blokland, Steinbusch, & Visser-Vandewalle, 2005). The details of the functional organisation within the STN are still a topic of discussion (Alkemade, 2013; Alkemade & Forstmann, 2014; Keuken et al., 2012; Lambert et al., 2015). Studies on protein expression in the human and non-human primate STN provide evidence for glutamatergic, GABA-ergic, dopaminergic, as well as serotonergic signalling (for a review, see Alkemade, Schnitzler, & Forstmann, 2015).

Striatum The striatum represents the largest structure within the BG. It is named after its characteristic striated (striped) appearance on brain sections. The striatum has a complex organisation and consists of a number of substructures. The striatum is a main input nucleus in addition to the STN (Figure 16.2). The cortex signals to the rest of the BG via the striatum. A number of midbrain and thalamic nuclei modulate subsequent signalling to the output nuclei of the BG. The output structures project to the midbrain in addition to other subcortical targets. Alternatively, signals are sent to the cortex via the thalamus through the cortico-BG thalamocortical circuitry. Cortical signals to the BG striatum and STN are conveyed using glutamate. The STN provides an excitatory signal to the output nuclei of the BG. In addition, glutamate is the main neurotransmitter signalling via the thalamostriatal routes, which regulate thalamocortical circuits.

In humans and non-human primates, the striatum comprises the caudate and putamen, as well as the nucleus accumbens (NAcb), which is often referred to as the ventral striatum.

Structures within the striatum have the same telencephalic origin and are not fully separated in the anterior region. The caudate and putamen are largely separated by the internal capsule, a white matter tract consisting of projections from the cortex as well as other brain areas. The NAcb is located in the rostroventral part of the striatum. Distinctions between the subparts can be made on the basis of the excitatory cortical, thalamic, and amygdaloid inputs. However, borders between the substructures of the striatum cannot be distinguished on MRI images, and even in post-mortem histological sections, this is challenging. The subparts of the striatum have the same basic structure and origin; therefore, clear boundaries are absent (Voorn, Vanderschuren, Groenewegen, Robbins, & Pennartz, 2004), although the NAcb has a higher cell and basophil density as compared to the rest of the striatum (Lauer & Heinsen, 1996). Over 95% of the striatal neurons are medium spiny neurons, which have inhibitory properties and use GABA as their main neurotransmitter (Beckstead, 1985; Kanazawa, Emson, & Cuello, 1977; Kemp & Powell, 1971; S. Vincent, Hokfelt, Christensson, & Terenius, 1982; S. R. Vincent, Hokfelt, Christensson, & Terenius, 1982). These neurons receive dense dendritic innervation, including glutamatergic innervation from the thalamus, as well as modulatory input from the amygdala, hippocampus and brain stem, dorsal raphe nucleus, ventral tegmental area, and LC (Groenewegen, Berendse, & Haber, 1993; Kelley, Domesick, & Nauta, 1982; Nieuwenhuis, 1985). The actions of spiny striatal neurons are modulated by dopaminergic input from the substantia nigra and VTA (Nieuwenhuis, 1985). In addition, striatal activity is regulated via serotonergic signals originating from the dorsal nucleus of the raphe and caudal linear nucleus, which show dense projections to the striatum, SN, and GP (Imai, Park, Steindler, & Kitai, 1986; Soghomonian, Descarries, & Watkins, 1989; Ternaux et al., 1977; van der Kooy & Hattori, 1980; Vertes, 1991). Other neurons in the striatum are interneurons, which may be acetylcholinergic or GABA-ergic (Kawaguchi, Wilson, Augood, & Emson, 1995). In turn, the striatum projects to the output structures of the BG using GABA, substance P, and dynorphin (Bolam & Izzo, 1988; Reid et al., 1990; Yung, Smith, Levey, & Bolam, 1996).

Substantia Nigra The SN is part of the mesencephalon and consists of a ventral zone, the reticular part, and the pars compacta, which contains the characteristic dopaminergic neurons containing neuromelanin, which is a dark-coloured dopamine-derived polymer from which the structure received its name (Zecca, Zucca, Wilms, & Sulzer, 2003).

Substantia Nigra Reticular Part Within the substantia nigra, two subparts can be distinguished: The dopamine-rich compact part, which contains a dense population of cells, and the reticular part, which is more diffuse and has a lower cell density (Hauber, 1998). The reticular part is considered a main output structure of the BG to the thalamus and midbrain. The striatum provides a GABA-ergic inhibitory signal to the SNr (Chevalier & Deniau, 1990; Deniau, Menetrey, & Thierry, 1994), which is part of the direct pathway (Figure 16.2). Additional GABA-ergic signals to the SNr originate from the GP and ventral pallidum (Groenewegen et al., 1993; Smith & Bolam, 1989, 1990). These signalling systems are part of the indirect pathway, and convergence of signalling takes place in this brain area. This means that GABA-ergic synaptic terminals from the striatum and globus pallidus project to single neurons within the SNr (Bevan, Smith, & Bolam, 1996; Bolam, Smith, Ingham, von Krosigk, & Smith, 1993). The converging neuroanatomy suggests that at this level, the direct and indirect pathway can interact, and these findings could imply that there is a convergence of signals that are related to limbic, sensorimotor, and cognitive function. The output of the SNr is dependent on dopamine (Cheramy, Leviel, & Glowinski, 1981; http://www.sciencedirect.com/science/article/pii/S0301008298000410 - BIB62). Dopamine can modulate the dopamine firing rate and release for the dopamine D2 receptor, and modulate GABA release from afferents through dopamine D1 receptors (Hauber, 1998; Kalivas, 1993; Robertson, 1992). The SNr is crucial for motor behavior (Robertson, 1992). The SNr receives a strong glutamatergic innervation from the STN and expresses a number of glutamatergic receptors (Albin et al., 1992; Hammond, Deniau, Rizk, & Feger, 1978; Nakanishi, Kita, & Kitai, 1987). The main targets of the SNr are the thalamus, superior colliculus, and the pedunculopontine tegmental nucleus (Beckstead, Domesick, & Nauta, 1979; Deniau & Chevalier, 1992; Nakanishi et al., 1987).

Substantia Nigra Compact Part The SNc and VTA provide dopaminergic input to the striatum, STN, and pallidum. These fibres often co-express cholecystokinin (CCK) or neurotensin (NT; Seroogy et al., 1988; Swanson, 1982). The SNc and VTA receive GABA-ergic feedback from the striatum and GP. The SNc additionally expresses tacykinin, substance P, dynorphin, and enkephalin receptors (Buck, Deshmukh, Burks, & Yamamura, 1981; Castel, Morino, Frey, Terenius, & Hokfelt, 1993). Also, the SN receives serotonergic afferent signals from the raphe nuclei, and both the VTA and SNpc receive achetylcholinergic input (Swanson, 1982).

Globus Pallidus The globus pallidus consists of an internal and an external segment (GPi and GPe). The GP expresses both GABA and enkephalin (Fonnum, Gottesfeld, & Grofova, 1978; Oertel & Mugnaini, 1984). The GP receives input from distinct neuronal populations within the striatum (Brog & Zahm, 1996). Additionally, it receives glumatergic innervation from the STN. Signalling from the nigrostriatal pathway provides a dopaminergic innervation of the GP (Lindvall & Bjorklund, 1979). Output from the GP is GABA-ergic, which signals to the striatum, STN, as well as the pedunculopontine tegmentum and the thalamus (Carter & Pycock, 1978; DeVito & Anderson, 1982; DeVito, Anderson, & Walsh, 1980; Fonnum et al., 1978; Grofova, 1975; Kincaid, Penney, Young, & Newman, 1991; Kita & Kitai, 1991; Parent & Hazrati, 1995a; Staines, Atmadja, & Fibiger, 1981; Staines & Fibiger, 1984; van der Kooy & Carter, 1981). Pallidal neurons provide signalling to multiple BG nuclei through axonal collaterals, allowing single neurons to signal to multiple structures. The globus pallidus has also prominent projections to the thalamus (Parent & Hazrati, 1995a; Shammah-Lagnado, Alheid, & Heimer, 1996). The GP controls BG output through direct projections to the SNr, indirect projections to the STN, and projections to the thalamus (Chesselet & Delfs, 1996).

Cortico-Subcortical Fibre Tracts Crucial for response inhibition are the white matter tracts between brain areas, which allow cross-talk and the integration of individual processes. The PFC connects to other parts of the brain via white matter projections (Schmahmann & Pandya, 2008): (a) association fibres to ipsilateral cortical areas, (b) striatal fibres to the BG, (c) commissural fibres to contralateral cerebral regions, (d) projection fibres to the thalamus, and (e) projection fibres to the brainstem and/or pontocerebellar system (Schmahmann & Pandya, 2008). The corticostriatal pathways are anatomically well organised for parallel processing. Projections from the cortex are organised according to a medial lateral and rostro-caudal gradient along the striatal nuclei, and integration can occur in specific brain regions, which show topographic overlap (Haber, 2003). Both hemispheres have a convergence zone where information comes together in the caudate nucleus as well as in the putamen. These convergence zones receive projections from the OFC, dorsolateral PFC, and parietal regions. Studies on functional connectivity on these convergence zones using resting state fMRI support the existence of these pathways. The described studies identified a network of converging corticostriatal projections. This network has been implicated in a large number of cognitive processes, including executive control, reward, as well as spatial attention (Jarbo & Verstynen, 2015).

The Striatal Hypothesis Understanding of the neuroanatomical structure underlying response inhibition is crucial, but even if the structure of a brain network has been elucidated, the question remains how this structure gives rise to function. The striatal hypothesis provides a model for understanding how changes in cortical signalling translate into a response. People decrease response thresholds by increasing activation from the cortex to the striatum, releasing the brain from inhibition (e.g., Bogacz, Wagenmakers, Forstmann, & Nieuwenhuis, 2010; Forstmann et al., 2008). From a functional point of view, the striatum is the main structure involved in the input of the BG through afferent signals from the entire cerebral cortex. The incoming signals are subsequently relayed via direct and indirect pathways to the SNr, which makes up the main output nucleus of the BG (Bogacz et al., 2010).

The STN Hypothesis As described above, the STN is considered a second input station of the BG in addition to the striatum, and the pathway running directly from the cortex, particularly the frontal lobe to the STN, is the hyperdirect pathway (Kita & Kitai, 1994; Kitai & Deniau, 1981; Levy et al., 1997; Mink, 1996; Mink & Thach, 1993; Monakow, Akert, & Kunzle, 1978; Nambu, Takada, Inase, & Tokuno, 1996; Nambu et al., 2000; Nambu, Tokuno, Inase, & Takada, 1997). During response inhibition, the IFG modulates the excitatory influence of the pre-SMA on the STN. This results in a downstream further inhibition from the STN to the motor cortex (Rae, Hughes, Anderson, & Rowe, 2015). The strength of the interaction between pre-SMA and STN and the degree of modulation by the IFG predicts interindividual variation in the stop-signal reaction time (Aron & Poldrack, 2006), although a later study did not replicate this finding (Congdon et al., 2010). The finding from Aron and Poldrack (2006) was corroborated by tractography data showing mean diffusivity in tracts between pre-SMA and STN, as well as IFG and STN, which also predicted interindividual differences in stopping efficiency (Rae et al., 2015). In addition, the white matter structure connecting pre-SMA and STN was related to the effective connectivity. In sum, these results demonstrated that the network dynamics and the modulatory function of the PFC can explain individual differences in inhibitory control (Rae et al., 2015).

Summary of the Brain Circuitries Involved in Response Inhibition

As described above, the brain networks involved in response inhibition include the cortex, thalamus, and the BG. Responses are driven from the cortex and the thalamus, which have extensive projections to the BG. The output nuclei signal to the motor nuclei of the brain stem as well as the motor cortical areas through the thalamus. From a functional point of view, the striatum and the STN form the main BG input structures for afferent signals from the entire cerebral cortex. The incoming signals are subsequently relayed via direct and indirect pathways to the SNr, which makes up the main output nucleus of the BG. The GP and STN make up integral parts of the indirect pathway. It is likely that massive convergence of cortical signals occurs in the BG. This is supported by studies in rodents in which the neurons in the pallidal and nigral structures of the BG in rats are outnumbered 30 times by neurons in the caudate and putamen (Oorschot, 1996). These numbers support parallel as well as integrated signalling within the BG.

Conclusion

A combination of research approaches has greatly improved our understanding of the brain circuitries involved in response inhibition. fMRI allows non-invasive study of human brain function. It is, however, possible that MRI will not provide us with a complete overview of the brain circuitries involved in response inhibition. Limited visibility of smaller structures on MRI images may have contributed to their omission from current models of response inhibition (de Hollander, Keuken, & Forstmann, 2015). The limitations in comparability between human and rodent neuroanatomy may also have contributed to our limited understanding of brain structures that differ between humans and other primates, such as the PFC. Comparability between rodents and primates cannot be improved, and therefore increased understanding of the brain circuitries involved in response inhibition will rely on further improvements of MRI techniques including higher MRI field strength such as 7T or higher. In addition, novel developments of MRI sequences and contrast calculations have dramatically improved the visibility of small subcortical brain structures. Still the majority of subcortical structures cannot be visualised using current state-of-the-art MR sequences, because MRI scan parameters need to be tailored individually on the basis of the tissue properties of individual brain structures (Keuken et al., 2014). As a first step, individual structures need to be visualised on anatomical scans, with gold-standard histological validation. Recent efforts using post-mortem brain samples have already resulted in the visualisation of the human subthalamic nucleus with unprecedented anatomical detail (Weiss et al., 2014).

It is clear that the neuroanatomical backbone underlying response inhibition is complex, and our understanding of the anatomy of brain structures continues to grow. A next logical step is to combine our knowledge on structural neuroanatomy with activation patterns obtained from fMRI studies. It is not always evident what is reflected by the observed BOLD signals. Even though we should be cautious when interpreting animal data and extrapolating findings to the human situations, anatomical studies in various species may provide insights on what an increased BOLD signal represents. Pharmacological studies have shown that within the stop-signal task, noradrenaline is associated primarily with stopping behaviour, whereas dopamine is highly involved in the go-process (Eagle, Tufft, Goodchild, & Robbins, 2007; Robinson et al., 2008). However, the functional neuroanatomy of stopping is much more complex than a simple model involving one neurotransmitter for stopping and another for going. Both human and rodent studies also indicate the involvement of serotonin in behavioural inhibition, although it is not involved in actual stopping behaviour, which was consistent with findings in humans (Chamberlain et al., 2006; Clark et al., 2005). Future technical developments are directed at detecting neurotransmitter concentrations in the in vivo brain in more detail by, for example, using MR spectroscopy, which will further increase our understanding of the biochemical processes in the human brain. In conclusion, classic neuroanatomical studies in combination with modern in vivo imaging techniques will further foster our understanding of the brain systems involved in response inhibition.

Acknowledgements

The authors would like to thank Dr. Max C Keuken for creating Figure 16.1. This research line is financially supported by the European Research Council (BUF), the Technology Foundation (STW) of the Netherlands Organisation for Scientific Research, and the Dutch Brain Foundation (AA and BUF).

References

Afsharpour, S. (1985a). Light microscopic analysis of Golgi-impregnated rat subthalamic neurons. *Journal of Comparative Neurology*, *236*(1), 1–13.

Afsharpour, S. (1985b). Topographical projections of the cerebral cortex to the subthalamic nucleus. *Journal of Comparative Neurology*, *236*(1), 14–28.

Albin, R. L., Makowiec, R. L., Hollingsworth, Z. R., Dure, L. S. T., Penney, J. B., & Young, A. B. (1992). Excitatory amino acid binding sites in the basal ganglia of the rat: A quantitative autoradiographic study. *Neuroscience*, *46*(1), 35–48. Retrieved from http://www.ncbi.nlm.nih.gov/pubmed/1317515

Alexander, G. E., & Crutcher, M. D. (1990). Functional architecture of basal ganglia circuits: Neural substrates of parallel processing. *Trends in Neurosciences*, *13*(7), 266–271. Retrieved from http://www.ncbi.nlm.nih.gov/pubmed/1695401

Alkemade, A. (2013). Subdivisions and anatomical boundaries of the subthalamic nucleus. *Journal of Neuroscience*, *33*(22), 9233–9234.

Alkemade, A., & Forstmann, B. U. (2014). Do we need to revise the tripartite subdivision hypothesis of the human subthalamic nucleus (STN)? *Neuroimage*, *95*, 326–329.

Alkemade, A., Schnitzler, A., & Forstmann, B. U. (2015). Topographic organization of the human and non-human primate subthalamic nucleus. *Brain Structure & Function*, *220*(6), 3075–3086.

Alluisi, E. A., & Warm, J. S. (1989). Things that go together: A review of stimulus-response compatibility and related effects. In R. W. Proctor & T. G. Reeve (Eds.), *Stimulus-response compatibility: An integrated perspective* (pp. 3–30). Amsterdam: Elsevier.

Aron, A. R. (2007). The neural basis of inhibition in cognitive control. *Neuroscientist*, *13*(3), 214–228.

Aron, A. R., & Poldrack, R. A. (2006). Cortical and subcortical contributions to Stop signal response inhibition: Role of the subthalamic nucleus. *Journal of Neuroscience*, *26*(9), 2424–2433.

Aron, A. R., Robbins, T. W., & Poldrack, R. A. (2004). Inhibition and the right inferior frontal cortex. *Trends in Cognitive Sciences*, *8*(4), 170–177.

Aron, A. R., Robbins, T. W., & Poldrack, R. A. (2014). Inhibition and the right inferior frontal cortex: One decade on. *Trends in Cognitive Sciences*, *18*(4), 177–185.

Aston-Jones, G., & Cohen, J. D. (2005). An integrative theory of locus coeruleus-norepinephrine function: Adaptive gain and optimal performance. *Annual Review of Neuroscience*, *28*, 403–450.

Beckstead, R. M. (1985). Complementary mosaic distributions of thalamic and nigral axons in the caudate nucleus of the cat: Double anterograde labeling combining autoradiography and wheat germ-HRP histochemistry. *Brain Research*, *335*(1), 153–159. Retrieved from http://www.ncbi.nlm.nih.gov/pubmed/4005540

Beckstead, R. M., Domesick, V. B., & Nauta, W. J. (1979). Efferent connections of the substantia nigra and ventral tegmental area in the rat. *Brain Research*, *175*(2), 191–217. Retrieved from http://www.ncbi.nlm.nih.gov/pubmed/314832

Berger, B., Trottier, S., Verney, C., Gaspar, P., & Alvarez, C. (1988). Regional and laminar distribution of the dopamine and serotonin innervation in the macaque cerebral cortex: A radioautographic study. *Journal of Comparative Neurology*, *273*(1), 99–119.

Bevan, M. D., Smith, A. D., & Bolam, J. P. (1996). The substantia nigra as a site of synaptic integration of functionally diverse information arising from the ventral pallidum and the globus pallidus in the rat. *Neuroscience*, *75*(1), 5–12. Retrieved from http://www.ncbi.nlm.nih.gov/pubmed/8923517

Bogacz, R., Wagenmakers, E. J., Forstmann, B. U., & Nieuwenhuis, S. (2010). The neural basis of the speed-accuracy tradeoff. *Trends in Neurosciences*, *33*(1), 10–16.

Bolam, J. P., & Izzo, P. N. (1988). The postsynaptic targets of substance P-immunoreactive terminals in the rat neostriatum with particular reference to identified spiny striatonigral neurons. *Experimental Brain Research*, *70*(2), 361–377. Retrieved from http://www.ncbi.nlm.nih.gov/pubmed/2454839

Bolam, J. P., Smith, Y., Ingham, C. A., von Krosigk, M., & Smith, A. D. (1993). Convergence of synaptic terminals from the striatum and the globus pallidus onto single neurones in the substantia nigra and the entopeduncular nucleus. *Progress in Brain Research*, *99*, 73–88. Retrieved from http://www.ncbi.nlm.nih.gov/pubmed/7509081

Brodmann, K. (1909). *Vergleichende Lokalisationslehre der Grosshirnrinde in ihren Prinzipien dargestellt auf Grund des Zellenbaues*. Leipzig: JA Barth.

Brog, J. S., & Zahm, D. S. (1996). Morphologically distinct subpopulations of neurotensin-immunoreactive striatal neurons observed in rat following dopamine depletions and D2 receptor blockade project to the globus pallidus. *Neuroscience*, *74*(3), 805–812. Retrieved from http://www.ncbi.nlm.nih.gov/pubmed/8884776

Brown, V. J., & Bowman, E. M. (2002). Rodent models of prefrontal cortical function. *Trends in Neurosciences, 25*(7), 340–343. Retrieved from http://www.ncbi.nlm.nih.gov/pubmed/12079756

Buck, S. H., Deshmukh, P. P., Burks, T. F., & Yamamura, H. I. (1981). A survey of substance P, somatostatin, and neurotensin levels in aging in the rat and human central nervous system. *Neurobiology of Aging, 2*(4), 257–264. Retrieved from http://www.ncbi.nlm.nih.gov/pubmed/6174878

Campbell, M. J., Lewis, D. A., Benoit, R., & Morrison, J. H. (1987). Regional heterogeneity in the distribution of somatostatin-28- and somatostatin-28(1-12)-immunoreactive profiles in monkey neocortex. *Journal of Neuroscience, 7*(4), 1133–1144. Retrieved from http://www.ncbi.nlm.nih.gov/pubmed/2883266

Carter, C. J., & Pycock, C. J. (1978). A study of the sites of interaction between dopamine and 5-hydroxytryptamine for the production of fluphenazine-induced catalepsy. *Naunyn-Schmiedeberg's Archives of Pharmacology, 304*(2), 135–139. Retrieved from http://www.ncbi.nlm.nih.gov/pubmed/568219

Casey, B. J., Trainor, R. J., Orendi, J. L., Schubert, A. B., Nystrom, L. E., Giedd, J. N.,...Rapoport, J. L. (1997). A developmental functional MRI study of prefrontal activation during performance of a go-no-go task. *Journal of Cognitive Neuroscience, 9*(6), 835–847.

Castel, M. N., Morino, P., Frey, P., Terenius, L., & Hokfelt, T. (1993). Immunohistochemical evidence for a neurotensin striatonigral pathway in the rat brain. *Neuroscience, 55*(3), 833–847. Retrieved from http://www.ncbi.nlm.nih.gov/pubmed/8105419

Chamberlain, S. R., Muller, U., Blackwell, A. D., Clark, L., Robbins, T. W., & Sahakian, B. J. (2006). Neurochemical modulation of response inhibition and probabilistic learning in humans. *Science, 311*(5762), 861–863.

Cheramy, A., Leviel, V., & Glowinski, J. (1981). Dendritic release of dopamine in the substantia nigra. *Nature, 289*(5798), 537–542. Retrieved from http://www.ncbi.nlm.nih.gov/pubmed/6258083

Chesselet, M. F., & Delfs, J. M. (1996). Basal ganglia and movement disorders: An update. *Trends in Neurosciences, 19*(10), 417–422. Retrieved from http://www.ncbi.nlm.nih.gov/pubmed/8888518

Chevalier, G., & Deniau, J. M. (1990). Disinhibition as a basic process in the expression of striatal functions. *Trends in Neurosciences, 13*(7), 277–280. Retrieved from http://www.ncbi.nlm.nih.gov/pubmed/1695403

Clark, L., Roiser, J. P., Cools, R., Rubinsztein, D. C., Sahakian, B. J., & Robbins, T. W. (2005). Stop signal response inhibition is not modulated by tryptophan depletion or the serotonin transporter polymorphism in healthy volunteers: Implications for the 5-HT theory of impulsivity. *Psychopharmacology (Berl), 182*(4), 570–578.

Congdon, E., Mumford, J. A., Cohen, J. R., Galvan, A., Aron, A. R., Xue, G.,...Poldrack, R. A. (2010). Engagement of large-scale networks is related to individual differences in inhibitory control. *Neuroimage, 53*(2), 653–663.

Crick, F. (1988). *What mad pursuit: A personal view of scientific discovery* (pp. 150). New York: Basic Books.

de Hollander, G., Keuken, M. C., & Forstmann, B. U. (2015). The subcortical cocktail problem; mixed signals from the subthalamic nucleus and substantia nigra. *PLoS ONE, 10*(3), e0120572.

Deniau, J. M., & Chevalier, G. (1992). The lamellar organization of the rat substantia nigra pars reticulata: Distribution of projection neurons. *Neuroscience, 46*(2), 361–377. Retrieved from http://www.ncbi.nlm.nih.gov/pubmed/1542412

Deniau, J. M., Menetrey, A., & Thierry, A. M. (1994). Indirect nucleus accumbens input to the prefrontal cortex via the substantia nigra pars reticulata: A combined anatomical and electrophysiological study in the rat. *Neuroscience, 61*(3), 533–545. Retrieved from http://www.ncbi.nlm.nih.gov/pubmed/7526269

DeVito, J. L., & Anderson, M. E. (1982). An autoradiographic study of efferent connections of the globus pallidus in Macaca mulatta. *Experimental Brain Research, 46*(1), 107–117. Retrieved from http://www.ncbi.nlm.nih.gov/pubmed/7067782

DeVito, J. L., Anderson, M. E., & Walsh, K. E. (1980). A horseradish peroxidase study of afferent connections of the globus pallidus in Macaca mulatta. *Experimental Brain Research, 38*(1), 65–73. Retrieved from http://www.ncbi.nlm.nih.gov/pubmed/6766111

Donders, F. C. (1969). On the speed of mental processes. *Acta Psychologica (Amst), 30*, 412–431. Retrieved from http://www.ncbi.nlm.nih.gov/pubmed/5811531

Duncan, J., & Owen, A. M. (2000). Common regions of the human frontal lobe recruited by diverse cognitive demands. *Trends in Neurosciences*, *23*(10), 475–483. Retrieved from http://www.ncbi.nlm.nih.gov/pubmed/11006464

Eagle, D. M., Tufft, M. R., Goodchild, H. L., & Robbins, T. W. (2007). Differential effects of modafinil and methylphenidate on stop-signal reaction time task performance in the rat, and interactions with the dopamine receptor antagonist cis-flupenthixol. *Psychopharmacology (Berl)*, *192*(2), 193–206.

Eriksen., B. A., & Eriksen., C. W. (1974). Effect of noise letters upon the identification of a target letter in a nonsearch task. *Perception, & Psychophysics*, *16*, 143–149.

Fonnum, F., Gottesfeld, Z., & Grofova, I. (1978). Distribution of glutamate decarboxylase, choline acetyl-transferase and aromatic amino acid decarboxylase in the basal ganglia of normal and operated rats. Evidence for striatopallidal, striatoentopeduncular and striatonigral GABAergic fibres. *Brain Research*, *143*(1), 125–138. Retrieved from http://www.ncbi.nlm.nih.gov/pubmed/630396

Forstmann, B. U., Dutilh, G., Brown, S., Neumann, J., von Cramon, D. Y., Ridderinkhof, K. R., & Wagenmakers, E. J. (2008). Striatum and pre-SMA facilitate decision-making under time pressure. *Proceedings of the National Academy of Sciences of the United States of America*, *105*(45), 17538–17542.

Garavan, H., Ross, T. J., & Stein, E. A. (1999). Right hemispheric dominance of inhibitory control: An event-related functional MRI study. *Proceedings of the National Academy of Sciences of the United States of America*, *96*(14), 8301–8306. Retrieved from http://www.ncbi.nlm.nih.gov/pubmed/10393989

Garey, L. J. (1994). *Brodmann's 'Localisation in the cerebral cortex'*. London: Smith-Gordon & Co.

Gaspar, P., Berger, B., & Febvret, A. (1990). Neurotensin innervation of the human cerebral cortex: Lack of colocalization with catecholamines. *Brain Research*, *530*(2), 181–195. Retrieved from http://www.ncbi.nlm.nih.gov/pubmed/2265353

Gaspar, P., Berger, B., Febvret, A., Vigny, A., & Henry, J. P. (1989). Catecholamine innervation of the human cerebral cortex as revealed by comparative immunohistochemistry of tyrosine hydroxylase and dopamine-beta-hydroxylase. *Journal of Comparative Neurology*, *279*(2), 249–271.

Goldman-Rakic, P. S. (1987). Circuitry of primate prefrontal cortex and regulation of behaviour by representational memory. In *Handbook of physiology: The nervous system* (Vol. 5, pp 373–417). American Physiological Society.

Groenewegen, H. J., Berendse, H. W., & Haber, S. N. (1993). Organization of the output of the ventral striatopallidal system in the rat: Ventral pallidal efferents. *Neuroscience*, *57*(1), 113–142. Retrieved from http://www.ncbi.nlm.nih.gov/pubmed/8278047

Grofova, I. (1975). The identification of striatal and pallidal neurons projecting to substantia nigra. An experimental study by means of retrograde axonal transport of horseradish peroxidase. *Brain Research*, *91*(2), 286–291. Retrieved from http://www.ncbi.nlm.nih.gov/pubmed/51667

Haber, S. N. (2003). The primate basal ganglia: Parallel and integrative networks. *Journal of Chemical Neuroanatomy*, *26*(4), 317–330. Retrieved from http://www.ncbi.nlm.nih.gov/pubmed/14729134

Hammond, C., Deniau, J. M., Rizk, A., & Feger, J. (1978). Electrophysiological demonstration of an excitatory subthalamonigral pathway in the rat. *Brain Research*, *151*(2), 235–244. Retrieved from http://www.ncbi.nlm.nih.gov/pubmed/209862

Hampshire, A. (2015). Putting the brakes on inhibitory models of frontal lobe function. *Neuroimage*, *113*, 340–355.

Hauber, W. (1998). Involvement of basal ganglia transmitter systems in movement initiation. *Progress in Neurobiology*, *56*(5), 507–540. Retrieved from http://www.ncbi.nlm.nih.gov/pubmed/9775402

Haynes, W. I., & Haber, S. N. (2013). The organization of prefrontal-subthalamic inputs in primates provides an anatomical substrate for both functional specificity and integration: Implications for Basal Ganglia models and deep brain stimulation. *Journal of Neuroscience*, *33*(11), 4804–4814.

Imai, H., Park, M. R., Steindler, D. A., & Kitai, S. T. (1986). The morphology and divergent axonal organization of midbrain raphe projection neurons in the rat. *Brain and Development*, *8*(4), 343–354. Retrieved from http://www.ncbi.nlm.nih.gov/pubmed/2432796

Iritani, S., Fujii, M., & Satoh, K. (1989). The distribution of substance P in the cerebral cortex and hippocampal formation: An immunohistochemical study in the monkey and rat. *Brain Research Bulletin*, *22*(2), 295–303. Retrieved from http://www.ncbi.nlm.nih.gov/pubmed/2468400

Jarbo, K., & Verstynen, T. D. (2015). Converging structural and functional connectivity of orbitofrontal, dorsolateral prefrontal, and posterior parietal cortex in the human striatum. *Journal of Neuroscience*, *35*(9), 3865–3878.

Kalivas, P. W. (1993). Neurotransmitter regulation of dopamine neurons in the ventral tegmental area. *Brain Research. Brain Research Reviews*, *18*(1), 75–113. Retrieved from http://www.ncbi.nlm.nih.gov/pubmed/8096779

Kanazawa, I., Emson, P. C., & Cuello, A. C. (1977). Evidence for the existence of substance P-containing fibres in striato-nigral and pallido-nigral pathways in rat brain. *Brain Research*, *119*(2), 447–453. Retrieved from http://www.ncbi.nlm.nih.gov/pubmed/830395

Kawaguchi, Y., Wilson, C. J., Augood, S. J., & Emson, P. C. (1995). Striatal interneurones: Chemical, physiological and morphological characterization. *Trends in Neurosciences*, *18*(12), 527–535. Retrieved from http://www.ncbi.nlm.nih.gov/pubmed/8638293

Kawashima, R., Satoh, K., Itoh, H., Ono, S., Furumoto, S., Gotoh, R.,…Fukuda, H. (1996). Functional anatomy of GO/NO-GO discrimination and response selection—A PET study in man. *Brain Research*, *728*(1), 79–89. Retrieved from http://www.ncbi.nlm.nih.gov/pubmed/8864300

Kelley, A. E., Domesick, V. B., & Nauta, W. J. (1982). The amygdalostriatal projection in the rat—An anatomical study by anterograde and retrograde tracing methods. *Neuroscience*, *7*(3), 615–630. Retrieved from http://www.ncbi.nlm.nih.gov/pubmed/7070669

Kemp, J. M., & Powell, T. P. (1971). The structure of the caudate nucleus of the cat: Light and electron microscopy. *Philosophical Transactions of the Royal Society B: Biological Sciences*, *262*(845), 383–401. Retrieved from http://www.ncbi.nlm.nih.gov/pubmed/4107495

Keuken, M. C., Bazin, P. L., Crown, L., Hootsmans, J., Laufer, A., Muller-Axt, C.,…Forstmann, B. U. (2014). Quantifying inter-individual anatomical variability in the subcortex using 7T structural MRI. *Neuroimage*, *94*, 40–46.

Keuken, M. C., Uylings, H. B., Geyer, S., Schafer, A., Turner, R., & Forstmann, B. U. (2012). Are there three subdivisions in the primate subthalamic nucleus? *Frontiers in Neuroanatomy*, *6*, 14.

Kincaid, A. E., Penney, J. B., Jr., Young, A. B., & Newman, S. W. (1991). Evidence for a projection from the globus pallidus to the entopeduncular nucleus in the rat. *Neuroscience Letters*, *128*(1), 121–125. Retrieved from http://www.ncbi.nlm.nih.gov/pubmed/1656332

Kita, H., & Kitai, S. T. (1991). Intracellular study of rat globus pallidus neurons: Membrane properties and responses to neostriatal, subthalamic and nigral stimulation. *Brain Research*, *564*(2), 296–305. Retrieved from http://www.ncbi.nlm.nih.gov/pubmed/1810628

Kita, H., & Kitai, S. T. (1994). The morphology of globus pallidus projection neurons in the rat: An intracellular staining study. *Brain Research*, *636*(2), 308–319. Retrieved from http://www.ncbi.nlm.nih.gov/pubmed/8012814

Kitai, S. T., & Deniau, J. M. (1981). Cortical inputs to the subthalamus: Intracellular analysis. *Brain Research*, *214*(2), 411–415. Retrieved from http://www.ncbi.nlm.nih.gov/pubmed/7237177

Konishi, S., Nakajima, K., Uchida, I., Kikyo, H., Kameyama, M., & Miyashita, Y. (1999). Common inhibitory mechanism in human inferior prefrontal cortex revealed by event-related functional MRI. *Brain*, *122*(Pt. 5), 981–991. Retrieved from http://www.ncbi.nlm.nih.gov/pubmed/10355680

Konishi, S., Nakajima, K., Uchida, I., Sekihara, K., & Miyashita, Y. (1998). No-go dominant brain activity in human inferior prefrontal cortex revealed by functional magnetic resonance imaging. *European Journal of Neuroscience*, *10*(3), 1209–1213. Retrieved from http://www.ncbi.nlm.nih.gov/pubmed/9753190

Lambert, C., Zrinzo, L., Nagy, Z., Lutti, A., Hariz, M., Foltynie, T.,…Frackowiak, R. (2015). Do we need to revise the tripartite subdivision hypothesis of the human subthalamic nucleus (STN)? Response to Alkemade and Forstmann. *Neuroimage*, *110*, 1–2.

Lauer, M., & Heinsen, H. (1996). Cytoarchitectonics of the human nucleus accumbens. *Journal für Hirnforschung*, *37*(2), 243–254. Retrieved from http://www.ncbi.nlm.nih.gov/pubmed/8776506

Leung, H. C., & Cai, W. (2007). Common and differential ventrolateral prefrontal activity during inhibition of hand and eye movements. *Journal of Neuroscience*, *27*(37), 9893–9900.

Levy, R., Hazrati, L. N., Herrero, M. T., Vila, M., Hassani, O. K., Mouroux, M., . . .Hirsch, E. C. (1997). Re-evaluation of the functional anatomy of the basal ganglia in normal and Parkinsonian states. *Neuroscience*, *76*(2), 335–343. Retrieved from http://www.ncbi.nlm.nih.gov/pubmed/9015319

Lewis, D. A. (1992). The catecholaminergic innervation of primate prefrontal cortex. *Journal of Neural Transmission. Supplementum, 36*, 179–200. Retrieved from http://www.ncbi.nlm.nih.gov/pubmed/1527517

Lewis, D. A., Foote, S. L., & Cha, C. I. (1989). Corticotropin-releasing factor immunoreactivity in monkey neocortex: An immunohistochemical analysis. *Journal of Comparative Neurology, 290*(4), 599–613.

Liddle, P. F., Kiehl, K. A., & Smith, A. M. (2001). Event-related fMRI study of response inhibition. *Human Brain Mapping, 12*(2), 100–109. Retrieved from http://www.ncbi.nlm.nih.gov/pubmed/11169874

Lindvall, O., & Bjorklund, A. (1979). Dopaminergic innervation of the globus pallidus by collaterals from the nigrostriatal pathway. *Brain Research, 172*(1), 169–173. Retrieved from http://www.ncbi.nlm.nih.gov/pubmed/466461

Liu, X., Banich, M. T., Jacobson, B. L., & Tanabe, J. L. (2004). Common and distinct neural substrates of attentional control in an integrated Simon and spatial Stroop task as assessed by event-related fMRI. *Neuroimage, 22*(3), 1097–1106.

Logan, G. D. (1994). On the ability to inhibit thought and action: A users' guide to the stop signal paradigm. In D. Dagenbach & T. H. Carr (Eds.), *Inhibitory processes in attention, memory, and language* (pp. 189–239). San Diego, CA: Academic Press.

Menon, V., Adleman, N. E., White, C. D., Glover, G. H., & Reiss, A. L. (2001). Error-related brain activation during a Go/NoGo response inhibition task. *Human Brain Mapping, 12*(3), 131–143. Retrieved from http://www.ncbi.nlm.nih.gov/pubmed/11170305

Mink, J. W. (1996). The basal ganglia: Focused selection and inhibition of competing motor programs. *Progress in Neurobiology, 50*(4), 381–425. Retrieved from http://www.ncbi.nlm.nih.gov/pubmed/9004351

Mink, J. W., & Thach, W. T. (1993). Basal ganglia intrinsic circuits and their role in behavior. *Current Opinion in Neurobiology, 3*(6), 950–957. Retrieved from http://www.ncbi.nlm.nih.gov/pubmed/8124079

Monakow, K. H., Akert, K., & Kunzle, H. (1978). Projections of the precentral motor cortex and other cortical areas of the frontal lobe to the subthalamic nucleus in the monkey. *Experimental Brain Research, 33*(3–4), 395–403. Retrieved from http://www.ncbi.nlm.nih.gov/pubmed/83239

Nakanishi, H., Kita, H., & Kitai, S. T. (1987). Intracellular study of rat substantia nigra pars reticulata neurons in an in vitro slice preparation: Electrical membrane properties and response characteristics to subthalamic stimulation. *Brain Research, 437*(1), 45–55. Retrieved from http://www.ncbi.nlm.nih.gov/pubmed/3427482

Nambu, A., Takada, M., Inase, M., & Tokuno, H. (1996). Dual somatotopical representations in the primate subthalamic nucleus: Evidence for ordered but reversed body-map transformations from the primary motor cortex and the supplementary motor area. *Journal of Neuroscience, 16*(8), 2671–2683. Retrieved from http://www.ncbi.nlm.nih.gov/pubmed/8786443

Nambu, A., Tokuno, H., Hamada, I., Kita, H., Imanishi, M., Akazawa, T.,...Hasegawa, N. (2000). Excitatory cortical inputs to pallidal neurons via the subthalamic nucleus in the monkey. *Journal of Neurophysiology, 84*(1), 289–300. Retrieved from http://www.ncbi.nlm.nih.gov/pubmed/10899204

Nambu, A., Tokuno, H., Inase, M., & Takada, M. (1997). Corticosubthalamic input zones from forelimb representations of the dorsal and ventral divisions of the premotor cortex in the macaque monkey: Comparison with the input zones from the primary motor cortex and the supplementary motor area. *Neuroscience Letters, 239*(1), 13–16. Retrieved from http://www.ncbi.nlm.nih.gov/pubmed/9547160

Nambu, A., Tokuno, H., & Takada, M. (2002). Functional significance of the cortico-subthalamo-pallidal 'hyperdirect' pathway. *Neuroscience Research, 43*(2), 111–117. Retrieved from http://www.ncbi.nlm.nih.gov/pubmed/12067746

Nee, D. E., Wager, T. D., & Jonides, J. (2007). Interference resolution: Insights from a meta-analysis of neuroimaging tasks. *Cognitive, Affective, & Behavioral Neuroscience, 7*(1), 1–17.

Neubert, F. X., Mars, R. B., Sallet, J., & Rushworth, M. F. (2015). Connectivity reveals relationship of brain areas for reward-guided learning and decision making in human and monkey frontal cortex. *Proceedings of the National Academy of Sciences of the United States of America, 112*(20), E2695–2704.

Nieuwenhuis, R. (1985). *Chemoarchitecture of the brain.* Berlin: Springer Verlag.

O'Reilly, R. C., Noelle, D. C., Braver, T. S., & Cohen, J. D. (2002). Prefrontal cortex and dynamic categorization tasks: Representational organization and neuromodulatory control. *Cerebral Cortex, 12*(3), 246–257. Retrieved from http://www.ncbi.nlm.nih.gov/pubmed/11839599

Oertel, W. H., & Mugnaini, E. (1984). Immunocytochemical studies of GABAergic neurons in rat basal ganglia and their relations to other neuronal systems. *Neuroscience Letters, 47*(3), 233–238. Retrieved from http://www.ncbi.nlm.nih.gov/pubmed/6147799

Oorschot, D. E. (1996). Total number of neurons in the neostriatal, pallidal, subthalamic, and substantia nigral nuclei of the rat basal ganglia: A stereological study using the cavalieri and optical disector methods. *Journal of Comparative Neurology, 366*(4), 580–599.

Parent, A., & Hazrati, L. N. (1995a). Functional anatomy of the basal ganglia. *I. The cortico-basal ganglia-thalamo-cortical loop. Brain Research. Brain Research Reviews, 20*(1), 91–127. Retrieved from http://www.ncbi.nlm.nih.gov/pubmed/7711769

Parent, A., & Hazrati, L. N. (1995b). Functional anatomy of the basal ganglia. *II. The place of subthalamic nucleus and external pallidum in basal ganglia circuitry. Brain Research. Brain Research Reviews, 20*(1), 128–154. Retrieved from http://www.ncbi.nlm.nih.gov/pubmed/7711765

Patel, G. H., Yang, D., Jamerson, E. C., Snyder, L. H., Corbetta, M., & Ferrera, V. P. (2015). Functional evolution of new and expanded attention networks in humans. *Proceedings of the National Academy of Sciences of the United States of America, 112*(30), 9454–9459.

Paus, T., Castro-Alamancos, M. A., & Petrides, M. (2001). Cortico-cortical connectivity of the human mid-dorsolateral frontal cortex and its modulation by repetitive transcranial magnetic stimulation. *European Journal of Neuroscience, 14*(8), 1405–1411. Retrieved from http://www.ncbi.nlm.nih.gov/pubmed/11703468

Petersen, C. (2014). Cell-type specific function of GABAergic neurons in layers 2 and 3 of mouse barrel cortex. *Current Opinion in Neurobiology, 26*, 1–6.

Rae, C. L., Hughes, L. E., Anderson, M. C., & Rowe, J. B. (2015). The prefrontal cortex achieves inhibitory control by facilitating subcortical motor pathway connectivity. *Journal of Neuroscience, 35*(2), 786–794.

Rae, C. L., Hughes, L. E., Weaver, C., Anderson, M. C., & Rowe, J. B. (2014). Selection and stopping in voluntary action: A meta-analysis and combined fMRI study. *Neuroimage, 86*, 381–391.

Rafols, J. A., & Fox, C. A. (1976). The neurons in the primate subthalamic nucleus: A Golgi and electron microscopic study. *Journal of Comparative Neurology, 168*(1), 75–111.

Reid, M. S., Herrera-Marschitz, M., Hokfelt, T., Lindefors, N., Persson, H., & Ungerstedt, U. (1990). Striatonigral GABA, dynorphin, substance P and neurokinin A modulation of nigrostriatal dopamine release: Evidence for direct regulatory mechanisms. *Experimental Brain Research, 82*(2), 293–303. Retrieved from http://www.ncbi.nlm.nih.gov/pubmed/1704847

Robertson, H. A. (1992). Dopamine receptor interactions: Some implications for the treatment of Parkinson's disease. *Trends in Neurosciences, 15*(6), 201–206. Retrieved from http://www.ncbi.nlm.nih.gov/pubmed/1378662

Robinson, E. S., Eagle, D. M., Mar, A. C., Bari, A., Banerjee, G., Jiang, X.,...Robbins, T. W. (2008). Similar effects of the selective noradrenaline reuptake inhibitor atomoxetine on three distinct forms of impulsivity in the rat. *Neuropsychopharmacology, 33*(5), 1028–1037.

Rosenbloom, M. H., Schmahmann, J. D., & Price, B. H. (2012). The functional neuroanatomy of decision-making. *Journal of Neuropsychiatry and Clinical Neurosciences, 24*(3), 266–277.

Rubia, K., Cubillo, A., Smith, A. B., Woolley, J., Heyman, I., & Brammer, M. J. (2010). Disorder-specific dysfunction in right inferior prefrontal cortex during two inhibition tasks in boys with attention-deficit hyperactivity disorder compared to boys with obsessive-compulsive disorder. *Human Brain Mapping, 31*(2), 287–299.

Rubia, K., Russell, T., Overmeyer, S., Brammer, M. J., Bullmore, E. T., Sharma, T.,...Taylor, E. (2001). Mapping motor inhibition: Conjunctive brain activations across different versions of go/no-go and stop tasks. *Neuroimage, 13*(2), 250–261.

Satoh, K., & Matsumura, H. (1990). Distribution of neurotensin-containing fibers in the frontal cortex of the macaque monkey. *Journal of Comparative Neurology, 298*(2), 215–223.

Schmahmann, J. D., & Pandya, D. N. (2008). Disconnection syndromes of basal ganglia, thalamus, and cerebrocerebellar systems. *Cortex, 44*(8), 1037–1066.

Seroogy, K., Ceccatelli, S., Schalling, M., Hokfelt, T., Frey, P., Walsh, J.,...Goldstein, M. (1988). A subpopulation of dopaminergic neurons in rat ventral mesencephalon contains both neurotensin and cholecystokinin. *Brain Research, 455*(1), 88–98. Retrieved from http://www.ncbi.nlm.nih.gov/pubmed/3046712

Shammah-Lagnado, S. J., Alheid, G. F., & Heimer, L. (1996). Efferent connections of the caudal part of the globus pallidus in the rat. *Journal of Comparative Neurology, 376*(3), 489–507.

Simmonds, D. J., Pekar, J. J., & Mostofsky, S. H. (2008). Meta-analysis of Go/No-go tasks demonstrating that fMRI activation associated with response inhibition is task-dependent. *Neuropsychologia, 46*(1), 224–232.

Simon, J. R. (1989). The effects of an irrelevant directional que on human information processing. In R. W. Proctor & T. G. Reeve (Eds.), *Stimulus-response compatibility: An integrated perspective* (pp. 31–86). Amsterdam: Elsevier.

Smith, Y., & Bolam, J. P. (1989). Neurons of the substantia nigra reticulata receive a dense GABA-containing input from the globus pallidus in the rat. *Brain Research, 493*(1), 160–167. Retrieved from http://www.ncbi.nlm.nih.gov/pubmed/2476197

Smith, Y., & Bolam, J. P. (1990). The output neurones and the dopaminergic neurones of the substantia nigra receive a GABA-containing input from the globus pallidus in the rat. *Journal of Comparative Neurology, 296*(1), 47–64.

Soghomonian, J. J., Descarries, L., & Watkins, K. C. (1989). Serotonin innervation in adult rat neostriatum. *II. Ultrastructural features: A radioautographic and immunocytochemical study. Brain Research, 481*(1), 67–86. Retrieved from http://www.ncbi.nlm.nih.gov/pubmed/2706468

Staines, W. A., Atmadja, S., & Fibiger, H. C. (1981). Demonstration of a pallidostriatal pathway by retrograde transport of HRP-labeled lectin. *Brain Research, 206*(2), 446–450. Retrieved from http://www.ncbi.nlm.nih.gov/pubmed/7214143

Staines, W. A., & Fibiger, H. C. (1984). Collateral projections of neurons of the rat globus pallidus to the striatum and substantia nigra. *Experimental Brain Research, 56*(2), 217–220. Retrieved from http://www.ncbi.nlm.nih.gov/pubmed/6479259

Stroop, J. R. (1935). Studies of interference in serial verbal reactions. *Journal of Experimental Psychology, 18*(6), 643–662.

Swanson, L. W. (1982). The projections of the ventral tegmental area and adjacent regions: A combined fluorescent retrograde tracer and immunofluorescence study in the rat. *Brain Research Bulletin, 9*(1–6), 321–353. Retrieved from http://www.ncbi.nlm.nih.gov/pubmed/6816390

Swick, D., Ashley, V., & Turken, U. (2011). Are the neural correlates of stopping and not going identical? Quantitative meta-analysis of two response inhibition tasks. *Neuroimage, 56*(3), 1655–1665.

Temel, Y., Blokland, A., Steinbusch, H. W., & Visser-Vandewalle, V. (2005). The functional role of the subthalamic nucleus in cognitive and limbic circuits. *Progress in Neurobiology, 76*(6), 393–413.

Ternaux, J. P., Hery, F., Bourgoin, S., Adrien, J., Glowinski, J., & Hamon, M. (1977). The topographical distribution of serotoninergic terminals in the neostriatum of the rat and the caudate nucleus of the cat. *Brain Research, 121*(2), 311–326. Retrieved from http://www.ncbi.nlm.nih.gov/pubmed/832165

Trillet, M., Vighetto, A., Croisile, B., Charles, N., & Aimard, G. (1995). [Hemiballismus with logorrhea and thymo-affective disinhibition caused by hematoma of the left subthalamic nucleus]. *Revue Neurologique (Paris), 151*(6–7), 416–419. Retrieved from http://www.ncbi.nlm.nih.gov/pubmed/7481408

van der Kooy, D., & Carter, D. A. (1981). The organization of the efferent projections and striatal afferents of the entopeduncular nucleus and adjacent areas in the rat. *Brain Research, 211*(1), 15–36. Retrieved from http://www.ncbi.nlm.nih.gov/pubmed/6164450

van der Kooy, D., & Hattori, T. (1980). Dorsal raphe cells with collateral projections to the caudate-putamen and substantia nigra: A fluorescent retrograde double labeling study in the rat. *Brain Research, 186*(1), 1–7. Retrieved from http://www.ncbi.nlm.nih.gov/pubmed/7357440

Vertes, R. P. (1991). A PHA-L analysis of ascending projections of the dorsal raphe nucleus in the rat. *Journal of Comparative Neurology, 313*(4), 643–668.

Vincent, S., Hokfelt, T., Christensson, I., & Terenius, L. (1982). Immunohistochemical evidence for a dynorphin immunoreactive striato-nigral pathway. *European Journal of Pharmacology, 85*(2), 251–252. Retrieved from http://www.ncbi.nlm.nih.gov/pubmed/6129987

Vincent, S. R., Hokfelt, T., Christensson, I., & Terenius, L. (1982). Dynorphin-immunoreactive neurons in the central nervous system of the rat. *Neuroscience Letters, 33*(2), 185–190. Retrieved from http://www.ncbi.nlm.nih.gov/pubmed/6130499

Voorn, P., Vanderschuren, L. J., Groenewegen, H. J., Robbins, T. W., & Pennartz, C. M. (2004). Putting a spin on the dorsal-ventral divide of the striatum. *Trends in Neurosciences, 27*(8), 468–474.

Wager, T. D., Sylvester, C. Y., Lacey, S. C., Nee, D. E., Franklin, M., & Jonides, J. (2005). Common and unique components of response inhibition revealed by fMRI. *Neuroimage, 27*(2), 323–340.

Watanabe, J., Sugiura, M., Sato, K., Sato, Y., Maeda, Y., Matsue, Y.,...Kawashima, R. (2002). The human prefrontal and parietal association cortices are involved in NO-GO performances: An event-related fMRI study. *Neuroimage, 17*(3), 1207–1216. Retrieved from http://www.ncbi.nlm.nih.gov/pubmed/12414261

Weiss, M., Alkemade, A., Keuken, M. C., Muller-Axt, C., Geyer, S., Turner, R., & Forstmann, B. U. (2014). Spatial normalization of ultrahigh resolution 7T magnetic resonance imaging data of the postmortem human subthalamic nucleus: A multistage approach. *Brain Structure & Function, 220*(3), 1695–1703.

Whelan, R., Conrod, P. J., Poline, J. B., Lourdusamy, A., Banaschewski, T., Barker, G. J.,...Consortium, I. (2012). Adolescent impulsivity phenotypes characterized by distinct brain networks. *Nature Neuroscience, 15*(6), 920–925.

Yung, K. K., Smith, A. D., Levey, A. I., & Bolam, J. P. (1996). Synaptic connections between spiny neurons of the direct and indirect pathways in the neostriatum of the rat: Evidence from dopamine receptor and neuropeptide immunostaining. *European Journal of Neuroscience, 8*(5), 861–869. Retrieved from http://www.ncbi.nlm.nih.gov/pubmed/8743734

Zecca, L., Zucca, F. A., Wilms, H., & Sulzer, D. (2003). Neuromelanin of the substantia nigra: A neuronal black hole with protective and toxic characteristics. *Trends in Neurosciences, 26*(11), 578–580.

17

Neural Bases of Performance Monitoring

Markus Ullsperger

What Is Performance Monitoring?

How does an organism detect the necessity to change behaviour away from its current default activity? Behavioural, somatic, cognitive, and affective adaptation are needed when the current behaviour is unlikely to yield achievement of the intended goals, such as receiving a reward or avoiding harm. This is the case with inefficient or erroneous actions and unexpected changes in the environment or the organism's homeostasis. Thus, to ensure successful and efficient behaviour as well as compensation for errors and other unwanted events, the organism needs to monitor the environment and the body continuously for information indicating the necessity for adaptation. A general principle underlying this monitoring function seems to be the detection of discrepancies between expected states and the actual state. Whenever an event (e.g., sensory input that conveys information on an object in the environment) differs from the prediction (e.g., the object moves towards the body more quickly than expected), this discrepancy contains potentially relevant information (in this example, that a collision with the object can occur) that is useful to guide adaptive behaviour (in this example, an evasive movement). Formally, a ***prediction error*** reflects this difference between observed and predicted states. A prediction error conveys two pieces of information: (a) expectancy violation indicating importance or salience, reflected in the absolute magnitude of the prediction error (***unsigned prediction error***) and (b) the ***sign*** of the deviation from the expected state, which may indicate the direction of potentially needed adaptation. As the eventual goal of action is to maintain homeostasis, that is, to obtain rewards fulfilling needs and to avoid aversive events, its outcome can be evaluated as better or worse than expected. This information is contained in the ***reward prediction error***, a teaching signal for learning about stimulus and action values.

Whether or not an event associated with a prediction error signal should drive adaptation depends on its importance given previous experiences and prior knowledge of similar situations. For example, when a basketball champion misses the hoop once, this does not necessarily indicate that he has lost his fitness and should be replaced. Single pieces of evidence reflected in prediction errors for multiple events need to be accumulated and weighted appropriately to determine the necessity and strength of adaptation. On the basis of the individual's subjective beliefs and prior knowledge, the observed expectancy violation is attributed either to noise (i.e., inherent stochasticity) of the observation or to external or internal causes, which may convey information to the individual about

The Wiley Handbook of Cognitive Control, First Edition. Edited by Tobias Egner.

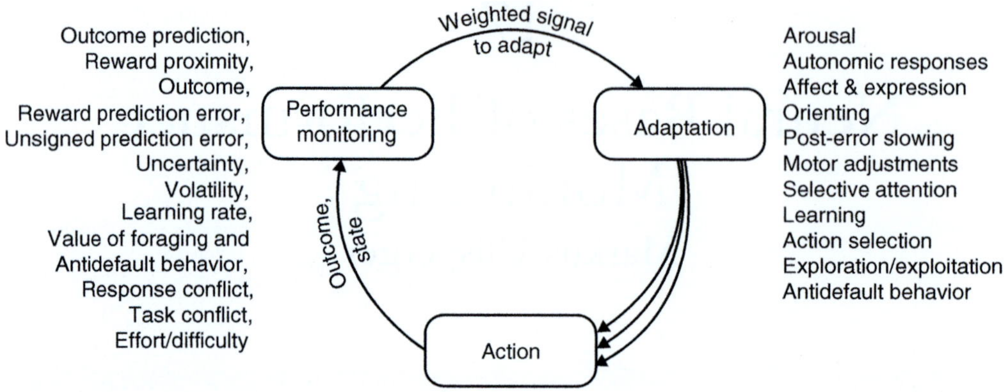

Figure 17.1 Schematic feedback loop of cognitive control. Performance monitoring evaluates motivationally salient events and determines the necessity, magnitude, and type of adaptation. A weighted adaptation signal is conveyed to other brain systems implementing adaptation. Left panel: Variables putatively represented and tracked in the performance monitoring system, on the basis of which the adaptation signal is elicited. Right panel: Various adaptations that can be initiated as a result of performance monitoring.

appropriate adaptations. Moreover, the costs of adaptations should also be taken into account (see Chapter 10 by Kool, Shenhav, & Botvinick in this volume). As a result, the performance monitoring system produces signals indicating the necessity, magnitude, and type of adaptation (Figure 17.1).

Notably, the principle of predictions, prediction errors, and resulting adjustments can be found at various levels of the nervous system. This chapter focuses on performance monitoring at the level of goal-directed behaviour at which action outcomes clearly have an emotional and motivational value.

Brain Correlates of Performance Monitoring

The discovery of the error-related negativity (ERN; also error negativity, Ne) in the early 1990s initiated a fast-growing new field of research (Falkenstein, Hohnsbein, Hoormann, & Blanke, 1990; Gehring, Goss, Coles, Meyer, & Donchin, 1993). In the following 25 years, numerous further electroencephalography (EEG) and fMRI correlates associated with different instances of performance monitoring have been described.

Event-related EEG analyses revealed a remarkably uniform sequence of deflections (Ullsperger, Danielmeier, & Jocham, 2014a; Ullsperger, Fischer, Nigbur, & Endrass, 2014b): A frontocentral negativity followed by a positivity of the same topography form a cycle of a theta band oscillation phase-locked to the eliciting event (see also Chapter 14 by M.X. Cohen in this volume), followed by a sustained centroparietal positivity. The classical example of a performance-monitoring-related frontocentral negativity, the ERN, occurs around 50–100 ms after errors in speeded reaction time tasks such as the Eriksen flanker, go/nogo, Simon, and Stroop tasks and is independent of stimulus and effector modality. Source localisation analyses suggest that it is generated in the posterior medial frontal cortex (pMFC; Dehaene, Posner, & Tucker, 1994; Gruendler, Ullsperger, & Huster, 2011), a brain region on the medial wall encompassing the anterior midcingulate cortex (aMCC), presupplementary motor area (pre-SMA), and adjacent dorsomedial prefrontal cortex (dmPFC). Simultaneous EEG and fMRI recordings during a flanker task revealed a within-subjects correlation of the single-trial ERN amplitudes, and the hemodynamic response in the aMCC: larger (more negative) ERN amplitudes were associated with stronger increases of the fMRI signal (Debener et al., 2005; Iannaccone et al., 2015). The ERN is followed by an early frontocentral error positivity (Pe) and a more sustained centroparietal late Pe (Falkenstein et al., 1990; Falkenstein, Hoormann, Christ, & Hohnsbein, 2000). The ERN is believed to reflect single pieces of objective evidence for an error, whereas the Pe appears to reflect the accumulation of subjective evidence leading to conscious perception of the error (Steinhauser & Yeung, 2010; Ullsperger, Harsay, Wessel, & Ridderinkhof, 2010).

In uncertain situations, external feedback is needed to disambiguate whether the response was correct or incorrect. Although in these cases no ERN is usually observable, the feedback elicits a frontocentral feedback-related negativity (FRN), which—depending on the feedback modality and complexity—has a latency of 200 to 400 ms (Miltner, Braun, & Coles, 1997). It is usually larger for negative outcomes, although this relationship may reverse depending on the information content and the unexpectedness of the outcome (Talmi, Atkinson, & El-Deredy, 2013; Walsh & Anderson, 2012). The FRN source has also been localised to the pMFC (Gehring & Willoughby, 2002) and is likely identical to the source of the ERN (Gentsch, Ullsperger, & Ullsperger, 2009; Gründler et al., 2009). The FRN is followed by an early, frontocentral P3a and a later centroparietal P3b (Ullsperger et al., 2014b).

When during an action difficulty turns out to be higher than expected and thus more effort is needed, a larger N2 locked to the imperative stimulus is found (Eichele, Juvodden, Ullsperger, & Eichele, 2010; Folstein & Van Petten, 2008; van Veen & Carter, 2002). Its latency varies substantially depending on the task at hand and the eliciting stimuli (Folstein & Van Petten, 2008). Again, a frontocentral source has been reported (Gründler et al., 2009), and again it is followed by a P3a and P3b (Ullsperger et al., 2014b).

FMRI results have consistently implicated the pMFC in performance monitoring. Response errors, negative feedback, increased difficulty, and response conflict have been shown to increase the fMRI signal in the aMCC, pre-SMA, and adjacent areas (Klein et al., 2007a; Nee, Kastner, & Brown, 2011; Ridderinkhof, Ullsperger, Crone, & Nieuwenhuis, 2004; Ullsperger & von Cramon, 2004). Although specialisations of pMFC subregions to specific performance monitoring functions have been repeatedly suggested (Desmet, Fias, Hartstra, & Brass, 2011; Garavan, Ross, Murphy, Roche, & Stein, 2002; Nachev, Rees, Parton, Kennard, & Husain, 2005; Nee et al., 2011; Ullsperger & von Cramon, 2001), often a large overlap of activations has obscured a putative functional dissociation of regions. Moreover, no clear criterion of functional separation has been found yet that would explain previously reported regional specialisations in an overarching framework. Also, no clear influence of motor effectors on the topographical distribution of pMFC activity has been found (Barch et al., 2001). However, the pMFC is a complex area consisting of subregions differing in their cytoarchitecture (Palomero-Gallagher, Zilles, Schleicher, & Vogt, 2013; Vogt, 2009) and connectivity (Neubert, Mars, Sallet, & Rushworth, 2015). Given the anatomical variability of this region (Huster, Westerhausen, Kreuder, Schweiger, & Wittling, 2007; Paus et al., 1996), systematic encoding patterns of performance monitoring and adaptation signals may be masked in group studies. Future neuroimaging research should take into account individual morphology and connectivity-based parcellation of the pMFC (Amiez et al., 2013; Johansen-Berg et al., 2004).

On the basis of anatomical findings and neuropsychological studies, a performance monitoring network of brain regions connected with the pMFC has been described. The ERN has proved to be a robust, sensitive, and specific measure of the functional integrity of this network. Lesions affecting the lateral frontal cortex or the basal ganglia circuits connected to the pMFC are associated with smaller or absent ERN deflections (Gehring & Knight, 2000; Peterburs et al., 2011; Seifert, von Cramon, Imperati, Tittgemeyer, & Ullsperger, 2011; Ullsperger, von Cramon, & Muller, 2002; Ullsperger & von Cramon, 2006a; Wessel, Klein, Ott, & Ullsperger, 2014). In the thalamus, a double dissociation was found: Lesions of the ventrolateral anterior and ventral anterior nuclei abolished the ERN and left the Pe unaffected, whereas lesions of the mediodorsal nucleus resulted in the reverse pattern (Klein, Ullsperger, & Danielmeier, 2013; Seifert et al., 2011). Lesions in the aMCC itself also affected the generation of the ERN, whereas the stimulus-locked N2 modulation by response conflict was less affected (Swick & Turken, 2002). Lesions of the pre- and subgenual anterior cingulate cortex and adjacent ventromedial prefrontal cortex also abolished the ERN (Stemmer, Segalowitz, Witzke, & Schonle, 2004; Turken & Swick, 2008), whereas orbitofrontal lesions confined to the anterior and frontopolar cortex had no effect on behavioural and EEG measures of performance monitoring (Ullsperger et al., 2002; see also Chapter 29 by Fellows in this volume). Functional neuroimaging has demonstrated consistent co-activations of the pMFC and the anterior insula and adjacent frontal operculum (Danielmeier, Eichele, Forstmann, Tittgemeyer, & Ullsperger, 2011; Danielmeier et al., 2015; Eichele et al., 2008; Klein et al., 2007a), suggesting very close interactions in monitoring and signalling the need for adaptation.

Performance Monitoring Signals and Theories

The first account put forward to explain the ERN and the underlying activity in the pMFC was that they reflect the detection of a mismatch between the representations of the actually performed motor action and of the intended (usually the correct) action (Coles, Scheffers, & Holroyd, 2001; Falkenstein et al., 2000; Gehring et al., 1993). This notion is roughly consistent with theories of performance monitoring put forward later, which suggest that the pMFC is part of an actor-critic reinforcement learning system and encodes its main parameters such as prediction error signals (see also Chapter 15 by Brown in this volume). These models will be discussed below. Another account that is not mutually exclusive with reinforcement learning theories, the response conflict monitoring theory, is discussed first as it has significantly influenced performance monitoring research over the last two decades.

Response Conflict Monitoring

In the late 1990, computational parallel distributed processing models of tasks eliciting the competition of simultaneously activated response tendencies, such as the Eriksen flanker task, the Simon task, and the Stroop task, have been extended by a control unit. The central idea is that errors as well as difficulties during ongoing actions requiring recruitment of control are associated with conflict between concurrently activated response tendencies. According to the theory, this response conflict is detected by the performance monitoring system and converted into an adaptation signal modulating motor threshold and attention (Botvinick, Braver, Barch, Carter, & Cohen, 2001; Botvinick, Nystrom, Fissell, Carter, & Cohen, 1999; Carter et al., 1998; see also Chapter 4 by Egner and Chapter 5 by Bugg in this volume). Numerous neuroimaging studies have demonstrated that the pMFC is engaged when response conflict occurs (Botvinick, Cohen, & Carter, 2004; Ridderinkhof et al., 2004). In EEG, the modulation of the stimulus-locked N2, ERN, and lateralised readiness potentials by differential stimulus conditions and attentional and pathological states can be explained by conflict monitoring models for the relevant tasks (Danielmeier, Wessel, Steinhauser,

& Ullsperger, 2009; Steinhauser, Maier, & Hubner, 2008; Steinhauser et al., 2012; van Veen & Carter, 2002; Yeung, Ralph, & Nieuwenhuis, 2007). The N2 is assumed to be modulated by pre-response conflict, that is, when an action unfolds and conflict is solved successfully, leading to a correct (but delayed) response. In contrast, the ERN reflects mostly post-response conflict between the executed response and the still-evolving alternative response tendency. Strong post-response conflict signals may serve as a reliable basis for internal error detection, rendering a specialised error detection mechanism unnecessary (Yeung, Cohen, & Botvinick, 2004).

The response conflict monitoring theory has been repeatedly challenged. Major points of criticism have been that it cannot account for all situations in which the performance monitoring system is active, such as processing of feedback in the absence of motor activity and during partial errors (Burle, Roger, Allain, Vidal, & Hasbroucq, 2008), and that the search for a response conflict signal at the single-neuron level in the pMFC of primates has not been successful up to now. A recent single-unit recording study in patients undergoing cingulotomy revealed neuronal activity in the aMCC that was stronger on trials involving high response conflict (Sheth et al., 2012). However, the investigation of conflict adaptation in trial sequences (cf. Chapter 4 by Egner in this volume) showed increased firing on trials after high response conflict, which is consistent with an adaptation signal rather than conflict magnitude (which, in contrast, should be reduced after high conflict as a consequence of increased control; Botvinick et al., 1999). In macaques, only firing rate increases related to errors as well as stimulus and task conflict have been found, but not to response conflict (Ebitz & Platt, 2015; Ito, Stuphorn, Brown, & Schall, 2003). Thus, although the response conflict monitoring model is well suited to predict reaction time and summed brain activity in EEG and fMRI, the relation to neurobiological mechanisms at the neuronal level still needs to be established.

Reinforcement Learning Signals

Most theories of performance monitoring suggest that a teaching signal that can not only drive learning but also other compensatory and adaptive adjustments is represented in the pMFC. Such teaching signals are the fundament of reinforcement learning (Rescorla & Wagner, 1972; Sutton & Barto, 1998) and can be formalised as weighted reward prediction errors (see above; see also Chapter 8 by Verguts and Chapter 15 by Brown in this volume). In 2002, Holroyd and Coles proposed the very influential reinforcement learning theory of performance monitoring (Holroyd & Coles, 2002). Briefly, it suggests that the ERN and FRN reflect a (negative) reward prediction error that influences a motor filter in the pMFC such that appropriate adjustments are initiated. Although early studies presented evidence in favour of the notion that ERN, FRN, and hemodynamic pMFC signals correlate with the reward prediction error (Holroyd & Coles, 2002; Holroyd & Krigolson, 2007; Holroyd, Nieuwenhuis, Yeung, & Cohen, 2003; Mars et al., 2005), others yielded contradictory results (de Bruijn, de Lange, von Cramon, & Ullsperger, 2009; Holroyd, Hajcak, & Larsen, 2006; Nieuwenhuis, Slagter, von Geusau, Heslenfeld, & Holroyd, 2005b; van Veen, Holroyd, Cohen, Stenger, & Carter, 2004). A recent meta-analysis suggests that the FRN is modulated by signed reward prediction errors (Walsh & Anderson, 2012). Model-based analyses showed that the frontocentral EEG in the latency range of the FRN correlates positively with the reward prediction error and with the outcome and negatively with reward prediction, suggesting that this activity is not merely outcome driven (Fischer & Ullsperger, 2013). In a combined EEG/MEG study, a rigorous axiomatic approach (Caplin & Dean, 2008; Rutledge, Dean, Caplin, & Glimcher, 2010) has been applied, again supporting the view that the FRN is strongly influenced by the signed reward prediction error (Talmi, Fuentemilla, Litvak,

Duzel, & Dolan, 2012). An inverse relation between the ERN and the FRN has been shown at different stages of reinforcement learning: as learning progresses, the error-related signal shifts forward in time from the feedback to the response, consistent with the principles of temporal difference error learning (Sutton & Barto, 1998).

Valence-Free Monitoring and Surprise Signals

Some neuroimaging studies failed to find stronger pMFC activity on negative feedback than on positive feedback during tasks in which context and negative feedback were insufficiently informative to determine how to adjust future behaviour (Nieuwenhuis et al., 2005b; van Veen et al., 2004). In a response switching experiment in which positive and negative feedback were equally informative for subsequent adjustments, monitoring-related pMFC activity increased, but did not differentiate outcome valence (Walton, Devlin, & Rushworth, 2004).

The N2, ERN, and even the FRN have also been shown to reflect unsigned PE or surprise. The N2 increases on unexpected, valence-free action outcomes and rare potentially action-relevant stimuli (Folstein & Van Petten, 2008; Wessel, Danielmeier, Morton, & Ullsperger, 2012). In tasks used to elicit the ERN, errors are always less frequent than correct responses and, thus, unexpected. The FRN and feedback-related pMFC activity have been shown to be elicited by unexpected positive outcomes, the omission of expected punishment, or to correlate with surprise (Amiez, Sallet, Procyk, & Petrides, 2012; Gentsch et al., 2009; Hauser et al., 2014; Jessup, Busemeyer, & Brown, 2010; Talmi et al., 2013; Ullsperger, Nittono, & von Cramon, 2007).

Origin and Destination of Performance Monitoring Signals

Invasive recordings in monkeys demonstrate coding of various parameters related to performance monitoring and principles of reinforcement learning in the pMFC, such as value (predicted outcome), obtained outcome, proximity of reward in action sequences, and signed and unsigned reward prediction errors (Hayden, Heilbronner, Pearson, & Platt, 2011; Kennerley, Behrens, & Wallis, 2011; Matsumoto, M., Matsumoto, K., Abe, & Tanaka, 2007; Scangos, Aronberg, & Stuphorn, 2013; Shidara & Richmond, 2002; Wang, Ulbert, Schomer, Marinkovic, & Halgren, 2005). It appears that, depending on task context, distribution of activity coding these variables may vary across neuronal ensembles in the pMFC, such that systemic signals in EEG, MEG, and fMRI can reflect each of them to a variable degree.

The origin of performance monitoring signals in the pMFC has been a matter of debate. The classical (Holroyd & Coles, 2002) and more recent reinforcement learning theories (Silvetti, Seurinck, & Verguts, 2011) of performance monitoring suggest that dopaminergic neurons projecting from the midbrain to the pMFC convey reward prediction error (specifically, temporal difference error) signals. Indeed, firing rates of mesencephalic neurons in the ventral tegmental area and substantia nigra have been shown to correlate with reward prediction errors (Matsumoto & Hikosaka, 2009; Schultz, Dayan, & Montague, 1997). However, the physiology and timing of the dopaminergic system make it unlikely that dopamine release can give rise to fast and short pMFC signals such as the ERN (Lapish, Kroener, Durstewitz, Lavin, & Seamans, 2007; Ullsperger et al., 2014b). In contrast, the predicted response-outcome theory (Alexander & Brown, 2011; Silvetti, Alexander, Verguts, & Brown, 2014) assumes an actor-critic architecture in the pMFC generating prediction errors for multiple possible action–outcome relations simultaneously. The net activation observable with EEG and fMRI correlates with the (un)expectedness of the observed outcome (see also Chapter 15 by Brown in this volume).

The transformation of encoded performance monitoring signals into adaptation, be it of future predictions or of behaviour, must be weighted with respect to expected uncertainty reflecting the stochastic distribution of possible outcomes and on reducible estimation uncertainty about the current beliefs regarding the task rules and context (Fischer & Ullsperger, 2014; McGuire, Nassar, Gold, & Kable, 2014; O'Reilly, 2013). Both forms of uncertainty can be based on prior knowledge and modified by experience, that is, reinforcement history. Lesions of the cingulate sulcus resulted in an inability of macaques to take into account reinforcement history for future decisions: Only the most recent outcome had a significant influence on action selection, whereas monkeys with intact cingulate cortex took into account 4–5 preceding outcomes (Kennerley, Walton, Behrens, Buckley, & Rushworth, 2006). In reinforcement learning models, the update of value representations is based on the prediction error and a weighting factor, the learning rate, reflecting the impact a single outcome has on beliefs and value representations. Neuroimaging studies suggest that, in addition to prediction errors, the learning rate is encoded in the pMFC (Behrens, Woolrich, Walton, & Rushworth, 2007; Jocham, Neumann, Klein, Danielmeier, & Ullsperger, 2009a). It was shown that the volatility of the context, which should determine the learning rate in optimal behaviour, is tracked in the pMFC (Behrens et al., 2007; Behrens, Hunt, Woolrich, & Rushworth, 2008). Factors influencing learning rate such as estimation uncertainty, the probability that an outcome indicates a change in context (which is related to surprise), and (non-informative) reward are coded in different networks of the brain, but appear to converge to a common adaptation system encompassing the pMFC, anterior insular, and parietal cortex (McGuire et al., 2014). In EEG, a centroparietal positivity in the time range of the P3b appears to reflect the activity of this common adaptive pathway (Fischer & Ullsperger, 2013). This is in line with the notion that this P3b-like positivity reflects the formation of decisions based on accumulating evidence (O'Connell, Dockree, & Kelly, 2012).

Recruitment of Control and Adaptation Based On Performance Monitoring

Weighted performance monitoring signals indicate the necessity of adaptation. General and specific adjustments can be distinguished.

General adjustments are largely unspecific to the task at hand, are part of the orienting reflex (Danielmeier & Ullsperger, 2011; Notebaert et al., 2009), and comprise changes in arousal, autonomic responses, slowing of subsequent reactions, and emotional expressions. They enable preparation of the organism to react to any unexpected event. Thus, under appropriate conditions, they may buy time and recruit resources for more task-specific adjustments.

Task-specific adjustments vary considerably with respect to their time course and level of processing depending on the task at hand and contextual conditions. For example, errors can be associated with adjustments in motor control, increase and redirection of selective attention, learning from errors, and changes from the previously performed default action, including switches between exploitation and exploratory behaviour.

In addition to these post-error adjustments, which are usually observed some time after error commission, there may be immediate attempts to stop or weaken the course of the ongoing erroneous action. An early finding was that skilled typists executed errors with less intense keystrokes than correct responses (Rabbitt, 1978), in line with the notion that errors were counteracted by pulling back. Such on-line within-trial inhibition of erroneous behaviour was also suggested by reduced and delayed electromyographic burst activity on errors compared with correct actions (Allain, Carbonnell, Burle, Hasbroucq, & Vidal, 2004).

In speeded choice reaction time tasks, participants often correct their errors by rapidly pressing the correct button even if they were not explicitly instructed to do so (Rabbitt, 1966a, 1966b). Usually, error corrections occur in tasks involving the competition between two response tendencies as a result of premature responding. It has therefore been argued that seemingly corrective responses may reflect delayed correct responses that are executed after the premature error rather than deliberately performed corrections (Ullsperger & von Cramon, 2006b). Compared with uncorrected errors, corrected errors are associated with higher pMFC activity and larger and earlier ERNs (Gehring et al., 1993; Fiehler, Ullsperger, & von Cramon, 2004, Fiehler, Ullsperger, & von Cramon, 2005; Rodriguez-Fornells, Kurzbuch, & Munte, 2002).

General Adjustments and the Orienting Reflex

Post-error slowing (PES) is a common finding that was first described by Patrick Rabbitt about 50 years ago (Rabbitt, 1966b). In many reaction time tasks, responses on trials after errors are slower than on trials following correct responses. Although under constant conditions PES has a rather high test-retest reliability (Danielmeier & Ullsperger, 2011), it is influenced by a number of factors. It appears to decay after an error: PES is higher when the interval to the next trial/response is shorter (Danielmeier & Ullsperger, 2011; Dudschig & Jentzsch, 2009; Jentzsch & Dudschig, 2009). Moreover, it appears to depend on the salience and subjective significance of the error (Notebaert et al., 2009; Nunez Castellar, Kuhn, Fias, & Notebaert, 2010; Ullsperger & Szymanowski, 2004). When the error rate becomes very high such that correct responses are more infrequent and salient, 'post-hit slowing' is found, which led to the hypothesis that PES is part of the orienting reflex and can be elicited by any infrequent or otherwise salient event (Notebaert et al., 2009). Indeed, valence-free and non-informative but salient action outcomes elicited identical slowing as errors. Moreover, those novel outcomes and errors were associated with overlapping brain activity, mainly in the pMFC (Wessel et al., 2012). Error-related activity in the pMFC, the amplitude of the ERN and early frontocentral Pe, and the power of error-related frontocentral theta oscillations are correlated with the magnitude of subsequent PES (Cavanagh, Cohen, & Allen, 2009; Cavanagh & Shackman, 2015; Danielmeier et al., 2011; Debener et al., 2005; Fischer, Endrass, Reuter, Kubisch, & Ullsperger, 2015; Gehring et al., 1993; Klein et al., 2007a; Wessel & Ullsperger, 2011; see also Chapter 14 by M.X. Cohen in this volume).

With respect to the underlying mechanism, it has been hypothesised that PES results from an increased motor threshold that, in neurophysiological terms, is equivalent to decreased corticospinal excitability (Botvinick et al., 2001; Dutilh et al., 2012). In fMRI studies, the hemodynamic signal in primary motor cortices was reduced on trials immediately following errors, and this activity reduction was correlated to PES (Danielmeier et al., 2011; King, Korb, von Cramon, & Ullsperger, 2010; Figure 17.2). EEG and MEG studies revealed a power increase in beta band oscillations over sensorimotor areas correlated with PES (Hartstra, Stolk, Mars, Toni, & Ullsperger, 2014; Marco-Pallares, Camara, Munte, & Rodriguez-Fornells, 2008). Transcranial magnetic stimulation (TMS) lends further support to the notion that decreased corticospinal excitability in the motor cortex underlies slowing after errors (Amengual et al., 2013) and other unexpected events of motivational salience (Wessel & Aron, 2013).

Diffusion-weighted imaging showed that the fractional anisotropy, a measure influenced by the directedness and integrity of fibre connections (Johansen-Berg, 2010), correlated with PES in the white matter beneath the pre-SMA and aMCC (Danielmeier et al., 2011). Probabilistic fibre tracking suggested that a network connecting the pre-SMA, right inferior

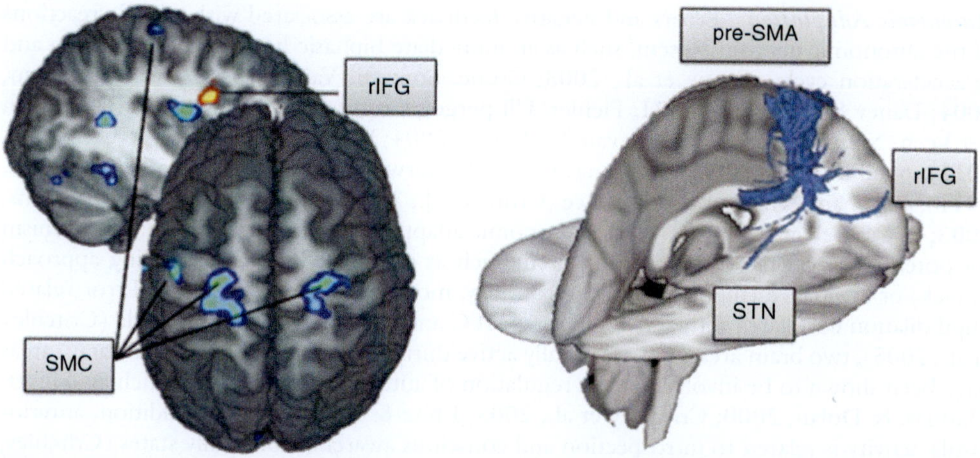

Figure 17.2 The neural basis of post-error slowing (PES). Left: Post-error fMRI signal in a face gender Simon task parametrically varying with PES. Larger PES is associated with signal reduction in the primary sensorimotor cortex (SMC), which may indicate reduced corticospinal excitability and increased activity in the right inferior frontal gyrus (rIFG). Right: Network reconstructed by probabilistic fibre tracking from a seed region in the white matter beneath the pre-SMA, where the fractional anisotropy determined with diffusion-tensor imaging correlated with individual PES. The reconstructed fibres connect structures involved in motor inhibition: pre-SMA, rIFG, and subthalamic nucleus (STN). Source: Adapted from King, 2010, and Danielmeier, 2011. Reproduced with permission of Journal of Neuroscience.

frontal gyrus (rIFG), and subthalamic nucleus (STN) is involved in PES (Figure 17.2). Interestingly, this network has been implicated in general motor inhibition (Aron, Behrens, Smith, Frank, & Poldrack, 2007; Swann et al., 2009; van Gaal, Ridderinkhof, Scholte, & Lamme, 2010). Recently, it has been shown that error-related fMRI signal increases in functionally connected areas overlapping with this network correlate with subsequent PES (Danielmeier et al., 2015), lending further support to the notion that PES is implemented via motor inhibition. The STN has been suggested to be crucially involved in regulating the speed of reactions, particularly motor inhibition (Bogacz, Wagenmakers, Forstmann, & Nieuwenhuis, 2010; Frank, Samanta, Moustafa, & Sherman, 2007; Swann et al., 2011). Direct recordings of local field potentials in the STN of Parkinson's patients showed error-related activity related to PES (Cavanagh, Sanguinetti, Allen, Sherman, & Frank, 2014; Siegert et al., 2014). Recently, it has been shown that genetic polymorphisms and pharmacological manipulations presumably associated with increased synaptic levels of serotonin enhance PES (Fischer et al., 2015), which is in line with the notion that serotonin is involved in motor inhibition in the context of aversive events (Crockett, Clark, & Robbins, 2009; Crockett, Clark, Smillie, & Robbins, 2012). Serotonergic modulation of PES could be mediated in the STN, which receives projections from the raphe nuclei (Xiang, Wang, & Kitai, 2005).

In sum, the accumulated evidence provides strong and converging evidence that a network consisting of pre-SMA, rIFG, and STN implements PES via mechanisms of general motor inhibition.

Autonomic Adaptation Errors and negative feedback are associated with specific reactions of the autonomic nervous system, such as an immediate biphasic heart rate deceleration and re-acceleration cycle (Crone et al., 2003; Crone, Somsen, Van Beek, & Van Der Molen, 2004; Danev & de Winter, 1971; Fiehler, Ullsperger, Grigutsch, & von Cramon, 2004; van der Veen, Nieuwenhuis, Crone, & van der Molen, 2004; Wessel, Danielmeier, & Ullsperger, 2011), pupil dilations (Critchley, Tang, Glaser, Butterworth, & Dolan, 2005; Wessel et al., 2011), and changes in skin conductance (Crone et al., 2004; Hajcak, McDonald, & Simons, 2003; O'Connell et al., 2007). These autonomic adaptations appear to prepare the organism for potential reactions to the eliciting event such as motor invigoration enabling approach (attack) or avoidance (flight) or, on the contrary, motor inhibition (freezing). Error-related pupil dilation correlates with activity in the pMFC and the anterior inferior insula (Critchley et al., 2005), two brain areas that are usually active during error commission. Both brain areas have been shown to be involved in the regulation of autonomic responses (Critchley, Elliott, Mathias, & Dolan, 2000; Critchley et al., 2003; Ebitz & Platt, 2015); in addition, anterior insula activity is related to introspection and conscious awareness of bodily states (Critchley, Wiens, Rotshtein, Ohman, & Dolan, 2004; Klein et al., 2013; Medford & Critchley, 2010; Ullsperger et al., 2010). It has been suggested that these phasic autonomic reactions co-vary with increases in norepinephrine release (Nieuwenhuis, Aston-Jones, & Cohen, 2005a; Sara & Bouret, 2012), which in turn may enable appropriate changes in cortical responses to the eliciting event (Aston-Jones & Cohen, 2005; Bouret & Sara, 2005). Notably, only errors associated with an autonomic response appear to be perceived consciously (O'Connell et al., 2007; Wessel et al., 2011).

Emotion Expression Unexpected and negative action outcomes are often accompanied by expressions of surprise and anger, such as changes in facial expression and cursing. Although this has not been studied extensively in the context of performance monitoring, the brain regions underlying expressions of emotions and performance monitoring appear to overlap, particularly in the pMFC (Etkin et al., 2011; Shackman et al., 2011). It may be speculated that these immediate emotional expressions have social functions supporting the ability of observers to detect errors made by the acting individual and use this information as a warning and to facilitate corresponding adaptations of their behaviour (de Bruijn et al., 2009; de Bruijn, Mars, Bekkering, & Coles, 2010; Nunez Castellar, Notebaert, Van den Bossche, & Fias, 2011; van Schie, Mars, Coles, & Bekkering, 2004; Yoshida, Saito, Iriki, & Isoda, 2011; Yoshida, Saito, Iriki, & Isoda, 2012).

Task-Specific Adjustments

Adaptive Increase of Selective Attention In speeded reaction time tasks, a substantial number of errors result from misdirection of selective attention towards distracting and misleading stimuli. In order to regain performance, attention must be redirected to task-relevant perceptual information, and processing of distractors must be suppressed. Thus, focus of selective attention is readjusted according to the task rules. At the behavioural level, this may be reflected in a general performance increase after errors, post-error improvement of accuracy (PIA), or a reduction in interference effects of distracting inputs on reaction times after errors (post-error reduction of interference, PERI) (Danielmeier & Ullsperger, 2011; see also Chapter 4 by Egner in this volume). These increases in selective attention are implemented by enhancing representations of task-relevant stimulus features and suppressing representations of distracting features in perceptual cortices. For example, in a face gender classification task, activity in face-sensitive extrastriate visual areas like the fusiform face area (FFA) was increased after errors (King et al., 2010). Participants with the highest post-error increase in FFA activity showed the highest PERI effects. At the same time, coding of task-irrelevant but distracting stimulus lateralisation was reduced on post-error trials (King et al., 2010). In a later study in which participants had to focus on stimulus colour and ignore its motion to achieve optimal performance, the amplitude of error-related pMFC activity correlated with subsequent enhancement of (task-relevant) colour processing in visual cortex V4 and suppression of (distracting) motion representations in the motion-sensitive cortex (Danielmeier et al., 2011). Animal research suggests that acetylcholine (ACh) released from basal forebrain neurons into perceptual cortices is involved in implementing selective attention (Golmayo, Nunez, & Zaborszky, 2003; Pinto et al., 2013). A recent pharmaco-fMRI study demonstrated that the top-down interaction between performance monitoring in the pMFC and perceptual areas is blocked by the antimuscarinic agent biperiden (Danielmeier et al., 2015). Although with placebo the previous findings—that the error-related signal increase in the pMFC is correlated with post-error enhancement in the visual cortex coding task-relevant stimulus features and post-error suppression in the visual cortices coding distracting information—were replicated, blocking muscarinic receptors reduces these interactions and abolishes the behavioural PERI effect (Figure 17.3). This provides strong support for the notion that performance-monitoring-driven top-down adjustments of selective attention are mediated by ACh. It appears that the performance monitoring system recruits activity of cholinergic neurons in the basal forebrain (in humans, the nucleus basalis of Meynert), which then modulate activity in the visual cortices according to the task demands. It could be speculated that the task rules represented in the dorsal attentional network encompassing lateral prefrontal and parietal cortices influence the visual cortices (Brass, Ullsperger, Knösche, von Cramon, & Phillips, 2005; King, Korb, & Egner, 2012; Zanto, Rubens, Thangavel, & Gazzaley, 2011; Zanto & Gazzaley, 2013) and that these top-down influences are phasically modulated by cholinergic inputs. The time course of these top-down effects is still poorly understood. Although suppression of irrelevant visual inputs seems to take effect immediately on the first post-error trial, processing of task-relevant inputs (which is reduced before and on the error trial itself) needs several trials to regain baseline and above-baseline activity (Danielmeier et al., 2011; Danielmeier et al., 2015). Methods with higher temporal resolution than fMRI are needed to address this question. Notably, not only errors but also increased difficulty and response conflict during an action can trigger adjustments in selective attentions. These conflict adaptation effects are discussed in Chapter 4 by Egner and Chapter 5 by Bugg in this volume.

Learning and Decision Making As discussed above, most theories of performance monitoring assume underlying reinforcement learning mechanisms. Thus, performance monitoring is used to appropriately update stimulus and action values. Indeed, it has been shown that performance-monitoring-related pMFC activity can predict adaptive switches in response selection (Cohen & Ranganath, 2007; Isoda & Hikosaka, 2007; Shima & Tanji, 1998; Williams, Bush, Rauch, Cosgrove, & Eskandar, 2004; Yoshida et al., 2011), learning (Klein et al., 2007b; Hester, Barre, Murphy, Silk, & Mattingley, 2008; Hester, Madeley, Murphy, & Mattingley, 2009), and the stability of learned task rules (Jocham, Neumann, et al., 2009; Jocham, Klein, et al., 2009). As discussed above, prediction errors alone may not be sufficient to predict future behaviour (Chase, Swainson, Durham, Benham, & Cools, 2011), but they must be weighted appropriately, which appears to involve a larger network encompassing the pMFC, anterior insula, and parietal cortex (McGuire et al., 2014) and may be reflected in centroparietal P3b-like EEG activity (Fischer & Ullsperger, 2013).

The pMFC appears to code prediction errors and value updates for actions rather than stimuli (Wunderlich, Rangel, & O'Doherty, 2009), which might explain why no

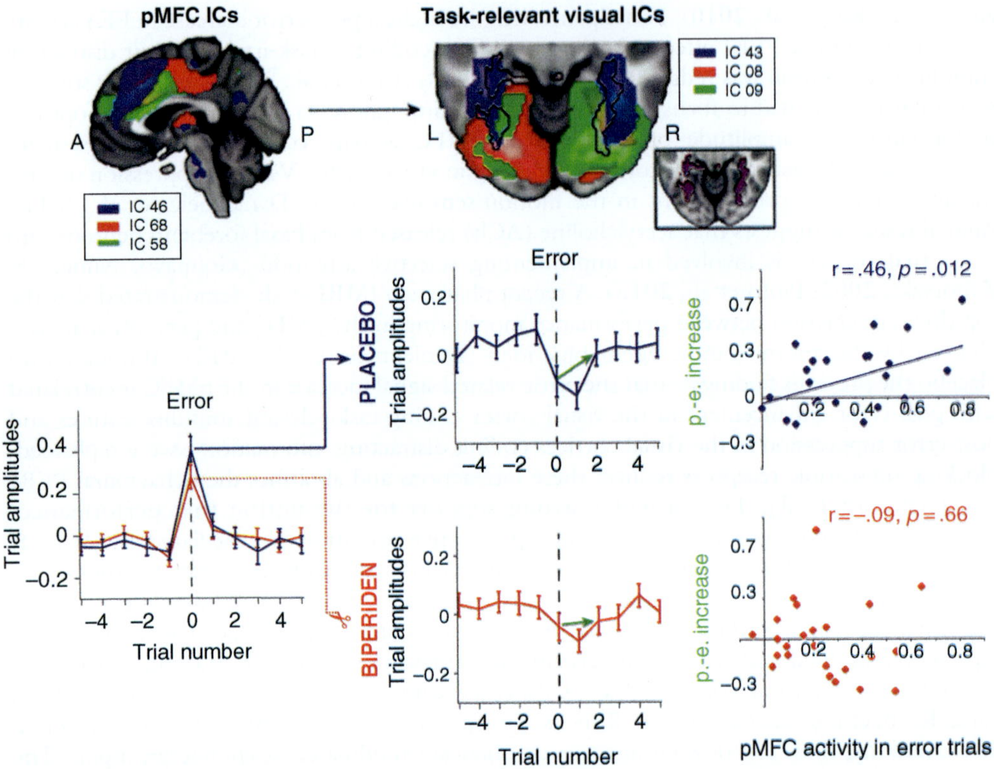

Figure 17.3 Post-error increase in selective attention is mediated by acetylcholine. Sagittal view of pMFC regions (independent components, ICs) showing increased activity on errors as determined by independent component analysis (pMFC ICs). Task-relevant visual ICs in a Simon interference task overlapping with activity in a separate colour localiser task (black outline and violet areas in inset). Below, mean activity of the averaged pMFC (left) and task-relevant visual (middle) ICs are shown for errors (±5 trials) with placebo (blue) and biperiden (red; note that the colour coding of time course data is independent of the colours used to visualise the ICs). Note that baseline activity here reflects brain activity during successfully solved trials. The green arrow indicates the post-error activity increase (p.-e. increase) in task-relevant visual areas, which is part of the correlation (right panel). The p.-e. increase was steeper with placebo than with the muscarinic antagonist biperiden. All y-axis units are z scores. Error bars represent the SEM. Right: Scatterplots for individual pMFC activities on error trials (x axes) and individual p.-e. increases over two trials (y axes) in task-relevant visual areas, shown separately for placebo (blue) and biperiden (red). The correlations demonstrate that pMFC activity in error trials predicts the subsequent activity increase in task-relevant visual areas under placebo, but not under biperiden. This relationship is indicated by the blue arrow and by the dotted line and scissors, respectively, in the left panel. A = anterior, P = posterior, L = left, R = right. Source: Danielmeier, 2015. Reproduced with permission of Elsevier.

prediction-error-related activity was found for fictive outcomes that help to update stimulus values in the absence of active choice (Fischer & Ullsperger, 2013).

In situations with more possible action strategies, necessitating a decision on whether to stay with a default behaviour or explore other non-default options, the pMFC appears to code the value of switching away from the current actions to an alternative strategy

(Boorman, Behrens, & Rushworth, 2011; Donoso, Collins, & Koechlin, 2014). Recent evidence suggests that shifts in general strategy are encoded by the somewhat more anteriorly located pregenual anterior cingulate cortex (Friedman et al., 2015; Schuck et al., 2015). The choice between continuing to exploit an established context/behaviour and shifting into an exploratory mode of behaviour that is costly but may promise higher rewards is analysed in foraging theory. Recent work suggests that the pMFC codes the value of foraging (Kolling, Behrens, Mars, & Rushworth, 2012; Kolling, Wittmann, & Rushworth, 2014).

Outlook

In sum, the performance monitoring system, a network centred on the pMFC, monitors for motivationally salient events that may indicate unexpected action outcomes and deviations of the observed states from the intended states of the organism and environment and uses this information to signal the necessity and magnitude of adaptive behaviour. To this end, the pMFC encodes in a context-dependent fashion various parameters and variables indicating outcome predictions, deviations from these predictions, and uncertainty. The directionality of interactions with other brain regions coding reward, prediction error, and salience signals such as the mesencephalon and the basal ganglia circuits still needs to be established. In particular, the neurochemical underpinnings and role of monoaminergic neuromodulators in these interactions needs further investigations across multiple species. A question for future research is how this performance monitoring system producing rather general alarm signals can initiate appropriate task- and context-specific adjustments. This question is related to the problem of credit assignment, which allows observed events and outcomes to be attributed to their putative cause. Although neuronal activity in the pMFC may be capable of supporting credit assignment (Scangos et al., 2013), substantial evidence suggests that the orbitofrontal cortex plays a major role in this function (Noonan et al., 2010; Walton, Behrens, Buckley, Rudebeck, & Rushworth, 2010). In general, the interactions between performance monitoring in the pMFC and anterior insula/frontal operculum and value coding in the ventromedial and orbitofrontal as well as frontopolar and parietal cortices requires further investigation and the development of overarching models of learning, decision making, and action selection.

References

Alexander, W. H., & Brown, J. W. (2011). Medial prefrontal cortex as an action-outcome predictor. *Nature Neuroscience*, *14*, 1338–1344.

Allain, S., Carbonnell, L., Burle, B., Hasbroucq, T., & Vidal, F. (2004). On-line executive control: An electromyographic study. *Psychophysiology*, *41*, 113–116.

Amengual, J. L., Marco-Pallares, J., Richter, L., Oung, S., Schweikard, A., Mohammadi, B., ... Munte, T. F. (2013). Tracking post-error adaptation in the motor system by transcranial magnetic stimulation. *Neuroscience*, *250*, 342–351.

Amiez, C., Neveu, R., Warrot, D., Petrides, M., Knoblauch, K., & Procyk, E. (2013). The location of feedback-related activity in the midcingulate cortex is predicted by local morphology. *Journal of Neuroscience*, *33*, 2217–2228.

Amiez, C., Sallet, J., Procyk, E., & Petrides, M. (2012). Modulation of feedback related activity in the rostral anterior cingulate cortex during trial and error exploration. *Neuroimage*, *63*, 1078–1090.

Aron, A. R., Behrens, T. E., Smith, S., Frank, M. J., & Poldrack, R. A. (2007). Triangulating a cognitive control network using diffusion-weighted magnetic resonance imaging (MRI) and functional MRI. *Journal of Neuroscience, 27*, 3743–3752.

Aston-Jones, G., & Cohen, J. D. (2005). Adaptive gain and the role of the locus coeruleus-norepinephrine system in optimal performance. *Journal of Comparative Neurology, 493*, 99–110.

Barch, D. M., Braver, T. S., Akbudak, E., Conturo, T., Ollinger, J., & Snyder, A. (2001). Anterior cingulate cortex and response conflict: Effects of response modality and processing domain. *Cerebral Cortex, 11*, 837–848.

Behrens, T. E., Hunt, L. T., Woolrich, M. W., & Rushworth, M. F. (2008). Associative learning of social value. *Nature, 456*, 245–249.

Behrens, T. E., Woolrich, M. W., Walton, M. E., & Rushworth, M. F. (2007). Learning the value of information in an uncertain world. *Nature Neuroscience, 10*, 1214–1221.

Bogacz, R., Wagenmakers, E. J., Forstmann, B. U., & Nieuwenhuis, S. (2010). The neural basis of the speed-accuracy tradeoff. *Trends in Neurosciences, 33*, 10–16.

Boorman, E. D., Behrens, T. E., & Rushworth, M. F. (2011). Counterfactual choice and learning in a neural network centered on human lateral frontopolar cortex. *PLoS Biology, 9*, e1001093.

Botvinick, M. M., Braver, T. S., Barch, D. M., Carter, C. S., & Cohen, J. D. (2001). Conflict monitoring and cognitive control. *Psychological Review, 108*, 624–652.

Botvinick, M. M., Cohen, J. D., & Carter, C. S. (2004). Conflict monitoring and anterior cingulate cortex: An update. *Trends in Cognitive Sciences, 8*, 539–546.

Botvinick, M., Nystrom, L. E., Fissell, K., Carter, C. S., & Cohen, J. D. (1999). Conflict monitoring versus selection-for-action in anterior cingulate cortex. *Nature, 402*, 179–181.

Bouret, S., & Sara, S. J. (2005). Network reset: A simplified overarching theory of locus coeruleus noradrenaline function. *Trends in Neurosciences, 28*, 574–582.

Brass, M., Ullsperger, M., Knösche, T. R., von Cramon, D. Y., & Phillips, N. A. (2005). Who comes first? The role of prefrontal and parietal cortex in cognitive control. *Journal of Cognitive Neuroscience, 17*, 1367–1375.

Burle, B., Roger, C., Allain, S., Vidal, F., & Hasbroucq, T. (2008). Error negativity does not reflect conflict: A reappraisal of conflict monitoring and anterior cingulate cortex activity. *Journal of Cognitive Neuroscience, 20*, 1637–1655.

Caplin, A., & Dean, M. (2008). Axiomatic methods, dopamine and reward prediction error. *Current Opinion in Neurobiology, 18*, 197–202.

Carter, C. S., Braver, T. S., Barch, D. M., Botvinick, M. M., Noll, D., & Cohen, J. D. (1998). Anterior cingulate cortex, error detection, and the online monitoring of performance. *Science, 280*, 747–749.

Cavanagh, J. F., Cohen, M. X., & Allen, J. J. (2009). Prelude to and resolution of an error: EEG phase synchrony reveals cognitive control dynamics during action monitoring. *Journal of Neuroscience, 29*, 98–105.

Cavanagh, J. F., Sanguinetti, J. L., Allen, J. J., Sherman, S. J., & Frank, M. J. (2014). The subthalamic nucleus contributes to post-error slowing. *Journal of Cognitive Neuroscience, 26*(11), 2637–2644.

Cavanagh, J. F., & Shackman, A. J. (2015). Frontal midline theta reflects anxiety and cognitive control: Meta-analytic evidence. *Journal of Physiology, Paris, 109*(1–3), 3–15.

Chase, H. W., Swainson, R., Durham, L., Benham, L., & Cools, R. (2011). Feedback-related negativity codes prediction error but not behavioral adjustment during probabilistic reversal learning. *Journal of Cognitive Neuroscience, 23*, 936–946.

Cohen, M. X., & Ranganath, C. (2007) Reinforcement learning signals predict future decisions. *Journal of Neuroscience, 27*, 371–378.

Coles, M. G., Scheffers, M. K., & Holroyd, C. B. (2001). Why is there an ERN/Ne on correct trials? Response representations, stimulus-related components, and the theory of error-processing. *Biological Psychology, 56*, 173–189.

Critchley, H. D., Elliott, R., Mathias, C. J., & Dolan, R. J. (2000). Neural activity relating to generation and representation of galvanic skin conductance responses: A functional magnetic resonance imaging study. *Journal of Neuroscience, 20*, 3033–3040.

Critchley, H. D., Mathias, C. J., Josephs, O., O'Doherty, J., Zanini, S., Dewar, B. K.,...Dolan, R. J. (2003). Human cingulate cortex and autonomic control: Converging neuroimaging and clinical evidence. *Brain, 126*, 2139–2152.

Critchley, H. D., Tang, J., Glaser, D., Butterworth, B., & Dolan, R. J. (2005). Anterior cingulate activity during error and autonomic response. *Neuroimage, 27*, 885–895.

Critchley, H. D., Wiens, S., Rotshtein, P., Ohman, A., & Dolan, R. J. (2004). Neural systems supporting interoceptive awareness. *Nature Neuroscience, 7*, 189–195.

Crockett, M. J., Clark, L., & Robbins, T. W. (2009). Reconciling the role of serotonin in behavioral inhibition and aversion: Acute tryptophan depletion abolishes punishment-induced inhibition in humans. *Journal of Neuroscience, 29*, 11993–11999.

Crockett, M. J., Clark, L., Smillie, L. D., & Robbins, T. W. (2012). The effects of acute tryptophan depletion on costly information sampling: Impulsivity or aversive processing? *Psychopharmacology (Berl), 219*, 587–597.

Crone, E. A., Somsen, R. J., Van Beek, B., & Van Der Molen, M. W. (2004). Heart rate and skin conductance analysis of antecendents and consequences of decision making. *Psychophysiology, 41*, 531–540.

Crone, E. A., van der Veen, F. M., van der Molen, M. W., Somsen, R. J. M., van Beek, B., & Jennings, J. R. (2003). Cardiac concomitants of feedback processing. *Biological Psychology, 64*(1–2), 143–156.

Danev, S. G., & de Winter, C. R. (1971). Heart rate deceleration after erroneous responses. A phenomenon complicating the use of heart rate variability for assessing mental load. *Psychologische Forschung, 35*, 27–34.

Danielmeier, C., Allen, E. A., Jocham, G., Onur, O. A., Eichele, T., & Ullsperger, M. (2015). Acetylcholine mediates behavioral and neural post-error control. *Current Biology, 25*, 1461–1468.

Danielmeier, C., Eichele, T., Forstmann, B. U., Tittgemeyer, M., & Ullsperger, M. (2011). Posterior medial frontal cortex activity predicts post-error adaptations in task-related visual and motor areas. *Journal of Neuroscience, 31*, 1780–1789.

Danielmeier, C., & Ullsperger, M. (2011). Post-error adjustments. *Front Psychology, 2*, 233.

Danielmeier, C., Wessel, J. R., Steinhauser, M., & Ullsperger, M. (2009). Modulation of the error-related negativity by response conflict. *Psychophysiology, 46*, 1288–1298.

de Bruijn, E. R. A., de Lange, F. P., von Cramon, D. Y., & Ullsperger, M. (2009). When errors are rewarding. *Journal of Neuroscience, 29*, 12183–12186.

de Bruijn, E. R. A., Mars, R. B., Bekkering, H., & Coles, M. G. H. (2010). *Your mistake is my mistake ... or is it?* Behavioral adjustments following own and observed actions in cooperative and competitive contexts. *Quarterly Journal of Experimental Psychology*, in press.

Debener, S., Ullsperger, M., Siegel, M., Fiehler, K., von Cramon, D. Y., & Engel, A. K. (2005). Trial-by-trial coupling of concurrent electroencephalogram and functional magnetic resonance imaging identifies the dynamics of performance monitoring. *Journal of Neuroscience, 25*, 11730–11737.

Dehaene, S., Posner, M. I., & Tucker, D. M. (1994). Localization of a neural system for error detection and compensation. *Psychological Science, 5*, 303–305.

Desmet, C., Fias, W., Hartstra, E., & Brass, M. (2011). Errors and conflict at the task level and the response level. *Journal of Neuroscience, 31*, 1366–1374.

Donoso, M., Collins, A. G., & Koechlin, E. (2014). Human cognition. *Foundations of human reasoning in the prefrontal cortex. Science, 344*, 1481–1486.

Dudschig, C., & Jentzsch, I. (2009). Speeding before and slowing after errors: Is it all just strategy? *Brain Research, 1296*, 56–62.

Dutilh, G., Vandekerckhove, J., Forstmann, B. U., Keuleers, E., Brysbaert, M., & Wagenmakers, E. J. (2012). Testing theories of post-error slowing. *Attention, Perception, & Psychophysics, 74*, 454–465.

Ebitz, R. B., & Platt, M. L. (2015). Neuronal activity in primate dorsal anterior cingulate cortex signals task conflict and predicts adjustments in pupil-linked arousal. *Neuron, 85*, 628–640.

Eichele, T., Debener, S., Calhoun, V. D., Specht, K., Engel, A. K., Hugdahl, K., ... Ullsperger, M. (2008). Prediction of human errors by maladaptive changes in event-related brain networks. *Proceedings of the National Academy of Sciences of the United States of America, 105*, 6173–6178.

Eichele, H., Juvodden, H. T., Ullsperger, M., & Eichele, T. (2010). Mal-adaptation of event-related EEG responses preceding performance errors. *Frontiers in Human Neuroscience*, 4.

Etkin, A., Egner, T., & Kalisch, R. (2011). Emotional processing in anterior cingulate and medial prefrontal cortex. *Trends in Cognitive Sciences*, *15*, 85–93.

Falkenstein, M., Hohnsbein, J., Hoormann, J., & Blanke, L. (1990). Effects of errors in choice reaction tasks on the ERP under focused and divided attention. In C. H. M. Brunia, A. W. K. Gaillard, & A. Kok (Eds.), *Psychophysiological brain research* (pp. 192–195). Tilburg: Tilburg University Press.

Falkenstein, M., Hoormann, J., Christ, S., & Hohnsbein, J. (2000). ERP components on reaction errors and their functional significance: A tutorial. *Biological Psychology*, *51*, 87–107.

Fiehler, K., Ullsperger, M., & von Cramon, D. Y. (2004). Neural correlates of error detection and error correction: Is there a common neuroanatomical substrate? *European Journal of Neuroscience*, *19*, 3081–3087.

Fiehler, K., Ullsperger, M., & von Cramon, D. Y. (2005). Electrophysiological correlates of error correction. *Psychophysiology*, *42*, 72–82.

Fiehler, K., Ullsperger, M., Grigutsch, M., & von Cramon, D. Y. (2004). Cardiac responses to error processing and response conflict. In M. Ullsperger & M. Falkenstein (Eds.), *Errors, conflicts, and the brain. Current opinions on performance monitoring* (pp. 135–140). Leipzig: MPI for Human Cognitive and Brain Sciences.

Fischer, A. G., Endrass, T., Reuter, M., Kubisch, C., & Ullsperger, M. (2015). Serotonin reuptake inhibitors and serotonin transporter genotype modulate performance monitoring functions but not their electrophysiological correlates. *Journal of Neuroscience*, *35*, 8181–8190.

Fischer, A. G., & Ullsperger, M. (2013). Real and fictive outcomes are processed differently but converge on a common adaptive mechanism. *Neuron*, *79*, 1243–1255.

Fischer, A. G., & Ullsperger, M. (2014). When is the time for a change? Decomposing dynamic learning rates. *Neuron*, *84*, 662–664.

Folstein, J. R., & Van Petten, C. (2008). Influence of cognitive control and mismatch on the N2 component of the ERP: A review. *Psychophysiology*, *45*, 152–170.

Frank, M. J., Samanta, J., Moustafa, A. A., & Sherman, S. J. (2007). Hold your horses: Impulsivity, deep brain stimulation, and medication in Parkinsonism. *Science*, *318*, 1309–1312.

Friedman, A., Homma, D., Gibb, L. G., Amemori, K., Rubin, S. J., Hood, A. S.,...Graybiel, A. M. (2015). A corticostriatal path targeting striosomes controls decision-making under conflict. *Cell*, *161*, 1320–1333.

Garavan, H., Ross, T. J., Murphy, K., Roche, R. A., & Stein, E. A. (2002). Dissociable executive functions in the dynamic control of behavior: Inhibition, error detection, and correction. *Neuroimage*, *17*, 1820–1829.

Gehring, W. J., Goss, B., Coles, M. G., Meyer, D. E., & Donchin, E. (1993). A neural system for error detection and compensation. *Psychological Science*, *4*, 385–390.

Gehring, W. J., & Knight, R. T. (2000). Prefrontal-cingulate interactions in action monitoring. *Nature Neuroscience*, *3*, 516–520.

Gehring, W. J., & Willoughby, A. R. (2002). The medial frontal cortex and the rapid processing of monetary gains and losses. *Science*, *295*, 2279–2282.

Gentsch, A., Ullsperger, P., & Ullsperger, M. (2009). Dissociable medial frontal negativities from a common monitoring system for self- and externally caused failure of goal achievement. *NeuroImage*, *47*, 2023–2030.

Golmayo, L., Nunez, A., & Zaborszky, L. (2003). Electrophysiological evidence for the existence of a posterior cortical-prefrontal-basal forebrain circuitry in modulating sensory responses in visual and somatosensory rat cortical areas. *Neuroscience*, *119*, 597–609.

Gruendler, T. O. J., Ullsperger, M., & Huster, R. J. (2011). Event-related potential correlates of performance-monitoring in a lateralized time-estimation task. *PLoS One*, *6*(10), e25591.

Hajcak, G., McDonald, N., & Simons, R. F. (2003). To err is autonomic: Error-related brain potentials, ANS activity, and post-error compensatory behavior. *Psychophysiology*, *40*(6), 895–903.

Hartstra, E., Stolk, A., Mars, R. B., Toni, I., & Ullsperger, M. (2014). *Temporal dynamics of post-error adaptations: Dissociation between early and late oscillatory effects.* In 20th Annual Meeting of the Organization for Human Brain Mapping (OHBM). Hamburg, Germany.

Hauser, T. U., Iannaccone, R., Stampfli, P., Drechsler, R., Brandeis, D., Walitza, S., & Brem, S. (2014). The feedback-related negativity (FRN) revisited: New insights into the localization, meaning and network organization. *Neuroimage, 84*, 159–168.

Hayden, B. Y., Heilbronner, S. R., Pearson, J. M., & Platt, M. L. (2011). Surprise signals in anterior cingulate cortex: Neuronal encoding of unsigned reward prediction errors driving adjustment in behavior. *Journal of Neuroscience, 31*, 4178–4187.

Hester, R., Barre, N., Murphy, K., Silk, T. J., & Mattingley, J. B. (2008). Human medial frontal cortex activity predicts learning from errors. *Cerebral Cortex, 18*, 1933–1940.

Hester, R., Madeley, J., Murphy, K., & Mattingley, J. B. (2009). Learning from errors: Error-related neural activity predicts improvements in future inhibitory control performance. *Journal of Neuroscience, 29*, 7158–7165.

Holroyd, C. B., & Coles, M. G. (2002). The neural basis of human error processing: Reinforcement learning, dopamine, and the error-related negativity. *Psychological Review, 109*, 679–709.

Holroyd, C. B., Hajcak, G., & Larsen, J. T. (2006). The good, the bad and the neutral: Electrophysiological responses to feedback stimuli. *Brain Research, 1105*, 93–101.

Holroyd, C. B., & Krigolson, O. E. (2007). Reward prediction error signals associated with a modified time estimation task. *Psychophysiology, 44*, 913–917.

Holroyd, C. B., Nieuwenhuis, S., Yeung, N., & Cohen, J. D. (2003). Errors in reward prediction are reflected in the event-related brain potential. *Neuroreport, 14*, 2481–2484.

Huster, R. J., Westerhausen, R., Kreuder, F., Schweiger, E., & Wittling, W. (2007). Morphologic asymmetry of the human anterior cingulate cortex. *NeuroImage, 34*, 888–895.

Iannaccone, R., Hauser, T. U., Staempfli, P., Walitza, S., Brandeis, D., & Brem, S. (2015). Conflict monitoring and error processing: New insights from simultaneous EEG-fMRI. *Neuroimage, 105*, 395–407.

Isoda, M., & Hikosaka, O. (2007). Switching from automatic to controlled action by monkey medial frontal cortex. *Nature Neuroscience, 10*, 240–248.

Ito, S., Stuphorn, V., Brown, J. W., & Schall, J. D. (2003). Performance monitoring by the anterior cingulate cortex during saccade countermanding. *Science, 302*, 120–122.

Jentzsch, I., & Dudschig, C. (2009). Why do we slow down after an error? Mechanisms underlying the effects of posterror slowing. *Quarterly Journal of Experimental Psychology (Colchester), 62*, 209–218.

Jessup, R. K., Busemeyer, J. R., & Brown, J. W. (2010). Error effects in anterior cingulate cortex reverse when error likelihood is high. *Journal of Neuroscience, 30*, 3467–3472.

Jocham, G., Klein, T. A., Neumann, J., von Cramon, D. Y., Reuter, M., & Ullsperger, M. (2009). Dopamine DRD2 polymorphism alters reversal learning and associated neural activity. *Journal of Neuroscience, 29*, 3695–3704.

Jocham, G., Neumann, J., Klein, T. A., Danielmeier, C., & Ullsperger, M. (2009). Adaptive coding of action values in the human rostral cingulate zone. *Journal of Neuroscience, 29*, 7489–7496.

Johansen-Berg, H. (2010). Behavioural relevance of variation in white matter microstructure. *Current Opinion in Neurology, 23*, 351–358.

Johansen-Berg, H., Behrens, T. E., Robson, M. D., Drobnjak, I., Rushworth, M. F., Brady, J. M.,... Matthews, P. M. (2004). Changes in connectivity profiles define functionally distinct regions in human medial frontal cortex. *Proceedings of the National Academy of Sciences of the United States of America, 101*(36), 13335–13340.

Kennerley, S. W., Behrens, T. E., & Wallis, J. D. (2011). Double dissociation of value computations in orbitofrontal and anterior cingulate neurons. *Nature Neuroscience, 14*, 1581–1589.

Kennerley, S. W., Walton, M. E., Behrens, T. E., Buckley, M. J., & Rushworth, M. F. (2006). Optimal decision making and the anterior cingulate cortex. *Nature Neuroscience, 9*, 940–947.

King, J. A., Korb, F. M., & Egner, T. (2012). Priming of control: Implicit contextual cuing of top-down attentional set. *Journal of Neuroscience, 32*, 8192–8200.

King, J. A., Korb, F. M., von Cramon, D. Y., & Ullsperger, M. (2010). Post-error behavioral adjustments are facilitated by activation and suppression of task-relevant and task-irrelevant information processing. *Journal of Neuroscience, 30*, 12759–12769.

Klein, T. A., Endrass, T., Kathmann, N., Neumann, J., von Cramon, D. Y., & Ullsperger, M. (2007a). Neural correlates of error awareness. *NeuroImage*, *34*, 1774–1781.

Klein, T. A., Neumann, J., Reuter, M., Hennig, J., von Cramon, D. Y., & Ullsperger, M. (2007b). Genetically determined differences in learning from errors. *Science*, *318*, 1642–1645.

Klein, T. A., Ullsperger, M., & Danielmeier, C. (2013). Error awareness and the insula: Links to neurological and psychiatric diseases. *Frontiers in Human Neuroscience*, *7*, 14.

Kolling, N., Behrens, T. E., Mars, R. B., & Rushworth, M. F. (2012). Neural mechanisms of foraging. *Science*, *336*, 95–98.

Kolling, N., Wittmann, M., & Rushworth, M. F. (2014). Multiple neural mechanisms of decision making and their competition under changing risk pressure. *Neuron*, *81*, 1190–1202.

Lapish, C. C., Kroener, S., Durstewitz, D., Lavin, A., & Seamans, J. K. (2007). The ability of the mesocortical dopamine system to operate in distinct temporal modes. *Psychopharmacology (Berl)*, *191*, 609–625.

Marco-Pallares, J., Camara, E., Munte, T. F., & Rodriguez-Fornells, A. (2008). Neural mechanisms underlying adaptive actions after slips. *Journal of Cognitive Neuroscience*, *20*, 1595–1610.

Mars, R. B., Coles, M. G., Grol, M. J., Holroyd, C. B., Nieuwenhuis, S., Hulstijn, W., & Toni, I. (2005). Neural dynamics of error processing in medial frontal cortex. *Neuroimage*, *28*, 1007–1013.

Matsumoto, M., & Hikosaka, O. (2009). Two types of dopamine neuron distinctly convey positive and negative motivational signals. *Nature*, *459*, 837–841.

Matsumoto, M., Matsumoto, K., Abe, H., & Tanaka, K. (2007). Medial prefrontal cell activity signaling prediction errors of action values. *Nature Neuroscience*, *10*, 647–656.

McGuire, J. T., Nassar, M. R., Gold, J. I., & Kable, J. W. (2014). Functionally dissociable influences on learning rate in a dynamic environment. *Neuron*, *84*, 870–881.

Medford, N., & Critchley, H. D. (2010). Conjoint activity of anterior insular and anterior cingulate cortex: Awareness and response. *Brain Structure & Function*, *214*, 535–549.

Miltner, W. H. R., Braun, C. H., & Coles, M. G. H. (1997). Event-related brain potentials following incorrect feedback in a time-estimation task: Evidence for a 'generic' neural system for error detection. *Journal of Cognitive Neuroscience*, *9*, 788–798.

Nachev, P., Rees, G., Parton, A., Kennard, C., & Husain, M. (2005). Volition and conflict in human medial frontal cortex. *Current Biology*, *15*, 122–128.

Nee, D. E., Kastner, S., & Brown, J. W. (2011). Functional heterogeneity of conflict, error, task-switching, and unexpectedness effects within medial prefrontal cortex. *Neuroimage*, *54*, 528–540.

Neubert, F. X., Mars, R. B., Sallet, J., & Rushworth, M. F. (2015). Connectivity reveals relationship of brain areas for reward-guided learning and decision making in human and monkey frontal cortex. *Proceedings of the National Academy of Sciences of the United States of America*, *112*, E2695–2704.

Nieuwenhuis, S., Aston-Jones, G., & Cohen, J. D. (2005a). Decision making, the P3, and the locus coeruleus-norepinephrine system. *Psychological Bulletin*, *131*, 510–532.

Nieuwenhuis, S., Slagter, H. A., von Geusau, N. J., Heslenfeld, D. J., & Holroyd, C. B. (2005b). Knowing good from bad: Differential activation of human cortical areas by positive and negative outcomes. *European Journal of Neuroscience*, *21*, 3161–3168.

Noonan, M. P., Walton, M. E., Behrens, T. E., Sallet, J., Buckley, M. J., & Rushworth, M. F. (2010). Separate value comparison and learning mechanisms in macaque medial and lateral orbitofrontal cortex. *Proceedings of the National Academy of Sciences of the United States of America*, *107*, 20547–20552.

Notebaert, W., Houtman, F., Opstal, F. V., Gevers, W., Fias, W., & Verguts, T. (2009). Post-error slowing: An orienting account. *Cognition*, *111*, 275–279.

Nunez Castellar, E., Kuhn, S., Fias, W., & Notebaert, W. (2010). Outcome expectancy and not accuracy determines posterror slowing: ERP support. *Cognitive, Affective, & Behavioral Neuroscience*, *10*, 270–278.

Nunez Castellar, E., Notebaert, W., Van den Bossche, L., & Fias, W. (2011). How monitoring other's actions influences one's own performance. *Experimental Psychology*, *58*, 499–508.

O'Connell, R. G., Dockree, P. M., Bellgrove, M. A., Kelly, S. P., Hester, R., Garavan, H.,...Foxe, J. J. (2007). The role of cingulate cortex in the detection of errors with and without awareness: A high-density electrical mapping study. *European Journal of Neuroscience*, *25*, 2571–2579.

O'Connell, R. G., Dockree, P. M., & Kelly, S. P. (2012). A supramodal accumulation-to-bound signal that determines perceptual decisions in humans. *Nature Neuroscience*, *15*, 1729–1735.

O'Reilly, J. X. (2013). Making predictions in a changing world-inference, uncertainty, and learning. *Frontiers in Neuroscience*, *7*, 105.

Palomero-Gallagher, N., Zilles, K., Schleicher, A., & Vogt, B. A. (2013). Cyto- and receptor architecture of area 32 in human and macaque brains. *Journal of Comparative Neurology*, *521*(14), 3272–3286.

Paus, T., Tomaiuolo, F., Otaky, N., MacDonald, D., Petrides, M., Atlas, J.,...Evans, A. C. (1996). Human cingulate and paracingulate sulci: Pattern, variability, asymmetry, and probabilistic map. *Cerebral Cortex*, *6*, 207–214.

Peterburs, J., Pergola, G., Koch, B., Schwarz, M., Hoffmann, K. P., Daum, I., & Bellebaum, C. (2011). Altered error processing following vascular thalamic damage: Evidence from an antisaccade task. *PLoS ONE*, *6*, e21517.

Pinto, L., Goard, M. J., Estandian, D., Xu, M., Kwan, A. C., Lee, S. H.,...Dan, Y. (2013). Fast modulation of visual perception by basal forebrain cholinergic neurons. *Nature Neuroscience*, *16*, 1857–1863.

Rabbitt, P. M. (1978). Detection of errors by skilled typists. *Ergonomics*, *21*, 945–958.

Rabbitt, P. M. (1966a). Error correction time without external error signals. *Nature*, *212*, 438.

Rabbitt, P. M. (1966b). Errors and error correction in choice-response tasks. *Journal of Experimental Psychology*, *71*, 264–272.

Rescorla, R. A., & Wagner, A. R. (1972). A theory of Pavlovian conditioning: Variations in the effectiveness of reinforcement and nonreinforcement. In A. H. Black & W. F. Prokrasy (Eds.), *Classical conditioning II: Current research and theory* (pp. 64–99). New York: Appleton-Century-Crofts.

Ridderinkhof, K. R., Ullsperger, M., Crone, E. A., & Nieuwenhuis, S. (2004). The role of the medial frontal cortex in cognitive control. *Science*, *306*, 443–447.

Rodriguez-Fornells, A., Kurzbuch, A. R., & Munte, T. F. (2002). Time course of error detection and correction in humans: Neurophysiological evidence. *Journal of Neuroscience*, *22*, 9990–9996.

Rutledge, R. B., Dean, M., Caplin, A., & Glimcher, P. W. (2010). Testing the reward prediction error hypothesis with an axiomatic model. *Journal of Neuroscience*, *30*, 13525–13536.

Sara, S. J., & Bouret, S. (2012). Orienting and reorienting: The locus coeruleus mediates cognition through arousal. *Neuron*, *76*, 130–141.

Scangos, K. W., Aronberg, R., & Stuphorn, V. (2013). Performance monitoring by presupplementary and supplementary motor area during an arm movement countermanding task. *Journal of Neurophysiology*, *109*, 1928–1939.

Schuck, N. W., Gaschler, R., Wenke, D., Heinzle, J., Frensch, P. A., Haynes, J. D., & Reverberi, C. (2015). Medial prefrontal cortex predicts internally driven strategy shifts. *Neuron*, *86*, 331–340.

Schultz, W., Dayan, P., & Montague, P. R. (1997). A neural substrate of prediction and reward. *Science*, *275*, 1593–1599.

Seifert, S., von Cramon, D. Y., Imperati, D., Tittgemeyer, M., & Ullsperger, M. (2011). Thalamocingulate interactions in performance monitoring. *Journal of Neuroscience*, *31*, 3375–3383.

Shackman, A. J., Salomons, T. V., Slagter, H. A., Fox, A. S., Winter, J. J., & Davidson, R. J. (2011). The integration of negative affect, pain and cognitive control in the cingulate cortex. *Nature Reviews Neuroscience*, *12*, 154–167.

Sheth, S. A., Mian, M. K., Patel, S. R., Asaad, W. F., Williams, Z. M., Dougherty, D. D.,...Eskandar, E. N. (2012). Human dorsal anterior cingulate cortex neurons mediate ongoing behavioural adaptation. *Nature*, *488*(7410), 218–221.

Shidara, M., & Richmond, B. J. (2002). Anterior cingulate: Single neuronal signals related to degree of reward expectancy. *Science*, *296*, 1709–1711.

Shima, K., & Tanji, J. (1998). Role for cingulate motor area cells in voluntary movement selection based on reward. *Science*, *282*, 1335–1338.

Siegert, S., Herrojo Ruiz, M., Brucke, C., Huebl, J., Schneider, G. H., Ullsperger, M., & Kuhn, A. A. (2014). Error signals in the subthalamic nucleus are related to post-error slowing in patients with Parkinson's disease. *Cortex*, *60*, 103–120.

Silvetti, M., Alexander, W., Verguts, T., & Brown, J. W. (2014). From conflict management to reward-based decision making: Actors and critics in primate medial frontal cortex. *Neuroscience & Biobehavioral Reviews, 46*(Pt. 1), 44–57.

Silvetti, M., Seurinck, R., & Verguts, T. (2011). Value and prediction error in medial frontal cortex: Integrating the single-unit and systems levels of analysis. *Frontiers in Human Neuroscience, 5*, 75.

Steinhauser, M., Eichele, H., Juvodden, H. T., Huster, R. J., Ullsperger, M., & Eichele, T. (2012). Error-preceding brain activity reflects (mal-)adaptive adjustments of cognitive control: A modeling study. *Frontiers in Human Neuroscience, 6*, 97.

Steinhauser, M., Maier, M., & Hubner, R. (2008). Modeling behavioral measures of error detection in choice tasks: Response monitoring versus conflict monitoring. *Journal of Experimental Psychology: Human Perception and Performance, 34*, 158–176.

Steinhauser, M., & Yeung, N. (2010). Decision processes in human performance monitoring. *Journal of Neuroscience, 30*, 15643–15653.

Stemmer, B., Segalowitz, S. J., Witzke, W., & Schonle, P. W. (2004). Error detection in patients with lesions to the medial prefrontal cortex: An ERP study. *Neuropsychologia, 42*, 118–130.

Sutton, R. S., & Barto, A. G. (1998). *Reinforcement learning: An introduction*. Cambridge, MA: The MIT Press.

Swann, N., Poizner, H., Houser, M., Gould, S., Greenhouse, I., Cai, W.,...Aron, A. R. (2011). Deep brain stimulation of the subthalamic nucleus alters the cortical profile of response inhibition in the beta frequency band: A scalp EEG study in Parkinson's disease. *Journal of Neuroscience, 31*, 5721–5729.

Swann, N., Tandon, N., Canolty, R., Ellmore, T. M., McEvoy, L. K., Dreyer, S.,...Aron, A. R. (2009). Intracranial EEG reveals a time- and frequency-specific role for the right inferior frontal gyrus and primary motor cortex in stopping initiated responses. *Journal of Neuroscience, 29*, 12675–12685.

Swick, D., & Turken, A. U. (2002). Dissociation between conflict detection and error monitoring in the human anterior cingulate cortex. *Proceedings of the National Academy of Sciences of the United States of America, 99*, 16354–16359.

Talmi, D., Atkinson, R., & El-Deredy, W. (2013). The feedback-related negativity signals salience prediction errors, not reward prediction errors. *Journal of Neuroscience, 33*, 8264–8269.

Talmi, D., Fuentemilla, L., Litvak, V., Duzel, E., & Dolan, R. J. (2012). An MEG signature corresponding to an axiomatic model of reward prediction error. *Neuroimage, 59*, 635–645.

Turken, A. U., & Swick, D. (2008). The effect of orbitofrontal lesions on the error-related negativity. *Neuroscience Letters, 441*, 7–10.

Ullsperger, M., & von Cramon, D. Y. (2001). Subprocesses of performance monitoring: A dissociation of error processing and response competition revealed by event-related fMRI and ERPs. *Neuroimage, 14*, 1387–1401.

Ullsperger, M., & von Cramon, D. Y. (2004). Neuroimaging of performance monitoring: Error detection and beyond. *Cortex, 40*, 593–604.

Ullsperger, M., & von Cramon, D. Y. (2006a). The role of intact frontostriatal circuits in error processing. *Journal of Cognitive Neuroscience, 18*, 651–664.

Ullsperger, M., & von Cramon, D. Y. (2006b). How does error correction differ from error signaling? An event-related potential study. *Brain Research, 1105*, 102–109.

Ullsperger, M., von Cramon, D. Y., & Muller, N. G. (2002). Interactions of focal cortical lesions with error processing: Evidence from event-related brain potentials. *Neuropsychology, 16*, 548–561.

Ullsperger, M., Danielmeier, C., & Jocham, G. (2014a). Neurophysiology of performance monitoring and adaptive behavior. *Physiological Reviews, 94*, 35–79.

Ullsperger, M., Fischer, A. G., Nigbur, R., & Endrass, T. (2014b). Neural mechanisms and temporal dynamics of performance monitoring. *Trends in Cognitive Sciences, 18*, 259–267.

Ullsperger, M., Harsay, H. A., Wessel, J. R., & Ridderinkhof, K. R. (2010). Conscious perception of errors and its relation to the anterior insula. *Brain Structure & Function, 214*, 629–643.

Ullsperger, M., Nittono, H., & von Cramon, D. Y. (2007). When goals are missed: Dealing with self-generated and externally induced failure. *NeuroImage, 35*, 1356–1364.

Ullsperger, M., & Szymanowski, F. (2004). ERP Correlates of error relevance. In M. Ullsperger & M. Falkenstein (Eds.), *Errors, conflicts, and the brain: Current opinions on performance monitoring* (pp. 171–184). Leipzig: MPI for Human Cognitive and Brain Sciences.

van der Veen, F. M., Nieuwenhuis, S., Crone, E. A., & van der Molen, M. W. (2004). Cardiac and electro-cortical responses to performance feedback reflect different aspects of feedback processing. In M. Ullsperger & M. Falkenstein (Eds.), *Errors, conflicts, and the brain. Current opinions on performance monitoring* (pp 140–147). Leipzig: MPI for Human Cognitive and Brain Sciences.

van Gaal, S., Ridderinkhof, K. R., Scholte, H. S., & Lamme, V. A. (2010). Unconscious activation of the prefrontal no-go network. *Journal of Neuroscience, 30*, 4143–4150.

van Schie, H. T., Mars, R. B., Coles, M. G., & Bekkering, H. (2004). Modulation of activity in medial frontal and motor cortices during error observation. *Nature Neuroscience, 7*, 549–554.

van Veen, V., & Carter, C. S. (2002). The timing of action-monitoring processes in the anterior cingulate cortex. *Journal of Cognitive Neuroscience, 14*, 593–602.

van Veen, V., Holroyd, C. B., Cohen, J. D., Stenger, V. A., & Carter, C. S. (2004). Errors without conflict: Implications for performance monitoring theories of anterior cingulate cortex. *Brain and Cognition, 56*, 267–276.

Vogt, B. A. (2009). Regions and subregions of the cingulate cortex. In B. A. Vogt (Ed.), *Cingulate neurobiology and disease* (pp. 3–30). New York: Oxford University Press.

Walsh, M. M., & Anderson, J. R. (2012). Learning from experience: Event-related potential correlates of reward processing, neural adaptation, and behavioral choice. *Neuroscience & Biobehavioral Reviews, 36*, 1870–1884.

Walton, M. E., Behrens, T. E., Buckley, M. J., Rudebeck, P. H., & Rushworth, M. F. (2010). Separable learning systems in the macaque brain and the role of orbitofrontal cortex in contingent learning. *Neuron, 65*, 927–939.

Walton, M. E., Devlin, J. T., & Rushworth, M. F. (2004). Interactions between decision making and performance monitoring within prefrontal cortex. *Nature Neuroscience, 7*, 1259–1265.

Wang, C., Ulbert, I., Schomer, D. L., Marinkovic, K., & Halgren, E. (2005). Responses of human anterior cingulate cortex microdomains to error detection, conflict monitoring, stimulus-response mapping, familiarity, and orienting. *Journal of Neuroscience, 25*, 604–613.

Wessel, J. R., & Aron, A. R. (2013). Unexpected events induce motor slowing via a brain mechanism for action-stopping with global suppressive effects. *Journal of Neuroscience, 33*, 18481–18491.

Wessel, J. R., Danielmeier, C., Morton, J. B., & Ullsperger, M. (2012). Surprise and error: Common neuronal architecture for the processing of errors and novelty. *Journal of Neuroscience, 32*, 7528–7537.

Wessel, J. R., Danielmeier, C., & Ullsperger, M. (2011). Error awareness revisited: Accumulation of multimodal evidence from central and autonomic nervous systems. *Journal of Cognitive Neuroscience, 23*, 3021–3036.

Wessel, J. R., Klein, T. A., Ott, D. V., & Ullsperger, M. (2014). Lesions to the prefrontal performance-monitoring network disrupt neural processing and adaptive behaviors after both errors and novelty. *Cortex, 50*, 45–54.

Wessel, J. R., & Ullsperger, M. (2011). Selection of independent components representing event-related brain potentials: A data-driven approach for greater objectivity. *Neuroimage, 54*, 2105–2115.

Williams, Z. M., Bush, G., Rauch, S. L., Cosgrove, G. R., & Eskandar, E. N. (2004). Human anterior cingulate neurons and the integration of monetary reward with motor responses. *Nature Neuroscience, 7*, 1370–1375.

Wunderlich, K., Rangel, A., & O'Doherty, J. P. (2009). Neural computations underlying action-based decision making in the human brain. *Proceedings of the National Academy of Sciences of the United States of America, 106*, 17199–17204.

Xiang, Z., Wang, L., & Kitai, S. T. (2005). Modulation of spontaneous firing in rat subthalamic neurons by 5-HT receptor subtypes. *Journal of Neurophysiology, 93*, 1145–1157.

Yeung, N., Cohen, J. D., & Botvinick, M. M. (2004). The neural basis of error detection: Conflict monitoring and the error-related negativity. *Psychological Review, 111*, 931–959.

Yeung, N., Ralph, J., & Nieuwenhuis, S. (2007). Drink alcohol and dim the lights: The impact of cognitive deficits on medial frontal cortex function. *Cognitive, Affective, & Behavioral Neuroscience, 7*, 347–355.

Yoshida, K., Saito, N., Iriki, A., & Isoda, M. (2011). Representation of others' action by neurons in monkey medial frontal cortex. *Current Biology, 21*, 249–253.

Yoshida, K., Saito, N., Iriki, A., & Isoda, M. (2012). Social error monitoring in macaque frontal cortex. *Nature Neuroscience, 15*, 1307–1312.

Zanto, T. P., & Gazzaley, A. (2013). Fronto-parietal network: Flexible hub of cognitive control. *Trends in Cognitive Sciences, 17*(12), 602–603.

Zanto, T. P., Rubens, M. T., Thangavel, A., & Gazzaley, A. (2011). Causal role of the prefrontal cortex in top-down modulation of visual processing and working memory. *Nature Neuroscience, 14*, 656–661.

18

Control and Connectivity
Dynamic Networks in the Human Brain

Michael W. Cole

The human brain is an extremely complex network consisting of billions of nodes and trillions of connections (Azevedo et al., 2009). Somehow, the complex spatiotemporal dynamics that play out on this network architecture produce cognitive control—a broad domain composed of goal-directed thoughts and behaviours. Until recently, however, even those who conceptualised cognitive control in terms of brain network mechanisms nonetheless continued to exclusively use localisation tools such as single-unit electrophysiology and functional MRI (fMRI) general linear models to investigate its neural basis. Here I will cover recent developments that are allowing for proper characterisation of the brain network basis of cognitive control. This includes methodological advances in characterising human brain connectivity, such as resting-state and task-state functional connectivity MRI and diffusion weighted imaging (DWI), as well as advances in identifying more descriptive network components such as brain hubs. Finally, I will cover theoretical insights gained from these advances, such as the nested organisation of cognitive control brain networks and the role of hub dynamics in implementing cognitive control functionality.

Applying Network Science to Neuroscience

Recent advances in network science have begun to facilitate our ability to characterise complex brain network properties underlying cognitive control. Network science is an interdisciplinary field built on graph theory (a branch of mathematics) to investigate complex interactive processes such as occur within the Internet and social networks (Bullmore & Bassett, 2011; Newman, 2008; Watts & Strogatz, 1998). For example, highly connected nodes in a network are called hubs (Berlingerio, Coscia, Giannotti, Monreale, & Pedreschi, 2011; Guimera, Mossa, Turtschi, & Amaral, 2005), with a variety of graph theoretical measures estimating different properties thought to characterise hubs (Figure 18.1). Google.com is an example of a hub website on the Internet, while a successful politician might be considered a social network hub. Network science also develops algorithms for identifying sub-networks (called modules) within a larger network, such as clusters of websites on related topics or social cliques (Figure 18.1).

These and other graphical approaches have been applied to the brain (Buckner et al., 2009; Bullmore & Bassett, 2011; Sporns, Honey, & Kötter, 2007), as part of ongoing developments in network science being adapted for neuroscience. A major challenge for applying

The Wiley Handbook of Cognitive Control, First Edition. Edited by Tobias Egner.

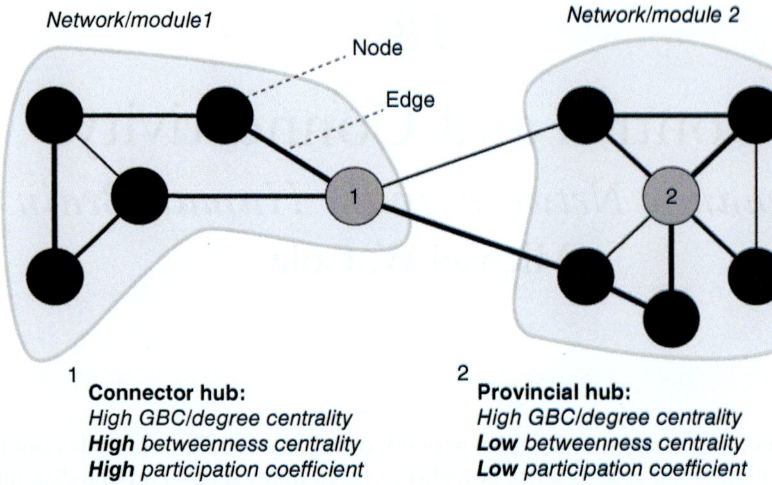

Figure 18.1 Hubs and modules. There are many graph theoretical constructs (and associated measures) that can be used to characterise networks. Particularly useful constructs include hubs, which are highly connected nodes, and modules, which are clusters of interconnected nodes. A hub that is primarily highly connected outside its own module is a connector hub (grey node 1 above). A hub that is primarily highly connected within its own module is a provincial hub (grey node 2 above). Graph theoretical measures that can (and cannot) be used to distinguish these hub subtypes are listed. Degree centrality simply counts the number of connections with a node, whereas global brain connectivity (GBC) averages the weighted connectivity (giving that node's average connection strength). Participation coefficient counts the number of between-module connections relative to within-module connections, therefore reflecting how well distributed a node's connections are across modules. Betweenness centrality counts the number of shortest paths between all nodes that pass through the node of interest, such that high betweenness centrality indicates that a node is on many of the most efficient/direct routes in a network. In practice, nodes with high betweenness centrality tend to have many between-module connections. See Cole, Ito, & Braver, 2015, for more details.

network science to the brain, however, is obtaining accurate and interpretable data regarding brain networks. Initial interest in this area focused on the brain's structural network—the comprehensive collection of white matter tracts connecting neurons termed the structural 'connectome' (Sporns, Tononi, & Kötter, 2005). An initial source of comprehensive large-scale structural connectivity data for the mammalian brain came from the CoCoMac database (Kötter, 2004), which was built from published tract tracing studies of the macaque monkey brain. This database was used along with graph theoretical measures to identify brain hubs (Sporns et al., 2007) and other network properties (Passingham, Stephan, & Kötter, 2002; Rubinov & Sporns, 2010) of the primate brain.[1]

Two primary approaches are used to estimate network structure in the living human brain. The first—DWI—uses MRI to detect white matter tracts via estimation of the diffusion of water along those tracts (Johansen-Berg & Rushworth, 2009). The second major method estimates functional connectivity—the statistical associations between brain activity time series—typically via correlation of activity across distinct brain locations (B. Biswal, Yetkin, Haughton, & Hyde, 1995; Friston, 1994). A variety of (primarily noninvasive) neuroimaging methods have been used to estimate functional connectivity in humans, such as fMRI, electroencephalography, and magnetoencephalography.

DWI involves the collection of multiple MRI images, each sensitive to a different direction of water diffusion (Figure 18.2). White matter tract directions are estimated on the basis of

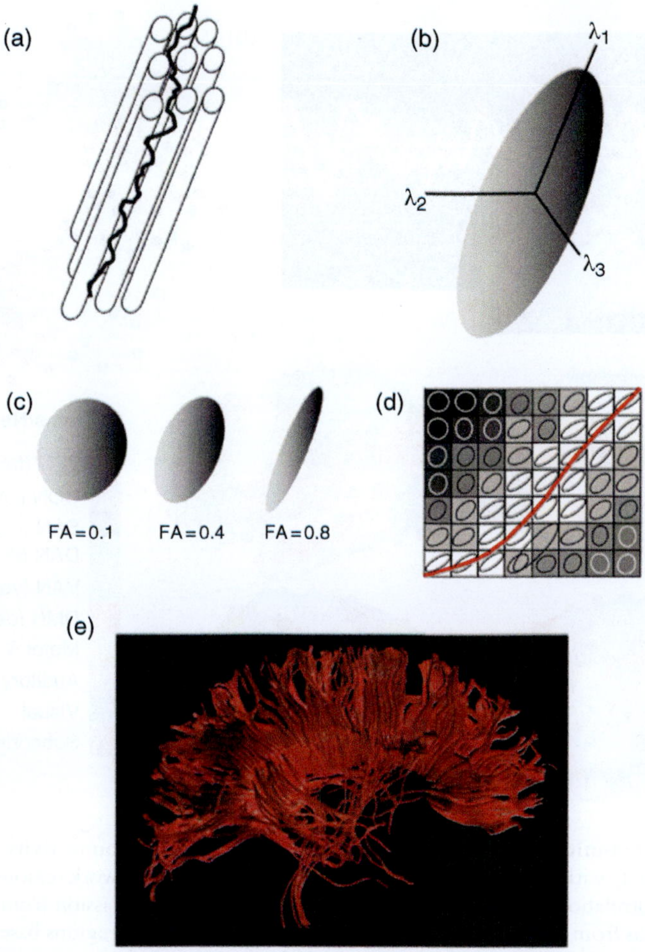

Figure 18.2 Structural connectivity estimated via diffusion weighted imaging (DWI). a) Diffusion weighted imaging (DWI) is sensitive to the direction of water diffusion within a measured brain volume. Water moves along an axon, producing a dominant diffusion direction as measured by DWI. b) The dominant direction of diffusion can be measured using a tensor with three axes, as shown here. More sophisticated models are available as well. These more sophisticated models are recommended, given the likelihood of fibre crossings (multiple dominant directions) within a given volume of measurement. c) As a given direction of diffusion increases in dominance, so does fractional anisotropy (FA). d) Each volume's FA (or equivalent) can be linked together to infer tracts over multiple volumes. e) These fibre tracts can then be analysed statistically and visualised. Figure adapted from (Johansen-Berg & Rushworth, 2009).

the principle that water is more likely to diffuse in the direction of tracts than other directions (Johansen-Berg & Rushworth, 2009). At least six DWI directions are necessary to estimate tracts, though many more directions (e.g., 64 or 264) are typically used to increase the accuracy of tract tracing. Even with many more directions than necessary, however, there are issues with movement during data collection that changes estimated tract directions (Mukherjee, Chung, Berman, Hess, & Henry, 2008; Yendiki, Koldewyn, Kakunoori, Kanwisher, & Fischl, 2013). The best way to model tracts is also unclear (Dell'Acqua & Catani, 2012)—especially

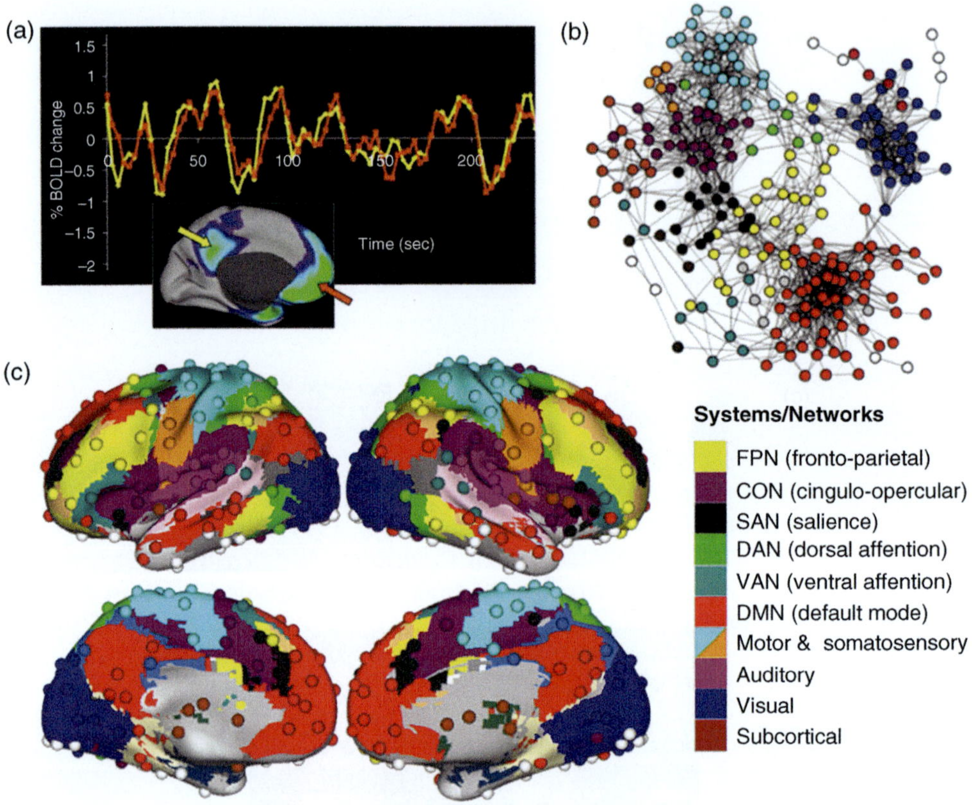

Figure 18.3 Community detection with resting-state functional connectivity. **a**) Resting-state functional connectivity with fMRI. Time series from two default-mode network regions are shown, indicating their high correlation (functional connectivity). Adapted with permission from (Raichle, 2010). **b**) A graph of regions from throughout the brain, with spatial proximity of regions based on the strength of their resting-state functional connectivity (closer regions = stronger connectivity). They are colour-coded on the basis of their network/module affiliation. Adapted with permission from (Power et al., 2011). **c**) The locations of the regions from panel B, along with their network/module affiliations, and equivalent network assignments based on a voxel-wise analysis. Source: Power 2011. Reproduced with permission of Elsevier.

at fibre crossings in single voxels. Indeed, there may be some fundamental limit to the accuracy of DWI data (C. Thomas et al., 2014; Van Essen et al., 2013). For instance, several tracts known to exist within the macaque monkey brain are not detectable using DWI even with especially high-resolution images (C. Thomas et al., 2014).

A common functional connectivity approach infers network structure on the basis of coordinated spontaneous fMRI activations while subjects are at rest (B. B. Biswal et al., 2010; B. Biswal et al., 1995; Raichle, 2010; Figure 18.3a). This resting-state functional connectivity method was developed after high correlations between each hemisphere's somatomotor cortices were observed even though subjects were at rest (B. Biswal et al., 1995). This paralleled these subjects' somatomotor activations during a motor task. Subsequent research has verified and extended those initial observations to the rest of the brain, determining that resting-state functional connectivity is functionally meaningful (Cole, Yarkoni, Repovs,

Anticevic, & Braver, 2012; Laird et al., 2013; S. M. Smith et al., 2009; Waites, Stanislavsky, Abbott, & Jackson, 2005), highly reproducible (Shehzad et al., 2009), and related (but not identical) to structural connectivity (as measured using DWI tractography; Honey et al., 2009). These studies have recently culminated in the identification of all major networks in the human brain via network partitioning—clustering of highly connected brain regions into communities (Power et al., 2011; Yeo et al., 2011; Figures 18.3b & 18.3c). However, there is uncertainty regarding exactly how to interpret resting-state functional connectivity results. One possibility consistent with existing data is that resting-state functional connectivity reflects spontaneous activity dynamics playing out on the brain's structural network (Deco et al., 2013; Goñi et al., 2014), perhaps modulated by synaptic strengths (Lewis, Baldassarre, Committeri, Romani, & Corbetta, 2009; Cole et al., 2016). Thus, this approach may reveal the likelihood of activity flowing across structural connections (and their synapses), even if multiple connections are involved in a chain—a kind of network pattern that is difficult to identify using structural data alone (Dum & Strick, 2013).

Cognitive Control as a Network Phenomenon

The above connectivity methods have been used to identify network properties related to cognitive control. Prior to this, however, a set of brain regions distributed across prefrontal and parietal cortices was found to be co-activated across various cognitive control demands (Duncan & Owen, 2000; Schneider & Chein, 2003; Wager & Smith, 2003; Figure 18.4a). These regions include the dorsolateral prefrontal cortex (DLPFC), anterior cingulate cortex (ACC; also called anterior mid-cingulate cortex), anterior insula cortex (AIC), posterior parietal cortex (PPC), and pre-motor cortex (PMC). Some additional regions may be included such as the pre-supplementary motor area (pre-SMA) and the rostrolateral prefrontal cortex,

Figure 18.4 The cognitive control network and nested sub-networks. **a**) A meta-analysis of the fMRI literature using the term 'cognitive control' with Neurosynth (Yarkoni, Poldrack, Nichols, Van Essen, & Wager, 2011). The activation pattern is consistent, with many individual studies of cognitive control assessing a variety of cognitive control demands. The statistical map ($p < 0.01$, false discovery rate corrected for multiple comparisons; forward inference) was generated by Neurosynth (which is periodically updated with new studies) on June 9, 2015. As many as 377 studies including the term 'cognitive control' contributed. **b**) Resting-state functional connectivity fMRI was used to parcellate the brain into seven networks (Yeo et al., 2011), of which three corresponding to the cognitive control co-activation pattern identified in panel A are shown. The orange network roughly corresponds to the frontoparietal control network, the green network roughly corresponds to the dorsal attention network, and the magenta roughly corresponds to the cingulo-opercular network. **c**) Nested within the coarse partition of the cognitive control network (three networks), seven networks are identified (based on a whole-brain 17-network solution; Yeo et al., 2011). B and C adapted with permission from Yeo et al. (2011).

depending on one's means of defining the set of co-active regions. Several groups speculated at the time that this set of regions forms a distinct brain network, yet it might have been that each region was no more connected with this set of regions as with any others.

Initial hints of a coherent cognitive control network had been identified over a decade earlier in non-human primates (Friedman & Goldman-Rakic, 1994; Pandya, Van Hoesen, & Mesulam, 1981; Quintana & Fuster, 1999; Selemon & Goldman-Rakic, 1988). These studies used structural connectivity and multi-site electrophysiological recordings to identify a distributed network across lateral frontal, medial frontal, and parietal cortices in macaque monkeys. Most of this work had been in the context of eye movement control and associated attentional shifts (Colby & Goldberg, 1999), leaving open the possibility that this frontoparietal network was really a specific eye movement control system rather than a general cognitive control system. Further, it might have been that this frontoparietal system—if not simply an eye movement system—was specifically an attentional system rather than something more general. Thus, even if a non-human primate investigators knew of this frontoparietal network, it was unknown if an analogue also existed in humans and if it corresponded to the co-activation patterns seen in humans across diverse cognitive control demands.

Subsequent studies have used resting-state functional connectivity to determine that these frontoparietal regions are not only highly interconnected in humans, but that their high connectivity continues during rest, when control processes are at a minimum (Cole & Schneider, 2007; Dosenbach et al., 2007; Fox et al., 2005; Vincent, Kahn, Snyder, Raichle, & Buckner, 2008). This demonstrated the existence of the frontoparietal cognitive control network in humans, as well as its intrinsic (non-task-dependent) nature.

Note that very little work has been done to investigate cognitive control network connectivity using DWI (relative to resting-state functional connectivity). One such study investigated the structural connectivity organisation of the prefrontal cortex (Croxson, 2005). Another study combined fMRI with DWI, linking a prefrontal region sensitive to inhibitory control with the subthalamic nucleus (Aron, Behrens, Smith, Frank, & Poldrack, 2007). A more recent study used probabilistic tractography along with individual difference correlations with behaviour to identify differential contributions of tracts to cognitive control (Strenziok, Greenwood, Santa Cruz, Thompson, & Parasuraman, 2013). The limited number of published DWI studies of cognitive control suggests there is much to learn from more such studies.

Dissociations within the Cognitive Control Network

Parallel to (and perhaps in theoretical opposition to) the discovery of an intrinsic cognitive control network, the dominant paradigm for understanding the neural basis of cognitive control was in terms of specialisation (i.e., dissociations) among cognitive-control-related regions (Botvinick, Braver, Barch, Carter, & Cohen, 2001; Henson, 2005; Saxe, Brett, & Kanwisher, 2006). This followed from the traditional assumption that all brain regions constitute functionally specialised 'modules', such that each cognitive-control-related region should have a unique function if we could just find the right cognitive manipulation to reveal it. This was effective for some portions of the cognitive control network, but not (so far) for others. Note, however, that both of these perspectives can be subsumed into a single framework characterising the nested organisation of cognitive control networks.

One particularly important task dissociation was hypothesised between the DLPFC the ACC (cf. Chapter 10 by Kool et al. in this volume; Chapter 4 by Egner in this volume). The difficulty in demonstrating this dissociation was made apparent in a meta-analysis showing the co-activation of these two distant prefrontal regions across a wide variety of tasks

(Duncan & Owen, 2000). This is in contrast to several other studies that did find dissociations between these regions (Barch et al., 1997; Braver et al., 1997; Macdonald, Cohen, Stenger, & Carter, 2000). These studies suggested that the DLPFC is involved in both sustained and transient working memory processes, whereas the ACC is involved in transient conflict monitoring/resolution processes (Botvinick, Cohen, & Carter, 2004; Braver, Reynolds, & Donaldson, 2003). Because of their emphasis on working memory tests, we will refer to this as the 'working memory' theoretical perspective on these regions.

Soon after these distinct roles for the DLPFC the ACC were posited, however, another study came to essentially the opposite conclusion (Dosenbach et al., 2006). They used a unique activity modelling approach to separate sustained from transient task-evoked activations (a mixed blocked/event-related fMRI design; Visscher et al., 2003), finding that ACC activity was sustained (and possibly transient as well), whereas DLPFC activity was only transient across a variety of tasks. This suggested that the ACC implements sustained processing for each task set, whereas the DLPFC implements transient changes in the task set. Because of their emphasis on changes in task set, we will refer to this as the 'task set' theoretical perspective on these regions.

These two views are yet to be resolved. One important consideration is that both accounts suggest a role for the DLPFC in highly dynamic cognitive control, such as task switches or possibly working memory manipulation. This is in contrast to, for example, simple working memory maintenance, which the accounts disagree on. This still leaves disagreement regarding the role of the ACC. One possibility supported by a subsequent study is that the ACC is involved in transient attentional processes as well as anticipatory attentional processes, such that ACC activity is not truly sustained but rather 'fills the gap' between trials in anticipation for the need for attention (Cole & Schneider, 2007). This anticipatory role for the ACC is supported by studies emphasising a role for the ACC in predicting the need for cognitive control (Alexander & Brown, 2011; Brown & Braver, 2005; Shenhav, Botvinick, & Cohen, 2013; Chapter 15 by Brown in this volume; Chapter 10 by Kool et al. in this volume). This role is also consistent with evidence that ACC activity is associated with arousal/vigilance, rather than maintaining task information per se (Sadaghiani & D'Esposito, 2014). Further research is necessary to resolve these theoretical and empirical discrepancies regarding the ACC, as well as the DLPFC.

Nested Cognitive Control Networks: From One Network to Many Sub-Networks

The popular 'working memory' theoretical perspective on the DLPFC and ACC emphasises their highly interactive nature (J. Cohen, Botvinick, & Carter, 2000), suggesting they should be part of the same network. For example, a now classic model suggests that the ACC detects decision conflict and signals the DLPFC to increase cognitive control during decision making (Botvinick et al., 2001). In contrast, however, the 'task set' account suggests that the DLPFC and ACC should be part of separate networks (Dosenbach et al., 2006). Determining whether these regions are in separate networks could therefore serve to adjudicate between these competing theoretical perspectives.

The resting-state functional connectivity studies above provided evidence that the DLPFC and ACC are nested within the same large-scale network (Cole & Schneider, 2007; Vincent et al., 2008). This was based on strongly above-chance correlations between the resting-state fMRI time series of these regions. However, advocates for the 'task set' account used graph theory to find that these regions segregate into separate networks once higher thresholds are applied (Dosenbach et al., 2007). This approach places the PPC along with the DLPFC, and the AIC separately along with the ACC. Thus, there is evidence for both perspectives.

These findings can be reconciled by considering a nested organisation: Portions of the ACC and DLPFC are part of separate sub-networks (Figure 18.4c) within a single higher-order frontoparietal control network (Figure 18.4b). This reflects the strongly above-chance functional connectivity between the ACC and DLPFC (Cole & Schneider, 2007; Vincent et al., 2008), while acknowledging that each of these regions has even stronger functional connectivity with other regions (Dosenbach et al., 2007). This also appears to reflect the frequency of observed task activation-based dissociations in the literature; the DLPFC and ACC appear to be more frequently dissociated than the DLPFC and the PPC or the ACC and the anterior insula (Dosenbach et al., 2006; Yeo et al., 2015). This tentative characterisation of the co-activation patterns of the DLPFC and ACC in the literature is consistent with a large meta-analysis of the fMRI literature (Figure 18.5), which finds that a co-activation component

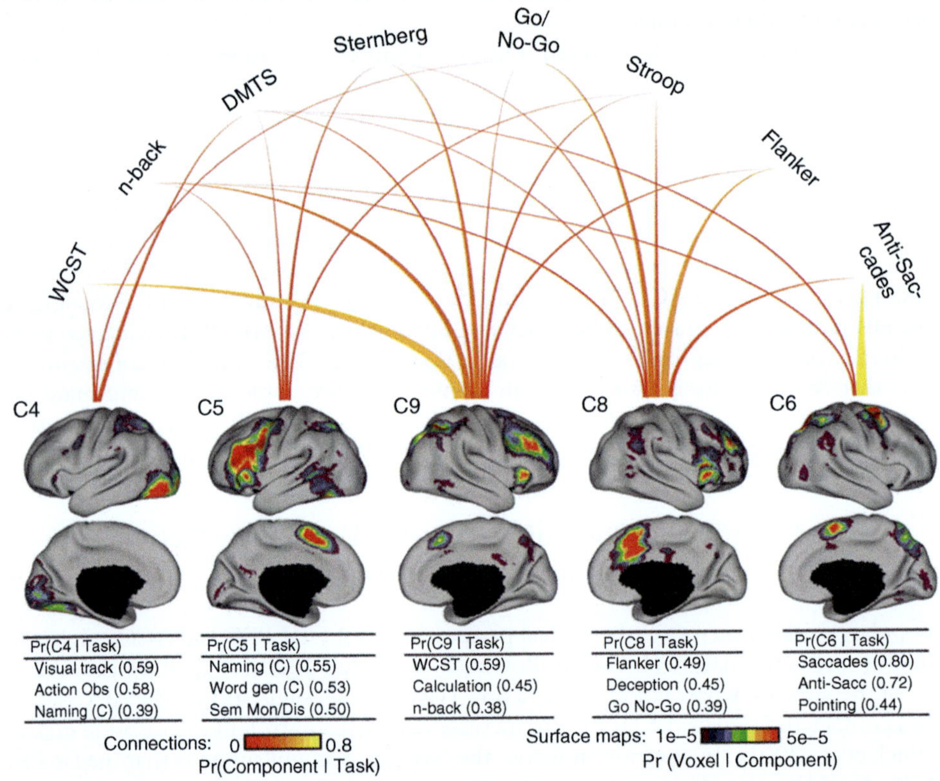

Figure 18.5 Task fMRI meta-analysis reveals co-activation pattern components related to cognitive control. A 13-component model was used to identify separate co-activation patterns across task demands (Yeo et al., 2015). Five components are shown that contributed prominently to activation patterns during cognitive control task performance. Of note, components 8 and 9 (C8 and C9) each contributed to all but one cognitive control task. C8 appears to correspond to the cingulo-opercular or salience resting-state functional connectivity network, whereas C9 appears to correspond to the fronto-parietal control network. When a 12-component model was used (rather than a 13-component model), C8 and C9 collapsed into a single component, suggesting that these components/networks have a nested organisation. The values along the top are names of cognitive control tasks. WCST = Wisconsin Card Sorting Test. DMTS = delayed match to sample. Adapted with permission from Yeo et al. (2015).

dominated by the ACC and anterior insula (component 8 in Figure 18.5) is separate from a co-activation component dominated by the DLPFC and PPC (component 9 in Figure 18.5; Yeo et al., 2015). Simultaneously, however, the ACC is present in the DLPFC component (component 9) in the same meta-analysis. Further, when a 12-component model was used instead of a 13-component model, the DLPFC and ACC components (components 8 and 9) were collapsed into a single component. Thus, the sub-network structure apparent in resting-state functional connectivity is reflected in co-activation patterns: regions in separate sub-networks are more readily dissociable, yet (reflecting their relationship in the higher-order network structure) are also often co-activated.

Identification of Cognitive Control Sub-Networks

Multiple cognitive control sub-networks have been identified (Figure 18.3c). First, a dorsal attention network (PMC and anterior PPC) and a ventral attention network (ventrolateral prefrontal cortex and temperoparietal junction) were identified using task co-activation patterns (Corbetta & Shulman, 2002) and verified by subsequent resting-state functional connectivity (Fox, Corbetta, Snyder, Vincent, & Raichle, 2006). The dorsal attention system appears to be specialised for eye movement control and goal-directed, or 'top-down', attention to external stimuli, whereas the ventral attention system appears to be specialised for stimulus-driven, or 'bottom-up', shifts in attention. Subsequently, as already mentioned, resting-state functional connectivity was used to separate the frontoparietal control network (the DLPFC and PPC) from the cingulo-opercular network (the ACC and AIC; Dosenbach et al., 2007). Simultaneously, another group dissociated these regions into essentially identical sets, but chose to label the cingulo-opercular network the 'salience' network (Seeley et al., 2007; but see below for evidence that they may be nearby separate networks).

At this point, with at least four cognitive control sub-networks, it became unclear how they all related to one another and to other networks in the brain. Further, it was unclear if each of the networks had been mapped comprehensively, or if extraneous regions had been added to one or more networks because of theoretical biases. What was necessary was a comprehensive map of all of the brain's networks at once, applying a single method uniformly to all networks to identify network affiliations throughout the brain.

As mentioned previously (see Figure 18.3), two groups of researchers took on this challenge (Power et al., 2011; Yeo et al., 2011). Power et al. (2011) applied a graph theoretical community detection algorithm to a large whole-brain resting-state functional connectivity dataset, identifying a variable but largely consistent set of brain networks across multiple thresholds (Figure 18.3c). In the same year, Yeo et al. (2011) used a similar dataset with an alternate clustering algorithm (involving projecting the data into a higher-dimensional space to facilitate segmentation into networks), identifying nearly identical networks as Power et al. (Figure 18.4c). Importantly, both studies threw out all but the very highest functional connections and applied multiple strict thresholds in order to force their algorithms to segment the data into a variety of separate networks. This suggests that it remains possible for higher-order network structures to exist (combining the identified networks at more liberal thresholds), as suggested above, to form an overarching cognitive control network.

Both the Power and Yeo studies found evidence for at least five cognitive control sub-networks. This included a dorsal attention network, a ventral attention network, a frontoparietal control network, a cingulo-opercular network and, perhaps surprisingly, a separate salience network. The cingulo-opercular and salience networks have a very similar connectivity pattern and are nearby spatially, yet they appear to be separate networks. This is especially illustrative of the importance of characterising all brain networks simultaneously: even hard-to-distinguish networks can thus be identified. Note, however, that both studies reported multiple network sets because of the need to use multiple thresholds for segmenting network affiliations. This presents a current limitation in our understanding, and leaves open the possibility of a nested organisation of cognitive control networks, as described above. Consistent with this, Yeo et al. (2011) found that several control networks collapsed between their 17-network and 7-network solutions (Figures 18.4b and 18.4c).

Beyond Networks to Network Mechanisms: Biased Competition and Guided Activation

If we are ever going to say we truly understand the neural basis of cognitive control, we will need to move beyond identifying the component parts underlying cognitive control in the brain. This means moving beyond identifying cognitive control-related brain regions and their networks. Instead, we need to map these and other constituent components to *mechanisms.*

Mechanisms are sets of causal relations explaining a phenomenon of interest. Mechanisms can be at any level of organisation, but in neuroscience they typically refer to a relationship between levels (Craver, 2007). Mechanism is inherent in our desire to map neural properties at one level to mental properties at another level. However, there are many levels between these fundamental two levels. One might begin at the ion channel level, move to synapses and axons, then entire neurons. Neurons then form local circuits such as cortical columns or cell assemblies, which form brain areas, which form sub-networks that in turn constitute entire distributed networks/systems, which then compose the entire brain. From interactions among these parts, mental phenomena emerge at yet another higher level of organisation. A mechanistic focus with the goal of eventual understanding of the neural basis of cognition should therefore focus on all of these levels and how the constituent parts produce the higher-order phenomena of interest—in this case, cognitive control.

Identifying the cognitive control network and the sub-networks that compose that network, then, is only the first step towards mechanistically understanding the neural basis of cognitive control. The ideal mechanistic explanation would involve a particular set of interactions among neural components that produce behavioural output consistent with a cognitive control process of interest.[2]

An excellent example of such a mechanism is the biased competition theory of visual attention (Desimone & Duncan, 1995; Desimone, Wessinger, Thomas, & Schneider, 1990). This theory suggests that potential attentional foci constantly compete for dominance, with both top-down and bottom-up properties influencing the eventual outcome of what is attended to. This involves distributed visual representations throughout the dorsal and ventral streams of the visual system, such that both spatial locations and objects compete for attention. Further, the theory speculates that the DLPFC is involved in biasing attentional selection based on current goals (Stokes et al., this volume). This goal-directedness is thought to be implemented via connectivity between the DLPFC and the visual system.

Simultaneous with the development of the biased competition theory, theoretical work applying connectionist modelling to the Stroop task revealed important related insights (J. D. Cohen, Dunbar, & McClelland, 1990). In this model, key nodes were able to overcome response conflict by biasing activity according to current goal-like activity patterns in those nodes. These key nodes were subsequently considered a model of the prefrontal cortex (Botvinick et al., 2001). Combining these insights with the biased competition theory led to a new theoretical account of the role of the prefrontal cortex in cognitive control called the guided activation theory (Miller & Cohen, 2001). Rather than being restricted to visual attention (or even attention at all), the account suggests that the prefrontal cortex maintains goal representations and uses its extensive connectivity to interact with a variety of task-relevant regions to bias processing towards goal-relevant outcomes (Figure 18.6a). Thus, the guided activation theory was a substantial extension of the biased competition theory. Further, this theory represents something close to an ideal in network-level neuroscience (as well as cognitive neuroscience) research: going beyond simply elucidating the regions and networks associated with cognitive phenomena to specify network interaction mechanisms underlying those phenomena.

Flexible Hubs as a Network Mechanism for Cognitive Control

With the discovery of the cognitive control network—with its strong integration between the prefrontal cortex and other distributed regions—the guided activation theory's sole focus on the prefrontal cortex may no longer be warranted. Further, recent studies have found that regions within the cognitive control network serve as 'hubs' with extensive functional (Cole, Pathak, & Schneider, 2010b; Cole & Schneider, 2007; Power et al., 2011) and structural (Modha & Singh, 2010; Van Den Heuvel & Sporns, 2011) connectivity throughout the brain. This both supports the guided activation theory and calls for further refinement of the theory with regard to how a distributed network of hub regions could implement cognitive control processes.

This line of reasoning led to the recently formulated flexible hub theory (Cole et al., 2012, 2013b; Cole, Laurent, & Stocco, 2013a; Cole, Repovs, & Anticevic, 2014; Zanto & Gazzaley, 2013). This theory posits that cognitive control network regions are hubs that shift their global functional connections depending on current task goals, biasing distributed goal-relevant processes (consistent with the guided activation theory). This allows for a balance between distributed and localised control: A network of highly integrated regions work together to implement control processes in a domain-general fashion. The domain generality of the network is made possible by its hub status, because it can shift its influences by biasing any of a wide variety of specialised brain networks.

A recent study carried out by my group provided strong evidence for this theory (Cole et al., 2013b). We tested the theory by looking at rapid functional connectivity updates across dozens of tasks, using a recently developed paradigm in which tasks are learned from instructions during fMRI scanning (Cole, Bagic, Kass, & Schneider, 2010a). We found that frontoparietal control network regions (one of the sub-networks within the cognitive control network) on average had the highest global variability coefficients among all major brain networks (Figure 18.7). The global variability coefficient is a graph theoretical measure created to test the flexible hub theory. It involves calculating the standard deviation of the strength of each functional connection for a region across tasks, followed by averaging all of that region's standard deviations—thereby summarising each region's overall tendency for global functional connectivity changes across tasks.

This supports a role for the frontoparietal control network in implementing the hypothesised global functional connectivity changes. However, it remained unclear if these changes were systematically related to the currently implemented task. We therefore used an approach related to representational similarity analysis (Kriegeskorte et al., 2008)—in which the relationship between connectivity patterns are matched to the relationship between mental representations—finding that the similarity of the frontoparietal control network's functional connections across tasks was related to the rule similarity across tasks. This suggests that

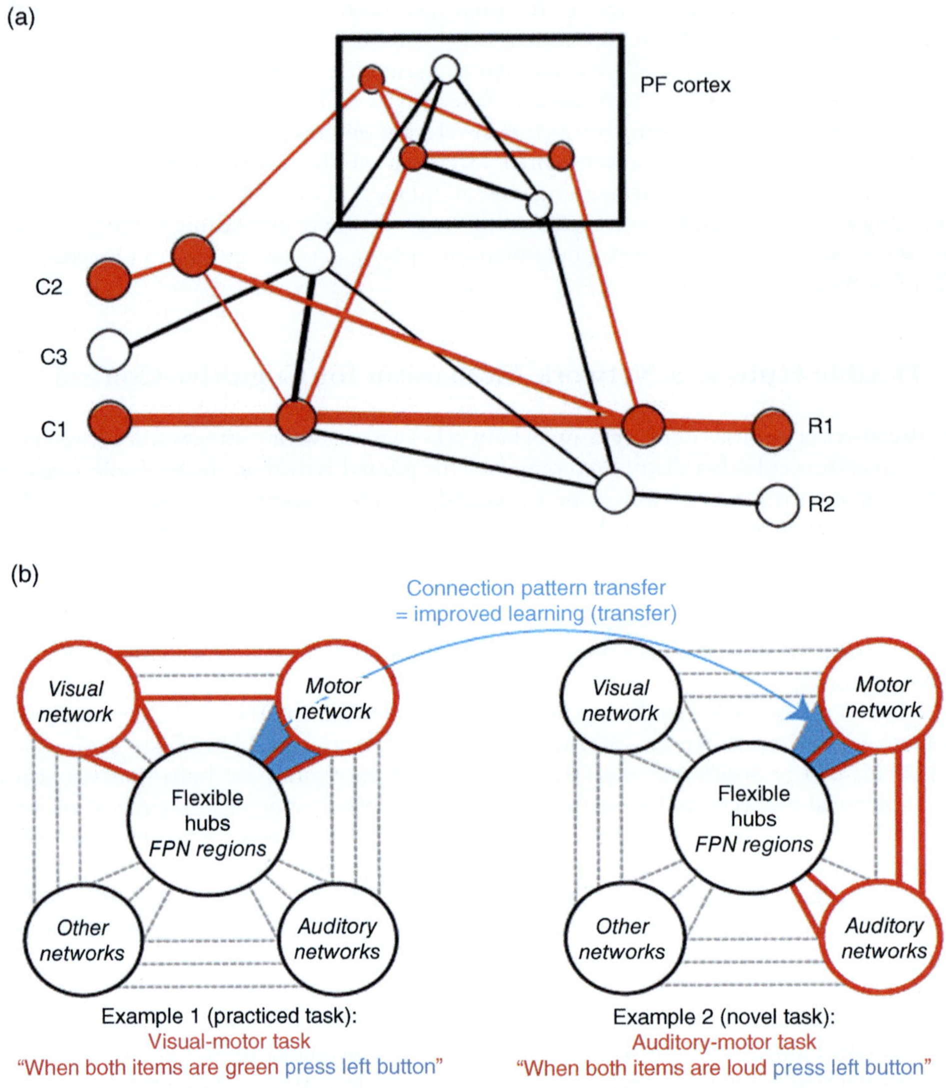

Figure 18.6 The guided activation and flexible hub theories as dynamic brain network accounts of cognitive control. a) The primary mechanisms of the guided activation theory are shown (Miller & Cohen, 2001). Activity and connectivity patterns within the prefrontal cortex are associated with large-scale influences on hidden units/neurons between sensory inputs (C1, C2, and C3) and motor outputs (R1 and R2). These enhancements are thought to specify particular behaviours required by current task demands. Figure adapted from Miller & Cohen (2001). b) A depiction of mechanisms of the flexible hub theory (Cole et al., 2013b). This theory is highly compatible with the guided activation theory, but expands functionality to the entire frontoparietal network (FPN), emphasises hub connectivity as a primary property allowing for these mechanisms, and provides a possible explanation for rapid instructed task learning via connection pattern transfer (see Cole et al., 2013b).

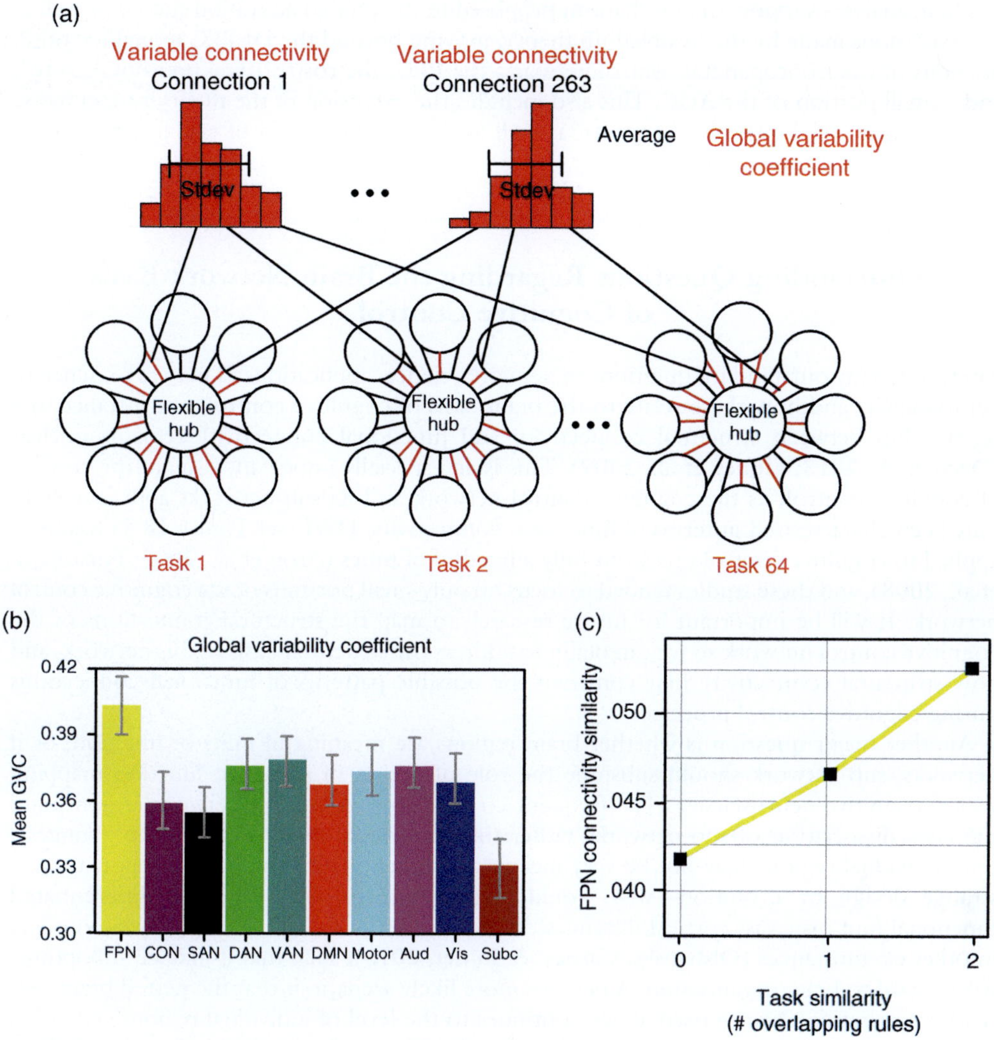

Figure 18.7 Evidence for flexible hub dynamics and task representation via the frontoparietal control network. The PRO (permuted rule operations) cognitive paradigm was used along with a new graph theoretical measure and multivariate pattern analysis (machine learning) to investigate novel task learning across 64 distinct cognitive tasks. a) A new dynamic graph theoretical measure for estimating global brain variability hypothesised to underlie flexible hub dynamics during learning (see text). b) The frontoparietal network (FPN) had the highest global variability of all major brain networks. See Figure 18.3c for brain network definitions used here. c) Multivariate pattern analysis (Kriegeskorte, Mur, & Bandettini, 2008) was used to determine that fine-grained task information (task rule identity) was contained in FPN functional connectivity patterns, consistent with the flexible hub hypothesis. Source: Adapted Cole 2013b.

functional connections between the frontoparietal control network and the rest of the brain represent the task rules composing the tested tasks. Further, we used support vector machines to successfully predict which task was being performed on the basis of the pattern of frontoparietal control network functional connections.

These analyses support the mechanism proposed in the guided activation theory as well as the extensions made by the flexible hub theory, moving beyond the DLPFC to include other portions of the frontoparietal control network: the PPC, the rostrolateral prefrontal cortex, and a small portion of the ACC. This also includes the extension of the theory to emphasise the hub status of the involved regions, as well as the representation of task information in global functional connections.

Outstanding Questions Regarding the Brain Network Basis of Cognitive Control

There are many outstanding questions in network science, applications of network science to neuroscience, and how these relate to the brain basis of cognitive control. For instance, the relationship between structural connectivity and functional connectivity remains unclear (Deco et al., 2013; Honey et al., 2009). This is an especially important issue in the context of cognitive control, as the cognitive control network and its sub-networks appear to have only been characterised in terms of functional connectivity. DWI (see Figure 18.2) has been applied to cognitive control questions only a handful of times (Aron et al., 2007; Forstmann et al., 2008), and these studies tended to focus on only small portions of the cognitive control network. It will be important for future research to map the structural connections of the cognitive control network so we can begin to address the structural basis of this network, and how structural connectivity may constrain the possible patterns of functional connections during cognitive control processes.

Another major question is whether brain regions are meaningful units of function, or if networks/sub-network should subsume the role of region in structure–function mapping within cognitive neuroscience. The frequent co-activation of control network regions, and the clear dissociation of sub-networks rather than individual regions (see above), suggests that individual regions may not be very meaningful functionally. This would appear to be a strange design by evolution—why would it distribute portions of an undifferentiated functional unit over vast cortical distances? Yet, evolution is known to be highly suboptimal in other circumstances (Olshansky, Carnes, & Butler, 2015), such that it may be suboptimal with regard to brain organisation. A perhaps more likely scenario is that the nested brain network organisation characterised above continues to the level of individual regions (i.e., each region has a unique function within a sub-network). This would imply that functional dissociations will be possible but much more difficult (and perhaps less functionally relevant) among regions within the same sub-networks.

Considering this issue with respect to the frontoparietal control network, it remains unclear if each frontoparietal region is equally involved in a wide variety of functions (domain general). Alternatively, frontoparietal regions could pool their (relatively broad) functionalities together such that the network as a whole is domain general. This would imply a nested organisation with an emergent property—domain generality—due to the coordinated interactions among regions within a network. It will be difficult to test this possibility, owing to the need to test a wide variety of functions in many individual brains. Nonetheless, some clues from large meta-analyses of fMRI activation data suggest that this may be true (Niendam et al., 2012; Yeo et al., 2015).

In other cases—such as the ACC/pre-SMA—there may be questions regarding how many functional regions comprise a portion of the cortex (Cole, Yeung, Freiwald, & Botvinick, 2009). ACC/pre-SMA is often considered to be a single region in the literature, with activations in that general portion of the cortex used to constrain inferences regarding possible functions of that region (Ridderinkhof, Nieuwenhuis, & Braver, 2007). This has led to particularly broad functional characterisations of the ACC/pre-SMA (Botvinick et al., 2001; Brown & Braver, 2005; Grinband, Hirsch, & Ferrera, 2006; Shenhav et al., 2013; Chapter 15 by Brown in this volume; Chapter 10 by Kool et al. in this volume), as well as some heated debates with relatively little resolution over time. However, cytoarchitecture (B. A. Vogt, Nimchinsky, Vogt, & Hof, 1995), receptor densities (Palomero-Gallagher, Mohlberg, Zilles, & Vogt, 2008), and other criteria (Picard & Strick, 2001; Power et al., 2011; Yeo et al., 2015) have identified a variety of distinct regions within the ACC/pre-SMA. It is especially noteworthy that the ACC/pre-SMA sits at the convergence of at least five large-scale functional networks (Power et al., 2011; Power, Schlaggar, Lessov-Schlaggar, & Petersen, 2013), with portions of each in the vicinity of what many in the literature assume to be a single region. Proper characterisation of functional subdivisions within the ACC/pre-SMA could substantially improve our understanding of the regions and the functional network associations that comprise this portion of the cortex.

Moving up a level from cognitive-control-related brain regions, the distinct and overlapping functions of the cognitive control sub-networks remain unclear. As outlined above, there is evidence for top-down control of attention via the dorsal attention network, and sensitivity to task salience in the salience network. There is also evidence for especially dynamic top-down cognitive control processes occurring in the frontoparietal network, whereas the cingulo-opercular network appears to maintain activation levels over the entire duration of a task (Dosenbach et al., 2006; Dosenbach, Fair, Cohen, Schlaggar, & Petersen, 2008). However, it remains unclear whether task information is actually being maintained by these sustained activations, or if these activations are rather maintaining arousal or sustained vigilance (Sadaghiani & D'Esposito, 2014). This is an example of the kind of questions that remain to be answered regarding the particular functions of each cognitive control sub-network.

Even after answering these kinds of questions, however, we would not have a complete theory of the neural basis of cognitive control. A set of interrelated functional mechanisms involving (in part) highly distributed network dynamics would also be necessary, which would bind the properties discovered in the process of answering these questions. Frameworks such as the guided activation theory and the flexible hub theory can act as these types of binders, though even these frameworks currently lack the detail necessary for a comprehensive explanation of cognitive control in the brain. What may be necessary is a kind of feedback loop between these types of theories and pragmatic empirically driven approaches—perhaps led by the novel data collection and network science methods described in the first section above—to answer outstanding questions. These questions would be driven in part by these theoretical frameworks (and possibly associated computational models; O'Reilly, Herd, & Pauli, 2010), while answering these questions would lead to refinement of the theoretical frameworks to incorporate that new information. Ultimately this process would help identify key mechanisms binding levels of organisation between lower-order brain processing (such as within-region dynamics) to higher-order brain processes (such as large-scale network dynamics) to cognition and behaviour.

Acknowledgements

I would like to acknowledge Ravi Mill for his assistance with preparation of this manuscript. This work was supported by the U.S. National Institutes of Health under award K99-R00 MH096801.

Notes

1 Note that while impressively comprehensive and accurate, this large-scale connectome lacks the detail of micro-connectomes characterizing neuron-level connectivity, such as is available in the aplysia sea slug and (for a subset of brain regions) the mouse.

2 Note that it may be impractical to start at certain levels of organization for explaining a given phenomenon of interest (e.g., the molecular level, or quantum mechanics level), because features often shift independently of the phenomenon of interest at certain levels (e.g., potassium molecules constantly being replaced in the prefrontal cortex).

References

Alexander, W. H., & Brown, J. W. (2011). Medial prefrontal cortex as an action-outcome predictor. *Nature Neuroscience, 14*(10), 1338–1344.

Aron, A. R., Behrens, T. E., Smith, S., Frank, M. J., & Poldrack, R. A. (2007). Triangulating a cognitive control network using diffusion-weighted magnetic resonance imaging (MRI) and functional MRI. *Journal of Neuroscience, 27*(14), 3743–3752.

Azevedo, F. A. C., Carvalho, L. R. B., Grinberg, L. T., Farfel, J. M., Ferretti, R. E. L., Leite, R. E. P.,… Herculano-Houzel, S. et al. (2009). Equal numbers of neuronal and nonneuronal cells make the human brain an isometrically scaled-up primate brain. *The Journal of Comparative Neurology, 513*(5), 532–541.

Barch, D., Braver, T., Nystrom, L., Forman, S., Noll, D., & Cohen, J. (1997). Dissociating working memory from task difficulty in human prefrontal cortex. *Neuropsychologia, 35*(10), 1373–1380.

Berlingerio, M., Coscia, M., Giannotti, F., Monreale, A., & Pedreschi, D. (2011). The pursuit of hubbiness: Analysis of hubs in large multidimensional networks. *Journal of Computational Science, 2*(3), 223–237.

Biswal, B. B., Mennes, M., Zuo, X.-N., Gohel, S., Kelly, C., Smith, S. M.,…Milham, M. P. (2010). Toward discovery science of human brain function. *Proceedings of the National Academy of Sciences of the United States of America, 107*(10), 4734–4739.

Biswal, B., Yetkin, F. Z., Haughton, V. M., & Hyde, J. S. (1995). Functional connectivity in the motor cortex of resting human brain using echo-planar MRI. *Magnetic Resonance in Medicine : Official Journal of the Society of Magnetic Resonance in Medicine/Society of Magnetic Resonance in Medicine, 34*(4), 537–541.

Botvinick, M., Braver, T., Barch, D., Carter, C., & Cohen, J. (2001). Conflict monitoring and cognitive control. *Psychological Review, 108*(3), 624–652.

Botvinick, M., Cohen, J., & Carter, C. (2004). Conflict monitoring and anterior cingulate cortex: An update. *Trends in Cognitive Sciences, 8*(12), 539–546.

Braver, T. S., Cohen, J. D., Nystrom, L. E., Jonides, J., Smith, E. E., & Noll, D. C. (1997). A parametric study of prefrontal cortex involvement in human working memory. *NeuroImage, 5*(1), 49–62.

Braver, T. S., Reynolds, J. R., & Donaldson, D. I. (2003). Neural mechanisms of transient and sustained cognitive control during task switching. *Neuron, 39*(4), 713–726.

Brown, J., & Braver, T. (2005). Learned predictions of error likelihood in the anterior cingulate cortex. *Science (New York, NY), 307*(5712), 1118–1121.

Buckner, R. L., Sepulcre, J., Talukdar, T., Krienen, F. M., Liu, H., Hedden, T.,…Johnson, K. A. (2009). Cortical hubs revealed by intrinsic functional connectivity: Mapping, assessment of stability, and relation to Alzheimer's disease. *Journal of Neuroscience, 29*(6), 1860–1873.

Bullmore, E. T., & Bassett, D. S. (2011). Brain graphs: Graphical models of the human brain connectome. *Annual Review of Clinical Psychology, 7*(1), 113–140.

Cohen, J. D., Dunbar, K., & McClelland, J. L. (1990). On the control of automatic processes: A parallel distributed processing account of the Stroop effect. *Psychological Review, 97*(3), 332–361.

Cohen, J., Botvinick, M., & Carter, C. (2000). Anterior cingulate and prefrontal cortex: Who's in control? *Nature Neuroscience, 3*(5), 421–423.

Colby, C., & Goldberg, M. (1999). Space and attention in parietal cortex. *Annual Review of Neuroscience*, *22*, 319–349.

Cole, M. W., & Schneider, W. (2007). The cognitive control network: Integrated cortical regions with dissociable functions. *NeuroImage*, *37*(1), 343–360.

Cole, M. W., Bagic, A., Kass, R., & Schneider, W. (2010a). Prefrontal dynamics underlying rapid instructed task learning reverse with practice. *Journal of Neuroscience*, *30*(42), 14245–14254.

Cole, M. W., Ito, T., & Braver, T. S. (2015). Lateral prefrontal cortex contributes to fluid intelligence through multinetwork connectivity. *Brain Connectivity*, *5*(8), 497–504.

Cole M. W., Ito, T., Bassett, D. S., & Schultz, D. H. (2016). Activity flow over resting-state networks shapes cognitive task activations, bioRxiv. *Cold Spring Harbor Labs Journals*. http://doi.org/10.1101/055194

Cole, M. W., Laurent, P., & Stocco, A. (2013a). Rapid instructed task learning: A new window into the human brain's unique capacity for flexible cognitive control. *Cognitive, Affective & Behavioral Neuroscience*, *13*(1), 1–22.

Cole, M. W., Pathak, S., & Schneider, W. (2010b). Identifying the brain's most globally connected regions. *NeuroImage*, *49*(4), 3132–3148.

Cole, M. W., Repovs, G., & Anticevic, A. (2014). The frontoparietal control system: A central role in mental health. *The Neuroscientist*, *20*(6), 652–664.

Cole, M. W., Reynolds, J. R., Power, J. D., Repovs, G., Anticevic, A., & Braver, T. S. (2013b). Multi-task connectivity reveals flexible hubs for adaptive task control. *Nature Neuroscience*, *16*(9), 1348–1355.

Cole, M. W., Yarkoni, T., Repovs, G., Anticevic, A., & Braver, T. S. (2012). Global connectivity of prefrontal cortex predicts cognitive control and intelligence. *Journal of Neuroscience*, *32*(26), 8988–8999.

Cole, M. W., Yeung, N., Freiwald, W. A., & Botvinick, M. (2009). Cingulate cortex: Diverging data from humans and monkeys. *Trends in Neurosciences*, *32*(11), 566–574.

Corbetta, M., & Shulman, G. L. (2002). Control of goal-directed and stimulus-driven attention in the brain. *Nature Reviews Neuroscience*, *3*(3), 201–215.

Craver, C. (2007). *Explaining the brain*. Oxford: Oxford University Press.

Croxson, P. L. (2005). Quantitative investigation of connections of the prefrontal cortex in the human and macaque using probabilistic diffusion tractography. *Journal of Neuroscience*, *25*(39), 8854–8866.

Deco, G., Ponce-Alvarez, A., Mantini, D., Romani, G. L., Hagmann, P., & Corbetta, M. (2013). Resting-state functional connectivity emerges from structurally and dynamically shaped slow linear fluctuations. *Journal of Neuroscience*, *33*(27), 11239–11252.

Dell'Acqua, F., & Catani, M. (2012). Structural human brain networks: Hot topics in diffusion tractography. *Current Opinion in Neurology*, *25*(4), 375–383.

Desimone, R., & Duncan, J. (1995). Neural mechanisms of selective visual attention. *Annual Review of Neuroscience*, *18*, 193–222.

Desimone, R., Wessinger, M., Thomas, L., & Schneider, W. (1990). Attentional control of visual perception: Cortical and subcortical mechanisms. *Cold Spring Harbor Symposia on Quantitative Biology*, *55*(0), 963–971.

Dosenbach, N. U. F., Fair, D. A., Cohen, A. L., Schlaggar, B. L., & Petersen, S. E. (2008). A dual-networks architecture of top-down control. *Trends in Cognitive Sciences*, *12*(3), 99–105.

Dosenbach, N. U. F., Fair, D. A., Miezin, F. M., Cohen, A. L., Wenger, K. K., Dosenbach, R. A. T.,… Petersen, S. E. (2007). Distinct brain networks for adaptive and stable task control in humans. *Proceedings of the National Academy of Sciences of the United States of America*, *104*(26), 11073–11078.

Dosenbach, N. U. F., Visscher, K. M., Palmer, E. D., Miezin, F. M., Wenger, K. K., Kang, H. C.,… Petersen, S. E. (2006). A core system for the implementation of task sets. *Neuron*, *50*(5), 799–812.

Dum, R. P., & Strick, P. L. (2013). Transneuronal tracing with neurotropic viruses reveals network macroarchitecture. *Current Opinion in Neurobiology*, *23*(2), 245–249.

Duncan, J., & Owen, A. (2000). Common regions of the human frontal lobe recruited by diverse cognitive demands. *Trends in Neurosciences*, *23*(10), 475–483.

Forstmann, B. U., Jahfari, S., Scholte, H. S., Wolfensteller, U., van den Wildenberg, W. P. M., & Ridderinkhof, K. R. (2008). Function and structure of the right inferior frontal cortex predict

individual differences in response inhibition: A model-based approach. *Journal of Neuroscience, 28*(39), 9790–9796.
Fox, M. D., Corbetta, M., Snyder, A. Z., Vincent, J. L., & Raichle, M. E. (2006). Spontaneous neuronal activity distinguishes human dorsal and ventral attention systems. *Proceedings of the National Academy of Sciences of the United States of America, 103*(26), 10046–10051.
Fox, M. D., Snyder, A. Z., Vincent, J. L., Corbetta, M., Van Essen, D. C., & Raichle, M. E. (2005). The human brain is intrinsically organized into dynamic, anticorrelated functional networks. *Proceedings of the National Academy of Sciences of the United States of America, 102*(27), 9673–9678.
Friedman, H. R., & Goldman-Rakic, P. S. (1994). Coactivation of prefrontal cortex and inferior parietal cortex in working memory tasks revealed by by 2DG functional mapping in the rhesus monkey. *Journal of Neuroscience, 14*(5 Pt. 1), 2775–2788.
Friston, K. J. (1994). Functional and effective connectivity in neuroimaging: A synthesis. *Human Brain Mapping, 2*, 56–78.
Goñi, J., Van Den Heuvel, M. P., Avena-Koenigsberger, A., Velez de Mendizabal, N., Betzel, R. F., Griffa, A.,…Sporns, O. (2014). Resting-brain functional connectivity predicted by analytic measures of network communication. *Proceedings of the National Academy of Sciences of the United States of America, 111*(2), 833–838.
Grinband, J., Hirsch, J., & Ferrera, V. (2006). A neural representation of categorization uncertainty in the human brain. *Neuron, 49*(5), 757–763.
Guimera, R., Mossa, S., Turtschi, A., & Amaral, L. A. N. (2005). The worldwide air transportation network: Anomalous centrality, community structure, and cities' global roles. *Proceedings of the National Academy of Sciences of the United States of America, 102*(22), 7794.
Henson, R. (2005). What can functional neuroimaging tell the experimental psychologist? *The Quarterly Journal of Experimental Psychology Section A, 58*(2), 193–233.
Honey, C. J., Sporns, O., Cammoun, L., Gigandet, X., Thiran, J. P., Meuli, R., & Hagmann, P. (2009). Predicting human resting-state functional connectivity from structural connectivity. *Proceedings of the National Academy of Sciences of the United States of America, 106*(6), 2035–2040.
Johansen-Berg, H., & Rushworth, M. F. S. (2009). Using diffusion imaging to study human connectional anatomy. *Annual Review of Neuroscience, 32*(1), 75–94.
Kötter, R. (2004). Online retrieval, processing, and visualization of primate connectivity data from the CoCoMac database. *Neuroinformatics, 2*(2), 127–144.
Kriegeskorte, N., Mur, M., & Bandettini, P. (2008). Representational similarity analysis—Connecting the branches of systems neuroscience. *Frontiers in Systems Neuroscience, 2*, 4.
Laird, A. R., Eickhoff, S. B., Rottschy, C., Bzdok, D., Ray, K. L., & Fox, P. T. (2013). Networks of task co-activations. *NeuroImage, 80*(C), 505–514.
Lewis, C. M., Baldassarre, A., Committeri, G., Romani, G. L., & Corbetta, M. (2009). Learning sculpts the spontaneous activity of the resting human brain. *Proceedings of the National Academy of Sciences of the United States of America, 106*(41), 17558–17563.
Macdonald, A., Cohen, J., Stenger, V., & Carter, C. (2000). Dissociating the role of the dorsolateral prefrontal and anterior cingulate cortex in cognitive control. *Science (New York, NY), 288*(5472), 1835–1838.
Miller, E., & Cohen, J. (2001). An integrative theory of prefrontal cortex function. *Annual Review of Neuroscience, 24*, 167–202.
Modha, D. S., & Singh, R. (2010). Network architecture of the long-distance pathways in the macaque brain. *Proceedings of the National Academy of Sciences of the United States of America, 107*(30), 13485–13490.
Mukherjee, P., Chung, S. W., Berman, J. I., Hess, C. P., & Henry, R. G. (2008). Diffusion tensor MR imaging and fiber tractography: Technical considerations. *AJNR American Journal of Neuroradiology, 29*(5), 843–852.
Newman, M. (2008). The physics of networks. *Physics Today, 61*(11), 33–38.
Niendam, T. A., Laird, A. R., Ray, K. L., Dean, Y. M., Glahn, D. C., & Carter, C. S. (2012). Meta-analytic evidence for a superordinate cognitive control network subserving diverse executive functions. *Cognitive, Affective & Behavioral Neuroscience, 12*(2), 241–268.
O'Reilly, R. C., Herd, S. A., & Pauli, W. M. (2010). Computational models of cognitive control. *Current Opinion in Neurobiology, 20*(2), 257–261.

Olshansky, S. J., Carnes, B. A., & Butler, R. N. (2015). If humans were built to last. *Scientific American*, *24*, 106–111.

Palomero-Gallagher, N., Mohlberg, H., Zilles, K., & Vogt, B. (2008). Cytology and receptor architecture of human anterior cingulate cortex. *The Journal of Comparative Neurology*, *508*(6), 906–926.

Pandya, D., Van Hoesen, G., & Mesulam, M. (1981). Efferent connections of the cingulate gyrus in the rhesus monkey. *Experimental Brain Research*, *42*(3–4), 319–330.

Passingham, R. E., Stephan, K. E., & Kötter, R. (2002). The anatomical basis of functional localization in the cortex. *Nature Reviews Neuroscience*, *3*(8), 606–616.

Picard, N., & Strick, P. (2001). Imaging the premotor areas. *Current Opinion in Neurobiology*, *11*(6), 663–672.

Power, J. D., Cohen, A. L., Nelson, S. M., Wig, G. S., Barnes, K. A., Church, J. A.,...Petersen, S. E. (2011). Functional network organization of the human brain. *Neuron*, *72*(4), 665–678.

Power, J. D., Schlaggar, B. L., Lessov-Schlaggar, C. N., & Petersen, S. E. (2013). Evidence for hubs in human functional brain networks. *Neuron*, *79*(4), 798–813.

Quintana, J., & Fuster, J. (1999). From perception to action: Temporal integrative functions of prefrontal and parietal neurons. *Cerebral Cortex (New York, NY: 1991)*, *9*(3), 213–221.

Raichle, M. E. (2010). Two views of brain function. *Trends in Cognitive Sciences*, *14*(4), 180–190.

Ridderinkhof, K. R., Nieuwenhuis, S., & Braver, T. S. (2007). Medial frontal cortex function: An introduction and overview. *Cognitive, Affective & Behavioral Neuroscience*, *7*(4), 261–265.

Rubinov, M., & Sporns, O. (2010). Complex network measures of brain connectivity: Uses and interpretations. *NeuroImage*, *52*(3), 1059–1069.

Sadaghiani, S., & D'Esposito, M. (2014). Functional characterization of the cingulo-opercular network in the maintenance of tonic alertness. *Cerebral Cortex*, *25*(9), 2763–2773.

Saxe, R., Brett, M., & Kanwisher, N. (2006). Divide and conquer: A defense of functional localizers. *NeuroImage*, *30*(4), 1088–1096; discussion 1097–9.

Schneider, W., & Chein, J. (2003). Controlled & automatic processing: Behavior, theory, and biological mechanisms. *Cognitive Science*, *27*(3), 525–559.

Seeley, W. W., Menon, V., Schatzberg, A. F., Keller, J., Glover, G. H., Kenna, H.,...Greicius, M. D. (2007). Dissociable intrinsic connectivity networks for salience processing and executive control. *Journal of Neuroscience*, *27*(9), 2349–2356.

Selemon, L., & Goldman-Rakic, P. (1988). Common cortical and subcortical targets of the dorsolateral prefrontal and posterior parietal cortices in the rhesus monkey: Evidence for a distributed neural network subserving spatially guided behavior. *The Journal of Neuroscience: The Official Journal of the Society for Neuroscience*, *8*(11), 4049–4068.

Shehzad, Z., Kelly, A. M. C., Reiss, P. T., Gee, D. G., Gotimer, K., Uddin, L. Q.,...Milham, M. P. (2009). The resting brain: Unconstrained yet reliable. *Cerebral Cortex (New York, NY: 1991)*, *19*(10), 2209–2229.

Shenhav, A., Botvinick, M. M., & Cohen, J. D. (2013). The expected value of control: An integrative theory of anterior cingulate cortex function. *Neuron*, *79*(2), 217–240.

Smith, S. M., Fox, P. T., Miller, K. L., Glahn, D. C., Fox, P. M., Mackay, C. E.,...Beckmann, C. F. (2009). Correspondence of the brain's functional architecture during activation and rest. *Proceedings of the National Academy of Sciences of the United States of America*, *106*(31), 13040–13045.

Sporns, O., Honey, C. J., & Kötter, R. (2007). Identification and classification of hubs in brain networks. *PLoS ONE*, *2*(10), e1049.

Sporns, O., Tononi, G., & Kötter, R. (2005). The human connectome: A structural description of the human brain. *PLoS Computational Biology*, *1*(4), e42.

Strenziok, M., Greenwood, P. M., Santa Cruz, S. A., Thompson, J. C., & Parasuraman, R. (2013). Differential contributions of dorso-ventral and rostro-caudal prefrontal white matter tracts to cognitive control in healthy older adults. *PLoS ONE*, *8*(12), e81410.

Thomas, C., Ye, F. Q., Irfanoglu, M. O., Modi, P., Saleem, K. S., Leopold, D. A., & Pierpaoli, C. (2014). Anatomical accuracy of brain connections derived from diffusion MRI tractography is inherently limited. *Proceedings of the National Academy of Sciences of the United States of America*, *111*(46), 16574–16579.

Van Den Heuvel, M. P., & Sporns, O. (2011). Rich-club organization of the human connectome. *Journal of Neuroscience, 31*(44), 15775–15786.

Van Essen, D. C., Jbabdi, S., Sotiropoulos, S. N., Chen, C., Dikranian, K., Coalson, T., Glasser, M. F. (2013). Mapping connections in humans and non-human primates: Aspirations and challenges for diffusion imaging. In *Diffusion MRI, 2nd: From quantitative measurement to in vivo neuroanatomy.* San Diego: Academic Press.

Vincent, J. L., Kahn, I., Snyder, A. Z., Raichle, M. E., & Buckner, R. L. (2008). Evidence for a fronto-parietal control system revealed by intrinsic functional connectivity. *Journal of Neurophysiology, 100*(6), 3328–3342.

Visscher, K., Miezin, F., Kelly, J., Buckner, R., Donaldson, D., McAvoy, M.,...Petersen, S. E. (2003). Mixed blocked/event-related designs separate transient and sustained activity in fMRI. *NeuroImage, 19*(4), 1694–1708.

Vogt, B. A., Nimchinsky, E. A., Vogt, L. J., & Hof, P. R. (1995). Human cingulate cortex: Surface features, flat maps, and cytoarchitecture. *The Journal of Comparative Neurology, 359*(3), 490–506.

Wager, T., & Smith, E. (2003). Neuroimaging studies of working memory: A meta-analysis. *Cognitive, Affective & Behavioral Neuroscience, 3*(4), 255–274.

Waites, A. B., Stanislavsky, A., Abbott, D. F., & Jackson, G. D. (2005). Effect of prior cognitive state on resting state networks measured with functional connectivity. *Human Brain Mapping, 24*(1), 59–68.

Watts, D. J., & Strogatz, S. H. (1998). Collective dynamics of 'small world' networks. *Nature, 393*(6684), 440–442.

Yarkoni, T., Poldrack, R. A., Nichols, T. E., Van Essen, D. C., & Wager, T. D. (2011). Large-scale automated synthesis of human functional neuroimaging data. *Nature Methods, 8*(8), 665–670.

Yendiki, A., Koldewyn, K., Kakunoori, S., Kanwisher, N., & Fischl, B. (2013). Spurious group differences due to head motion in a diffusion MRI study. *NeuroImage, 88C*, 79–90.

Yeo, B. T. T., Krienen, F. M., Eickhoff, S. B., Yaakub, S. N., Fox, P. T., Buckner, R. L.,...Chee, M. W. (2015). Functional specialization and flexibility in human association cortex. *Cerebral Cortex, 25*(10), 3654–3672.

Yeo, B. T. T., Krienen, F. M., Sepulcre, J., Sabuncu, M. R., Lashkari, D., Hollinshead, M.,...Buckner, R. L. (2011). The organization of the human cerebral cortex estimated by intrinsic functional connectivity. *Journal of Neurophysiology, 106*(3), 1125–1165.

Zanto, T. P., & Gazzaley, A. (2013). Fronto-parietal network: Flexible hub of cognitive control. *Trends in Cognitive Sciences, 17*(12), 602–603.

19

The Neurochemical Modulation of Prefrontal Control Processes

Trevor W. Robbins and Angie A. Kehagia

Introduction

Cognitive control has been defined in various ways, but includes a reference to the coordination of thoughts or actions with respect to goals or intentions (Miller & Cohen, 2001), which presumably subsumes the ability to flexibly adapt behaviour with respect to particular task demands, in order to optimise performance. This flexible control includes the capacity to inhibit actions or thoughts that are pre-potent or no longer goal directed when contingencies change (Chapter 6 by Verbruggen & Logan in this volume; Chapter 16 by Forstmann & Alkemade in this volume). In a broader context, such control includes the capacity to search strategically through memory and also, when appropriate, to exert emotional regulation (Chapter 22 by Pessoa in this volume; Chapter 30 by McTeague & Etkin in this volume). The capacities for inhibition and flexible responding are two of the prime components of what Friedman et al. (2006) have termed *executive control*, the updating of working memory being their third major component. This updating of working memory appears to be the best correlate of general intelligence or *g*, whereas the other components of cognitive control appear less directly related to this construct. Moreover, other aspects of prefrontal cortex (PFC) function that relate, for example, to decision making and aspects of social cognition, may provide yet other components of PFC function that contribute to behavioural and cognitive output, requiring additional integrative control.

These concepts and considerations are useful starting points, but fail to do justice to the complexities of executive control, primarily because they do not consider the various states under which the executive system has to operate. These states include different stages of the sleep–wake arousal cycle, fatigue, stress, fluctuations in mood and motivation, and environmental (e.g., reinforcing) feedback, while also being dependent on inherent individual variation in many of these states, which may also be trait dependent. Clues concerning the mediation of these states come from what we know about the innervation of the PFC and related structures by the ascending neurotransmitter systems, including notably the catecholamine (i.e., noradrenaline, NA; and dopamine, DA) and indoleamine (i.e., serotonin, 5-hydroxytryptamine, 5-HT) systems, as well as the acetylcholine (ACh) and less well-studied histaminergic and gamma-aminobutyric acid (GABA)-ergic ascending systems (Cooper, Bloom, & Roth, 2002). These relatively slow, 'modulatory' systems contrast with the fast signalling amino acid neurotransmitters glutamate and GABA, which comprise most of the neural networks in the cerebral cortex including the PFC, but will not be the main subject of this chapter (Table 19.1).

The Wiley Handbook of Cognitive Control, First Edition. Edited by Tobias Egner.

Table 19.1 Major Chemical Neurotransmitters and Drugs Commonly Used in Human Psychopharmacological Studies With Their Modes of Action (see Cooper et al., 2002, for more information on mechanisms).

Classical neurotransmitters	*Drug plus mode of action*
'Fast signalling'	
Glutamate (excitatory)	memantine (NMDA-R antagonist) ketamine (NMDA-R antagonist)
Gamma-aminobutyric acid (inhibitory) (GABA)	diazepam (benzodiazepine; GABA-R potentiator)
'Slow modulatory'	
Acetylcholine (ACh):	donepezil (ACh-esterase), nicotine, scopolamine (muscarinic-R antagonist); mecamylamine (nicotine-R antagonist)
Dopamine(DA)	bromocriptine (D2-R agonist) pergolide (mixed D1-D2-R agonist) sulpiride, haloperidol (D2-R antagonists) amphetamine, methylphenidate (indirect R-agonists)
Noradrenaline (NA) (norepinephrine)	amphetamine, methylphenidate (indirect R-agonists) atomoxetine (noradrenaline reuptake inhibitor, NRI) clonidine (mixed α 1-2 adrenergic (A) R agonist) guanfacine (α 2 A-R agonist) prazosin (α -1A-R- antagonist) idazoxan (α -2A-R- antagonist propranolol (ß-A R antagonist)_
Serotonin (5-hydroxytryptamine, 5-HT)	Acute tryptophan depletion (reduces central 5-HT levels) citalopram (selective serotonin reuptake inhibitor, SSRI)

The landmark study demonstrating the importance of this chemical neuromodulation was one in which the PFC monoamine systems were manipulated pharmacologically in rhesus monkeys to demonstrate a profound contribution of prefrontal DA to working memory function (Brozoski, Brown, Rosvold, & Goldman, 1979). Specifically, depletion of DA in the principal sulcus of the PFC using the locally infused neurotoxin 6-hydroxydopamine produced impairments in spatial working memory that were as profound as following removal of the tissue of this region itself. Moreover, subsequent studies showed a degree of amelioration of this deficit by systemically administered DA agonist drugs such as L-Dopa. This study clearly demonstrated that one of the major cognitive functions of the PFC, working memory, was crucially dependent on its modulation by PFC DA.

Subsequent preclinical studies have highlighted several other important principles that have emerged in the study of the neuromodulation of executive control. First, it is evident that the neuromodulation exerted by the ascending neurotransmitter systems has an optimal level; that is to say, too much as well as too little neuromodulation may lead to suboptimal performance. In other words, excessive PFC DA activity may lead to deficits in spatial working memory, just as clearly as deficient levels of PFC DA (Williams & Goldman-Rakic, 1995; Zahrt, Taylor, Mathew, & Arnsten, 1997)—although possibly for different reasons and with distinct forms of impairment. This is tantamount to the well-known inverted-U-shaped function linking cognitive performance to levels of PFC DA activity. This has been alluded to in other contexts as the Yerkes–Dodson function (although this carries with it an additional principle that will be discussed below). The Yerkes–Dodson (1908) principle was originally couched not specifically in terms of 'PFC DA function' but as a 'general arousal' or drive-like construct. In general, it has not been welcomed by human experimental psychologists studying cognitive function, who have critically identified several shortcomings, which include the difficulty of testing the function rigorously with the appropriate levels of 'arousal'. Moreover, when predictions are made of the various traditional ways of manipulating 'arousal', including stress, fatigue, drugs, and motivation, the outcomes cannot always be made to fit a single unidimensional arousal construct (see Robbins, 1986).

These criticisms are perhaps less critical from a neuroscientific perspective. First, it is relatively easy to manipulate different levels of neuromodulation, via the determination of pharmacological dose–response curves—just as one can manipulate the level of cognitive 'load', which also impinges on this function. Second, there is no reason to suppose that there is a single curvilinear function that captures all of the sources of neuromodulation that may derive from the contributions of the other monoaminergic or non-monoaminergic influences impinging on the PFC. Thus, as will become apparent, there are probably also optimal levels of NA and 5-HT function for efficient PFC function and adaptive behaviour, and these may vary somewhat independently of the function governing PFC DA, although they will obviously interact in their combined influences on PFC processing.

A further consideration is the interaction of the nature and level of cognitive processing itself with level of neuromodulation. The original Yerkes–Dodson principle (Yerkes & Dodson, 1908) provided the earliest evidence that the influence of different levels of motivation in mice escaping aversive outcomes depended essentially on the complexity of the required discrimination or cognitive load; easy discriminations were best negotiated with relatively high levels of motivation, whereas the reverse was true for more complex levels of discrimination. Easy tasks may then be performed optimally at levels of arousal that actually impair 'difficult' tasks, and vice versa. There is increasing evidence that some form of this principle may apply to the influences of the ascending neurotransmitter systems.

One caveat might be that each of the ascending neurotransmitter systems may function in at least two modes; fast phasic and slower tonic (see Rajkowski, Majczynski, Clayton, & Aston-Jones, 2004; Sarter, Lustig, Howe, Gritton, & Berry, 2014; and Schultz, Dayan, & Montague, 1997), which may have very different implications for the function and nature of neuromodulation; indeed, in discussing such findings for ACh, Sarter et al. consider that this neurotransmitter can play a *mediating* role within cortical circuitry at least as much as a modulatory influence upon it.

Given this degree of complexity, it is perhaps no surprise that each of the neuromodulatory systems is governed by a variety of intrinsic control systems; for example, the effects of a particular neurotransmitter are mediated by several different receptors that may have different affinities for the neurotransmitter and thus come into play under different circumstances. This is the case, for example, for each of the main modulatory systems of the PFC: for DA, there are D1 receptors, which outnumber D2 and D4 receptors (Lidow, Goldman-Rakic, Gallager, & Rakic, 1991); for NA, there are α-1 adrenergic (-A) and α-2-A, as well as, ß-receptors (Ramos & Arnsten, 2007; see Table 19.1). Moreover, given the complex cytoarchitectonic heterogeneity of the PFC, probably associated with equivalent complexity in terms of PFC functioning, it also appears that the different monoaminergic neurotransmitter innervations exhibit some specificity. Thus, rather than innervating the entire cortical column uniformly, such innervation may preferentially be to specific laminae, for example, deep layers targeting pyramidal output neurons (Lidow et al., 1991; Lidow, Wang, Cao, & Goldman-Rakic, 1998). Finally, there is evidence that the neuromodulation itself is under top-down control, for example, via descending neuronal tracts to the brain stem nuclei from which the ascending modulatory systems arise (Amat et al., 2005). This evidence shows that the neuromodulation of processes such as cognitive control is far from being passive, but is shaped by PFC processing itself.

Much of the evidence for these organisational principles has arisen from behavioural neuroscience studies employing rodents or non-human primates as experimental subjects, and this will become evident in our review of human studies of cognitive control. However, it is also instructive to match the conceptual complexity of human cognitive neuroscience to these principles, which in turn serve to refine them further.

Neurochemistry of Cognitive Control: Basic Neuroscience Studies

Initial Studies on the Roles of the Catecholamines in Working Memory

Following the initial breakthrough study by Brozoski et al., the main focus of further studies of DA modulation of PFC function was initially on working memory processes. Most of the focus was on the spatial delayed response paradigm, including the related delayed saccade procedure, which allows a greater degree of experimental control. Combined with the use of single-cell electrophysiology and the local application to the PFC of DA-ergic drugs via microiontophoresis, it was quickly established that DA D1 receptors were importantly implicated in the modulatory effects of PFC DA in rhesus monkeys, and that this modulatory effect did indeed appear to follow the curvilinear function consistent with the Yerkes–Dodson principle (Vijayraghavan et al., 2007; Williams & Goldman-Rakic, 1995). In rodent studies, it was similarly found that D1 agonist treatment could lead to impaired spatial delayed alternation performance, mimicking deficits also produced by stress that were both reversed by treatment with a D1 receptor antagonist, further bolstering the validity of the Yerkes–Dodson function (Zahrt et al., 1997). Furthermore, other neuromodulators were investigated; thus, at least in aged and catecholamine-depleted monkeys, systemically administered α-2- adrenergic (-A) agonists produced dose-dependent improvements in delayed response performance (Ramos & Arnsten, 2007), although the likelihood that these effects were mediated by the PFC could not be confirmed by these studies alone. However, infusion of an α-2-A antagonist directly into the monkey dorsolateral PFC did impair performance (Li & Mei, 1994). Both behavioural and electrophysiological studies of 5-HT revealed a complex possible relationship with working memory performance (Williams, Rao, & Goldman-Rakic, 2002), whereas a recent investigation has suggested that the involvement of ACh mainly manifests at the acquisition stage of this behaviour (Croxson, Kyriazis, & Baxter, 2011).

Working memory is, of course, a complex process and one likely comprising several sub-components (see, e.g., Chapter 3 by Meier & Kane in this volume). In the behavioural version of the rhesus delayed response task, in addition to 'maintaining stimuli on-line', there is an element of requiring response inhibitory control required to suppress pre-potent reaching responses as well as to resist interference from preceding trials, whereas other control processes regulate efficient attention during the encoding and delay phases. The precise role of the various neuromodulators in each of these sub-processes is not known, although there are some important clues, especially from electrophysiological studies in the delayed saccade paradigm.

For example, NA stimulation of postsynaptic α2a-A receptors strengthens directional signals within the neuronal network, whereas optimal DA stimulation of D1 receptors has been shown to weaken network connections from contra-directional inputs (i.e., reduces 'noise' within the network; Arnsten, Wang, & Paspalas, 2015). These gating actions are thought to occur in dendritic spines via the regulation of cAMP influences on hyperpolarisation-activated cyclic nucleotide gated cation (HCN) channels, which pass the h-current when in the open state. Electrophysiological studies have further shown that α2a-A receptor stimulation increases network firing for preferred directions, and this improvement can be reversed by manipulations that increase or mimic cAMP (Wang et al. 2007). Conversely, blockade of α2a-A receptors induces a breakdown of activity in the network that can be rescued by blocking HCN channels. Thus, it appears likely that DA D1 and NA α2a-A receptors contribute to working memory performance in complementary ways. In addition, in terms of PFC NA receptors, the α1 subtype exerts opposing, detrimental effects on working memory function to its α2 partner, hypothetically as a function of stress leading to excessive PFC activity that engages the relatively low-affinity α1-A receptor (Ramos & Arnsten, 2007).

Beyond Working Memory Modulation: Neurochemical Dissociations Among Executive Functions

An extension of this neuropharmacological approach has embraced neurocomputational modelling to provide crucial perspectives on the different roles of PFC D1 and D2 receptors. Although the former greatly outnumber the latter in the PFC, the D2 receptor clearly has an important and distinctive role. Durstewitz and Seamans (2008) have developed a model based on data from electrophysiological studies that suggest that the D1 receptor is especially well placed via its prominent position on dendritic spines in synapses and its interactions with the NMDA and GABA-a receptors to preserve the working memory 'trace' and to protect it from distraction. By contrast, the D2 receptor is suggested to be more influential during high levels of DA release owing to its lower affinity for the neurotransmitter, but when engaged, it enhances attentional lability, enabling switching between dominant modes of processing. The D2 receptor is primarily localised on the dendrites of pyramidal cells in layer V, including in glutamatergic projections to the striatum, and electrophysiological evidence suggests it is especially brought into play during activity immediately prior to or following the performance of the motor response (Arnsten et al., 2015). Presumably, the conjoint activity of DA at these receptors determines the characteristic inverted 'U' modulation of PFC executive function. It remains a major challenge, as we will see, to predict how DA affects cognitive control functions in humans, in health and disease. In order to meet this challenge, it may prove necessary to test the neuromodulatory roles of DA and NA in a number of cognitive test paradigms.

There are some indications that working memory and other PFC functions are not modulated in precisely the same way, reflecting either functional anatomical differences within the PFC or the possibility that different tasks may be differentially configured by neural network ensembles within a given PFC region. For example, PFC DA depletion in the marmoset monkey has differential effects on performance of a spatial delayed response task (impaired; Roberts et al., 1994), a visuospatial sequencing task (unaffected; Collins, Roberts, Dias, Everitt, & Robbins, 1998), and on a visual attentional set-shifting battery (improved, Roberts et al., 1994), despite the fact that each of these tasks is sensitive to damage to the lateral PFC. This pattern of findings is not at all inconsistent with a version of the Yerkes–Dodson principle. Further experimental analysis revealed that enhanced extra-dimensional set shifting was associated with impaired serial intra-dimensional performance, suggesting a form of attentional lability that hypothetically also contributed to distractibility during the delay phase of spatial delayed response performance (Crofts et al., 2001; Robbins & Roberts, 2007; Roberts et al., 1994).

This emphasis on control of attention is consistent with some rodent studies; for example, intra-PFC infusions of D1 agonists to poorly performing rats enhanced the accuracy of performance on a continuous performance task that requires attentional monitoring in rodents (Granon et al., 2000); moreover, such effects appeared stronger than on working memory performance per se (Chudasama & Robbins, 2004). Indeed, the indirect DA and NA agonist d-amphetamine restored attentional performance on this task following lesions of the rat medial PFC to a greater extent than working memory performance (Chudasama, Nathwani, & Robbins, 2005). Clearly, a systematic analysis of different task performance is required to understand the significance of these findings. Other rodent evidence supports this view: Floresco, Magyar, Ghods-Sharifi, Vexelman, and Tse (2006) found that intra-PFC D1 agonist failed to affect attentional set-shifting performance, whereas a D4 receptor antagonist improved it.

There may be similar dissociations of PFC NA mechanisms, as much basic neuroscience evidence supports an important role for the coeruleal-cortical noradrenergic projection in attentional selection and response control (Aston-Jones, Rajkowski, Kubiak, & Alexinsky, 1994; Rajkowski et al., 2004). This evidence includes findings of significant influences of PFC NA manipulations on attentional set shifting (Lapiz & Morilak, 2006; Tait et al., 2007), continuous (5-choice) performance (Carli et al., 1983), and response inhibitory control (Bari et al., 2011) based on effects in rodents of selective NA depletion (Carli et al., 2006; Tait et al., 2007) and adrenergic drugs such as the NA reuptake blocker atomoxetine (Bari et al., 2011) and the α2 receptor antagonist atipamezole (Lapiz & Morilak, 2006). Ma, Qi, Peng, and Li (2003) found selective deficits in go/no-go performance induced by blockade of α2a-A receptors in rhesus monkeys, but whether this occurs at doses that also impair working memory performance is unclear. Of particular relevance may be recent evidence of effects of atomoxetine in rats to speed up stop-signal reaction time performance, suggestive of specific effects on inhibitory control (Robinson et al., 2008). Such effects are also reproduced when atomoxetine is infused into two regions of the PFC, the anterior cingulate cortex and the lateral orbitofrontal cortex, but not more medial regions (Bari et al., 2011). Although this drug has been shown to block NA reuptake, it is also known to increase DA levels in the PFC. However, the beneficial effects on stop-signal response time (SSRT) were shown more likely to depend on its NA action by additional pharmacological evidence of NA but not DA involvement. This evidence of neurochemical specificity is further bolstered by evidence of a relative lack of 5-HT involvement in rodent SSRT performance (Eagle et al., 2009).

The Complexities of Multiple PFC Regions and Neuromodulators: Roles of 5-HT and ACh

Studies such as Bari et al. (2011) highlight another major dimension to be considered in the issue of neuromodulation of cognitive control mediated by the PFC, given the heterogeneity of this structure. An early, albeit controversial, distinction had been made between spatial working memory and visual working memory (i.e., for features or objects), the former requiring dorsolateral PFC and the latter ventrolateral PFC (Walker's area 12/45; Levy & Goldman-Rakic, 2000). Recent studies have focused on visuo-associative learning mediated by the PFC, highlighting comparative effects of D1 and D2 receptor antagonists (Puig & Miller, 2012, 2015). Thus, both DA antagonists affected new learning and cognitive flexibility, but only the D2 receptor antagonist affected motivation (findings not neatly fitting with the Durstewitz–Seamans model of dopaminergic modulation of the PFC). However, a more obvious functional division of the cortex is between the lateral and orbitofrontal PFC, the latter being implicated in emotional processing and the representation of value. Although considerable electrophysiological evidence supports the concept that the OFC is implicated in processing rewarding outcomes, there has been relatively little specific evidence of DA-ergic modulation at the level of the OFC, despite the well-known existence of neurons encoding reward prediction errors within the origins of the ascending DA systems at the level of the midbrain ventral tegmental area (Schultz et al., 1997). Additional evidence is provided by studies of comparative effects of DA and 5-HT depletion specifically within OFC regions of the marmoset. Monkeys with dopamine depletion were insensitive to conditioned reinforcers and displayed persistent responding in the absence of reward in extinction, a pattern of deficits that may reflect basic deficits in the associative processing of reward (Walker, Robbins, & Roberts, 2009). In contrast, marmosets with 5-HT depletion of the OFC displayed stimulus-bound responding on both tests of conditioned reinforcement and discrimination extinction, suggesting that orbitofrontal serotonin plays a specific role in preventing competing, task-irrelevant, salient stimuli from biasing responding. Thus, there is a demonstrable double dissociation of neuromodulatory effects of the monoamines within a relatively restricted region of the PFC. However, it is interesting to speculate whether this indeed reflects differential neurochemical control within the same region, or alternatively if the independent functions of two adjacent areas have been isolated, each being sensitive to a dominant neuromodulation of that discrete region. In either case, the specificity of chemical neuromodulation is apparent, raising the question of how these individual components of neuromodulation normally interact in the control of cognitive function (see Robbins & Roberts, 2007).

One evident possible consequence of the effects of profound 5-HT depletion of the OFC is to reduce 'exploration' or possible alternative contingencies during extinction of learnt discrimination performance, thus producing excessive 'exploitation' of the original contingency, expressed as maladaptive perseverative responding (Clarke, Walker, Dalley, Robbins, & Roberts, 2007). This possible deficit in processing seems likely to contribute to the robust deficits in both simple and serial visual reversal learning following OFC depletion (Clarke, Dalley, Crofts, Robbins, & Roberts, 2004; Clarke et al., 2005, 2007), which indeed appear to be perseverative in nature and are neurochemically specific in that OFC DA depletion had no comparable effect. The dramatic specificity of this involvement is further highlighted by the fact that extra-dimensional set shifting ([EDS], that is, shifting between superordinate stimulus dimensions or categories such as shape and colour) and response sequencing are completely spared by the same loss of PFC 5-HT, suggesting that its effects are largely restricted to the OFC, possibly reflecting the greater innervation of the ventral PFC by serotonergic projections (Way, Lacan, Fairbanks, & Melega, 2007). The significance of this result is further highlighted by the fact that attentional set-shifting performance is affected by manipulations of the catecholamine systems DA and NA, although the latter result has only been shown to date in rats (see Robbins & Roberts, 2007, for a review). Attentional set shifting comprises both extra-dimensional set shifting and intra-dimensional set shifting (for a discrimination within a dimension such as colour, shifting to new exemplars of colours—hence establishing a 'learning set' or more abstract rule-based discrimination rather than one applying simply to particular exemplars such as 'red' or 'green').

Thus, the neurochemical modulation of these two forms of cognitive flexibility or plasticity (i.e., attentional set shifting and reversal learning) is highly selective and respects the apparently hierarchical distinctions in processing at the level of reward-based processing of specific objects versus the 'higher-order' processing of perceptual dimensions, or abstract sets or rules, which appear to be mediated dissociably by the OFC and the lateral PFC respectively in the non-human primate brain (Dias, Robbins, & Roberts, 1996). Findings in both rats and rhesus monkeys are also consistent with the special role for OFC 5-HT in reversal learning. Recent evidence shows that individual differences in rat reversal learning performance, including those arising from stress, can be treated with sub-chronic treatment with selective serotonin reuptake inhibitors and 5-HT2 receptor agents (Barlow et al., 2015). Moreover, the effects contrast with behavioural disinhibition leading to premature responding in a rodent continuous performance task, which appear to be mediated in part by descending influences of the medial PFC, and by different 5-HT receptors (e.g., Winstanley et al., 2003). Such findings are inconsistent with a simple, global role for 5-HT in exerting top-down 'behavioural inhibition'.

These findings further suggest that there is much to be learnt about the chemical neuromodulatory systems of cognitive control mechanisms in the PFC—and analyses clearly have to be extended to other paradigms and neuromodulators. Given the likely involvement of acetylcholine in cognitive symptoms of many neurodegenerative disorders, this is an obvious further target for comparative analysis. Cortical acetylcholine is strongly implicated in attentional and working memory processes from studies in both rodents and monkeys (Everitt & Robbins, 1997). However, despite the apparent relevance of such observations to neuromodulatory roles of basal forebrain acetylcholine on PFC function, recent evidence and arguments have suggested a more deterministic role on cortical processing per se. Thus, Sarter and colleagues (2014) have pointed out that thalamocortical and corticocortical projections can evoke brief cholinergic release events, through presynaptic receptors expressed on cholinergic terminals. These ACh release events occur on a rapid, sub-second to seconds-long time scale ('transients'). In rats performing an attentional task requiring both the detection of cues and the report of non-cue events, cholinergic transients mediate the detection of cues specifically in trials involving a shift from the state of monitoring possible cues to cue-directed responding. Thus, poorly timed cholinergic transients, generated using optogenetic methods, prompt false detections in trials without cues. This evidence is consistent with the hypothesis that cholinergic transients reduce detection uncertainty. Clearly, comparative evidence of parallel functioning of the other neuromodulatory neurotransmitters is now required, using similar techniques for measuring fast and transient neurotransmitter release events in these systems in order to gauge the full significance of these findings. Possible translational implications of such results are considered further below. In achieving this translation, however, we have to bear in mind that the tools for studying these functions, including selective compounds, are not available in humans, and that the anatomical localisation of drug effects in causal terms is not feasible—all that is usually possible is the combination of correlative neuroimaging methodology, such as fMRI and positron emission tomography (PET) with behavioural observations following systemic drug treatments. In addition, the sophisticated cognitive paradigms available for measuring control generally cannot be implemented in experimental animals, although there are some apparent exceptions that are potentially informative.

Neurochemical Modulation of PFC During Cognitive Control in Humans

Human psychopharmacological studies will now be discussed in the context of clusters of experimental paradigms (Figure 19.1) with common neurochemical sensitivities. Studies in humans employ systemic drug administration rather than targeted microinfusion, and lead to widespread central effects. Hence, inferences concerning prefrontally mediated effects on cognition rely on a good understanding of the mechanism of action of the agents employed, which are rarely selective, and known distribution of receptor densities in the brain. Functional magnetic resonance imaging (fMRI) addresses blood oxygenation level dependent (BOLD) changes in a given region and offers an indirect measure of drug effects, whereas PET offers a direct and quantifiable measure of receptor occupancy. Together, these techniques offer the means to delineate and confirm locus and mechanism. Within a coherent framework of translation, placebo-controlled studies in healthy humans mirror many of the findings in experimental animals, whereas neurogenetic studies are key in accounting for individual variation in response to pharmacological manipulations. Studies of psychopharmacology in patients with neuropsychiatric disorders that impinge on prefrontal function, such as attention deficit hyperactivity disorder (ADHD), schizophrenia, depression, and Parkinson's disease, complete the final stage of translation. We consider the main neuromodulators in a sequence that

Task	Description	Figure
1. Stroop	Attentionally resolved response conflict Ink naming and colour word reading	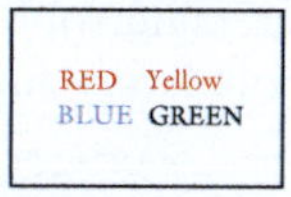
2. Probabilistic reversal learning	Behavioural flexibility in the context of probabilistic feedback Can be decomposed into processing the incentive value of stimuli, and adapting behaviour toward these consequent to changes in this value	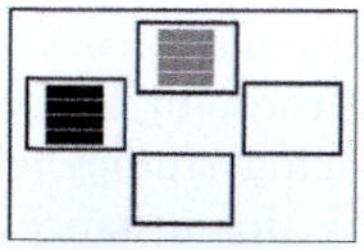
3. Decision making	Choice behaviour in a framework of probabilistic outcomes, often presented in the context of gambling tasks (CANTAB battery)	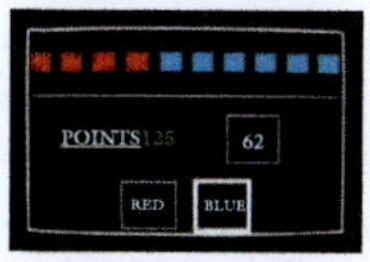
4. Task switching	Behavioural flexibility unconfounded by the processing of feedback. Switches between S-R mappings governed by rules (task sets) Concrete rules: 1:1 S-R mappings Abstract rules: categorical responses (n:1 S-R mappings)	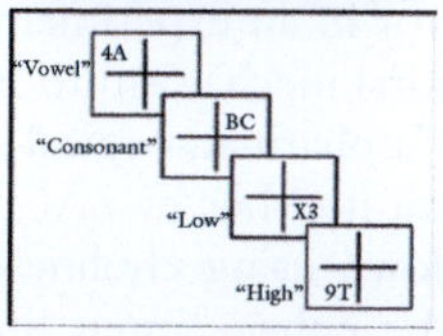
5. Wisconsin Card Sorting Test	Rule formation following feedback and flexible shifting to new rules	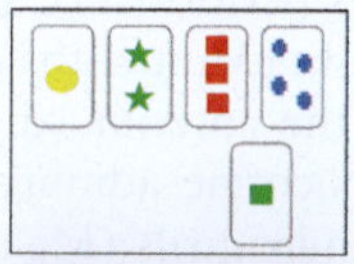
6. Tower of London	Planning and problem solving The One-Touch Stockings of Cambridge is a computer-based analog (CANTAB battery)	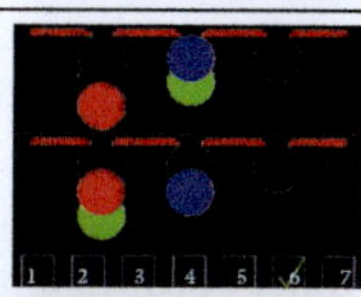
7. Rapid Visual information processing	Sustained attention, involves the presentation of infrequent targets (CANTAB battery)	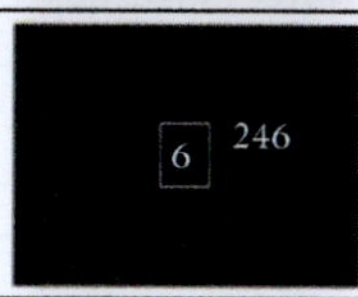
8. Stop Signal task (SST)	Response inhibition,with an attentional component (CANTAB battery)	
9. Intra and extra-dimensional set shifting	Set formation and reversal on the basis of feedback at within (ID) and between (ED) stimulus categories (CANTAB battery)	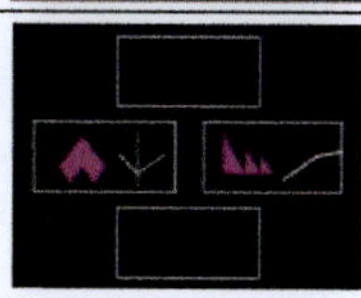

Figure 19.1 Neuropsychological tasks employed in human cognitive neuroscience and psychopharmacological studies. Figure 2 (in row 2) is adapted from Chamberlain et al., 2006. Figure 4 (in row 4) is adapted from Kehagia, Cools, Barker, & Robbins, 2009 and Figure 5 (in row 5) is adapted from Grant, & Berg, 1948. Figures 3, 6–9 (in rows 3, 6–9) are drawn from the CANTAB test battery. Reproduced with permission from Cambridge Cognition.

implies increasing sophistication of processing; whether this implies some form of hierarchy in neuromodulatory function is, however, less clear.

Cholinergic Modulation of Attention as a Necessary Condition for Control

ACh plays a major role in the regulation of attention, the sine qua non of cognitive control, primarily through its action at the ionotropic nicotinic receptor (nAChR). In humans, neuroimaging studies implicate right (R) PFC regions, especially BA9 in cue detection, employing, for instance, arterial spin labelling (ASL), which enables direct BOLD signal quantification. Increased perfusion in R BA9 was shown during a sustained attention task (Demeter, Hernandez-Garcia, Sarter, & Lustig, 2011). Evidence for genetic variation in cholinergic PFC function stems from recent studies addressing the functional polymorphism (Ile89Val) in the choline transporter (SLC5A7) gene. Compared to Ile89Val homozygotes, heterozygotic individuals with limited cholinergic capacity reported greater subjective vulnerability to distraction in everyday life and were more susceptible to distraction by a video distractor in an experimental context (Berry et al., 2014). A follow-up fMRI study located this neural mechanism to R BA9 (Berry, Blakely, Sarter, & Lustig, 2015).

Psychopharmacological manipulations targeting cholinergic function broadly aim to augment it either by nAChR agonism with agents such as nicotine, or inhibiting enzymatic breakdown using cholinesterase inhibitors (AChEIs) used clinically in dementia. Evidence from the enhancement literature addressed the effects of both of these agents on flight-simulated performance in pilots, reporting pronounced effects on the most attentionally taxing tasks (Mumenthaler et al., 2003). The prototypical attentional conflict paradigm that engages the anterior cingulate cortex is the Stroop task (Pardo, Pardo, Janer, & Raichle, 1990). Nicotine administration confers general improvements on Stroop performance in healthy individuals (Mancuso, Warburton, Melen, Sherwood, & Tirelli, 1999) and ameliorates deficits in ADHD patients (Potter & Newhouse, 2004) as well as patients with schizophrenia, known to have compromised nAChR function (Barr et al., 2008). Interactions between nicotinic receptors and glutamate are potent (Knott et al., 2011), and highly relevant to understanding phenomena such as the high incidence of smoking in individuals with schizophrenia (Hughes, Hatsukami, Mitchell, & Dahlgren, 1986) leading to nicotinic self-medication hypotheses in this group (Kumari & Postma, 2005). Given that nicotine also stimulates catecholamine release, cholinesterase inhibition offers a more selective manipulation and yields commensurately perhaps more convincing results. For example, in patients with mild Alzheimer's dementia, 12-week administration of the AChEI donepezil conferred specific improvements on the Stroop that correlated with the degree of AChE inhibition in the ACC (Bohnen et al., 2005).

Serotonin Modulates Cognitive Control in the Face of Uncertainty, Loss, and Delay

Evidence for the functional contribution of prefrontal serotonin to cognitive control comes from probabilistic reversal learning and decision-making paradigms. Probabilistic reversal learning offers a rudimentary, and thus specific, framework in which to investigate ventromedial prefrontal function (Fellows & Farah, 2003). Conversely, decision-making tasks (Bechara, Damasio, Damasio, & Anderson, 1994) are more complex, and thus less specific, but may offer greater sensitivity, and capture the common denominators of a cluster of serotonergically sensitive paradigms: the delivery of feedback (negative is salient in this context), as well as the probabilistic context that creates uncertainty regarding the outcome of a given choice, rendering choices into bets.

Early work with acute tryptophan depletion (ATD), which aims to reduce synthesis globally in the brain (Young, Smith, Pihl, & Ervin, 1985), showed significant effects on reversal learning, dependent on OFC function, with no effects on the Tower of London (ToL) test of planning (Park et al., 1994). In agreement with the animal literature (e.g., Walker et al., 2009), this established a specific role for 5-HT in maintaining a flexible response set, particularly following non-reward, which plays out over longer temporal windows compared to phasic DA (Daw, Kakade, & Dayan, 2002). ATD improved learning from negative cues (Cools, Robinson, & Sahakian, 2008; Rogers, Blackshaw, et al., 1999), but may disrupt learning from negative outcomes when these are delayed (Tanaka et al., 2009). Moreover, the selective serotonin reuptake inhibitor (SSRI) citalopram, but not the noradrenaline reuptake inhibitor (NRI) atomoxetine, impaired reversal learning in healthy individuals, presumably by perturbing the serotonergic optimum required for this task (Chamberlain et al., 2006). A double dissociation was observed in that atomoxetine but not citalopram improved response inhibition, discounting the possibility that serotonin is generally involved in behavioural inhibition. Furthermore, allelic variation in the 5-HT transporter polymorphism (5-HTTLPR) dictated subjects' ability to adapt behaviour after punishment (den Ouden et al., 2013).

Similar effects are seen on decision making assessed by gambling paradigms. Early ATD studies demonstrated general deficits in this probabilistic reward context (Rogers, Blackwell, et al., 1999). Manipulating 5-HT through tryptophan loading, the procedural obverse of ATD, shifted risky choice for large but improbable reward towards smaller negative outcomes, further supporting a role of 5-HT in loss aversion (Murphy et al., 2009) through modulation of orbitofrontal function (Rogers, Owen, et al., 1999). Decision-making studies involving the familiar 5-HTTLPR polymorphism consistently implicate serotonin in processing probability cues for risky choices. Carriers of the short allele exhibited aversion to risky choice (Kuhnen & Chiao, 2009) in an experimental context, and were less likely to invest in equities and maintain credit lines in daily life (Kuhnen, Samanez-Larkin, & Knutson, 2013).

In the last decade, neuroscientific research on social decision making has borrowed paradigms from economics, allowing the investigation of choice in the context of exchange with other agents. Paradigms such as the Ultimatum Game and Prisoner's Dilemma extend notions of reward beyond that implied by the actual financial gain in a transaction to reputation and the intrinsically rewarding effects of prosociality. In these respective paradigms, ATD increased subjects' tendency to accept unfair offers (Crockett, Clark, Tabibnia, Lieberman, & Robbins, 2008) and increased impulsive choice as well as altruistic punishment, signifying engagement in punishing unfair behaviour at, or despite, personal cost (Crockett, Clark, Lieberman, Tabibnia, & Robbins, 2010). These findings are complex in cognitive terms, but their implications for understanding social withdrawal caused by disorders such as depression are of immense clinical value given the role of SSRIs in enhancing affiliative behaviour (Knutson et al., 1998).

Dopamine at the Interface of Reward, Learning, and Flexibility

In building inferences from human psychopharmacology concerning DA-ergic function in PFC, the task of localising drug effects to this region is necessarily complicated by extensive interactions with subcortical regions mediated by parallel corticostriatal loops (see Cools & D'Esposito, 2011). However, the effects of dopaminergic agents on the PFC can be understood according to the model of Durstewitz and Seamans (2008) described above, in which D2 versus D1 dominant states underlie respectively greater and less representational perturbability.

Although the prediction error signal that underlies learning (Schultz et al., 1997; Sutton & Barto, 1998) originates in the ventral striatum, orbitofrontal regions likely integrate these signals into behavioural control at the cortical level (Gottfried, O'Doherty, & Dolan, 2003), consistent with strong reciprocal connectivity. In healthy humans, however, the D2/D3 antagonist sulpiride had no effects on learning rate even at a high dose (800 mg) while impairing discrimination performance (Eisenegger et al., 2014) suggesting a more important role for the D1 receptor in reinforcement learning. However, sulpiride increased the discriminability of reward-related BOLD activity in the OFC (Kahnt, Weber, Haker, Robbins, & Tobler, 2015), presumably by reducing its perturbability through D2 receptor blockade.

Performance on paradigms that index flexibility in the absence of reinforcement is also disrupted by D2 receptor antagonism. In a task-switching paradigm where switches between stimuli are neutrally cued, sulpiride led to switching deficits (Mehta, Manes, Magnolfi, Sahakian, & Robbins, 2004), which were reversed by the D2 agonist bromocriptine (van Holstein et al., 2011). Bromocriptine was also shown to improve flexible updating (switching) of relevant information in working memory in high-impulsive subjects, but not in low-impulsive subjects (Cools, Sheridan, Jacobs, & D'Esposito, 2007). These behavioural effects in high-impulsive subjects were accompanied by dissociable effects on frontostriatal activity. Specifically, the lateral frontal cortex was modulated by bromocriptine during distraction but not during switching. By contrast, bromocriptine modulated the striatum during switching but not during distraction from relevant information in working memory. These findings thus provided a key link between dopamine D2 receptor function, impulsivity, and frontostriatal activity during component processes of working memory.

A compelling, albeit less specific, source of evidence on the role of DA comes from dopaminergic manipulations in patients with Parkinson's disease (PD), in whom neurodegeneration leads to extensive dorsal striatal deficits that progressively encroach on ventral regions and the PFC. Overnight withdrawal from the DA precursor L-DOPA as well as DA agonists impaired task switching at the stimulus level (e.g., Cools, Barker, Sahakian, & Robbins, 2001), but not between abstract rules (Kehagia, Cools, Barker, & Robbins, 2009; resembling extra-dimensional set shifting). Notably, DA-ergic medication in PD also impaired probabilistic reversal learning (Swainson et al., 2000), reliant on ventral corticostriatal circuitry to the OFC, possibly through 'overdosing' of intact circuits at early disease stages (Gotham, Brown, & Marsden, 1988).

Dopaminergic effects are also seen on more complex tasks of rule formation and planning. On the Wisconsin Card Sorting Test, bromocriptine improved performance in patients with frontal lobe damage (McDowell, Whyte, & D'Esposito, 1998). On the ToL planning task, sulpiride impaired performance (Mehta, Sahakian, McKenna, & Robbins, 1999), whereas methylphenidate, which increases extracellular DA via DA transporter (DAT) inhibition, improved performance, both in healthy subjects (Elliott et al., 1997) and adults with ADHD (Tucha et al., 2011). Dopaminergic medication also improved ToL, as well as spatial working memory performance in PD patients (Lange et al., 1992). Interestingly, these D2-mediated effects do not extend to pergolide, a D1/D2 agonist (Kimberg & D'Esposito, 2003), which only improved working memory, suggesting possible pharmacological dissociation between dopaminergic functions of control and working memory at the receptor level.

Extensive research into risk of dopaminergic dysfunction in schizophrenia and heterogeneity of response to dopaminergic stimulation by amphetamine (Egan et al., 2001; Mattay et al., 2003) has identified the catechol-O-methyl-transferase (COMT) polymorphism, which determines the limits of catecholaminergic action, as a powerful modulator of prefrontal DA. The number of val (compared to met) alleles carried renders the transferase more efficient at limiting the action of catecholamines, an effect that overshadows that of DAT owing to its relative paucity in the cortex. Val homozygotes underperform compared to met homozygotes on the ToL, within an inverted U distribution. This nonlinear response to dopaminergic stimulation extends to understanding medication effects in PD, which are additionally difficult to parse given that early on in the disease, prefrontal DA is actually upregulated in a compensatory response to subcortical loss, but eventually also declines as the disease progresses. In this case, mildly affected, dopaminergically medicated met homozygotic patients perform worse compared to their val homozygote counterparts (Foltynie et al., 2004; Williams-Gray et al., 2009), once more reflecting deleterious DA-ergic overdosing of upregulated PFC circuitry (Figure 19.2).

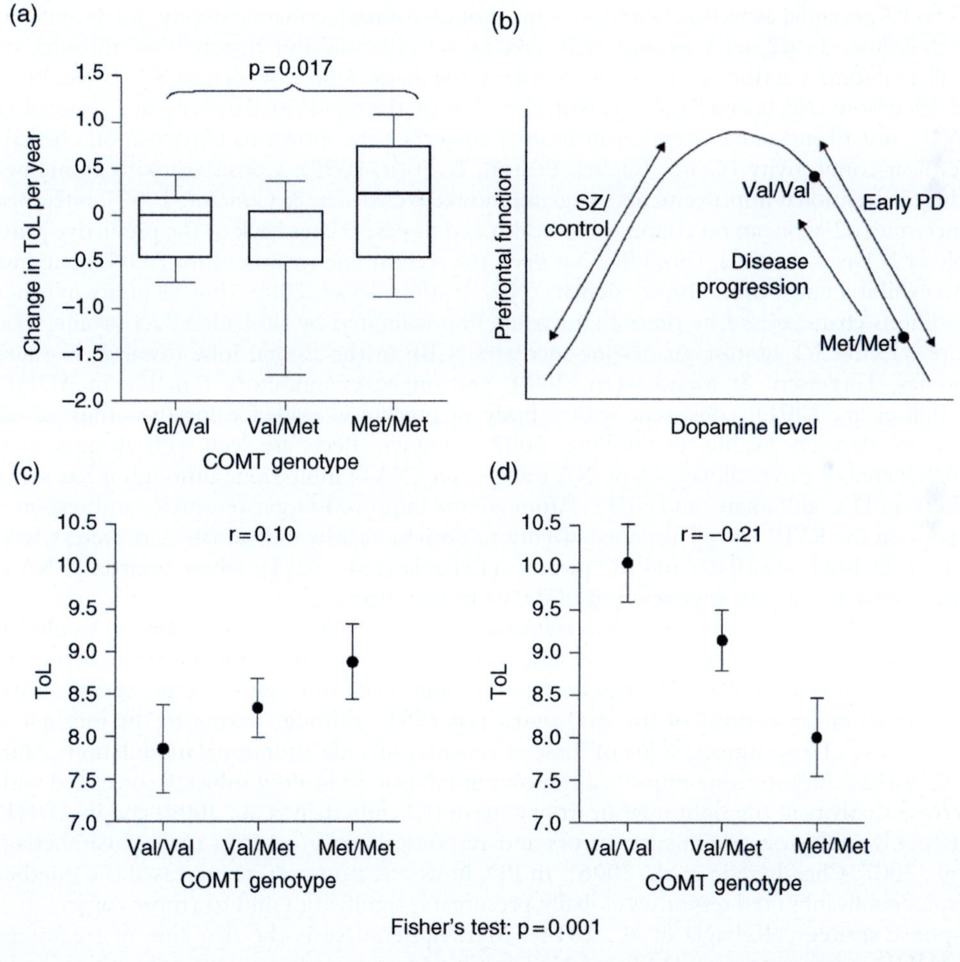

Figure 19.2 **a**. ToL performance as a function of COMT genotype of a group of 70 PD patients. **b**. Cross-sectional analysis of PD patients with relatively short (<1.6 years) and longer (>1.6 years) disease duration, and patients with schizophrenia (SZ), illustrating the inverted-U-shaped distribution. **c**. and **d**. ToL performance in respectively early and later stage PD patients stratified by COMT genotype. From Williams-Gray et al., 2009. Reproduced with permission from the publishers.

Noradrenergic Effects on Attention, Inhibition and Set Shifting

Beyond stress and arousal, cortical noradrenaline modulates sustained attention, response inhibition, and higher forms of flexibility pertaining to stimulus classes embodied in extra-dimensional set shifting. Given the variable engagement of α1 and α2 receptors, as well as COMT, predicting the cognitive effects of adrenergic stimulation will depend on receptor selectivity and on whether the manipulation effects a shift towards or away from the optimal level of NA as a function of task demands.

Sustained attention is linked to vigilance, a necessary precondition for effective control in situations where relevant targets, and consequently the required behaviour, appear infrequently. Animal studies reviewed above provide considerable evidence for a role of NA in attention as well as working memory, and these modulatory effects of the locus coeruleus system in humans can probably be measured from scalp electrodes as the P3 component of the event-related brain potential (Nieuwenhuis, Aston-Jones, & Cohen, 2005).

The P3 potential as well as behaviour can (though somewhat inconsistently) be modulated by clonidine, an α2/α1A agonist that also causes sedation and hypotension through its parallel inhibitory action on the locus coeruleus, the major source of cortical NA. Using PET, the deleterious effects of α2 agonism with clonidine on the rapid visual information processing (RVIP) test of sustained attention in healthy subjects were shown to impact upon cortico-coerulear connectivity (Coull, Buchel, Friston, & Frith, 1999). Consistent with clonidine-induced attentional improvements in ageing monkeys (Arnsten & Contant, 1992), inference concerning α2 agonism on cognition in health and disease relates back to the predictive principle of NA-ergic optima. Clonidine has been the second-line treatment for ADHD patients who exhibit a range of inhibitory deficits (Nair & Mahadevan, 2009), but its utility in other conditions characterised by prefrontal dysfunction is limited by this side-effect profile. The more selective α2 agonist guanfacine increases rCBF in the frontal lobe (Swartz, Kovalik, Thomas, Torgersen, & Mandelkern, 2000) and improves inhibitory function in ADHD (Scahill et al., 2001), consistent with a body of preclinical evidence for the utility of α2 agonism (Arnsten, Scahill, & Findling, 2007). Simpler effects are seen with atomoxetine, which increases extracellular NA by NA transporter (NAT) inhibition, although it has some effects on DA, glutamate, and 5-HT. Atomoxetine improved target sensitivity and response latency on the RVIP test of sustained attention, both in healthy individuals (Crockett, Clark, Hauser, & Robbins, 2010) and PD patients (Kehagia et al., 2014), where prefrontal NA is compromised owing to degeneration of the locus coeruleus.

In ADHD patients, however, atomoxetine leads to fewer commission errors on the RVIP task, indicating that in this group the effect signifies an improvement in response control (Chamberlain et al., 2007). The amelioration of impulsivity through NA-ergic agents is best exemplified in the context of the stop-signal task (SST) although owing to the infrequent presentation of stop signals, some of these effects may include attentional modulation (Pauls et al., 2012). Atomoxetine improved response inhibition in healthy subjects correlated with increased activity in the right inferior frontal gyrus (Chamberlain et al., 2009) and in ADHD patients by reducing commission errors and response latency on stop trials (Chamberlain et al., 2007; Chamberlain et al., 2006). In PD, however, atomoxetine increased the number of successfully inhibited responses globally, presumably signifying a shift to a more conservative response strategy (Kehagia et al., 2014). Methylphenidate is the first line of treatment in ADHD, conferring both DAT and NAT inhibition and improving response inhibition in ADHD patients (Aron, Dowson, Sahakian, & Robbins, 2003) by modulating activity in inferior PFC regions (Cubillo et al., 2014).

Prefrontal NA modulates cognitive control at higher levels of a notional 'control hierarchy', namely, the ability to flexibly reorient between classes of stimuli during extra-dimensional set shifting. Following on from demonstrations of attentional set-shifting deficits after lesions of the DNAB in experimental animals (Tait et al., 2007) and the restorative effects of NA-ergic agents, isolating receptor-specific effects in healthy humans has proved challenging. For example, the detrimental effect of clonidine on extra-dimensional shifting is not mitigated but in fact potentiated by the α2-A receptor antagonist idazoxan (Middleton, Sharma, Agouzoul, Sahakian, & Robbins, 1999), likely reflecting complexities of the pre- and post-synaptic inhibitory effects. Clear improvement on extra-dimensional shifting deficits in schizophrenic patients, as well as in sleep-deprived surgeons (Sugden, Housden, Aggarwal, Sahakian, & Darzi, 2012), are however conferred by modafinil (Turner et al., 2004), though not in healthy individuals or ADHD patients. Its wide-ranging action includes NAT and DAT inhibition, yet its noradrenergic effects appear to be especially relevant. Modafinil lowers baseline firing in the locus coeruleus, but increases it during trials that index cognitive control and enhances coeruleal connectivity with regions such as the dorsolateral PFC (Minzenberg, Watrous, Yoon, Ursu, & Carter, 2008). Moreover, the improvement of ToL performance by modafinil is antagonised by the α-1 antagonist prazosin (Winder-Rhodes et al., 2010).

Conclusions

Accumulating evidence of pharmacological modulation of cognitive control functions both in experimental animals and humans has clinical implications for the treatment of patients with control deficits, for example, in schizophrenia, ADHD, and PD, as well as revealing the essentially state-dependent nature of cognitive control. The evidence reviewed here, which is necessarily incomplete owing to limitations of space, as indicated in Table 19.2, richly indicates a degree of convergence between basic studies using experimental animals and humans, despite the obvious difficulties of measuring higher-order control functions in infra-human species. However, we have seen here clear cross-species translation of findings, often within the same research groups, not only, for example, in the role of DA in working memory, but also in that of NA in attention and inhibitory control, and 5-HT and ACh in forms of

Table 19.2 Commonalities of Neurochemical Modulation of Prefrontal Cognitive Control Across Animal and Human Studies.

	Rodent/non-human primate	*Human*
Dopamine		
	Spatial delayed response/ delayed alternation (working memory)	Tower of London planning Working memory
	Attentional set shifting,	Task-set switching
	Reversal learning	Probabilistic reversal
	Reinforcement learning	Reinforcement learning
Noradrenaline		
	Stop-signal inhibition	Stop-signal inhibition
	Continuous (5-choice) performance RVIP performance	
	Attentional set shifting	Attentional set shifting
Serotonin		
	Reversal learning	Reversal learning
Acetylcholine		
	Attentional regulation	Attentional regulation

response and attentional control, respectively. We can expect to see further examples of these dissociable influences in future investigations as well as a more complete and sophisticated specification of neurochemical mechanisms within the PFC itself.

References

Amat, J., Baratta, M. V., Paul, E., Bland, S. T., Watkins, L. R., & Maier, S. F. (2005). Medial prefrontal cortex determines how stressor controllability affects behavior and dorsal raphe nucleus. *Nature Neuroscience, 8*, 365–371.

Arnsten, A. F. T., & Contant, T. A. (1992). Alpha-2 adrenergic agonists decrease distractibility in aged monkeys performing the delayed response task. *Psychopharmacology, 108*, 159–169.

Arnsten, A. F. T., Scahill, L., & Findling, R. L. (2007). alpha2-Adrenergic receptor agonists for the treatment of attention-deficit/hyperactivity disorder: Emerging concepts from new data. *Journal of Child and Adolescent Psychopharmacology, 17*, 393–406.

Arnsten, A. F. T., Wang, M., & Paspalas, C. D. (2015). Dopamine's action in primate prefrontal cortex: Challenges for treating cognitive disorders. *Pharmacological Reviews*, in press.

Aron, A. R., Dowson, J. H., Sahakian, B. J., & Robbins, T. W. (2003). Methylphenidate improves response inhibition in adults with attention-deficit/hyperactivity disorder. *Biological Psychiatry, 54*, 1465–1468.

Aston-Jones, G., Rajkowski, J., Kubiak, P., Alexinsky, T. (1994). Locus coeruleus neurons in monkey are selectively activated by attended cues in a vigilance task. *Journal of Neuroscience, 14*, 4467–4480.

Bari, A., Mar, A. C., Theobald, D. E. H., Elands, S. A., Oganya, K. C., Eagle, D. M., & Robbins, T. W. (2011). Prefrontal and monoaminergic contributions to top-signal performance in rats. *Journal of Neuroscience, 31*, 9254–9263.

Barlow, R., Alsio, J., Jupp, B., Rabinovitch, R., Lui,G., Ho, V.,...Dalley, J. W. (2015). Markers of serotonergic function in the orbitofrontal cortex and dorsal raphé nucleus predict individual variation in spatial-discrimination serial reversal learning. *Neuropsychopharmacology, 40*(7), 1619–1630.

Barr, R. S., Culhane, M. A., Jubelt, L. E., Mufti,R. S., Dyer, M. A., Weiss, A. P.,...Evins, A. E. (2008). The effects of transdermal nicotine on cognition in nonsmokers with schizophrenia and nonpsychiatric controls. *Neuropsychopharmacology, 33*(3), 480–490.

Bechara, A., Damasio, A. R., Damasio, H., & Anderson, S. W. (1994). Insensitivity to future consequences following damage to human prefrontal cortex. *Cognition, 50*, 7–15.

Berry, A. S., Blakely, R. D., Sarter, M., & Lustig, C. (2015). Cholinergic capacity mediates prefrontal engagement during challenges to attention: Evidence from imaging genetics. *Neuroimage, 108*, 386–395.

Berry, A. S., Demeter, E,. Sabhapathy, S., English, B. A., Blakely, R. D., Sarter, M., & Lustig, C. (2014). Disposed to distraction: Genetic variation in the cholinergic system influences distractibility but not time-on-task effects. *Journal of Cognitive Neuroscience, 26*, 1981–91.

Bohnen, N. I., Kaufer, D. I., Hendrickson, R., Ivanco, L. S., Lopresti, B. J., Koeppe, R. A.,...Moore, R. Y. (2005). Degree of inhibition of cortical acetylcholinesterase activity and cognitive effects by donepezil treatment in Alzheimer's disease. *Journal of Neurology, Neurosurgery, & Psychiatry, 76*(3), 315–319.

Brozoski, T., Brown, R. M., Rosvold, H. E., & Goldman, P. S. (1979). Cognitive deficit caused by regional depletion of dopamine in prefrontal cortex of rhesus monkey. *Science, 205*, 929–931.

Carli, M., Baviera, M., Invernizzi, R. W., & Balducci, C. (2006). Dissociable contribution of 5-HT1A and 5-HT2A receptors in the medial prefrontal cortex to different aspects of executive control such as impulsivity and compulsive perseveration in rats. *Neuropsychopharmacology, 31*, 757–767.

Carli, M., Robbins, T. W., Evenden, J. L., & Everitt, B. J. (1983). Effects of lesions to ascending noradrenergic neurones on performance of a 5-choice serial reaction task in rats; implications for theories of dorsal noradrenergic bundle function based on selective attention and arousal. *Behavioural Brain Research, 9*, 361–380.

Chamberlain, S. R., Muller, U., Blackwell, A. D., Clark, L., Robbins, T. W., & Sahakian, B. J. (2006). Neurochemical modulation of response inhibition and probabilistic learning in humans. *Science, 311*, 861–863.

Chamberlain, S. R., Del Campo, N., Dowson, J., Muller, U., Clark, L., Robbins, T. W., & Sahakian, B. J. (2007). Atomoxetine improved response inhibition in adults with attention deficit/hyperactivity disorder. *Biological Psychiatry, 62*, 977–984.

Chamberlain, S. R., Hampshire, A., Muller, U., Rubia, K., Del Campo, N., Craig, K.,...Sahakian, B. J. (2009). Atomoxetine modulates right inferior frontal activation during inhibitory control: A pharmacological functional magnetic resonance imaging study. *Biological Psychiatry, 65*(7), 550–555.

Chudasama, Y., & Robbins, T. W. (2004). Dopaminergic modulation of visual attention and working memory in the rodent prefrontal cortex. *Neuropsychopharmacology, 29*, 1628–1636.

Chudasama, Y., Nathwani, F., & Robbins, T. W. (2005). D-amphetamine remediates attentional performance in rats with dorsal prefrontal lesions. *Behavioural Brain Research, 158*, 97–107.

Clarke, H. F., Dalley, J. W., Crofts, H. S., Robbins, T. W., & Roberts, A. C. (2004). Cognitive inflexibility after prefrontal serotonin depletion. *Science, 304*, 878–880.

Clarke, H. F., Walker, S. C., Crofts, H. S., Dalley, J. W., Robbins, T. W., & Roberts, A. C. (2005). Prefrontal serotonin depletion affects reversal learning but not attentional set shifting. *Journal of Neuroscience, 25*, 532–538.

Clarke, H. F., Walker, S. C., Dalley, J. W., Robbins, T. W., & Roberts, A. C. (2007). Cognitive inflexibility after prefrontal serotonin depletion is behaviourally and neurochemically specific. *Cerebral Cortex, 17*, 18–27.

Collins, P., Roberts, A. C., Dias, R., Everitt, B. J., & Robbins, T. W. (1998). Perseveration and strategy in a novel spatial self-ordered sequencing task for non-human primates: Effects of excitotoxic lesions and dopamine depletions of the prefrontal cortex. *Journal of Cognitive Neuroscience, 10*, 332–354.

Cools, R., Barker, R. A., Sahakian, B. J., & Robbins, T. W. (2001). Enhanced or impaired cognitive function in Parkinson's disease as a function of dopaminergic medication and task demands. *Cerebral Cortex, 11*, 1136–1143.

Cools, R., Sheridan, M., Jacobs, E., & D'Esposito, M. (2007). Impulsive personality predicts dopamine-dependent changes in frontostriatal activity during component processes of working memory. *Journal of Neuroscience, 27*, 5506–5514.

Cools, R,. Robinson, O. J., & Sahakian, B. (2008). Acute tryptophan depletion in healthy volunteers enhances punishment prediction but does not affect reward prediction. *Neuropsychopharmacology, 33*, 2291–2299.

Cools, R., & D'Esposito, M. (2011). Inverted U-shaped dopamine actions on human working memory and cognitive control. *Biological Psychiatry, 69*, 113–125.

Cooper, J. R., Bloom, F. E., & Roth, R. H. (2002). *The biochemical basis of neuropharmacology.* New York: Oxford University Press.

Coull, J. T., Buchel, C., Friston, K. J., & Frith, C. D. (1999). Noradrenergically mediated plasticity in a human attentional neuronal network. *Neuroimage, 10*, 705–715.

Crockett, M. J., Clark, L., Hauser, M. D., & Robbins, T. W. (2010). Serotonin selectively influences moral judgment and behavior through effects on harm aversion. *Proceedings of the National Academy of Sciences of the United States of America, 107*, 17433–17438.

Crockett, M. J., Clark, L., Lieberman, M. D., Tabibnia, G., & Robbins, T. W. (2010). Impulsive choice and altruistic punishment are correlated and increase in tandem with serotonin depletion. *Emotion, 10*, 855–862.

Crockett, M. J., Clark, L., Tabibnia, G., Lieberman, M. D., & Robbins, T. W. (2008). Serotonin modulates behavioral reactions to unfairness. *Science, 320*, 1739.

Crofts, H. S., Dalley, J. W., Collins, P., Van Denderen, J. C. M., Everitt, B. J., Robbins, T. W., & Roberts, A. C. (2001). Differential effects of 6-OHDA lesions of the prefrontal cortex and caudate nucleus on the ability to acquire an attentional set. *Cerebral Cortex, 11*, 1015–1026.

Croxson, P. L., Kyriazis, D. A., & Baxter, M. G. (2011). Cholinergic modulation of a specific memory function of prefrontal cortex. *Nature Neuroscience, 14*, 1510–1512.

Cubillo, A., Smith, A. B., Barrett, N., Giiampetro, V., Brammer, M. J., Simmons, A., & Rubia, K. (2014). Shared and drug-specific effects of atomoxetine and methylphenidate on inhibitory brain dysfunction in medication-naive ADHD boys. *Cerebral Cortex, 24*, 174–185.

Daw, N. D., Kakade, S., & Dayan, P. (2002). Opponent interactions between serotonin and dopamine. *Neural Networks, 15*, 603–616.

Demeter, E., Hernandez-Garcia, L., Sarter, M., & Lustig, C. (2011). Challenges to attention: A continuous arterial spin labeling (ASL) study of the effects of distraction on sustained attention. *Neuroimage, 54,* 1518–1529.

den Ouden, H. E., Daw, N. D., Fernandez, G., Elshout, J. A., Rijpkema, M., Hoogman, M.,...Cools, R. (2013). Dissociable effects of dopamine and serotonin on reversal learning. *Neuron, 80,* 1090–1100.

Dias, R., Robbins, T. W., & Roberts, A. C. (1996). Dissociation in prefrontal cortex of affective and attentional shifts. *Nature, 380,* 69–72.

Durstewitz D., & Seamans, J. K. (2008). The dual-state theory of prefrontal cortex dopamine function with relevance to catechol-o-methyltransferase genotypes and schizophrenia. *Biological Psychiatry, 64,* 739–749.

Eagle, D. M., Lehmann, O., Theobald, D. E. H., Peña, Y., Zakaria, R., Ghosh, R.,...Robbins, T. W. (2009). Serotonin depletion impairs waiting but not stop-signal reaction time in rats: Implications for theories of the role of the 5-HT in behavioural inhibition. *Neuropsychopharmacology, 34*(5), 1311–1321.

Egan, M. F., Goldberg, T. E., Kolachana, B. S., Callicott, J. H., Mazzanti, C. M., Straub, R. E.,... Weinberger, D. R. (2001). Effect of COMT Val108/158 Met genotype on frontal lobe function and risk for schizophrenia. *Proceedings of the National Academy of Sciences of the United States of America, 98*(12), 6917–6922.

Eisenegger, C., Naef, M., Linssen, A., Clark, L., Gandamaneni, P. K., Muller, U., & Robbins, T. W. (2014). Role of dopamine D2 receptors in human reinforcement learning. *Neuropsychopharmacology, 39,* 2366–2375.

Elliott, R., Sahakian, B. J., Matthews, K., Bannerjea, A., Rimmer, J., & Robbins, T. W. (1997). Effects of methylphenidate on spatial working memory and planning in healthy young adults. *Psychopharmacology, 131,* 196–206.

Everitt, B. J., & Robbins, T. W. (1997) Central cholinergic systems and cognition. *Annual Review of Psychology, 48,* 649–684.

Fellows, L. K., &, Farah, M. J. (2003). Ventromedial frontal cortex mediates affective shifting in humans: Evidence from a reversal learning paradigm. *Brain, 126,* 1830–1837.

Floresco, S. B., Magyar, O., Ghods-Sharifi, S., Vexelman, C., & Tse, M. T. (2006). Multiple dopamine receptor subtypes in the medial prefrontal cortex of the rat regulate set-shifting. *Neuropsychopharmacology, 31,* 297–309.

Foltynie, T., Goldberg, T. E., Lewis, S. G., Blackwell, A. D., Kolachana, B. S., Weinberger, D. R.,... Barker, R. A. (2004). Planning ability in Parkinson's disease is influenced by the COMT val158met polymorphism. *Movement Disorders, 19*(8), 885–891

Friedman, N. P., Miyake, A., Corley, R. P., Young, S. E., DeFries, J. C., & Hewitt, J. K. (2006). Not all executive functions are related to intelligence. *Psychological Science, 17,* 172–179.

Gotham, A. M., Brown, R. G., & Marsden, C. D. (1988). 'Frontal' cognitive function in patients with Parkinson's disease 'on' and 'off' levodopa. *Brain, 111*(Pt. 2), 299–321.

Gottfried, J. A., O'Doherty, J., & Dolan, R. J. (2003). Encoding predictive reward value in human amygdala and orbitofrontal cortex. *Science, 301,* 1104–1107.

Granon, S., Passetti, F., Thomas, K. L., Dalley, J. W., Everitt, B. J., & Robbins, T. W. (2000). Enhanced and impaired attentional performance after infusion of D1 dopaminergic receptor agents into rat prefrontal cortex. *Journal of Neuroscience, 20,* 1208–1215.

Grant, D. A., & Berg, E. A. (1948). A behavioral analysis of degree of reinforcement and ease of shifting to new responses in a Weigl- type card-sorting problem. *Journal of Experimental Psychology, 38,* 404–411.

Hughes, J. R., Hatsukami, D. K., Mitchell, J. E., & Dahlgren, L. A. (1986). Prevalence of smoking among psychiatric outpatients. *American Journal of Psychiatry, 143,* 993–997.

Kahnt, T., Weber, S. C., Haker, H., Robbins, T. W., & Tobler, P. N. (2015). Dopamine D2-receptor blockade enhances decoding of prefrontal signals in humans. *Journal of Neuroscience, 35,* 4104–4111.

Kehagia, A. A., Cools, R., Barker, R. A., & Robbins, T. W. (2009). Switching between abstract rules reflects disease severity but not dopaminergic status in Parkinson's disease. *Neuropsychologia, 47,* 1117–1127.

Kehagia, A. A., Housden, C. R., Regenthal, R., Barker, R. A., Muller, U., Rowe, J.,...Robbins, T. W. (2014). Targeting impulsivity in Parkinson's disease using atomoxetine. *Brain, 137*(Pt. 7), 1986–1997.

Kimberg, D. Y., & D'Esposito, M. (2003). Cognitive effects of the dopamine receptor agonist pergolide. *Neuropsychologia, 41*, 1020–1027.
Knott, V. J., Millar, A. M., McIntosh, J. F., Shah, D. K., Fisher, D. J., Blais, C. M.,...Horn, E. (2011). Separate and combined effects of low dose ketamine and nicotine on behavioural and neural correlates of sustained attention. *Biological Psychology, 88*(1), 83–93.
Knutson, B., Wolkowitz, O. M., Cole, S. W., Chan, T., Moore, E. A., Johnson, R. C.,...Reus, V. I. (1998). Selective alteration of personality and social behavior by serotonergic intervention. *American Journal of Psychiatry, 155*(3), 373–379.
Kuhnen, C. M., & Chiao, J. Y. (2009). Genetic determinants of financial risk taking. *PLoS ONE, 4*, e4362.
Kuhnen, C. M., Samanez-Larkin, G. R., & Knutson, B. (2013). Serotonergic genotypes, neuroticism, and financial choices. *PLoS ONE, 8*, e54632.
Kumari, V., & Postma, P. (2005). Nicotine use in schizophrenia: The self medication hypotheses. *Neuroscience & Biobehavioral Reviews, 29*, 1021–1034.
Lange, K. W., Robbins, T. W., Marsden, C. D., James, M., Owen, A. M., & Paul, G. M. (1992). L-dopa withdrawal in Parkinson's disease selectively impairs cognitive performance in tests sensitive to frontal lobe dysfunction. *Psychopharmacology, 107*, 394–404.
Lapiz, M. D., & Morilak, D. A. (2006). Noradrenergic modulation of cognitive function in rat medial prefrontal cortex as measured by attentional set shifting capability. *Neuroscience, 137*, 1039–1049.
Levy, R., & Goldman-Rakic, P. S. (2000). Segregation of working memory functions within the dorsolateral prefrontal cortex. *Experimental Brain Research, 133*, 23–32.
Li, B. M., & Mei, Z. T. (1994). Delayed-response deficit induced by local injection of the alpha 2-adrenergic antagonist yohimbine into the dorsolateral prefrontal cortex in young adult monkeys. *Behavioral and Neural Biology, 62*(2), 134–139.
Lidow, M. S., Goldman-Rakic, P. S., Gallager, D. W., & Rakic, P. (1991). Distribution of dopaminergic receptors in the primate cerebral cortex: Quantitative autoradiographic analysis using [3H]raclopride, [3H]spiperone, and [3H]SCH 23390. *Neuroscience, 40*, 657–671.
Lidow, M. S., Wang, F., Cao, Y., & Goldman-Rakic, P. S. (1998). Layer V pyramidal neurons bear the majority of mRNAs encoding the five distinct dopamine receptor subtypes in the primate prefrontal cortex. *Synapse, 28*, 10–20.
Ma, C. L., Qi, X. L., Peng, J. Y., & Li, B. M. (2003). Selective deficit in no-go performance induced by blockade of prefrontal cortical alpha 2-adrenoceptors in monkeys. *Neuroreport, 14*, 1013–1016.
Mancuso, G., Warburton, D. M., Melen, M., Sherwood, N., & Tirelli, E. (1999). Selective effects of nicotine on attentional processes. *Psychopharmacology, 146*(2), 199–204.
Mattay, V. S., Goldberg, T. E., Fera, F., Hariri, A. R., Tessitore, A., Egan, M. F.,...Weinberger, D. R. (2003). Catechol O-methyltransferase val158-met genotype and individual variation in the brain response to amphetamine. *Proceedings of the National Academy of Sciences of the United States of America, 100*, 6186–6191.
McDowell, S., Whyte, J., & D'Esposito, M. (1998). Differential effect of a dopaminergic agonist on prefrontal function in traumatic brain injury patients. *Brain, 121*(Pt. 6), 1155–1164.
Mehta, M. A., Manes, F. F., Magnolfi, G., Sahakian, B. J., & Robbins, T. W. (2004). Impaired set-shifting and dissociable effects on tests of spatial working memory following the dopamine D2 receptor antagonist sulpiride in human volunteers. *Psychopharmacology, 176*, 331–342.
Mehta, M. A., Sahakian, B. J., McKenna, P. J., & Robbins, T. W. (1999). Systemic sulpiride in young adult volunteers simulates the profile of cognitive deficits in Parkinson's disease. *Psychopharmacology, 146*, 162–174.
Middleton, H. C., Sharma, A., Agouzoul, D., Sahakian, B. J., & Robbins, T. W. (1999). Idazoxan potentiates rather than antagonizes some of the cognitive effects of clonidine. *Psychopharmacology, 145*, 401–411.
Miller, E. K., & Cohen, J. D. (2001). An integrative theory of prefrontal cortex function. *Annual Review of Neuroscience, 24*, 167–202.
Minzenberg, M. J., Watrous, A. J., Yoon, J. H., Ursu, S., & Carter, C. S. (2008). Modafinil shifts human locus coeruleus to low-tonic, high-phasic activity during functional MRI. *Science, 322*, 1700–1702.
Mumenthaler, M. S., Yesavage, J. A., Taylor, J. L., O'Hara, R., Friedman, L., Lee, H., & Kraemer, H. C. (2003). Psychoactive drugs and pilot performance: A comparison of nicotine, donepezil, and alcohol effects. *Neuropsychopharmacology, 28*, 1366–1373.

Murphy, S. E., Longhitano, C., Ayres, R. E., Cowen, P. J., Harmer, C. J., & Rogers, R. D. (2009). The role of serotonin in nonnormative risky choice: The effects of tryptophan supplements on the 'reflection effect' in healthy adult volunteers. *Journal of Cognitive Neuroscience*, *21*, 1709–1719.

Nair, V., & Mahadevan, S. (2009). Randomised controlled study-efficacy of clonidine versus carbamazepine in children with ADHD. *Journal of Tropical Pediatrics*, *55*, 116–121.

Nieuwenhuis, S., Aston-Jones, G., & Cohen, J. D. (2005). Decision-making, the P3 and the locus-coeruleus-norepinephrine system. *Psychological Bulletin*, *131*, 510–532.

Pardo, J. V., Pardo, P. J., Janer, K. W., & Raichle, M. E. (1990). The anterior cingulate cortex mediates processing selection in the Stroop attentional conflict paradigm. *Proceedings of the National Academy of Sciences of the United States of America*, *87*, 256–259.

Park, S. B., Coull, J. T., McShane, R. H., Young, A. H., Sahakian, B. J., Robbins, T. W., & Cowen, P. J. (1994). Tryptophan depletion in normal volunteers produces selective impairments in learning and memory. *Neuropharmacology*, *33*, 575–588.

Pauls, A. M., O'Daly, O. G., Rubia, K., Riedel, W. J., Williams, S. C., & Mehta, M. A. (2012). Methylphenidate effects on prefrontal functioning during attentional-capture and response inhibition. *Biological Psychiatry*, *72*, 142–149.

Potter, A. S., & Newhouse, P. A. (2004). Effects of acute nicotine administration on behavioral inhibition in adolescents with attention-deficit/hyperactivity disorder. *Psychopharmacology*, *176*, 182–194.

Puig, M. V., & Miller, E. K. (2012). The role of prefrontal dopamine D1 receptors in the neural mechanisms of associative memory. *Neuron*, *74*, 874–886.

Puig, M. V., & Miller, E. K. (2015). Neural substrates of dopamine D2 receptor modulated executive functions in the monkey prefrontal cortex. *Cerebral Cortex*, *25*(9), 2980–2987.

Rajkowski, J., Majczynski, H., Clayton, E., & Aston-Jones, G. (2004). Activation of monkey locus coeruleus neurons varies with difficulty and performance in a target detection task. *Journal of Neurophysiology*, *92*, 361–371.

Ramos, B. P., & Arnsten, A. F. T. (2007). Adrenergic pharmacology and cognition: Focus on the prefrontal cortex. *Pharmacology and Therapeutics*, *113*, 523–536.

Robbins, T. W. (1986). Psychopharmacological and neurobiological aspects of energetics of information processing. In R. Hockey, A. Gaillard, & M. Cole, (Eds.), *Energetics and human information processing* (pp. 71–90). Dordrecht, The Netherlands: Martinus Nijhoff.

Robbins, T. W., & Roberts, A. C. (2007). Differential regulation of fronto-executive function by the monoamines and acetylcholine. *Cerebral Cortex*, *17*, i151–i160.

Roberts, A. C., De Salvia, M. A., Wilkinson, L. S., Collins, P., Muir, J. L.,...& Robbins, T. W. (1994). 6-hydroxydopamine lesions of the prefrontal cortex in monkeys enhance performance on an analog of the Wisconsin Card Sort Test: Possible interactions with subcortical dopamine. *Journal of Neuroscience*, *14*(5 Pt. 1), 2531–2544.

Robinson, E. S. J., Eagle, D. M., Mar, A. C., Bari, A., Banerjee, G., Jiang, X.,...Robbins, T. W. (2008). Similar effects of the selective noradrenaline reuptake inhibitor atomoxetine on three distinct forms of impulsivity in the rat. *Neuropsychopharmacology*, *33*(5), 1028–1037.

Rogers, R. D., Blackshaw, A. J., Middleton, H. C., Matthews, K., Hawtin, K., Crowley, C.,...Robbins, T. W. (1999). Tryptophan depletion impairs stimulus-reward learning while methylphenidate disrupts attentional control in healthy young adults: Implications for the monoaminergic basis of impulsive behaviour. *Psychopharmacology (Berl)*, *146*(4), 482–491.

Rogers, R. D., Owen, A. M., Middleton, H. C., Williams, E. J., Pickard, J. D., & Robbins, T. W. (1999). Choosing between small, likely rewards and large, unlikely rewards activates inferior and orbital prefrontal cortex. *Journal of Neuroscience*, *19*(20), 9029–9038.

Sarter, M., Lustig, C., Howe, W. M., Gritton, H., & Berry, A. S. (2014). Deterministic functions of cortical acetylcholine. *European Journal of Neuroscience*, *39*, 1912–20

Scahill, L., Chappell, P. B., Kim, Y. S., Schultz, R. T., Katsovich, L., Shepherd, E.,...Leckman, J. F. (2001). A placebo-controlled study of guanfacine in the treatment of children with tic disorders and attention deficit hyperactivity disorder. *American Journal of Psychiatry*, *158*(7), 1067–1074.

Schultz, W., Dayan, P., & Montague, P. R. (1997). A neural substrate of prediction and reward. *Science*, *275*, 1593–1599.

Sugden, C., Housden, C. R., Aggarwal, R., Sahakian, B. J., & Darzi, A. (2012). Effect of pharmacological enhancement on the cognitive and clinical psychomotor performance of sleep-deprived doctors: A randomized controlled trial. *Annals of Surgery, 255*, 222–227.

Sutton, R. S., & Barto, A. G. (1998). *Reinforcement learning: An introduction*. Cambridge, MA: MIT Press.

Swainson, R., Rogers, R. D., Sahakian, B. J., Summers, B. A., Polkey, C. E., & Robbins, T. W. (2000). Probabilistic learning and reversal deficits in patients with Parkinson's disease or frontal or temporal lobe lesions: Possible adverse effects of dopaminergic medication. *Neuropsychologia, 38*, 596–612.

Swartz, B. E., Kovalik, E., Thomas, K., Torgersen, D., & Mandelkern, M. A. (2000). The effects of an alpha-2 adrenergic agonist, guanfacine, on rCBF in human cortex in normal controls and subjects with focal epilepsy. *Neuropsychopharmacology, 23*, 263–275.

Tait, D. S., Brown, V. J., Farovik, A., Theobald, D. E. H., Dalley, J. W., & Robbins, T. W. (2007). Lesions of the dorsal noradrenergic bundle impair attentional set-shifting in the rat. *European Journal of Neuroscience, 25*, 3719–3724.

Tanaka, S. C., Shishida, K., Schweighofer, N., Okamoto, Y., Yamawaki, S., & Doya, K. (2009). Serotonin affects association of aversive outcomes to past actions. *Journal of Neuroscience, 29*, 15669–15674.

Tucha, L., Tucha, O., Sontag, T. A., Stasik, D., Laufkotter, R., & Lange, K. W. (2011). Differential effects of methylphenidate on problem solving in adults with ADHD. *Journal of Attention Disorders, 15*, 161–173.

Turner, D. C., Clark, L., Pomarol-Clotet, E., McKenna, P., Robbins, T. W., & Sahakian, B. J. (2004). Modafinil improves cognition and attentional set shifting in patients with chronic schizophrenia. *Neuropsychopharmacology, 29*, 1363–1373.

van Holstein, M., Aarts, E., van der Schaaf, M. E., Geurts, D. E., Verkes, R. J., Franke, B.,...Cools, R. (2011). Human cognitive flexibility depends on dopamine D2 receptor signaling. *Psychopharmacology, 218*(3), 567–578.

Vijayraghavan, S., Wang, M., Birnbaum, S. G., Bruce, C. J., Williams, G. V., & Arnsten, A. F. T. (2007). Inverted-U dopamine D1 receptor actions on prefrontal neurons engaged in working memory. *Nature Neuroscience, 10*, 376–384.

Walker, S. C., Robbins, T. W., & Roberts, A. C. (2009). Differential contributions of dopamine and serotonin to orbitofrontal cortex function in the marmoset. *Cerebral Cortex, 19*, 889–898.

Wang, M., Ramos, B., Paspalas, C., Shu, Y., Simen, A., Duque, A.,...Arnsten, A. F. T. (2007). Alpha2A-adrenoceptor stimulation strengthens working memory networks by inhibiting cAMP-HCN channel signaling in prefrontal cortex. *Cell, 129*(2), 397–410.

Way, B. M., Lacan, G., Fairbanks, L. A., & Melega, W. P. (2007). Architectonic distribution of the serotonin transporter within the orbitofrontal cortex. *Neuroscience, 148*, 937–948.

Williams, G. V., & Goldman-Rakic, P. S. (1995). Blockade of dopamine D1 receptors enhances memory fields of prefrontal neurons in primate cerebral cortex. *Nature, 376*, 572–575.

Williams, G. V., Rao, S. G., & Goldman-Rakic, P. S. (2002). The physiological role of 5-HT2A receptors in working memory. *Journal of Neuroscience, 22*, 2843–2854.

Williams-Gray, C. H., Evans, J. R., Goris, A., Foltynie, T., Ban, M., Robbins, T. W.,...Barker, R. A. (2009). The distinct cognitive syndromes of Parkinson's disease: 5 year follow-up of the CamPaIGN cohort. *Brain, 132*(Pt. 11), 2958–2969.

Winder-Rhodes, S. E., Chamberlain, S. R., Idris, M. I., Robbins, T. W., Sahakian, B. J., & Muller, U. (2010). Effects of modafinil and prazosin on cognitive and physiological functions in healthy volunteers. *Journal of Psychopharmacology, 24*, 1649–1657.

Winstanley, C. A., Chudasama, Y., Dalley, J. W., Theobold, D. E., Glennon, J. C., & Robbins, T. W. (2003). Intra-prefrontal 8-OH-DPAT and M100907 improve visuospatial attention and decrease impulsivity on the five-choice serial reaction time task in rats. *Psychopharmacology, 167*, 304–314.

Young, S. N., Smith, S. E., Pihl, R. O., & Ervin, F. R. (1985). Tryptophan depletion causes a rapid lowering of mood in normal males. *Psychopharmacology, 87*, 173–177.

Yerkes, R. M., & Dodson, J. D. (1908). The relation of strength of stimulus to rapidity of habit formation. *Journal of Comparative Neurology and Psychology, 18*, 459–482.

Zahrt, J., Taylor, J. R., Mathew, R. G., & Arnsten, A. F. (1997). Supranormal stimulation of D1 dopamine receptors in the rodent prefrontal cortex impairs spatial working memory performance. *Journal of Neuroscience, 17*, 8528–8535.

Section III

Cognitive Control in Context

Interactions with Other Cognitive and Affective Domains

20

Cognitive Control in Memory Encoding and Retrieval

Nicole M. Dudukovic and Brice A. Kuhl

Introduction

A critical feature of human memory is that it can be utilised in a goal-directed manner. Aligning our memories with our goals requires cognitive control, both during the formation of memories as well as during subsequent retrieval of memories. At any given moment, our environment contains more information than we can process and, accordingly, more information than we can encode into memory (Chun, Golomb, & Turk-Browne, 2011). Cognitive control allows for goal-relevant information to be selectively encoded while distracting or irrelevant information is ignored. Cognitive control is similarly essential for goal-directed retrieval of memories. Just as we are constrained in terms of the amount of information in the external environment that we can encode, we are also constrained in terms of the amount of information we can—or want to—retrieve at one time. Cognitive control of memory retrieval enables selective processing of, and strategic access to, those memories that are relevant to current goals and, in some cases, deliberately stopping the retrieval process to keep unwanted memories out of mind. In this chapter, we consider how cognitive control mechanisms permit the selective encoding and retrieval of information from long-term memory.

Cognitive Control and Encoding

Given the virtually nonstop stream of new information in our environment, what determines which information we ultimately remember and how well we remember it? First, we will discuss encoding processes that contribute to successful remembering and examine the role that cognitive control plays in these processes. Then, we consider how cognitive control mechanisms allow us to selectively encode goal-relevant information and filter out goal-irrelevant information.

Successful Encoding

What neural mechanisms contribute to the effective encoding of new events or information? It has long been known that the medial temporal lobe (MTL) is critical for encoding conscious memories for facts and events, and that MTL damage produces an amnesic condition (e.g., Squire, 1992). However, clinicians have also observed memory deficits in humans with lesions to the prefrontal cortex (PFC) that are qualitatively different from the deficits associated with MTL damage (e.g., Shimamura, 1995). Lateral PFC damage produces

The Wiley Handbook of Cognitive Control, First Edition. Edited by Tobias Egner.

20

Cognitive Control in Memory Encoding and Retrieval

Nicole M. Dudukovic and Brice A. Kuhl

Introduction

A critical feature of human memory is that it can be utilised in a goal-directed manner. Aligning our memories with our goals requires cognitive control, both during the formation of memories as well as during subsequent retrieval of memories. At any given moment, our environment contains more information than we can process and, accordingly, more information than we can encode into memory (Chun, Golomb, & Turk-Browne, 2011). Cognitive control allows for goal-relevant information to be selectively encoded while distracting or irrelevant information is ignored. Cognitive control is similarly essential for goal-directed retrieval of memories. Just as we are constrained in terms of the amount of information in the external environment that we can encode, we are also constrained in terms of the amount of information we can—*or want to*—retrieve at one time. Cognitive control of memory retrieval enables selective processing of, and strategic access to, those memories that are relevant to current goals and, in some cases, deliberately stopping the retrieval process to keep unwanted memories out of mind. In this chapter, we consider how cognitive control mechanisms permit the selective encoding and retrieval of information from long-term memory.

Cognitive Control and Encoding

Given the virtually nonstop stream of new information in our environment, what determines which information we ultimately remember and how well we remember it? First, we will discuss encoding processes that contribute to successful remembering and examine the role that cognitive control plays in those processes. Then, we consider how cognitive control mechanisms allow us to selectively encode goal-relevant information and filter out goal-irrelevant information.

Successful Encoding

What neural mechanisms contribute to the effective encoding of new events or information? It has long been known that the medial temporal lobe (MTL) is critical for encoding conscious memories for facts and events and that MTL damage produces an amnesic condition (e.g., Squire, 1992). However, clinicians have also observed memory deficits in humans with lesions to the prefrontal cortex (PFC) that are qualitatively different from the deficits associated with MTL damage (e.g., Shimamura, 1995). Lateral PFC damage produces

---

The Wiley Handbook of Cognitive Control, First Edition. Edited by Tobias Egner.

impairments in memory tasks that involve the strategic control of memory, such as remembering the temporal order of events (e.g., Milner, Petrides, & Smith, 1985), attributing an item to a particular learning context (e.g., Janowsky, Shimamura, & Squire, 1989), and learning overlapping associations (e.g., Shimamura, Jurica, Mangels, Gershberg, & Knight, 1995). Although these impairments may be due to deficits in strategic control that partly occur during memory retrieval, it has become increasingly clear that the lateral PFC plays an important role during encoding. Specifically, lateral PFC control processes interact with the MTL to shape mnemonic processing on the basis of current goals or task requirements (Simons & Spiers, 2003). For example, PFC control mechanisms direct attention to goal-relevant stimulus features and bias the processing and maintenance of goal-relevant representations in the posterior cortex (Desimone & Duncan, 1995; Miller & Cohen, 2001). These representations may then act as inputs to the MTL memory system (Wagner, 2002).

A wealth of evidence supporting the role of PFC processes at encoding comes from studies using the *subsequent memory paradigm*. In this paradigm, neuroimaging data are collected during the encoding of events, such as pictures or words. These neural data are then sorted on the basis of whether an event is subsequently remembered or forgotten (Brewer, Zhao, Desmond, Glover, & Gabrieli, 1998; Wagner et al., 1998). Positive subsequent memory effects, or greater encoding activation for events that are later remembered as opposed to forgotten, have consistently been observed in the lateral PFC, along with medial temporal and posterior cortical regions (Kim, 2011; Spaniol et al., 2009). The specific neural regions that are activated during successful encoding vary not only with stimulus content (e.g., verbal vs. visuospatial stimuli), but also with task demands (e.g., phonological vs. semantic processing of word stimuli), suggesting that the neural responses reflect the goal-oriented focusing of attention (Paller & Wagner, 2002). Thus, encoding may be a by-product of processing that is dependent on a number of factors, including attentional modulations, task demands, and stimulus features. A specific event may recruit PFC control processes to a different extent depending on these factors, with greater PFC recruitment associated with enhanced subsequent memory (Wagner, 2002).

Within the PFC, different regions may execute different control processes that play a role in memory encoding. Whereas a wealth of studies have observed robust subsequent memory effects in prefrontal regions, these effects tend to be localised within the ventrolateral PFC (VLPFC; Broadmann's areas [BA] 44, 45, and 47) and are much less likely to be found within the dorsolateral PFC (DLPFC; BA 46 and 9; Blumenfeld & Ranganath, 2007; Figure 20.1). Given existing evidence that the VLPFC plays a key role in the selection of task-relevant information (e.g., Badre, Poldrack, Paré-Blagoev, Insler, & Wagner, 2005; Jonides & Nee, 2006; Thompson-Schill, Bedny, & Goldberg, 2005), the VLPFC may be involved in selecting and strengthening the representation of goal-relevant features during encoding, which would explain the frequency of subsequent memory effects observed in the VLPFC. The DLPFC, on the other hand, has been shown to be involved in the organisation of items in working memory (e.g., D'Esposito, Postle, Jonides, & Smith, 1999; Postle, Berger, & D'Esposito, 1999; Wagner, Maril, Bjork, & Schacter, 2001), and it may promote the encoding of associations between items. Subsequent memory effects may be less frequently observed in the DLPFC because the types of encoding tasks that are typically used do not encourage the formation of associations between items. When studies utilise encoding tasks that promote the formation of inter-item associations, DLPFC activation is predictive of encoding success (Blumenfeld & Ranganath, 2007). For example, in one study that required participants to actively reorder a set of three words on the basis of a stimulus dimension, DLPFC activation during reordering was positively associated with subsequent memory performance (Blumenfeld & Ranganath, 2006). Similarly, enhanced DLPFC activation during encoding has been associated with subsequent free recall performance, but

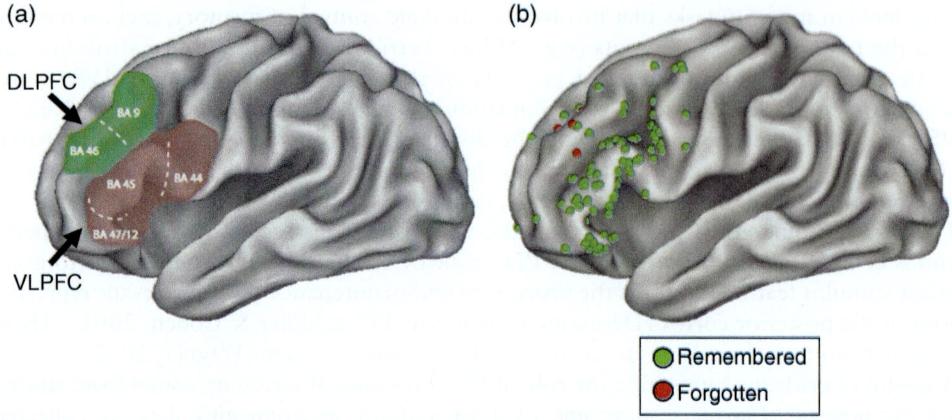

Figure 20.1 Prefrontal contributions to successful encoding. Source: Adapted from Blumenfeld and Ranganath 2007. (a) Dorsolateral prefrontal cortex (DLPFC, green) and ventrolateral prefrontal cortex (VLPFC, red) and corresponding Broadmann's areas (BA). (b) Meta-analysis of fMRI subsequent memory effects. Each green dot corresponds to a published report of a positive subsequent memory effect: greater activity for information that is subsequently remembered compared to information subsequently forgotten. Each red dot corresponds to a published report of a negative subsequent memory effect: greater activity for information subsequently forgotten compared to information subsequently remembered. As can be seen, positive subsequent memory effects have overwhelmingly been localised to the VLPFC.

not recognition (Staresina & Davachi, 2006). Free recall tests are thought to be more sensitive to organisational strategies and inter-item associations in memory than recognition tests (Tulving, 1968).

The existing evidence provides strong support that PFC cognitive control processes promote successful encoding. Prefrontal mechanisms contribute to encoding by selecting relevant item or feature information and by forming associations among items. The role that the PFC plays in encoding is dependent on a number of factors, including stimulus attributes, task demands, and the context in which an item or event is encountered.

Encoding in the Face of Distraction

Much of real-life encoding occurs among distractions. For example, while driving your car, you listen to and encode the juicy gossip your passenger is telling you; or while giving a lecture, you notice and encode that much of your audience is asleep. Encoding is resource demanding, as it involves the goal-dependent allocation of attention to the to-be-remembered event. Furthermore, memory benefits from more elaborate encoding, such that the target event is processed for meaning, actively incorporated into ongoing mental activities, and associated with existing memories (Craik & Tulving, 1975). A classic finding in the memory literature is that limiting attentional resources at the time of encoding results in diminished subsequent memory for an event relative to when attentional resources are fully dedicated during encoding (e.g., Baddeley, Lewis, Eldridge, & Thomson, 1984; Craik, Govoni, Naveh-Benjamin, & Anderson, 1996). Reducing the availability of attentional resources disrupts not only intentional learning, but also incidental encoding (e.g., Chiu & Egner, 2015; Jacoby, Woloshyn, & Kelley, 1989; Kensinger, Clarke, & Corkin, 2003), highlighting the importance of the availability of attention for encoding. Although memory retrieval tends to be less encumbered by dual-task situations, successful retrieval of an item in a dual-task situation reduces the likelihood of remembering that item again in the future relative to successful retrieval under full attention, suggesting that the power of retrieval as an encoding event is dependent on attentional resources (Dudukovic, DuBrow, & Wagner, 2009). Early PET studies observed reduced encoding activation in the left PFC when simultaneously engaging in a challenging secondary task in comparison to an easy secondary task (e.g., Shallice et al., 1994). Electrophysiological data indicate that engaging in a dual task at encoding results in a reduction in negative event-related potentials (ERP) at left inferior PFC sites and that enhanced negative ERPs in these sites are associated with later memory (Mangels, Picton, & Craik, 2001). Similarly, an fMRI study found that encoding activations in the left PFC are reduced and less likely to predict subsequent memory under demanding dual-task conditions (Kensinger et al., 2003). Thus, these data provide evidence that performing a dual task during encoding reduces activation in the left PFC regions that are essential for long-term memory formation.

Distractions can disrupt encoding not only in the context of performing dual tasks but also in situations where attention could be captured by other task-irrelevant perceptual stimuli in the environment. When listening to a lecture, your attention might be automatically drawn to the flashing images on the laptop screen of the person sitting next to you, and top-down control is required to focus on and encode the lecture material. In these situations, top-down control can be exerted not only through the enhancement of task-relevant processing in posterior sensory cortices, but also through the active suppression of the sensory processing of task-irrelevant stimuli. For example, when face and scene stimuli are presented together, neural responses in face-sensitive and scene-sensitive regions of ventral temporal cortex are enhanced or attenuated depending on which stimulus category is actively attended (Gazzaley, Cooney, McEvoy, Knight, & D'Esposito, 2005; O'Craven, Downing, & Kanwisher, 1999; Yi & Chun, 2005). Attended stimuli are associated with not only enhanced neural responses but also superior subsequent memory relative to unattended, or ignored, stimuli (Dudukovic, Preston, Archie, Glover, & Wagner, 2011; Gazzaley, Cooney, McEvoy, et al., 2005; Yi & Chun, 2005). More direct evidence for a link between top-down attentional modulation and encoding success comes from a study demonstrating that selective attention to specific contextual features (e.g., colour or location) was associated with the modulation of visual cortical activation, and the strength of this modulation was associated with hippocampal activation and with subsequent remembering (Uncapher & Rugg, 2009).

If memory is a product of the top-down modulation of attention, what mechanisms support the control of top-down attention? Evidence from animal and human studies suggests that the PFC modulates neural activity in sensory processing regions via long-range cortico-cortical projections (Kastner & Ungerleider, 2000). Functional connectivity analyses support the idea that top-down signals from the PFC exert modulatory control on activation in the visual cortical regions (Gazzaley et al., 2007). A recent study using transcranial magnetic stimulation to temporarily disrupt PFC function demonstrated a causal link between PFC activity and top-down modulation in the posterior cortical regions (Zanto, Rubens, Thangavel, & Gazzaley, 2011). Importantly, top-down modulation in visual processing regions was also causally linked to subsequent working memory performance, providing direct evidence that the PFC drives the top-down attentional control that is essential for encoding success.

In addition to, and in concert with, PFC mechanisms, the parietal cortex also plays an important role in the top-down control of attention. According to the dual-attention model, there are two dissociable frontoparietal attentional systems, such that the dorsal and ventral regions of the lateral posterior parietal cortex (PPC) play distinct roles (e.g., Corbetta & Shulman, 2002). Specifically, the dorsal PPC is involved in the top-down, or goal-directed, allocation of attention, whereas ventral PPC engagement reflects bottom-up, or stimulus-driven, attention. Recent evidence suggests that these two types of attention may have opposite effects on encoding success, such that goal-directed attention contributes to successful encoding of a target item, whereas stimulus-driven attention that is captured by salient but task-irrelevant information contributes to encoding failures (Uncapher & Wagner, 2009). For example, in meeting someone new at a party, if you successfully direct your attention to their introduction, you will likely encode their name, but if your attention is automatically drawn to the food table, you may later offend your new acquaintance by failing to remember his or her name. Goal-directed attention can bias cortical processing to attended information (Beck & Kastner, 2009), and there is evidence that this biased enhancement may impact encoding activity in the MTL (e.g., Uncapher & Rugg, 2009). The ***dual-attention encoding hypothesis*** accounts for the fact that positive subsequent memory effects are frequently observed in the dorsal PPC, whereas negative subsequent memory effects (i.e., greater activation during the encoding of items that are later forgotten compared to those that are later remembered) have been associated with the ventral PPC (Uncapher & Wagner, 2009).

The relationship between PPC attention mechanisms and memory encoding was directly tested in a study using a spatial cueing paradigm in which object pictures appeared at validly cued or invalidly cued locations (Uncapher, Hutchinson, & Wagner, 2011; Figure 20.2). Goal-directed orientation of attention to cued locations was associated with dorsal PPC activity, but these shifts of attention were only associated with successful encoding when the objects appeared in validly cued locations. When objects appeared at uncued locations, these shifts of attention *negatively* impacted encoding success. This trade-off was also paralleled by functional connectivity data: Stronger connectivity between the dorsal PPC (intraparietal sulcus) and perceptual cortices (fusiform/lateral occipital cortex) immediately prior to stimulus presentation was associated with a higher probability of successful encoding of the stimulus, whereas stronger connectivity between the ventral PPC (temporoparietal junction) and perceptual cortices during the same period was associated with a lower probability of successful stimulus encoding. These data provide evidence that parietal attention mechanisms play a role in episodic encoding and, specifically, that dorsal parietal mechanisms promote successful encoding via top-down orienting of attention.

In sum, frontoparietal cortical mechanisms drive the top-down control of attention, modulating activation in posterior representational areas and ultimately influencing which stimuli and which stimulus attributes are encoded into memory. These acts of attentional control are essential in an environment that is rich in goal-relevant and goal-irrelevant information. In fact, failures of attentional control can be characterised by increased learning of *goal-irrelevant* information (e.g., Gazzaley, Cooney, Rissman, & D'Esposito, 2005; Uncapher et al., 2011), highlighting the importance of cognitive control for adaptive encoding.

Cognitive Control and Retrieval

Once information is encoded into memory, what determines whether and in what context the information will be retrieved? In some instances, information is automatically retrieved because a cue triggers the reactivation of associated information. For example, when you see a close friend approaching, you may automatically retrieve the friend's name and basic information about him or her. This automatic retrieval can be contrasted with the many contexts in which effort is required to exert control over the retrieval process (Miyashita, 2004). Here, we review several distinct ways in which cognitive control may be recruited during memory retrieval.

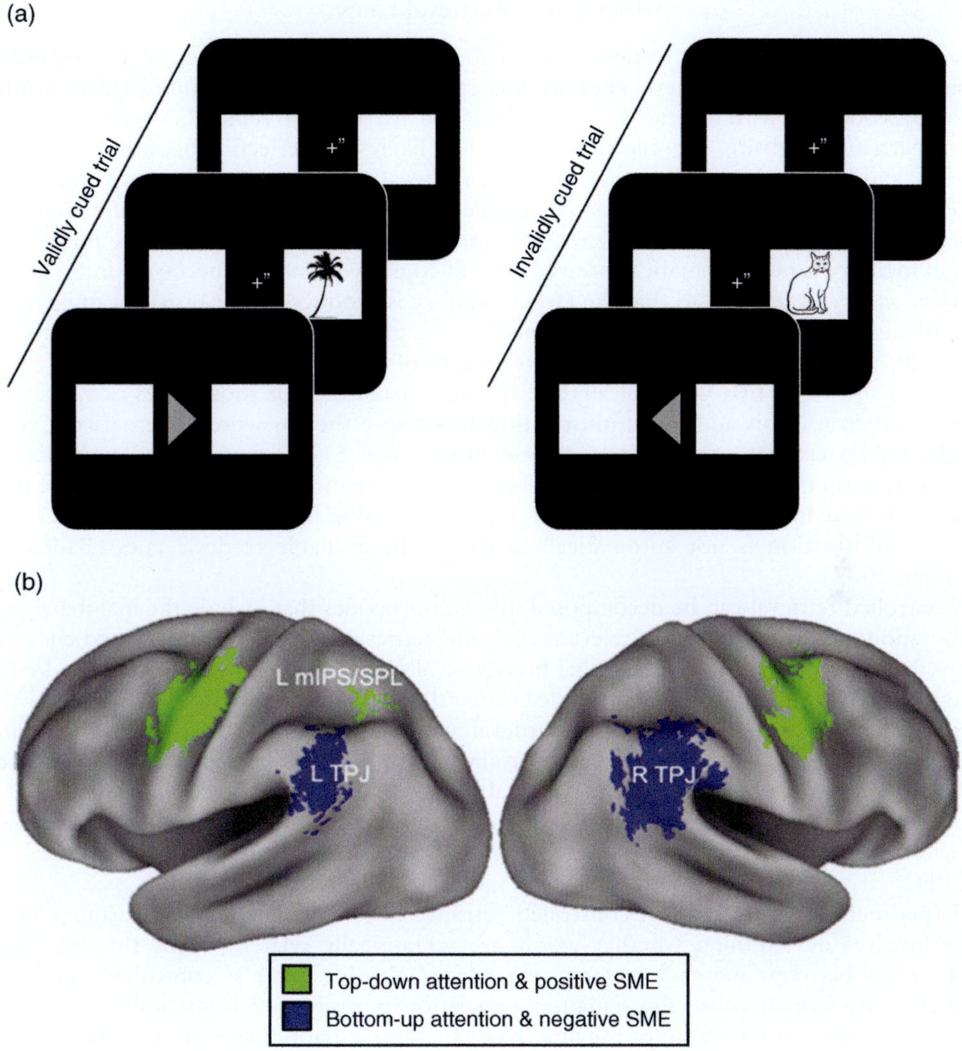

Figure 20.2 Attention and encoding. Source: Adapted from Uncapher 2011. (a) During encoding trials, a spatial cue (left- vs. right-facing arrow) cued a spatial location for participants to attend. The to-be-encoded image appeared either at the cued location (validly cued trials; left panel) or opposite the cued location (invalidly cued trials; right panel) following a variable-duration delay. (b) fMRI data were separately analysed to identify regions that exhibited top-down versus bottom-up attention as well as regions that exhibited positive or negative subsequent memory effects. Top-down attention effects were defined as cue-related activation that was modulated by the duration of the delay between the spatial cue and the image. Bottom-up attention effects were defined as greater activation in response to objects at expected locations (valid cues) versus unexpected locations (invalid cues). Bottom-up attention effects overlapped with negative subsequent memory effects for validly cued images in the bilateral temporoparietal junction (TPJ), whereas top-down attention effects overlapped with positive subsequent memory effects for validly cued images in the left medial intraparietal sulcus (mIPS)/superior parietal lobule (SPL). Thus, dorsal parietal regions that contributed to top-down attention (mIPS/SPL) positively influenced memory for images that appeared at attended locations, but activity in ventral parietal regions that reflected bottom-up capture of attention (TPJ) signalled failed encoding of images at attended locations.

Ineffective Retrieval Cues

On some occasions, available retrieval cues fail to elicit retrieval of goal-relevant information. This may happen when retrieval cues are non-specific, or if the desired information is not strongly associated with the cue. For example, if you are trying to remember your password for a particular website, the cue 'password?' will likely be ineffective because it is non-specific and is unlikely to automatically trigger retrieval of your password. Similarly, even a more detailed password hint such as 'It's a street you have lived on' may be ineffective if it refers to a former address that you have since moved away from. In instances like these, the target memory is not automatically retrieved, and cognitive control is necessary in order to retrieve *more information* so that the target memory is produced. This process can involve identifying additional retrieval cues that may help single out the sought-after memory. For example, you might note that the website is part of a larger category of websites (e.g., banking) and recall that you typically use the same password for those types of websites. You can then use this additional information to retrieve the password. Alternatively, you might try to estimate the period of time during which you first created your login information on this particular website and use that information to retrieve the name of the street you were living on at the time. This type of *controlled retrieval* is required when the desired information is not automatically activated by available retrieval cues (Badre & Wagner, 2007).

Controlled retrieval can be decomposed into sub-processes that include the maintenance, elaboration, and specification of relevant cues and retrieval plans. These elaborated cues or plans then serve as top-down signals that bias retrieval towards relevant information (Badre & Wagner, 2007). The demand for controlled retrieval is inversely proportional to the effectiveness of retrieval cues: Less effective retrieval cues require greater controlled retrieval. Controlled retrieval has been extensively studied in the context of semantic and episodic memory, and this work has led to the proposal that the VLPFC plays a common role in controlled retrieval. In particular, controlled retrieval of information that is semantic or conceptual in nature has consistently been associated with engagement of the anterior aspect of the left VLPFC.

Experimentally, demands on controlled retrieval have been manipulated using a task that involves determining whether words are semantically related. When the semantic association between words is strong (e.g., 'candle' and 'flame'), controlled retrieval demands are low. Because the semantic association is easily or automatically retrieved, there is no demand to search for the link. But when the semantic association is weak (e.g., 'candle' and 'halo'), the semantic association may not be automatically retrieved, and it may require a more effortful search to establish the link between the words. Indeed, activation in the left anterior VLPFC is greater when retrieving or establishing relatively weak semantic relationships compared to relatively strong semantic relationships (e.g., Badre et al., 2005; Figure 20.3). Although the left anterior VLPFC is thought to *drive* the retrieval search process, the search requires coordination with structures that store the relevant information. For example, the left anterior VLPFC activity functionally couples with activity in the lateral temporal cortex during controlled retrieval of semantic or conceptual information (e.g., Badre et al., 2005; Dobbins & Wagner, 2005; Gold et al., 2006), consistent with the idea that the left anterior VLPFC supplies a top-down signal that biases processing within the lateral temporal cortex. Similarly, functional interactions between the right VLPFC and the occipito-temporal cortex have been observed when perceptual information is goal relevant (e.g., Dobbins & Wagner, 2005), suggesting that left versus right lateralisation of controlled retrieval processes may depend on whether target information is semantic or perceptual, respectively.

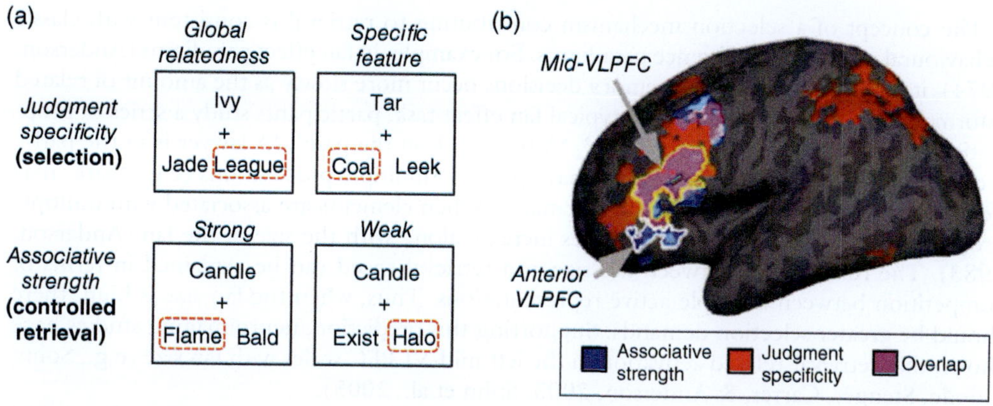

Figure 20.3 Controlled retrieval and selection. Source: Adapted from Badre et al. (2005). (a) Participants were presented with cue words and were instructed to select semantically related associates. On some trials, participants were instructed to select associates based on *global relatedness*, whereas on other trials, participants were given a *specific feature* on which to base their selection of the associate ('colour', 'shape', 'size', or 'texture'). Decisions based on specific features were intended to entail greater selection demands than decisions based on global relatedness because feature decisions required favouring goal-relevant information (the cued feature) and ignoring competing information (the non-cued features). Among the global relatedness trials, the associative strength between the cue and the target associate was varied from weak (e.g., 'candle' and 'halo') to strong (e.g., 'candle' and 'flame'). Low associative strength trials were intended to require higher controlled retrieval demands compared to high associative strength trials because the relevant semantic link between the words was unlikely to automatically come to mind. (b) Judgement specificity (specific > global), the selection manipulation, was associated with increased activity within several frontoparietal regions, including the left mid-VLPFC. Associative strength (weak > strong), the controlled retrieval manipulation, was associated with increased activity in the left anterior VLPFC.

Selection

Cognitive control is required not only when retrieval cues do not automatically elicit retrieval of the target memory, but also in situations in which more information is retrieved than is needed. In these instances, a target memory must be *selected* from competing alternatives. For example, in trying to remember where you were at 11:00 a.m. yesterday, you may initially retrieve several episodes that occurred yesterday—coffee with a friend, attending a meeting, parking your car—before you are able to 'select' the event that occurred at 11:00 a.m. Selection involves a form of cognitive control that is qualitatively distinct from controlled retrieval. Specifically, whereas controlled retrieval is a pre-retrieval mechanism that is necessary when initial retrieval cues fail to activate the target memory, selection is a post-retrieval mechanism that sorts through active memories in order to identify the target information (Badre & Wagner, 2007).

As with controlled retrieval, selection is a mechanism that generalises across different memory domains. Neuroimaging studies of semantic memory (e.g., Badre et al., 2005; Gold et al., 2006; Thompson-Schill, D'Esposito, Aguirre, & Farah, 1997), working memory (e.g., Badre & Wagner, 2005), and episodic memory (e.g., Dobbins & Wagner, 2005) have provided converging evidence that the mid to posterior extent of the left VLPFC supports selection from memory (Figure 20.3). Selection demands can be manipulated by varying the number and/or strength of competing (non-target) memories that are active during retrieval. As the number and/or strength of competitors increase, so do the demands on selection.

The concept of a selection mechanism contributing to retrieval is consistent with classic behavioural memory interference paradigms. For example, in fan effect paradigms (Anderson, 1974), it has been shown that memory decisions occur more slowly as the amount of related information in memory grows. In a typical fan effect task, participants study a series of propositions (e.g., 'A doctor is in the bank', 'A fireman is in the park', 'A lawyer is in the park', etc.). However, some of the elements may appear in multiple propositions (e.g., 'park' may be associated with both 'lawyer' and 'fireman'). When elements are associated with multiple propositions (i.e., a 'fan'), retrieval times increase along with the size of the fan (Anderson, 1983). The relationship between fan size and retrieval speed can be explained in terms of competition between multiple active representations. Thus, when the fan size is high, there should be greater selection demands. Supporting this prediction, neuroimaging studies have shown that retrieval-related activation in the left mid-VLPFC scales with fan size (e.g., Sohn, Goode, Stenger, Carter, & Anderson, 2003; Sohn et al., 2005).

More generally, selection demands are thought to be present whenever a specific criterial detail is required during retrieval. For example, when encountering a friend, if the retrieval goal is to recall the last place you saw this friend, selection demands will likely be present because many previous encounters may come to mind. However, if the retrieval goal is simply to determine whether you know this person, any prior encounter would constitute positive evidence and demands for selective retrieval would be absent. Consistent with this perspective, the left mid-VLPFC is particularly engaged when retrieval tasks require recollection of specific event details or the 'source' of an event, relative to simple decisions of whether an item has been encountered before (e.g., Dobbins, Foley, Schacter, & Wagner, 2002; Dobbins & Wagner, 2005; Kostopoulos & Petrides, 2003).

Although the left mid-VLPFC may play a critical role in enabling selection among competing memories, VLPFC-mediated selection is likely only one of several mechanisms that contribute to post-retrieval cognitive control. For example, the anterior cingulate cortex (ACC) is thought to play a domain-general role in *detecting* competition or conflict between active representations (Botvinick, Braver, Barch, Carter, & Cohen, 2001). ACC conflict detection is thought to play an initial role in signalling the need for control, which may ultimately be implemented by the lateral PFC (Norman, Newman, & Detre, 2007). The idea that the ACC detects conflict among active memories has received support from several neuroimaging studies (Badre & Wagner, 2004; Kuhl, Dudukovic, Kahn, & Wagner, 2007; Kuhl, Kahn, Dudukovic, & Wagner, 2008). In one study, multi-voxel pattern analysis (MVPA) of fMRI data was used to index the *reactivation* of memories within temporal lobe regions during retrieval (Kuhl, Bainbridge, & Chun, 2012). When a retrieval cue had been paired with more than one associate—similar to the fan effect paradigms—retrieval of one of the associates (the target) was often accompanied by reactivation of the non-target or competing association. Moreover, the relative strength with which the target versus competing memories were reactivated was related to engagement of the ACC (Figure 20.4). Specifically, when the competing memory was weak relative to the target memory, ACC activation was low, but when the competing memory was strong relative to the target memory, ACC activation was high. Thus, this study directly linked trial-by-trial variations in ACC activity to competition between memories in posterior cortical areas.

The ultimate goal of post-retrieval selection is to favour the memory that matches current behavioural goals. To the extent that selection is successful, goal-relevant memories should therefore be afforded stronger neural representations than non-target events or irrelevant features of a retrieved event. In an fMRI study that tested this idea, participants studied various word–picture pairs that varied according to two dimensions: Images consisted of either faces or scenes, and they appeared either on the left- or right-hand side of the computer screen (Kuhl, Johnson, & Chun, 2013) (Figure 20.5). At retrieval, participants were presented

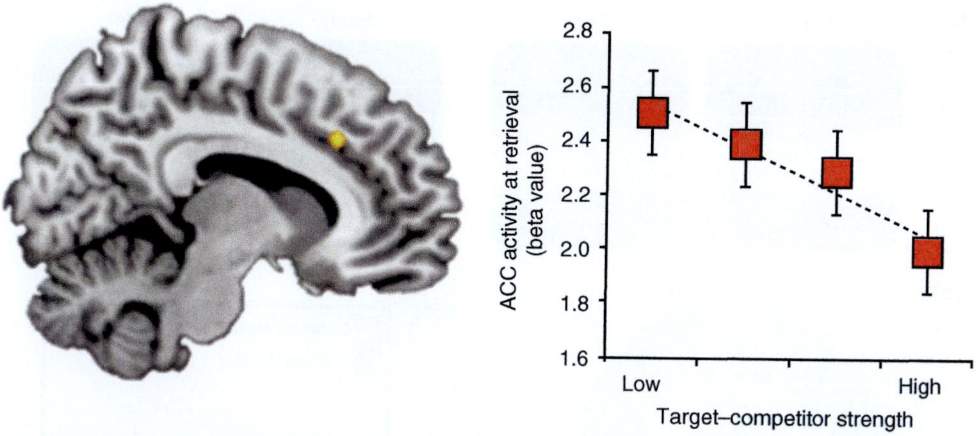

Figure 20.4 ACC and retrieval competition. Source: Adapted from Kuhl 2012. Multi-voxel pattern analysis (MVPA) was applied to fMRI data collected during cued retrieval to separately measure the strength with which target and competing memories were reactivated. Trial-by-trial variability in the strength of target reactivation *relative to* competitor reactivation was correlated with activity in the ACC. When the competing memory was relatively strong (resulting in a relatively small difference in strength between target and competitor memories), ACC activity was high; when the competing memory was relatively weak (resulting in a relatively large difference in strength between target and competitor memories), ACC activity was low.

with word cues and were asked to retrieve either the location (left vs. right) or the visual category (face vs. scene) of the corresponding picture. MVPA was used to test for reactivation of visual category information as a function of whether that information was relevant at retrieval. Within the temporal lobe regions, visual category information was robustly reactivated regardless of whether it was behaviourally relevant or irrelevant, indicating relatively automatic reactivation within temporal lobe regions. In contrast, reactivation of visual category information within the lateral parietal cortex was more robust when that information was relevant (category trials) than irrelevant (location trials). Thus, the lateral parietal cortex preferentially represented information that matched the specific retrieval goal, consistent with the desired outcome of post-retrieval selection.

Consequences of Selective Retrieval

As described above, when retrieval cues elicit activation of multiple memories, selection is required to enable a target memory to 'win' the competition. However, such acts of competitive remembering can carry lasting consequences for the memories that 'lose' the competition. The consequences of competitive retrieval have been explored using the *retrieval practice paradigm* (Anderson, Bjork, & Bjork, 1994). In the standard retrieval practice paradigm, participants study category–exemplar word pairs (e.g., 'fruit–apple', 'fruit–banana', 'drink–whiskey', 'drink–scotch') and then engage in selective retrieval (retrieval practice) of some of the exemplars from some of the categories (e.g., 'fruit-a___' given as a probe to retrieve 'apple'). Later, participants are asked to recall all of the exemplars they studied earlier. Not surprisingly, practised items (e.g., 'apple') are recalled more often than baseline items, which were exemplars from categories that were not practised at all during the selective retrieval practice phase (e.g., 'whiskey' or 'rum'). Critically, however, items that come from practised categories but were not actually practised themselves (e.g., 'banana') are recalled

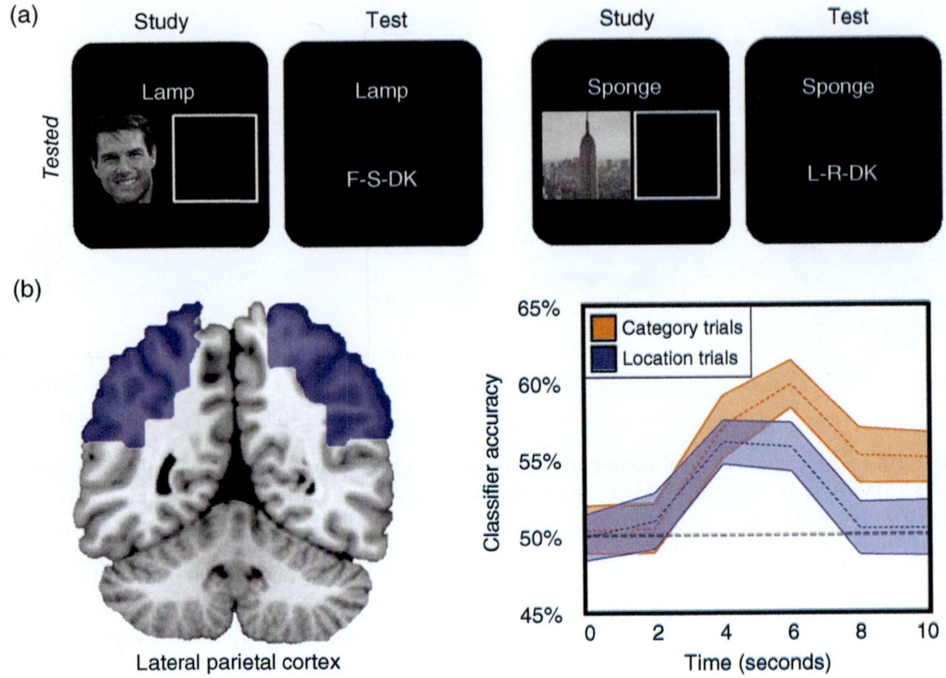

Figure 20.5 Favouring goal-relevant information in memory. Source: Adapted from Kuhl 2013. (a) During fMRI scanning, participants alternated between study and test blocks. During study, words were paired with images of either faces or scenes; each image appeared either on the left- or right-hand side of the screen. Participants were instructed to encode the word–image association as well as the location information. Each test block began with a cue indicating whether image category (face vs. scene) or location (left vs. right) was relevant for that block. On each test trial, a cue word was presented, and participants retrieved the relevant information: On category blocks, participants indicated the category of the associated image (face – F; scene – S, or don't know – DK); or a scene (S); on location blocks, participants indicated which side of the screen the associated image had appeared (left – L; right – R, or don't know – DK). (b) MVPA was applied to fMRI data from the test trials to measure reactivation of visual category information. Specifically, pattern classifiers decoded whether each cue had been paired with a face or a scene, with the accuracy of the classifier indexing the strength of category reactivation. Reactivation of category information (classifier accuracy) was then compared as a function of whether that information was behaviourally relevant (category trials) or irrelevant (location trials). Within the lateral parietal cortex, category information was robustly reactivated when it was behaviourally relevant; however, when category information was not relevant, reactivation in the lateral parietal cortex was markedly lower. In contrast, within the ventral temporal cortex (not shown), category reactivation was equally robust regardless of whether it was behaviourally relevant. Thus, whereas reactivation of category information was relatively automatic in the ventral temporal cortex, reactivation in the lateral parietal cortex was clearly modulated by retrieval goals.

less often than the baseline items (Levy & Anderson, 2002). In other words, practice retrieving 'apple' can make it more difficult to remember 'banana' compared to a situation in which fruits were not practised at all. The forgetting of unpractised items from practised categories is referred to as retrieval-induced forgetting. Retrieval-induced forgetting has been explained in terms of competing memories being *inhibited* specifically because they interfere with target retrieval (Levy & Anderson, 2002; Norman et al., 2007).

If retrieval-induced forgetting reduces competition, then successive acts of selective retrieval should require less cognitive control than initial retrieval attempts. Consistent with this prediction, an fMRI study using the retrieval practice paradigm found robust decreases in lateral and medial PFC engagement across repeated retrieval practice attempts (Kuhl et al., 2007). Moreover, if the ACC signals competition (as discussed previously), then ACC engagement should be particularly sensitive to the strength of competing memories (Norman et al., 2007). This prediction was supported by the finding that reductions in ACC activation across successive retrieval practice attempts were correlated with the amount of forgetting that competing memories experienced (Kuhl et al., 2007). In other words, disengagement of the ACC during target retrieval was a clue that competing memories had become weaker.

A more recent fMRI study provided even more direct insight into the phenomenon of retrieval-induced forgetting (Wimber, Alink, Charest, Kriegeskorte, & Anderson, 2015) (Figure 20.6). MVPA was used to identify patterns of neural activity within high-level visual cortical regions that corresponded to individual images. During retrieval of a target image, the evoked pattern of neural activity was compared to a 'template' representing the competing image (i.e., the image that shared the same retrieval cue). With repeated retrieval of the target, the evoked neural activity pattern became increasingly dissimilar to the competing image template. Critically, this reduction in competitor activation predicted competitor forgetting. Moreover, VLPFC activity also decreased with repeated target retrieval, and VLPFC activation was correlated with the reductions in competitor activation observed in high-level visual cortex. These data suggest an active role for the VLPFC in reducing retrieval competition by weakening competing memories.

During initial attempts to retrieve a target memory, the probability that a competing memory will become simultaneously active is thought to be a function of the overlap of the target and competitor hippocampal representations. If the hippocampal representations are highly overlapping, simultaneous activation of both memories—and corresponding engagement of cognitive control mechanisms—is more likely. From this perspective, one interpretation of the reduced cognitive control demands that are observed with retrieval practice is that underlying hippocampal representations become modified—specifically, that representations of target and competing memories overlap less and less. Indeed, there is recent evidence that retrieval practice results in differentiation of hippocampal activity patterns corresponding to target versus competing memories (Hulbert & Norman, 2014), though a direct link between the overlap of hippocampal activity patterns and PFC control demands has not yet been established.

Stopping Retrieval

Thus far, we have considered how cognitive control enables goal-relevant memories to be retrieved. However, in some cases our goal is to stop the retrieval process altogether. The motivations for stopping retrieval range from the mundane (e.g., keeping attention focused on encoding a current conversation) to the dramatic (e.g., preventing painful, emotional memories from coming to mind). Research using the Think/No-Think paradigm specifically addresses how retrieval is stopped and the consequences of these actions (Anderson & Green, 2001). In the standard Think/No-Think paradigm, participants study a series of cue–target word pairs (e.g., 'ordeal–roach') and are then presented with the cue words (e.g., 'ordeal') and asked to either recall the associated target word (e.g., 'roach'; *Think* trials) or prevent the associated target word from coming to mind (*No-Think* trials). After repeatedly retrieving or preventing the retrieval of these associates, participants are asked to retrieve all of the target words that they studied earlier. When participants practice preventing target words from coming to mind, they remember those words less well than not only the words they practised

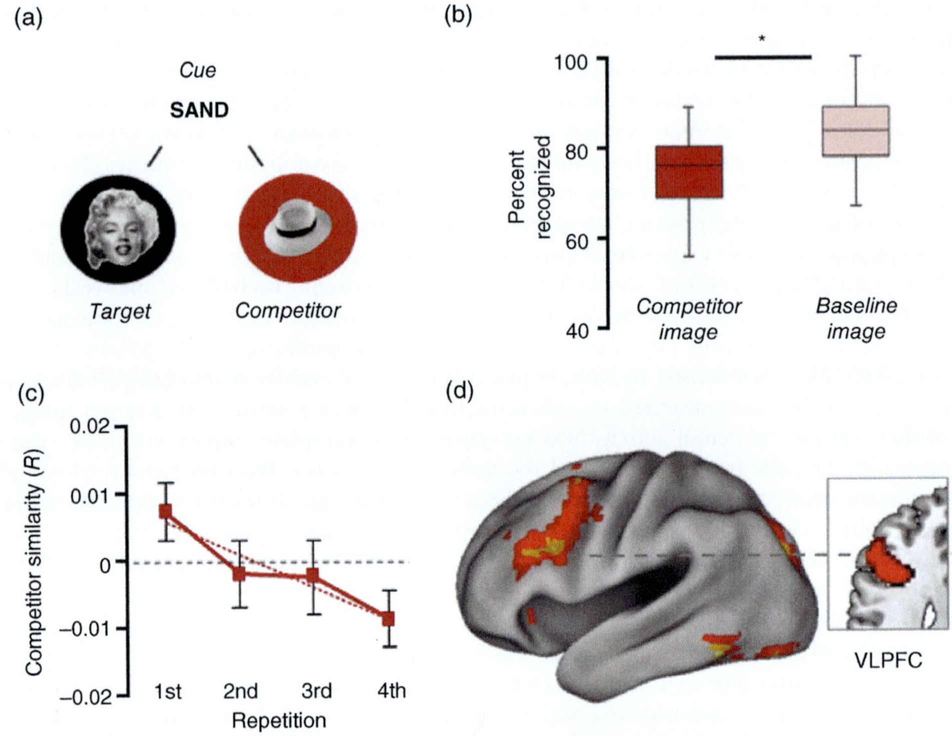

Figure 20.6 Retrieval-induced forgetting. Source: Adapted from Wimber 2015. Reproduced with permission of Nature Publishing Group. (a) Participants initially learned associations between words (cues) and images, with individual cues paired with two images. During a subsequent retrieval phase, participants were shown cues and practised recalling *one* of the associates (the target), with the other associate (the competitor) remaining unpractised. (b) Selective retrieval of target images produced *retrieval-induced forgetting* of the competitor image, as reflected by poorer recognition memory for competitor images relative to baseline images that were associated with an unpractised cue. (c) Visual cortical activity patterns elicited during retrieval of a target memory were compared with the activity pattern corresponding to the competitor memory. With repeated target retrieval, evidence for competitor activation decreased. This decrease in competitor activation correlated with forgetting of the competitor memory (not shown). (d) Left VLPFC activity was higher during early target retrieval attempts compared to later target retrieval attempts, indicating that demands on the VLPFC decreased with retrieval practice. VLPFC activity also correlated with reductions in competitor activation (not shown).

retrieving but also baseline words that they studied but that they did not practice either retrieving or suppressing (Anderson & Green, 2001). The forgetting of No-Think items is reminiscent of retrieval-induced forgetting; however, whereas retrieval-induced forgetting is a by-product of intentional retrieval, the forgetting of No-Think items is a result of deliberate attempts to not retrieve. In this way, the Think/No-Think paradigm is a better model for real-life situations in which we have a motivation to keep a particular memory out of mind.

Mechanistically, what happens when one tries to stop retrieval? Neuroimaging studies of the Think/No-Think paradigm have provided evidence that stopping the retrieval process is associated with increased activation in the lateral PFC (Anderson et al., 2004; Depue, Curran, & Banich, 2007; Levy & Anderson, 2012), and more specifically, the strength of DLPFC engagement during intentional attempts to keep memories out of mind has been shown to predict the amount of forgetting that excluded memories suffer (Anderson et al., 2004; Depue et al., 2007). The strong response in PFC control regions during memory suppression suggests that stopping retrieval does not reflect a passive failure to engage retrieval processes but instead is an active process of keeping memories out of mind (Anderson & Huddleston, 2012).

The increases in prefrontal cortex activity that are associated with No-Think trials are also paralleled by ***decreased*** activity in the MTL regions, including the hippocampus, suggesting that core memory regions are downregulated during attempts to stop retrieval (Anderson et al., 2004). Activation in the hippocampus during No-Think trials has also been related to participants' self-reports of the degree to which No-Think items accidentally ***intrude*** (Levy & Anderson, 2012; Figure 20.7). Although self-reported intrusions tend to decrease with repeated No-Think attempts—and these decreases predict forgetting—the relationship between intrusions and hippocampal activity is somewhat surprising: Intrusion trials are associated with ***lower*** hippocampal activity than non-intrusion trials. That is, when the to-be-avoided memory comes to mind, downregulation of the hippocampus is most apparent. This observation suggests that hippocampal downregulation is an active reaction to intrusions. In fact, downregulation of the hippocampus during intrusion trials is a strong predictor of the amount of forgetting of No-Think items, establishing a functional link between memory intrusions, hippocampal activity, and forgetting.

Conclusions

Cognitive control allows memories to be encoded and retrieved strategically, in line with current goals. The ability to selectively process and retain relevant information contributes to an adaptive memory system. Here, we have considered how cognitive control mechanisms influence long-term memory encoding and retrieval. During encoding, cognitive control mechanisms supported by the PFC drive the selection of goal-relevant information and promote the formation of associations among items (Blumenfeld & Ranganath, 2007), thereby shaping what we ultimately remember (Fletcher, Shallice, & Dolan, 2000). These PFC control mechanisms interact with MTL encoding processes, directing and modifying MTL representations in accordance with current goals (Simons & Spiers, 2003). Successful encoding is also supported by parietal control mechanisms that direct attention, in a top-down manner, towards relevant sources of information (Uncapher et al., 2011). Top-down attentional signals generated by the parietal cortex directly interact with and modulate processing within sensory regions, thereby prioritising specific sources of information for encoding.

Although memory retrieval can occur as an automatic response to an environmental cue, more often than not cognitive control mechanisms are necessary to guide and shape retrieval so that goal-relevant information can be produced (Miyashita, 2004). The PFC plays a critical role in representing retrieval goals (Miller & Cohen, 2001) and in generating the top-down signals that bias the retrieval process (Tomita, Ohbayashi, Nakahara, Hasegawa, & Miyashita, 1999). Controlled retrieval is necessary when available retrieval cues do not elicit retrieval of goal-relevant information and is supported by the anterior VLPFC, whereas post-retrieval selection is required when a target memory must be chosen among competing alternatives and is supported by the mid-VLPFC (Badre & Wagner, 2007).

Recruiting cognitive control mechanisms to guide memory retrieval in the present also influences what we will remember in the future. In particular, when competing memories are selected ***against***, they become weaker and are more likely to be forgotten (Anderson et al., 1994). This weakening of competing memories is closely related to the engagement of PFC control mechanisms during target memory retrieval (Kuhl et al., 2007; Wimber et al., 2015).

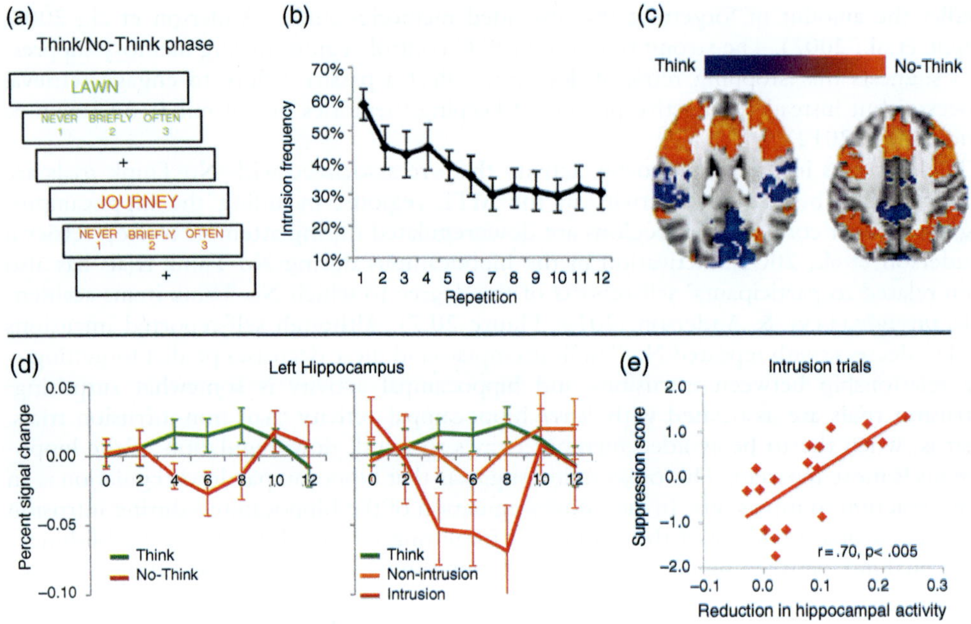

Figure 20.7 Think/No-Think paradigm. Source: Adapted from Levy 2012. (a) Participants first studied a series of word pairs. During the critical Think/No-Think phase, participants were shown individual word cues and were either instructed to 'Think' of the associated word (green font = Think) or to 'Not Think' of the associated word (red font = No Think). After each trial, participants indicated how strongly the corresponding word came to mind. (b) With repeated practice, participants were better able to volitionally keep 'No-Think' items out of mind, as reflected by a reduced rate of intrusions. (c) Actively keeping No-Think items out of mind was associated with greater recruitment of prefrontal cortex (warm colours), compared to Think trials (cool colours). (d) No-Think trials were associated with reduced activity in the Hippocampus compared to Think trials (left panel), consistent with the idea that No-Think trials 'downregulated' hippocampal retrieval processes. This downregulation of hippocampal activity was *more apparent* when participants reported intrusions from No-Think items relative to when No-Think items did not intrude (right panel). (e) Individual differences in downregulation of hippocampal activity during intrusion trials were correlated with forgetting of No-Think items (suppression score).

Similarly, volitionally halting the retrieval process in an effort to keep memories out of mind contributes to the forgetting of excluded memories (Anderson & Green, 2001). Stopping retrieval is an effortful process that is associated with increased PFC engagement and corresponding downregulation of MTL regions, including the hippocampus. In fact, the forgetting of excluded memories is predicted by increases in prefrontal activity (Anderson et al., 2004; Depue et al., 2007) and decreases in hippocampal activity (Levy & Anderson, 2012). Thus, cognitive control processes not only contribute to retrieving goal-relevant information in the present, but they actively weaken those memories that are irrelevant, interfering, or undesired.

Although our review has given separate consideration to cognitive control processes engaged during encoding versus retrieval, it should be emphasised that these processes are interrelated. For example, to the extent that the PFC supports selective encoding of goal-relevant information (Fletcher et al., 2000), the demands for prefrontal-mediated selection at retrieval (Badre & Wagner, 2007) should be reduced. Likewise, recent methodological advances in neuroimaging studies that map individual memories to specific profiles of activation within the hippocampus (Hulbert & Norman, 2014) offer the opportunity to predict cognitive control engagement at retrieval on the basis of how individual memories are encoded in relation to other (competing) memories. Although direct investigations of this kind have been limited to date, they represent an exciting avenue for future research and synthesis.

References

Anderson, J. R. (1974). Retrieval of propositional information from long-term memory. *Cognitive Psychology, 6*, 451–474.

Anderson, J. R. (1983). Retrieval of information from long-term memory. *Science, 220*, 25–30.

Anderson, M. C., Bjork, R. A., & Bjork, E. L. (1994). Remembering can cause forgetting: Retrieval dynamics in long-term memory. *Journal of Experimental Psychology: Learning, Memory, and Cognition, 20*, 1063–1087.

Anderson, M. C., & Green, C. (2001). Suppressing unwanted memories by executive control. *Nature, 410*, 366–369.

Anderson, M. C., & Huddleston, E. (2012). Towards a cognitive and neurobiological model of motivated forgetting. In R. F. Belli (Ed.), *True and false recovered memories: Toward a reconciliation of the debate. Vol. 58: Nebraska symposium on motivation* (pp. 53–120), New York: Springer.

Anderson, M. C., Ochsner, K., Kuhl, B., Cooper, J., Robertson, E., Gabrieli, S. W., & Gabrieli, J. D. E. (2004). Neural systems underlying the suppression of unwanted memories. *Science, 303*, 232–235.

Baddeley, A., Lewis, V., Eldridge, M., & Thomson, N. (1984). Attention and retrieval from long-term memory. *Journal of Experimental Psychology: General, 113*, 518–540.

Badre, D., Poldrack, R. A., Paré-Blagoev, E. J., Insler, R. Z., & Wagner, A. D. (2005). Dissociable controlled retrieval and generalized selection mechanisms in ventrolateral prefrontal cortex. *Neuron, 47*, 907–918.

Badre, D., & Wagner, A. D. (2004). Selection, integration, and conflict monitoring: Assessing the nature and generality of prefrontal cognitive control mechanisms. *Neuron, 41*, 473–487.

Badre, D., & Wagner, A. D. (2005). Frontal lobe mechanisms that resolve proactive interference. *Cerebral Cortex, 15*, 2003–2012.

Badre, D., & Wagner, A. D. (2007). Left ventrolateral prefrontal cortex and the cognitive control of memory. *Neuropsychologia, 45*, 2883–2901.

Beck, D. M., & Kastner, S. (2009). Top-down and bottom-up mechanisms in biasing competition in the human brain. *Vision Research, 49*, 1154–1165.

Blumenfeld, R. S., & Ranganath, C. (2006). Dorsolateral prefrontal cortex promotes long-term memory formation through its role in working memory organization. *Journal of Neuroscience, 26*, 916–925.

Blumenfeld, R. S., & Ranganath, C. (2007). Prefrontal cortex and long-term memory encoding: An integrative review of findings from neuropsychology and neuroimaging. *The Neuroscientist, 13*, 280–291.

Botvinick, M. M., Braver, T. S., Barch, D. M., Carter, C. S., & Cohen, J. D. (2001). Conflict monitoring and cognitive control. *Psychological Reviews, 108*, 624–652.

Brewer, J. B., Zhao, Z., Desmond, J. E., Glover, G. H., & Gabrieli, J. D. E. (1998). Making memories: Brain activity that predicts how well visual experience will be remembered. *Science, 281*, 1185–1187.

Chiu, Y.-C., & Egner, T. (2015). Inhibition-induced forgetting: When more control leads to less memory. *Psychological Science, 26*, 27–38.

Chun, M. M., Golomb, J. D., & Turk-Browne, N. B. (2011). A taxonomy of external and internal attention. *Annual Reviews of Psychology, 62*, 73–101.

Corbetta, M., & Shulman, G. (2002). Control of goal-directed and stimulus-driven attention in the brain. *Nature Reviews Neuroscience, 3*, 201–215.

Craik, F. I. M., Govoni, R., Naveh-Benjamin, M., & Anderson, N. D. (1996). The effects of divided attention on encoding and retrieval processes in human memory. *Journal of Experimental Psychology: General, 125*, 159–180.

Craik, F. I. M., & Tulving, E. (1975). Depth of processing and the retention of words in episodic memory. *Journal of Experimental Psychology: General, 104*, 268–294.

D'Esposito, M., Postle, B. R., Jonides, J., & Smith, E. E. (1999). The neural substrate and temporal dynamics of interference effects in working memory as revealed by event-related functional MRI. *Proceedings of the National Academy of Sciences of the United States of America, 96*, 7514–7519.

Depue, B. E., Curran, T., & Banich, M. T. (2007). Prefrontal regions orchestrate suppression of emotional memories via a two-phase process. *Science, 37*, 215–219.

Desimone, R., & Duncan, J. (1995). Neural mechanisms of selective visual attention. *Annual Review of Neuroscience, 18*, 193–222.

Dobbins, I. G., Foley, H., Schacter, D. L., & Wagner, A. D. (2002). Executive control during episodic retrieval: Multiple prefrontal processes subserve source memory. *Neuron, 35*, 989–996.

Dobbins, I. G., & Wagner, A. D. (2005). Domain-general and domain-sensitive prefrontal mechanisms for recollecting events and detecting novelty. *Cerebral Cortex, 15*, 1768–1778.

Dudukovic, N. M., DuBrow, S., & Wagner, A. D. (2009). Attention during memory retrieval enhances future remembering. *Memory & Cognition, 37*, 953–961.

Dudukovic, N. M., Preston, A. R., Archie, J. J., Glover, G. H., & Wagner, A. D. (2011). High-resolution fMRI reveals match enhancement and attentional modulation in the human medial temporal lobe. *Journal of Cognitive Neuroscience, 23*, 670–682.

Fletcher, P. C., Shallice, T., & Dolan, R. J. (2000). "Sculpting the response space"—An account of left prefrontal activation at encoding. *Neuroimage, 12*, 404–417.

Gazzaley, A., Cooney, J. W., McEvoy, K., Knight, R. T., & D'Esposito, M. (2005). Top-down enhancement and suppression of the magnitude and speed of neural activity. *Journal of Cognitive Neuroscience, 17*, 507–517.

Gazzaley, A., Cooney, J. W., Rissman, J., & D'Esposito, M. (2005). Top-down suppression deficit underlies working memory impairment in normal aging. *Nature Neuroscience, 8*, 1298–1300.

Gazzaley, A., Rissman, J., Cooney, J., Rutman, A., Seibert, T., Clapp, W., & D'Esposito, M. (2007). Functional interactions between prefrontal and visual association cortex contribute to top-down modulation of visual processing. *Cerebral Cortex, 17*, i125–i135.

Gold, B. T., Balota, D. A., Jones, S. J., Powell, D. K., Smith, C. D., & Andersen, A. H. (2006). Dissociation of automatic and strategic lexical-semantics: Functional magnetic resonance imaging evidence for differing roles of multiple frontotemporal regions. *Journal of Neuroscience, 26*(24), 6523–6532.

Hulbert, J. C., & Norman, K. A. (2014). Neural differentiation tracks improved recall of competing memories following interleaved study and retrieval practice. *Cerebral Cortex, 25*(10), 3994–4008.

Jacoby, L. L., Woloshyn, V., & Kelley, C. (1989). Becoming famous without being recognized: Unconscious influences of memory produced by dividing attention. *Journal of Experimental Psychology: General, 118*, 115–125.

Janowsky, J. S., Shimamura, A. P., & Squire, L. R. (1989). Source memory impairment in patients with frontal lobe lesions. *Neuropsychologia, 27*, 1043–1056.

Jonides, J., & Nee, D. E. (2006). Brain mechanisms of proactive interference in working memory. *Neuroscience, 139*, 181–193.

Kastner, S., & Ungerleider, L. G. (2000). Mechanisms of visual attention in the human cortex. *Annual Reviews of Neuroscience, 23*, 315–341.

Kensinger, E. A., Clarke, R. J., & Corkin, S. (2003). What neural correlates underlie successful encoding and retrieval? A functional magnetic resonance imaging study using a divided attention paradigm. *Journal of Neuroscience, 23*, 2407–2415.

Kim, H. (2011). Neural activity that predicts subsequent memory and forgetting: A metananalysis of 74 fMRI studies. *Neuroimage, 54*, 2446–2461.

Kostopoulos, P., & Petrides, M. (2003). The mid-ventrolateral prefrontal cortex: Insights into its role in memory retrieval. *European Journal of Neuroscience, 17*, 1489–1497.

Kuhl, B. A., Bainbridge, W. A., & Chun, M. M. (2012). Neural reactivation reveals mechanisms for updating memory. *Journal of Neuroscience, 32*, 3453–3461.

Kuhl, B. A., Dudukovic, N. M., Kahn, I., & Wagner, A. D. (2007). Decreased demands on cognitive control following memory suppression reveal benefits of forgetting. *Nature Neuroscience, 10,* 908–914.

Kuhl, B. A., Johnson, M. K., & Chun, M. M. (2013). Dissociable neural mechanisms for goal-directed versus incidental memory reactivation. *Journal of Neuroscience, 33,* 16099–16109.

Kuhl, B. A., Kahn, I., Dudukovic, N. M., & Wagner, A. D. (2008). Overcoming suppression in order to remember: Contributions from anterior cingulate and ventrolateral prefrontal cortex. *Cognitive, Affective, & Behavioral Neuroscience, 8,* 211–221.

Levy, B. J., & Anderson, M. C. (2002). Inhibitory processes and the control of memory retrieval. *Trends in Cognitive Sciences, 6,* 299–305.

Levy, B. J., & Anderson, M. C. (2012). Purging of memories from conscious awareness tracked in the human brain. *Journal of Neuroscience, 32,* 16785–16794.

Mangels, J. A., Picton, T. W., & Craik, F. I. M. (2001). Attention and successful episodic encoding: An event-related potential study. *Cognitive Brain Research, 11,* 77–95.

Miller, E. K., & Cohen, J. D. (2001). An integrative theory of prefrontal cortex function. *Annual Review of Neuroscience, 24,* 167–202.

Milner, B., Petrides, M., & Smith, M. L. (1985). Frontal lobes and the temporal organization of memory. *Human Neurobiology, 4,* 137–142.

Miyashita, Y. (2004). Cognitive memory: Cellular and network machineries and their top-down control. *Science, 306,* 435–440.

Norman, K. A., Newman, E. L., & Detre, G. (2007). A neural network model of retrieval-induced forgetting. *Psychological Reviews, 114,* 887–953.

O'Craven, K. M., Downing, P. E., & Kanwisher, N. (1999). fMRI evidence for objects as the units of attentional selection. *Nature, 401,* 584–587.

Paller, K. A., & Wagner, A. D. (2002). Observing the transformation of experience into memory. *Trends in Cognitive Sciences, 6,* 93–102.

Postle, B. R., Berger, J. S., & D'Esposito, M. (1999). Functional neuroanatomical double dissociation of mnemonic and executive control processes contributing to working memory performance. *Proceedings of the National Academy of Sciences of the United States of America, 96,* 12959–12964.

Shallice, T., Fletcher, P., Frith, C. D., Grasby, P., Frackowiak, R. S., & Dolan, R. J. (1994). Brain regions associated with acquisition and retrieval of verbal episodic memory. *Nature, 368,* 633–635.

Shimamura, A. P. (1995). Memory and the prefrontal cortex. *Annals of the New York Academy of Sciences, 769,* 151–159.

Shimamura A. P., Jurica, P. J., Mangels, J. A., Gershberg, F. B., & Knight, R. T. (1995). Susceptibility to memory interference effects following frontal lobe damage: Findings from tests of paired-associate learning. *Journal of Cognitive Neuroscience, 7,* 144–152.

Simons, J. S., & Spiers, H. J. (2003). Prefrontal and medial temporal lobe interactions in long-term memory. *Nature Reviews Neuroscience, 4,* 637–648.

Sohn, M. H., Goode, A., Stenger, V. A., Carter, C. S., & Anderson, J. R. (2003). Competition and representation during memory retrieval: Roles of the prefrontal cortex and the posterior parietal cortex. *Proceedings of the National Academy of Sciences of the United States of America, 100,* 7412–7417.

Sohn, M. H., Goode, A., Stenger, V. A., Jung, K. J., Carter, C. S., & Anderson, J. R. (2005). An information-processing model of three cortical regions: Evidence in episodic memory retrieval. *Neuroimage, 25,* 21–33.

Spaniol, J., Davidson, P. S., Kim, A. S., Han, H., Moscovitch, M., & Grady, C. L. (2009). Event-related fMRI studies of episodic encoding and retrieval: Meta-analyses using activation likelihood estimation. *Neuropsychologia, 47,* 1765–1779.

Squire, L. R. (1992). Memory and the hippocampus: A synthesis from findings with rats, monkeys, and humans. *Psychological Review, 99,* 195–231.

Staresina, B. P., & Davachi, L. (2006). Differential encoding mechanisms for subsequent associative recognition and free recall. *Journal of Neuroscience, 26,* 9162–9172.

Thompson-Schill, S. L., Bedny, M., & Goldberg, R. F. (2005). The frontal lobes and the regulation of mental activity. *Current Opinion in Neurobiology, 15,* 219–224.

Thompson-Schill, S. L., D'Esposito, M., Aguirre, G. K., & Farah, M. J. (1997). Role of left inferior prefrontal cortex in retrieval of semantic knowledge: A reevaluation. *Proceedings of the National Academy of Sciences of the United States of America, 94,* 14792–14797.

Tomita, H., Ohbayashi, M., Nakahara, K., Hasegawa, I., & Miyashita, Y. (1999). Top-down signal from prefrontal cortex in executive control of memory retrieval. *Nature, 401,* 699–703.

Tulving, E. (1968). Theoretical issues in free recall. In T. R. Dixon and D. L. Horton (Eds.), *Verbal behavior and general behavior theory* (pp. 2–36). Englewood Cliffs, NJ: Prentice Hall.

Uncapher, M. R., Hutchinson, J. B., & Wagner, A. D. (2011). Dissociable effects of top-down and bottom-up attention during episodic encoding. *Journal of Neuroscience, 31,* 12613–12628.

Uncapher, M. R., & Rugg, M. D. (2009). Selecting for memory: The influence of selective attention on the mnemonic binding of contextual information. *Journal of Neuroscience, 29,* 8270–8279.

Uncapher, M. R., & Wagner, A. D. (2009). Posterior parietal cortex and episodic encoding: Insights from fMRI subsequent memory effects and dual-attention theory. *Neurobiology of Learning and Memory, 91,* 139–154.

Wagner, A. D. (2002). Cognitive control and episodic memory: Contributions from prefrontal cortex. In L. R. Squire & D. L. Schacter (Eds.), *Neuropsychology of memory* (3rd ed., pp. 174–192). New York: Guilford Press.

Wagner, A. D., Maril, A., Bjork, R. A., & Schacter, D. L. (2001). Prefrontal contributions to executive control: fMRI evidence for functional distinctions within lateral prefrontal cortex. *Neuroimage, 14,* 1337–1347.

Wagner, A. D., Schacter, D. L., Rotte, M., Koutstaal, W., Maril, A., Dale, A. M.,...Buckner, R. L. (1998). Building memories: Remember and forgetting of verbal experiences as predicted by brain activity. *Science, 281,* 1188–1191.

Wimber, M., Alink, A., Charest, I., Kriegeskorte, N., & Anderson, M. C. (2015). Retrieval induces adaptive forgetting of competing memories via cortical pattern suppression. *Nature Neuroscience, 18,* 582–589.

Yi, D. J., & Chun, M. M. (2005). Attentional modulation of learning-related attenuation effects in human parahippocampal cortex. *Journal of Neuroscience, 25,* 3593–3600.

Zanto, T. P., Rubens, M. T., Thangavel, A., & Gazzaley, A. (2011). Causal role of the prefrontal cortex in top-down modulation of visual processing and working memory. *Nature Neuroscience, 14,* 656–661.

21

Learning Cognitive Control

Apoorva Bhandari, David Badre, and Michael J. Frank

Introduction

Human behaviour is remarkably flexible. In a wide variety of settings, we actively engage in behaviour that is consistent with our task goals, rather than just being reflexively driven by stimuli in the immediate environment. As our task environment changes, or our task goals evolve, we rapidly adjust our behaviour to the new constraints. Such flexibility is believed to emerge from the operation of a putative cognitive control system, consisting of a set of *controllers* that configure our cognitive processes for the task at hand. By adopting different controller states, our cognitive processes can be rapidly reconfigured for different tasks or different operations within a complex task (Norman & Shallice, 1986). To achieve flexible control, therefore, we require *control policies* that dictate what state the controllers adopt as task requirements change. In addition, we must develop useful *control representations* that efficiently code for a repertoire of controller states that can be invoked in familiar tasks and generalised to novel ones.

The past two decades have seen significant strides in unpacking these abstract ideas and identifying the neural machinery that implements cognitive control. Human imaging and patient studies have identified a network of brain regions, including areas in the prefrontal and parietal cortex, as well as sub-cortical nuclei in the basal ganglia, that are associated with control functions (Badre & D'Esposito, 2007; Bechara, Damasio, Damasio, & Anderson, 1994; Cole & Schneider, 2007; Dosenbach et al., 2006; Duncan & Owen, 2000; Jahfari et al., 2012; Lewis, Dove, Robbins, Barker, & Owen, 2004; Monchi, Petrides, Petre, Worsley, & Dagher, 2001; Nomura et al., 2010; Owen, Downes, Sahakian, Polkey, & Robbins, 1990) and may, therefore, serve as the neural instantiation of controllers (Power & Petersen, 2013). Neurophysiological studies in non-human primates have revealed the rich, distractor-resistant coding of task-relevant information by neurons in the prefrontal cortex (Asaad, Rainer, & Miller, 1998; Mansouri, Tanaka, & Buckley, 2009; Sakagami & Niki, 1994; White & Wise, 1999), implicating the region as the neural substrate of a working memory function that supports flexible task performance (Duncan, 2001; Miller & Cohen, 2001; Chapter 13 by Stokes, Buschman, & Miller in this volume). Meanwhile, detailed computational models have helped build explicit links between cognitive constructs and specific biological mechanisms in the prefrontal cortex and basal ganglia, thus providing an integrated theoretical account of how control might operate mechanistically (Collins & Frank, 2013; Frank & Badre, 2012; Frank, Loughry, & O'Reilly, 2001; Kriete, Noelle, Cohen, & O'Reilly, 2013; O'Reilly & Frank, 2006; O'Reilly, Herd, & Pauli, 2010 for review; Wiecki & Frank, 2013; Chapter 8 by Verguts in this volume).

Central to the problem of cognitive control are issues of learning. How do we come to acquire task control policies that can drive appropriate transitions between controller states in

The Wiley Handbook of Cognitive Control, First Edition. Edited by Tobias Egner.

accordance with changing task demands? How do we come to discover control representations that are suitable for a given task and can be generalised to novel tasks? Answering these questions is critical for building accounts of cognitive control that do not rely on a hidden homunculus, which 'just knows' how to behave and does all the heavy lifting of control behind the scenes (Monsell, 2003).

The past decade has seen key developments in this respect, as our growing understanding of learning mechanisms in the brain has been incorporated into models of cognitive control. One fruitful development has been the seminal discovery of 'reward prediction error' signals in the mid-brain dopaminergic nuclei (Montague, Dayan, & Sejnowski, 1996; Schultz, Dayan, & Montague, 1997) that are widely and diffusely distributed to the frontal cortex and the striatum. These signals have been shown to support reinforcement learning in motor action selection (Graybiel, 1995), and it has been proposed that they may also support the learning of control policies (Braver & Cohen, 2000; O'Reilly & Frank, 2006). Similarly, recent studies have begun to reveal how an interplay of the architecture of the cognitive control system, dopaminergic reinforcement learning, and domain-general cortical learning mechanisms may support the discovery of abstract control representations, which are critical for generalisation and rapid learning in novel situations (Collins & Frank, 2013; Frank & Badre, 2012; Reynolds & O'Reilly, 2009; Rougier, Noelle, Braver, Cohen, & O'Reilly, 2005).

Simultaneously, the intersection of research in these two areas has also foregrounded the influence of cognitive control on learning. It has been recognised that the control mechanisms of working memory and gating provide a solution to learning problems involving temporal delays and partially observable environments (Hochreiter & Schmidhuber, 1997; Todd, Niv, & Cohen, 2009). Similarly, computational models and model-based analyses of fMRI data have shown that a putative hierarchy of abstract representations in the prefrontal cortex can support a form of hierarchical reinforcement learning (Badre & Frank, 2012; Frank & Badre, 2012). Indeed, recent work has demonstrated that even simple learning problems engage working memory, showing its ubiquitous influence in learning processes (Collins & Frank, 2012, 2013), and that the extent to which participants recruit prefrontal hierarchical control is related to their speeded learning and transfer (Collins, Cavanagh, & Frank, 2014; Frank & Badre, 2012).

In this chapter, we examine both sides of this reciprocal interaction between cognitive control and learning. We begin by reviewing the key functional and mechanistic components of cognitive control, and, in the process, highlight the primary loci where learning interacts with cognitive control. We then consider the specific learning mechanisms at play in the learning of control policies and control representations. Finally, we examine how cognitive control processes influence learning.

Cognitive Control and the Gating Framework

Working Memory and Top-Down Control

Early models of cognitive control identified two key biological components of a control system. The first is working memory, a function associated with pyramidal neurons in the lateral prefrontal cortex (reviewed in Goldman-Rakic, 1995). These neurons encode information over delays (Fuster, 1973; Fuster & Alexander, 1971), and in the face of distracting events in a task episode (Miller, Erickson, & Desimone, 1996). Biophysical models have demonstrated that these properties can emerge from recurrent neural activity, intrinsic bistability, or synaptic mechanisms (Durstewitz & Seamans, 2002; Mongillo, Barak, & Tsodyks, 2008; Wang, 1999; Zipser, 1991). In models of cognitive control, these neurons are envisaged as representing the task 'context' including information about task goals, rules, or other information necessary for carrying out a task (Chapter 9 by Chiew & Braver in this volume; Cohen, Dunbar, & McClelland, 1990; Dehaene, Kerszberg, & Changeux, 1998). The second component of the control system is 'top-down bias', implemented via projections from lateral prefrontal cortex neurons to other brain regions (Cohen et al., 1990; Cohen & Servan-Schreiber, 1992; Desimone & Duncan, 1995). These connections allow the contents of working memory to influence attention and decision-making processes in the rest of the brain. Collectively, these two components recapitulate the architecture of a controller and its influence on processors in biological terms, with the pattern of activity in lateral prefrontal cortex neurons reflecting the state of the controller, and projections from these neurons configuring processing in the rest of the brain for the task at hand.

Dynamic Gating Mechanisms

Subsequent work recognised that flexible control places conflicting functional demands on working memory (Braver & Cohen, 2000; Frank et al., 2001). On the one hand, previously encountered task-relevant information must be stably maintained in working memory in the face of irrelevant, distracting events. On the other hand, one must sometimes rapidly and selectively update the contents of working memory with new relevant information. Mechanistically, these two goals are in conflict: Strengthening robust maintenance involves making the working memory system less sensitive to inputs, whereas efficient updating involves making the system more sensitive to inputs.

In order to solve this 'flexibility versus stability' paradox, a number of computational models have incorporated a computationally separate 'input gating' mechanism (Braver & Cohen, 2000; Cohen, Braver, & O'Reilly, 1996; Frank et al., 2001; O'Reilly, Braver, & Cohen, 1999), which provides a means of modulating the influence of external inputs onto prefrontal neurons. Such a separate gating mechanism is computationally efficient (Hochreiter & Schmidhuber, 1997). In such a system, if the gate is 'open', prefrontal neurons are sensitive to the external input, leading to an update of the contents of working memory. If the gate is 'closed', the neurons remain robust to the external input and continue to actively maintain the contents of working memory. In other words, input gating serves as a selection mechanism for determining when the content of working memory should be updated.

In a similar vein, one may also view the influence of stored working memory on attention and decision making as being subject to selection, especially when multiple items are simultaneously stored in working memory (Chatham, Frank, & Badre, 2014; Hazy, Frank, & O'Reilly, R., 2007). In this case, the state of the 'output gate' determines whether and which information being maintained in working memory has an influence on processing in the rest of the brain. Output gating can also be used to select which higher-order, abstract PFC action representation should be used to contextualise gating of lower-order actions (Collins & Frank, 2013; Frank & Badre, 2012; Kriete et al., 2013).

This mechanistic perspective highlights two key learning problems relevant to control. First, the system must learn what information is relevant, and should be stored in working memory for successful control of future behaviour. Second, the system must learn which information currently in working memory is allowed to influence cognitive processing in the rest of the brain. The input and output 'gates' discussed above are key loci at the heart of these learning problems as they serve as the selection mechanisms for working memory update and the influence of working memory, respectively. On this view, the control system must be able to learn a 'gating policy' that specifies when, and under what conditions, the gates to working memory are opened or closed. Such a gating policy would shape the contents of working memory, thus enabling the adoption of task-appropriate controller states and cognitive configurations.

Biological Substrates of Gating

Different proposals have emerged for how gating and the learning of a gating policy are biologically implemented. The specifics of these proposals are important to discuss, because they make different predictions about where in the brain learning must take place. One class of models implements gating via the actions of dopaminergic inputs from the ventral tegmental area (VTA) and substantia nigra pars compacta (SNc) into the prefrontal cortex. Dopamine has long been implicated in working memory functions (Arnsten, Cai, Murphy, & Goldman-Rakic, 1994; Brozoski, Brown, Rosvold, & Goldman, 1979; Luciana & Collins, 1997; Luciana, Depue, Arbisi, & Leon, 1992). Mid-brain dopaminergic neurons show tonic firing, and brief, phasic departures from this tonic firing in response to specific events in the environment. In a novel task environment, these phasic responses occur when the organism receives unexpected reinforcement—bursts for positive reinforcement, dips for negative reinforcement. With more experience with the task environment, these phasic responses start occurring in response to unexpected cues that reliably predict future reinforcement (reviewed in Schultz, 2013). Neuroimaging studies have also found evidence for a similar phasic dopamine signal in humans (D'Ardenne, McClure, Nystrom, & Cohen, 2008). This 'reward prediction error' signal, because it occurs in response to unexpected contextual cues that reliably predict future reinforcement, is appropriately timed to gate the storage of the contextual cue in working memory. Indeed, biophysically, dopamine can shift the balance in prefrontal networks from stable to flexible via different modes of processing dependent on D1 and D2 receptors (Durstewitz & Seamans, 2002). D'Ardenne et al. (2012) recently demonstrated using fMRI that activation in the SNc/VTA correlated with behaviour and activity in a prefrontal region causally related to context updating. Though this study did not measure DA activity directly, the phasic nature of the VTA–PFC interaction implied by these results is consistent with this view.

However, there are at least two problems with dopamine prediction error signals acting as the sole gating mechanism. First, dopamine signals are relatively homogeneous, and hence a DA-based gating mechanism is sufficient when only one item must be stored, but is incapable of providing a selective gating mechanism that would allow updating of one piece of information while continuing to support robust maintenance of other information, as is required in several tasks (Frank et al., 2001). Second, and relatedly, phasic DA signals are sufficient when a stimulus is presented that could elicit a reward prediction error (because its maintenance is predictive of subsequent success), but this mechanism would fail if a task requires updating one stimulus and then another before making a final response. In this case, once learning has occurred, the first stimulus could elicit a reward prediction error, but the second would not (because the reward was already expected), and hence it would not be updated.

A more flexible mechanism was proposed by Frank et al. (2001), who hypothesised that although dopamine might be important for learning when to gate, the gating itself is carried out by the basal ganglia (BG), consistent with an analogous role they play in motion action selection (Doya, 1999; Graybiel, 1995; Gurney, Prescott, & Redgrave, 2001; Mink, 1996; Redgrave, Prescott, & Gurney, 1999), and where dopamine is also relevant for learning which action to select (Frank, 2005). The lateral prefrontal cortex and the BG are connected via multiple, parallel, re-entrant loops (Alexander, DeLong, & Strick, 1986), each connecting one region in the prefrontal cortex to a corresponding patch of the dorsal striatum, and back to the prefrontal cortex via the thalamus. These loops are further subdivided at the micro level, where individual prefrontal 'stripes' of relatively isolated, interconnected neurons can be selectively targeted by distinct circuits in the BG and thalamus (Frank et al., 2001). In the model, these loops implement input gating through the contrasting effects of 'go' and 'no-go' neurons in the dorsal striatum that are respectively part of the 'direct' and 'indirect' pathways. 'Go' neuron firing has a disinhibitory effect on thalamo-cortical communication, leading to a loop of excitation in the corresponding prefrontal stripe that transiently increases its sensitivity to external inputs, thus allowing update. The model also incorporates a mechanism for output gating via similar prefrontal-BG loops (Hazy et al., 2007). This scheme has received support from patient studies (Baier et al., 2010), as well as recent results from neuroimaging (Chatham et al., 2014; McNab & Klingberg, 2008; van Schouwenburg, den Ouden, & Cools, 2010). Unlike the dopamine-based gating mechanism, this architecture supports the storage of multiple items in working memory, and allows selective updating or output gating of each item independently (Hazy et al., 2007). Dopamine reward prediction errors are still relevant, however, for learning when and what to gate (see below).

Learning Gating Policies

The Problem

Biological learning mechanisms supporting the acquisition of a gating policy must rely only on reinforcement signals, given that, often, no explicit signal indicates what items one should hold in mind to solve a problem. Any such mechanism faces a temporal credit assignment problem (O'Reilly & Frank, 2006). In any complex task, the reinforcement associated with a adopting a controller state (via gating) is usually available only later in time, often after several intervening actions or events. Therefore, the learning mechanism must be able to bridge the gap between the adoption of a relevant controller state and later reinforcement, and increase the likelihood of the gating signal that led to the adoption of controller states associated with positive outcomes (note that positive outcomes do not need to imply primary reward, but can be internally generated based on successful achievement of a goal). Models that rely on multiple controllers also face a spatial credit assignment problem: When a positive outcome occurs, which of the multiple gating signals should receive credit for having obtained this outcome?

Learning Gating Policies with Reinforcement Learning

In both classes of gating models discussed earlier, the acquisition of gating policies relies on dopamine 'reward prediction error' signals. Such prediction error signals are often utilised to drive learning in computational reinforcement learning algorithms (Glimcher, 2011; Schultz et al., 1997). The effect of dopamine on neural circuits is well established. At synapses, dopamine has been shown to enhance long-term potentiation (LTP) through its action via the D1-receptor (Calabresi, Picconi, Tozzi, & Di Filippo, 2007; Reymann & Frey, 2007; Shen, Flajolet, Greengard, & Surmeier, 2008). Indeed, evidence from pharmacogenetic and optogenetic manipulations of dopamine across species support its role in reinforcement learning (reviewed in Collins & Frank, 2014).

The properties of dopamine neurons make them well suited to solving the temporal credit assignment problem. As described above, these neurons initially produce phasic responses to unexpected reinforcement that, with experience, move to the onset of stimuli that reliably predict the reinforcement. In other words, a prediction error signal bridges the temporal gap between the cue and the reinforcing outcome, and hence, can be employed to train a gating signal. This training is based on dopamine modulating activity in units that generate the gating signal and reinforcing learning at synapses between cortical neurons representing external events or internal states, and the neurons that are part of the gating machinery.

In the model proposed by Braver and Cohen (2000), where dopamine also serves as the gating signal, learning proceeds via the temporal-difference (TD) algorithm. Dopamine activity is computed as the difference between the total expected current and (discounted) future reward at an earlier moment in time, and the actual current reinforcement (Sutton & Barto, 1998). When such a prediction error signal occurs, it is used to modify the expectation of reward at earlier moments in time via the updating of the weights of the inputs into the dopaminergic units themselves. If an earlier cue reliably predicts rewards, with experience, it will drive reward expectation (and thus dopamine bursts). As this dopamine activity is also used for gating, this will drive the gating of the cue into working memory. As long as such a gating operation results in positive outcomes, the TD mechanism will support its learning. This mechanism, though simple and elegant, suffers from a key limitation. It can only support the acquisition of a single, global updating policy and cannot support the learning of a complex gating policy that controls the selective gating of multiple items.

How are the dopamine mechanisms extended to support selective gating when multiple items must be maintained separately? O'Reilly and Frank (2006) also employ dopamine-based reinforcement learning, with the phasic dopamine signal being generated by associative mechanisms (O'Reilly, Frank, Hazy, & Watz, 2007) rather than by TD mechanisms per se. But rather than implementing the gating signal itself, dopamine is used to train striatal gating. More specifically, the phasic dopamine burst reinforces firing of 'go' units in the dorsal striatum via the action of simulated D1 receptors, while a phasic dopamine dip reinforces 'no-go' firing via the action of D2 receptors. Thus, when a prefrontal representation is associated with a reward prediction error (because its maintenance has been predictive of success in the past), the phasic burst reinforces the gating of the particular striatal loop that recently elicited a gating signal, and hence was likely to have caused that prediction error. In this way, the learning is applied selectively to some loops, while others can learn to not gate information but to continue maintaining it. Conversely, gating of irrelevant stimuli that lead to poor performance results in dopamine dips. This reinforces 'no-go' firing, and thereby reduces the likelihood of a future gating signal. As the learning and gating mechanisms are separate, this model can support the learning of selective gating signals.

In summary, these models demonstrate that gating policies can be acquired using reinforcement learning mechanisms. Indeed, the learning of gating policies is readily situated into the wider action selection and actor-critic reinforcement learning framework, with gating being a kind of 'cognitive action' (O'Reilly & Frank, 2006). Empirically, systemic administration of D2 receptor agonists and antagonists in healthy participants produces analogous effects on reinforcement learning and working memory task performance (Frank & O'Reilly, 2006). On the other hand, Parkinson disease patients on and off dopaminergic medication show contrasting deficits in working memory tasks, with patients off medication (low dopamine levels) more resistant to distractors, but impaired in learning a new gating policy (Cools, Miyakawa, Sheridan, & D'Esposito, 2010; Moustafa, Sherman, & Frank, 2008). However, these studies cannot easily separate the effects of dopamine in the striatum and prefrontal cortex. D'Ardenne et al. (2012) reported phasic activity in the VTA/SNc during updating of context stimuli, but these results cannot speak to whether this activity reflects a gating or a training signal, or both. More direct and specific tests of the hypothesis that phasic dopamine trains gating policies are necessary to differentiate these putative roles.

Learning Abstract, Hierarchical Action Policies

Tasks that involve a large number of stimulus–response contingencies can sometimes be simplified if an abstract rule is available that allows the complex contingencies to be construed hierarchically, with a higher-level context constraining contingencies at a lower level. When such a structure is available in a task, it can be exploited to aid learning by simplifying the problem to be learnt. Frank and Badre (2012) investigated how such abstract, hierarchical action policies could be discovered using PFC-BG gating circuits and dopaminergic reinforcement learning.

Building on the O'Reilly and Frank (2006) model, Frank and Badre (2012) incorporated hierarchically nested PFC-BG loops into their model. More specifically, they posited two PFC-BG loops, one higher-order loop, located more rostral (corresponding to dorsal pre premotor cortex—prePMd), and one lower-order loop, located more caudal (corresponding to dorsal premotor cortex—PMd). The rostral loop input gates higher-level contextual representations into working memory, which constrain the output gating of the lower-order working memory representation in the caudal loop. The output-gated caudal representation then contextualises action selection in a motor loop. These nested, interacting loops allow the testing of hierarchical, conditional hypotheses and thus support the learning of abstract, hierarchical action policies.

The model was deployed to explain data from an fMRI study (Badre, Kayser, & D'Esposito, 2010) in which subjects were tasked with learning a set of 18 stimulus–action mappings. Each stimulus was defined by a conjunction of three stimulus dimensions—shape, colour, and orientation—and was mapped onto one of three response keys. Each subject learnt two different sets of mappings: a 'flat' mapping in which the 18 mappings were independent, and a 'hierarchical' mapping, which could also be learnt as an abstract policy, where one stimulus dimension indicated which of the other two stimulus dimensions was relevant.

The hierarchical corticostriatal model recapitulated the behaviour of human subjects by learning faster in the hierarchical condition by learning an abstract policy, and this learning critically depended on dopaminergic reinforcement of striatal gating. The model also displayed similar activity dynamics in the prePMd with activity declining with learning in the flat, but not in the hierarchical condition. In the model, this activity decline was explained by the fact that, in the absence of hierarchical structure in the flat condition, prePMD activity patterns are counterproductive because they force attention on a single dimension, thus leading to negative prediction errors. Over time, the BG module learns to reduce the likelihood that any stimuli will be gated into the prePMd. An empirical analysis of the imaging data was consistent with this mechanism: those subjects exhibiting the strongest striatal response to reward prediction errors also showed the greatest decline in prePMd activity in the flat condition (Badre & Frank, 2012).

Learning Control Representations

The Problem

In the gating framework, inputs into the prefrontal cortex, when accompanied by an enabling gating signal, drive an update of working memory, thus transitioning controllers to a new task-appropriate state. This account presupposes that neurons in the lateral prefrontal cortex are able to represent a repertoire of controller states and that inputs into these neurons have well-established connectivity patterns and weights that can trigger appropriate transitions between these states. Indeed, in computational models of control, including those that are able to learn gating policies from experience, these prefrontal representations are often hand-coded (Braver & Cohen, 2000; Cohen et al., 1990; Dehaene et al., 1998; O'Reilly et al., 1999). However, just like gating policies, these control representations must emerge from a learning process that integrates training experience and prior biases. Therefore, a critical question concerns the learning mechanisms that support the development of useful control representations. Indeed, the nature of these control representations is critical to how a control system performs in novel task environments.

Real-world task environments are often highly complex, and learning a new control policy with reinforcement learning algorithms can require very large amounts of training (as, indeed, models do). On the other hand, humans are able to adapt to novel task environments much more quickly. It has been suggested that we accomplish this feat by exploiting *shared structure* in real-world task environments (Botvinick & Cohen, 2014; Collins & Frank, 2013; Gershman & Niv, 2010; Kemp & Tenenbaum, 2008). Such structure affords opportunities for abstraction and generalisation, allowing us to make powerful inductive inferences about a new task environment with very little experience (Tenenbaum, Kemp, Griffiths, & Goodman, 2011). The use of abstract, componential control representations may aid such generalisation in cognitive control (Botvinick & Cohen, 2014). Indeed, the prefrontal cortex has long been known to represent abstract rules (Wallis, Anderson, & Miller, 2001). It has also been suggested that these abstract representations and the ability to use them emerges slowly through development and that flexible behaviour in adults depends on such representations (Bunge & Zelazo, 2006; Jacques & Zelazo, 2005; Kharitonova & Munakata, 2011). More recently, multivariate pattern analysis of human fMRI data from the prefrontal cortex has provided evidence for abstract, componential representations in the prefrontal cortex (Cole, Etzel, Zacks, Schneider, & Braver, 2011; Reverberi, Görgen, & Haynes, 2012; Woolgar, Thompson, Bor, & Duncan, 2011).

Learning to Represent Abstract Stimulus Dimensions

A common feature of several tasks is that often only one of several stimulus dimensions is relevant to a particular task's performance. For instance, when packing clothes in a travel bag, shape is much more relevant than colour. On the other hand, for sorting clothes for the laundry, colour becomes much more relevant. At the same time, these dimensions are usually relevant across multiple tasks—for instance, colour information is important at the traffic light, while watching a game of football, or while looking for ripe bananas at the grocery store. It would be useful, therefore, for a control system involved in performing all these tasks to develop the ability to represent the colour of the stimulus.

Rougier et al. (2005) asked whether such representations could emerge in a model using domain-general cortical learning mechanisms and a gating architecture for cognitive control, if it was exposed to a training environment in which the stimulus dimension was a shared structure feature of several tasks. To test this, they created a model with a gating architecture. The model incorporates a hidden layer, a 'PFC' layer, and 'dopamine'-based gating. Critically, the weights of the inputs into the 'PFC' layer were randomly initialised. These weights were then learnt during training using the general-purpose Leabra algorithm that combines Hebbian and error-driven learning, capturing in abstract form several cortical learning mechanisms. Initially, the 'PFC' units had no particular selectivity. They created a training environment made of four different tasks involving multidimensional stimuli. For each task, only one of the stimulus dimensions was relevant, and the relevant feature in that dimension switched every few trials. Therefore, the key control demand in all these tasks was the maintenance of the relevant stimulus dimension over the task.

Rougier et al. (2005) compared two instantiations of the model, one that was trained on all four tasks, and another that was trained on only two of the tasks, but with twice the number of trials. Both the models were then tested on previously unseen stimuli. Both models learnt to perform well on the tasks and stimuli they were trained on, but the model trained on four tasks was better at generalising to novel stimuli. Therefore, the breadth of experience (rather than the amount, which was equated) promoted generalisation. Moreover, an examination of the trained weights into the context layer showed that each unit was selective to a single stimulus dimension and all features in that dimension. Therefore, the PFC layer had discovered a componential representation comprising units that formed a full basis set for coding any stimulus dimension. Rougier et al. (2005) also showed that such an abstract, componential representation did not emerge in other networks that either lacked a PFC layer, or lacked adaptive gating, despite the same training. In particular, the gating mechanism facilitated a stable PFC representation during blocks of trials in which a particular stimulus dimension was relevant, which would be destabilised owing to negative reward prediction errors when the task changed.

Learning to Represent Abstract, Generalisable Task Sets with Arbitrary Dynamic Task Shifts

Another important structural feature of real-world task environments is stable relationships between stimuli, actions, and outcomes (task sets), which recur in multiple contexts. For instance, answering your mobile phone is an appropriate response to receiving a phone call in multiple different contexts, like when you are at home, or outdoors, or at a party. But, it may be inappropriate in other contexts, like in a business meeting, at a funeral, or in a classroom. Moreover, any given task set cued by a context (home vs. classroom) might define not only one action (answering the mobile phone when it rings) but also which action to take in response to other events (whether and how to greet a friend or acquaintance). Task sets are, therefore, abstract features of a task that can constrain action selection across a variety of situations. From the perspective of a control system, an abstract task-set would be a useful feature that can serve as a control representation, with contextual cues then serving to activate such a representation. Indeed, humans are capable of learning and reusing such task sets (Collins & Koechlin, 2012). In addition, evidence from human imaging studies using multivariate pattern analysis supports the idea that the prefrontal cortex represents such abstract task sets that are invariant to the contextual cues that activated them (Woolgar et al., 2011).

How are such abstract task-set representations learnt and used? Collins and Frank (2013) built a neural network model based on the gating architecture employed by O'Reilly and Frank (2006) and its hierarchical extension by Frank and Badre (2012) to examine how abstract task-set representations emerge in the biologically realistic cortico-BG network model. In their model, context inputs into the PFC layer were fully connected and had random initial weights, such that any given PFC pattern could become active in response to any or multiple contexts, subject to BG gating. The PFC layer projected to a parietal cortex layer, which also received input from lower-level stimulus information, so that stimulus representations were contextualised by PFC input, before projecting to a second premotor cortex-BG action selection circuit. Reinforcement learning signals operated at both circuits, so that with learning to gate, the PFC representations came to represent abstract task sets that guide lower-level action selection, and the lower-level circuit learnt which response to make given the selected task set and the input stimulus. Unlike the Rougier et al. (2005) model, the task context input could change arbitrarily (e.g., on every trial).

Critically, the model learnt when to reuse existing task-set representations across multiple contexts, if those contexts shared the same set of rules. In that case, the gating of similar task representations is reinforced, and the lower-level circuit automatically inherits all of the previous learning attributed to that task state. Furthermore, these abstract representations allowed the model to append any new knowledge (new stimulus–action associations) to an existing structure (much as one can append a new word to a known language), which can then be immediately generalised to other contexts indicative of that structure. The hierarchical mechanisms proposed have also been supported by EEG studies showing early anterior frontal signals that precede later posterior frontal signals during task-set learning, which are predictive of subjects' ability to transfer knowledge to novel contexts (Collins et al., 2014). EEG correlates of hierarchically structured reward prediction errors are also predictive of subjects' ability to append new knowledge to existing structures (Collins & Frank, 2016).

Collectively, these results powerfully demonstrate how the combination of architectural constraints, reinforcement learning mechanisms, general-purpose cortical learning principles, and a structured training environment can support the emergence of abstract control representations that support generalisation to novel environments.

Cognitive Control Over Learning

In the previous sections, we examined how reinforcement learning and cortical learning mechanisms are deployed in the service of cognitive control, and how they support the acquisition of control policies and the discovery of useful control representations. In this section, we examine the reciprocal question of how cognitive control mechanisms influence learning. In particular, we focus on instrumental learning, which has been extensively analysed in terms of reinforcement learning mechanisms, and review studies that show how control mechanisms complement reinforcement learning mechanisms in a variety of learning situations.

Reinforcement learning (RL) models treat decision making in a task environment as a Markov decision process (MDP), which consists of a set of *states* that the environment can exist in, and a set of *actions* that the agent can take in each of these states. The state of the environment determines the consequences of taking a specific action, both in terms of the reward and in terms of transitions to a different state. The agent's job is to learn a *policy*—a mapping between states and actions—that maximises long-term reward (Sutton & Barto, 1998). This usually involves estimating *values* (expected future reward) associated with each state–action pair using either model-free methods like TD learning, or model-based methods that involve building and searching a world model. In the MDP framework, the state space is assumed to be fully observable, and the agent is assumed to have complete knowledge of the current state. There are, however, a number of learning problems in which the state is not fully observable. Examples include situations in which contingencies depend not just on the current stimulus, but also on past observations, or when they are constrained by a temporally distant goal or a rule. In such *partially observable* environments, a cognitive control system involving a working memory that can maintain past observations may provide a solution to the learning problem. Indeed, Todd et al. (2009) has analysed the psychological constructs of working memory, cognitive control, and gating as comprising an approximate solution to partially observable Markov decision problems.

Control mechanisms may also play an important role in learning problems with a hierarchical structure. Badre et al. (2010) found that participants learnt a set of stimulus–response mappings faster when an abstract rule was available that enabled generalisation across several individual stimulus–response pairs. As discussed earlier, the discovery of such an abstract rule depended on a hierarchical interaction between anterior and posterior PFC-BG loops (Badre & Frank, 2012; Frank & Badre, 2012) and the treatment of one stimulus feature as a higher-level context. In other words, the architecture supported the testing of the hierarchical hypothesis of the underlying rule.

An even more striking demonstration of the a priori consideration of a hierarchical structure comes from Collins and Frank (2013), who asked participants to learn four stimulus–response mappings where the four stimuli were defined by two colours and two shapes. Crucially, the two feature dimensions were symmetric—both colour and shape are required for making a correct response, and indeed, it was possible to learn the mappings as a conjunction of the two stimulus features. However, after learning, subjects displayed differential reaction time switch costs for switching shape or colour, suggesting that they were treating one of the two dimensions as a higher-order context signal. The transfer phase in the experiment provided strong evidence that participants built the proposed structure. Subjects who had learnt to treat colour as a contextual cue were given stimuli with two new colours (but the same shapes). For one of these colours, stimulus–response–outcome contingencies matched with those of a previously encountered colour. For the other colour, novel stimulus–response–outcome contingencies were used. Consistent with the predictions of the model, subjects showed positive transfer for the matched contingencies and negative transfer for the novel contingencies. In other words, subjects had successfully generalised a previously learnt task set to a novel context. On the other hand, when a truly novel context appeared, signalling a new task set, they learnt more slowly. Moreover, the generalisation benefits were predicted by the switch costs in the earlier learning phase.

These findings are consistent with the above-described neural network model, in which cognitive control mechanisms prevent action selection in the lower response circuit until conflict is resolved at the level of task-set selection. Simulations showed that such control mechanisms induce switch costs but prevent interference in learnt mappings across task sets, and hence are more efficient across learning trials. Moreover, the empirical findings provide striking evidence for humans spontaneously using abstract task sets even in a simple learning task, when using such a structure provided no initial benefit. As Collins and Frank (2013) point out, this preference might be motivated by the long-term benefits of creating abstract task sets that affords generalisation in novel situations that may be encountered in the future. Indeed, even 8-month-old infants show the tendency to build a priori structure in the same task, suggesting that control functions may play a key role in the development of abstract representations (Werchan, Collins, Frank, & Amso, 2015). Collectively, these results reveal the rich involvement of cognitive control mechanisms in simple and complex learning problems that complement slow, RL mechanisms. They also point to interactions between these mechanisms that will require careful consideration when studying either process.

Conclusions

The flexibility of human behaviour is often laid at the feet of our cognitive control system. Yet, learning processes are essential partners in the enterprise. Our cognitive control system makes several demands on our learning systems, continually relying on them for shaping control policies and building useful control representations. In the past decade, as our understanding of cortical and dopaminergic learning mechanisms has grown, they have been firmly implanted into accounts of cognitive control. The use of detailed, explicit computational modelling at different levels of abstraction has also allowed us to understand the complex interplay of different learning mechanisms, the architecture of cognitive control, and the structure of task environments. In turn, we are beginning to discover that cognitive control may be a critical player in both simple and complex learning tasks. Therefore, it will become more and more crucial in the future to study how these processes interact to generate flexible, adaptive cognition and behaviour. In particular, a key open area of investigation relates to the learning mechanisms that support the remarkable human ability to rapidly adopt appropriate control structures in novel tasks with arbitrary structure and complexity, relying only a few lines of verbal instruction.

References

Alexander, G. E., DeLong, M. R., & Strick, P. L. (1986). Parallel organization of functionally segregated circuits linking basal ganglia and cortex. *Annual Review of Neuroscience*, *9*, 357–381.

Arnsten, A. F., Cai, J. X., Murphy, B. L., & Goldman-Rakic, P. S. (1994). Dopamine D1 receptor mechanisms in the cognitive performance of young adult and aged monkeys. *Psychopharmacology (Berl)*, *116*(2), 143–151. Retrieved from http://www.ncbi.nlm.nih.gov/pubmed/7862943

Asaad, W. F., Rainer, G., & Miller, E. K. (1998). Neural activity in the primate prefrontal cortex during associative learning. *Neuron*, *21*(6), 1399–1407.

Badre, D., & D'Esposito, M. (2007). Functional magnetic resonance imaging evidence for a hierarchical organization of the prefrontal cortex. *Journal of Cognitive Neuroscience*, *19*(12), 2082–2099.

Badre, D., & Frank, M. J. (2012). Mechanisms of hierarchical reinforcement learning in Cortico–Striatal Circuits 2: Evidence from fMRI. *Cerebral Cortex*, *22*(3), 527–536.

Badre, D., Kayser, A. S., & D'Esposito, M. (2010). Frontal cortex and the discovery of abstract action rules. *Neuron*, *66*(2), 315–326.

Baier, B., Karnath, H. O., Dieterich, M., Birklein, F., Heinze, C., & Muller, N. G. (2010). Keeping memory clear and stable—The contribution of human basal ganglia and prefrontal cortex to working memory. *Journal of Neuroscience*, *30*(29), 9788–9792.

Bechara, A., Damasio, A. R., Damasio, H., & Anderson, S. W. (1994). Insensitivity to future consequences following damage to human prefrontal cortex. *Cognition*, *50*(1–3), 7–15. Retrieved from http://www.ncbi.nlm.nih.gov/pubmed/8039375

Botvinick, M. M., & Cohen, J. D. (2014). The computational and neural basis of cognitive control: Charted territory and new frontiers. *Cognitive Science*, *38*(6), 1249–1285.

Braver, T. S., & Cohen, J. D. (2000). On the control of control: The role of dopamine in regulating prefrontal function and working memory. *Control of cognitive processes: Attention and performance XVIII* (pp. 713–737). MIT Press: Cambridge, MA.

Brozoski, T. J., Brown, R. M., Rosvold, H. E., & Goldman, P. S. (1979). Cognitive deficit caused by regional depletion of dopamine in prefrontal cortex of rhesus monkey. *Science*, *205*(4409), 929–932. Retrieved from http://www.ncbi.nlm.nih.gov/pubmed/112679

Bunge, S. A., & Zelazo, P. D. (2006). A brain-based account of the development of rule use in childhood. *Current Directions in Psychological Science*, *15*(3), 118–121.

Calabresi, P., Picconi, B., Tozzi, A., & Di Filippo, M. (2007). Dopamine-mediated regulation of corticostriatal synaptic plasticity. *Trends in Neurosciences*, *30*(5), 211–219.

Chatham, C. H., Frank, M. J., & Badre, D. (2014). Corticostriatal output gating during selection from working memory. *Neuron*, *81*(4), 930–942.

Cohen, J. D., Braver, T. S., & O'Reilly, R. C. (1996). A computational approach to prefrontal cortex, cognitive control and schizophrenia: Recent developments and current challenges. *Philosophical Transactions of the Royal Society of London. Series B: Biological Sciences*, *351*(1346), 1515–1527.

Cohen, J. D., Dunbar, K., & McClelland, J. L. (1990). On the control of automatic processes: A parallel distributed processing account of the Stroop effect. *Psychological Review*, *97*(3), 332.

Cohen, J. D., & Servan-Schreiber, D. (1992). Context, cortex, and dopamine: A connectionist approach to behavior and biology in schizophrenia. *Psychological Review*, *99*(1), 45–77. Retrieved from http://www.ncbi.nlm.nih.gov/pubmed/1546118

Cole, M. W., Etzel, J. A., Zacks, J. M., Schneider, W., & Braver, T. S. (2011). Rapid transfer of abstract rules to novel contexts in human lateral prefrontal cortex. *Frontiers in Human Neuroscience*, *5*, 142.

Cole, M. W., & Schneider, W. (2007). The cognitive control network: Integrated cortical regions with dissociable functions. *Neuroimage*, *37*(1), 343–360.

Collins, A. G. E., Cavanagh, J. F., & Frank, M. J. (2014). Human EEG uncovers latent generalizable rule structure during learning. *Journal of Neuroscience*, *34*(13), 4677–4685.

Collins, A. G. E., & Frank, M. J. (2012). How much of reinforcement learning is working memory, not reinforcement learning? A behavioral, computational, and neurogenetic analysis. *European Journal of Neuroscience*, *35*(7), 1024–1035.

Collins, A. G. E., & Frank, M. J. (2013). Cognitive control over learning: Creating, clustering, and generalizing task-set structure. *Psychological Review*, *120*(1), 190.

Collins, A. G. E., & Frank, M. J. (2014). Opponent actor learning (OpAL): Modeling interactive effects of striatal dopamine on reinforcement learning and choice incentive. *Psychological Review*, *121*(3), 337–366.

Collins, A. G. E., & Frank, M. J. (2016). Neural signature of hierarchically structured expectations predicts clustering and transfer of rule sets in reinforcement learning. *Cognition*, *152*, 160–169.

Collins, A. G. E., & Koechlin, E. (2012). Reasoning, learning, and creativity: Frontal lobe function and human decision-making. *PLoS Biology*, *10*(3), e1001293.

Cools, R., Miyakawa, A., Sheridan, M., & D'Esposito, M. (2010). Enhanced frontal function in Parkinson's disease. *Brain*, *133*(Pt 1), 225–233.

D'Ardenne, K., Eshel, N., Luka, J., Lenartowicz, A., Nystrom, L. E., & Cohen, J. D. (2012). Role of prefrontal cortex and the midbrain dopamine system in working memory updating. *Proceedings of the National Academy of Sciences of the United States of America*, *109*(49), 19900–19909.

D'Ardenne, K., McClure, S. M., Nystrom, L. E., & Cohen, J. D. (2008). BOLD responses reflecting dopaminergic signals in the human ventral tegmental area. *Science*, *319*(5867), 1264–1267.

Dehaene, S., Kerszberg, M., & Changeux, J. P. (1998). A neuronal model of a global workspace in effortful cognitive tasks. *Proceedings of the National Academy of Sciences of the United States of America*, *95*(24), 14529–14534.

Desimone, R., & Duncan, J. (1995). Neural mechanisms of selective visual attention. *Annual Review of Neuroscience*, *18*(1), 193–222.

Dosenbach, N. U. F., Visscher, K. M., Palmer, E. D., Miezin, F. M., Wenger, K. K., Kang, H. C.,... Petersen, S. E. (2006). A core system for the implementation of task sets. *Neuron*, *50*(5), 799.

Doya, K. (1999). What are the computations of the cerebellum, the basal ganglia and the cerebral cortex? *Neural Networks*, *12*(7–8), 961–974. Retrieved from http://www.ncbi.nlm.nih.gov/pubmed/12662639

Duncan, J. (2001). An adaptive coding model of neural function in prefrontal cortex. *Nature Reviews Neuroscience*, *2*(11), 820–829.

Duncan, J., & Owen, A. M. (2000). Common regions of the human frontal lobe recruited by diverse cognitive demands. *Trends in Neurosciences*, *23*(10), 475–483.

Durstewitz, D., & Seamans, J. K. (2002). The computational role of dopamine D1 receptors in working memory. *Neural Networks*, *15*(4–6), 561–572. Retrieved from http://www.ncbi.nlm.nih.gov/pubmed/12371512

Frank, M. J. (2005). Dynamic dopamine modulation in the basal ganglia: A neurocomputational account of cognitive deficits in medicated and nonmedicated Parkinsonism. *Journal of Cognitive Neuroscience*, *17*(1), 51–72.

Frank, M. J., & Badre, D. (2012). Mechanisms of hierarchical reinforcement learning in corticostriatal circuits 1: Computational analysis. *Cerebral Cortex*, *22*(3), 509–526.

Frank, M. J., Loughry, B., & O'Reilly, R. C. (2001). Interactions between frontal cortex and basal ganglia in working memory: A computational model. *Cognitive, Affective, & Behavioral Neuroscience*, *1*(2), 137–160.

Frank, M. J., & O'Reilly, R. C. (2006). A mechanistic account of striatal dopamine function in human cognition: Psychopharmacological studies with cabergoline and haloperidol. *Behavioral Neuroscience*, *120*(3), 497–517.

Fuster, J. M. (1973). Unit activity in prefrontal cortex during delayed-response performance: Neuronal correlates of transient memory. *Journal of Neurophysiology*, *36*(1), 61–78.

Fuster, J. M., & Alexander, G. E. (1971). Neuron activity related to short-term memory. *Science*, *173*(997), 652–654.

Gershman, S. J., & Niv, Y. (2010). Learning latent structure: Carving nature at its joints. *Current Opinion in Neurobiology*, *20*(2), 251–256.

Glimcher, P. W. (2011). Understanding dopamine and reinforcement learning: The dopamine reward prediction error hypothesis. *Proceedings of the National Academy of Sciences of the United States of America*, *108*(Suppl. 3), 15647–15654.

Goldman-Rakic, P. S. (1995). Cellular basis of working memory. *Neuron*, *14*(3), 477–485. Retrieved from http://www.ncbi.nlm.nih.gov/pubmed/7695894

Graybiel, A. M. (1995). Building action repertoires: Memory and learning functions of the basal ganglia. *Current Opinion in Neurobiology*, *5*(6), 733–741. Retrieved from http://www.ncbi.nlm.nih.gov/pubmed/8805417

Gurney, K., Prescott, T. J., & Redgrave, P. (2001). A computational model of action selection in the basal ganglia. II. Analysis and simulation of behaviour. *Biological Cybernetics*, *84*(6), 411–423. Retrieved from http://www.ncbi.nlm.nih.gov/pubmed/11417053

Hazy, T. E., Frank, M. J., & O'Reilly, R. C. (2007). Towards an executive without a homunculus: Computational models of the prefrontal cortex/basal ganglia system. *Philosophical Transactions of the Royal Society of London. Series B: Biological Sciences*, *362*(1485), 1601–1613.

Hochreiter, S., & Schmidhuber, J. (1997). Long short-term memory. *Neural Computation*, *9*(8), 1735–1780.

Jacques, S., & Zelazo, P. D. (2005). On the possible roots of cognitive flexibility. In B. Homer & C. Tamis-Lemonda (Eds.), *The development of social understanding and communication* (pp. 53–81). Mahwah, NJ: Lawrence Erlbaum Associates.

Jahfari, S., Verbruggen, F., Frank, M. J., Waldorp, L. J., Colzato, L., Ridderinkhof, K. R., & Forstmann, B. U. (2012). How preparation changes the need for top-down control of the basal ganglia when inhibiting premature actions. *Journal of Neuroscience*, *32*(32), 10870–10878.

Kemp, C., & Tenenbaum, J. B. (2008). The discovery of structural form. *Proceedings of the National Academy of Sciences of the United States of America*, *105*(31), 10687–10692.

Kharitonova, M., & Munakata, Y. (2011). The role of representations in executive function: Investigating a developmental link between flexibility and abstraction. *Frontiers in Psychology*, *2*, 347.

Kriete, T., Noelle, D. C., Cohen, J. D., & O'Reilly, R. C. (2013). Indirection and symbol-like processing in the prefrontal cortex and basal ganglia. *Proceedings of the National Academy of Sciences of the United States of America*, *110*(41), 16390–16395.

Lewis, S. J., Dove, A., Robbins, T. W., Barker, R. A., & Owen, A. M. (2004). Striatal contributions to working memory: A functional magnetic resonance imaging study in humans. *European Journal of Neuroscience*, *19*(3), 755–760. Retrieved from http://www.ncbi.nlm.nih.gov/pubmed/14984425

Luciana, M., & Collins, P. F. (1997). Dopaminergic modulation of working memory for spatial but not object cues in normal humans. *Journal of Cognitive Neuroscience*, *9*(3), 330–347.

Luciana, M., Depue, R. A., Arbisi, P., & Leon, A. (1992). Facilitation of working memory in humans by a d2 dopamine receptor agonist. *Journal of Cognitive Neuroscience*, *4*(1), 58–68.

Mansouri, F. A., Tanaka, K., & Buckley, M. J. (2009). Conflict-induced behavioural adjustment: A clue to the executive functions of the prefrontal cortex. *Nature Reviews Neuroscience*, *10*(2), 141–152.

McNab, F., & Klingberg, T. (2008). Prefrontal cortex and basal ganglia control access to working memory. *Nature Neuroscience*, *11*(1), 103–107.

Miller, E. K., & Cohen, J. D. (2001). An integrative theory of prefrontal cortex function. *Annual Review of Neuroscience*, *24*, 167–202.

Miller, E. K., Erickson, C. A., & Desimone, R. (1996). Neural mechanisms of visual working memory in prefrontal cortex of the macaque. *The Journal of Neuroscience*, *16*(16), 5154–5167.

Mink, J. W. (1996). The basal ganglia: Focused selection and inhibition of competing motor programs. *Progress in Neurobiology*, *50*(4), 381–425. Retrieved from http://www.ncbi.nlm.nih.gov/pubmed/9004351

Monchi, O., Petrides, M., Petre, V., Worsley, K., & Dagher, A. (2001). Wisconsin Card Sorting revisited: Distinct neural circuits participating in different stages of the task identified by event-related functional magnetic resonance imaging. *Journal of Neuroscience*, *21*(19), 7733–7741. Retrieved from http://www.ncbi.nlm.nih.gov/pubmed/11567063

Mongillo, G., Barak, O., & Tsodyks, M. (2008). Synaptic theory of working memory. *Science*, *319*(5869), 1543–1546.

Monsell, S. (2003). Task switching. *Trends in Cognitive Sciences*, *7*(3), 134–140.

Montague, P. R., Dayan, P., & Sejnowski, T. J. (1996). A framework for mesencephalic dopamine systems based on predictive Hebbian learning. *Journal of Neuroscience*, *16*(5), 1936–1947. Retrieved from http://www.ncbi.nlm.nih.gov/pubmed/8774460

Moustafa, A. A., Sherman, S. J., & Frank, M. J. (2008). A dopaminergic basis for working memory, learning and attentional shifting in Parkinsonism. *Neuropsychologia*, *46*(13), 3144–3156.

Nomura, E. M., Gratton, C., Visser, R. M., Kayser, A., Perez, F., & D'Esposito, M. (2010). Double dissociation of two cognitive control networks in patients with focal brain lesions. *Proceedings of the National Academy of Sciences of the United States of America, 107*(26), 12017–12022.

Norman, D. A., & Shallice, T. (1986). Attention to action: Willed and automatic control of behavior. In R. J. Davidson, G. E. Schwartz, & D. Shapiro (Eds.), *Consciousness and self-regulation* (pp. 1–18). New York: Plenum Press.

O'Reilly, R. C., Braver, T. S., & Cohen, J. D. (1999). A biologically based computational model of working memory. In A. Miyake & P. Shah (Eds.), *Models of working memory: Mechanisms of active maintenance and executive control* (pp. 375–411). New York: Cambridge University Press.

O'Reilly, R. C., & Frank, M. J. (2006). Making working memory work: A computational model of learning in the prefrontal cortex and basal ganglia. *Neural Computation, 18*(2), 283–328.

O'Reilly, R. C., Frank, M. J., Hazy, T. E., & Watz, B. (2007). PVLV: The primary value and learned value Pavlovian learning algorithm. *Behavioral Neuroscience, 121*(1), 31–49.

O'Reilly, R. C., Herd, S. A., & Pauli, W. M. (2010). Computational models of cognitive control. *Current Opinion in Neurobiology, 20*(2), 257–261.

Owen, A. M., Downes, J. J., Sahakian, B. J., Polkey, C. E., & Robbins, T. W. (1990). Planning and spatial working memory following frontal lobe lesions in man. *Neuropsychologia, 28*(10), 1021–1034. Retrieved from http://www.ncbi.nlm.nih.gov/pubmed/2267054

Power, J. D., & Petersen, S. E. (2013). Control-related systems in the human brain. *Current Opinion in Neurobiology, 23*(2), 223–228.

Redgrave, P., Prescott, T. J., & Gurney, K. (1999). The basal ganglia: A vertebrate solution to the selection problem? *Neuroscience, 89*(4), 1009–1023. Retrieved from http://www.ncbi.nlm.nih.gov/pubmed/10362291

Reverberi, C., Görgen, K., & Haynes, J. D. (2012). Compositionality of rule representations in human prefrontal cortex. *Cerebral Cortex, 22*(6), 1237–1246.

Reymann, K. G., & Frey, J. U. (2007). The late maintenance of hippocampal LTP: Requirements, phases, 'synaptic tagging', 'late-associativity' and implications. *Neuropharmacology, 52*(1), 24–40.

Reynolds, J. R., & O'Reilly, R. C. (2009). Developing PFC representations using reinforcement learning. *Cognition, 113*(3), 281–292.

Rougier, N. P., Noelle, D. C., Braver, T. S., Cohen, J. D., & O'Reilly, R. C. (2005). Prefrontal cortex and flexible cognitive control: Rules without symbols. *Proceedings of the National Academy of Sciences of the United States of America, 102*(20), 7338–7343.

Sakagami, M., & Niki, H. (1994). Spatial selectivity of go/no-go neurons in monkey prefrontal cortex. *Experimental Brain Research, 100*(1), 165–169.

Schultz, W. (2013). Updating dopamine reward signals. *Current Opinion in Neurobiology, 23*(2), 229–238.

Schultz, W., Dayan, P., & Montague, P. R. (1997). A neural substrate of prediction and reward. *Science, 275*(5306), 1593–1599. Retrieved from http://www.ncbi.nlm.nih.gov/pubmed/9054347

Shen, W., Flajolet, M., Greengard, P., & Surmeier, D. J. (2008). Dichotomous dopaminergic control of striatal synaptic plasticity. *Science, 321*(5890), 848–851.

Sutton, R. S., & Barto, A. G. (1998). *Reinforcement learning: An introduction* (Vol. *1*). Cambridge: MIT Press.

Tenenbaum, J. B., Kemp, C., Griffiths, T. L., & Goodman, N. D. (2011). How to grow a mind: Statistics, structure, and abstraction. *Science, 331*(6022), 1279–1285.

Todd, M. T., Niv, Y., & Cohen, J. D. (2009). *Learning to use working memory in partially observable environments through dopaminergic reinforcement.* Paper presented at the Advances in neural information processing systems.

van Schouwenburg, M. R., den Ouden, H. E., & Cools, R. (2010). The human basal ganglia modulate frontal-posterior connectivity during attention shifting. *Journal of Neuroscience, 30*(29), 9910–9918.

Wallis, J. D., Anderson, K. C., & Miller, E. K. (2001). Single neurons in prefrontal cortex encode abstract rules. *Nature, 411*(6840), 953–956.

Wang, X. J. (1999). Synaptic basis of cortical persistent activity: The importance of NMDA receptors to working memory. *Journal of Neuroscience, 19*(21), 9587–9603. Retrieved from http://www.ncbi.nlm.nih.gov/pubmed/10531461

Werchan, D. M., Collins, A. G., Frank, M. J., & Amso, D. (2015). 8-month-old infants spontaneously learn and generalize hierarchical rules. *Psychological Science*, *26*(6), 805–815.
White, I. M., & Wise, S. P. (1999). Rule-dependent neuronal activity in the prefrontal cortex. *Experimental Brain Research*, *126*(3), 315–335. Retrieved from http://www.ncbi.nlm.nih.gov/pubmed/10382618
Wiecki, T. V., & Frank, M. J. (2013). A computational model of inhibitory control in frontal cortex and basal ganglia. *Psychological Review*, *120*(2), 329–355.
Woolgar, A., Thompson, R., Bor, D., & Duncan, J. (2011). Multi-voxel coding of stimuli, rules, and responses in human frontoparietal cortex. *Neuroimage*, *56*(2), 744–752.
Zipser, D. (1991). Recurrent network model of the neural mechanism of short-term active memory. *Neural Computation*, *3*(2), 179–193.

22

Cognitive Control and Emotional Processing

Luiz Pessoa

Thorny Conceptual Issues

How should we understand the status of mental terms such as 'perception', 'attention', 'motivation', 'cognition', and 'emotion'? These terms, which are often the titles of textbook chapters, are frequently conceptualised in a fairly independent fashion. Thus, for example, perception and cognition are distinct *entities* (though it is accepted that they interact). Although this approach is useful in organising a textbook (or sessions of a large conference), it is poorly suited to understanding the brain. That is because the mapping of structure and function in the brain is highly complex and does not respect linguistic boundaries ('perception', 'attention', etc.). Indeed, I argue that *reification* (literally, 'thing-ification') of these linguistic terms leads to an impoverished conceptualisation of the brain.

Let us examine an example involving motivation. Methodologically, disentangling the contributions of cognition and motivation to neural signals is far from easy (see Hickey, Chelazzi, & Theeuwes, 2010; Peck, Jangraw, Suzuki, Efem, & Gottlieb, 2009). For example, participants may be instructed via a cue stimulus that a potential reward will result if their performance is both fast and accurate. In such cases, increased brain signals may reflect enhanced attention because subjects are more likely to engage attention when a reward is at stake. Maunsell (2004, pp. 262–263) raised this important point in his discussion of monkey physiology studies of attention:

> When the effects of spatial attention are examined, subjects are motivated to direct attention to one location or another only by expectations about which location is more likely to be associated with a reward. ... Such reward manipulations reliably lead to shifts in attention. ... However, these experiments typically provide no basis for assigning changes preferentially to attention or to expectations about reward. In most cases, attention-related modulation could equally well be described in terms of expectation about rewards because the two are inextricably confounded.

The Wiley Handbook of Cognitive Control, First Edition. Edited by Tobias Egner.

What was at stake in his discussion was whether increased cell responses *actually* reflected greater attention (something well understood at the time) or whether they reflected reward (something that was novel).

The same set of issues is at stake when considering interactions between emotion and perception/cognition. Consider a study (further discussed below) by my colleagues and I to understand the role of affective significance during the attentional blink (Lim, Padmala, & Pessoa, 2009). As expected, when affectively significant items were presented, potential amygdala contributions to visual cortex responses were observed (and emotional items were better detected). Furthermore, statistical analysis revealed an indirect contribution of the amygdala via the prefrontal cortex. How are we to interpret these findings? It is tempting to describe the direct amygdala contribution as involving 'emotion' and the one via the prefrontal cortex as reflecting increased 'attention'—both leading to improved detection performance when affectively significant items were presented.

It is thus instructive to consider the relationship between affective signals (emotion, motivation) and cognition more abstractly (keep in mind here that cognition also involves perception in these examples). One possibility is that affective signals have effects that take place independently of cognition (Figure 22.1a). A second is that affective signals modulate behaviour by engaging the same set of functions that are used by cognition, in which case, the impact of affective signals on behaviour could be described as 'mediated by cognition' (Figure 22.1b). This mediation could be partial only, such that both direct (motivation-to-behaviour) and indirect (motivation-via-cognition-to-behaviour) effects take place. A third possibility is that cognition and affect are more intertwined, such that they *jointly* guide behaviour (Figure 22.1c), in which case, although certain processes could be described as 'cognitive' and others as 'affective', the interactions between them are sufficiently high that their strict separation is more semantic than real. In terms of these scenarios, the situation described by Maunsell could be portrayed by the mediation model. In other words, *mechanistically*, effects of reward are obtained via 'attentional circuits'.

More recently, Chelazzi, Perlato, Santandrea, and Della Libera (2013) described multiple classes of effects of reward on visual processing (Figure 22.2). Interestingly, each class was backed up by a subset of the growing literature on reward and perception/attention. Although not described by Chelazzi and colleagues, parallel scenarios exist in the case of emotion, too.

To a great extent, the mission of neuroscience is to understand the nature of signals observed in different parts of the brain, and to attempt to *disentangle* the potential contributions to

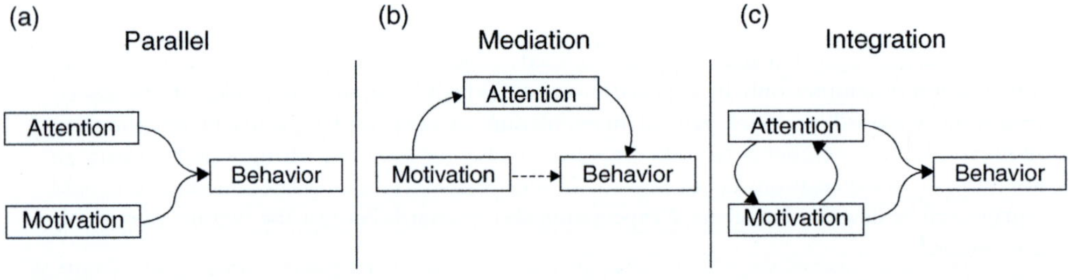

Figure 22.1 Three models of the relationships between cognition and affective signals (emotion, motivation). (a) In the parallel model, cognition and affect have independent effects on behaviour. (b) In the mediation model, the influence of affective signals on behaviour is mediated via cognitive systems (such as attention). (c) In the integration model, cognitive and affective systems interact so strongly that they cannot be decomposed. Source: Adapted from Pessoa 2010. Frontiers in Neuroscience.

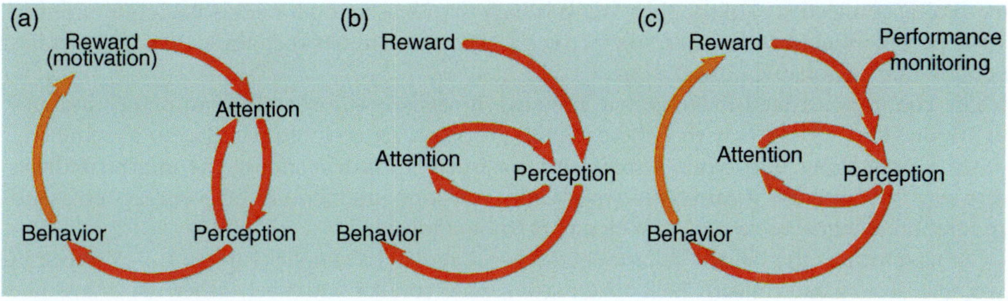

Figure 22.2 Additional representations of the interactions between cognition and affective signals (emotion, motivation). Source: Chelazzi 2013. Reproduced with permission of Elsevier.

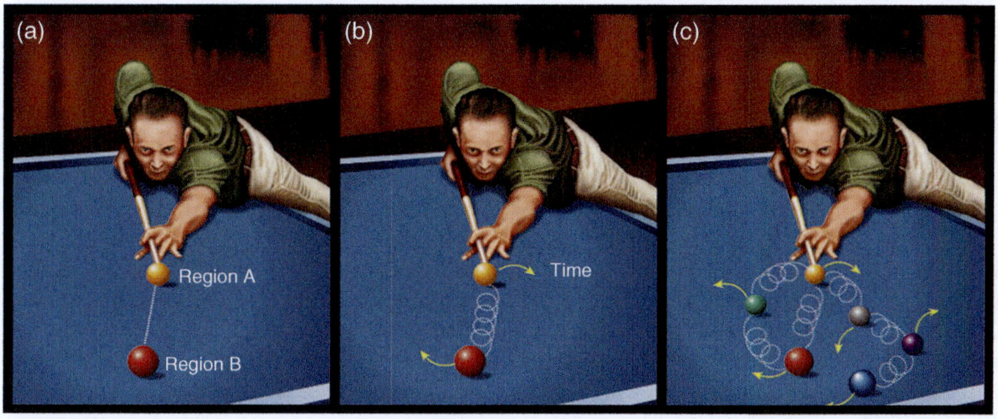

Figure 22.3 Schematic diagrams of causal frameworks. (a) Billiard ball scheme. Complex system scheme with two (b) or many particles (c). Source: Pessoa 2015. Reproduced with permission of Elsevier.

those signals. But here lies the problem, too. The problem is not so much with diagrams such as in Figure 22.1, but when these, implicitly perhaps, are associated with what I call the *billiard ball* causal model (Figure 22.3a). In this model, force applied to a ball leads to its movement on the table until it hits the target ball. In this case, the reason the target ball moves is obvious; the first ball hits it, and via the force applied to the target ball, the target ball moves. But this mode of thinking, which has been very productive in the history of science, is too impoverished when complex systems—the brain for one—are considered.

To illustrate why, let us consider two scenarios of what the diagram in Figure 22.1c might represent. The first case can be called *decomposable.* Attention processes (and brain circuits) operate separately from affective processes (and brain circuits). The separation in this case is relative, as evidenced by the arrows linking the two boxes. But the overall system with attention and affect is decomposable in that each subsystem is modular and operates according to its own intrinsic principles, independently of the other (Bechtel & Richardson, 1993). The second case can be called *non-decomposable*, because the interrelatedness (in the brain, via anatomical connectivity) of the subsystems is such that they are no longer isolable—interfering with one will influence the other, and vice versa. Of course, in between these two extremes lies a continuum of possible organisations.

The contention that I make is that affective signals are *not* isolable from perception and cognition. This is *not* to say that these mental processes are so interrelated as to become one and the same thing. But when systems are not isolable, understanding the interrelatedness between 'subsystems' means that we should consider interactions between systems and integration of signals as the central elements to be unravelled.

Let us return to the billiard ball model discussed above. Its simplicity lies in the existence of two spatially separate billiard balls that make simple contact with each other. Diagrams like those in Figure 22.1 place mental processes such as emotion and attention in separate boxes (like billiard balls) that can affect each other in direct, simple ways (like a ball hitting another). But this analogy will not be helpful in non-decomposable systems—like the brain. Although thinking of causation in complex systems is much more challenging, consider the modification shown in Figure 22.3b. Here, the two balls are connected by a spring, and the goal of explanation is not to explain where $ball_2$ ends up. Instead, when the initial force is applied to $ball_1$, the goal is to understand the evolution of the $ball_1$–$ball_2$ *system* as the two balls interact with each other. More generally, a series of springs with different coupling properties links the multiple elements in the system (Figure 22.3c). In a related vein, an important goal for brain scientists should be to understand the interactions between affect and perception/cognition.

Control and Emotional Processing

Both emotion (discussed here) and motivation (see Chapter 24 by Krebs & Woldorff in this volume) influence competition at both the perceptual and the executive levels (Pessoa, 2009, 2013; Figure 22.4). *Perceptual competition* refers to processes that shape the ways visual items (or items from other sensory modalities) compete with one another. In essence, it speaks to how stimulus features or objects are selected so that they will affect behaviour. Because these mechanisms involve more 'central' processing aspects, including attention and control, they should not be viewed as simply perceptual. But we leave the term *executive competition* to refer to situations that deal less with sensory aspects of competition and more with functions such as *inhibition*, *shifting*, and *updating*, with the understanding that perceptual and executive competition are interdependent.

Perceptual Competition

Perceptual competition is affected by emotional content. Such influence has been extensively studied by using behavioural and cognitive neuroscience techniques to illustrate how emotion-laden items are highlighted in perception (Pourtois, Schettino, & Vuilleumier, 2012) What are the mechanisms by which affective significance influences perceptual processing and hence competition? Here the focus will be on visual processing.

An important mechanism involves the frontoparietal 'attentional network' (Figure 22.5), which provides top-down signals capable of biasing visual responses (Corbetta & Shulman, 2002; Kastner & Ungerleider, 2000). I suggest that regions in the lateral frontal cortex and parietal cortex influence visual processing according to an item's affective significance. In particular, both the frontal eye field and parietal cortex contain a *priority map*, namely, a representation of spatial locations that are salient (e.g., high-contrast stimuli) or relevant (e.g., stimuli connected to current goals; Awh, Lane, Hawken, Zanke, & Kim, 2013; Fecteau & Munoz, 2006; Serences & Yantis, 2006). To embed affective significance into priority maps,

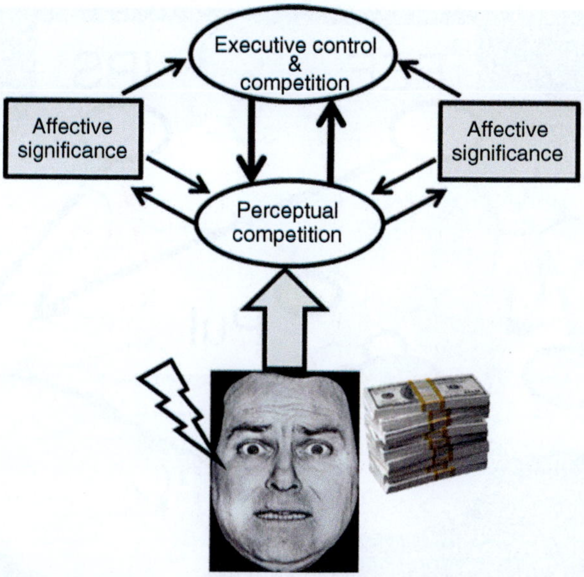

Figure 22.4 Dual competition model. Affective significance impacts the flow of information processing on the basis of emotional and motivational factors, for both of which competition occurs at the perceptual and executive levels. Arrows denote functional pathways that do not necessarily map to specific anatomical connections. Individual differences in state or trait anxiety and sensitivity to reward are predicted to modulate the impact of affective significance on information processing. Source: Pessoa 2013. Reproduced with permission of Cambridge University Press.

I suggest that frontoparietal regions work closely with evaluative sites, such as the amygdala, orbitofrontal cortex, and anterior insula, and even hypothalamus, to prioritise processing based on the emotional value of a sensory stimulus (Figure 22.5).

Evidence for emotion–attention interactions involving the frontoparietal network was provided in a study by Mohanty, Egner, Monti, and Mesulam (2009). In a variant of an endogenous cueing task, a central cue indicated both the expected location of an upcoming target face stimulus and its valence (negative or neutral). Responses in frontoparietal regions were modulated by spatial *and* emotional cue information. When the cue signalled an emotional target, Mohanty and colleagues observed increased functional connectivity (that is, correlated signals) between these areas and the amygdala, which, they suggested, facilitated detection of the affectively significant item.

Still other evidence regarding frontoparietal interactions with the visual cortex comes from an attentional blink study (Lim et al., 2009). In the attentional blink paradigm, participants are asked to report the occurrence of two targets, T1 and T2, among a rapid stream of visual stimuli. When T2 follows T1 after a brief delay (200–500 ms), participants are more likely to miss it, as if they had blinked (hence the name). The attentional blink is believed to reflect capacity-limited mechanisms associated with the processing of the first target. We investigated how affective significance influences visual perception during an attentional blink task that was combined with aversive conditioning. During the initial phase of the experimental session, houses or buildings (counterbalanced) were paired with mild shock. Behaviourally, following aversive learning, affectively significant scenes (CS+; houses or buildings, but not both) were better detected than unpaired scenes (CS-) ones (72% vs. 62% correct, respectively) during the blink period.

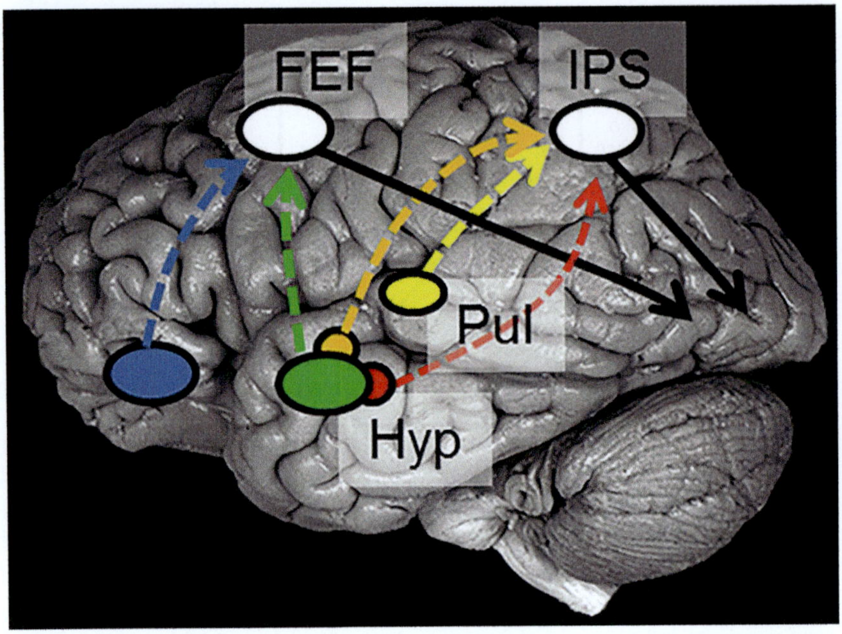

Figure 22.5 Interactions between emotion and perception in the brain. Visual processing is modulated by affective value via several mechanisms, including those involving projections from the amygdala (not shown). Other mechanisms rely on interactions between evaluative sites (coloured ellipses and circles) and attentional control sites (white ellipses). Located in the frontoparietal cortex, the latter sites are known to exert top-down influences on visual processing. Dotted lines indicate possible indirect connections. All locations are approximate, and the positions of the amygdala, basal forebrain, hypothalamus, and pulvinar are shown on the lateral surface for schematic purposes. Labels from panel A were omitted to reduce clutter. A, amygdala; BF, basal forebrain; FEF, frontal eye field; Hyp, hypothalamus; IPS, intraparietal sulcus; OFC, orbitofrontal cortex; Pul, pulvinar. Source: Adapted from Pessoa 2010. Dialogues in Clinical Neuroscience with the permission of the publisher (Les Laboratories Servier, Suresnes, France).

To elucidate how emotional content counters the blink, my colleagues and I evaluated potential contributions from the amygdala to the visual cortex by performing statistical path analysis and structural equation modelling. Given that areas in the lateral prefrontal cortex had been suggested to be critical 'bottleneck' regions during the attentional blink (Marois & Ivanoff, 2005; see also Tombu et al., 2011), we were interested in evaluating whether the impact of the amygdala on the visual cortex was mediated by the prefrontal cortex. The results of our analysis were consistent with both direct and indirect (via the prefrontal cortex) influences on visual cortical responses, such that both influences supported the behavioural advantage of affectively significant items during the attentional blink task. These results indicate that interactions between frontoparietal and valuation regions provide a mechanism by which emotion participates in perceptual competition—in this case, a temporal competition mechanism as probed in the attentional blink paradigm.

The interactions between the amygdala and components of the frontoparietal network may rely on indirect anatomical connections because direct ones are sparse. But, amygdala signals are capable of being rapidly broadcast across the entirety of the prefrontal cortex, for instance (e.g., Averbeck & Seo, 2008; Pessoa, 2013, ch. 9). Thus, a strong *functional* relationship does not have to depend on robust direct connections (Pessoa, 2014). The posterior cingulate cortex, which is both extensively interconnected with the amygdala and connected with the lateral frontal areas, may also bridge the two systems (Mohanty et al., 2009).

Another emotion–attention mechanism involves the pulvinar complex of the thalamus (Figure 22.5). As proposed elsewhere (Pessoa & Adolphs, 2010), the importance of the pulvinar for affective processing is due not to its putative role as part of a subcortical pathway but, instead, to its connectivity with other cortical regions. Thus, the role of the pulvinar extends beyond standard attentional functions (see also Shipp, 2004) to include mechanisms sensitive to affective significance. This proposal is consistent with findings by Ward and colleagues showing the impact of pulvinar lesions on the processing of emotion-laden visual stimuli in humans (Ward, Calder, Parker, & Arend, 2007; Ward, Danziger, & Bamford, 2005), as well as with the findings of our attentional blink study (Padmala, Lim, & Pessoa, 2010). Having observed a significant relationship between pulvinar responses and the ability to detect emotion-laden targets on a trial-by-trial basis, we proposed that the pulvinar amplifies responses to stimuli of potential value to the organism during challenging sensory conditions.

To summarise, during the past decade, researchers have described how, via projections to much of the ventral visual cortex, the amygdala influences visual processing according to affective significance. Although this type of mechanism is important, as briefly illustrated here, other circuits need to be considered, too. These circuits, which include both cortical and subcortical structures, generate complex network interactions that sculpt how visual signals evolve in response to the behavioural and affective significance of sensory stimuli (for discussion of additional mechanisms, including those involving the amygdala-to-visual cortex connections, see Pessoa, 2013, ch. 7).

Executive Competition

Executive control is involved in maintaining and updating information in working memory, monitoring conflict and errors, resisting distracting information, inhibiting prepotent responses, and shifting mental sets. Behavioural research indicates that executive control is not unitary and that different functions may have their own limited processing capacities, or *resources* (Kahneman, 1973; Norman & Bobrow, 1975). The standard strategy to study this question is by devising dual-task interference tasks. When performing function *X* does not interfere with function *Y* and vice versa (a 'double dissociation'), it suggests that their resources do not overlap and that their resource pool is not unitary (Kahneman, 1973). Neuropsychological research also supports the dissociation of cognitive operations, consistent with the *fractionation* of the central executive (Norman & Shallice, 1986; Stuss & Knight, 2002). Here, again, findings of double dissociations are used to argue for the separation of the processes in question (for an extended discussion, see Chapter 29 by Fellows in this volume; Shallice, 1988; Shallice & Cooper, 2011).

Yet, ample evidence also suggests at least some *unity* of executive functions—certain mechanisms are shared across functions (Duncan, Emslie, Williams, Johnson, & Freer, 1996; Miyake et al., 2000). For example, Duncan and Owen (2000) proposed that the processes underlying fluid intelligence are shared in key ways and overlap considerably in both the lateral and dorsal-medial prefrontal cortex, as well as in the parietal cortex. Capacity sharing has implications for human information processing because it implies *executive competition*: Subcomponents of executive control are mutually interacting, such that multiple functions cannot be independently executed simultaneously. For example, handling conflict may make it especially challenging to deal with inhibition at the same time. This competition can be cast in terms of resources. Accordingly, even though some executive processes rely on partly independent mechanisms, they share a *common pool* of resources. Therefore, when a given function is needed, resources devoted to one operation will not be available for other operations, and behavioural interference will ensue. More generally, the question of the unity or fragmentation of executive control relates to the notion of cognitive architecture and modularity (Shallice & Cooper, 2011).

Because emotion can either enhance or impair cognitive performance, to see how emotional content interacts with executive control, we must consider at least two key factors: the strength or *arousal* of the stimulus (or manipulation) and *task relevance* (see also Mather & Sutherland, 2011). When arousal is 'low' and affective significance is task *irrelevant*, interference with the main task may be observed, and the behavioural effect will be typically small.

When, however, arousal is 'high' and the stimulus/manipulation is task irrelevant, resources are more fully *diverted* toward the processing of the emotional item and, because the mobilisation of resources is more pronounced, the effects on behaviour are greater. For example, in our investigation of cognitive–emotional interactions, Choi, Padmala, and Pessoa observed that response conflict increased on trials with the possibility of shock, suggesting that the impact of emotion on behaviour comes in part from the more vigorous recruitment of attentional/effortful control required to prioritise the processing of high-arousal items (Choi, Padmala, & Pessoa, 2012). Attentional/effortful control involves resources that are shared across executive functions and, because situations associated with high levels of arousal are expected to recruit some of these resources (see also Bishop, 2007; Eysenck, Derakshan, Santos, & Calvo, 2007; Mathews & Mackinstosh, 1998), interference with other executive functions will ensue (Figure 22.6a). The impact of emotion on performance thus occurs because of

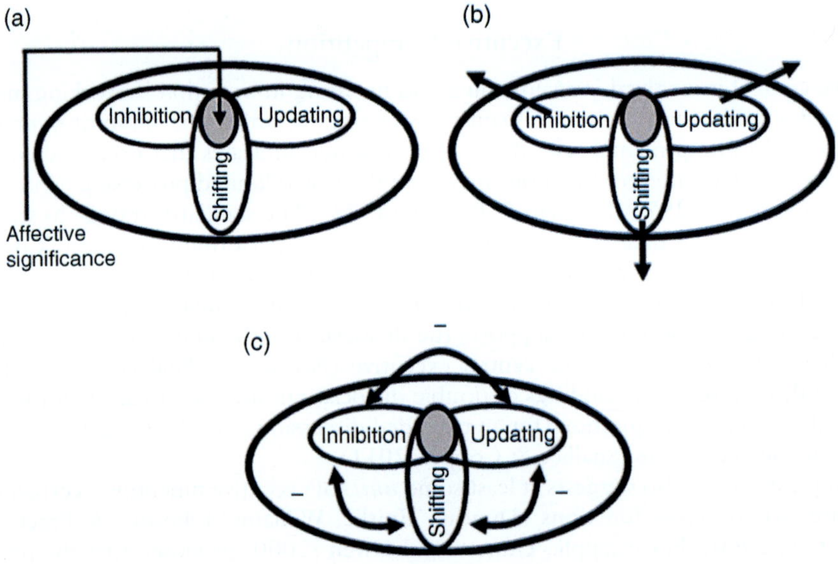

Figure 22.6 Executive control, competition, and processing resources. (a–c) Processes are proposed to share resources called 'common-pool resources' (smaller ellipses in grey), such that the engagement of one will detract from the processing of the other. Common-pool resources are necessary for general functions of attentional/effortful control. (a) High-arousal emotional stimuli recruit common-pool resources that allow their processing to be prioritised, thus detracting from other mechanisms sharing those resources. (b) These stimuli also trigger executive functions, such as updating, shifting, and inhibition, to handle the challenges to the organism, as indicated by the arrows emanating from attentional/effortful control. Source: Pessoa 2009. Reproduced with permission of Elsevier.

limited processing capacity and competition for 'common-pool resources'. For example, performance during a response inhibition task that required subjects to withhold responses to 'stop' stimuli was compromised when they viewed high-arousal pictures (Verbruggen & De Houwer, 2007).

What about the situation when the emotional stimulus is task *relevant*? Two outcomes are possible. If the affective intensity is 'low', task performance might improve because control will be mobilised in the service of handling the task at hand, and the executive functions needed for task completion will more effectively compete for resources. In all, task performance will be enhanced. If, however, the affective intensity is sufficiently high, task performance might be compromised. Thus, in a study of response inhibition, my colleagues and I employed emotional stimuli as stop signals (Pessoa, Padmala, Kenzer, & Bauer, 2012). Subjects had to perform a simple discrimination task but to withhold responding when they saw the stop signal. We found that, when we used both fearful and happy faces as low-arousal stop signals, response inhibition was enhanced relative to neutral faces, but when we employed high-arousal emotional stimuli (previously paired with mild shock) as stop signals, response inhibition was impaired relative to neutral stimuli. Thus, inhibition performance was degraded even though emotional content was task *relevant*. We conjectured that processing the emotional stimulus consumed resources needed for inhibition.

In summary, dealing with an emotional stimulus requires the types of behavioural adjustments that characterise executive functions. For example, to refresh the contents of working memory, to switch the current task set, and to cancel previously planned actions might require *updating*, *shifting*, and *inhibition*, respectively. Such adjustments recruit specific resources in the service of emotional processing (Figure 22.6b) and, if these resources are temporarily unavailable for the task at hand, behavioural performance will be compromised—the more so, the stronger the emotional manipulation (see below). An example may help to illustrate. Suppose a subject is performing a cognitive task and a change in background colour signals that the subject will receive a shock sometime in the next 30 s. The subject might *update* the contents of working memory to include the 'shock possible' information, *shift* between the execution of the cognitive task and 'monitoring for shock' every few seconds, and, if another cue indicated that the shock would be delivered in the next second, *inhibit* a response to the cognitive task to prepare for the shock. In other words, dealing with the emotional situation necessitates the same types of executive functions that are considered to be the hallmark of cognition.

Processing Resources

In the discussion above, I repeatedly used the concept of resources. But what does the concept mean?

Although the concept of resources so often invoked in accounts of the limits of human information processing has been criticised in the past (e.g., Logan, 1988; Navon, 1984; Neisser, 1976) and has not been mechanistically specified, further insight into it can be gained by examining brain regions sensitive to changes in the distribution of resources. Researchers have identified a set of regions consistently involved in attentionally demanding tasks; this 'attentional network' includes sites in the lateral and dorsal-medial prefrontal cortex, anterior insula, and parietal cortex (Corbetta & Shulman, 2002; Kastner & Ungerleider, 2000). These regions help generate and maintain endogenous signals on the basis of current goals and task-relevant contingencies, and send top-down signals that bias the processing of appropriate stimulus features and locations in the sensory cortex (Corbetta, Patel, & Shulman, 2008; see also Chapter 13 by Stokes, Buschman, & Miller and Chapter 18 by Cole in this volume).

To investigate the neural bases of capacity limits more directly, researchers have probed attentional *bottlenecks* observed during tasks such as the attentional blink, as well as the phenomenon known as the 'psychological refractory period'. When subjects perform two speeded sensorimotor tasks separated by a variable delay, the refractory period refers to the increase in reaction time to the second stimulus as the interval between the two tasks decreases. The effect may result from a central attentional bottleneck responsible for operations such as response selection (Pashler, 1994). Indeed, Tombu, Marois, and colleagues (2011) have proposed the existence of a 'unified' attentional bottleneck that involves several regions of the frontoparietal attentional network.

If robust emotional manipulations indeed consume processing resources, then they should engage sites implicated as 'bottleneck areas'. An informal meta-analysis of activation peaks in aversive conditioning functional MRI studies revealed sites throughout the lateral and medial prefrontal cortex, in addition to the anterior insula (Pessoa, 2009). Formal meta-analyses also have implicated these regions during aversive conditioning (Etkin & Wager, 2007; Mechias, Etkin, & Kalisch, 2010). Thus, attentional bottleneck regions are consistently recruited during emotion processing. If this recruitment prevents them from being adequately engaged when neutral task-related processing is required, we should expect to see behavioural impairments.

Still, what is meant by 'resources'? A precise description could be accomplished by specifying the mechanisms involved computationally (Grossberg, 1980). More informally, we can think of resources as directly linked to how specific neural circuits are engaged by task conditions. For concreteness, consider a population of neurons across multiple sites in the prefrontal cortex. Suppose one such population is consistently engaged when ***updating*** is called for, and call it P_U; likewise, call distributed populations of neurons engaged by ***inhibition*** and ***shifting***, P_I and P_S, respectively. If these neuronal populations are shared between these three processes, it is possible to imagine that if a given population is engaged at a particular moment in time, that it will be temporarily unable to participate as effectively in other operations. The extent of interference will depend on the precise mechanisms, of course, but here the general idea suffices.

Now, consider two ways in which emotional processing might recruit the populations of neurons above. First, as mentioned, handling the triggering emotional event might lead to the types of 'cognitive' readjustments that are the staple of executive control. This case is relatively straightforward, as it is similar to a situation in which two non-emotional conditions engage the same population (possibly in close temporal succession). Second, it is also possible that the populations P_U, P_I, and P_S participate, at least in part, in processes that are closer to what is commonly attributed to affective processing, for example, assessing the costs associated with an action, an essential component of monitoring and adjusting levels of control.

Thus, although the use of the concept of resources is admittedly vague, it can (and should) be made more precise as we advance our understanding of executive control—as summarised in the chapters of this book. Greater rigour can be obtained with computational models, of course, but also empirically, as modern recording techniques start revealing properties of distributed neuronal populations across brain regions (see Kiani et al., 2015, for an example involving populations spanning a 4 × 4 mm patch of the prefrontal cortex)

Neural Interactions

What kind of architecture supports the types of interactions and intertwining between emotion and cognition that we discussed above?

Cognitive–emotional interactions rely on the communication between 'task networks' (e.g., the attentional network during attention tasks) and 'valuation networks', which involve both subcortical regions, such as the amygdala and the hypothalamus, and cortical ones, such as the orbitofrontal cortex, anterior insula, and medial prefrontal cortex (see also Chapter 24 by Krebs & Woldorff and Chapter 23 by Winecoff & Huettel in this volume). These interactions are suggested to take place via multiple *modes of communication* (Figure 22.7).

The first mode involves *direct* pathways between task and valuation networks. One example is the pathway between the orbitofrontal and lateral prefrontal cortex (Barbas & Pandya, 1989). The orbitofrontal cortex is important for assessing the value of a stimulus, and the lateral prefrontal cortex, as we know, is heavily engaged during cognitive tasks. Other examples involve the extensive interconnections between the lateral surface of the prefrontal frontal cortex (including the dorsal-lateral PFC) and all cingulate regions (Morecraft & Tanji, 2009). Thus, direct pathways provide a substrate for cognitive–emotional interactions.

A second mode of communication relies on *hub* regions at the intersection of task and valuation networks. Hubs are highly connected and central regions that play a key role in information communication between different parts of a network (Guimera & Nunes Amaral, 2005; Sporns, 2011; Pessoa, 2014; Chapter 18 by Cole in this volume).

The dorsal-medial prefrontal cortex plays a prominent role as a hub region, or *common node*, of executive and emotional networks because of its participation in integrating inputs from diverse sources, notably cognitive and affective ones (e.g., Devinsky, Morrell, & Vogt, 1995; Figure 22.7). This region is involved in multiple executive functions, such as conflict

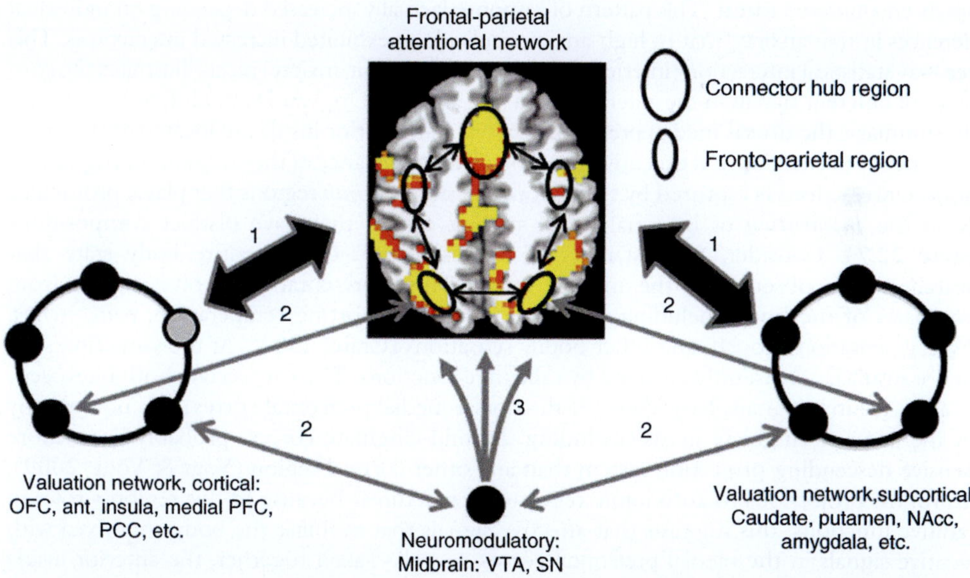

Figure 22.7 Modes of interaction between cognitive and affective networks. (1) Interactions rely on *hub* regions, such as those in the dorsal-medial prefrontal cortex, which are part of both attentional and motivational networks (hub region in the slice and grey node in the cortical valuation network). (2) In addition, specific regions may link the two networks, either directly or via the thalamus. (3) Finally, motivational signals are further embedded within cognitive mechanisms through the action of diffuse neuromodulatory systems. Ant., anterior; NAcc, nucleus accumbens; OFC, orbitofrontal cortex; PCC, posterior cingulate cortex; PFC, prefrontal cortex; SN, substantia nigra; VTA, ventral tegmental area. Source: Pessoa 2010. Frontiers in Neuroscience. Reproduced with permission of Pessoa and Engelmann.

detection, error likelihood processing, and error monitoring (Alexander & Brown, 2011; Chapter 15 by Brown in this volume). It is also important for attentional processing more generally, including spatial attention (Pessoa & Ungerleider, 2004). The dorsal-medial prefrontal cortex is also reliably engaged during conditions involving negative affect, as are all sectors of the anterior-medial prefrontal cortex (Pessoa, 2013; Shackman et al., 2011; Chapter 30 by McTeague & Etkin in this volume).

A second hub region, the anterior insula, is important for interoception, which involves monitoring the sensations needed for integrity of the state not just of the viscera (Craig, 2002, 2009) but of the entire body (Paulus & Stein, 2006). Moreover, threat, uncertainty, and risk are all factors that engage the anterior insula (Singer, Critchley, & Preuschoff, 2009), which is also reliably recruited by long-term memory, working memory, task switching, attention, and many other cognitive processes (Van Snellenberg & Wager, 2010). Indeed, Craig (2009, p. 65) observed that 'no other region of the brain is activated in all of these tasks'. And, in a recent analysis of the *functional diversity* of brain regions, the anterior insula emerged as one of the most diverse (Anderson, Kinnison, & Pessoa, 2013; see also Uddin, Kinnison, Pessoa, & Anderson, 2013).

The anterior insula is thus robustly engaged by both cognitive and emotional tasks, and it provides an interface between the two, as exemplified by the findings of the response-conflict study (Choi et al., 2012), where subjects were asked to indicate whether a picture was of a house or of some other building while ignoring task-irrelevant words. My colleagues and I observed interference-related (i.e., larger) responses in the anterior insula when subjects encountered an incongruent stimulus. And we also observed an interaction between the cognitive and emotional dimensions of the task in the anterior insula, namely, larger conflict responses when subjects encountered threat. This pattern of responses actually increased depending on individual differences in trait anxiety, that is, high-anxious individuals exhibited increased interactions. This three-way statistical interaction (interference by emotion by trait anxiety) nicely illustrates the confluence of different signals in the anterior insula (see also Gu, Liu, Van Dam, Hof, & Fan, 2012).

In summary, the dorsal-medial prefrontal cortex and anterior insula are located at the intersection between emotion and executive function. The importance of these regions in cognition–emotion interactions is captured by their proposed status as *hub* regions that play a prominent role in the *integration* of information by mixing signals that have distinct compositions (Figure 22.7). Consider, for instance, the representation of the entire body state that researchers have observed in the anterior insula. This representation involves signals from many parts of the body, including the viscera, signals reporting temperature, pain, itches, muscular sensations, touch, and other bodily sensations (Craig, 2002). At the same time, the anterior insula is consistently engaged by executive functions. Thus, it receives both interoceptive and executive signals. Now consider the dorsal-medial prefrontal cortex. It is noteworthy that the anterior cingulate gyrus (including the mid-cingulate cortex) probably has a more extensive descending projection system than any other cortical region (Vogt & Vogt, 2009), with robust projections to autonomic regulatory structures. Because of this region's roles in cognitive functions, this suggests that affective signals that mobilise the body are mixed with executive signals in the medial prefrontal cortex as well. Taken together, the anterior insula and dorsal-medial prefrontal cortex provide the substrate for ample cognitive–emotional integration that, in broad terms, includes both bodily 'input' and 'output' signals (see Chapter 9 in Pessoa, 2013, for a further discussion). Moreover, these regions do not work in isolation. During cognitive–emotional interactions, they interact with the lateral prefrontal cortex and parietal cortex, among others (Figure 22.7).

A third mode of communication depends on the diffuse action of neuromodulatory systems, including the action of dopamine and norepinephrine. Widespread connections originating from these systems reach large portions of the cortical surface and several subcortical areas, from which they are able to rapidly influence brain responses. Although neuromodulation can be viewed as another aspect of the network interactions described above, it is worth separating it from other network interactions because neuromodulatory interactions are often characterised as 'regulatory' functions. For example, neuromodulatory signals in the isocortex (i.e., the cortex containing six layers) target superficial (I–III) and deeper (V–VI) cortical layers, but tend to avoid layer IV (e.g., Raghanti et al., 2008), thus appearing to provide less of a 'driving input' and more of a '*modulatory* input' to ongoing information processing. Although neuromodulation will not be reviewed further here (but see Chapter 19 by Robbins & Kehagia in this volume), its effects have been documented in stress studies (Arnsten, 2009; Panksepp, 1998). For example, exposing animals to acute stress leads to release of norepinephrine across a widely distributed brain network, including the prefrontal cortex (Arnsten & Li, 2005; Hermans et al., 2011; Joels, Pu, Wiegert, Oitzl, & Krugers, 2006). Acute stress also results in activation of the dopamine system, which has considerable effects on prefrontal cortex function and associated cognitive tasks, particularly working memory (Arnsten, 2009; Deutch & Roth, 1990).

Conclusions

In discussing the relationship between cognitive control and emotional processing, naturally, the central question pertains to the broader relationship between emotion and cognition. My approach, briefly described here, advocates characterising the interactions between these systems, and studying how cognition and emotion are integrated in the brain. Interactions and integration are supported by a large-scale architecture that provides ample opportunities for signal exchange and refinement.

At an even broader level, the architecture described here speaks to the types of causation schemes discussed previously (Figure 22.3). A segregationist view of emotion and cognition is closer to the billiard ball scheme of causation: When emotion is invoked, how does it impact cognition? When emotion hits cognition, where does cognition end up, so to speak? The integration framework described here cannot be as easily diagrammed, but comes closer in spirit to the complex systems in Figure 22.3 b–c, where emotion and cognition co-evolve and co-determine one another. (For further discussion on segregation and integration, see the commentaries to my target article: Pessoa, 2015).

In conclusion—and it goes without saying—the literature of the interactions between cognitive control and emotional processing is rather extensive. The goal of the present chapter was to review the framework developed in my laboratory in the past decade, and so other contributions were not highlighted here. From the perspective advocated in this chapter, a rather interesting proposal is the one by Inzlicht, Bartholow, and Hirsh (2015), who suggest that control itself can be understood as an emotional process. They suggest that cognitive control is initiated when goal conflicts evoke changes to 'emotional primitives' that both focus attention on the presence of goal conflicts and energise conflict resolution to support goal-directed behaviour. Finally, the approach advocated here has also been used to describe interactions between motivation and perception/cognition (Pessoa, 2009, 2013).

References

Alexander, W. H., & Brown, J. W. (2011). Medial prefrontal cortex as an action-outcome predictor. *Nature Neuroscience*, *14*, 1338–1344.

Anderson, M. L., Kinnison, J., & Pessoa, L. (2013). Describing functional diversity of brain regions and brain networks. *NeuroImage*, *73*, 50–58.

Arnsten, A. F. (2009). Stress signalling pathways that impair prefrontal cortex structure and function. *Nature Reviews Neuroscience, 10,* 410–422.

Arnsten, A. F., & Li, B. M. (2005). Neurobiology of executive functions: Catecholamine influences on prefrontal cortical functions. *Biological Psychiatry, 57,* 1377–1384.

Averbeck, B. B., & Seo, M. (2008). The statistical neuroanatomy of frontal networks in the macaque. *PLoS Computational Biology, 4, 4*(4), e1000050.

Awh, C. C., Lane, A.-M., Hawken, S., Zanke, B., & Kim, I. K. (2013). CFH and ARMS2 genetic polymorphisms predict response to antioxidants and zinc in patients with age-related macular degeneration. *Ophthalmology, 120,* 2317–2323.

Barbas, H., & Pandya, D. N. (1989). Architecture and intrinsic connections of the prefrontal cortex in the rhesus monkey. *Journal of Comparative Neurology, 286,* 353–375.

Bechtel, W., & Richardson, R. C. (1993). *Discovering complexity: Decomposition and localization as scientific research strategies.* Princeton, NJ: Princeton University Press.

Bishop, S. J. (2007). Neurocognitive mechanisms of anxiety: An integrative account. *Trends in Cognitive Sciences, 11,* 307–316.

Chelazzi, L., Perlato, A., Santandrea, E., & Della Libera, C. (2013). Rewards teach visual selective attention. *Vision Research, 85,* 58–72.

Choi, J. M., Padmala, S., & Pessoa, L. (2012). Impact of state anxiety on the interaction between threat monitoring and cognition. *NeuroImage, 59,* 1912–1923.

Corbetta, M., Patel, G., & Shulman, G. L. (2008). The reorienting system of the human brain: From environment to theory of mind. *Neuron, 58,* 306–324.

Corbetta, M., & Shulman, G. L. (2002). Control of goal-directed and stimulus-driven attention in the brain. *Nature Reviews Neuroscience, 3,* 201–215.

Craig, A. D. (2002). How do you feel? Interoception: The sense of the physiological condition of the body. *Nature Reviews Neuroscience, 3,* 655–666.

Craig, A. D. (2009). How do you feel—Now? The anterior insula and human awareness. *Nature Reviews Neuroscience, 10,* 59–70.

Deutch, A. Y., & Roth, R. H. (1990). The determinants of stress-induced activation of the prefrontal cortical dopamine system. *Progress in Brain Research, 85,* 367–402; discussion 402–363.

Devinsky, O., Morrell, M. J., & Vogt, B. A. (1995). Contributions of anterior cingulate cortex to behaviour. *Brain, 118*(Pt. 1), 279–306.

Duncan, J., Emslie, H., Williams, P., Johnson, R., & Freer, C. (1996). Intelligence and the frontal lobe: The organization of goal-directed behavior. *Cognitive Psychology, 30,* 257–303.

Duncan, J., & Owen, A. M. (2000). Common regions of the human frontal lobe recruited by diverse cognitive demands. *Trends in Neurosciences, 23,* 475–483.

Etkin, A., & Wager, T. D. (2007). Functional neuroimaging of anxiety: A meta-analysis of emotional processing in PTSD, social anxiety disorder, and specific phobia. *American Journal of Psychiatry, 164,* 1476–1488.

Eysenck, M. W., Derakshan, N., Santos, R., & Calvo, M. G. (2007). Anxiety and cognitive performance: Attentional control theory. *Emotion, 7,* 336–353.

Fecteau, J. H., & Munoz, D. P. (2006). Salience, relevance, and firing: A priority map for target selection. *Trends in Cognitive Sciences, 10,* 382–390.

Grossberg, S. (1980). How does a brain build a cognitive code? *Psychological Review, 87,* 1–51.

Gu, X., Liu, X., Van Dam, N. T., Hof, P. R., & Fan, J. (2012). Cognition-emotion integration in the anterior insular cortex. *Cerebral Cortex, 23*(1), 20–27.

Guimera, R., & Nunes Amaral, L. A. (2005). Functional cartography of complex metabolic networks. *Nature, 433,* 895–900.

Hermans, E. J., van Marle, H. J., Ossewaarde, L., Henckens, M. J., Qin, S., van Kesteren, M. T.,… Fernandez, G (2011). Stress-related noradrenergic activity prompts large-scale neural network reconfiguration. *Science, 334,* 1151–1153.

Hickey, C., Chelazzi, L., & Theeuwes, J. (2010). Reward changes salience in human vision via the anterior cingulate. *Journal of Neuroscience, 30,* 11096–11103.

Inzlicht, M., Bartholow, B. D., & Hirsh, J. B. (2015). Emotional foundations of cognitive control. *Trends in Cognitive Sciences, 19,* 126–132.

Joels, M., Pu, Z., Wiegert, O., Oitzl, M. S., & Krugers, H. J. (2006). Learning under stress: How does it work? *Trends in Cognitive Sciences, 10,* 152–158.

Kahneman, D. (1973). *Attention and effort.* Englewood Cliffs, NJ: Prentice-Hall.

Kastner, S., & Ungerleider, L. G. (2000). Mechanisms of visual attention in the human cortex. *Annual Review of Neuroscience, 23,* 315–341.

Kiani, R., Cueva, C. J., Reppas, J. B., Peixoto, D., Ryu, S. I., & Newsome, W. T. (2015). Natural grouping of neural responses reveals spatially segregated clusters in prearcuate cortex. *Neuron, 85,* 1359–1373.

Lim, S. L., Padmala, S., & Pessoa, L. (2009). Segregating the significant from the mundane on a moment-to-moment basis via direct and indirect amygdala contributions. *Proceedings of the National Academy of Sciences of the United States of America, 106,* 16841–16846.

Logan, G. D. (1988). Automaticity, resources, and memory: Theoretical controversies and practical implications. *Human Factors, 30,* 583–598.

Marois, R., & Ivanoff, J. (2005). Capacity limits of information processing in the brain. *Trends in Cognitive Sciences, 9,* 296–305.

Mather, M., & Sutherland, M. R. (2011). Arousal-biased competition in perception and memory. *Perspectives on Psychological Science, 6,* 114–133.

Mathews, A., & Mackinstosh, B. (1998). A cognitive model of selective processing in anxiety. *Cognitive Therapy and Research, 22,* 539–560.

Maunsell, J. H. (2004). Neuronal representations of cognitive state: Reward or attention? *Trends in Cognitive Sciences, 8,* 261–265.

Mechias, M.-L., Etkin, A., & Kalisch, R. (2010). A meta-analysis of instructed fear studies: Implications for conscious appraisal of threat. *NeuroImage, 49,* 1760–1768.

Miyake, A., Friedman, N. P., Emerson, M. J., Witzki, A. H., Howerter, A., & Wager, T. D. (2000). The unity and diversity of executive functions and their contributions to complex 'Frontal Lobe' tasks: A latent variable analysis. *Cognitive Psychology, 41,* 49–100.

Mohanty, A., Egner, T., Monti, J. M., & Mesulam, M. M. (2009). Search for a threatening target triggers limbic guidance of spatial attention. *Journal of Neuroscience, 29,* 10563–10572.

Morecraft, R. J., & Tanji, J. (2009). Cingulofrontal Interactions and the cingulate motor areas. In B. A. Vogt (Ed.), *Cingulate neurobiology and disease.* New York: Oxford University Press.

Navon, D. (1984). Resources—A theoretical soup stone? *Psychological Review, 91,* 216–234.

Neisser, U. (1976). *Cognition and reality.* San Francisco: Freeman.

Norman, D. A., & Bobrow, D. G. (1975). On data-limited and resource-limited processes. *Cognitive Psychology, 7,* 44–64.

Norman, D. A., & Shallice, T. (1986). Attention to action: Willed and automatic control of behavior. In R. J. Davidson, G. E. Schwartz, & D. Shapiro (Eds.), *Consciousness and self-regulation.* New York: Plenum.

Padmala, S., Lim, S.-L., & Pessoa, L. (2010). Pulvinar and affective significance: Responses track moment-to-moment visibility. *Frontiers in Human Neuroscience, 4,* 1–9.

Panksepp, J. (1998). *Affective neuroscience: The foundations of human and animal emotions.* New York: Oxford University Press.

Pashler, H. (1994). Dual-task interference in simple tasks: Data and theory. *Psychological Bulletin, 116,* 220–244.

Paulus, M. P., & Stein, M. B. (2006). An insular view of anxiety. *Biological Psychiatry, 60,* 383–387.

Peck, C. J., Jangraw, D. C., Suzuki, M., Efem, R., & Gottlieb, J. (2009). Reward modulates attention independently of action value in posterior parietal cortex. *Journal of Neuroscience, 29,* 11182–11191.

Pessoa, L. (2009). How do emotion and motivation direct executive control? *Trends in Cognitive Sciences, 13,* 160–166.

Pessoa, L. (2010). Emergent processes in cognitive-emotional interactions. *Dialogues in Clinical Neuroscience, 12,* 433–448.

Pessoa, L. (2013). *The cognitive-emotional brain: From interactions to integration.* Cambridge: MIT Press.

Pessoa, L. (2014). Understanding brain networks and brain organization. *Physics of Life Reviews, 11,* 400–435.

Pessoa, L. (2015). Precis of the cognitive-emotional brain. *Behavioral and Brain Sciences, 38*, e71.

Pessoa, L., & Adolphs, R. (2010). Emotion processing and the amygdala: From a 'low road' to 'many roads' of evaluating biological significance. *Nature Reviews Neuroscience, 11*, 773–783.

Pessoa, L., & Engelmann, J. B. (2010). Embedding reward signals into perception and cognition. *Frontiers in Neuroscience, 4*, pii: 17.

Pessoa, L., & Najafi, M. (2015). Complex-system causality in large-scale brain networks: Comment on 'Foundational perspectives on causality in large-scale brain networks by M. Mannino and S.L. Bressler'. *Physics of Life Reviews, 15*, 124–127.

Pessoa, L., Padmala, S., Kenzer, A., & Bauer, A. (2012). Interactions between cognition and emotion during response inhibition. *Emotion, 12*, 192–197.

Pessoa, L., & Ungerleider, L. G. (2004). Top-down mechanisms for working memory and attentional processes. In M. S. Gazzaniga (Ed.), *The new cognitive neurosciences* (3rd ed., pp. 919–930). Cambridge, MA: MIT Press.

Pourtois, G., Schettino, A., & Vuilleumier, P. (2012). Brain mechanisms for emotional influences on perception and attention: What is magic and what is not. *Biological Psychology, 92*(3), 492–512.

Raghanti, M. A., Stimpson, C. D., Marcinkiewicz, J. L., Erwin, J. M., Hof, P. R., & Sherwood, C. C. (2008). Cholinergic innervation of the frontal cortex: Differences among humans, chimpanzees, and macaque monkeys. *Journal of Comparative Neurology, 506*, 409–424.

Serences, J. T., & Yantis, S. (2006). Selective visual attention and perceptual coherence. *Trends in Cognitive Sciences, 10*, 38–45.

Shackman, A. J., Salomons, T. V., Slagter, H. A., Fox, A. S., Winter, J. J., & Davidson, R. J. (2011). The integration of negative affect, pain and cognitive control in the cingulate cortex. *Nature Reviews Neuroscience, 12*, 154–167.

Shallice, T. (1988). *From neuropsychology to mental structure*. New York: Cambridge University Press.

Shallice, T., & Cooper, R. P. (2011). *The organisation of mind*. Oxford: Oxford University Press.

Shipp, S. (2004). The brain circuitry of attention. *Trends in Cognitive Sciences, 8*, 223–230.

Singer, T., Critchley, H. D., & Preuschoff, K. (2009). A common role of insula in feelings, empathy and uncertainty. *Trends in Cognitive Sciences, 13*, 334–340.

Sporns, O. (2011). *Networks of the brain*. Cambridge, MA: MIT Press.

Stuss, D. T., & Knight, R. T. (Eds.) (2002) *Principles of frontal lobe function*. New York: Oxford University Press.

Tombu, M. N., Asplund, C. L., Dux, P. E., Godwin, D., Martin, J. W., & Marois, R. (2011). A unified attentional bottleneck in the human brain. *Proceedings of the National Academy of Sciences of the United States of America, 108*(33), 13426–13431.

Uddin, L. Q., Kinnison, J., Pessoa, L., & Anderson, M. L. (2013). Beyond the tripartite cognition-emotion-interoception model of the human insular cortex. *Journal of Cognitive Neuroscience, 26*, 16–27.

Van Snellenberg, J. X., & Wager, T. D. (2010). Cognitive and motivational functions of the human prefrontal cortex. In A.-L. Christensen, E. Goldberg, D. Bougakov (Eds.), *Luria's legacy in the 21st century* (pp. 30–60). Oxford: Oxford University Press.

Verbruggen, F., & De Houwer, J. (2007). Do emotional stimuli interfere with response inhibition? Evidence from the stop signal paradigm. *Cognition & Emotion, 21*, 391–403.

Vogt, B. A., & Vogt, L. J. (2009). Mu-opiod receptors, placebo map, descending systems, and cingulate-mediated control of vocalization and pain. In B. A. Vogt (Ed.), *Cingulate neurobiology and disease* (pp. 339–364). Oxford: Oxford University Press.

Ward, R., Calder, A. J., Parker, M., & Arend, I. (2007). Emotion recognition following human pulvinar damage. *Neuropsychologia, 45*, 1973–1978.

Ward, R., Danziger, S., & Bamford, S. (2005). Response to visual threat following damage to the pulvinar. *Current Biology, 15*, 571–573.

23

Cognitive Control and Neuroeconomics

Amy A. Winecoff and Scott A. Huettel

Introduction

Over the past 15 years, an explosion of research has begun to elucidate the neural mechanisms that support decision making. Neuroscientific methods have been applied to many of the key questions previously studied using traditional psychological and economic methods: What changes objective qualities of a reward into subjective utility (Rangel, Camerer, & Montague, 2008)? How do uncertainty and risk shape decisions (Platt & Huettel, 2008)? How are others' rewards and risks incorporated into social decisions (Ruff & Fehr, 2014)? Collectively, studies addressing these and related questions have coalesced into a new interdisciplinary domain now called 'decision neuroscience' or 'neuroeconomics' (Levallois, Clithero, Wouters, Smidts, & Huettel, 2012).

To date, neuroeconomics represents a prototypic example of how concepts in one discipline cross over into other disciplines. The behaviours investigated in neuroeconomics—risk aversion, temporal discounting, competition in games, etc.—have been well characterised through previous laboratory and field experiments, often as part of the growing interest in 'behavioural economics'. As such, there is a growing recognition that the processes of cognition shape decision-making behaviour. Conversely, the methods of neuroeconomics are those of cognitive neuroscience: human neuroimaging and electrophysiology, single-unit recording in monkeys and rats, disruption of function through lesions and magnetic stimulation, and administration of drugs that alter brain activity (Ruff & Huettel, 2013). Given the strong connection to behaviour and the use of neuroscience methods that measure and manipulate brain function, neuroeconomics can be conceptualised as a special case of cognitive neuroscience. As such, advances within cognitive neuroscience (e.g., in the understanding of cognitive control) have shaped and will continue to shape research on neuroeconomics.

Despite the natural parallels between cognitive neuroscience and neuroeconomics, research on the neural basis of decision making has largely proceeded apace from research on other aspects of cognition. As evidence, analysis of the semantic structure of articles published in major neuroscience journals revealed a remarkable pattern: The topics most associated with neuroeconomics (e.g., reward, risk) were clustered in an 'island' that was only weakly connected to the main network of traditional terms like vision and memory (Beam, Applebaum, Jack, Moody, & Huettel, 2014). This disconnection likely reflects both the extensive influences upon neuroeconomics from the non-psychological social sciences

The Wiley Handbook of Cognitive Control, First Edition. Edited by Tobias Egner.

(most notably economics) and the neuroeconomic emphasis on very careful measurement of behaviour rather than inferences about mental states (Levallois et al., 2012). In effect, even those concepts within the cognitive neuroscience literature that might be most connected to goal-directed behaviour—like *cognitive control*—have been mostly absent from neuroeconomic research.

Yet, there exist promising avenues for research on cognitive control to influence neuroeconomic studies, and vice versa. For example, most research on the neural basis of decision making when people are faced with choices between safe and risky options has pointed to the important roles played by regions of the prefrontal cortex (PFC), notably parts of the lateral PFC (Mohr, Biele, & Heekeren, 2010). Improved understanding of these regions' function could change how neuroeconomic research characterises risky choices. And, even those aspects of decision making that seem most distant from traditional cognitive control research—such as the basic processes supported by subcortical regions like the ventral striatum (VS)—have been shown by recent research to be influenced by control processes. In the following sections, we highlight four potential links whereby cognitive control can (a) alter the process of valuation, (b) shape how people process risk and uncertainty, (c) change the strategies people apply in complex decisions, and (d) alter both short-term and long-term behaviours that underlie natural decisions (e.g., eating).

Cognitive Control of Motivation and Value

The influence upon our behaviour of motivationally relevant stimuli—those that spark our emotions and drive our decisions—can be altered by processes of cognitive control. Scientific understanding of the psychological and neural mechanisms of how cognitive control shapes the processes of emotion and affect (Chapter 30 by McTeague & Etkin in this volume; Chapter 22 by Pessoa in this volume), and reward valuation (Chapter 24 by Krebs & Woldorff in this volume) has advanced considerably in the last few decades, with concomitant connections to neuroeconomics. Though emotion and reward valuation are often studied separately, these processes share many overlapping neural mechanisms (Lindquist, 2013; Lindquist & Barrett, 2012; Wager, Davidson, Hughes, Lindquist, & Ochsner, 2008; Winecoff et al., 2013). Thus, an evaluation of research on emotion regulation (i.e., how cognitive control can impact emotion) can lend insight into how reward information can be consciously regulated. Moreover, by importing methods from studies of the neural mechanisms of emotion regulation, researchers interested in value regulation can identify similarities and differences between the cognitive control of potentially distinct sorts of affective stimuli.

The process model of emotion regulation (Gross, 1998b) has been a highly influential framework for understanding the temporal progression and nature of emotion regulation. In this model, emotion regulation strategies are classified as either antecedent-focused (i.e., occurring before an emotion has been fully generated) or response-focused (i.e., occurring after the emotional response has already taken place). Whereas response-focused strategies alter the behavioural expression of emotion, antecedent-focused strategies either alter the context in which emotions are experienced or the way in which emotions are felt (Gross, 1998a). Because antecedent-focused strategies alter emotion at the 'input' phase rather than at the 'output' phase (Gross, 1998a), antecedent-focused strategies are more effective than response-focused strategies in terms of neural, psychological, and physiological effects (Goldin, McRae, Ramel, & Gross, 2008; Gross, 1998a; John & Gross, 2004).

Of all the strategies detailed in the process model of emotion regulation (i.e., situation selection, situation modification, selective attention, cognitive reappraisal, and suppression), cognitive reappraisal has been studied most extensively within the field of cognitive neuroscience. Cognitive reappraisal entails thinking about the emotion-evoking stimulus in a way that alters its emotional impact. For example, when reading a news story about children whose parents are in the midst of a messy divorce, the reader could imagine that the children's well-being will ultimately be served by the divorce as both parents will likely be happier when separated. Cognitive reappraisal can also operate by increasing one's emotional distance from a stimulus to weaken its effect. Using the same example, the reader could focus on the fact that the children in the news story are not her own nor the children of any of her friends or family. Regardless of strategy type, the goal of cognitive reappraisal is to use a cognitive manipulation to alter either the specific emotion that is experienced or to alter the intensity with which it is experienced (Gross & Thompson, 2007).

As of 2013, nearly 50 studies have investigated the neural mechanisms of cognitive reappraisal (Buhle et al., 2013). Though peak coordinates and regions of activation vary somewhat between studies (Buhle et al., 2013), on the whole, cognitive reappraisal has been shown to engage many of the same regions of the brain as are involved in attention, working memory, and executive functioning (Ochsner & Gross, 2008). That is, cognitive reappraisal is thought to rely on the ventrolateral PFC (vlPFC), the dorsolateral PFC (dlPFC), the dorsomedial PFC (dmPFC), and the posterior parietal cortex (pPAR), which interact to modulate activation in subcortical areas such as the amygdala (Buhle et al., 2013). These results converge with other psychophysiological studies showing that reappraisal modulates indices of physiological arousal such as skin conductance (Giuliani, McRae, & Gross, 2008) and startle eye blink (Jackson, Malmstadt, Larson, & Davidson, 2000).

Ideally, emotion regulation should not only alter subjective affective responses in the moment, but also promote adaptive subsequent behaviour. Though the typical reappraisal paradigm is well suited to studying regulation effects *while* an emotion is elicited, that same paradigm provides little information about whether regulation alters emotional responses to those same stimuli in the future (i.e., whether current regulation also attenuates later emotional responses). Most reappraisal paradigms employ stimuli such as emotional pictures and videos, which would likely still induce bottom-up affective responses whether they had been previously regulated or not. Another limitation of the most frequently employed cognitive reappraisal paradigms is that they typically focus only on negative emotions. Thus, it is not possible to conclude whether emotion regulation alters arousal, valence, or both. Lastly, given the limited range of emotional stimuli used in reappraisal studies, they have failed to characterise how responses to a wider array of stimuli can be altered via cognitive control.

Neuroeconomics has posed a solution to these problems by importing the methods of cognitive reappraisal investigations and applying them to ideas and processes important for decision making. Several investigations have studied the effect of reappraisal on responses to neutral stimuli that have been repeatedly paired with rewards (i.e., reward-conditioned responses). In this way, these studies focus on affective responses to stimuli that do not inherently generate bottom-up affect—and show the extent to which the mechanisms of reappraisal are shared when the target response is appetitive rather than aversive.

Early research on regulation of reward-conditioned cues showed clear parallels to prior research on emotion reappraisal; that is, regulating the subjective value of a stimulus alters brain systems for valuation just as regulating emotional affect alters brain systems for emotion. In one study (Delgado, Gillis, & Phelps, 2008), functional magnetic resonance imaging (fMRI) and skin conductance data were collected from participants while they viewed coloured squares that had previously been paired with monetary rewards (CS+) or no monetary rewards (CS−). Participants were instructed to either focus on the meaning of coloured squares (i.e., what outcomes they were paired with) or to think of something calming in nature that was associated with the colour of the square (e.g., the ocean for a blue square). When participants regulated anticipatory responses to the CS+, they showed reduced skin conductance responses compared to when they focused on the meaning of the CS+. Similarly, response in the VS was attenuated when regulating reward anticipation to the CS+ than when experiencing the CS+ without regulation. This decrease in the VS happened concurrently with an increase in activation in the left middle frontal gyrus (MFG), a region that has been implicated in the reappraisal of negative and positive emotions (Ochsner, Bunge, Gross, & Gabrieli, 2002; Ochsner et al., 2004; Winecoff et al., 2013; Winecoff, LaBar, Madden, Cabeza, & Huettel, 2011). Collectively, these results suggest that both arousal, as indexed by skin conductance response, and incentive salience, as indexed by decreased activation in the VS are altered by cognitive regulation.

That anticipation of rewards can be affected by conscious regulation suggests that reward-conditioned responses can be modulated through similar mechanisms as cognitive reappraisal of emotion. Yet reward conditioning involves affective responses not only during anticipation, but also during the receipt of reward. Thus, understanding how reward responses can be altered via emotion regulation strategies requires an examination of affective modulation during both the anticipation and outcome phase of reinforcement learning paradigms. Using a modified monetary incentive delay paradigm (Knutson, Westdorp, Kaiser, & Hommer, 2000), Staudinger, Erk, Abler, and Walter (2009) scanned participants while they either won or did not win variable sums of money. On half of all trials, participants were instructed to allow their emotional responses to the cues to unfold naturally, whereas on the other half of trials, participants were instructed to distance themselves from their feelings by thinking about the cues as a neutral observer would. This design enabled the authors to disentangle the effects of regulation on expected value (i.e., anticipation of reward) from the effects of regulation on experienced outcomes (i.e., receipt of reward). On trials in which participants experienced their responses naturally, greater expected value was associated with greater response in the VS. In contrast, in the regulation condition, VS response did not differ between large and small rewards, suggesting that regulation reduced the subjective affective impact of larger rewards to that of smaller rewards. Similarly, responses in the left VS scaled with prediction error when participants experienced their affective responses, but did not show this pattern during regulation. Paralleling emotion reappraisal studies, regulation across both expectation and outcome periods activated the inferior parietal cortex, the lateral orbitofrontal cortex (OFC), the rostral cingulate cortex, and the dlPFC.

A follow-up study showed converging results (Staudinger, Erk, & Walter, 2011). When employing distancing-type reappraisal, participants showed reduced activation in the putamen compared to when they naturally experienced their affective reactions to winning money. This regulation-related reduction in striatal activation was accompanied by an increase in activation in the dlPFC, the intraparietal lobule (IPL), and the temporoparietal junction (TPJ). In contrast to the inverse coupling between the dlPFC and putamen that might be expected during regulation, a psychophysiological interaction analysis (PPI) demonstrated greater functional connectivity between the dlPFC and the putamen during receipt of small rewards than during receipt of large results. Though at first blush these results seem counterintuitive, the authors suggested that when participants are experiencing their affective reactions naturally, they might experience winning comparatively smaller rewards as subjectively aversive. Thus, the positive coupling between the dlPFC and the putamen might indicate a regulation-related increase in the appetitive response to small monetary gains. Collectively, studies of the effects of cognitive regulation of conditioned reward responses reveal a network of regulatory regions similar to that found for emotion regulation (i.e., lateral PFC, parietal cortex) but different affective target regions (i.e., striatum).

Cognitive Control During Risky Decision Making

Reappraisal of value signals alters the way in which rewards are processed in the brain. Furthermore, the use of reappraisal is associated with attenuated self-reports of affective intensity as well as increased activation in regions of the brain thought to be responsible for cognitive control (Delgado et al., 2008; Staudinger et al., 2009, 2011). These data imply that emotion regulation acts through cognitive control circuitry to modulate both neural and subjective affective responses to rewarding stimuli. Yet these studies alone do not indicate whether a change in affective responding is associated with changes in subsequent behaviour. That is, they leave open the question of whether reappraisal of reward responses leads to shifts in goal-directed behaviour.

Investigations of the effects of reappraisal on risky decision making have shown that the regulation of risk relies on many of the same processes as the regulation of reward. In an investigation by Martin and Delgado (2011), participants underwent fMRI while they viewed stimuli that were associated with a monetary reward (CS+) or no reward (CS–). Participants were instructed to 'look' (to experience their emotions naturally), 'relax' (to downregulate their emotional response by thinking of something relaxing), or 'excite'(to think of something exciting). After the presentation of the CS+ and the instruction cue, participants were asked to choose between a safe option (100% probability of winning) and a risky option (20%–80% probability of winning) of equal expected value. In addition to undergoing fMRI, participants also rated how successful they were implementing the regulation strategies. Breaking participants into 'regulators' and 'non-regulators' based on their own self-reported regulation success, regulators showed an interaction in the VS between choice type and instruction type. For the 'look' condition, regulators showed more VS activation when making risky decisions than when making safe decisions. In contrast, during the 'relax' trials, they showed no difference between safe and risky decisions in striatal activation. That is, within individuals who are successful at cognitive regulation, implementing reappraisal effectively reduces the striatal signal typically evoked by processing risky gambles to the level of activation associated with making a safe decision. Accompanying these neural changes, regulators also chose the safe option more frequently when regulating than when naturally experiencing their affective reactions. Reappraisal of risky decision making, therefore, can alter both the neural response to decision parameters as well as subsequent decisions.

A limitation of the investigation by Martin and Delgado (2011) is that participants made decisions between gambles with equivalent expected value. Consequently, it is not possible to determine whether the observed shift from risky to safe decisions during emotion regulation represents a shift towards a more adaptive pattern of decision making, a change in risk preferences, or both. To disentangle these possibilities, a follow-up investigation was performed to examine the effect of regulation on risk and expected value separately (Martin Braunstein, Herrera, & Delgado, 2014). In a design similar to the 2011 study, participants were asked to 'look', 'emphasise' (regard the upcoming decisions as very important), or 'deemphasise' (think of the upcoming decision as just a single decision within a portfolio of many choices). Following the instruction presentation, participants where then asked to choose between a risky and safe option where the expected value of each varied across trials. Analyses of participants' behaviour indicated that there was a significant main effect of strategy whereby risk taking was highest when deemphasising affective responses and lowest when emphasising. Moreover, there was a trend for regulation strategy to interact with decision type. When emphasising the importance of decisions in which the risky option was disadvantageous, risk taking was reduced, whereas when deemphasising the importance of decisions in which the risky option was advantageous, risk taking was increased. Plainly stated, the use of reappraisal shifted participants' behaviour towards more advantageous decision making irrespective of whether the decision involved risk.

These findings suggest that reappraisal can alter and even improve decision making when implemented in specific decision contexts; however, the positive effects of cognitive reappraisal on decision making may be more widespread. In a gambling card task in which participants could either win or lose depending on which cards are selected, habitual use of reappraisal was correlated with risk taking. Follow-up analyses indicated that regular use of reappraisal was associated with both less sensitivity to changes in the probabilities of outcomes as well as less sensitivity to loss (Panno, Lauriola, & Figner, 2013). Similarly, the use of cognitive reappraisal on induced negative emotion has also been associated with reduced risk aversion as indexed by the Balloon Analogue Risk Task (Heilman, Crisan, Houser, Miclea, & Miu, 2010).

As with risk, regions of the brain involved in reward and motivation can also be altered by reappraisal of loss aversion. In one fMRI investigation (Sokol-Hessner, Camerer, & Phelps, 2012), participants were asked to make decisions in which they selected between a guaranteed outcome and a risky outcome. Similar to emotional distancing forms of cognitive reappraisal, participants were asked to either 'regulate' by thinking of each choice as only one choice within a portfolio of possible decisions or 'attend' by thinking of each choice on its own. Subjects varied in how much their choices were influenced by information about gains compared to information about losses; a typical ratio (denoted by λ) is about 2:1 (Kahneman & Tversky, 1979). The researchers explored both how behavioural estimates of λ were altered by regulation and how neural responses changed between the regulate and attend conditions. Results indicated that regulation significantly reduced λ (i.e., diminished loss aversion). In terms of neural response, when participants were effectively able to alter loss aversion through regulation, they showed decreased activation in the amygdala to loss outcomes compared to those who were unsuccessful at regulating loss aversion. This reappraisal-related decrease in amydalar engagement was accompanied by increased activation in the right dlPFC, ventromedial PFC (vmPFC), and striatum, structures that have all been implicated in the regulation of negative emotion (Diekhof, Geier, Falkai, & Gruber, 2011; Ochsner et al., 2002; Ochsner et al., 2004; Wager et al., 2008).

While activation in the dlPFC during the regulation of reward, risk, and loss aversion likely reflects the implementation of cognitive control, the interpretation of increased activation in canonical reward regions in the PFC (e.g., the medial OFC and vmPFC) is less straightforward. A small meta-analysis indicated that the vmPFC is the only region commonly activated across a range of negative emotion regulation strategies (Diekhof et al., 2011), suggesting that this region is critical for control of affective reactions. Yet, recently we demonstrated that reappraisal of positive emotion led to *decreased* activation in the vmPFC (Winecoff et al., 2013)—a pattern opposite of what would be expected if the vmPFC were the locus of emotional control. Therefore, it is possible that increases in the vmPFC during the regulation of negative affect reflect a shift from a more negative stimulus valuation to a more neutral and thereby more positive valuation. In essence, the vmPFC does not implement emotion regulation per se, but reflects the consequences of emotion regulation in an integrated domain-general value signal. Increased activation in the vmPFC during the regulation of loss aversion could represent a shift from an emotionally aversive state to a more neutral—and therefore comparatively more positive—affective state (see also Chapter 30 by McTeague & Etkin in this volume).

Cognitive Control of Decision Strategies

Reward value and risk are among the elemental features of decision making. Much of neuroeconomic research has employed simple tasks with well-controlled manipulations of these features—along with other features like the delay until reward delivery—to isolate their effects on brain function. And, where the effects of cognitive control have been studied, most research has examined how the exertion of control alters how these features influence reward valuation and decision making, as described in the previous sections.

Another, more recent connection between cognitive control and neuroeconomics involves decisions that have a more complex structure. Suppose, for the moment, that you seek to purchase a new car. You walk onto the lot of a large dealership, and you see hundreds of cars comprising dozens of distinct models each with different sets of options. It would be impractical to evaluate the subjective value of every car on that lot in order to choose the highest value—and the decision space would get even larger if one were considering all dealerships in one city or all cars available online. In essence, for many decisions we face in daily life (and not in laboratories), there are many, many elemental features that combine in very complex and computationally intensive ways. How can we simplify very complex decisions with many potential choice options into something more manageable? This question was one of the drivers of early research on bounded rationality (Simon, 1957), the idea that decision-making processes are rational within specific computational and environmental constraints. In the example of car shopping, a typical consumer might narrow down their choice options to a very small set of cars (e.g., two models) on the basis of subjective criteria like looks or feel—and then proceed to compare those cars' features and prices in a more analytic manner.

Effective simplification of decision problems often involves the application of rules appropriate to the current decision context. Some such general rules have been very well established through decades of research in behavioural economics and psychology; these are applied to a wide variety of decision problems and have come to be known through named *heuristics*. For example, 'the availability heuristic' entails making decisions on the basis of information that comes to mind most easily. Other rules may be specific to a particular decision, may be the result of conscious planning or instructions from others, or may be learned through experience with a particular task. In either case, application of the right rule to the right decision context can be critical for effective decision making (Payne, Bettman, & Johnson, 1993) both by reducing the computational demands associated with a complex decision problem and by taking advantage of the structure of that problem to actually improve choices (Gigerenzer & Goldstein, 1999).

Some heuristics have become so endemic to decision making that they have been objects of study in neuroeconomics. Illustrative examples come from studies of framing effects, most notably that of gain/loss framing (Tversky & Kahneman, 1981). In a landmark neuroeconomic experiment, De Martino, Kumaran, Seymour, and Dolan (2006) showed that the anterior dorsomedial PFC (aDMPFC) exhibited greater activity when people made choices counter to the typical framing effect (e.g., being risk seeking for gambles described as gains) than for choices consistent with typical framing effects (e.g., being risk averse in that same scenario). The authors explained this activation as reflecting cognitive control that was exerted when the emotional pull of the safe option needed to be overcome to make a riskier but higher choice between two systems for decision making, such that the dorsal ACC detects conflict between analytic, rational signals and emotional signals from the amygdala. That explanation is consistent with more recent theories that suggest conflict between decision options suffices for engaging the same sort of control processes as seen in other domains of cognition (Brown & Braver, 2008; Pochon, Riis, Sanfey, Nystrom, & Cohen, 2008; Chapter 10 by Kool, Shenhav, & Botvinick in this volume).

Yet, this dual-system idea—and what it implies about cognitive control in complex decision making—cannot account for other sorts of heuristic behaviour. Across several studies, Venkatraman and colleagues explored how people simplify complex gambles into something more manageable (Venkatraman, Payne, Bettman, Luce, & Huettel, 2009; Venkatraman, Payne, & Huettel, 2011). Previous work had shown that when choosing between five-outcome mixed gambles (i.e., those containing both gains and losses), people frequently adopted a 'probability of loss heuristic' that simply chooses the gamble with the highest ratio of good outcomes to bad outcomes, ignoring the magnitude of each outcome (Payne, 2005). Taking advantage of both inter- and intra-subject variability in the expression of this heuristic, Venkatraman, Payne, Bettman, Luce, and Huettel (2009) argued that the aDMPFC was specifically engaged for *strategic* control of behaviour; specifically, making a choice counter to one's own dominant choice tendency (i.e., 'with- or against-heuristic'). That same region also showed choice-related patterns of functional connectivity with the regions most engaged by with- or against-heuristic choices, further supporting the idea that it codes for strategic tendencies independent of choices. Subsequent studies from the same group demonstrated a topographic organisation of control processing in the medial PFC—posterior *response control*, middle *decision control*, and anterior *strategic control*—that was also present in functional connectivity between the medial and lateral PFC (Taren, Venkatraman, & Huettel, 2011; Venkatraman, Rosati, Taren, & Huettel, 2009). This pattern echoes the topographies previously postulated in the lateral PFC from evidence in non-motivated (Badre & D'Esposito, 2007; Koechlin, Ody, & Kouneiher, 2003) and the DMPFC in motivated executive function tasks (Kouneiher, Charron, & Koechlin, 2009).

A synthetic model of the relationship between motivational information and allocation of cognitive control has been proffered by Shenhav, Botvinick, and Cohen (2013; see also Chapter 10 by Kool, Shenhav, & Botvinick in this volume), who suggest that the core functional role of the aDMPFC is to balance information about potential rewards against estimates of the control-related cost associated with receiving those rewards. Their model argues that the aDMPFC seeks to maximise the 'expected value of control' by engaging control processes in which such engagement would improve future reward outcomes, but avoiding engagement in circumstances when control could be ineffective or counterproductive. The strengths of this model are that it can account both for the original observations that linked the aDMPFC to cognitive control—most of which were obtained using paradigms outside of the reward context—as well as for the effects of incentives on expression of control processes in both behaviour and brain. Moreover, analogous computations could be applied at different levels of task complexity, from very simple motivated behaviour to complex decision tasks that involve competing strategic biases (De Martino et al., 2006; Venkatraman, Rosati, et al., 2009). This model does not, however, make express predictions about the organisation of control within the aDMPFC, and thus it remains agnostic on the specific claims that the aDMPFC contributes preferentially to control in strategic decision making (see Coutlee & Huettel, 2012, for a related account that posits that the dorsal aDMPFC engages the dlPFC topographically on the basis of task demands).

Cognitive Control in Natural Decisions

Cognitive control influences both how heuristic strategies are implemented as well as how value signals and decision-making parameters can be altered to optimise behaviour. Neuroeconomics has provided evidence as to the neural mechanisms responsible for such processes through rigorous laboratory studies. Nevertheless, understanding the neural substrates of cognitive control of reward and decision making has implications not only for decisions made in artificial laboratory settings, but also for the kinds of self-controlled decisions made in everyday life. (for a review of the closely related literature on 'self-regulation', as defined in social psychology, see Chapter 25 by Davisson & Hoyle in this volume). Decisions in everyday life are often more complex than those encountered in controlled neuroeconomic investigations. For example, decision makers are often weighing the relative value of two different reward types against each other while simultaneously deciding between immediate but small rewards and larger but delayed rewards.

Forgoing immediate gains in favour of longer-term goal acquisition has obvious relevance for everyday behaviours such as retirement planning; however, this process is also important for dietary choices. The human brain likely evolved in circumstances in which periods of food scarcity were commonplace. Thus, a physiological system that drives consumption of palatable (i.e., high-calorie) foods whenever they are available has evolutionary advantages. For example, the majority of pigeons will peck a key to receive a smaller, immediate reward over refraining from pecking the key to receive a delayed, but larger reward (Ainslie, 1974). Similarly, once endowed with a smaller reward, carrion crows will not exchange their initial endowment for larger quantities of the same reward (Wascher, Dufour, & Bugnyar, 2012). Though non-human primates such as chimpanzees and bonobos have been shown to be capable of forgoing immediate rewards for larger gains (Rosati, Stevens, Hare, & Hauser, 2007), results in other animal species suggest that evolutionary pressures may have shaped humans' relative preference for immediate gratification. In the current Western obesogenic environment (Lake & Townshend, 2006), however, the evolved motivational system is put at odds with the health consequences of overconsumption of high-fat, high-sugar foods. From 2011 to 2012, the rate of obesity was nearly 17% for children aged 2 to 19 and almost 35% for adults (Ogden, Carroll, Kit, & Flegal, 2014). These rising rates of obesity are projected to have staggering health and financial consequences within the near future (Wang, McPherson, Marsh, Gortmaker, & Brown, 2011). Research on the neural mechanisms of self-control over drive for food consumption can elucidate the process by which reward and motivational systems are modulated by cognitive control and could potentially lend insight into new avenues for the prevention and treatment of overweight and obesity.

Early investigations of the neural mechanisms of self-control over diet reveal a pattern of results that is similar to research on the neural mechanisms of reappraisal of reward responses. In a seminal experiment, Hare, Camerer, and Rangel (2009) showed participants images of healthy and unhealthy food items while they underwent fMRI. Participants made decisions about whether they would prefer to eat the item shown or a reference item that was rated as intermediate on both healthfulness and tastiness. At the end of the experiment, one of the participant's decisions was selected, and she was offered the food item to consume. Based on participants' decisions, participants were classified as either 'self-controllers' (those who chose to reject tasty but unhealthy items) or 'non-self-controllers' (those who made decisions based on taste alone). The goal value of each food item—a measure of the value of the item that incorporates both taste and health information—was associated with activation in the vmPFC. As would be predicted by participants' behaviour, activation in the vmPFC for self-controllers was modulated both by taste as well as health value, whereas for non-self-controllers, vmPFC activation was modulated by taste value alone. During trials in which participants rejected liked but unhealthy food items (i.e., successful self-control), activation in the left dlPFC/inferior frontal gyrus (IFG) increased. Moreover, activation in this region during successful self-control was negatively associated with activation in the vmPFC when participants viewed palatable but unhealthy food items.

These results imply that the prefrontal control regions often activated by emotion or value reappraisal (see earlier in this chapter) are likewise responsible for implementing self-control over diet by downregulating reward signals in the brain. To test this prediction further, Hare and colleagues employed a PPI analysis to look at functional connectivity between their dlPFC/IFG seed region and other areas of the brain. Results indicated that activation in the seed region was associated with activation in the vmPFC indirectly via functional connectivity with the IFG/MFG. Thus, paralleling findings from delay discounting experiments (e.g., Kable & Glimcher, 2007), the vmPFC integrates information about the immediate hedonic impact of food items with longer-term health goals, a process that is facilitated by a network of interconnected regions in the dorsolateral and ventrolateral PFC.

Hare et al. (2009) investigated the neural mechanisms of incidental self-control; however, often self-control is employed in situations in which individuals set specific, conscious goals to control dietary behaviour. Providing insight into this process, results explicitly instructing participants to deploy self-regulation over their desire for pleasurable foods have found results similar to those of Hare and colleagues. When instructed to focus on the long-term consequences of consuming unhealthy food options, healthy, normal-weight participants showed increased activation in the bilateral IFG, the anterior insula (aINS), the lateral OFC, and the dlPFC. In contrast, when instructed not to deploy self-control, participants showed increased activation in the VS and central insular cortex in response to viewing tasty versus non-tasty food items (Hollmann et al., 2012).

Individual differences in both hedonic drive as well as self-control may account not only for neural responses to food, but also health behaviour. For example, in one investigation, women underwent fMRI scanning while they consumed either a chocolate milkshake or a tasteless solution. They were then scanned again at a six-month follow-up using the same paradigm. Women who showed an increase in body mass index (BMI) of 2.5% or greater showed a reduction from baseline measurement in right caudate activation when contrasting milkshake versus tasteless solution trials (Stice, Yokum, Blum, & Bohon, 2010).

In addition to associations with long-term health outcomes, individual differences in reward-related brain activation also correlate with short-term behaviour. Following a 6-hr fast, activation in response to food images in the VS was associated with subsequent consumption of snack chips. Moreover, participants who reported having high control over food and high levels of VS activation to food cues had lower BMIs. In contrast, participants who reported having low self-control over food showed a positive correlation between VS activation to food items and BMI (Lawrence, Hinton, Parkinson, & Lawrence, 2012).

That those with high striatal response and low self-control tended to have higher BMIs is not surprising; an unregulated but highly responsive motivational system would likely drive further consumption. Yet, it is not altogether unexpected that those with high striatal response and high self-control would have lower BMIs. A pattern of strong implicit preferences for unhealthy snack food paired with high response inhibition has been shown to predict subsequent weight loss (Nederkoorn, Houben, Hofmann, Roefs, & Jansen, 2010). Furthermore, individuals with anorexia nervosa, who are high in behavioural inhibition (Harrison, O'Brien, Lopez, & Treasure, 2010; Winecoff, Ngo, Moskovich, Merwin, & Zucker, 2015), also often show heightened response in the striatum in response to canonically rewarding stimuli (Bailer et al., 2012; Frank et al., 2005; Frank et al., 2012). Thus, as with studies of conscious regulation of reward responses, studies of self-control over diet implicate interactions between cognitive control regions of the brain and regions typically associated with reward processing.

Summary of Regulation of Reward and Decision Making

In this chapter, we have reviewed the ways in which the neural mechanisms of cognitive control affect reward valuation, the evaluation of risk and uncertainty, strategic decision making, and everyday dietary decisions. Though the nature of each of these processes affects the specific neural mechanisms through which cognitive control operates, in general, cognitive control within neuroeconomic tasks engages aspects of the PFC including the aDMPFC, MFG, IFG, and dlPFC. These regions shape processing in other brain regions such as the parietal cortex, vmPFC, or subcortical regions such as the VS or amygdala depending on the demands of the task.

As with the regulation of emotion, reappraisal of value signals both in the anticipation and receipt of reward increases activation in the dlPFC, MFG, IPL, and TPJ while decreasing activation in reward regions such as the dorsal and VS. A similar subset of prefrontal regions are engaged when risk and loss aversion are regulated, with analogous changes in the target regions of affect regulation (e.g., the striatum and amygdala). Increases in activity in the prefrontal regions coupled with decreases in the subcortical affective regions are associated with changes in decision making that likely shift choice towards more advantageous behaviour. Cognitive control can also be used to alter prepotent tendencies that bias decisions based on emotional responses (e.g., framing effects). Furthermore, control processes in the brain can be used to modify the use of heuristic decision making, both at simpler as well as more complex levels of the decision process. Overall, these forms of cognitive control tend to engage the dlPFC and aspects of the aDMPFC and are topographically arranged such that more simple forms of cognitive control are implemented more posteriorly while more complex forms are implemented more anteriorly (see also Chapter 12 by Koechlin in this volume). Lastly, as with conscious regulation of affective signals such as reward valuation, prefrontal control regions interact with striatal regions to shape everyday dietary choices.

Though neuroeconomics has functioned largely apart from other subdisciplines in cognitive neuroscience (Beam et al., 2014), such independence can be counterproductive. The research highlighted in this chapter demonstrates that cognitive control alters the processes of decision making in a very similar manner to what has been observed for executive function, attention, and other processes. In the future, neuroeconomics can benefit from making more direct connections to other fields of cognitive and affective neuroscience, particularly with respect to how these fields conceptualise cognitive control. Likewise, many of the ideas and methods of neuroeconomics could be useful in characterising precise effects of cognitive control upon brain and behaviour.

References

Ainslie, G. W. (1974). Impulse control in pigeons. *Journal of the Experimental Analysis of Behavior*, *21*(3), 485–489.

Badre, D., & D'Esposito, M. (2007). Functional magnetic resonance imaging evidence for a hierarchical organization of the prefrontal cortex. *Journal of Cognitive Neuroscience*, *19*(12), 2082–2099.

Bailer, U. F., Narendran, R., Frankle, W. G., Himes, M. L., Duvvuri, V., Mathis, C. A., & Kaye, W. H. (2012). Amphetamine induced dopamine release increases anxiety in individuals recovered from anorexia nervosa. *The International Journal of Eating Disorders*, *45*(2), 263–271.

Beam, E., Applebaum, L., Jack, J., Moody, J., & Huettel, S. A. (2014). Mapping the semantic structure of cognitive neuroscience. *Journal of Cognitive Neuroscience*, *26*(9), 1949–1965.

Brown, J. W., & Braver, T. S. (2008). A computational model of risk, conflict, and individual difference effects in the anterior cingulate cortex. *Brain Research*, *1202*, 99–108.

Buhle, J. T., Silvers, J. A., Wager, T. D., Lopez, R., Onyemekwu, C., Kober, H.,...Ochsner, K. N. (2013). Cognitive reappraisal of emotion: A meta-analysis of human neuroimaging studies. *Cerebral Cortex*, *24*(11), 2981–2990.

Coutlee, C. G., & Huettel, S. A. (2012). The functional neuroanatomy of decision making: Prefrontal control of thought and action. *Brain Research*, *1428*, 3–12.

De Martino, B., Kumaran, D., Seymour, B., & Dolan, R. J. (2006). Frames, biases, and rational decision-making in the human brain. *Science*, *313*(5787), 684–687.

Delgado, M. R., Gillis, M. M., & Phelps, E. A. (2008). Regulating the expectation of reward via cognitive strategies. *Nature Neuroscience, 11*(8), 880–881.

Diekhof, E. K., Geier, K., Falkai, P., & Gruber, O. (2011). Fear is only as deep as the mind allows: A coordinate-based meta-analysis of neuroimaging studies on the regulation of negative affect. *NeuroImage, 58*(1), 275–285.

Frank, G. K., Bailer, U. F., Henry, S. E., Drevets, W., Meltzer, C. C., Price, J. C.,...Kaye, W. H. (2005). Increased dopamine D2/D3 receptor binding after recovery from anorexia nervosa measured by positron emission tomography and [11c]raclopride. *Biological Psychiatry, 58*(11), 908–912.

Frank, G. K. W., Reynolds, J. R., Shott, M. E., Jappe, L., Yang, T. T., Tregellas, J. R., & O'Reilly, R. C. (2012). Anorexia nervosa and obesity are associated with opposite brain reward response. *Neuropsychopharmacology, 37*(9), 2031–2046.

Gigerenzer, G., & Goldstein, D. G. (1999). Betting on one good reason: Take the best heuristic. In G. Gigerenzer & P. T. Todd (Eds.), *Simple heuristics that make us smart* (1st ed., pp. 75–95). New York: Oxford University Press.

Giuliani, N. R., McRae, K., & Gross, J. J. (2008). The up- and down-regulation of amusement: Experiential, behavioral, and autonomic consequences. *Emotion, 8*(5), 714–719.

Goldin, P. R., McRae, K., Ramel, W., & Gross, J. J. (2008). The neural bases of emotion regulation: Reappraisal and suppression of negative emotion. *Biological Psychiatry, 63*(6), 577–586.

Gross, J. J. (1998a). Antecedent-and response-focused emotion regulation: Divergent consequences for experience, expression, and physiology. *Journal of Personality and Social Psychology, 74*, 224–237.

Gross, J. J. (1998b). The emerging field of emotion regulation: An integrative review. *Review of General Psychology, 2*(3), 271–299.

Gross, J., & Thompson, R. (2007). Handbook of emotion regulation. In J. J. Gross (Ed.), *Handbook of emotion regulation*. New York: Guilford Press.

Hare, T. A., Camerer, C. F., & Rangel, A. (2009). Self-control in decision-making involves modulation of the vmPFC valuation system. *Science, 324*(5927), 646–648.

Harrison, A., O'Brien, N., Lopez, C., & Treasure, J. (2010). Sensitivity to reward and punishment in eating disorders. *Psychiatry Research, 177*(1–2), 1–11.

Heilman, R. M., Crisan, L. G., Houser, D., Miclea, M., & Miu, A. C. (2010). Emotion regulation and decision making under risk and uncertainty. *Emotion, 10*(2), 257–265.

Hollmann, M., Hellrung, L., Pleger, B., Schlögl, H., Kabisch, S., Stumvoll, M.,...Horstmann, A. (2012). Neural correlates of the volitional regulation of the desire for food. *International Journal of Obesity, 36*(5), 648–655.

Jackson, D. C., Malmstadt, J. R., Larson, C. L., & Davidson, R. J. (2000). Suppression and enhancement of emotional responses to unpleasant pictures. *Psychophysiology, 37*, 515–522.

John, O. P., & Gross, J. J. (2004). Healthy and unhealthy emotion regulation: Personality processes, individual differences, and life span development. *Journal of Personality, 72*(6), 1301–33.

Kable, J. W., & Glimcher, P. W. (2007). The neural correlates of subjective value during intertemporal choice. *Nature Neuroscience, 10*(12), 1625–1633.

Kahneman, D., & Tversky, A. (1979). Prospect theory: An analysis of decision under risk. *Econometrica, 47*(2), 263–292.

Knutson, B., Westdorp, A., Kaiser, E., & Hommer, D. (2000). FMRI visualization of brain activity during a monetary incentive delay task. *NeuroImage, 12*(1), 20–7.

Koechlin, E., Ody, C., & Kouneiher, F. (2003). The architecture of cognitive control in the human prefrontal cortex. *Science (New York, N.Y.), 302*(5648), 1181–1185.

Kouneiher, F., Charron, S., & Koechlin, E. (2009). Motivation and cognitive control in the human prefrontal cortex. *Nature Neuroscience, 12*(7), 939–945.

Lake, A., & Townshend, T. (2006). Obesogenic environments: Exploring the built and food environments. *The Journal of the Royal Society for the Promotion of Health, 126*(6), 262–267.

Lawrence, N. S., Hinton, E. C., Parkinson, J. A., & Lawrence, A. D. (2012). Nucleus accumbens response to food cues predicts subsequent snack consumption in women and increased body mass index in those with reduced self-control. *NeuroImage, 63*(1), 415–422.

Levallois, C., Clithero, J. A., Wouters, P., Smidts, A., & Huettel, S. A. (2012). Translating upwards: Linking the neural and social sciences via neuroeconomics. *Nature Reviews Neuroscience, 13*(11), 789–797.

Lindquist, K. A. (2013). Emotions emerge from more basic psychological ingredients: A modern psychological constructionist model. *Emotion Review*, *5*(4), 356–368.

Lindquist, K. A., & Barrett, L. F. (2012). A functional architecture of the human brain: Emerging insights from the science of emotion. *Trends in Cognitive Sciences*, *16*(11), 533–540.

Martin Braunstein, L., Herrera, S. J., & Delgado, M. R. (2014). Reappraisal and expected value modulate risk taking. *Cognition & Emotion*, *28*(1), 172–181.

Martin, L. N., & Delgado, M. R. (2011). The influence of emotion regulation on decision-making under risk. *Journal of Cognitive Neuroscience*, *23*(9), 2569–2581.

Mohr, P. N. C., Biele, G., & Heekeren, H. R. (2010). Neural processing of risk. *The Journal of Neuroscience*, *30*(19), 6613–6619.

Nederkoorn, C., Houben, K., Hofmann, W., Roefs, A., & Jansen, A. (2010). Control yourself or just eat what you like? Weight gain over a year is predicted by an interactive effect of response inhibition and implicit preference for snack foods. *Health Psychology: Official Journal of the Division of Health Psychology, American Psychological Association*, *29*(4), 389–393.

Ochsner, K. N., Bunge, S. A., Gross, J. J., & Gabrieli, J. D. E. (2002). Rethinking feelings: An fMRI study of the cognitive regulation of emotion. *Journal of Cognitive Neuroscience*, *14*(8), 1215–1229.

Ochsner, K. N., & Gross, J. J. (2008). Cognitive emotion regulation. *Current Directions in Psychological Science*, *17*(2), 153–158.

Ochsner, K. N., Ray, R. D., Cooper, J. C., Robertson, E. R., Chopra, S., Gabrieli, J. D. E., & Gross, J. J. (2004). For better or for worse: Neural systems supporting the cognitive down-and up-regulation of negative emotion. *Neuroimage*, *23*(2), 483–499.

Ogden, C. L., Carroll, M. D., Kit, B. K., & Flegal, K. M. (2014). Prevalence of childhood and adult obesity in the United States, 2011–2012. *The Journal of the American Medical Association*, *311*(8), 806–814.

Panno, A., Lauriola, M., & Figner, B. (2013). Emotion regulation and risk taking: Predicting risky choice in deliberative decision making. *Cognition & Emotion*, *27*(April 2015), 326–334.

Payne, J. W. (2005). It is whether you win or lose: The importance of the overall probabilities of winning or losing in risky choice. *Journal of Risk and Uncertainty*, *30*(1), 5–19.

Payne, J. W., Bettman, J. R., & Johnson, E. J. (1993). *The adaptive decision maker*. Cambridge: Cambridge University Press.

Platt, M. L., & Huettel, S. A. (2008). Risky business: The neuroeconomics of decision making under uncertainty. *Nature Neuroscience*, *11*(4), 398–403.

Pochon, J.-B., Riis, J., Sanfey, A. G., Nystrom, L. E., & Cohen, J. D. (2008). Functional imaging of decision conflict. *The Journal of Neuroscience*, *28*(13), 3468–3473.

Rangel, A., Camerer, C., & Montague, P. R. (2008). A framework for studying the neurobiology of value-based decision making. *Nature Reviews Neuroscience*, *9*(7), 545–56.

Rosati, A. G., Stevens, J. R., Hare, B., & Hauser, M. D. (2007). The evolutionary origins of human patience: Temporal preferences in chimpanzees, bonobos, and human adults. *Current Biology*, *17*, 1663–1668.

Ruff, C. C., & Fehr, E. (2014). The neurobiology of rewards and values in social decision making. *Nature Reviews Neuroscience*, *15*(8), 549–562.

Ruff, C. C., & Huettel, S. A. (2013). Experimental methods in cognitive neuroscience. In P. W. Glimcher & E. Fehr (Eds.), *Neuroeconomics: Decision making and the brain* (2nd ed., pp. 77–108). London: Academic Press.

Shenhav, A., Botvinick, M., & Cohen, J. (2013). The expected value of control: An integrative theory of anterior cingulate cortex function. *Neuron*, *79*(2), 217–240.

Simon, H. A. (1957). *Models of man: Social and rational*. New York: Wiley.

Sokol-Hessner, P., Camerer, C. F., & Phelps, E. A. (2012). Emotion regulation reduces loss aversion and decreases amygdala responses to losses. *Social Cognitive and Affective Neuroscience*, *8*(3), 341–350.

Staudinger, M. R., Erk, S., Abler, B., & Walter, H. (2009). Cognitive reappraisal modulates expected value and prediction error encoding in the ventral striatum. *Neuroimage*, *47*(2), 713–721.

Staudinger, M. R., Erk, S., & Walter, H. (2011). Dorsolateral prefrontal cortex modulates striatal reward encoding during reappraisal of reward anticipation. *Cerebral Cortex*, *21*(11), 2578–2588.

Stice, E., Yokum, S., Blum, K., & Bohon, C. (2010). Weight gain is associated with reduced striatal response to palatable food. *The Journal of Neuroscience*, *30*(39), 13105–13109.

Taren, A. A., Venkatraman, V., & Huettel, S. A. (2011). A parallel functional topography between medial and lateral prefrontal cortex: Evidence and implications for cognitive control. *The Journal of Neuroscience*, *31*(13), 5026–5031.

Tversky, A., & Kahneman, D. (1981). The framing of decisions and the psychology of choice. *Science*, *211*(4481), 453–458.

Venkatraman, V., Payne, J. W., Bettman, J. R., Luce, M. F., & Huettel, S. A. (2009). Separate neural mechanisms underlie choices and strategic preferences in risky decision making. *Neuron*, *62*(4), 593–602.

Venkatraman, V., Payne, J. W., & Huettel, S. A. (2011). Neuroeconomics of risky decisions. In M. Delgado, E. A. Phelps, & T. W. Robbins (Eds.), *Decision making, affect, and learning: Attention and performance XXIII* (pp. 153–172). Oxford: Oxford University Press.

Venkatraman, V., Rosati, A. G., Taren, A. A., & Huettel, S. A. (2009). Resolving response, decision, and strategic control: Evidence for a functional topography in dorsomedial prefrontal cortex. *The Journal of Neuroscience*, *29*(42), 13158–13164.

Wager, T. D., Davidson, M. L., Hughes, B. L., Lindquist, M. A., & Ochsner, K. N. (2008). Prefrontal-subcortical pathways mediating successful emotion regulation. *Neuron*, *59*(6), 1037–1050.

Wang, Y. C., McPherson, K., Marsh, T., Gortmaker, S. L., & Brown, M. (2011). Health and economic burden of the projected obesity trends in the USA and the UK. *Lancet*, *378*(9793), 815–825.

Wascher, C. A. F., Dufour, V., & Bugnyar, T. (2012). Carrion crows cannot overcome impulsive choice in a quantitative exchange task. *Frontiers in Psychology*, *3*(April), 1–6.

Winecoff, A. A., Ngo, L., Moskovich, A., Merwin, R., & Zucker, N. (2015). The functional significance of shyness in anorexia nervosa. *European Eating Disorders Review*, *23*, 327–332.

Winecoff, A., Clithero, J. A, Carter, R. M., Bergman, S. R., Wang, L., & Huettel, S. A. (2013). Ventromedial prefrontal cortex encodes emotional value. *The Journal of Neuroscience*, *33*(27), 11032–11039.

Winecoff, A., LaBar, K. S., Madden, D. J., Cabeza, R., & Huettel, S. A. (2011). Cognitive and neural contributors to emotion regulation in aging. *Social Cognitive and Affective Neuroscience*, *6*(2), 165–176.

24

Cognitive Control and Reward

Ruth M. Krebs and Marty G. Woldorff

Introduction

Cognitive control can be considered as a set of functions that enable us to adapt to changing demands in our environment and to optimise our behaviour accordingly. As such, our behaviour is guided by internal and external goals, which are in turn influenced by motivational factors. For example, our ability to focus attention on a certain task at work, without being distracted by co-workers or our own unrelated thoughts, will not only depend on our skills but just as much on our current level of motivation. Although intrinsic factors of motivation (e.g., the joy of solving a task) are difficult to manipulate and measure in an experimental setting, we are able to investigate how extrinsic motivation, for example, in the form of monetary incentives (rewards), can modulate cognitive control processes and ultimately change behavioural output. For example, it has long been known, from research on instrumental learning, that rewarding an animal (or human) for a particular action will increase the likelihood of this action's repetition in the future (Wise, 2004). More recently, however, researchers have started to investigate how the anticipation of reward can modulate complex cognitive functions, including visual attention, conflict resolution, response inhibition, and episodic memory encoding (for a review, see Braver et al., 2014; Botvinick & Braver, 2015).

In this chapter, we take the view that a broad distinction can be made between two main ways by which the prospect of reward can be brought to bear on task performance and behaviour, namely, by pre-task reward cueing and by target–reward associations (Figure 24.1). In the first case, the availability of reward versus no reward for a specific task, or for a high versus low level of reward, is indicated ahead of time. This pre-task reward information can either be set in a sustained way across time, such as for a block of task trials, or in a more rapidly varying way, such as by a cue providing reward information on a trial-by-trial basis. Both of these possibilities can lead to a differential state of preparation for the subsequent task and for the processing of the task-relevant stimuli, which can then in turn lead to a modulation of task performance.

The second major way of using reward to manipulate task performance is by associating it directly with certain stimuli or stimulus features that are part of the task itself. Such target–reward associations may have been set by previous experience (e.g., reward has been previously signalled by these stimulus types or features), or because one has been instructed regarding their reward-related properties; that is, when such stimulus types or features occur

The Wiley Handbook of Cognitive Control, First Edition. Edited by Tobias Egner.

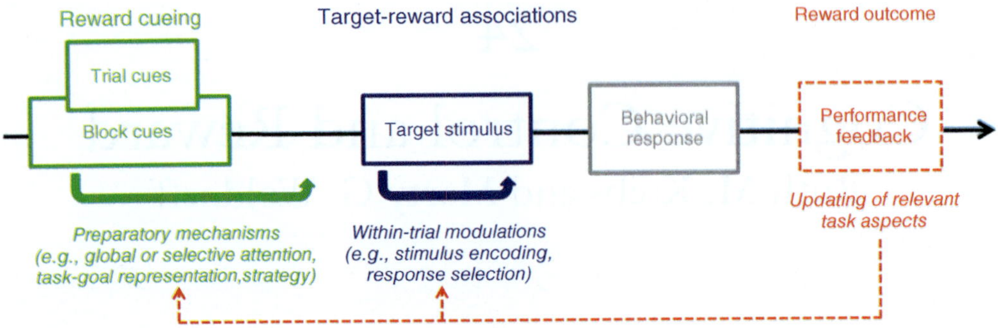

Figure 24.1 Key ways in which reward information can influence cognitive control. The availability of reward can be indicated in advance of a trial or a block of trials (Reward cueing, green), or it can be directly embedded in the target stimulus itself in form of reward-associated stimulus types or features (Target–reward associations, blue). Whereas reward cueing is thought to act at the level of preparatory mechanisms that in turn modulate subsequent target processing, target–reward associations likely trigger rapid within-trial modulations. Both manipulations have been shown to modulate the behavioural response to the target (grey). Note that though combined in one figure for illustration purposes here, reward cuing and target–reward associations are typically employed in independent paradigms. In both cases, if performance feedback is provided on a trial-by-trial basis, it can be used to update the relevant aspects of a given task (Reward outcome, orange; not the focus of this chapter).

in a trial, participants are rewarded for accurate task performance. In contrast to cue or block manipulations, which allow for differential preparation in advance of the target, within-trial target–reward associations can only start to affect stimulus encoding and response selection once the target appears.[1]

The following sections are organised along these two main ways by which reward can influence cognitive control. We first discuss the putative pro-active effects that occur when reward information about a task trial (or a block of trials) is provided prior to the task, and we then move on to with in-trial processes triggered by target–reward associations that seem to be rather re-active (or even automatic) in nature. Within this general framework, we will focus on reward-related effects on a range of cognitive control functions. In this regard, we note that we consider attentional control to be a subset of the broader category of cognitive control, which also includes processes related to conflict resolution (Chapter 4 by Egner in this volume), working memory (Chapter 3 by Meier & Kane in this volume), task switching and task-goal maintenance (Chapter 2 by Monsell in this volume), and response inhibition[2] (Chapter 6 by Verbruggen & Logan in this volume). As part of this discussion, we will also consider the putative underlying neural mechanisms of the behavioural modulations due to reward cuing and target–reward associations, and highlight ongoing discussions and outstanding questions that remain to be tackled in future studies.

Modulations of Preparatory Attentional Control via Reward Cues

Consider a situation in which you have a high expectation to be rewarded for performing a certain task well or quickly, before you actually start. You could endeavour to enhance your performance on the task by taking measures that help you to concentrate better, such as putting on headphones to limit distracting noise, or to try to bring yourself into a focused mental state. These can be considered ‘pro-active’ or global preparatory strategies (Braver, 2012; see also Chapter 9 by Chiew & Braver in this volume) that may promote your performance and in turn increase the chances to earn the reward. In experimental research, reward manipulations of pro-active control can be implemented in a more sustained fashion by rewarding entire blocks of a task (e.g., Locke & Braver, 2008), or in a more transient, trial-to-trial fashion by using pre-task cues that indicate reward availability or no reward before each new trial (e.g., Krebs, Boehler, Roberts, Song, & Woldorff, 2012). In this section, we will consider the effects of both blocked and cued reward together because they both involve the engagement of preparatory processes, prior to the occurrence of the target stimulus. Moreover, they also are sometimes combined in one and the same study (e.g., Jimura, Locke, & Braver, 2010), and they seem likely to rely on partly overlapping mechanisms.

Probably the most commonly used reward manipulation in the cognitive neuroscience literature is the monetary incentive delay (MID) task (Knutson & Cooper, 2005; Knutson, Westdorp, Kaiser, & Hommer, 2000), which was designed to parallel animal models of reward anticipation and delivery (Schultz, 1997, 2001). In this task, a pre-task cue at the beginning of each trial indicates the chance to win a reward or no reward on that trial (or to indicate different reward probability levels). Performance in the cued upcoming task, which in the early versions of the MID paradigms was a simple target-detection task, is then rewarded according to the cue contingencies. After the response on each trial, given accurate performance, participants receive feedback about the actual reward they just earned, thereby facilitating the establishment of a stable association between reward prediction and actual outcome. When this task is performed while acquiring event-related measures of brain activity, such as functional magnetic resonance imaging (fMRI) or event-related potentials (ERPs), it enables the isolation of the preparatory neural processes related to reward anticipation from those of target-stimulus processing and reward feedback, thus paralleling the approaches used in analogous animal work. And indeed, in parallel to the animal findings, the anticipation of reward with this task was found to be associated with activity modulations in regions of the dopaminergic system, first and foremost the ventral striatum and the medial midbrain, which are now considered to be key regions of the human reward-processing circuit (Knutson, Adams, Fong, & Hommer, 2001a; Knutson, Fong, Adams, Varner, & Hommer, 2001b; Knutson, Taylor, Kaufman, Peterson, & Glover, 2005; Wittmann, Schiltz, Boehler, & Duzel, 2005). We note, however, that reward processing in the brain is not limited to these regions, as reward seems to modulate activity in networks that are relevant for a given cognitive function (Pessoa & Engelmann, 2010), and that regions of the dopaminergic system can also be recruited by other cognitive processes in the absence of reward (Boehler et al., 2011).

Enhancing Preparatory Attentional Control

Over the past decade, the MID logic has been adopted to probe for a range of specific cognitive control functions, including top-down attentional control and conflict resolution. One of the first MID-like paradigms in the attentional control domain was employed by Engelmann and Pessoa (2007, 2009). Using a rewarded exogenous attention-cuing paradigm, they reported that the participants' ability to detect a peripheral target improved with increasing incentive value, which was indicated by a pre-task cue for a set of trials (Engelmann & Pessoa, 2007). In a subsequent fMRI study (Engelmann, Damaraju, Padmala, & Pessoa, 2009), the authors demonstrated a similar reward benefit for detection sensitivity of noisy face targets in an endogenous attention-cuing paradigm. It should be noted that in both of these studies the cue indicating the reward level was given for a series of five (Engelmann et al., 2009) or eleven (Engelmann & Pessoa, 2007) upcoming trials rather than on a trial-by-trial basis as in the classic MID task, and thus these could be considered mini-block manipulations. Also, an earlier fMRI study already provided evidence for reward-enhanced attentional orienting by having participants perform an endogenous attention task in blocks with different incentive contexts, namely, having the potential to win, the potential to lose, or no monetary consequence (Small et al., 2005). Importantly, in both fMRI studies using endogenous attention tasks (Small et al., 2005; Engelmann et al., 2009), reward-induced performance benefits were associated with activity modulations in frontoparietal brain regions implicated by other work as being involved in attentional control (reviewed in Corbetta & Shulman, 2002), as well as regions of the reward circuit, suggesting that reward may improve performance by modulating task-relevant attentional processes, rather than acting on its own as an independent force (Pessoa & Engelmann, 2010).

More direct evidence for this notion has been provided by several studies specifically investigating the relationships between reward anticipation and attentional control. In an fMRI study from our group (Krebs et al., 2012), trial-by-trial cues for a visual discrimination task contained information about reward availability on the current trial as well as information about attentional task demands of the upcoming target stimulus. Importantly, activity in the brain networks associated with preparation for high-reward targets were largely overlapping with those for high-attentional-demand targets, supporting the above-mentioned notion that reward imparts its influence on task performance, at least in part, via attentional control networks. Beyond this overlap, there were a handful of regions, including the anterior cingulate cortex and the medial midbrain, in which activity was especially increased for anticipating attentionally demanding targets when reward was at stake. This latter finding highlights the flexible nature with which reward information seems to be integrated to meet the current task demands.

On the basis of this fMRI study, we employed the same paradigm, that is, crossing reward availability and attentional demand, while recording EEG to provide insights into the temporal dynamics of their integration (Schevernels, Krebs, Santens, Woldorff, & Boehler, 2014). In particular, while reward information modulated attention-related preparatory ERP activity (as indexed by the fronto-central negative-polarity contingent negative variation wave [CNV]) fairly early in the cue-target interval, the interaction with attentional-task-demand information was only observed late this interval, as reflected in a late enhancement of the pre-target CNV. In another recent EEG study, we investigated the effects of cued attention and cued reward on performance in a Stroop conflict task (van den Berg, Krebs, Lorist, & Woldorff, 2014). Relative to cues after which no target was to be expected (control cues), cues that indicated to attend to an upcoming target (attend cues) enhanced the CNV and reduced alpha power in the cue-target interval (both hallmark reflections of increased attention). Importantly, both of these effects were substantially enhanced when the attend cue signalled reward availability. Lastly, another study seeking to directly integrate spatial and temporal information by acquiring fMRI and EEG simultaneously during a classic MID task (Plichta et al., 2013) found that the attention-related preparatory CNV component was correlated with activity modulations in regions of the reward-processing circuit, including the medial prefrontal cortex, thalamus, and ventral striatum. Thus, all of these studies argue that the prospect of reward brings about its effects on behaviour and task performance, at least in part, by marshalling attentional control circuits.

Effects of Reward Cueing on Conflict Control

The preparatory effects of reward availability have also been studied in other cognitive control domains, such as in conflict processing. In an fMRI study by Padmala and Pessoa (2011), the effect of trial-by-trial reward cues on attentional control in a Stroop-like picture-word interference task was examined, in which participants were asked to discriminate images of houses and buildings while ignoring a matching or non-matching overlaid word (Padmala & Pessoa, 2011). These authors found that reward availability reduced the typical performance decrements on trials with incongruent irrelevant information (i.e., interference effects). Neurally, this reward-related performance benefit was accompanied by enhanced activity in the frontoparietal attention network and in subcortical reward-processing regions during cue processing, which was followed by diminished target-evoked activity in medial-frontal regions that have been implicated in conflict resolution. Together with the observation that reward availability reduced word-specific activity in the visual cortex, these results were interpreted in terms of improved attentional filtering of irrelevant information in reward trials (see also Soutschek, Stelzel, Paschke, Walter, & Schubert, 2015).

Although the picture-word interference task is considered to probe for conflict-control mechanisms similar to those in the classic colour-naming Stroop task (MacLeod & MacDonald, 2000), the results of two studies that included reward-predicting cues in a classic Stroop task suggest that there may be some key differences (van den Berg et al., 2014; Veling & Aarts, 2010). In particular, although reward availability facilitated responses globally in these other studies (faster and more accurate on reward trials), there was no reward-induced reduction of conflict interference, in contrast to the results by Padmala and Pessoa (2011). This discrepancy may be related to the nature of the stimuli in the picture-word interference task versus the classic Stroop task. Specifically, the reward-induced attentional filtering of the irrelevant word meaning, as suggested by Padmala and Pessoa (2011), may be more difficult in the classic Stroop paradigm because the to-be ignored dimension (word meaning) is embedded in the to-be attended stimulus (colour word), rather than being overlaid on a separable object. Moreover, the van den Berg et al. (2014) study also included ERP measures, and the results showed no reduction across subjects of the negative incongruency ERP wave (Ninc) following reward cues, paralleling the lack of a mean reduction in behavioural interference. However, this latter study found substantial variations between participants in the reduction of both the behavioural and neural conflict effects in the context of reward, which were correlated with participants' overall RT speeding on reward trials, as well as with the level of preparatory neural activity enhancement (larger CNVs, reduced alpha). Such a result suggests that some participants used the advance reward information to selectively enhance their attention to the relevant dimension of the Stroop task (font colour) and reduce interference, whereas others may not have utilised the reward information as effectively. In fact, some participants displayed larger interference effects on reward trials, which could be explained by enhanced attention to the entire Stroop stimulus, both the relevant and irrelevant dimensions, thereby actually increasing interference.

The above considerations become yet more complicated when irrelevant and relevant information of an interference task are separated in time. One of the first EEG studies that sought to investigate the influence of reward on interference processing used an MID-like version of a flanker task, in which the flanking distracters were presented about 100 ms before the central target letter (Seifert, Naumann, Hewig, Hagemann, & Bartussek, 2006). In that study, the prospect of reward enhanced overall response speed (similar to most of the above studies), but it also increased interference from incompatible flankers at the same time. This latter behavioural effect was paralleled by activity modulations in the lateralised readiness potentials (LRPs), reflecting motor-response preparation activity, indicating that reward cueing led to the motor system being initially biased to respond according to the distracter letters that were presented slightly earlier in time.

Promoting Cognitive Flexibility

The above studies seem to suggest that the exact nature of reward-based modulations in attentional control and conflict resolution can depend on specific differences in the task or stimulus structure, as well as in the reward mode. In addition, many of these studies suggest that increased attention to the target stimulus plays a crucial role in task performance. But what happens in tasks in which the task structure or task goals are more complex or change during the experiment? In this section, we discuss studies of reward effects on working memory maintenance, utilisation of contextual cues, and task switching.

In an early cued-reward fMRI study of such effects in working memory (Krawczyk, Gazzaley, & D'Esposito, 2007), participants had to memorise one stimulus category and ignore another, and were presented with a probe stimulus after a variable maintenance delay. The results indicated that during the encoding period, reward cues triggered activity increases in the lateral prefrontal cortex, an area implicated in working memory and task-goal maintenance, while also facilitating the behavioural responses to the subsequent probe stimulus. Moreover, activity in category-specific visual regions was selectively increased when stimuli were actively attended versus passively viewed, but only when a reward was at stake. These results were interpreted as reflecting enhanced top-down modulation of visual processing during working memory maintenance in the context of reward. A similar task structure was used by Jimura and colleagues in which trial-by-trial reward cuing was combined with a reward block manipulation in a working memory task (Jimura et al., 2010). The results indicated that a general reward context (blocks that included both reward and no-reward trials) benefited task performance, especially on those trials for which the trial cue indicated that there would not be a reward, suggesting a role of globally enhanced (sustained) pro-active control. Importantly, this context effect was associated with activity changes in lateral prefrontal regions typically implicated in working memory maintenance.

These results highlight the complex and flexible nature by which reward can shape our behaviour, depending on the exact task structure and context information. More evidence for this notion comes from an fMRI study that employed a rewarded stimulus–sequence contingency task (Locke & Braver, 2008), in which participants had to respond to a probe stimulus (X), but only when it was preceded by a specific cue stimulus (A). They found that performance in A-X trials was increased in reward blocks compared to baseline blocks. However, the possibility of reward also increased error rates on non-target trials (e.g., on A-Y trials), presumably reflecting increased response biasing by the cue stimulus (i.e., the 'A'). The fMRI data indicated these global reward-context effects were linked to increased activity in frontoparietal regions typically associated with attentional control.

In contrast to the above studies, in which task goals were complex but mostly stable throughout the experiment, other studies have required participants to flexibly switch between different task regimes from trial to trial in the context of reward (Aarts et al., 2010; Capa, Bouquet, Dreher, & Dufour, 2013; Kleinsorge & Rinkenauer, 2012), or to integrate information from multiple sources in order to make the correct response (Kouneiher, Charron, & Koechlin, 2009). The cued task-switching reward studies have produced diverse results, which may be partly explained by differences in task design and the focus of the analyses. In one fMRI study that used a Stroop-like cued task-switching paradigm while also assessing genetic variations (Aarts et al., 2010), reward cues improved task performance only on switch trials and only in participants with a certain genetic polymorphism (i.e., 9-repeat allele carriers of the dopamine transporter gene DAT1), which is typically (and was in this study) associated with increased activity in the ventral striatum during reward anticipation. Moreover, actual task execution was accompanied by increased dorsal-striatal activity in switch versus repeat trials. In another task-switching paradigm (Kleinsorge & Rinkenauer, 2012), if reward was only available for one of two tasks in an experiment (constant task-to-reward mapping), switch costs were significantly reduced across tasks. If, however, reward availability varied from trial to trial independent of the task (variable task-to-reward mapping), no reward benefit for switch trials was observed. Rather, reward anticipation facilitated responses on trials with a robust task–reward association, at the cost of performance detriments in the unrewarded task. The critical role of task structure in these processes is further underscored by a task-switching study that signalled reward prospect block-wise (Capa et al., 2013). In this case, no differential reduction of switch costs was observed above and beyond a general response facilitation in reward blocks. Although the comparability of these studies is limited owing to differences in task structure, general methods, and even participant selection, they indicate that reward can modulate the ability to flexibly employ cognitive control in response to changing task demands, and that these modulations seem to rely to some extent on activity changes in regions of the dopaminergic neural circuits.

Preparing for Inhibition

So far, we have discussed studies in which reward prospect was linked to the correct and sufficiently fast *execution* of a certain action. But what if the task requires the *withholding or stopping* of an already initiated or about to be initiated response (i.e., response inhibition)? This question seems especially relevant considering the common notion that approach behaviour is more easily triggered by reward-related or other positive stimuli, whereas withdrawal or avoidance behaviour tends to be associated with negative events (e.g., Guitart-Masip et al., 2011; Hare, Tottenham, Davidson, Glover, & Casey, 2005).

In the classic stop-signal task, the majority of trials require a fast response to a go stimulus. In a subset of trials, however, the go stimulus is rapidly followed by a stop stimulus, indicating that participants have to 'stop' (i.e., withhold their response to the go stimulus). The key finding in these studies is that the shorter the interval between the go and stop stimuli, the higher the likelihood of being able to withhold the response. By considering stopping success at different percentiles of the go-trial RT distribution, the so-called stop-signal response time (SSRT) can be calculated as an index of how fast stopping can be implemented by an individual (Verbruggen, Chambers, & Logan, 2013; Chapter 6 by Verbruggen & Logan in this volume).

One of the first studies that investigated motivational influences on response inhibition introduced specific monetary contingencies for go and stop trials across task blocks (Leotti & Wager, 2010). In this behavioural study, participants were rewarded for fast and accurate performance on go trials as well as for successful response inhibition on stop trials, whereas slow and inaccurate responses (go trials) and commission errors (stop trials) led to monetary loss. Importantly, the task was performed in blocks of different motivational contexts, in which either speed was favoured over accuracy or vice versa, or both were equally important. Specifically, reward and loss magnitudes were titrated adaptively throughout each of these blocks on the basis of the participants' performance, thereby prompting participants to make strategic adjustments to maximise their gain. And, indeed, in blocks that increasingly favoured stopping accuracy over speed, participants were more successful in inhibiting their responses on stop trials, displayed shorter SSRTs, and responded more slowly on go trials. These observations thus provided insights into dynamic behavioural adjustments in the context of extrinsic motivation and have contributed to contemporary models of response inhibition.

More recently, another group investigated whether reward can introduce global biases towards response execution and response inhibition while acquiring EEG measures of brain activity (Greenhouse & Wessel, 2013). In this study, participants were instructed that either response inhibition (stopping) would be more rewarding than response execution (responding) in a given block or vice versa (but in contrast to the study by Leotti and Wager [2010], these contingencies were fixed for each block). Behaviourally, it was found that stopping performance, as indexed by the SSRT, was facilitated when stopping was rewarded, but impaired when go responses were rewarded. These results are consistent with the study by Leotti and Wager (2010), as well as with an earlier report of impaired stopping performance when reward was exclusively associated with performance on go trials (Padmala & Pessoa,

2010). The analysis of the EEG data in the Greenhhouse and Wessel study (2013) furthermore revealed an enhanced P3 in rewarded stop trials: an ERP component that has been found to be sensitive to stopping success (Bekker, Kenemans, Hoeksma, Talsma, & Verbaten, 2005). Moreover, EEG phase coherence between fronto-central and occipito-parietal sites on go trials was greater when stopping rather than going was rewarded, suggesting that reward not only modulates response inhibition itself, but also leads to other preparatory changes.

A recent fRMI study used cues to signal the availability of reward in a stop-signal task before the go stimulus was presented (Rosell-Negre et al., 2014). Importantly, participants could earn bonuses for any correct performance after reward cues, that is, both on go trials and stop trials. The data showed that with these balanced reward probabilities, both the RTs on go trials as well as SSRTs on stop trials were significantly shorter after versus no-reward cues. Moreover, this fMRI study revealed increased activity in cognitive control regions in stop trials following reward cues compared to no-reward cues, suggesting that reward availability modulates the inhibitory process itself—at least to some extent (unfortunately, the neural response to the cue was not analysed in this study).

Together, the above findings highlight once again that reward availability can facilitate very specific cognitive functions, here response inhibition, by modulating neural processes involved in that particular function. Importantly, they further suggest that the natural approach tendencies for rewarding events can be overruled by explicitly rewarding the successful inhibition of responses.

Interim Summary

Together, the above studies support the notion that the prospect of reward, either instantiated in a block fashion or trial to trial, ramps up attention and cognitive resources in anticipation of the task at hand, doing so in a mostly preparatory or pro-active manner (see Figure 24.1). Neurally, this enhanced task preparation seems to be subserved by frontoparietal regions that have been widely implicated in attentional control, in concert with regions implicated in reward processing such as the ventral striatum, along with task-specific networks and components. Such activity changes are not only observed during the cue-target interval in MID-like paradigms, but also during block manipulations of reward, in which attention is modulated in a more sustained fashion across trials. In very general terms, it seems likely that in blocked reward manipulations, the contribution of global, strategic control is relatively high, whereas trial-by-trial reward cues promote a more transient (but still pro-active) control mode.

Despite these differences, both block and trial-by-trial reward cues signal reward availability ahead of time and can thus trigger top-down influences that are set up prior to the occurrence of the target stimulus. This is in contrast to circumstances in which previously established or instructed reward associations are embedded in the target stimulus itself in the absence of any pre-task cues. The effects of such target–reward associations are discussed in the next section.

Within-Trial and Post-Trial Modulations due to Stimulus–Reward Associations

In keeping with the example from the beginning of the chapter, not every task in real life or in the laboratory allows for strategic preparation in anticipation of an extra incentive. Rather, one often needs to react 'on the fly' to changing reward contingencies, without the benefit of any advance information. In particular, reward-related information can be introduced into a task or circumstance by associating reward with a certain target-stimulus type or feature, due, for example, to previous experience or by explicit instructions about such relationships upfront. Moreover, depending on the paradigm, reward associations appear to be able to affect performance not only on the current trial, but also on the subsequent trial, or even in a separate task phase in which reward is no longer available. Importantly, paradigms with stimulus–reward associations share two aspects that set them apart from cued or blocked reward manipulations. First, in the absence of any pre-task cues, reward effects in these paradigms rely more on re-active or perhaps even automatic processes rather than on preparatory mechanisms. Second, target–reward associations can both benefit and impair performance, depending on their compatibility with the task goal.

In this section, we discuss studies using different types of target–reward associations to examine reward effects on several key cognitive functions, in particular, effects on attentional orienting, conflict processing, and response inhibition.

Reward-Based Modulations of Visual Attention

One way to modulate task performance via reward without invoking preparatory attention is to directly link reward to distinct features of a target stimulus (e.g., colour). Kiss, Driver, and Eimer (2009) employed a visual-search paradigm, in which the target stimulus (a colour pop-out item in an array) could be associated with high or low reward, depending on its colour (Kiss et al., 2009). Explicit instructions about the reward–colour associations were given in advance. Typically, the attentional selection of the target stimulus among distractor stimuli in this task is indexed by a particular ERP component, the N2pc (posterior N2 wave contralateral to the target stimulus), peaking at about 220 ms after array onset (Luck, 1994). Kiss et al. (2009) report that this N2pc component was amplified for high- versus low-reward targets, which was paralleled by faster target discrimination, demonstrating a fast modulatory influence of stimulus–reward associations on attentional selection.

As mentioned above, reward information can also be delivered as post-target feedback, which can lead to incidental changes in the processing of subsequent target stimuli. Specifically, DellaLibera and Chelazzi (2006) introduced reward feedback in a negative priming experiment. In negative priming, if a particular stimulus type or feature is irrelevant on one trial and is to be ignored, then subjects are slower to identify that stimulus type or feature if it is relevant on the next trial (e.g., Tipper, 2001). In the study by DellaLibera and Chelazzi (2006), either high or low-reward feedback was given following the first trial of a pair of trials in a global–local number task (e.g., a large 5 made up of small 6's). Performance on the second trial of a pair was analysed depending on whether it contained the previously attended or ignored unique feature (e.g., a large 5 made up of small x's). The finding was that negative priming was enhanced when high-reward signals were presented in between the trials, but abolished after low-reward signals. These findings not only suggest that specific cognitive biases are augmented after high-reward feedback, but that target stimuli following low-reward feedback are not subject to typical priming effects, in turn suggesting that the relative-reward level (relative to the range of reward possibilities) modulates attention to a primed or relevant task feature.

A similar approach was employed in a rewarded visual-search task (Hickey, Chelazzi, & Theeuwes, 2010). Specifically, the circular search display contained a unique shape (i.e., target) of a certain colour, several distractors of a different shape but with the same colour, and one distractor of different shape and colour (i.e., salient distractor). Participants had to discriminate the orientation of a line within the target shape, and their response was followed by incidental reward feedback. Effects of incidental reward were assessed by sorting the current trials with regard to the reward feedback on the previous trial (high vs. low) and the colour mapping of target and salient distractor (same vs. swapped colours). This allowed testing for low-level attentional biasing towards features that had just been incidentally associated with reward. As the authors hypothesized, participants responded faster when the target colour was repeated in the next trial after high- compared to low-reward feedback, whereas the opposite pattern was observed when target and distractor colours were swapped on the next trial. Moreover, amplitude modulations of the N2pc ERP component as an index of attentional selection followed a similar pattern, supporting the notion that attention was more readily deployed to the previous target colour after high-reward feedback.

More recently, the same group adopted this logic to a more naturalistic context in a behavioural study by presenting reward feedback after participants had identified a target of a defined object category (e.g., a tree) on a natural scene image (Hickey, Kaiser, & Peelen, 2015). They could demonstrate that high-reward feedback resulted in a stronger attentional bias for the entire object category, as reflected in impaired performance if any object from the rewarded category was present as a distractor on the next trial. A companion fMRI study employing multivariate pattern analysis (Hickey & Peelen, 2015) further confirmed that the processing of reward-associated target and distracter objects in object-sensitive visual cortex regions was relatively enhanced and suppressed, respectively. These observations highlight how incidental reward signals can modulate object-specific visual attention and extends the previous finding that perceptual representations in visual sensory cortex are modulated by the reward history of a given item (Serences, 2008).

Such changes in visual saliency have been shown to linger even when reward is no longer at stake. Specifically, researchers have separated the acquisition of specific target–reward associations (acquisition phase) from a later experimental phase (test phase). Importantly, reward associations that were previously established, via either learning from feedback or by explicit instructions, were no longer task relevant in the test phase, so that any effects of reward could be attributed to lingering stimulus prioritisation rather than being a result of voluntary top-down control. Using this logic in a visual-search task, Anderson and colleagues demonstrated that attention is more readily captured by distractor colours that had been associated with reward during a previous independent study phase (Anderson, Laurent, & Yantis, 2011a, 2011b).

Attentional biasing effects due to previously learned reward associations have also been observed in other visual attention paradigms, including the attentional blink (Raymond & O'Brien, 2009), visual orienting (Failing and Theeuwes, 2014; Rutherford, O'Brien, & Raymond, 2010), and visual working memory tasks (Gong & Li, 2014). In addition, reward-related attentional biasing effects have been observed in economic decision-making tasks, where attention is drawn towards stimuli with previously learned associations for predicting gain likelihood in the task (San Martin, Appelbaum, Huettel, & Woldorff, 2016; see Gottlieb, 2012 for related work in monkeys). Moreover, the better an individual is at learning these associations, the greater this attentional biasing, and the better the economic decision-making performance (San Martin et al., 2016). Lastly, it is important to note that stimulus features that have been associated with reward from trial to trial (e.g., Hickey et al., 2010) or through repeated exposure during learning (e.g., Anderson et al., 2011b) can lead to performance *impairments* if they occur in irrelevant dimensions of a task or trigger incompatible response tendencies.

Reward Associations in Conflict Resolution

In addition to guiding or (misguiding) visual attention, target–reward associations have been shown to affect very specific cognitive-control functions, including conflict resolution. In a series of studies, we adopted the logic of explicitly associating certain target features with reward in a classic Stroop task in the absence of any pre-task cue (Krebs, Boehler, & Woldorff, 2010). In addition to having the standard Stroop task of responding to the ink colour of colour-word stimuli (task-relevant dimension) while ignoring the semantic meaning of the colour word (task-irrelevant dimension), participants were explicitly informed upfront that specific ink colours would be associated with reward upon accurate performance (e.g., stimuli in red font), whereas others were not. These explicit reward associations with the relevant stimulus dimension resulted in general response facilitation (faster RTs on those stimuli) and reduced interference from irrelevant word meanings on incongruent trials. Thus, these results show that reward prospect can benefit conflict resolution even in the absence of differential preparatory mechanisms (cf. Padmala & Pessoa, 2011). Moreover, when reward-related information occurred in the irrelevant stimulus dimension and was incongruent with the actual ink colour (e.g., 'RED' written in green ink when the red colour was associated with reward), we observed larger interference effects compared to trials with reward-unrelated irrelevant information. This latter observation is consistent with the distracting effects of irrelevant reward-related features in visual search discussed above, suggesting a shared underlying mechanism of reward-related saliency regardless of the cognitive function that is being probed by a specific paradigm.

In an fMRI version of the above-described rewarded Stroop task (Krebs, Boehler, Egner, & Woldorff, 2011) we found that both task-relevant and task-irrelevant reward associations recruited overlapping neural networks, including lateral prefrontal and visual processing regions, probably reflecting the general saliency of the reward-related features. However, only task-relevant reward information (e.g., actual red-coloured stimuli) increased activity in the ventral striatum, one key region of the reward-processing circuit, whereas task-irrelevant reward-associated information led to activity modulations at the level of response selection (i.e., in the pre-supplementary motor area), which may reflect the prevention of an erroneous response triggered by a salient but incongruent feature. Although fMRI studies can reveal the neural networks involved in a task, its low temporal resolution does not allow for the effects on different processing stages to be delineated, which is why we employed an EEG version of the same rewarded Stroop paradigm (Krebs, Boehler, Appelbaum, & Woldorff, 2013). The ERP data from this study revealed that reward associations were registered fairly early in the processing cascade, namely, by around 200 ms. Moreover, subsequent ERP components reflecting conflict processing (i.e., the Ninc and the late positive potential—see earlier section), were also modulated by the reward association, in particular, by occurring earlier in time. These results indicate that the behavioural effects of stimulus–reward associations would seem to rely on an early visual prioritisation of reward-related features, which in turn modulates subsequent conflict resolution processes. Interestingly, this parallels the effects of cued reward on conflict resolution (e.g., Padmala & Pessoa, 2011), but resulting from a more re-active mechanism, rather than differential pro-active control.

As with visual attention, reward information can also be introduced incidentally in cognitive control tasks, for example, by post-target feedback, which can in turn affect performance on the subsequent trial. One well-known phenomenon in the conflict-resolution literature is a decrease in behavioural interference on trials following incongruent trials, referred to as conflict adaptation (Chapter 4 by Egner in this volume), which is believed to reflect an enhancement of voluntary cognitive engagement after such trials or to be related to feature-binding processes and arousal (Botvinick, Braver, Barch, Carter, & Cohen, 2001; Hommel, 2004; Verguts & Notebaert, 2009). Several studies have examined whether incidental reward signals can modulate these conflict-triggered behavioural adaptations—with partly inconsistent findings. Van Steenbergen and colleagues (2009) introduced incidental reward and loss signals (i.e., smiley and sad faces) in a flanker task, and compared performance as a function of trial type (incongruent vs. congruent) and reward (reward vs. neutral vs. loss) on the previous trial. Intriguingly, whereas conflict adaptation was observed after neutral and loss signals, this was absent after reward signals, which led the authors to conclude that incidental positive events can cancel out the aversive experience of conflict and in turn counteract cognitive adaptation to this conflict. Another group presented bonus points as performance feedback in a conflict task (Braem, Verguts, Roggeman, & Notebaert, 2012) and found that conflict adaptation was actually enhanced in trials following these feedback signals, which is in contrast to Van Steenbergen et al. (2009). Critically, these two studies used different types of reward signals and different instructions regarding the contingency on task performance (see also Sturmer, Nigbur, Schacht, & Sommer, 2011).

Braem et al. (2013) further investigated how performance contingency can determine the effects of emotional (and reward-based) signals by introducing positive and negative emotional pictures between trials of a combined conflict/task-switching paradigm. Importantly, in one group of participants, the pictures were only presented following correct performance (performance-contingent), whereas they were always presented in the other group of subjects (non-contingent). The behavioural data showed that positive-valence signals, relative to negative ones, promoted a loosening of task focus (flexibility) in the performance-contingent context, as reflected by a reduction of the typical switch-cost increase after conflict trials (Goschke, 2000). In the non-contingent group, however, positive-valence signals lead to an increase in task focus (stability). Together, it seems that the effects of post-target reward signals strongly depend on the instructions (performance-contingent vs. non-contingent) and the stimuli used (affective vs. symbolic reward signals). Note, moreover, that these incidental or post-target reward effects are conceptually different from the instructed within-trial reward manipulations discussed above.

Reward-Based Response Inhibition

In the first part of this chapter, we laid out that reward cues can modulate the ability to withhold an initiated response process, which is already notable considering that response inhibition tends to be a quite stable feature within individuals (Cohen & Poldrack, 2008). But can such modulations occur even when participants are not given any chance to differentially prepare for a rewarded stop trial?

We investigated this issue by applying the same logic that we used in the rewarded Stroop task (Krebs et al., 2010) to a classic stop-signal task without cues (Boehler, Hopf, Stoppel, & Krebs, 2012). Specifically, reward availability on a given trial was indicated by the colour of the stop signal, which, as in classic stop-signal tasks, followed the go signal after a variable delay in a subset of trials. All go trials were therefore by definition unrewarded, whereas stop trials could be rewarded or unrewarded depending on the colour. Because stopping success rates in stop-signal tasks are usually kept stable at 50% by using a staircase procedure to vary the stop-signal delay, performance is typically assessed using the SSRT, which reflects the 'speed' with which an initiated response process can be cancelled (see also Chapter 6 by Verbruggen & Logan in this volume). The results showed that SSRTs were indeed shorter when the stop-signal colour was reward related compared to reward unrelated, indicating that the reward associations produced a faster response-cancellation process. Intriguingly, and in comparison with a blocked reward manipulation (Greenhouse & Wessel, 2013), reward must be able to affect behaviour in a rapid manner in this case, as the reward information was only provided after response initiation (i.e., when the stop signal occurred). When employing this task in the fMRI scanner, we also found that rewarded stop trials were associated with neural modulations in areas implicated in response inhibition, including the anterior insula and inferior frontal gyrus (Boehler, Schevernels, Hopf, Stoppel, & Krebs, 2014). Importantly, the observation that these modulations occurred for both successful and unsuccessful stop trials suggests that they were indeed related to inhibitory processes rather than being mere reflections of successful task performance.

Interim Summary

The above findings demonstrate that linking reward to certain types or features of a target stimulus can modulate performance and task-specific neural networks in diverse cognitive tasks. Unlike in cued or blocked reward paradigms, these modulations arise in the absence of classic preparatory (pro-active) attentional processes.[1] Rather, such target–reward associations seem to trigger changes at the time of the target-stimulus processing and/or response selection, which in turn affect task-specific processes, such as attentional orienting, conflict resolution, and response inhibition (see Figure 24.1). Importantly, these modulations can occur regardless of the relevance of the stimulus for the task and can hence lead either to behavioural facilitation or to distraction, depending on whether the reward feature is overlapping or in competition with the current task goal (see below). The occurrence of such detrimental effects suggest that target–reward associations can act, at least in part, on low-level processes and are less affected by voluntary control processes. However, several of the above-described neuroimaging studies support the notion that low-level effects are not the only driving force for behavioural modulations in these paradigms, and that (re-active) control mechanisms can be triggered on a fast scale within a trial.

Additional Considerations

The Role of Dopamine

Considering the origin of the MID task in the animal literature (Schultz, 1997, 2001), it seems worthwhile to discuss the potential role of the dopaminergic system in the different task regimes described above, that is, reward cuing and target–reward associations. On the one hand, the increased firing rate of dopaminergic neurons in animals to rewarding stimuli seems to reflect some sort of value signal, which strengthens the association between a certain action and the reward outcome (or between a reward-predictive cue and the reward outcome) via activity modulations in the striatal and prefrontal regions (Wise, 2004).[3] On the other hand, there is evidence that the dopaminergic system can be recruited in a top-down fashion without any explicit reward prospect, highlighting a more voluntary, invigorating role of dopamine which ensures optimal behaviour in demanding situations (Boehler et al., 2011; Salamone, Correa, Mingote, & Weber, 2005). Although there is evidence in the animal literature for these components occurring both in isolation and combined, depending on the paradigm, their experimental dissociation is much more challenging in humans owing to the low temporal resolution of fMRI. In the context of cued-reward studies, it seems possible to dissociate a pure value component from more invigorating signals to some extent by comparing fMRI responses in regions of the dopaminergic system between the respective phases of a trial (e.g., reward outcome vs. cue-target interval; see Figure 24.1). However, such temporal dissociations are much more difficult in paradigms using target–reward associations, because here a *single* event carries information about reward availability, response requirements, and implicitly also indicates reward outcome (in the case of accurate performance). Hence, observed modulations in dopaminergic regions in target–reward association paradigms could in theory reflect a combination of value and invigorating signals (or in other words, it is not possible to judge whether both components are actually present or not). One way to at least parcel out mere value signals arising from implicit reward outcome in correct trials is to compare activity in regions of the dopaminergic system between correct and incorrect trials (see, e.g., Boehler et al., 2014). Specifically, although an invigorating component would be expected to be present in both cases, a mere value component would be expected to be more emphasised on correct trials.

Detrimental Effects of Reward

In contrast to the long-standing global notion that motivation in general, and monetary incentives in particular, will benefit performance, several of the studies described above using target–reward associations revealed a potential flip side of reward-based performance enhancements, namely, performance detriments in the form of attentional distraction and erroneous response priming (see Anderson et al., 2011b; Krebs et al., 2010; Muhle-Karbe & Krebs, 2012; Rutherford et al., 2010; Seifert et al., 2006). These seem to occur in situations in which reward information is linked to an irrelevant stimulus, stimulus feature, or stimulus location, or is in some other way incompatible with the task goal at hand. Importantly, the same reward-associated feature can benefit performance if it is overlapping with the target stimulus, target feature, or target location. In contrast to these opposing effects of stimulus–reward associations, reward cues almost always have a beneficial impact on performance. More specifically, reward availability signalled by a cue further promotes preparatory attentional processes (and in turn benefit performance), by for example focusing attention on the task-relevant feature of the upcoming target. However, in exceptional cases, global reward manipulations (block cues) can impair performance in a subset of trials if one aspect of the task goal is emphasised over another through reward (e.g., going vs. stopping in Padmala and Pessoa, 2010), or if participants adopt a certain strategy that improves performance on most but not all trials (e.g., pro-active control strategy at the costs of a certain error type in Locke and Braver, 2008). From a more global perspective, a further investigation of situations that promote reward-triggered performance benefits versus costs seems warranted, given that in an applied context (e.g., education) the effects of motivation are often considered to be exclusively positive in nature.

Conclusions and Outlook

In this chapter, we have discussed the manifold influences of reward information in attention and cognitive control tasks. Despite the large variety of paradigms and manipulations, a useful distinction can be made between reward effects that rely on differential preparatory mechanisms (such as changes in global or selective attention, task goals, or task maintenance) and those that result from target–reward associations that trigger rapid changes in stimulus processing. From a general perspective, as mentioned above, these two approaches partly map on to classic cognitive control distinctions, in that the former allows for a more pro-active mode, whereas the latter captures rather re-active (or even automatic) processes. This is not to say, however, that in tasks employing target-reward associations no global preparatory influences can occur. It is, for example, conceivable that stimulus-reward associations affect the attentional set throughout the entire task, promoting a strategic search for certain stimuli or features (see Schevernels et al., 2015). Similarly, several cued-reward studies suggest that re-active processes upon target presentation can be altered in addition to preparatory modulations during the cue-target interval. Furthermore, within the target–reward association studies, no clear distinction has been made so far between low-level effects on stimulus encoding and response selection on the one hand, and effects that rely on re-active control mechanisms on the other hand. These potential interactions call for future studies that include systematic comparisons between the different ways in which reward factors can be introduced into a task or circumstance. Considering the importance of both reward processing and cognitive control for our ability to navigate through and adapt to our complex world, future studies of the multifaceted interactions between these two fundamental cognitive functions can be expected to be highly rewarding.

Notes

1 The above two ways by which reward can influence cognitive control parallel in some key ways a distinction that has been made in the cognitive-control literature more broadly, i.e., between pro-active control, resulting from sustained pre-task preparatory processes set in place before a task or trial (e.g., with cuing), and re-active control, resulting from differential within-trial processes due to particular stimulus or response characteristics (Braver et al., 2007; Braver, 2012; Chapter 9 by Chiew & Braver, this volume). While we generally consider reward cues and target-reward associations as key promoters of pro-active and re-active control processes, respectively, this mapping is not exclusive (see Conclusions and outlook).

2 We note that this chapter does not cover related cognitive domains that have shown to be sensitive to reward manipulations, including the exertion of mental and physical effort (e.g., Croxson, Walton, O'Reilly, Behrens, & Rushworth, 2009; Pessiglione et al., 2007; Schmidt, Lebreton, Clery-Melin, Daunizeau, & Pessiglione, 2012), value-based decision making (e.g., Glascher, Hampton, & O'Doherty, 2009; Larsen and O'Doherty, 2014; O'Doherty et al., 2004), memory formation (e.g., Adcock, Thangavel, Whitfield-Gabrieli, Knutson, & Gabrieli, 2006; Elward, Vilberg, & Rugg, 2015; Wittmann et al., 2005), and emotion processing (e.g., Choi, Padmala, Spechler, & Pessoa, 2014; Hu, Padmala, & Pessoa, 2013; Wittmann et al., 2008).

3 Note that there have been many hypotheses generated to describe the role of the reward anticipation response in the dopaminergic system. For a comprehensive review, see Wise, 2004.

References

Aarts, E., Roelofs, A., Franke, B., Rijpkema, M., Fernandez, G., Helmich, R. C., & Cools, R. (2010). Striatal dopamine mediates the interface between motivational and cognitive control in humans: Evidence from genetic imaging. *Neuropsychopharmacology*, *35*, 1943–1951.

Adcock, R. A., Thangavel, A., Whitfield-Gabrieli, S., Knutson, B., & Gabrieli, J. D. (2006). Reward-motivated learning: Mesolimbic activation precedes memory formation. *Neuron*, *50*, 507–517.

Anderson, B. A., Laurent, P. A., & Yantis, S. (2011a). Learned value magnifies salience-based attentional capture. *PLoS ONE*, *6*, e27926.

Anderson, B. A., Laurent, P. A., & Yantis, S. (2011b). Value-driven attentional capture. *Proceedings of the National Academy of Sciences of the United States of America*, *108*, 10367–10371.

Bekker, E. M., Kenemans, J. L., Hoeksma, M. R., Talsma, D., & Verbaten, M. N. (2005). The pure electrophysiology of stopping. *International Journal of Psychophysiology*, *55*, 191–198.

Boehler, C. N., Hopf, J. M., Stoppel, C. M., & Krebs, R. M. (2012). Motivating inhibition—Reward prospect speeds up response cancellation. *Cognition*, *125*, 498–503.

Boehler, C. N., Hopf, J. M., Krebs, R. M., Stoppel, C. M., Schoenfeld, M. A., Heinze, H. J., & Noesselt, T. (2011). Task-load-dependent activation of dopaminergic midbrain areas in the absence of reward. *Journal of Neuroscience*, *31*, 4955–4961.

Boehler, C. N., Schevernels, H., Hopf, J. M., Stoppel, C. M., & Krebs, R. M. (2014). Reward prospect rapidly speeds up response inhibition via reactive control. *Cognitive, Affective, & Behavioral Neuroscience*, *14*(2), 593–609.

Botvinick, M., & Braver, T. (2015). Motivation and cognitive control: From behavior to neural mechanism. *Annual Review of Psychology*, *66*, 83–113.

Botvinick, M. M., Braver, T. S., Barch, D. M., Carter, C. S., & Cohen, J. D. (2001). Conflict monitoring and cognitive control. *Psychological Review*, *108*, 624–652.

Braem, S., King, J. A., Korb, F. M., Krebs, R. M., Notebaert, W., & Egner, T. (2013). Affective modulation of cognitive control is determined by performance-contingency and mediated by ventromedial prefrontal and cingulate cortex. *Journal of Neuroscience*, *33*, 16961–16970.

Braem, S., Verguts, T., Roggeman, C., & Notebaert, W. (2012). Reward modulates adaptations to conflict. *Cognition*, *125*, 324–332.

Braver, T. S. (2012). The variable nature of cognitive control: A dual mechanisms framework. *Trends in Cognitive Sciences*, *16*, 106–113.

Braver, T. S., Gray, J. R., & Burgess, G. C. (2007). *Explaining the many varieties of working memory variation: Dual mechanisms of cognitive control.* New York: Oxford University Press.
Braver, T. S., Krug, M. K., Chiew, K. S., Kool, W., Westbrook, J. A., Clement, N. J.,...MOMCAI group (2014). Mechanisms of motivation-cognition interaction: Challenges and opportunities. *Cognitive, Affective, & Behavioral Neuroscience, 14*(2), 443–472.
Capa, R. L., Bouquet, C. A., Dreher, J. C., & Dufour, A. (2013). Long-lasting effects of performance-contingent unconscious and conscious reward incentives during cued task-switching. *Cortex, 49*, 1943–1954.
Choi, J. M., Padmala, S., Spechler, P., & Pessoa, L. (2014). Pervasive competition between threat and reward in the brain. *Social Cognitive and Affective Neuroscience, 9*, 737–750.
Cohen, J. R., & Poldrack, R. A. (2008). Automaticity in motor sequence learning does not impair response inhibition. *Psychonomic Bulletin & Review, 15*, 108–115.
Corbetta, M., & Shulman, G. L. (2002). Control of goal-directed and stimulus-driven attention in the brain. *Nature Reviews Neuroscience, 3*, 201–215.
Croxson, P. L., Walton, M. E., O'Reilly, J. X., Behrens, T. E., & Rushworth, M. F. (2009). Effort-based cost-benefit valuation and the human brain. *Journal of Neuroscience, 29*, 4531–4541.
Della Libera, C., & Chelazzi, L. (2006). Visual selective attention and the effects of monetary rewards. *Psychological Science, 17*, 222–227.
Elward, R. L., Vilberg, K. L., & Rugg, M. D. (2015). Motivated memories: Effects of reward and recollection in the core recollection network and beyond. *Cerebral Cortex, 25*, 3159–3166.
Engelmann, J. B., Damaraju, E., Padmala, S., & Pessoa, L. (2009). Combined effects of attention and motivation on visual task performance: Transient and sustained motivational effects. *Frontiers in Human Neuroscience, 3*, 4.
Engelmann, J. B., & Pessoa, L. (2007). Motivation sharpens exogenous spatial attention. *Emotion, 7*, 668–674.
Failing, M. F., & Theeuwes, J. (2014). Exogenous visual orienting by reward. *Journal of Vision, 14*, 6.
Glascher, J., Hampton, A. N., & O'Doherty, J. P. (2009). Determining a role for ventromedial prefrontal cortex in encoding action-based value signals during reward-related decision making. *Cerebral Cortex, 19*, 483–495.
Gong, M., & Li, S. (2014). Learned reward association improves visual working memory. *Journal of Experimental Psychology: Human Perception and Performance, 40*, 841–856.
Goschke, T. (2000). Involuntary persistence and intentional reconfiguration in task-set switching. In S. Monsell & J. Driver (Eds.), *Control of cognitive processes: Attention and performance XVIII* (pp. 331–355). Cambridge, MA: MIT Press.
Gottlieb, J. (2012). Attention, learning, and the value of information. *Neuron, 76*, 281–295.
Greenhouse, I., & Wessel, J. R. (2013). EEG signatures associated with stopping are sensitive to preparation. *Psychophysiology, 50*, 900–908.
Guitart-Masip, M., Fuentemilla, L., Bach, D. R., Huys, Q. J., Dayan, P., Dolan, R. J., & Duzel, E. (2011). Action dominates valence in anticipatory representations in the human striatum and dopaminergic midbrain. *Journal of Neuroscience, 31*, 7867–7875.
Hare, T. A., Tottenham, N., Davidson, M. C., Glover, G. H., & Casey, B. J. (2005). Contributions of amygdala and striatal activity in emotion regulation. *Biological Psychiatry, 57*, 624–632.
Hickey, C., Chelazzi, L., & Theeuwes, J. (2010). Reward changes salience in human vision via the anterior cingulate. *Journal of Neuroscience, 30*, 11096–11103.
Hickey, C., Kaiser, D., & Peelen, M. V. (2015). Reward guides attention to object categories in real-world scenes. *Journal of Experimental Psychology: General, 144*, 264–273.
Hickey, C., & Peelen, M. V. (2015). Neural mechanisms of incentive salience in naturalistic human vision. *Neuron, 85*, 512–518.
Hommel, B. (2004). Event files: Feature binding in and across perception and action. *Trends in Cognitive Sciences, 8*, 494–500.
Hu, K., Padmala, S., & Pessoa, L. (2013). Interactions between reward and threat during visual processing. *Neuropsychologia, 51*, 1763–1772.
Jimura, K., Locke, H. S., & Braver, T. S. (2010). Prefrontal cortex mediation of cognitive enhancement in rewarding motivational contexts. *Proceedings of the National Academy of Sciences of the United States of America, 107*, 8871–8876.

Kiss, M., Driver, J., & Eimer, M. (2009). Reward priority of visual target singletons modulates event-related potential signatures of attentional selection. *Psychological Science, 20*, 245–251.

Kleinsorge, T., & Rinkenauer, G. (2012). Effects of monetary incentives on task switching. *Experimental Psychology, 59*, 216–226.

Knutson, B., Adams, C. M., Fong, G. W., & Hommer, D. (2001a). Anticipation of increasing monetary reward selectively recruits nucleus accumbens. *Journal of Neuroscience, 21*, RC159.

Knutson, B., & Cooper, J. C. (2005). Functional magnetic resonance imaging of reward prediction. *Current Opinion in Neurology, 18*, 411–417.

Knutson, B., Fong, G. W., Adams, C. M., Varner, J. L., & Hommer, D. (2001b). Dissociation of reward anticipation and outcome with event-related fMRI. *Neuroreport, 12*, 3683–3687.

Knutson, B., Taylor, J., Kaufman, M., Peterson, R., & Glover, G. (2005). Distributed neural representation of expected value. *Journal of Neuroscience, 25*, 4806–4812.

Knutson, B., Westdorp, A., Kaiser, E., & Hommer, D. (2000). FMRI visualization of brain activity during a monetary incentive delay task. *NeuroImage, 12*, 20–27.

Kouneiher, F., Charron, S., & Koechlin, E. (2009). Motivation and cognitive control in the human prefrontal cortex. *Nature Neuroscience, 12*, 939–945.

Krawczyk, D. C., Gazzaley, A., & D'Esposito, M. (2007). Reward modulation of prefrontal and visual association cortex during an incentive working memory task. *Brain Research, 1141*, 168–177.

Krebs, R. M., Boehler, C. N., Appelbaum, L. G., & Woldorff, M. G. (2013). Reward associations reduce behavioral interference by changing the temporal dynamics of conflict processing. *PLoS ONE, 8*, e53894.

Krebs, R. M., Boehler, C. N., Egner, T., & Woldorff, M. G. (2011). The neural underpinnings of how reward associations can both guide and misguide attention. *Journal of Neuroscience, 31*, 9752–9759.

Krebs, R. M., Boehler, C. N., Roberts, K. C., Song, A. W., & Woldorff, M. G. (2012). The involvement of the dopaminergic midbrain and cortico-striatal-thalamic circuits in the integration of reward prospect and attentional task demands. *Cerebral Cortex, 22*, 607–615.

Krebs, R. M., Boehler, C. N., & Woldorff, M. G. (2010). The influence of reward associations on conflict processing in the Stroop task. *Cognition, 117*, 341–347.

Larsen, T., & O'Doherty, J. P. (2014). Uncovering the spatio-temporal dynamics of value-based decision-making in the human brain: A combined fMRI-EEG study. *Philosophical Transactions of the Royal Society of London. Series B: Biological Sciences, 369*(1655), pii: 20130473.

Leotti, L. A., & Wager, T. D. (2010). Motivational influences on response inhibition measures. *Journal of Experimental Psychology: Human Perception and Performance, 36*, 430–447.

Locke, H. S., & Braver, T. S. (2008). Motivational influences on cognitive control: Behavior, brain activation, and individual differences. *Cognitive, Affective, & Behavioral Neuroscience, 8*, 99–112.

Luck, S. J. (1994). Cognitive and neural mechanisms of visual search. *Current Opinion in Neurobiology, 4*, 183–188.

MacLeod, C. M., & MacDonald, P. A. (2000). Interdimensional interference in the Stroop effect: Uncovering the cognitive and neural anatomy of attention. *Trends in Cognitive Sciences, 4*, 383–391.

Muhle-Karbe, P. S., & Krebs, R. M. (2012). On the influence of reward on action-effect binding. *Frontiers in Psychology, 3*, 450.

O'Doherty, J., Dayan, P., Schultz, J., Deichmann, R., Friston, K., & Dolan, R. J. (2004). Dissociable roles of ventral and dorsal striatum in instrumental conditioning. *Science, 304*, 452–454.

Padmala, S., & Pessoa, L. (2010). Interactions between cognition and motivation during response inhibition. *Neuropsychologia, 48*, 558–565.

Padmala, S., & Pessoa, L. (2011). Reward reduces conflict by enhancing attentional control and biasing visual cortical processing. *Journal of Cognitive Neuroscience, 23*, 3419–3432.

Pessiglione, M., Schmidt, L., Draganski, B., Kalisch, R., Lau, H., Dolan, R. J., & Frith, C. D. (2007). How the brain translates money into force: A neuroimaging study of subliminal motivation. *Science, 316*, 904–906.

Pessoa, L., & Engelmann, J. B. (2010). Embedding reward signals into perception and cognition. *Frontiers in Neuroscience, 4*, pii: 17.

Plichta, M. M., Wolf, I., Hohmann, S., Baumeister, S., Boecker, R., Schwarz, A. J.,…Brandeis, D. (2013). Simultaneous EEG and fMRI reveals a causally connected subcortical-cortical network during reward anticipation. *Journal of Neuroscience, 33*(36), 14526–14533.

Raymond, J. E., & O'Brien, J. L. (2009). Selective visual attention and motivation: The consequences of value learning in an attentional blink task. *Psychological Science*, *20*, 981–988.

Rosell-Negre, P., Bustamante, J. C., Fuentes-Claramonte, P., Costumero, V., Benabarre, S., & Barros-Loscertales, A. (2014). Reward anticipation enhances brain activation during response inhibition. *Cognitive, Affective, & Behavioral Neuroscience*, *14*(2), 621–634.

Rutherford, H. J., O'Brien, J. L., & Raymond, J. E. (2010). Value associations of irrelevant stimuli modify rapid visual orienting. *Psychonomic Bulletin & Review*, *17*, 536–542.

Salamone, J. D., Correa, M., Mingote, S. M., & Weber, S. M. (2005). Beyond the reward hypothesis: Alternative functions of nucleus accumbens dopamine. *Current Opinion in Pharmacology*, *5*, 34–41.

San Martin, R., Appelbaum, L. G., Huettel, S. A., & Woldorff, M. G. (2016). Cortical brain activity reflecting attentional biasing toward reward-predicting cues covaries with economic decision-making performance. *Cerebral Cortex*, *26*, 1–11.

Schevernels, H., Bombeke, K., Van Der Borght, L., Hopf, J. -M., Krebs, R. M., & Boehler, C. N. (2015). Electrophysiological evidence for the involvement of proactive and reactive control in a rewarded stop-signal task. *Neuroimage*, *121*, 115–125.

Schevernels, H., Krebs, R. M., Santens, P., Woldorff, M. G., & Boehler, C. N. (2014). Task preparation processes related to reward prediction precede those related to task-difficulty expectation. *Neuroimage*, *84*, 639–647.

Schmidt, L., Lebreton, M., Clery-Melin, M. L., Daunizeau, J., & Pessiglione, M. (2012). Neural mechanisms underlying motivation of mental versus physical effort. *PLoS Biology*, *10*, e1001266.

Schultz, W. (1997). Dopamine neurons and their role in reward mechanisms. *Current Opinion in Neurobiology*, *7*, 191–197.

Schultz, W. (2001). Reward signaling by dopamine neurons. *Neuroscientist*, *7*, 293–302.

Seifert, J., Naumann, E., Hewig, J., Hagemann, D., & Bartussek, D. (2006). Motivated executive attention—Incentives and the noise-compatibility effect. *Biological Psychology*, *71*, 80–89.

Serences, J. T. (2008). Value-based modulations in human visual cortex. *Neuron*, *60*, 1169–1181.

Small, D. M., Gitelman, D., Simmons, K., Bloise, S. M., Parrish, T., & Mesulam, M. M. (2005). Monetary incentives enhance processing in brain regions mediating top-down control of attention. *Cerebral Cortex*, *15*, 1855–1865.

Soutschek, A., Stelzel, C., Paschke, L., Walter, H., & Schubert, T. (2015). Dissociable effects of motivation and expectancy on conflict processing: An fMRI study. *Journal of Cognitive Neuroscience*, *27*, 409–423.

Sturmer, B., Nigbur, R., Schacht, A., & Sommer, W. (2011). Reward and punishment effects on error processing and conflict control. *Frontiers in Psychology*, *2*, 335.

Tipper, S. P. (2001). Does negative priming reflect inhibitory mechanisms? A review and integration of conflicting views. *Quarterly Journal of Experimental Psychology A*, *54*, 321–343.

van den Berg, B., Krebs, R. M., Lorist, M. M., & Woldorff, M. G. (2014). Utilization of reward-prospect enhances preparatory attention and reduces stimulus conflict. *Cognitive, Affective, & Behavioral Neuroscience*, *14*, 561–577.

van Steenbergen, H., Band, G. P., & Hommel, B. (2009). Reward counteracts conflict adaptation. Evidence for a role of affect in executive control. *Psychological Science*, *20*, 1473–1477.

Veling, H., & Aarts, H. (2010). Cueing task goals and earning money: Relatively high monetary rewards reduce failures to act on goals in a Stroop task. *Motivation and Emotion*, *34*, 184–190.

Verbruggen, F., Chambers, C. D., & Logan, G. D. (2013). Fictitious inhibitory differences: How skewness and slowing distort the estimation of stopping latencies. *Psychological Science*, *24*, 352–362.

Verguts, T., & Notebaert, W. (2009). Adaptation by binding: A learning account of cognitive control. *Trends in Cognitive Sciences*, *13*, 252–257.

Wise, R. A. (2004). Dopamine, learning and motivation. *Nature Reviews Neuroscience*, *5*, 483–494.

Wittmann, B. C., Schiltz, K., Boehler, C. N., & Duzel, E. (2008). Mesolimbic interaction of emotional valence and reward improves memory formation. *Neuropsychologia*, *46*, 1000–1008.

Wittmann, B. C., Schott, B. H., Guderian, S., Frey, J. U., Heinze, H. J., & Duzel, E. (2005). Reward-related FMRI activation of dopaminergic midbrain is associated with enhanced hippocampus-dependent long-term memory formation. *Neuron*, *45*, 459–467.

25

The Social-Psychological Perspective on Self-Regulation

Erin K. Davisson and Rick H. Hoyle

From grand New Year's resolutions to train for a marathon to simple intentions like packing one's lunch, many of the everyday behaviours people engage in routinely involve the adoption and pursuit of goals. These goals may be short-term or long-term in nature. Regardless of the scope of the goal(s) in question, people frequently encounter threats to successful goal pursuit. They must ward off diverse social, environmental, and motivational challenges as they strive to achieve goals, big and small. The manner by which people grapple with these challenges in pursuit of their goals is termed self-regulation.

Self-regulation is a catchall term that refers to the general process by which people control their thoughts, feelings, and behaviour in the service of achieving their goals. As we present in this chapter, a large array of constructs, processes, and behaviours are relevant to self-regulation. Given this conceptual breadth, it is perhaps unsurprising that there is no one widely accepted definition assigned to self-regulation. Rather, behavioural scientists in fields from sociology (e.g., Smith-Lovin & Heise, 1988) to education (e.g., Pintrich, 2004) and psychology (e.g., Baumeister & Heatherton, 1996) have offered diverse descriptions of self-regulation. Even within social psychology—the field in which self-regulation has enjoyed a wealth of sustained and recent interest (cf. MacKenzie, Mezo, & Francis, 2012)[1]—researchers have not reached consensus on its definition. However, common to the varied definitions of self-regulation is a focus on the processes by which people manage goal pursuit. As would be expected from the large number of varying definitions of self-regulation, multiple social-psychological models offer an account of the process.

Self-regulation can be understood narrowly to refer only to those psychological processes directly implicated in the management of goal pursuit. For the present purposes, however, self-regulation is better understood in broad terms to include the basic capacities and more general propensities that enable it. This broader treatment facilitates our goal of highlighting connections between cognitive control constructs and processes and self-regulation. To that end, we cover an array of processes and constructs directly and indirectly implicated in self-regulation. We begin with *foundations*—the mental, motivational, and dispositional characteristics on which self-regulation draws. In the context of goal pursuit, these characteristics manifest as *processes*, the specific mental and emotional activities involved in self-regulation. These processes play out in the context of *strategies*, the specific means by which self-regulation is achieved. Although these processes and constructs might be presented with reference to specific models of self-regulation, our treatment of them here is more general, touching on multiple models that exemplify the social-psychological account of self-regulation.

The Wiley Handbook of Cognitive Control, First Edition. Edited by Tobias Egner.

Foundations of Self-Regulation

Self-regulation draws on a collection of basic mental processes and capacities in the management of goal pursuit. These are foundational because, without them, the processes required for self-regulation would not be possible. They also help account for the fact that people vary in their skill at self-regulation and preferences regarding how it is undertaken. Specifically, the foundations of self-regulation include those cognitive control processes commonly referred to as *executive functions*, as well as some components of *temperament* and *personality*. Though these basic processes and tendencies enable the management of goal pursuit (as well as other activities and processes), they themselves do not constitute self-regulation. Rather, they are necessary for self-regulation but not sufficient to fully account for the specific processes involved in self-regulation and the strategies by which they are enacted.

Executive Functions

Executive functions are those cognitive control processes and tendencies that implicate the prefrontal cortex (Best & Miller, 2010). Although a number of these functions have been identified, at the most basic level (and sufficient for present purposes), there are three—inhibition, updating, and shifting—that contribute to processes involved in self-regulation (Miyake et al., 2000). These functions do not serve self-regulation exclusively, but they are required for it. They are the bridge from neural activity to the cognitive, affective, and motivational processes involved in self-regulation.

Perhaps the most basic executive function is *inhibition*, the ability to restrain or override dominant or prepotent responses, including thoughts and behaviours, when needed (Miyake et al., 2000; Chapter 6 by Verbruggen & Logan in this volume). The active and deliberate inhibition of prepotent responses is commonly captured through executive functioning tasks like the Stroop task (Stroop, 1935) and the stop signal task (Logan, 1994). Inhibition may take the form of direct, global inhibition, which involves deliberately stopping prepotent responses, and indirect, competitive inhibition, which enables the inhibition of prepotent responses by activating goal-relevant processing that wards off these automatic response tendencies via competition (Munakata et al., 2011). This more nuanced characterisation of inhibition fits well with an emerging model of self-control (Hoyle & Davisson, 2016), one of the primary self-regulatory strategies.

The executive function of *updating* involves monitoring the contents of working memory and updating irrelevant information by replacing it with more relevant information (Miyake et al., 2000). Updating requires that information remain active in working memory, where it can be accessed quickly as dictated by task demands (Kane & Engle, 2002). Though the specific cognitive mechanism underlying the link between updating and working memory is not entirely clear (Miyake & Friedman, 2012), the involvement of working memory at a general level provides a conceptual link between updating and self-regulation. Working memory is critical for successful self-regulation insofar as active mental representations of goals are central in goal pursuit and, as such, has been implicated as a factor in self-regulatory failure (e.g., Schmeichel, Volokhov, & Demaree, 2008). However, as with inhibition, this executive function is foundational to rather than an instance of self-regulation, contributing to specific self-regulatory strategies such as self-control and goal shielding (Hofmann, Friese, Schmeichel, & Baddeley, 2011).

A third basic executive function relevant for self-regulatory process and strategies is *shifting*, the ability to move back and forth between tasks, rules, or mental states (Miyake et al., 2000). Although switching between tasks might appear to be largely driven by inhibition—requiring the abandonment of one task in favour of another—shifting differs from inhibition in that it requires the individual to move between mental sets that often contain several rules, as opposed to simply overriding a response as in inhibition (Crone, Somsen, Zanolie, & Van der Molen, 2006; Chapter 2 by Monsell in this volume). Because self-regulatory processes are fundamentally associated with goal pursuit, and because individuals rarely pursue one goal at a time using a single means (Louro, Pieters, & Zeelenberg, 2007), shifting is necessary for effective self-regulation. It enables the individual to flexibly switch between goals with varying time horizons and means of pursuing those goals, some of which may conflict with one another (Locke & Latham, 2002). Newer accounts of self-regulatory failures, which look to the cognitive control literature for potential explanations, suggest that such failures might be attributable in large measure to deficits in the ability to switch between goals effectively (Inzlicht, Schmeichel, & Macrae, 2014).

Underlying these basic executive functions—especially inhibition—is the 'common executive function', which allows individuals to actively maintain goals and goal-relevant information (Miyake & Friedman, 2012; Munakata et al., 2011). The core ability to keep active mental representations of goals is what enables executive functions to influence downstream processes that facilitate goal pursuit. This conceptualisation of the underlying executive factor argues for both unity in executive functions—the common factor that is especially critical to inhibition, but also to updating and shifting—and diversity, represented in the updating and shifting functions, which require both the active maintenance of mental representations common to all executive functions, as well as task-specific mechanisms involving flexibility and working memory operations (Friedman, Miyake, Robinson, & Hewitt, 2011; Miyake & Friedman, 2012). Alternative conceptualisations posit that attention is central to and unifies the executive functions (Kaplan & Berman, 2010; see also Chapter 3 by Meier & Kane in this volume). Attentional processes are core to several dimensions of temperament.

Temperament

A related category of constructs and process that are foundational for self-regulation is temperament, the 'affective, activational, and attentional core of personality' (Rothbart & Bates, 2006, p. 100). Although temperament is biological (and genetic) in origin, it is shaped by maturation and experience. Some aspects of temperament, like executive functioning, are necessary for self-regulation, whereas other aspects influence the way self-regulation is characteristically enacted. Especially during childhood and early adolescence, temperament underlies emerging individual differences in certain preferences, tendencies, deficits, and strengths related to self-regulation.

Although in-depth studies of infants (Thomas, Chess, & Birch, 1970) and young children (Rothbart, Ahadi, Hershey, & Fisher, 2001) have identified as many as 15 specific dimensions of temperament, two broader dimensions–reactive control and effortful control–are of most direct relevance to individual differences in self-regulation (Rothbart & Derryberry, 1981). (Note that the term 'reactive control' is used in a slightly different way in the self-regulation literature here than in the cognitive control literature; for comparison, see Chapter 9 by Chiew & Braver in this volume). Reactive control is automatic, encompassing reactive approach and avoidance responses. Either extreme–impulsive undercontrol or rigid overcontrol–is detrimental to self-regulation. Effortful control, a precursor to basic forms of self-regulation, is primarily characterised as an ability to inhibit a suboptimal dominant response in favour of a more desirable or adaptive secondary response (Rothbart & Rueda, 2005). Specific features of effortful control such as attentional shifting, attentional focusing, inhibitory control, and perceptual sensitivity support the cognitive effort of managing counterproductive reactivity. Variability in these core dimensions of temperament is expressed beginning in early adolescence in more proximal predictors of self-regulation such as reward sensitivity and self-control.

Personality

Temperament is the foundation from which personality emerges, linking personality to genetic and biological characteristics as expressed in executive functioning (Rothbart, Ahadi, & Evans, 2000). Similarly, personality provides a bridge between genetic and biological characteristics and self-regulation via executive functioning and temperament. In broad terms, personality refers to variability between people in characteristic patterns of cognition, emotion, and behaviour. General models of personality aim to capture the majority of that variability using a small number of relatively broad domains. The cost of parsimony is the conceptual and empirical explanatory power offered by a plethora of personality traits that aim to account for variability in specific cognitive, emotional, or behavioural preferences and tendencies. At the general and specific levels, personality can be understood as the intrapersonal backdrop against which self-regulation operates. Some domains and traits facilitate one or more phases of self-regulation, some are obstacles to effective self-regulation, and some reflect preferences or styles related to self-regulation.

The dominant comprehensive models of personality posit five broad personality domains, referred to as the Big Five, that account for the lion's share of the variance shared by specific traits (Costa & McCrae, 1992; John, 1990). The basis of these domains—neuroticism, agreeableness, conscientiousness, extraversion, and openness—is factor analysis of natural language descriptors of personality. These emergent factors offer a broad description of personality, but, when interpreted with reference to self-regulation, they imply processes that are causal and explanatory (DeYoung, 2015; Van Egeren, 2009). For example, extraversion reflects a general tendency towards approach and motivation with reference to specific rewards. Neuroticism reflects a tendency towards defensiveness and avoidance and motivation with reference to punishment. Conscientiousness, the domain most clearly foundational for self-regulatory processes, reflects a general tendency to remain on course despite disruptions to active goal pursuit. Whereas extraversion and neuroticism suggest different approaches to self-regulation, either of which might be effective in most situations, conscientiousness reflects tendencies that contribute directly to effective self-regulation regardless of how it is approached.

Beyond these broad personality domains is a range of specific traits that might be viewed as assets or liabilities for self-regulation or characteristic approaches to self-regulation. In terms of assets and liabilities, examples are optimism (Nes, Carlson, Crofford, de Leeuw, & Segerstrom, 2011) and impulsivity (Eisenberg et al., 2009), respectively. In terms of characteristic approaches to self-regulation, examples are regulatory focus, the preference for goal pursuit with reference to gains ('making good things happen') versus non-losses ('keeping bad things from happening'; Higgins, 1997) and regulatory mode, a principal concern with choosing goals and means of pursuing them (i.e., assessment) versus making progress in goal pursuit (i.e., locomotion; Higgins, Kruglanski, & Pierro, 2003). These traits and others (for a review, see Hoyle & Gallagher, 2015) serve as a proximal foundation for the processes that constitute self-regulation.

Self-Regulatory Processes

Having reviewed the array of processes and constructs that are foundational for self-regulation and established that they enable self-regulation but are not to be confused with it, we now turn to a review of those processes that, together, constitute self-regulation proper. Our focus is general, highlighting processes that are common to most models of self-regulation in social psychology.

Despite the lack of consensus regarding *the* theoretical model of self-regulation in social psychology, several influential models specify the processes involved in self-regulation with considerable overlap. These models can be categorised into two distinct but related sets (Karoly, 2010). *Type 1 self-regulation* involves the management of goal pursuit that spans time, occasion, and situation such as the pursuit of educational or health goals. These goals may become more or less salient based on contextual factors. When these goals are salient, their pursuit is managed by *type 2 self-regulation*, which is bound by circumstance (i.e., time and situation). Type 2 self-regulation can operate at non-conscious or conscious levels. Non-conscious self-regulation involves neither awareness nor conscious effort and is triggered by primes of goal-relevant environmental stimuli (Fitzsimons & Bargh, 2004). Conscious self-regulation involves both self-awareness (Duval & Wicklund, 1972) and concerted, deliberate effort towards goal pursuit on the part of the individual (Carver & Scheier, 1981). Though type 1 and type 2 self-regulation differ in the degree to which they are constrained by time and situation, each serves goal pursuit—often in an interdependent fashion.

Type 1 Self-Regulation

Accounts of type 1 self-regulation, which is oriented towards long-term goal pursuit, are often situated in models of self-regulated learning (e.g., Pintrich, 2004; Zimmerman & Campillo, 2003). The processes and behaviours that are relevant for the broad goal of acquiring skills and knowledge can be arrayed in a three-phase process (Zimmerman & Campillo, 2003). In the *forethought* phase, the individual must decide which goals should be pursued and create a plan that enables their achievement. Goal setting and planning require the activation of goal- and outcome-relevant behaviours and self-efficacy regarding their implementation. When plans are paired with adequate motivation for goal achievement, the individual shifts to the *performance* phase of self-regulation. Performance implicates self-control, a specific self-regulatory activity whereby the individual must eschew behaviours that threaten to derail goal pursuit, and instead initiate and engage in goal-relevant behaviours. Monitoring behaviour is crucial to successful performance; the individual must remain keenly aware of his or her own motivation, effort, and behaviour, in addition to monitoring the task, environment, and changes in each (Pintrich, 2004). After performance is complete, the individual transitions to the *self-reflection* phase, during which he or she reflects on and evaluates behaviours as they pertain to goal pursuit. A reaction follows self-reflection, by which future similar behaviours may be rewarded or discouraged based on their association with successful goal pursuit. At this point in the self-regulation process, the individual may revisit the goal and associated plans for its achievement and decide whether either needs adjustment for continued goal pursuit. This model—as is also the case with models of type 2 self-regulation—suggests that self-regulation is a cyclical process, assuming that the individual repeatedly moves through the phases while advancing towards goal attainment.

Type 2 Self-Regulation

Models of type 2 self-regulation emphasise conflicts between goal pursuit and those behaviours that threaten to derail it. Several prominent models of self-regulation are focused on the set of processes that enable an individual to recognise and ward off such threats. Common to these models is the concept of a feedback loop, wherein the current status is evaluated with reference to an active goal. If a significant discrepancy between the current state and the state that would reflect satisfactory goal pursuit is evident, then behavioural change is warranted. Behaviour is then altered with the goal of decreasing the discrepancy between the current and desired states, which is evaluated by a return to the first stage of the feedback loop, then a fresh cycle through it. Self-regulation terminates when the discrepancy is sufficiently small that the individual is not motivated to change goal-relevant behaviour. (Note the close parallels between this definition and the idea of 'conflict monitoring' in the cognitive control literature; for comparison, see Chapter 10 by Kool, Shenhav, & Botvinick and Chapter 4 by Egner in this volume).

Type 2 self-regulation unfolds in a set of phases not unlike those involved in type 1 self-regulation (Kanfer, 1970; cf. Carver & Scheier, 1981). The type 2 process begins with *self-monitoring*, which is triggered when roadblocks to goal pursuit are encountered—either because the course of action necessary for continued goal progress is unclear, or due to unexpected (or unwelcome) consequences resulting from prior action (Duval & Wicklund, 1972). Self-monitoring enables the individual to establish the reason for the interruption in goal pursuit, and construct a path to address the said interruption. Self-monitoring is followed by *self-evaluation*, which involves the determination of whether the current behaviour is relevant to a salient goal. If the behaviour is, in fact, goal relevant, then the active goal is used as a standard against which performance is measured. This phase of self-regulation enables the alteration of present behaviours to align with performance standards prescribed by the salient goal. Following self-evaluation is *self-reinforcement*, by which rewards and punishments are conferred as warranted by the match—or mismatch, as the case may be—between the standard offered by the goal and current behaviour. The self-reinforcement phase involves conditioning behaviours that sustain goal pursuit through rewards and positive reinforcement, and attempting to prevent future disruptions to goal pursuit through negative reinforcement and punishment (Kanfer & Karoly, 1972).

Though models of type 2 self-regulation implicate specific processes that permit testable hypotheses regarding how people alter their behaviours in the service of goals, they do not attend to the broader goal pursuit activities central to models of type 1 self-regulation. However, there is some empirical work that broadens the three-stage model of self-regulation to elaborate on how the process of self-regulation might unfold with respect to the pursuit of long-term goals (Miller & Brown, 1991). This more comprehensive model extends the self-regulation process to seven steps, which begin when the individual *receives* and *evaluates* relevant information, comparing it to norms and standards for behaviour. Next, the individual *triggers behavioural changes* oriented towards goal pursuit by *searching for options* that would lead to an alternative course of action. Once an appropriate course of action is identified, the individual then *formulates* and *implements a plan*, *assesses* its effectiveness, and cycles through the self-regulatory process as needed to make progress towards goal achievement.

The successful implementation of type 1 and type 2 self-regulatory processes requires flexible and appropriate use of specific self-regulatory strategies. Although the number of such strategies is likely large, a relatively small number, which we review in the next section, are routinely used to manage challenges to goal pursuit. Importantly, people vary widely in the skill with which they routinely apply these strategies, likely accounting for the variability in routine self-regulatory success and failure.

Self-Regulatory Strategies

Much of the social-psychological work on self-regulation centres on self-regulatory activity—the behavioural strategies that people employ in the service of goal pursuit. The behaviours—and people's tendencies to engage in them—that constitute the daily activity of self-regulation connect self-regulatory processes to outcomes. These self-regulatory strategies, which support self-regulatory processes, are enabled by the foundations reviewed earlier.

Delay of Gratification

Delay of gratification, a specific form of self-control, refers to the ability to defer an immediate reward and instead wait for a more preferred outcome. In a set of influential studies, Mischel and colleagues (Mischel, Ebbesen, & Zeiss, 1972; Mischel, Shoda, & Rodriguez, 1989) measured how long preschoolers could resist a small, immediate reward like one marshmallow in favour of a larger reward like two marshmallows after a waiting period. Children who were able to delay gratification effectively by waiting for the preferred reward evidenced better outcomes later in life including higher SAT scores, greater social competence, less aggressive behaviour, and lower obesity rates (Mischel et al., 2011).

Though delay of gratification is often associated with, and examined in, young children, the related delay discounting procedure (Green, Fry, & Myerson, 1994) has been used successfully as a valid marker of delay of gratification in adolescent and adult samples (Romer, Duckworth, Sznitman, & Park, 2010). Whatever the measure, the foundational executive functions are consequential for delay of gratification. In particular, updating of working memory is central to delay of gratification because it allows individuals to keep active information relevant to the delayed rewards as opposed to irrelevant information or information associated with immediate rewards (Berman et al., 2013). In addition, and as would be expected, inhibition is crucial for the successful delay of gratification insofar as this executive function enables individuals to override prepotent responses towards immediate rewards (Heatherton & Wagner, 2011; Volkow & Baler, 2015), as is also the case with respect to broader issues of self-control.

Self-Control

The self-regulatory strategy that has received the most attention from social psychologists is self-control. Indeed, many treatments of self-regulation and related processes use the terms interchangeably. However, as detailed earlier, the processes involved in self-regulation are numerous and more general than the specific strategy of self-control. The distinction is evident in our basic definition of self-control as a conscious decision by the individual to proceed, or to continue, in a manner that differs from that dictated by prevailing forces in the person (e.g., impulses, habits, desires) or situation (e.g., norms, requests/demands by others). This multiple-capacities conceptualisation of self-control is more nuanced than alternative definitions, highlighting the different forms that self-control might take when enacted in the service of self-regulation. Specifically, self-control might involve the inhibition of goal-inconsistent behaviour, the initiation of goal-consistent behaviour, or the continuation of goal-consistent behaviour that has been initiated (Davisson, 2013; Hoyle & Davisson, 2016; cf. Fujita, 2011). Other conceptualisations focus exclusively on the inhibition of impulses to behave in ways contrary to active goals (e.g., Friese & Hofmann, 2009; Maloney, Grawitch, & Barber, 2012) or the process by which self-control is enacted when immediately available behaviours or outcomes run counter to longer-term, goal-consistent interests (Trope & Fishbach, 2000). These conceptualisations share in common an assumption that self-control is an essential strategy for effective self-regulation.

The importance of self-control to self-regulation is evident in the many correlations between self-control and behaviours that reflect commonly held goals. For example, a widely used measure of self-control that emphasises inhibition of impulses predicts better academic performance, less binge eating and alcohol use, better interpersonal skills, and more secure attachments (Tangney, Baumeister, & Boone, 2004). An alternative measure focused specifically on poor self-control predicts risky behaviour (e.g., gambling, not wearing seat belts), general criminal involvement, and relatively low life satisfaction (Arneklev, Elis, & Medlicott, 2006). These behaviours generally reflect the achievement or failure to achieve goals that require self-regulation, underscoring the importance of self-control as a self-regulatory strategy.

A potential connection between self-control as understood by social psychologists and cognitive control is suggested by an influential body of research on the psychological resources that enable self-control. Much of this research is guided by the limited resource or strength model of self-control, which posits that self-control is made possible by a resource that is temporarily drawn down by acts of self-control (Baumeister, Bratslavsky, Muraven, & Tice, 1998). As such, the individual required to repeatedly enact self-control with relatively little time for recovery between occasions should do so with diminishing success as the resource is depleted without time for replenishment. A key feature of the model is the assumption that the same resource serves all acts of self-control. Given the centrality of some resource in the model, it is perhaps surprising that the precise nature of the resource is unknown (see, e.g., Beedie & Lane, 2012; Gailliot et al., 2007; Inzlicht & Schmeichel, 2012; Job, Dweck, & Walton, 2010; Kurzban, 2010). One of the more well-articulated accounts of the resource points to attention and motivation as key components (Inzlicht & Schmeichel, 2012). This account coupled with work showing that performing executive function tasks depletes the resource in the same manner as acts of self-control (Schmeichel, 2007) suggests a connection between cognitive control and the broader construct of self-control.

Grit

Whereas self-control generally concerns the management of specific, time-limited threats to effective goal pursuit, *grit*—passion and perseverance for long-term goals, particularly in the face of roadblocks—is relevant for managing threats to goal pursuits that play out over long periods of time (Duckworth, Peterson, Matthews, & Kelly, 2007; Eskreis-Winkler, Gross, & Duckworth, in press). Grit is associated with performance and achievement across diverse settings, including higher undergraduate grade point averages, cadet retention rates at West Point military academy, and performance at the Scripps National Spelling Bee (Duckworth et al., 2007). Though self-control is certainly relevant for achievement outcomes like academic performance, especially in the context of daily temptations, grit is unique in its prediction of outcomes in challenging situations that require dogged goal pursuit over time despite adversity (Duckworth et al., 2007; Duckworth & Gross, 2014). In other words, grit is a longer-term self-regulatory strategy that necessitates the use of other time-limited self-regulatory strategies such as self-control in the pursuit of major life goals.

Goal-Related Strategies

Given the primary focus of self-regulatory processes on the management of goal pursuit, it is not surprising that a subset of self-regulatory strategies concern the optimal setting, pursuit, and abandonment of goals. These strategies address one or more of several concerns related to goal pursuit: (1) the properties of goals influence their pursuit and the results of pursuit; (2) individuals are rarely pursuing a single goal and, therefore, must find ways to make progress on multiple, sometimes conflicting, goals; (3) motivation and attention related to goals wax and wane, requiring strategies that generate motivation and/or attention, or promote pursuit without it; (4) some goals concern the maintenance of a current state rather than the attainment of a new state; (5) sometimes the optimal goal-management strategy is to discontinue pursuit of one or more goals.

Although many goals are nonconsciously adopted as a result of exposure to family or societal priorities and others are imposed by other people or organisations, at least some goals are consciously set by the individual. When goals concern performance, as in work- or school-related goals, optimal goals have two important properties. They are challenging, requiring the commitment and effort necessary to perform at a level near the individual's maximum skill or ability. And they are specific, allowing for unambiguous evaluations of progress and determination of attainment (Locke & Latham, 2002). Although specificity is beneficial in goal setting, a more general construal of goal-related decisions and behaviours when pursuing the goal contributes to better self-control (Fujita, Trope, Liberman, & Levin-Sagim, 2006). This strategy suggests that goal properties are not fixed, but rather vary as a function of how they are construed at a particular point in time.

People rarely pursue goals in a singular fashion and, therefore, conflict between competing goals is a frequently encountered self-regulatory challenge. Given that goal conflict is so pervasive, individuals must make use of self-regulatory strategies that enable effective goal pursuit when alternative goals beckon. Some potentially appealing strategies may actually prove ineffective in practice (e.g., multifinality; Köpetz, Faber, Fishbach, & Kruglanski, 2011), but one effective strategy is the practice of goal shielding, by which commitment to an active goal results in the temporary inhibition of alternative goals (Dreisbach & Haider, 2009; Shah, Friedman, & Kruglanski, 2002; Monsell, this volume). This and other strategies for managing the demands of multiple interrelated goals implicate basic cognitive capacities such as shifting and updating.

The maintenance of attention and motivation are significant challenges in the management of goal pursuit. Relevant strategies include measures to increase attention and motivation or to automate goal pursuit so that conscious attention and motivation are not required. For example, motivation may be increased through self-generated measures that increase interest in a goal-related task (Sansone & Thoman, 2005). For repeated acts, the need for attention and motivation in goal pursuit can be diminished through the adoption of implementation intentions—if-then statements that associate desired behaviours (the 'then' component) with cues in the environment (the 'if' component; Gollwitzer, 1999). In terms of process, the use of implementation intentions induces inhibition (Veling & van Knippenberg, 2006) and can facilitate performance on shifting tasks (Cohen, Bayer, Jaudas, & Gollwitzer, 2008), particularly in situations in which executive control might be otherwise impaired (Scholz et al., 2009), suggesting that in addition to directly influencing goal pursuit by guiding behaviour, this strategy can contribute to self-regulation at the foundational level.

When these goal-related strategies do not produce satisfactory progress in goal pursuit, or when a goal is no longer relevant or feasible, the optimal self-regulatory strategy may be goal disengagement. Given the frustration and futility associated with continued pursuit of an unattainable goal, it is not surprising that effective goal disengagement is associated with lower rumination and perceived stress and higher subjective well-being (Wrosch, Scheier, Miller, Schulz, & Carver, 2003). In light of the fact that people typically are managing multiple goals, an important payoff associated with goal disengagement is the freeing of psychological resources that may contribute to improved progress on other goals.

These goal-related strategies, when coupled with more general strategies such as self-control, are the direct means by which goals and their pursuit are managed. Their focus on basic cognitive and motivational processes suggests a largely untapped connection between self-regulation and cognitive control.

Summary and Conclusions

We provided a broad overview of self-regulation with an emphasis on contributions from social and personality psychology. We highlighted a number of constructs, processes, and behaviours relevant for the ways in which people exert control over their emotions, cognitions, and actions in the service of goal attainment. Despite a lack of consensus around a singular definition or model of self-regulation, many treatments identify a common set of constructs relevant for the management of goal pursuit. We organised the wide array of relevant and overlapping constructs intro three broad categories: *foundations*, which include executive functions, temperament, and personality; *self-regulatory processes* such as self-monitoring and forethought; and *self-regulatory strategies*, including delay of gratification and self-control. Within each category, we highlighted constructs and processes that either directly or indirectly implicate basic cognitive and motivational processes characteristic of cognitive control.

Given the importance of effective self-regulation to many of life's important pursuits, a critical question is whether self-regulation might be improved through training or practice (see, e.g., Chapter 28 by Spencer-Smith & Klingberg in this volume). To date, researchers across fields have offered accounts of training programmes and interventions, many of which target the underlying foundations of self-regulation like executive functions (Berkman, Kahn, & Merchant, 2014; Houben & Jansen, 2011) and self-regulatory strategies like self-control (Davisson, 2013; Muraven, 2010). Executive function interventions involve practice of executive control tasks such as a go no-go task (Diamond, 2012) so that these foundational elements of self-regulation might become more efficient in daily use. Training programmes targeting self-control seem to function largely by giving people practice at enacting self-control strategies (e.g., avoiding tempting sweets) so that they gain experience acting appropriately when faced with self-control demands (e.g., Muraven, 2010), and perhaps through use of implementation intentions (Duckworth, Grant, Loew, Oettingen, & Gollwitzer, 2011) or by working towards the creation of habitual, automatic responses reflecting self-control (Rothman, Sheeran, & Wood, 2009).

Though well intentioned, most of these self-regulatory interventions evidence small effects that rarely transfer to other abilities and strategies implicated in successful self-regulation. For example, as Diamond (2012) notes, executive functioning training programmes targeting inhibition do not result in improvements to working memory. Similarly, practice effects involving self-control appear to be short-lived in nature (Davisson, 2013). Despite these shortcomings, attempts to improve self-regulation at a broader level—that is, simultaneously targeting the foundations and strategies that serve self-regulation—would likely be beneficial for a wide variety of consequential outcomes (e.g., Moffitt et al., 2011).

Acknowledgement

The preparation of this chapter was supported in part by grant P30DA023026 from the National Institute on Drug Abuse.

Note

1 Although social and personality psychology are separate fields, the boundary between them is not always clear. Many theories and models, including the treatments of self-regulation we cover in this chapter, draw on constructs and processes from both fields. Thus, references to social-psychological models and perspectives imply an accounting for relevant personality constructs and processes.

References

Arneklev, B. J., Elis, L., & Medlicott, S. (2006). Testing the general theory of crime: Comparing the effects of 'imprudent behavior' and an attitudinal indicator of 'low self-control'. *Western Criminology Review, 7,* 41–55.

Baumeister, R. F., Bratslavsky, E., Muraven, M., & Tice, D. M. (1998). Ego depletion: Is the active self a limited resource? *Journal of Personality and Social Psychology, 74,* 1252–1265.

Baumeister, R. F., & Heatherton, T. F. (1996). Self-regulation failure: An overview. *Psychological Inquiry, 7,* 1–15.

Beedie, C. J., & Lane, A. M. (2012). The role of glucose in self-control: Another look at the evidence and an alternative conceptualization. *Personality and Social Psychology Review, 16,* 143–153.

Berkman, E. T., Kahn, L. E., & Merchant, J. S. (2014). Training-induced changes in inhibitory control network activity. *The Journal of Neuroscience, 34,* 149–157.

Berman, M. G., Yourganov, G., Askren, M. K., Ayduk, O., Casey, B. J., Gotlib, I. H.,...Jonides, J. (2013). Dimensionality of brain networks linked to life-long individual differences in self-control. *Nature Communications, 4,* 1373–1379.

Best, J. R., & Miller, P. H. (2010). A developmental perspective on executive function. *Child Development, 81,* 1641–1660.

Carver, C. S., & Scheier, M. F. (1981). *Attention and self-regulation: A control-theory approach to human behavior.* New York: Springer.

Cohen, A. L., Bayer, U. C., Jaudas, A., & Gollwitzer, P. M. (2008). Self-regulatory strategy and executive control: Implementation intentions modulate task switching and Simon task performance. *Psychological Research, 72,* 12–26.

Costa, P. T., Jr., & McCrae, R. R. (1992). Normal personality assessment in clinical practice: The NEO Personality Inventory. *Psychological Assessment, 4,* 5–13.

Crone, E. A., Somsen, R. J. M., Zanolie, K.,& Van der Molen, M. W. (2006). A heart rate analysis of developmental change in feedback processing and rule shifting from childhood to early adulthood. *Journal of Experimental Child Psychology, 95,* 99–116.

Davisson, E. K. (2013). *Strengthening self-control by practicing inhibition and initiation.* (Unpublished doctoral dissertation). Duke University, Durham, NC.

DeYoung, C. G. (2015). Cybernetic Big Five Theory. *Journal of Research in Personality, 56,* 33–58.

Diamond, A. (2012). Activities and programs that improve children's executive functions. *Current Directions in Psychological Science, 21,* 335–341.

Dreisbach, G., & Haider, H. (2009). How task representations guide attention: Further evidence for the shielding function of task sets. *Journal of Experimental Psychology: Learning, Memory, and Cognition, 35,* 477–486.

Duckworth, A. L., Grant, H., Loew, B., Oettingen, G., & Gollwitzer, P. M. (2011). Self-regulation strategies improve self-discipline in adolescents: Benefits of mental contrasting and implementation intentions. *Educational Psychology, 31,* 17–26.

Duckworth, A. L., & Gross, J. J. (2014). Self-control and grit related but separable determinants of success. *Current Directions in Psychological Science, 23,* 319–325.

Duckworth, A. L., Peterson, C., Matthews, M. D., & Kelly, D. R. (2007). Grit: Perseverance and passion for long-term goals. *Journal of Personality and Social Psychology, 92,* 1087–1111.

Duval, S., & Wicklund, R. A. (1972). *A theory of objective self-awareness.* New York: Academic Press.

Eisenberg, N., Valiente, C., Spinrad, T. L., Liew, J., Zhou, Q., Losoya, S. H.,...Cumberland, A. (2009). Longitudinal relations of children's effortful control, impulsivity, and negative emotionality to their externalizing, internalizing, and co-occurring behavior problems. *Developmental Psychology, 45,* 988–1008.

Eskreis-Winkler, L., Gross, J. J., & Duckworth, A. L. (in press). Grit: Sustained self-regulation in the service of superordinate goals. In K. D. Vohs & R. F. Baumeister (Eds.), *Handbook of self-regulation* (3rd ed.). New York: Guilford Press.

Fitzsimons, G. M., & Bargh, J. A. (2004). Automatic self-regulation. In R. F. Baumeister & K. D. Vohs (Eds.), *Handbook of self-regulation: Research, theory, and applications* (pp. 151–170). New York: Guilford Press.

Friedman, N. P., Miyake, A., Robinson, J. L., & Hewitt, J. K. (2011). Developmental trajectories in toddlers' self-restraint predict individual differences in executive functions 14 years later: A behavioral genetic analysis. *Developmental Psychology, 47*, 1410–1430.

Friese, M., & Hofmann, W. (2009). Control me or I will control you: Impulses, trait self-control, and the guidance of behavior. *Journal of Research in Personality, 43*, 795–805.

Fujita, K. (2011). On conceptualizing self-control as more than the effortful inhibition of impulses. *Personality and Social Psychology Review, 15*, 352–366.

Fujita, K., Trope, Y., Liberman, N., & Levin-Sagim, M. (2006). Construal levels and self-control. *Journal of Personality and Social Psychology, 90*, 351–367.

Gailliot, M. T., Baumeister, R. F., DeWall, C. N., Maner, J. K., Plant, E. A., Tice, D. M.,...Schmeichel, B. J. (2007). Self-control relies on glucose as a limited energy source: Willpower is more than a metaphor. *Journal of Personality and Social Psychology, 92*, 325–336.

Gollwitzer, P. M. (1999). Implementation intentions: Strong effects of simple plans. *American Psychologist, 54*, 493–503.

Green, L., Fry, A. F., & Myerson, J. (1994). Discounting of delay rewards: A lifespan comparison. *Psychological Science, 5*, 33–36.

Heatherton, T. F., & Wagner, D. D. (2011). Cognitive neuroscience of self-regulation failure. *Trends in Cognitive Sciences, 15*, 132–139.

Higgins, E. T. (1997). Beyond pleasure and pain. *American Psychologist, 52*, 2180–1300.

Higgins, E. T., Kruglanski, A. W., & Pierro, A. (2003). Regulatory mode: Locomotion and assessment as distinct orientations. *Advances in Experimental Social Psychology, 35*, 293–344.

Hofmann, W., Friese, M., Schmeichel, B. J., & Baddeley, A. D. (2011). Working memory and self-regulation. In R. F. Baumeister & K. D. Vohs (Eds.), *Handbook of self-regulation: Research, theory, and applications*, (2nd ed., pp. 204–226). New York: Guilford Press.

Houben, K., & Jansen, A. (2011). Training inhibitory control: A recipe for resisting sweet temptations. *Appetite, 56*, 345–349.

Hoyle, R. H., & Davisson, E. K. (2016). Varieties of self-control and their personality correlates. In K. D. Vohs & R. F. Baumeister (Eds.), *Handbook of self-regulation* (3rd ed., pp. 396–413.). New York: Guilford Press.

Hoyle, R. H., & Gallagher, P. (2015). The interplay of personality and self-regulation. In M. Mikulincer, P. R. Shaver, M. L. Cooper, & R. J. Larsen (Eds.), *APA handbook of personality and social psychology: Vol. 4, Personality processes and individual differences* (pp. 189–207). Washington, DC: American Psychological Association.

Inzlicht, M., & Schmeichel, B. J. (2012). What is ego depletion? Toward a mechanistic revision of the resource model of self-control. *Perspectives on Psychological Science, 7*, 450–463.

Inzlicht, M., Schmeichel, B. J., & Macrae, C. N. (2014). Why self-control seems (but may not be) limited. *Trends in Cognitive Sciences, 18*, 127–133.

Job, V., Dweck, C. S., & Walton, G. M. (2010). Ego depletion: Is it all in your head? Implicit theories about willpower affect self-regulation. *Psychological Science, 21*, 1686–1693.

John, O. P. (1990). The 'Big Five' factor taxonomy: Dimensions of personality in the natural language and in questionnaires. In L. Pervin (Ed.), *Handbook of personality: Theory and research* (pp. 66–100). New York: Guilford Press.

Kane, M. J., & Engle, R. W. (2002). The role of prefrontal cortex in working-memory capacity, executive attention, and general fluid intelligence: An individual-differences perspective. *Psychonomic Bulletin and Review, 9*, 637–671.

Kanfer, F. H. (1970). Self-regulation: Research, issues, and speculation. In C. Neuringer & J. L. Michaels (Eds.), *Behavior modification in clinical psychology* (pp. 178–220). New York, NY: Appleton-Century-Crofts.

Kanfer, F. H., & Karoly, P. (1972). Self-control: A behavioristic excursion into the lion's den. *Behavior Therapy, 3*, 389–416.

Kaplan, S., & Berman, M. G. (2010). Directed attention as a common resource for executive functioning and self-regulation. *Perspectives on Psychological Science, 5*, 43–57.

Karoly, P. (2010). Goal systems and self-regulation: An individual differences perspective. In R. H. Hoyle (Ed.), *Handbook of personality and self-regulation* (pp. 218–242). Malden, MA: Blackwell.

Köpetz, C., Faber, T., Fishbach, A., & Kruglanski, A. W. (2011). The multifinality constraints effect: How goal multiplicity narrows the means set to a focal end. *Journal of Personality and Social Psychology, 100*, 810–826.

Kurzban, R. (2010). Does the brain consume additional glucose during self-control tasks? *Evolutionary Psychology, 8*, 244–259.

Locke, E. A., & Latham, G. P. (2002). Building a practically useful theory of goal setting and task motivation: A 35-year odyssey. *American Psychologist, 57*, 705–717.

Logan, G. D. (1994). On the ability to inhibit thought and action: A user's guide to the stop signal paradigm. In D. Dagenbach & T. H. Carr (Eds.), *Inhibitory processes in attention, memory, and language* (pp. 189–239). San Diego, CA: Academic Press.

Louro, M. J., Pieters, R., & Zeelenberg, M. (2007). Dynamics of multiple-goal pursuit. *Journal of Personality and Social Psychology, 93*, 174–193.

MacKenzie, M. B., Mezo, P. G., & Francis, S. E. (2012). A conceptual framework for understanding self-regulation in adults. *New Ideas in Psychology, 30*, 155–165.

Maloney, P. W., Grawitch, M. J., & Barber, L. K. (2012). The multi-factor structure of the Brief Self-Control Scale: Discriminant validity of restraint and impulsivity. *Journal of Research in Personality, 46*, 111–115.

Miller, W. R., & Brown, J. M. (1991). Self-regulation as a conceptual basis for the prevention and treatment of addictive behaviours. In N. Heather, W. R. Miller, & J. Greeley (Eds.), *Self-control and the addictive behaviours* (pp. 3–79). Sydney: Maxwell Macmillan Publishing Australia.

Mischel, W., Ayduk, O., Berman, M. G., Casey, B. J., Gotlib, I. H., Jonides, J.,...Shoda, Y. (2011). 'Willpower' over the life span: Decomposing self-regulation. *Social Cognitive and Affective Neuroscience, 6*, 252–256.

Mischel, W., Ebbesen, E. B., & Zeiss, A. R. (1972). Cognitive and attentional mechanisms in delay of gratification. *Journal of Personality and Social Psychology, 21*, 204–218.

Mischel, W., Shoda, Y., & Rodriguez, M. L. (1989). Delay of gratification in children. *Science, 244*, 933–938.

Miyake, A., & Friedman, N. P. (2012). The nature of organization of individual differences in executive functions: Four general conclusions. *Current Directions in Psychological Science, 21*, 8–14.

Miyake, A., Friedman, N. P., Emerson, M. J., Witzki, A. J., Howerter, A., & Wager, T. D. (2000). The unity and diversity of executive functions and their contributions to complex 'frontal lobe' tasks: A latent variable analysis. *Cognitive Psychology, 41*, 49–100.

Moffitt, T. E., Arseneault, L., Belsky, D., Dickson, N., Hancox, R. J., Harrington, H., ... Caspi, A. (2011). A gradient of childhood self-control predicts health, wealth, and public safety. *Proceedings of the National Academy of Sciences of the United States of America, 108*, 2693–2698.

Munakata, Y., Herd, S. A., Chatham, C. H., Depue, B. E., Banich, M. T., & O'Reilly, R. C. (2011). A unified framework for inhibitory control. *Trends in Cognitive Sciences, 15*, 453–459.

Muraven, M. (2010). Building self-control strength: Practicing self-control leads to improved self-control performance. *Journal of Experimental Social Psychology, 46*, 465–468.

Nes, L. S., Carlson, C. R., Crofford, L. J., de Leeuw, R., & Segerstrom, S. C. (2011). Individual differences and self-regulatory fatigue: Optimism, conscientiousness, and self-consciousness. *Personality and Individual Differences, 50*, 475–480.

Pintrich, P. R. (2004). A conceptual framework for assessing motivation and self-regulated learning in college students. *Educational Psychology Review, 16*, 385–407.

Romer, D., Duckworth, A. L., Sznitman, S., & Park, S. (2010). Can adolescents learn self-control? Delay of gratification in the development of control over risk taking. *Prevention Science, 11*, 319–330.

Rothbart, M. K., Ahadi, S. A., & Evans, D. E. (2000). Temperament and personality: Origins and outcomes. *Journal of Personality and Social Psychology, 78*, 122–135.

Rothbart, M. K., Ahadi, S. A., Hershey, K., & Fisher, P. (2001). Investigations of temperament at three to seven years: The Children's Behavior Questionnaire. *Child Development, 72*, 1394–1408.

Rothbart, M. K., & Bates, J. E. (2006). Temperament. In W. Damon, R. L. Lerner (Series Eds.), & N. Eisenberg (Vol. Ed.), *Handbook of child psychology: Vol. 3: Social, emotional, and personality development* (6th ed., pp. 99–166). New York: Wiley.

Rothbart, M. K., & Derryberry, D. (1981). Development of individual differences in temperament. In M. E. Lamb & A. L. Brown (Eds.), *Advances in developmental psychology* (Vol. *1*, pp. 37–86). Hillsdale, NJ: Erlbaum.

Rothbart, M. K., & Rueda, M. R. (2005). The development of effortful control. In U. Mayr, E. Awh, & S. Keele (Eds.), *Developing individuality in the human brain: A tribute to Michael I. Posner* (pp. 167–188). Washington, DC: American Psychological Association.

Rothman, A. J., Sheeran, P., & Wood, W. (2009). Reflective and automatic processes in the initiation and maintenance of dietary change. *Annals of Behavioral Medicine*, *38*, 4–17.

Sansone, C., & Thoman, D. B. (2005). Interest as the missing motivator in self-regulation. *European Psychologist*, *10*, 175–186.

Schmeichel, B. J. (2007). Attention control, memory updating, and emotion regulation temporarily reduce the capacity for executive control. *Journal of Experimental Psychology: General*, *136*, 241–255.

Schmeichel, B. J., Volokhov, R. N., & Demaree, H. A. (2008). Working memory capacity and the self-regulation of emotional expression and experience. *Journal of Personality and Social Psychology*, *95*, 1526–1540.

Scholz, U., La Marca, R., Nater, U. M., Aberle, I., Ehlert, U., Hornung, R.,...Kliegel, M. (2009). Go no-go performance under psychosocial stress: Beneficial effects of implementation intentions. *Neurobiology of Learning and Memory*, *91*, 89–92.

Shah, J. Y., Friedman, R., & Kruglanski, A. W. (2002). Forgetting all else: On the antecedents and consequences of goal shielding. *Journal of Personality and Social Psychology*, *83*, 1261–1280.

Smith-Lovin, L., & Heise, D. R. (Eds.) (1988). *Analyzing social interaction: Advances in affect control theory*. New York: Gordon & Breach.

Stroop, J. R. (1935). Studies of interference in serial verbal reactions. *Journal of Experimental Psychology*, *18*, 643–662.

Tangney, J. P., Baumeister, R. F., & Boone, A. L. (2004). High self-control predicts good adjustment, less pathology, better grades, and interpersonal success. *Journal of Personality*, *72*, 271–324.

Thomas, A., Chess, S., & Birch, S. G. (1970). The origin of personality. *Scientific American*, *223*, 102–109.

Trope, Y., & Fishbach, A. (2000). Counteractive self-control in overcoming temptation. *Journal of Personality and Social Psychology*, *79*, 493–506.

Van Egeren, L. F. (2009). A cybernetic model of global personality traits. *Personality and Social Psychology Review*, *13*, 92–108.

Veling, H., & van Knippenberg, A. (2006). Shielding intentions from distraction: Forming an intention induces inhibition of distracting stimuli. *Social Cognition*, *4*, 409–425.

Volkow, N. D., & Baler, R. D. (2015). NOW vs LATER brain circuits: Implications for obesity and addiction. *Trends in Neuroscience*, *38*, 345–352.

Wrosch, C., Scheier, M. F., Miller, G. E., Schulz, R., & Carver, C. S. (2003). Adaptive self-regulation of unattainable goals: Goal disengagement, goal reengagement, and subjective well-being. *Personality and Social Psychology Bulletin*, *29*, 1494–1508.

Zimmerman, B. J., & Campillo, M. (2003). Motivating self-regulated problem solvers. In J. E. Davidson & R. J. Sternberg (Eds.), *The nature of problem solving* (pp. 233–262). New York, NY: Cambridge University Press.

Section IV

Cognitive Control in Practice

Development, Training, and Clinical Perspectives

26

The Neurobiology of Adolescent Self-Control

Alexandra O. Cohen and B. J. Casey

You see us as you want to see us ... in the simplest terms ... as a brain, an athlete, a basket case, a princess and a criminal.

The Breakfast Club

Introduction

Too often adolescents are characterised as having no minds of their own – a brain with no brake or steering wheel, only an accelerator (Bell & McBride, 2010). Yet, this disparaging description of adolescence is not used to characterise other developmental stages. For instance, when an infant is incapable of walking or talking at birth, we describe this as part of typical development (Steinberg, 2010), but when a teen makes a bad decision in the heat of the moment, we characterise that behaviour as aberrant. This chapter highlights changes in neural and behavioural correlates of cognitive control in both emotionally charged and relatively benign situations, to better understand when and how self-control may lapse during adolescence and why.

Adolescence is a typical developmental transition between childhood and adulthood, when the individual must learn to be relatively independent of the parent. This developmental period is a time when the brain is more "plastic" than it will ever be again, capable of remarkable adaptability in light of the many social, physical, sexual, and intellectual challenges that this developmental phase brings (Casey, Galván, & Lee, 2015). With these challenges comes the need for cognitive control: the ability to suppress irrelevant thoughts, desires, and actions in favour of adaptive or goal-oriented ones (Casey, 2015). During adolescence, the capacity for cognitive control is influenced by a heightened sensitivity to rewards (Galvan et al., 2006; Geier, Terwilliger, Teslovich, Velanova, & Luna, 2010; Van Leijenhorst et al., 2010), potential threats (Cohen Gilbert & Thomas, 2013; Dreyfuss et al., 2014; Grose-Fifer, Rodrigues, Hoover, & Zottoli, 2013; Hare et al., 2008), and peers (Chein et al., 2011; Steinberg et al., 2009).

In this chapter, we first describe the development and neural correlates of cognitive control and then how motivational and socioemotional information alters this capacity specifically during adolescence. We provide a neurobiological model of adolescence based on the existing animal, imaging, and post mortem studies that attempt to move beyond

The Wiley Handbook of Cognitive Control, First Edition. Edited by Tobias Egner.

26

The Neurobiology of Adolescent Self-Control

Alexandra O. Cohen and B. J. Casey

You see us as you want to see us—in the simplest terms ... as a brain, an athlete, a basket case, a princess and a criminal.

The Breakfast Club

Introduction

Too often adolescents are characterised as having no minds of their own—a brain with no brake or steering wheel, only an accelerator (Bell & McBride, 2010). Yet, this disparaging description of adolescence is not used to characterise other developmental stages. For instance, when an infant is incapable of walking or talking at birth, we describe this as part of typical development (Steinberg, 2012), but when a teen makes a bad decision in the heat of the moment, we characterise that behaviour as aberrant. This chapter highlights changes in neural and behavioural correlates of cognitive control in both emotionally charged and relatively benign situations, to better understand when and how self-control may lapse during adolescence and why.

Adolescence is a typical developmental transition between childhood and adulthood, when the individual must learn to be relatively independent of the parent. This developmental period is a time when the brain is more 'plastic' than it will ever be again, capable of remarkable adaptability in light of the many social, physical, sexual, and intellectual challenges that this developmental phase brings (Casey, Glatt, & Lee, 2015). With these challenges comes the need for cognitive control: the ability to suppress irrelevant thoughts, desires, and actions in favour of adaptive or goal-oriented ones (Casey, 2015). During adolescence, the capacity for cognitive control is influenced by a heightened sensitivity to rewards (Galván et al., 2006; Geier, Terwilliger, Teslovich, Velanova, & Luna, 2010; Van Leijenhorst et al., 2010), potential threats (Cohen-Gilbert & Thomas, 2013; Dreyfuss et al., 2014; Grose-fifer, Rodrigues, Hoover, & Zottoli, 2013; Hare et al., 2008), and peers (Chein et al., 2011; Steinberg et al., 2009).

In this chapter, we first describe the development and neural correlates of cognitive control and then how motivational and socioemotional information alters this capacity specifically during adolescence. We provide a neurobiological model of adolescence based on the existing animal, imaging, and post-mortem studies that attempts to move beyond

The Wiley Handbook of Cognitive Control, First Edition. Edited by Tobias Egner.

simple accounts of self-control to circuit-based accounts of adolescent behaviour (Casey, 2015). We conclude by speculating on why the developing brain may be programmed this way (Casey, Getz, & Galván, 2008).

Development of Cognitive Control

Cognitive control is a theoretical construct at the heart of numerous theories of attention and memory (Baddeley, 1986; Cohen, Servan-Schreiber, & McClelland, 1992; Desimone & Duncan, 1995; Shallice, 1988) and has many different labels (e.g., 'central executive', 'attentional bias', 'supervisory system'). These terms all suggest a mechanism that is required to inhibit reflexive actions (Shallice, 1988; Chapter 6 by Verbruggen & Logan in this volume) and to redirect attention (Desimone & Duncan, 1995) and actions (Miller & Cohen, 2001) in favour of adaptive ones (Chapter 13 by Stokes, Buschman, & Miller in this volume). Thus, cognitive control refers to the capacity to suppress irrelevant thoughts and actions in favour of adaptive or goal-oriented ones (Allport, 1987; Casey et al. 2000; Casey, Thomas, Davidson, Kunz, & Franzen, 2002; Cohen & Servan-Schreiber, 1992; Kahneman, Treisman, & Burkell, 1983). Typically, when this ability involves the suppression of inappropriate emotions and desires, we refer to it as self-control (Casey, 2015; Chapter 25 by Davisson & Hoyle in this volume). The neural circuitry underlying control differs by the type of information or stage of processing being suppressed (Casey et al., 2000; Casey, Galván, & Hare, 2005; Jonides & Nee, 2006). In other words, different circuitry may come into play when we suppress unwanted information by paying attention to something else; suppress unwanted thoughts by thinking about something else; or suppress an action, impulse or desire by engaging in an alternative action or withholding an action (Casey, Jones, et al., 2011; Mischel et al., 2011).

Imaging studies of cognitive control are consistent with this idea in that these studies show non-overlapping patterns of brain activity for a number of tasks including the flanker, go/no-go, and Stroop tasks (Nee, Wager, & Jonides, 2007). These tasks and others that recruit cognitive control reveal modest, if any, correlations with one another in behavioural performance as well (Earles et al., 1997; Grant & Dagenbach, 2000; Kramer, Humphrey, Larish, Logan, & Strayer, 1994; Shilling, Chetwynd, & Rabbitt, 2002; Tipper & Baylis, 1987). Together, these behavioural and imaging findings suggest that control processes involved in resolving interference may be due to multiple mechanisms rather than a single mechanism, brain region, or circuit (see also Chapter 4 by Egner in this volume). Nonetheless, various lateral prefrontal circuits and the posterior parietal cortex have been implicated fairly reliably in imaging studies of cognitive control (Vincent et al., 2008).

Developmental studies of cognitive control processes such as the suppression of unwanted information in memory in favour of relevant information (i.e., working memory) or suppressing an action in favour of an alternative one (i.e., response inhibition) show steady improvement during childhood but plateau by adolescence (Casey, 2013; Davidson, Amso, Anderson, & Diamond, 2006; Luna, Padmanabhan, & O'Hearn, 2010; Paus, 2005). The improvement in performance with age is paralleled by changes in patterns of activity in frontoparietal and frontostriatal circuitry across development (Adleman et al., 2002; Bunge et al., 2002; Casey et al., 1995, 1997; Crone, Wendelken, Donohue, van Leijenhorst, & Bunge, 2006; Darki & Klingberg, 2014; Durston, Thomas, Worden, Yang, & Casey, 2002; Durston et al., 2006; Kwon, Reiss, & Menon, 2002; Luna et al., 2001; Rubia et al., 2000, 2006; Thomas et al., 1999) and in white matter tracts within and across these circuits (Giedd, Keshavean, & Paus, 2008; Liston et al 2006; Paus, 2005; Stevens, Skudlarski, Pearlson, & Vince, 2009), consistent with the basic circuitry implicated in adult neuroimaging studies of cognitive control. However, this development occurs in the context of significant reorganisation of brain circuitry, especially during adolescence, when circuits become more hardwired with experience and maturation (Casey, 2015; Casey et al., 2015).

Reorganisation of the Brain During Adolescence

Cognitive, emotional, and behavioural changes during adolescence are paralleled by hormonal and neurobiological changes that are regional in nature (Casey et al., 2015). Non-human primate and human post-mortem studies show that this period is associated with regional elimination and stabilisation of synapses that are predominantly excitatory (Huttenlocher & Dabholkar, 1997; Rakic, Bourgeois, & Goldman-Rakic, 1994). Sensorimotor regions are pruned before higher-order association cortices. These regional changes in synaptic connections are paralleled by regional changes in patterns of cell firing and in the availability neurotrophins, neurochemicals, gonadal hormones, and their receptors (Lee et al., 2014). For example, dopamine modulates regional changes in excitatory and inhibitory balance within this circuitry (Floresco & Tse, 2007) with peaks in dopamine receptor expression within the accumbens (Brenhouse, Sonntag, & Andersen, 2008) occurring before peak dopamine signalling within the prefrontal cortex (Tseng & O'Donnell, 2007).

Human imaging studies reveal similar regional changes in brain development with changes in the phylogentically older sensorimotor cortex and the subcortical limbic regions developing before association cortices, such as regions of the prefrontal cortex that do not reach maturity until the early twenties (Chugani, Phelps, & Mazziotta, 1987; Gogtay et al., 2004; Mills, Goddings, Clasen, Giedd, & Blakemore, 2014; Sowell et al., 2004). Increases in white matter are concurrent with decreases in grey matter (Giedd et al., 1999; Giorgio et al., 2008; Simmonds, Hallquist, Asato, & Luna, 2014). The ubiquitous regional progressive and regressive changes in grey and white matter volumes and brain activity throughout the brain are thought to reflect refinement of functional circuits (Brown et al., 2005; Brown, Petersen, & Schlaggar, 2006; Durston et al., 2006).

Fundamental changes in patterns of functional connectivity throughout the brain have been shown with resting state functional MRI. These studies reveal that multiple functional networks, including frontoparietal and cingulo-opercular networks implicated in cognitive control processes, all undergo reorganisation from childhood to adulthood. Over the course of development, short-range, local functional connectivity decreases, while long-range connections among brain regions increase (Fair et al., 2007; 2008; 2009). These findings suggest that there is a developmental shift towards integration of various brain networks into distributed functional networks with connectivity rapidly increasing with the onset of adolescence and plateauing by the mid-twenties (Dosenbach et al., 2010).

Circuits or networks are the product of both experience and evolutionary biological constraints (Karmiloff-Smith, 2009). Experiences help to shape a network that can then be modulated by another network. Presumably the effectiveness of modulated networks relies upon maturation within a network and between networks, with within-circuit development occurring before between-network development (Silvers et al., 2016).

In the context of adolescent brain development, significant structural and functional changes occur in limbic subcortical regions before the prefrontal cortex (Galván et al., 2006; Mills et al., 2014). These regional structural and functional changes have been hypothesised to lead to transient imbalances in functional brain circuitry, leading to dysregulation of emotions and actions during adolescence (Casey et al., 2008). Figure 26.1 highlights development of neurocircuitry implicated in cognitive and affective control. Functional connectivity between these circuits depends on the relative maturity within each circuit (Cohen et al., 2015; Silvers et al., 2016; Van den Bos, Cohen, Kahnt, & Crone, 2012; Van den Bos, Guroglu, van den Bulk, Rombouts, & Crone, 2009). Refinement of subcortical connections

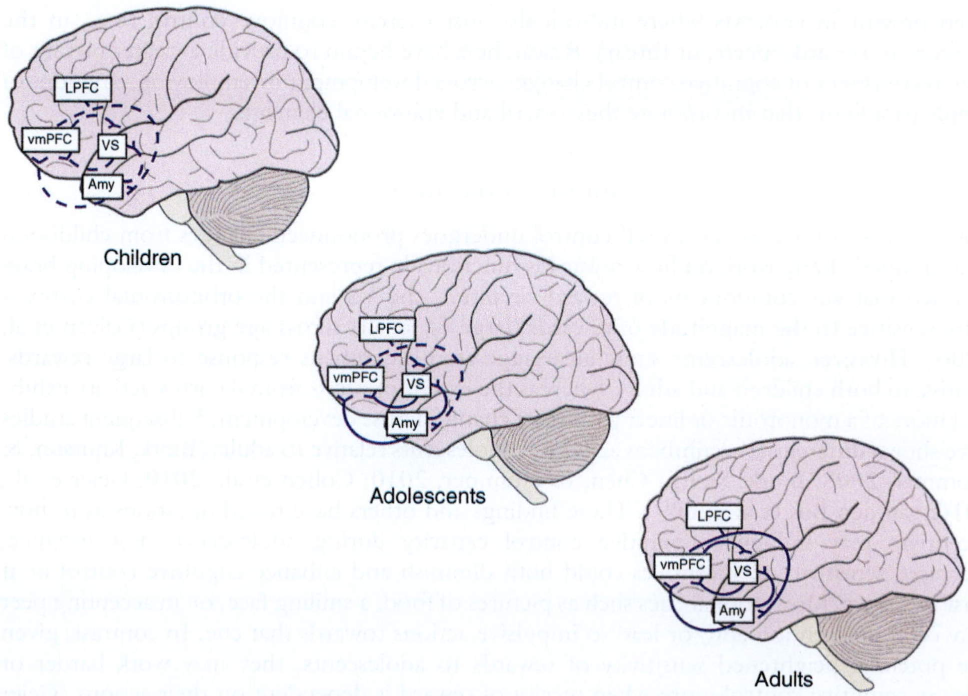

Figure 26.1 Schematic of cognitive and motivational circuitry development. The ventromedial prefrontal cortex (vmPFC) has reciprocal connections with both the amygdala (Amy) and ventral striatum (VS). The VS also receives direct projections from the basolateral amygdala. Subcortical connections from the amygdala to the VS show changes before cortico-subcortical projections from the prefrontal cortex. Subsequently, cortico-cortico connections from the lateral prefrontal cortex (lPFC) to vmPFC implicated in cognitive control mature. Dotted lines represent immature projections.

such as those from the amygdala to the ventral striatum that lead to motivated behaviour (Stuber et al., 2011) are thought to mature before cortico-subcortical projections such as those from the ventromedial PFC to GABA-ergic inhibitory neurons within the amygdala and ventral striatum that help modulate motivated and emotive behaviours (Gee et al., 2013). Subsequently, refinement of cortico-cortico connections or distributed functional networks occur that help modulate complex thoughts, emotions, and actions. For the sake of simplicity, we illustrate these changes in distributed functional networks across development in Figure 26.1 by singling out connections between lateral and medial prefrontal circuits thought to support cognitive and affective computations, respectively (Cohen et al., 2016). Recent developmental imaging studies have examined the organisation and function of these circuits by examining cognitive control capacity in the context of motivational and socioemotional contexts (i.e., self-control).

Motivational and Socioemotional Influences on Cognitive Control Development

The palette on which cognitive and neural development take place becomes more colourful and complex when we consider how emotional or motivational processes modulate or impact cognitive control processes, especially during adolescence. Socioemotional information is often present in contexts where individuals must exercise cognitive control (e.g., in the presence of rewards, peers, or threat). Researchers have begun to provide a clearer picture of how recruitment of cognitive control changes across development by employing variations of simple paradigms that incorporate these social and emotional elements.

Impact of Incentives

The influence of incentives on self-control undergoes pronounced changes from childhood to adulthood. Early work on how reward is functionally represented in the developing brain revealed that key components of reward circuitry—the VS and the orbitofrontal cortex—were sensitive to the magnitude of rewards (large vs. small) across age groups (Galván et al. 2006). However, adolescents show an exaggerated accumbens response to large rewards, relative to both children and adults, whereas the extent of orbitofrontal cortex activity exhibited more of a monotonic or linear pattern of change across development. Subsequent studies have shown differential accumbens activity in adolescents relative to adults (Bjork, Knutson, & Hommer, 2008; Bjork, Smith, Chen, & Hommer, 2010; Cohen et al., 2010; Geier et al., 2010; van den Bos et al., 2009). These findings and others have raised questions as to how incentives may influence cognitive control capacity during adolescence. For instance, increased sensitivity to incentives could both diminish and enhance cognitive control in at least two ways. Motivational cues such as pictures of food, a smiling face, or an accepting peer may capture attention and/or lead to impulsive actions towards that cue. In contrast, given the potential heightened sensitivity of rewards to adolescents, they may work harder or engage cognitive control more when receipt of reward is dependent on their actions (Geier et al., 2010; Hardin et al., 2009; Teslovich et al., 2014). We highlight work examining each of these potential outcomes below.

A number of behavioural studies show marked performance differences across development when reward cues or incentives are used. Across these tasks, adolescents, as compared to adults, show increased risk taking when making decisions following immediate or positive reward outcomes as opposed to delayed ones (Cauffman et al., 2010; Figner, Mackinlay, Wilkening, & Weber, 2009). This increase in risky behaviour appears to peak in mid-adolescence to late adolescence. Delay discounting of reward values also varies with age, with younger adolescents showing a preference for smaller rewards sooner over larger rewards later, indicating diminished future orientation (Steinberg et al., 2009). These studies suggest that sensitivity to rewards peaks in mid-adolescence and that the salience of rewards may lead to breakdowns in performance on cognitive control tasks.

An imaging study of a monetary gambling task revealed that increasing the expected value of gambles more strongly influences gambling choices in adolescents than in adults (Barkley-Levenson & Galván, 2014). This behavioural effect was paralleled by increased VS activity, which persisted even when groups were matched for performance, indicating that the neural developmental effect was likely due to a developmental difference in recruitment of reward circuitry.

In contrast to appetitive cues diminishing cognitive control, heightened sensitivity to rewards during adolescence has been shown to improve performance on cognitive control tasks, but only when the rewards are contingent upon how well individuals perform. Variations of impulse control tasks like the antisaccade task, in which a participant must look in the opposite direction of a cue, have been used to test this. Adolescents showed enhanced performance relative to adults when instructed that the amount of money they could win was contingent on their performance on the task (Geier et al., 2010; Hardin et al., 2009). This improved performance was paralleled by increased ventral striatal activity in adolescents relative to adults when preparing to make a response in anticipation of a reward (Geier et al.,

2010). Adolescents also showed increased activity in frontal eye field regions when preparing to make an eye movement, suggesting that rewards modulated or upregulated this control circuitry more than in adults.

Reward-based perceptual discrimination tasks show similar results. When small and large rewards are paired with correct detection of the direction of moving dots, adolescents appear to be less impulsive with slower reaction times than adults for trials associated with the largest rewards, as if letting evidence accumulate until they are confident in their choice before making a decision to maximise their gains. These behavioural findings are paralleled by ventral striatal activity on large reward trials across ages but enhanced recruitment of frontoparietal control regions by adolescents relative to adults (Teslovich et al., 2014). Together these findings suggest that adolescents' sensitivity to rewards is associated with changes in activation of striatal reward circuitry that influences the recruitment of cognitive control circuitry differentially from adults.

Influence of Social Cues and Peers

During adolescence there is a heightened orientation and sensitivity to peer influences relative to either childhood or adulthood (Guyer, Choate, Pine, & Nelson, 2012; Somerville, Jones, Ruberry, & Dyke, 2013; Steinberg & Monahan, 2007). Social cues such as a simple smiling face can lead to impulsive behaviour as measured on a go/no-go task in teens (Somerville, Hare, & Casey, 2011). Specifically, when participants were told not to respond to rare smiling faces, only to neutral ones, adolescents were less able to withhold responses to the positive social cues as compared to both children and adults (Somerville et al., 2011). This impulsive, adolescent-specific behaviour was paralleled by increased ventral striatal activity in adolescents relative to children and adults (Figure 26.2). A functional connectivity analysis showed increased corticostriatal coupling in adults, suggesting that immaturity of connections between control frontostriatal circuitry and reward circuitry may underlie diminished cognitive control in response to positive social cues during adolescence.

Like positive social cues, the mere presence of a peer can have a striking effect on cognitive control capacity. Performance in decision-making tasks, such as whether to drive through a yellow light at an intersection or not, show adolescent-specific deterioration in performance when a peer is present versus when performing the task alone (Chein, Albert, O'Brien, Uckert, & Steinberg, 2011; Gardner & Steinberg, 2005; Weigard, Chein, Albert, Smith, & Steinberg, 2014). This pattern of performance is paralleled by increased activation of brain areas implicated in reward circuitry, including the VS by adolescents, when choosing to go through the yellow light.

The influence of peers on adolescent behaviour is not specific to humans. Heightened sensitivity to peers is also observed in rodents. For example, adolescent mice consume more alcohol in the presence of another juvenile mouse than when alone, whereas adult mice show no difference in alcohol consumption with or without another mouse present (Logue, Chein, Gould, Holliday, & Steinberg, 2014). These findings suggest that the increase in risky behaviours when in the presence of peers during adolescence may be an evolutionarily conserved process across species.

Understanding how adolescents interpret and learn from social signals from their peers can provide insight into the observed shift in social sensitivity during this period. Jones and colleagues (2014) recently examined the effects of positive social feedback on brain and behaviour across development. Modelling trial-by-trial learning, children and adults showed higher positive learning rates than adolescents when receiving variable reinforcement from peers, suggesting that adolescents show less differentiation in their actions for peers who provided more or less positive feedback. Forming expectations about receiving positive social

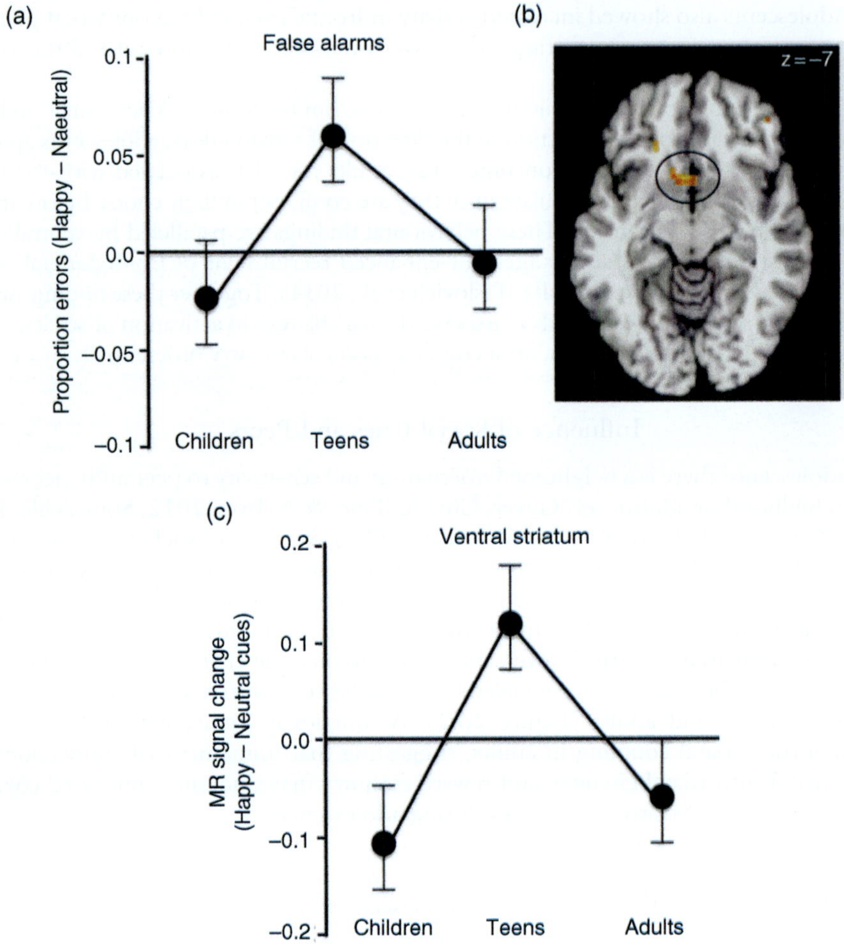

Figure 26.2 Sensitivity to positive social cues. Adolescents make more errors to rare happy face cues (a) relative to both children and young adults. This behaviour is paralleled by increased ventral striatum (b) activity (c). Source: Casey, 2015. Reproduced with permission of MIT Press; Adapted from Somerville 2011.

reinforcement correlated with neural activity within the medial prefrontal cortex and VS across age. Adolescents, unlike children and adults, showed greater insular activity during positive prediction error learning and increased activity in the supplementary motor cortex and the putamen when receiving positive social feedback regardless of the expected outcome. These results suggest that peer approval may motivate adolescents towards action. Whereas different amounts of positive social reinforcement enhanced learning in children and adults, all positive social reinforcement equally motivated adolescents. Together, these findings indicate that sensitivity to peer approval during adolescence goes beyond simple reinforcement theory accounts and suggests possible explanations for how peers may motivate action in adolescents.

Influence of Potential Threat

Not all experiences during adolescence are positive. As adolescents begin to negotiate the world on their own without the protection of a caregiver, potential threats may arise. Potential cues of threat or danger are believed to increase arousal and attention to prepare the organism for action (fight or flight). Life experiences with fearful facial expressions lead to the interpretation of these cues as signals of potential imminent danger in the environment (Ohman & Mineka, 2001). Such stimuli can lead to behavioural inhibition as measured by slower responses times on cognitive tasks (Hare, Tottenham, Davidson, Glover, & Casey, 2005; Macleod, 1991; Simpson et al., 2000) and changes in psychophysiological responses, such as increased skin conductance and fluctuations in heart rate (Bradley, Codispoti, Cuthbert, & Lang, 2001; Ohman & Soares, 1998), consistent with the physiology of heightened arousal and increases in amygdala activity (Hare et al., 2005; Hariri, Tessitore, Mattay, Fera, & Weinberger, 2002; Scherf, Smyth, & Delgado, 2013; Whalen et al., 2001; see also Chapter 30 by McTeague & Etkin in this volume).

Recent developmental studies suggest that the characteristic behavioural and neural responses to negative emotional stimuli shown in adults are altered during adolescence, a time when these socially relevant stimuli are particularly salient (Cohen-Gilbert & Thomas, 2013; Grose-Fifer et al., 2013; Monk et al., 2003). Adolescents, unlike children and adults, show more impulsive responses as measured by a go/no-go task when non-targets (letters) are superimposed on task-irrelevant negative scenes (Cohen-Gilbert & Thomas, 2013). They also show greater attentional capture by fearful faces when used as task-irrelevant flankers (i.e., greater interference effects, Grose-Fifer et al., 2013). These behavioural studies indicate that adolescents have difficulty suppressing attention and action towards cues of potential threat, even when the emotional cues are irrelevant to the task.

Imaging studies parallel these findings showing enhanced amygdala activity to fear faces in adolescents relative to adults (Baird et al., 1999; Guyer et al., 2008; Monk et al., 2003) and children (Hare et al., 2008, see Figure 26.3). Activity in this region is positively correlated with reaction time, such that the greater the amygdala activity the slower reaction times in detection of these cues. Recently, we examined the effect of threat cues on cognitive control. Adolescents were found to impulsively react to (i.e., approach rather than withdraw from) potential threat cues in a go/no-go task more than children or adults (Dreyfuss et al., 2014). This adolescent-specific peak in impulsivity to threat cues was paralleled by increased activity in medial prefrontal regions implicated in emotion regulation when successfully suppressing the impulse to respond to rare fearful face cues, whereas greater lateral prefrontal activity, implicated in cognitive control, was associated with overall behavioural performance across age. These results together with evidence of increased amygdala reactivity to emotional cues in adolescents (Guyer et al., 2008; Hare et al., 2008) suggest that greater exertion of top-down regulation of subcortical structures may be necessary for effective recruitment of cognitive control circuitry in the presence of cues of potential danger. Taken together, these studies of cognitive control suggest that adolescent reactivity to emotional cues and less mature cortico-subcortical circuitry may lead to less efficient modulation by lateral prefrontal control circuitry, resulting in diminished cognitive control capacity in emotional situations.

Towards a Circuit-Based Neurobiological Model of Adolescence

At least three neurobiological models of adolescence have emerged over the past decade to explain the pronounced changes in cognitive control during this developmental period (Blakemore & Robbins, 2012; Casey et al., 2008; Ernst, Romeo, & Andersen, 2006;

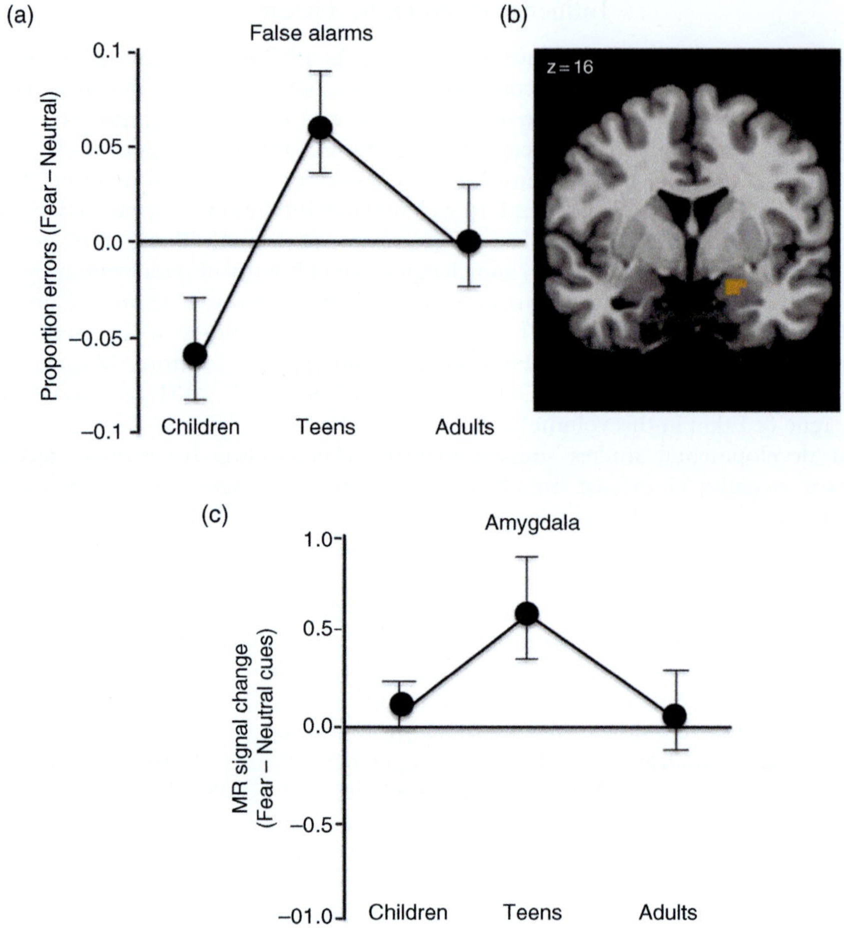

Figure 26.3 Sensitivity to cues of potential threat. Adolescents show greater false alarms (a) and amygdala (b) reactivity (c), relative to both children and adults, in response to fearful faces relative to neutral ones. Source: Adapted from Dreyfuss 2014. Reproduced with permission of S. Karger AG, Basel; Hare, 2008. Reproduced with permission of Elsevier.

Steinberg, 2005; Figure 26.4 adapted from Casey, 2015). Of these, the dual system model has received the most attention, serving as a simple heuristic for linking brain and behaviour. According to this model, there are orthogonal limbic 'hot' and prefrontal 'cold' systems that are in competition with one another (McClure, Laibson, Loewenstein, & Cohen, 2004; Metcalfe & Mischel, 1999). One analogy for characterising these two systems builds on the 1960s television series *Star Trek* and its two main characters: James T. Kirk, the passionate captain of the *Starship Enterprise* from planet Earth, who sometimes runs with his emotions; and his first officer, Mr. Spock, a logical and rational native of the planet Vulcan. According to this analogy, the cool prefrontal system is rational like Mr. Spock, whereas the hot limbic system is like the passionate Captain Kirk, who has a tendency to ignore Starfleet regulations. This model provides a simple heuristic for how brain systems may compete, focusing largely on reward-based processes. Accordingly, the value-based dominance of the

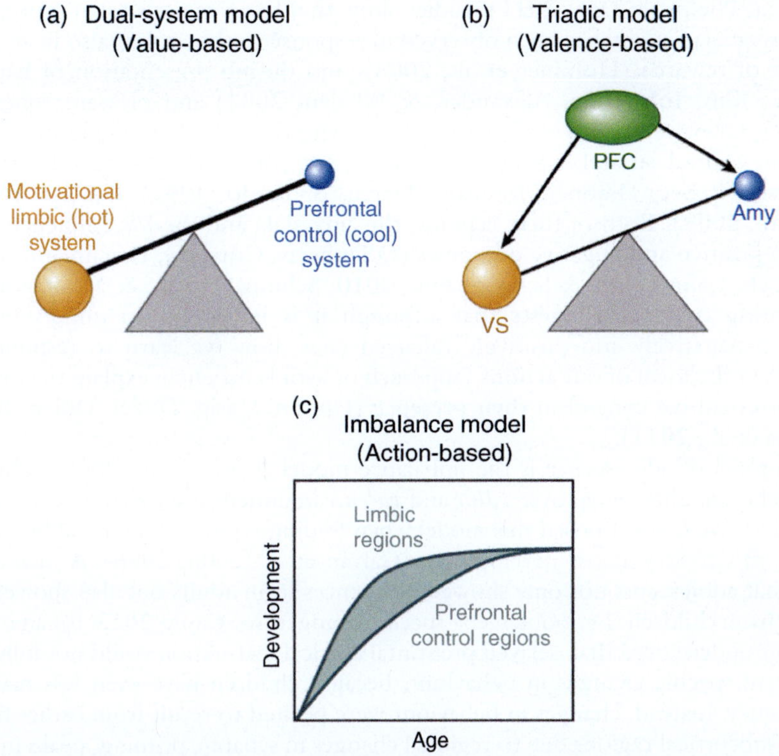

Figure 26.4 Neurobiological Models of Adolescence. Abbreviations: Amy, amygdala; PFC, prefrontal cortex; VS, ventral striatum. Source: Adapted from Casey, 2015. Reproduced with permission of Annual Review of Psychology.

limbic system during adolescence is thought to result in diminished cognitive control and increased risky decision making. With development of the prefrontal cortex, the brain becomes more 'vulcanised', which helps with the delay of gratification and more logical decisions (Cohen, 2005; McClure et al., 2004). Steinberg suggests that during adolescence the cold system of the prefrontal cortex is less mature than the hot system, and so the latter wins out when making decisions in motivated situations (among peers or rewards; Steinberg et al., 2008).

A second model, proposed by Ernst and colleagues (2006; 2009), is the 'triadic' model. This model further subdivides the hot system from a reward system into both positively and negatively valenced subsystems, and it localises reward and threat to the VS and amygdala, respectively. The prefrontal cortex serves as the modulator of these systems. Motivated behaviour is proposed to stem from changes in gonadal hormonal and social reorientation that weight the reward system more heavily than threat during adolescence (Ernst et al., 2006; Ernst et al., 2009). This heavier weighting towards rewards over threats is thought to lead to thrill seeking and suboptimal choices during adolescence (Helfinstein & Casey, 2014; Spielberger, Olino, Forbes, & Dahl, 2014). On the surface, thinking of adolescence as one big roller coaster ride of thrills seems tenable (Spielberger et al., 2014). However, converging animal (Paton, Belova, Morrison, & Salzman, 2006), imaging (Delgado, Schotter, Ozbay, & Phelps, 2008; Levita et al., 2009), and computational (Li, Schiller,

Schoenbaum, Phelps, & Daw, 2011) studies show that brain regions are not organised by valence. Amygdala activity has been observed in response to threat, but also in anticipation and receipt of rewards (Hommer et al., 2003), and during presentation of happy faces (Somerville, Kim, Johnstone, Alexander, & Whalen, 2004) and pleasant tones (Levita et al., 2009). Likewise, the VS has been associated with reward processing (anticipation and receipt), but ventral striatal activation has also been observed to social (Britton et al., 2006), novel (Rebec, Grabner, Johnson, Pierce, & Bardo, 1997) and aversive stimuli (Levita et al., 2009). Both of these regions, the amygdala and the VS, influence action in response to positive and negative outcomes (O'Doherty, Critchley, Deichmann, & Dolan, 2003; Roesch, Calu, Esber, & Schoenbaum, 2010; Schultz, Dayan, & Montague, 1997). This mounting evidence suggests that although it is important to understand neural responses to negatively and positively valenced cues, how we learn to respond to this information in the form of our actions (approach or avoid) may help explain the diminished capacity for cognitive control in their presence (Guitart-Masip, Duzel, Dolan, & Dayan, 2014; Huys et al., 2011).

A third model of adolescence is the imbalance model (Casey et al., 2008). This circuit-based model highlights *action* over *value* and *valence* in understanding adolescent behaviour (Casey, 2015). We first proposed this model when finding regional functional brain changes within reward circuitry across development (Galván et al., 2005, 2006). A nuance of this work was that adolescents not only showed differences from adults but also showed distinct differences from children (i.e., adolescent-specific changes; see Casey, 2013, for an overview). The findings underscored that delayed prefrontal cortical maturation could not fully account for adolescent-specific changes in behaviour, because children have even less mature prefrontal circuitry. Instead, changes in behaviour were posited to result from earlier functional changes in subcortical regions due to regional changes in synaptic pruning, peaks in available neurochemicals, and increased local connectivity relative to prefrontal circuitry implicated in self-control.

A subtle difference between the dual system and the imbalance model may be best conceptualised by calling upon the *Star Trek* analogy again. Captain Kirk and Mr. Spock had a dynamic relationship on the Starship Enterprise. Mr. Spock interjected his logical reasoning on the bridge when the captain was being irrational or illogical in his behaviour or decisions. There was no need to impose his logic when rational decisions were being made. Similarly, in development, there would be no need to strengthen connections between the prefrontal and limbic regions if there were very little limbic activity. In fact, children show little subcortical limbic activity, specifically in the amygdala, in the presence of a parent, even in emotionally arousing situations; the parent serves to buffer and regulate the child's emotions (Gee et al. 2014, Moriceau & Sullivan, 2006). During adolescence, this effect is not evidenced (Gee et al., 2014). Instead, there appears to be an enhanced signal in the amygdala during this time (e.g., Gee et al., 2014; Hare et al 2008; Monk et al., 2003) that may help the adolescent learn to suppress their own emotions and actions by signalling the prefrontal cortex for help. Signalling of the prefrontal cortex could facilitate the hard wiring back from this region to inhibitory GABA-ergic cells within limbic structures, to enhance the capacity for regulation of emotions and actions within social bounds, and without the parent, in preparing the adolescents for adult roles. This view of the adolescent brain underscores the need for circuit-level change within local and ascending connections of subcortical regions before distal descending prefrontal connections (see Figure 26.1). More recently, work to understand how cortico-cortical connections help modulate this circuitry in cognitive control tasks has begun to emerge. These studies suggest that between-circuit modulation may be dependent on within-circuit maturity (Silvers et al., 2016; Teslovich et al., 2014; van den Bos et al., 2009).

All three neurobiological models underscore the importance of understanding the impact of social and emotional processes on cognitive control during adolescence. Each model is in part consistent with human and non-human animal studies supporting the self-control circuitry depicted in Figure 26.1. Early work in rodents began to map out the connectivity as well as unique and complementary functions of each of these brain regions (Groenewegen & Berendse, 1990; Krettek & Price, 1977; Wright & Groenewegen, 1995). The prefrontal cortex is implicated broadly in behavioural regulation and can be roughly grouped into medial and lateral subdivisions. The medial PFC has reciprocal connections with both the amygdala and the VS, and the lateral PFC projects directly to the amygdala and indirectly to the VS, with returning projections primarily going through the medial PFC (Miller & Cohen, 2001; Tanji & Hoshi, 2008). The striatum, implicated in computations related to prediction of outcomes (Haber & Knutson, 2010), and the amygdala, a region involved in associative learning (Janak & Tye, 2015), are directly connected via ascending projections from the basolateral amygdala to the nucleus accumbens (Stuber et al., 2011). Considering these regions as interacting networks rather than unitary regions with orthogonal functions may drive developmental cognitive neuroscience research more towards a mechanistic understanding of why adolescents engage in the behaviours they do (Casey et al., 2015).

Conclusions

Although cognitive control ability emerges during the adolescence years, this capacity is compromised in emotionally charged situations. Converging evidence from human imaging, post-mortem, and animal studies indicates that this is a time where brain structure and connectivity are dynamically changing. Brain circuitry implicated in emotional and motivational behaviour shows earlier neurochemical, neurotrophin, and connectivity changes relative to more distributed cortical networks implicated in recruitment of cognitive control, suggesting that imbalances in these circuitries exist during adolescence. We have shown how age-related differences in functional connectivity in a simplified model of cognitive control and motivational circuitry may underlie adolescent cognitive, emotional, and behavioural changes. These changes parallel heightened sensitivities to emotional and motivational cues that can lead to diminished cognitive control (Casey, Duhoux, & Malter Cohen, 2010), but why would the brain be programmed this way, especially given that this diminished self-control corresponds with a 200% increase in mortality rates due largely to preventable causes (accidental death, suicide, and homicide) rather than disease?

Speculation would suggest that this developmental pattern is an evolutionary feature. Individuals need to engage in high-risk behaviour to leave their families and villages to find a mate, and this occurs at the time when gonadal hormones are elevated and sexual behaviours increase. In today's society, however, adolescence seems to extend indefinitely, with children living with parents or having financial dependence on them and choosing mates later in life. Further, increased peer interactions and fighting with parents are not specific to humans but observed across species (Spear, 2013). Thus, from an evolutionary perspective, adolescence is the period in which the very skills one needs to increase the chances of survival upon separation from a caregiver (Casey et al., 2008) may depend on a heightened awareness of social incentives and threats. This heightened sensitivity to rewards and threats can lead to diminished cognitive control but also provides a tension in brain circuitry that must be resolved, and thus may serve as a precursor to brain maturation. A biologically driven imbalance in brain circuits that process emotional inputs that drive behaviour and cognitive control circuitry that regulates behaviour may, in fact, enhance wiring of the brain. In other words, enhanced signalling of the prefrontal control regions from the limbic regions could lead to strengthening of connections from the prefrontal regions back to those limbic structures to enhance the capacity for regulation of emotions and actions within social bounds, as a function of experience and social learning.

The brain is sculpted by evolutionary biological constraints and experiences as it adapts to the unique intellectual, physical, sexual, and social challenges of adolescence. Accumulation of a variety of social and emotional experiences with varying outcomes may be important in maturing into a healthy, socially competent adult. Describing adolescents as how we see them '... *in the simplest terms*' does not do justice to the complexities and challenges of this period. It is important to remember the unique challenges that this period brings and that, as the Breakfast Club highlights, 'each one of [them] is a brain' in progress, ultimately taking on different identities that will be realised in adulthood.

References

Adleman, N. E., Menon, V., Blasey, C. M., White, C. D., Warsofsky, I. S., Glover, G. H., & Reiss, A. L. (2002). A developmental fMRI study of the Stroop color-word task. *NeuroImage*, *16*(1), 61–75.

Allport, A. (1987). Selection for action: Some behavioral and neurophysiological considerations of attention and action. In H. Heuer & H. F. Sanders (Eds.), *Perspectives on perception and action* (pp. 395–419). Hillsdale, NJ: Lawrence Erlbaum.

Baddeley, A. (1986). *Working memory*. Oxford, England: Clarendon Press.

Baird, A. A., Gruber, S. A., Fein, D. A., Maas, L. C., Steingard, R. J., Renshaw, P. F.,...Yurgelun-Todd, D. A. (1999). Functional magnetic resonance imaging of facial affect recognition in children and adolescents. *Journal of American Academy of Child Adolescent Psychiatry*, *38*(2), 195–199.

Barkley-Levenson, E., & Galván, A. (2014). Neural representation of expected value in the adolescent brain. *Proceedings of the National Academy of Sciences of the United States of America*, *111*(4), 1646–51.

Bell, C. C., & McBride, D. F. (2010). Affect regulation and prevention of risky behaviors. *JAMA: The Journal of the American Medical Association*, *304*(5), 565–566.

Bjork, J. M., Knutson, B., & Hommer, D. W. (2008). Incentive-elicited striatal activation in adolescent children of alcoholics. *Addiction*, *103*(8), 1308–1319.

Bjork, J. M., Smith, A. R., Chen, G., & Hommer, D. W. (2010). Adolescents, adults and rewards: Comparing motivational neurocircuitry recruitment using fMRI. *PLoS ONE*, *5*(7): e11440.

Blakemore, S.-J., & Robbins, T. W. (2012). Decision-making in the adolescent brain. *Nature Neuroscience*, *15*(9), 1184–1191.

Bradley, M. M., Codispoti, M., Cuthbert, B. N., & Lang, P. J. (2001). Emotion and motivation I: Defensive and appetitive reactions in picture processing. *Emotion*, *1*(3), 276–298.

Brenhouse, H. C., Sonntag, K. C., & Andersen, S. L. (2008). Transient D1 dopamine receptor expression on prefrontal cortex projection neurons: Relationship to enhanced motivational salience of drug cues in adolescence. *The Journal of Neuroscience: The Official Journal of the Society for Neuroscience*, *28*(10), 2375–82.

Britton, J. C., Phan, K. L., Taylor, S. F., Welsh, R. C., Berridge, K. C., & Liberzon, I. (2006). Neural correlates of social and nonsocial emotions: An fMRI study. *NeuroImage*, *31*(1), 397–409.

Brown, T. T., Lugar, H. M., Coalson, R. S., Miezin, F. M., Petersen, S. E., & Schlaggar, B. L. (2005). Developmental changes in human cerebral functional organization for word generation. *Cerebral Cortex*, *15*(3), 275–290.

Brown, T. T., Petersen, S. E., & Schlaggar, B. L. (2006). Commentary: Does human functional brain organization shift from diffuse to focal with development? *Developmental Science*, *9*(1), 9–11.

Bunge, S. A., Dudukovic, N. M., Thomason, M. E., Vaidya, C. J., & Gabrieli, J. D. (2002). Immature frontal lobe contributions to cognitive control in children: Evidence from fMRI. *Neuron*, *33*(2), 301–311.

Casey, B., Getz, S., & Galvan, A. (2008). The adolescent brain. *Developmental Reviews, 28*(1), 62–77.

Casey, B. J. (2015). Beyond simple models of self-control to circuit-based accounts of adolescent behavior. *Annual Review of Psychology, 66*, 295–319.

Casey, B. J., Cohen, J. D., Jezzard, P., Turner, R., Noll, D. C., Trainor, R. J.,...Rapoport, J. L. (1995). Activation of prefrontal cortex in children during a nonspatial working memory task with functional MRI. *NeuroImage, 2*(3), 221–229.

Casey, B. J., Duhoux, S., & Malter Cohen, M. (2010). Adolescence: What do transmission, transition, and translation have to do with it? *Neuron, 67*(5), 749–760.

Casey, B. J., Jones, R. M., & Somerville, L. H. (2011). Braking and accelerating of the adolescent brain. *Journal of Research on Adolescence, 21*(1), 21–33.

Casey, B. J., Somerville, L. H., Gotlib, I. H., Ayduk, O., Franklin, N. T., Askren, M. K.,...Shoda, Y. (2011). Behavioral and neural correlates of delay of gratification 40 years later. *Proceedings of the National Academy of Sciences of the United States of America, 108*(36), 14998–15003.

Casey, B. J. (2013). The teenage brain: An overview. *Current Directions in Psychological Science, 22*(2), 80–81.

Casey, B. J., Galvan, A., & Hare, T. A. (2005). Changes in cerebral functional organization during cognitive development. *Current Opinion in Neurobiology, 15*(2), 239–244.

Casey, B. J., Glatt, C. E., & Lee, F. S. (2015). Treating the developing versus developed brain: Translating preclinical mouse and human studies. *Neuron, 86*(6), 1358–1368.

Casey, B. J., Thomas, K. M., Davidson, M. C., Kunz, K., & Franzen, P. L. (2002). Dissociating striatal and hippocampal function developmentally with a stimulus-response compatibility task. *The Journal of Neuroscience, 22*(19), 8647–8652.

Casey, B. J., Thomas, K. M., Welsh, T. F., Badgaiyan, R. D., Eccard, C. H., Jennings, J. R., & Crone, E. A. (2000). Dissociation of response conflict, attentional selection, and expectancy with functional magnetic resonance imaging. *Proceedings of the National Academy of Sciences of the United States of America, 97*(15), 8728–8733.

Casey, B. J., Trainor, R. J., Orendi, J. L., Schubert, A. B., Nystrom, L. E., Giedd, J. N.,...Rapoport, J. L. (1997). A developmental functional MRI study of prefrontal activation during performance of a go-no-go task. *Journal of Cognitive Neuroscience, 9*, 835–847.

Cauffman, E., Shulman, E. P., Steinberg, L., Claus, E., Banich, M. T., Graham, S. J., & Woolard, J. (2010). Age differences in affective decision making as indexed by performance on the Iowa Gambling Task. *Developmental Psychology, 46*(1), 193–207.

Chein, J., Albert, D., O'Brien, L., Uckert, K., & Steinberg, L. (2011). Peers increase adolescent risk taking by enhancing activity in the brain's reward circuitry. *Developmental Science, 14*(2), F1–F10.

Chugani, H. T., Phelps, M. E., & Mazziotta, J. C. (1987). Positron emission tomography study of human brain functional development. *Annals of Neurology, 22*(4), 487–497.

Cohen, A. O., Breiner, K., Steinberg, L., Bonnie, R. J., Scott, E. S., Taylor-Thompson, K. A., ... Casey, B. J. (2016). When is an adolescent an adult? Assessing cognitive control in emotional and non-emotional contexts. *Psychological Science, 27*(4), 549–562.

Cohen, J. D. (2005). The vulcanization of the human brain: A neural perspective on interactions between cognition and emotion. *Journal of Economic Perspectives, 19*, 3–24.

Cohen, J. D., & Servan-Schreiber, D. (1992). Context, cortex, and dopamine: A connectionist approach to behavior and biology in schizophrenia. *Psychological Review, 99*(1), 45–77.

Cohen, J. D., Servan-Schreiber, D., & McClelland, J. L. (1992). A parallel distributed processing approach to automaticity. *American Journal of Psychology, 105*(2), 239–269.

Cohen, J. R., Asarnow, R. F., Sabb, F. W., Bilder, R. M., Bookheimer, S. Y., Knowlton, B. J., & Poldrack, R. A. (2010). A unique adolescent response to reward prediction errors. *Nature Neuroscience, 13*(6), 669–671.

Cohen-Gilbert, J. E., & Thomas, K. M. (2013). Inhibitory control during emotional distraction across adolescence and early adulthood. *Child Development, 84*(6), 1954–1966.

Crone, E. A., Wendelken, C., Donohue, S., van Leijenhorst, L., & Bunge, S. A. (2006). Neurocognitive development of the ability to manipulate information in working memory. *Proceedings of the National Academy of Sciences of the United States of America, 103*(24), 9315–9320.

Darki, F., & Klingberg, T. (2014). The role of fronto-parietal and fronto-striatal networks in the development of working memory: A longitudinal study. *Cerebral Cortex (New York, NY: 1991)*, 1–9.

Davidson, M. C., Amso, D., Anderson, L. C., & Diamond, A. (2006). Development of cognitive control and executive functions from 4 to 13 years: Evidence from manipulations of memory, inhibition, and task switching. *Neuropsychologia, 44*(11), 2037–2078.

Delgado, M. R., Schotter, A., Ozbay, E. Y., & Phelps, E. A. (2008). Understanding overbidding: Using the neural circuitry of reward to design economic auctions. *Science, 321*, 1849–1852.

Desimone, R., & Duncan, J. (1995). Neural mechanisms of selective visual attention. *Annual Review of Neuroscience, 18*, 193–222.

Dosenbach, N. U. F., Nardos, B., Cohen, A. L., Fair, D. A., Power, D., Church, J. A.,...Schlaggar, B. L. (2010). Prediction of individual brain maturity using fMRI. *Science, 329*(5997), 1358–1361.

Dreyfuss, M., Caudle, K., Drysdale, A. T., Johnston, N. E., Cohen, A. O., Somerville, L. H.,...Casey, B. J. (2014). Teens impulsively react rather than retreat from threat. *Developmental Neuroscience, 36*, 220–227.

Durston, S., Davidson, M. C., Tottenham, N., Galvan, A., Spicer, J., Fossella, J. A., & Casey, B. J. (2006). A shift from diffuse to focal cortical activity with development. *Developmental Science, 9*(1), 1–8.

Durston, S., Thomas, K. M., Worden, M. S., Yang, Y., & Casey, B. J. (2002). The effect of preceding context on inhibition: An event-related fMRI study. *NeuroImage, 16*(2), 449–453.

Earles, J. L., Connor, L. T., Frieske, D., Park, D. C., Smith, A. D., & Zwahr, M. (1997). Age differences in inhibition: Possible causes and consequences. *Aging, Neuropsychology, and Cognition, 4*(1), 45–57.

Ernst, M., Pine, D. S., & Hardin, M. (2006). Triadic model of the neurobiology of motivated behavior in adolescence. *Psychological Medicine, 36*(3), 299–312.

Ernst, M., Romeo, R. D., & Andersen, S. L. (2009). Neurobiology of the development of motivated behaviors in adolescence: A window into a neural systems model. *Pharmacology, Biochemistry, and Behavior, 93*(3), 199–211.

Fair, D. A., Cohen, A. L., Dosenbach, N. U. F., Church, J. A., Miezin, F. M., Barch, D. M.,...Schlaggar, B. L. (2008). The maturing architecture of the brain's default network. *Proceedings of the National Academy of Sciences of the United States of America, 105*(10), 4028–4032.

Fair, D. A., Cohen, A. L., Power, J. D., Dosenbach, N. U., Church, J. A., Miezin, F. M.,...Petersen, S. E. (2009). Functional brain networks develop from a 'local to distributed' organization. *PLoS Computational Biology, 5*(5), 1–14.

Fair, D. A., Dosenbach, N. U. F., Church, J. A., Cohen, A. L., Brahmbhatt, S., Miezin, F. M.,... Schlaggar, B. L. (2007). Development of distinct control networks through segregation and integration. *Proceedings of the National Academy of Sciences of the United States of America, 104*(33), 13507–13512.

Figner, B., Mackinlay, R. J., Wilkening, F., & Weber, E. U. (2009). Affective and deliberative processes in risky choice: Age differences in risk taking in the Columbia Card Task. *Journal of Experimental Psychology: Learning, Memory, and Cognition, 35*(3), 709–730.

Floresco, S. B., & Tse, M. T. (2007). Dopaminergic regulation of inhibitory and excitatory transmission in the basolateral amygdala-prefrontal cortical pathway. *Journal of Neuroscience, 27*(8), 2045–2057.

Galvan, A., Hare, T. A., Parra, C. E., Penn, J., Voss, H., Glover, G., & Casey, B. J. (2006). Earlier development of the accumbens relative to orbitofrontal cortex might underlie risk-taking behavior in adolescents. *Journal of Neuroscience, 26*(25), 6885–6892.

Galvan, A., Hare, T. A., Davidson, M., Spicer, J., Glover, G., & Casey, B. J. (2005). The role of ventral frontostriatal circuitry in reward-based learning in humans. *Journal of Neuroscience, 25*(38), 8650–8656.

Gardner, M., & Steinberg, L. (2005). Peer influence on risk taking, risk preference, and risky decision making in adolescence and adulthood: An experimental study. *Developmental Psychology, 41*, 625–635.

Gee, D. G., Gabard-Durnam, L., Telzer, E. H., Humphreys, K. L., Goff, B., Shapiro, M.,...Tottenham, N. (2014). Maternal buffering of human amygdala-prefrontal circuitry during childhood but not during adolescence. *Psychological Science, 25*(11), 2067–2078.

Gee, D. G., Humphreys, K. L., Flannery, J., Goff, B., Telzer, E. H., Shapiro, M.,...Tottenham, N. (2013). A developmental shift from positive to negative connectivity in human amygdala—Prefrontal circuitry. *Journal of Neuroscience, 33*(10), 4584–4593.

Geier, C. F., Terwilliger, R., Teslovich, T., Velanova, K., & Luna, B. (2010). Immaturities in reward processing and its influence on inhibitory control in adolescence. *Cerebral Cortex*, *20*(7), 1613–1629.

Giedd, J. N., Blumenthal, J., Jeffries, N. O., Castellanos, F. X., Liu, H., Zijdenbos, A.,...Rapoport, J. (1999). Brain development during childhood and adolescence: A longitudinal MRI study. *Nature Neuroscience*, *2*, 861–863.

Giedd, J. N., Keshavean, M. S., & Paus, T. (2008). Why do many psychiatric disorders emerge during adolescence? *Nature Reviews Neuroscience*, *9*(12), 947–957.

Giorgio, A., Watkins, K. E., Douaud, G., James, A. C., James, S., De Stefano, N.,...Johansen-Berg, H. (2008). Changes in white matter microstructure during adolescence. *NeuroImage*, *39*(1), 52–61.

Gogtay, N., Giedd, J. N., Lusk, L., Hayashi, K. M., Greenstein, D., Vaituzis, A. C.,...Thompson, P. M. (2004). Dynamic mapping of human cortical development during childhood through early adulthood. *Proceedings of the National Academy of Sciences of the United States of America*, *101*(21), 8174–8179.

Grant, J. D., & Dagenbach, D. (2000). Further considerations regarding inhibitory processes, working memory, and cognitive aging. *American Journal of Psychology*, *113*(1), 69–94.

Groenewegen, H. J., & Berendse, H. W. (1990). Connections of the subthalamic nucleus with ventral striatopallidal parts of the basal ganglia in the rat. *The Journal of Comparative Neurology*, *294*(4), 607–622.

Grose-fifer, J., Rodrigues, A., Hoover, S., & Zottoli, T. (2013). Attentional capture by emotional faces in adolescence. *Advances in Cognitive Psychology*, *9*(2), 81–91.

Guitart-Masip, M., Duzel, E., Dolan, R., & Dayan, P. (2014). Action versus valence in decision making. *Trends in Cognitive Sciences*, *18*(4), 194–202.

Guyer, A. E., Choate, V. R., Pine, D. S., & Nelson, E. E. (2012). Neural circuitry underlying affective response to peer feedback in adolescence. *Social Cognitive and Affective Neuroscience*, *7*(1), 81–92.

Guyer, A. E., Monk, C. S., Mcclure-tone, E. B., Nelson, E. E., Adler, A. D., Fromm, S. J.,...Ernst, M. (2008). A developmental examination of amygdala response to facial expressions. *Journal of Cognitive Neuroscience*, *20*(9), 1565–1582.

Haber, S. N., & Knutson, B. (2010). The reward circuit: Linking primate anatomy and human imaging. *Neuropsychopharmacology*, *35*(1), 4–26.

Hardin, M. G., Mandell, D., Mueller, S. C., Dahl, R. E., Pine, D. S., & Ernst, M. (2009). Inhibitory Control in anxious and healthy adolescents is modulated by incentive and incidental affective stimuli. *Journal of Child Psychology & Psychiatry*, *50*(12), 1550–1558.

Hare, T. A., Tottenham, N., Davidson, M. C., Glover, G. H., & Casey, B. J. (2005). Contributions of amygdala and striatal activity in emotion regulation. *Biological Psychiatry*, *57*(6), 624–632. Retrieved from http://www.ncbi.nlm.nih.gov/entrez/query.fcgi?cmd=Retrieve&db=PubMed&dopt=Citation&list_uids=15780849

Hare, T. A., Tottenham, N., Galvan, A., Voss, H. U., Glover, G. H., & Casey, B. J. (2008). Biological substrates of emotional reactivity and regulation in adolescence during an emotional go-nogo task. *Biological Psychiatry*, *63*(10), 927–934.

Hariri, A. R., Tessitore, A., Mattay, V. S., Fera, F., & Weinberger, D. R. (2002). The amygdala response to emotional stimuli: A comparison of faces and scenes. *NeuroImage*, *17*, 317–323.

Hommer, D. W., Knutson, B., Fong, G. W., Bennett, S., Adams, C. M., & Varner, J. L. (2003). Amygdalar recruitment during anticipant of monetary rewards: An event-related fMRI study. *Annals of the New York Academy of Sciences*, *985*, 476–478.

Helfinstein, S. M., & Casey, B. J. (2014). Commentary on Spielberg at al., 'Exciting fear in adolescence: Does pubertal development alter threat processing?' *Developmental Cognitive Neuroscience*, *8*, 96–97.

Huttenlocher, P. R., & Dabholkar, A. S. (1997). Regional differences in synaptogenesis in human cerebral cortex. *Journal of Comparative Neurology*, *387*(2), 167–178.

Huys, Q. J. M., Cools, R., Gölzer, M., Friedel, E., Heinz, A., Dolan, R. J., & Dayan, P. (2011). Disentangling the roles of approach, activation and valence in instrumental and Pavlovian responding. *PLoS Computational Biology*, *7*(4), e1002028.

Janak, P. H., & Tye, K. M. (2015). From circuits to behaviour in the amygdala. *Nature*, *517*(7534), 284–292.

Jones, R. M., Somerville, L. H., Li, J., Ruberry, E. J., Powers, A., Mehta, N.,…Casey, B. J. (2014). Adolescent-specific patterns of behavior and neural activity during social reinforcement learning. *Cognitive, Affective & Behavioral Neuroscience*, *14*(2), 683–697.

Jonides, J., & Nee, D. E. (2006). Brain mechanisms of proactive interference in working memory. *Neuroscience*, *139*(1), 181–193.

Kahneman, D., Treisman, A., & Burkell, J. (1983). The cost of visual filtering. *Journal of Experimental Psychology: Human Perception and Performance*, *9*, 510–522.

Karmiloff-Smith, A. (2009). Nativism versus neuroconstructivism: Rethinking the study of developmental disorders. *Developmental Psychology*, *45*(1), 56–63.

Kramer, A. F., Humphrey, D. G., Larish, J. F., Logan, G. D., & Strayer, D. L. (1994). Aging and inhibition: Beyond a unitary view of inhibitory processing in attention. *Psychology and Aging*, *9*(4), 491–512.

Krettek, J. E., & Price, J. L. (1977). The cortical projections of the mediodorsal nucleus and adjacent thalamic nuclei in the rat. *The Journal of Comparative Neurology*, *171*(2), 157–191.

Kwon, H., Reiss, A. L., & Menon, V. (2002). Neural basis of protracted developmental changes in visuo-spatial working memory. *Proceedings of the National Academy of Sciences of the United States of America*, *99*(20), 13336–13341.

Lee, F. S., Heimer, H., Giedd, J. N., Lein, E. S., Šestan, N., Weinberger, D. R., & Casey, B. J. (2014). Adolescent mental health—Opportunity and obligation. *Science*, *346*(6209), 547–549.

Levita, L., Hare, T. A., Voss, H. U., Glover, G., Ballon, D. J., & Casey, B. J. (2009). The bivalent side of the nucleus accumbens. *NeuroImage*, *44*(3), 1178–1187.

Li, J., Schiller, D., Schoenbaum, G., Phelps, E. A., & Daw, N. D. (2011). Differential roles of human striatum and amygdala in associative learning. *Nature Neuroscience*, *14*(10), 1250–1252.

Liston, C., Watts, R., Tottenham, N., Davidson, M. C., Niogi, S., Ulug, A. M., & Casey, B. J. (2006). Frontostriatal microstructure modulates efficient recruitment of cognitive control. *Cerebral Cortex*, *16*(4), 553–560. Retrieved from http://www.ncbi.nlm.nih.gov/entrez/query.fcgi?cmd=Retrieve&db=PubMed&dopt=Citation&list_uids=16033925

Logue, S., Chein, J., Gould, T., Holliday, E., & Steinberg, L. (2014). Adolescent mice, unlike adults, consume more alcohol in the presence of peers than alone. *Developmental Science*, *17*(1), 79–85.

Luna, B., Padmanabhan, A., & Hearn, K. O. (2010). What has fMRI told us about the development of cognitive control. *Brain and Cognition*, *72*(1), 1–28.

Luna, B., Thulborn, K. R., Munoz, D. P., Merriam, E. P., Garver, K. E., Minshew, N. J.,…Sweeney, J. A. (2001). Maturation of widely distributed brain function subserves cognitive development. *NeuroImage*, *13*(5), 786–793. Retrieved from http://www.ncbi.nlm.nih.gov/entrez/query.fcgi?cmd=Retrieve&db=PubMed&dopt=Citation&list_uids=11304075

Macleod, C. M. (1991). Half a century of research on the Stroop effect: An integrative review. *Psychological Bulletin*, *109*(2), 163–203.

McClure, S. M., Laibson, D. I., Loewenstein, G., & Cohen, J. D. (2004). Separate neural systems value immediate and delayed monetary rewards. *Science*, *306*(5695), 503–507.

Metcalfe, J., & Mischel, W. (1999). A hot/cool-system analysis of delay of gratification: Dynamics of willpower. *Psychological Review*, *106*(1), 3–19.

Miller, E. K., & Cohen, J. D. (2001). An integrative theory of prefrontal cortex function. *Annual Reviews in Neuroscience*, *24*, 167–202.

Mills, K. L., Goddings, A.-L., Clasen, L. S., Giedd, J. N., & Blakemore, S.-J. (2014). The developmental mismatch in structural brain maturation during adolescence. *Developmental Neuroscience*, *36*(3–4), 147–160.

Mischel, W., Ayduk, O., Berman, M. G., Casey, B. J., Gotlib, I. H., Jonides, J.,…Shoda, Y. (2011). 'Willpower' over the life span: Decomposing self-regulation. *Social Cognitive and Affective Neuroscience*, *6*(2), 252–256.

Monk, C. S., McClure, E. B., Nelson, E. E., Zarahn, E., Bilder, R. M., Leibenluff, E.,…Pine, D. S. (2003). Adolescent immaturity in attention-related brain engagement to emotional facial expressions. *NeuroImage*, *20*, 420–428.

Moriceau, S., & Sullivan, R. M. (2006). Maternal presence serves as a switch between learning fear and attraction in infancy. *Nature Neuroscience, 9*(8), 1004–1006.

Nee, D. E., Wager, T. D., & Jonides, J. (2007). Interference resolution: Insights from a meta-analysis of neuroimaging tasks. *Cognitive, Affective, & Behavioral Neuroscience, 7*(1), 1–17.

O'Doherty, J., Critchley, H., Deichmann, R., & Dolan, R. J. (2003). Dissociating valence of outcome from behavioral control in human orbital and ventral prefrontal cortices. *Journal of Neuroscience, 23*(21), 7931–7939.

Ohman, A., & Mineka, S. (2001). Fears, phobias, and preparedness: Toward an evolved module of fear and fear learning. *Psychological Review, 108*(3), 483–522.

Ohman, A., & Soares, J. J. (1998). Emotional conditioning to masked stimuli: Expectancies for aversive outcomes following nonrecognized fear-relevant stimuli. *Journal of Experimental Psychology. General, 127*(1), 69–82.

Paton, J. J., Belova, M. A., Morrison, S. E., & Salzman, C. D. (2006). The primate amygdala represents the positive and negative value of visual stimuli during learning. *Nature, 439*, 865–870.

Paus, T. (2005). Mapping brain maturation and cognitive development during adolescence. *Trends in Cognitive Sciences, 9*(2), 60–69.

Rakic, P., Bourgeois, J. P., & Goldman-Rakic, P. S. (1994). Synaptic development of the cerebral cortex: Implications for learning, memory and mental illness. *Progress in Brain Research, 102*, 227–243.

Rebec, G. V., Grabner, C. P., Johnson, M., Pierce, R. C., & Bardo, M. T. (1997). Transient increases in catecholaminergic activity in medial prefrontal cortex and nucleus accumbens shell during novelty. *Neuroscience, 76*(3), 707–714.

Roesch, M. R., Calu, D. J., Esber, G. R., & Schoenbaum, G. (2010). All that glitters … dissociating attention and outcome expectancy from prediction errors signals. *Journal of Neurophysiology, 104*(2), 587–595.

Rubia, K., Overmeyer, S., Taylor, E., Brammer, M., Williams, S. C., Simmons, A.,…Bullmore, E. T. (2000). Functional frontalisation with age: Mapping neurodevelopmental trajectories with fMRI. *Neuroscience & Biobehavioral Reviews, 24*(1), 13–19.

Rubia, K., Smith, A. B., Woolley, J., Nosarti, C., Heyman, I., Taylor, E., & Brammer, M. J. (2006). Progressive increase of frontostriatal brain activation from childhood to adulthood during event-related tasks of cognitive control. *Human Brain Mapping, 27*, 973–993.

Scherf, K. S., Smyth, J. M., & Delgado, M. R. (2013). The amygdala: An agent of change in adolescent neural networks. *Hormones and Behavior, 64*(2), 298–313.

Schultz, W., Dayan, P., & Montague, P. R. (1997). A neural substrate of prediction and reward. *Science, 275*(5306), 1593–1599.

Shallice, T. (1988). *From neuropsychology to mental structure*. New York, NY: Cambridge University Press.

Shilling, V. M., Chetwynd, A., & Rabbitt, P. M. A. (2002). Individual inconsistency across measures of inhibition: An investigation of the construct validity of inhibition in older adults. *Neuropsychologia, 40*(6), 605–619.

Silvers, J. A., Insel, C., Powers, A., Franz, P., Helion, C., Martin, R. E.,…Ochsner, K. N. (2016). vlpfc-vmpfc-amygdala interactions underlie age-related differences in cognitive regulation of emotion. *Cerebral Cortex*, pii: bhw073.

Simmonds, D. J., Hallquist, M. N., Asato, M., & Luna, B. (2014). Developmental stages and sex differences of white matter and behavioral development through adolescence: A longitudinal diffusion tensor imaging (DTI) study. *NeuroImage, 92*, 356–368.

Simpson, J. R., Ongür, D., Akbudak, E., Conturo, T. E., Ollinger, J. M., Snyder, A. Z.,…Raichle, M. E. (2000). The emotional modulation of cognitive processing: An fMRI study. *Journal of Cognitive Neuroscience, 12*(Suppl. 2), 157–170.

Somerville, L. H., Kim, H., Johnstone, T., Alexander, A. L., & Whalen, P. J. (2004). Human amygdala responses during presentation of happy and neutral faces: Correlations with state anxiety. *Biological Psychiatry, 55*(9), 897–903.

Somerville, L. H., Hare, T., & Casey, B. J. (2011). Frontostriatal maturation predicts cognitive control failure to appetitive cues in adolescents. *Journal of Cognitive Neuroscience, 23*(9), 2123–2134.

Somerville, L. H., Jones, R. M., Ruberry, E. J., & Dyke, J. P. (2013). Medial prefrontal cortex and the emergence of self-conscious emotion in adolescence. *Psychological Science, 24*(8), 1554–1562.

Sowell, E. R., Thompson, P. M., Leonard, C. M., Welcome, S. E., Kan, E., & Toga, A. W. (2004). Longitudinal mapping of cortical thickness and brain growth in normal children. *Journal of Neuroscience, 24*(38), 8223–8231.

Spear, L. P. (2013). Adolescent neurodevelopment. *Journal of Adolescent Health, 52*(2), S7–S13.

Spielberger, J. M., Olino, T. M., Forbes, E. E., & Dahl, R. E. (2014). Exciting fear in adolescence: Does pubertal development alter threat processing? *Developmental Cognitive Neuroscience, 8*, 86–95.

Steinberg, L. (2005). Cognitive and affective development in adolescence. *Trends in Cognitive Sciences, 9*(2), 69–74.

Steinberg, L., Albert, D., Cauffman, E., Banich, M., Graham, S., & Woolard, J. (2008). Age differences in sensation seeking and impulsivity as indexed by behavior and self-report: Evidence for a dual systems model. *Developmental Psychology, 44*(6), 1764–1778.

Steinberg, L., Graham, S. J., O'Brien, L., Woolard, J., Cauffman, E., & Banich, M. (2009). Age differences in future orientation and delay discounting. *Child Development, 80*(1), 28–44.

Steinberg, L., & Monahan, K. C. (2007). Age differences in resistance to peer influence. *Developmental Psychology, 43*(6), 1531–1543.

Steinberg, L. (2012). Should the science of adolescent brain development inform public policy? *Issues in Science and Technology, 28*(3), 67–78.

Stevens, M. C., Skudlarski, P., Pearlson, G. D., & Vince, D. (2009). Age-related cognitive gains are mediated by the effects of white matter development on brain network integration. *NeuroImage, 48*(4), 738–746.

Stuber, G. D., Sparta, D. R., Stamatakis, A. M., van Leeuwen, W. A., Hardjoprajitno, J. E., Cho, S.,... Bonci, A. (2011). Excitatory transmission from the amygdala to nucleus accumbens facilitates reward-seeking. *Nature, 475*, 377–380.

Tanji, J., & Hoshi, E. (2008). Role of the lateral prefrontal cortex in executive behavioral control. *Physiological Reviews, 88*(1), 37–57.

Teslovich, T., Mulder, M., Franklin, N. T., Ruberry, E. J., Millner, A., Somerville, L. H.,...Casey, B. J. (2014). Adolescents let sufficient evidence accumulate before making a decision when large incentives are at stake. *Developmental Science, 17*(1), 59–70.

Thomas, K. M., King, S. W., Franzen, P. L., Welsh, T. F., Berkowitz, A. L., Noll, D. C.,...Casey, B. J. (1999). A developmental functional MRI study of spatial working memory. *NeuroImage, 10*(3 Pt. 1), 327–338.

Tipper, S. P., & Baylis, G. C. (1987). Individual differences in selective attention: The relation of priming and interference to cognitive failure. *Personality and Individual Differences, 8*(5), 667–675.

Tseng, K., & O'Donnell, P. (2007). Dopamine modulation of prefrontal cortical interneurons changes during adolescence. *Cerebral Cortex, 17*(5), 1235–1240.

Van den Bos, W., Cohen, M. X., Kahnt, T., & Crone, E. A. (2012). Striatum-medial prefrontal cortex connectivity predicts developmental changes in reinforcement learning. *Cerebral Cortex, 22*(6), 1247–1255.

Van den Bos, W., Guroglu, B., van den Bulk, B. G., Rombouts, S. A., & Crone, E. A. (2009). Better than expected or as bad as you thought? The neurocognitive development of probabilistic feedback processing. *Frontiers in Human Neuroscience, 3*, 52.

Van Leijenhorst, L., Zanolie, K., Van Meel, C. S., Westenberg, P. M., Rombouts, S. A., & Crone, E. A. (2010). What motivates the adolescent? Brain regions mediating reward sensitivity across adolescence. *Cerebral Cortex, 20*(1), 61–69.

Vincent, J. L., Kahn, I., Snyder, A. Z., Raichle, M. E., Buckner, R. L., Philippi, C. L.,...Raichle, M. E. (2008). Evidence for a frontoparietal control system revealed by intrinsic functional connectivity. *Journal of Neurophysiology, 100*, 3328–3342.

Weigard, A., Chein, J., Albert, D., Smith, A., & Steinberg, L. (2014). Effects of anonymous peer observation on adolescents' preference for immediate rewards. *Developmental Science, 17*(1), 71–78.

Whalen, P. J., Shin, L. M., McInerney, S. C., Fischer, H., Wright, C. I., & Rauch, S. L. (2001). A functional MRI study of human amygdala responses to facial expressions of fear versus anger. *Emotion, 1*(1), 70–83. Retrieved from http://www.ncbi.nlm.nih.gov/pubmed/12894812

Wright, C. I., & Groenewegen, H. J. (1995). Patterns of convergence and segregation in the medial nucleus accumbens of the rat: Relationships of prefrontal cortical, midline thalamic, and basal amygdaloid afferents. *Journal of Comparative Neurology, 361*(3), 383–403.

27

Cognitive Control and the Ageing Brain

Theodore P. Zanto and Adam Gazzaley

Introduction

The expected human lifespan is now longer than ever before, resulting in an expanding population of older (i.e., >65 years of age) adults. As this population continues to grow, it becomes imperative to understand the changes that occur in fundamental cognitive control abilities and its underpinnings in alterations of the ageing brain. Cognitive control refers to high-level executive functions of the brain such as attention, working memory, and task management (e.g., task switching, multitasking). Cognitive control abilities enable most activities of daily living, such as safely and effectively navigating our environment, and the consequence of cognitive control decline can severely alter the quality of life. This chapter will focus exclusively on research conducted on healthy older adults with an emphasis on neuroimaging research incorporating functional magnetic resonance imaging (fMRI) and electroencephalography (EEG). To better understand cognitive control and the aging brain, the chapter is organised by functions: selective attention, inhibitory control, working memory, multitasking, task switching, and finally what we can do to enhance cognitive control abilities in older adults.

It is thought that by 90 years of age, brain weight has decreased by 11–14% of the maximum achieved weight in early adulthood (Dekaban, 1978; Jernigan et al., 2001). Compared to younger adults, older adults exhibit lower grey and white matter volumes, particularly in the prefrontal cortex (PFC) and parietal cortex (Gordon et al., 2008; Raz, Ghisletta, Rodrigue, Kennedy, & Lindenberger, 2010). Given the prominent role of the PFC in cognitive control (Cole et al., 2013; Corbetta & Shulman, 2002), many theories of cognitive ageing consider alterations of the PFC as a primary culprit for age-related declines in cognitive control (Dempster, 1992; Hasher, Lustig, & Zacks, 2007; Hasher & Zacks, 1988; Park & Reuter-Lorenz, 2009; Reuter-Lorenz & Cappell, 2008; Reuter-Lorenz & Lustig, 2005; West, 1996). In response to these anatomical declines, compensatory neural activity may serve to maximise inefficient neural resources, especially in the PFC, in order to uphold behavioural performance. Moreover, once age-related peripheral sensory changes (e.g., visual acuity and hearing loss) are accounted for, perceptual declines in older adults remain (Strouse, Ashmead, Ohde, & Grantham, 1998; Zanto, Toy, & Gazzaley, 2010). Thus, diminished perceptual abilities may also require compensation by PFC-mediated cognitive control to retain performance abilities. Evidence now exists for age-related compensation of both perceptual

The Wiley Handbook of Cognitive Control, First Edition. Edited by Tobias Egner.

decline and inefficient/reduced neural resources, placing greater demands on the cognitive control abilities of older adults. However, compensatory mechanisms are not always engaged, or they may be engaged but fail to sufficiently compensate, in which case age-related deficits become noticeable in multiple cognitive control domains, such as inhibitory control, working memory, task switching, and multitasking. Fortunately, the brain remains plastic throughout the lifespan, and many age-related declines in cognitive control may be reversed through physical exercise and cognitive training.

Selective Attention and Inhibitory Control

Selective attention refers to goal-directed focus on task-relevant information and the ignoring (or inhibition) of irrelevant information. A common approach to studying selective attention and inhibitory control is via the use of target detection paradigms. Target detection often requires participants to detect a target with or without the presence of simultaneous distracting non-targets and involves either open searching or the presentation of cues to orient attention toward relevant, and away from irrelevant, spatial locations, features, or objects. Note that we here employ the term *inhibitory control* to refer to the attentional suppression of the processing of task-irrelevant stimuli, which is distinct from inhibitory control in the domain of action, or *response inhibition* (see Chapter 6 by Verbruggen & Logan in this volume).

When presented with a visual cue that predicts the subsequent location of a target stimulus, younger and older adults exhibit comparable benefits in target detection abilities compared to uncued target detection (Gottlob & Madden, 1998; Hartley, Kieley, & Slabach, 1990; Nissen & Corkin, 1985). Similarly, in the auditory domain, comparable cue-based facilitation of target detection is observed between younger and older adults (Singh, Pichora-Fuller, & Schneider, 2008). Although this suggests that selectively orienting attention to a location is preserved in ageing, older adults utilise different brain regions to accomplish the same levels of performance. Notably, older adults exhibit declines in attentional modulation of sensory cortical activity, but exhibit increased prefrontal cortical (PFC) activity during spatial (Geerligs, Saliasi, Maurits, Renken, & Lorist, 2014; Li, Gratton, Fabiani, & Knight, 2013; Madden & Hoffman, 1997; Madden et al., 2007) and feature (Alperin, Mott, Rentz, Holcomb, & Daffner, 2014; Cashdollar et al., 2013; O'connell et al., 2012; Zhuravleva et al., 2014) target detection. Whereas optimal visual target detection performance in younger adults is associated with increased activity in the visual cortex, older adults who exhibit increased PFC activity show improved target detection performance (Allen & Payne, 2012; Madden et al., 2007; Madden, Turkington, Provenzale, Hawk, & Hoffman, 1997). The relationship between increased PFC activity and improved behaviour suggests that this PFC recruitment serves as a compensatory mechanism to offset sensory and/or perceptual deficits in ageing. The differential networks utilised by older adults support a posterior-to-anterior shift in ageing (PASA) model (Davis, Dennis, Daselaar, Fleck, & Cabeza, 2008), which proposes that age-related sensory cortical (posterior) declines may be compensated by the recruitment of PFC (anterior) regions to retain performance abilities.

Despite the ability of older adults to utilise compensatory mechanisms under certain conditions, there are many scenarios in which age-related performance in cognitive control abilities in the selective attention domain declines. For example, when target stimuli share features or spatial locations as distracting stimuli, older adults are slower and less accurate in detecting targets in both visual (Hommel, Li, & Li, 2004; Madden & Whiting, 2004; McDowd & Shaw, 2000) and auditory (Duquesnoy, 1983; Singh et al., 2008) domains. Age-related performance declines in target detection occur concomitantly with impaired (attenuated and/or slowed) neural signatures of expectation, attentional orienting, and selection in the PFC and occipital regions (Kenemans, Smulders, & Kok, 1995; Lorenzo-Lopez et al., 2011; Lorenzo-Lopez, Amenedo, & Cadaveira, 2008; Quigley, Andersen, Schulze, Grunwald, & Muller, 2010; Quigley & Muller, 2014; Zanto et al., 2011). Thus, age-related declines in target detection are associated with alterations in multiple brain regions that span diverse cognitive processes, including expectation, attentional orientation, and selection. Moreover, older adults do not appear to successfully engage compensatory mechanisms when distractors are similar to targets.

When distractors share similar qualities as a target stimulus, this may be considered to be a more difficult task than target detection with more distinct distractors. As such, task difficulty may account for why compensatory mechanisms in ageing are not consistently engaged to uphold performance abilities comparable to younger adults. Indeed, the cognitive reserve hypothesis (Stern, 2002), the compensation-related utilisation of neural circuit hypothesis (CRUNCH; Reuter-Lorenz & Cappell, 2008; Reuter-Lorenz & Lustig, 2005), and the scaffolding theory of ageing and cognition (STAC; Park & Reuter-Lorenz, 2009) all suggest that compensatory mechanisms may be successfully engaged only until a capacity limit is reached. Therefore, it is proposed that older adults utilise and benefit from compensatory mechanisms during tasks with low demands, but exhibit performance declines as task demands increase. In support of this view, compared to younger adults, older adults exhibit a disproportionate decline in performance with increased task difficulty during target detection (Prakash et al., 2009) and discrimination (Wang, Fu, Greenwood, Luo, & Parasuraman, 2012). In accordance with performance declines during high task demands, older adults exhibit slowed neural processing associated with attentional selection (Wang et al., 2012). Whereas younger adults recruit additional PFC and parietal regions under high task demands, older adults do not exhibit such recruitment (Prakash et al., 2009). Because older adults display similar network activity during low task demands as younger adults during high task demands (Prakash et al., 2009), this provides evidence that older adults utilise all available resources during a low-demand task, which prevents the recruitment of additional compensatory mechanisms during high-demand tasks. Moreover, these age-related declines due to increased task difficulty have been associated with reduced grey matter volume (Muller-Oehring, Schulte, Rohlfing, Pfefferbaum, & Sullivan, 2013) and declines in white matter integrity (Bennett, Motes, Rao, & Rypma, 2012) within the PFC and parietal cortical regions that are associated with attentional control. Together, these results support theories and hypotheses that suggest the use of compensatory mechanisms in ageing is limited by cognitive resources, which may become unavailable with increased task demands.

The observed age-related slowing of both neural processes and target detection performance, as described above, reinforces theories that suggest age-related declines in cognitive control stem from slowed processing speed (Salthouse, 1996; Salthouse & Madden, 2007). Indeed, age-related declines in white matter integrity and grey matter volume are known to contribute to decreased processing speed (Rabbitt, Lunn, et al., 2007; Rabbitt, Mogapi, et al., 2007; Turken et al., 2008) and provide an anatomical basis for widespread cognitive control declines in ageing. The consequence of such generalised slowing of neural processing is the increased computational time required for a particular process, making less information available for higher-level functions in a set amount of time. Therefore, older adults require more time to orient attention to a target (Amenedo, Lorenzo-Lopez, & Pazo-Alvarez, 2012), away from a target (Cona, Bisiacchi, Amodio, & Schiff, 2013), and away from distractors (Cashdollar et al., 2013).

Alternatively, the disproportionate negative impact of distractors in ageing also supports theories attributing age-related declines in cognitive control to deficient inhibitory abilities that arise from anatomical and functional decline of the PFC (Dempster, 1992; Hasher et al.,

2007; Hasher & Zacks, 1988; West, 1996). Deficient target detection performance has been observed in ageing concomitantly with a selective decline in suppressing neural activity to irrelevant stimuli. This age-related suppression deficit has been observed in the PFC, parietal, and sensory regions during both visual (Alperin et al., 2013; Daffner, Alperin, Mott, & Holcomb, 2014; Haring et al., 2013) and auditory target detection tasks (Passow et al., 2014). Similar results have been observed with target discrimination tasks such that older adults exhibit declines in performance as well as increased distractor processing in sensory cortical regions (de Fockert, Ramchurn, van Velzen, Bergstrom, & Bunce, 2009; Wais & Gazzaley, 2014a). A common target discrimination task, the Stroop task, requires participants to report the colour of a word and not the word itself. During this task, older adults exhibit diminished ability in inhibiting the irrelevant word and increased interference-related activity as well as increased interference-related extent of activity in the PFC and parietal regions (Langenecker, Nielson, & Rao, 2004; Nielson et al., 2004; Nielson, Langenecker, & Garavan, 2002), providing additional evidence for an age-related decline in suppressing irrelevant information. These age-related deficits in inhibitory control during target discrimination have been associated with declines in white matter integrity in the anterior corpus callosum (Sullivan, Adalsteinsson, & Pfefferbaum, 2006) as well as within the parietal and occipital regions (Kennedy & Raz, 2009). Importantly, the relationship between age-based declines in inhibitory control and decreased white matter integrity may not be fully attributed to general slowing with age (Wolf et al., 2014). Thus, anatomical declines associated with ageing contribute to inhibitory deficits and slowed processing speed, both of which contribute to deficient cognitive control in ageing.

Overall, research on selective attention and inhibitory control in ageing has shown scenarios in which older and younger adults exhibit comparable task performance, but differential neural activity profiles. Compared to younger adults, older adults often exhibit decreased posterior (sensory-related) activity concomitantly with increased PFC activity, which serves to compensate for deficient sensory and/or perceptual processes in ageing. However, older adults' use of compensatory neural regions is limited by the amount of cognitive resources, which may become unavailable during increased task demands. The consequence of unavailable or failed compensatory mechanisms reveals age-related slowing of cognitive processing as well as a selective decline in inhibitory control, both of which contribute to deficient task performance. Importantly, the inability to utilise compensatory mechanisms as well as age-related declines in processing speed and inhibition are related to lower grey matter volume and decreased white matter integrity within the PFC and parietal cortex.

Working Memory

Change detection tasks are often used to assess working memory capacity and its underlying neural mechanisms. Generally, the goal of a change detection task is to determine whether previously presented stimuli that are held for a brief time mentally have changed in some way, such as a shift in features, spatial location, or orientation. Thus, change detection tasks may be referred to as a delayed match to sample task, location matching, N-back, or a same/different task. Given the amount of overlap between selective attention and working memory (Cowan, 1995; Gazzaley & Nobre, 2012), similar age-related declines in working memory may be observed (see also Chapter 3 by Meier & Kane in this volume). Indeed, in line with the target detection paradigms discussed above, older adults may recruit additional PFC and parietal regions to compensate for visual processing declines and retain performance abilities during change detection (Burianova, Lee, Grady, & Moscovitch, 2013; Lee, Grady, Habak, Wilson, & Moscovitch, 2011; Macpherson et al., 2014; Rypma & D'Esposito, 2000) and verbal working memory (Grossman et al., 2002) tasks, which provide additional support for the PASA model (Davis et al., 2008).

However, older adults may also exhibit increased PFC and parietal activity during change detection tasks concomitantly with performance declines (Ansado et al., 2013; Ansado, Monchi, Ennabil, Faure, & Joanette, 2012; Grady, Yu, & Alain, 2008; Piefke, Onur, & Fink, 2012). This neural recruitment in ageing without a performance benefit is considered failed compensation (Grady, 2012) and may arise when task demands exceed available cognitive resources (Park & Reuter-Lorenz, 2009; Reuter-Lorenz & Cappell, 2008; Reuter-Lorenz & Lustig, 2005; Stern, 2002). Indeed, age-related performance declines in change detection (i.e., working memory) abilities are most pronounced under increased task difficulty (Ansado et al., 2012, 2013; Piefke et al., 2012; Prakash, Heo, Voss, Patterson, & Kramer, 2012; Sander, Werkle-Bergner, & Lindenberger, 2012). Interestingly, high-performing older adults exhibit more 'youth-like' neural activity modulation in the PFC and parietal cortex with increased task difficulty, suggesting that increased cognitive reserve in high-performing older adults may serve to uphold task performance in lieu of compensatory recruitment of regions (Nagel et al., 2009), at least until a resource ceiling is reached (Toepper et al., 2014).

Much research and theory, as described above, indicate that many age-related performance declines in working memory and inhibitory control stem from a lack of cognitive resources to compensate for sensory and/or perceptual deficits. However, age-related recruitment of PFC regions may not always be used to compensate for sensory and/or perceptual declines, but rather, compensate for reduced PFC resources and/or inefficient PFC processing. Evidence for this exists in the well-documented hemispheric asymmetry reduction in older adults (HAROLD) model (Cabeza, 2002), where younger adults exhibit unilateral PFC activity to achieve task goals and older adults recruit contralateral (i.e., bilateral) PFC activity to achieve the same goals. Notably, this reduced asymmetry in ageing has been linked to an age-related compensation for inefficient use of PFC resources such that low-performing older adults exhibit unilateral PFC activity similar to younger adults, whereas high-performing older adults utilise bilateral PFC regions (Cabeza, Anderson, Locantore, & McIntosh, 2002). This reduced asymmetry in ageing has been observed during working memory (Dixit et al., 2000; Piefke et al., 2012; Reuter-Lorenz et al., 2000) and inhibitory control processes (Langenecker et al., 2004; Nielson et al., 2002), and as such, may reflect a general compensatory response for cognitive control declines in ageing.

In order to better understand cognitive control changes in ageing, it is important to distinguish whether age-related declines stem from deficient higher-order processes or an inability to fully compensate for lower-level (i.e., sensory/perceptual) deficits. When age-related sensory and perceptual differences are equated and general age-related slowing has been accounted for, older adults still exhibit deficient working memory performance along with decreases in attentional modulation of visual cortical activity (Zanto et al., 2010). Changes in attentional modulation of the visual cortex have been attributed to declines in functional interactions with the PFC (Zanto, Chadick, Satris, & Gazzaley, 2013; Zanto, Rubens, Thangavel, & Gazzaley, 2011), supporting models that posit age-based declines in working memory stem from deficient communication between the PFC and the sensory cortex (Gazzaley, 2012). This deficit in working memory in ageing extends to a related cognitive control ability, mental imagery, which is marked by an age-related decline in the selectivity of representations in the visual cortex during imagery and a corresponding reduction in selectivity of functional connectivity between the PFC and the visual cortex (Kalkstein, Checksfield, Bollinger, & Gazzaley, 2011).

Older adults also exhibit a disproportionate decline in working memory performance in the presence of distractors during the encoding period, which is manifest through a selective decline in inhibiting sensory cortical activity to irrelevant stimuli (Clapp & Gazzaley, 2012;

Deiber et al., 2010; Gazzaley et al., 2008; Gazzaley, Cooney, Rissman, & D'Esposito, 2005). Importantly, this deficit in suppressing irrelevant sensory information not only predicts working memory decline due to distraction, but is also related to decreased activity in the PFC as well as decreased functional connectivity between the sensory cortex and the PFC (Campbell, Grady, Ng, & Hasher, 2012; Chadick, Zanto, & Gazzaley, 2014). Because age-related declines in sensory and/or perceptual processes would equally affect processing of attended and ignored stimuli, it is unlikely that such changes would underlie a selective age-related decline in suppressing irrelevant information. Thus, deficient top-down communication between the PFC and the sensory cortex appears to be independent of sensory and/or perceptual decline and plays a role in decreased working memory and inhibitory control in ageing.

Suppression deficits in ageing during working memory encoding are observed as alterations in both amplitude and latency of neural activity to irrelevant information (Clapp & Gazzaley, 2012; Deiber et al., 2010; Gazzaley et al., 2008). Therefore, these findings help reconcile models that suggest age-related declines in cognition result from deficient inhibition (Hasher et al., 2007; Hasher & Zacks, 1988) or slowed processing speed (Salthouse, 1996; Salthouse & Madden, 2007). In line with other models of cognitive ageing that emphasise declines in PFC structure, age-related declines in working memory and top-down suppression signals between the PFC and the visual cortex have been associated with decreased grey matter volume and white matter integrity within the PFC (Chadick et al., 2014). Thus, age-related declines in the PFC structure leads to deficient top-down modulation of sensory cortical activity, which in turn results in lowered working memory performance.

It is interesting to note that age-related declines in top-down modulation of sensory cortical activity occur not only during attentional orientation and selection processes (i.e., 100–500 ms post stimulus onset; Deiber et al., 2010; Gazzaley et al., 2008), but also occur in anticipation of stimulus onset to impact both working memory and perceptual performance, suggesting the presence of an expectation deficit associated with ageing (Bollinger, Rubens, Masangkay, & Gazzaley, 2011; Sander et al., 2012; Zanto et al., 2011). Accordingly, predictive information does not help older adults suppress neural activity to expected irrelevant distractions, nor does it help improve working memory performance (Clapp & Gazzaley, 2012; Zanto, Hennigan, Ostberg, Clapp, & Gazzaley, 2010). Similarly, older adults, unlike younger adults, do not benefit from retrospective cues that would otherwise enhance working memory abilities by orienting attention away from irrelevant working memory representations (Rizio & Dennis, 2014). This age-related decline in suppressing working memory representations result from an inability to recruit the PFC and parietal regions that are selectively involved in suppressing irrelevant working memory representations (Rizio & Dennis, 2014). Together, the failure to ignore impending distractions or release irrelevant information from working memory may overload limited memory stores and hinder the ability to remember relevant information (Gazzaley et al., 2005; Schmitz, Cheng, & De Rosa, 2010). It is noteworthy that the negative impact of reduced inhibitory control in ageing extends beyond the brief time periods involved in working memory, and also results in deficient long-term memory retrieval (Wais, Martin, & Gazzaley, 2012), which is associated with impaired communication between the PFC, sensory cortex, and medial temporal lobe (Wais & Gazzaley, 2014a).

Overall, research on working memory in ageing has shown remarkable overlap with findings from inhibitory control research. Although there is less evidence for age-equivalent working memory performance compared to inhibitory control during target detection, when it does occur, older adults often exhibit increased PFC activity. This increased activity serves to compensate for deficient sensory/perceptual processes, reduced PFC resources, and/or inefficient PFC processes in ageing. However, the age-related recruitment of compensatory neural regions is limited by the amount of cognitive resources available, which decline across the lifespan in response to anatomical changes, especially in the PFC. Therefore, older adults have less resources available for compensation during increased task demands. The consequences of unavailable or failed compensatory mechanisms reveal age-related declines in PFC modulation of sensory cortical activity, which would normally serve working memory performance by enhancing representations of relevant stimuli and suppressing irrelevant stimuli. These declines in top-down modulation of sensory processes are also associated with anatomical changes in the PFC, notably, decreased grey matter volume and white matter integrity.

Multitasking and Task Switching

Multitasking refers to an aspect of cognitive control that is characterised by dividing attention between multiple skills, tasks, or cognitive sets, whereas task switching refers to a type of multitasking that involves shifting between multiple skills, tasks, or cognitive sets (see Chapter 2 by Monsell in this volume). In line with research from the single-task paradigms discussed above, older adults may recruit the PFC and parietal regions as a compensatory mechanism to support multitasking performance abilities (Fernandes, Pacurar, Moscovitch, & Grady, 2006; Hartley, Jonides, & Sylvester, 2011; Wild-Wall, Hahn, & Falkenstein, 2011), especially during task switching (DiGirolamo et al., 2001; Goffaux, Phillips, Sinai, & Pushkar, 2008; Smith et al., 2001). Yet, age-related performance declines in multitasking become pronounced with increased task difficulty (McDowd & Craik, 1988). There is evidence that these age-related declines may stem from generalised slowing because older adults respond slower to a second task when the onset between tasks decreases (Allen, Smith, Vires-Collins, & Sperry, 1998; Glass et al., 2000). However, when generalised slowing is accounted for, age-based multitasking performance declines persist (Verhaeghen, Steitz, Sliwinski, & Cerella, 2003), which may reflect deficient task switching processes (Glass et al., 2000; Hartley & Little, 1999) that arise from reduced PFC engagement (Anguera et al., 2013) as well as reduced functional connectivity within the PFC and parietal cortex networks (Madden et al., 2010). Moreover, when older adults are required to disengage a secondary task and re-engage a primary task, they fail to dynamically allocate attention and switch between functional brain networks involving the PFC and the sensory cortex (Clapp, Rubens, Sabharwal, & Gazzaley, 2011). Thus, multitasking declines in ageing support models of cognitive ageing that attribute cognitive deficits to slowed processing speed (Salthouse & Madden, 2007; Salthouse, 1996) and deficient attentional orienting/task switching through diminished communication abilities between the PFC and the sensory cortex (Gazzaley, 2012).

Enhancing Cognitive Control in Ageing

Although many aspects of cognitive control decline with age, there is mounting evidence that the brain remains plastic throughout our lifespan, and cognitive control abilities may be maintained or recovered (see also Chapter 28 by Spencer-Smith & Klingberg in this volume). One of the most well-documented means of retaining or regaining cognitive control in ageing is through physical exercise. Physical exercise, notably cardiorespiratory and resistance training (Kelly et al., 2014), has been associated with increased grey and white matter volume, particularly in the PFC (Colcombe et al., 2006). As such, physical exercise improves cognitive function in ageing, and most prominently, cognitive control functions (Colcombe & Kramer, 2003). This includes improved inhibitory control (Dustman et al., 1984), working memory (Fabre, Chamari, Mucci, Massé-Biron, & Préfaut, 2002; but also see

Smith et al., 2010), and multitasking/task switching ability (Hawkins, Kramer, & Capaldi, 1992), which may arise from increased activity in the PFC and parietal cortex in older adults who exercise (Colcombe et al., 2004). It is interesting to note that physical exercise appears to benefit the brain regions and corresponding cognitive functions that are most affected in ageing.

In addition to physical exercise, cognitive exercise (also referred to as cognitive training or exergaming) may also help counteract age-related declines in various cognitive control abilities (for reviews, see Brehmer, Kalpouzos, Wenger, & Lövdén, 2014; Lampit, Hallock, & Valenzuela, 2014; Lustig, Shah, Seidler, & Reuter-Lorenz, 2009). As discussed throughout this chapter, perceptual declines in ageing may contribute to cognitive performance declines, as limited cognitive resources may be usurped to compensate for perceptual deficits. Fortunately, perceptual training in older adults can improve perceptual abilities such that after training, older adults get better at discriminating two stimuli with similar features (Berry et al., 2010). The consequence of improved perceptual performance in older adults results in a transfer of benefit to improved visual working memory performance due to more efficient sensory processing in the visual cortex (Berry et al., 2010).

Aside from perceptual declines in ageing, another prominent source of age-related declines in cognitive control has been attributed to slow processing speed. Targeting this decline, cognitive speed of processing training in older adults not only speeds cognitive processing, but translates to improved everyday abilities such as driving performance (Ball, Edwards, & Ross, 2007), which has been attributed to speeded multitasking abilities (Edwards, Ruva, O'Brien, Haley, & Lister, 2013). Interestingly, auditory cognitive training that emphasises speed of processing enables faster and less variable neural responses within older adults' brainstem, and these benefits extend to improved working memory performance (Anderson, White-Schwoch, Parbery-Clark, & Kraus, 2013).

Cognitive training appears to be potentially beneficial for whatever cognitive domain is targeted, and in older adults, declines in inhibitory control are often in need of improvement. It was recently shown that distraction training in aged humans and aged rats shows improved inhibitory control concomitantly with sharpened sensory receptive fields (in aged rats) and decreased PFC activity to distractors (in aged humans ; Mishra, de Villers-Sidani, Merzenich, & Gazzaley, 2014). Interestingly, distraction training also resulted in a transfer of benefit to working memory and sustained attention abilities (Mishra et al., 2014). Similarly, age-related declines in multitasking performance have also been targeted by cognitive training in older adults, which demonstrated that improved performance occurs concomitantly with PFC activity profiles that become more similar to younger adults (Anguera et al., 2013; Erickson et al., 2007). Moreover, multitasking training has been shown to yield a transfer of benefit to working memory and attention abilities (Anguera et al., 2013; Basak, Boot, Voss, & Kramer, 2008).

Although more research will be needed to fully understand the how long cognitive training benefits may last after training has ceased, research thus far has reported post-training benefits that last from at least 6 months (Anderson, White-Schwoch, Choi, & Kraus, 2014; Anguera et al., 2013) to 2 years (Ball et al., 2007). Furthermore, it is unclear in which scenarios cognitive training will enhance untrained domains, as a transfer of benefit is not always observed, or is minimal (Gross et al., 2012; Verhaeghen, Marcoen, & Goossens, 1992). Finally, additional research will be required to understand any synergistic effects that may be gained by combining both physical and cognitive exercise. It may be that the combination of both physical and cognitive exercise will result in optimal cognitive control gains such that physical exercise increases the potential for neurogenesis/synaptogenesis, while cognitive exercise guides it to induce more targeted changes (Bamidis et al., 2014; Fissler, Küster, Schlee, & Kolassa, 2013).

Summary

Whereas behavioural performance measures do not always identify age-related declines in cognitive control abilities, neuroimaging data often uncover differential neural activity profiles between younger and older adults when engaged in demanding tasks. Owing to decreased grey and white matter volume in ageing, older adults may recruit additional neural regions, typically in the PFC, to support performance by compensating for declines in perception or inefficient/reduced cognitive resources. Because cognitive resources are limited, compensatory mechanisms are not always available under increased task demands. Thus, cognitively demanding tasks are among the most likely to uncover age-related declines in cognitive control. Fortunately, the brain remains plastic throughout the lifespan, and many age-related declines in cognitive control may be reversed through physical exercise and/or cognitive training.

References

Allen, H. A., & Payne, H. (2012). Similar behaviour, different brain patterns: Age-related changes in neural signatures of ignoring. *Neuroimage, 59*(4), 4113–4125.

Allen, P. A., Smith, A. F., Vires-Collins, H., & Sperry, S. (1998). The psychological refractory period: Evidence for age differences in attentional time-sharing. *Psychology and Aging, 13*(2), 218–229.

Alperin, B. R., Haring, A. E., Zhuravleva, T. Y., Holcomb, P. J., Rentz, D. M., & Daffner, K. R. (2013). The dissociation between early and late selection in older adults. *Journal of Cognitive Neuroscience, 25*(12), 2189–2206.

Alperin, B. R., Mott, K. K., Rentz, D. M., Holcomb, P. J., & Daffner, K. R. (2014). Investigating the age-related 'anterior shift' in the scalp distribution of the P3b component using principal component analysis. *Psychophysiology, 51*(7), 620–633.

Amenedo, E., Lorenzo-Lopez, L., & Pazo-Alvarez, P. (2012). Response processing during visual search in normal aging: The need for more time to prevent cross talk between spatial attention and manual response selection. *Biological Psychology, 91*(2), 201–211.

Anderson, S., White-Schwoch, T., Choi, H. J., & Kraus, N. (2014). Partial maintenance of auditory-based cognitive training benefits in older adults. *Neuropsychologia, 62*, 286–296.

Anderson, S., White-Schwoch, T., Parbery-Clark, A., & Kraus, N. (2013). Reversal of age-related neural timing delays with training. *Proceedings of the National Academy of Sciences of the United States of America, 110*(11), 4357–62.

Anguera, J. A., Boccanfuso, J., Rintoul, J. L., Al-Hashimi, O., Faraji, F., Janowich, J., … Gazzaley, A. (2013). Video game training enhances cognitive control in older adults. *Nature, 501*, 97–101.

Ansado, J., Monchi, O., Ennabil, N., Deslauriers, J., Jubault, T., Faure, S., & Joanette, Y. (2013). Coping with task demand in aging using neural compensation and neural reserve triggers primarily intra-hemispheric-based neurofunctional reorganization. *Neuroscience Research, 75*(4), 295–304.

Ansado, J., Monchi, O., Ennabil, N., Faure, S., & Joanette, Y. (2012). Load-dependent posterior-anterior shift in aging in complex visual selective attention situations. *Brain Research, 1454*, 14–22.

Ball, K. K., Edwards, J. D., & Ross, L. A. (2007). The impact of speed of processing training on cognitive and everyday functions. *Journals of Gerontology Series B-Psychological Sciences and Social Sciences, 62B*, 19–31.

Bamidis, P. D., Vivas, A. B., Styliadis, C., Frantzidis, C., Klados, M., Schlee, W., … Papageorgiou, S. G. (2014). A review of physical and cognitive interventions in aging. *Neuroscience and Biobehavioral Reviews, 44*, 206–220.

Basak, C., Boot, W. R., Voss, M. W., & Kramer, A. F. (2008). Can training in a real-time strategy video game attenuate cognitive decline in older adults? *Psychology and Aging, 23*(4), 765–777.

Bennett, I. J., Motes, M. A., Rao, N. K., & Rypma, B. (2012). White matter tract integrity predicts visual search performance in young and older adults. *Neurobiology of Aging, 33*(2), 433.e21-31.

Berry, A. S., Zanto, T. P., Clapp, W. C., Hardy, J. L., Delahunt, P. B., Mahncke, H. W., & Gazzaley, A. (2010). The influence of perceptual training on working memory in older adults. *PLoS ONE*, *5*(7), e11537.

Bollinger, J., Rubens, M. T., Masangkay, E., & Gazzaley, A. (2011). An expectation-based memory deficit in aging. *Neuropsychologia*, *49*, 1466–1475.

Brehmer, Y., Kalpouzos, G., Wenger, E., & Lövdén, M. (2014). Plasticity of brain and cognition in older adults. *Psychological Research*, 790–802.

Burianova, H., Lee, Y., Grady, C. L., & Moscovitch, M. (2013). Age-related dedifferentiation and compensatory changes in the functional network underlying face processing. *Neurobiology of Aging*, *34*(12), 2759–2767.

Cabeza, R. (2002). Hemispheric asymmetry reduction in older adults: The HAROLD model. *Psychology and Aging*, *17*(1), 85–100.

Cabeza, R., Anderson, N. D., Locantore, J. K., & McIntosh, A. R. (2002). Aging gracefully: Compensatory brain activity in high-performing older adults. *Neuroimage*, *17*(3), 1394–1402.

Campbell, K. L., Grady, C. L., Ng, C., & Hasher, L. (2012). Age differences in the frontoparietal cognitive control network: Implications for distractibility. *Neuropsychologia*, *50*(9), 2212–2223.

Cashdollar, N., Fukuda, K., Bocklage, A., Aurtenetxe, S., Vogel, E. K., & Gazzaley, A. (2013). Prolonged disengagement from attentional capture in normal aging. *Psychology and Aging*, *28*(1), 77–86.

Chadick, J. Z., Zanto, T. P., & Gazzaley, A. (2014). Structural and functional differences in medial prefrontal cortex underlie distractibility and suppression deficits in ageing. *Nature Communications*, *5*, 4223.

Clapp, W. C., Rubens, M. T., Sabharwal, J., & Gazzaley, A. (2011). Deficit in switching between functional brain networks underlies the impact of multitasking on working memory in older adults. *Proceedings of the National Academy of Sciences of the United States of America*, *108*(17), 7212–7217.

Clapp, W., & Gazzaley, A. (2012). Distinct mechanisms for the impact of distraction and interruption on working memory in aging. *Neurobiology of Aging*, *33*(1), 134–148.

Colcombe, S. J., Erickson, K. I., Scalf, P. E., Kim, J. S., Prakash, R., McAuley, E., ... Kramer, A. F. (2006). Aerobic exercise training increases brain volume in aging humans. *Journals of Gerontology Series A-Biological Sciences and Medical Sciences*, *61*(11), 1166–1170.

Colcombe, S. J., Kramer, A. F., Erickson, K. I., Scalf, P., McAuley, E., Cohen, N. J., ... Elavsky, S. (2004). Cardiovascular fitness, cortical plasticity, and aging. *Proceedings of the National Academy of Sciences of the United States of America*, *101*(9), 3316–3321.

Colcombe, S., & Kramer, A. F. (2003). Fitness effects on the cognitive function of older adults: A meta-analytic study. *Psychological Science: A Journal of the American Psychological Society/APS*, *14*(2), 125–130.

Cole, M. W., Reynolds, J. R., Power, J. D., Repovs, G., Anticevic, A., & Braver, T. S. (2013). Multi-task connectivity reveals flexible hubs for adaptive task control. *Nature Neuroscience*, *16*(9), 1348–1355.

Cona, G., Bisiacchi, P. S., Amodio, P., & Schiff, S. (2013). Age-related decline in attentional shifting: Evidence from ERPs. *Neuroscience Letters*, *556*, 129–134.

Corbetta, M., & Shulman, G. L. (2002). Control of goal-directed and stimulus-driven attention in the brain. *Nature Reviews Neuroscience*, *3*(3), 201–215.

Cowan, N. (1995). *Attention and memory: An integrated framework. Oxford psychology series* (Vol. 26). New York: Oxford University Press.

Daffner, K. R., Alperin, B. R., Mott, K. K., & Holcomb, P. J. (2014). Age-related differences in the automatic processing of single letters: Implications for selective attention. *Neuroreport*, *25*(2), 77–82.

Davis, S. W., Dennis, N. A., Daselaar, S. M., Fleck, M. S., & Cabeza, R. (2008). Que PASA? The posterior-anterior shift in aging. *Cerebral Cortex*, *18*(5), 1201–1209.

De Fockert, J. W., Ramchurn, A., van Velzen, J., Bergstrom, Z., & Bunce, D. (2009). Behavioral and ERP evidence of greater distractor processing in old age. *Brain Research*, *1282*, 67–73.

Deiber, M. P., Rodriguez, C., Jaques, D., Missonnier, P., Emch, J., Millet, P., ... Ibanez, V. (2010). Aging effects on selective attention-related electroencephalographic patterns during face encoding. *Neuroscience*, *171*(1), 173–186.

Dekaban, A. S. (1978). Changes in brain weights during the span of human life: Relation of brain weights to body heights and body weights. *Annals of Neurology, 4*(4), 345–356.

Dempster, F. N. (1992). The rise and fall of the inhibitory mechanism: Toward a unified theory of cognitive development and aging. *Developmental Review, 12*, 45–75.

DiGirolamo, G. J., Kramer, A. F., Barad, V., Cepeda, N. J., Weissman, D. H., Milham, M. P., ... McAuley, E. (2001). General and task-specific frontal lobe recruitment in older adults during executive processes: A fMRI investigation of task-switching. *Neuroreport, 12*(9), 2065–2071.

Dixit, N. K., Gerton, B. K., Kohn, P., Meyer-Lindenberg, A., Berman, K. F., Dahle, C. L., ... Raz, N. (2000). Age-related changes in rCBF activation during an N-back working memory paradigm occur prior to age 50. *NeuroImage, 11*(5), S94–162.

Duquesnoy, A. J. (1983). Effect of single interfering noise or speech source upon the binaural sentence intelligibility of aged persons. *Journal of the Acoustical Society of America, 74*(3), 739–743.

Dustman, R. E., Ruhling, R. O., Russell, E. M., Shearer, D. E., Bonekat, H. W., Shigeoka, J. W., ... Bradford, D. C. (1984). Aerobic exercise training and improved neuropsychological function of older individuals. *Neurobiology of Aging, 5*(1), 35–42.

Edwards, J. D., Ruva, C. L., O'Brien, J. L., Haley, C. B., & Lister, J. J. (2013). An examination of mediators of the transfer of cognitive speed of processing training to everyday functional performance. *Psychology and Aging, 28*(2), 314–321.

Erickson, K. I., Colcombe, S. J., Wadhwa, R., Bherer, L., Peterson, M. S., Scalf, P. E., ... Kramer, A. F. (2007). Training-induced functional activation changes in dual-task processing: An fMRI study. *Cerebral Cortex, 17*(1), 192–204.

Fabre, C., Chamari, K., Mucci, P., Massé-Biron, J., & Préfaut, C. (2002). Improvement of cognitive function by mental and/or individualized aerobic training in healthy elderly subjects. *International Journal of Sports Medicine, 23*(6), 415–421.

Fernandes, M. A., Pacurar, A., Moscovitch, M., & Grady, C. (2006). Neural correlates of auditory recognition under full and divided attention in younger and older adults. *Neuropsychologia, 44*(12), 2452–2464.

Fissler, P., Küster, O., Schlee, W., & Kolassa, I. T. (2013). Novelty interventions to enhance broad cognitive abilities and prevent dementia: Synergistic approaches for the facilitation of positive plastic change. *Progress in Brain Research, 207*, 403–434.

Gazzaley, A. (2012). Top-down modulation deficit in the aging brain: An emerging theory of cognitive aging. In D. T. Stuss & R. T. Knight (Eds.), *Principles of frontal lobe function* (2nd ed., pp. 593–608). New York: Oxford University Press.

Gazzaley, A., Clapp, W., Kelley, J., McEvoy, K., Knight, R., & D'Esposito, M. (2008). Age-related top-down suppression deficit in the early stages of cortical visual memory processing. *Proceedings of the National Academy of Sciences of the United States of America, 105*(35), 13122–13126.

Gazzaley, A., Cooney, J. W., Rissman, J., & D'Esposito, M. (2005). Top-down suppression deficit underlies working memory impairment in normal aging. *Nature Neuroscience, 8*(10), 1298–1300.

Gazzaley, A., & Nobre, A. C. (2012). Top-down modulation: Bridging selective attention and working memory. *Trends in Cognitive Sciences, 16*(2), 129–135.

Geerligs, L., Saliasi, E., Maurits, N. M., Renken, R. J., & Lorist, M. M. (2014). Brain mechanisms underlying the effects of aging on different aspects of selective attention. *Neuroimage, 91*, 52–62.

Glass, J. M., Schumacher, E. H., Lauber, E. J., Zurbriggen, E. L., Gmeindl, L., Kieras, D. E., & Meyer, D. E. (2000). Aging and the psychological refractory period: Task-coordination strategies in young and old adults. *Psychology and Aging, 15*(4), 571–595.

Goffaux, P., Phillips, N. A., Sinai, M., & Pushkar, D. (2008). Neurophysiological measures of task-set switching: Effects of working memory and aging. *Journals of Gerontology Series B: Psychological Sciences and Social Sciences, 63*(2), 57–66.

Gordon, B. A., Rykhlevskaia, E. I., Brumback, C. R., Lee, Y., Elavsky, S., Konopack, J. F., ... Fabiani, M. (2008). Neuroanatomical correlates of aging, cardiopulmonary fitness level, and education. *Psychophysiology, 45*(5), 825–838.

Gottlob, L. R., & Madden, D. J. (1998). Time course of allocation of visual attention after equating for sensory differences: An age-related perspective. *Psychology and Aging, 13*(1), 138–149.

Grady, C. (2012). The cognitive neuroscience of ageing. *Nature Reviews Neuroscience*, *13*(7), 491–505.

Grady, C. L., Yu, H., & Alain, C. (2008). Age-related differences in brain activity underlying working memory for spatial and nonspatial auditory information. *Cerebral Cortex*, *18*(1), 189–199.

Gross, A. L., Parisi, J. M., Spira, A. P., Kueider, A. M., Ko, J. Y., Saczynski, J. S., ... Rebok, G. W. (2012). Memory training interventions for older adults: A meta-analysis. *Aging & Mental Health*, *16*(6), 722–734.

Grossman, M., Cooke, A., DeVita, C., Alsop, D., Detre, J., Chen, W., & Gee, J. (2002). Age-related changes in working memory during sentence comprehension: An fMRI study. *Neuroimage*, *15*(2), 302–317.

Haring, A. E., Zhuravleva, T. Y., Alperin, B. R., Rentz, D. M., Holcomb, P. J., & Daffner, K. R. (2013). Age-related differences in enhancement and suppression of neural activity underlying selective attention in matched young and old adults. *Brain Research*, *1499*, 69–79.

Hartley, A. A., Jonides, J., & Sylvester, C. Y. C. (2011). Dual-task processing in younger and older adults: Similarities and differences revealed by fMRI. *Brain and Cognition*, *75*(3), 281–291.

Hartley, A. A., Kieley, J. M., & Slabach, E. H. (1990). Age differences and similarities in the effects of cues and prompts. *Journal of Experimental Psychology-Human Perception and Performance*, *16*(3), 523–537.

Hartley, A. A., & Little, D. M. (1999). Age-related differences and similarities in dual-task interference. *Journal of Experimental Psychology: General*, *128*(4), 416–449.

Hasher, L., Lustig, C., & Zacks, J. M. (2007). Inhibitory mechanisms and the control of attention. In A. Conway, C. Jarrold, M. Kane, A. Miyake, & J. Towse (Eds.), *Variation in working memory* (pp. 227–249). New York: Oxford University Press.

Hasher, L., & Zacks, R. T. (1988). Working memory, comprehension and aging: A review and a new view. In G. H. Bower (Ed.), *The psychology of learning and motivation* (Vol. *22*, pp. 193–225). New York, NY: Academic Press.

Hawkins, H. L., Kramer, A. F., & Capaldi, D. (1992). Aging, exercise, and attention. *Psychology and Aging*, *7*(4), 643–653.

Hommel, B., Li, K. Z. H., & Li, S. C. (2004). Visual search across the life span. *Developmental Psychology*, *40*(4), 545–558.

Jernigan, T. L., Archibald, S. L., Fennema-Notestine, C., Gamst, A. C., Stout, J. C., Bonner, J., & Hesselink, J. R. (2001). Effects of age on tissues and regions of the cerebrum and cerebellum. *Neurobiology of Aging*, *22*(4), 581–594.

Kalkstein, J., Checksfield, K., Bollinger, J., & Gazzaley, A. (2011). Diminished top-down control underlies a visual imagery deficit in normal aging. *Journal of Neuroscience*, *31*(44), 15768–15774.

Kelly, M. E., Loughrey, D., Lawlor, B. A., Robertson, I. H., Walsh, C., & Brennan, S. (2014). The impact of exercise on the cognitive functioning of healthy older adults: A systematic review and meta-analysis. *Ageing Research Reviews*, *16*(1), 12–31.

Kenemans, J. L., Smulders, F. T. Y., & Kok, A. (1995). Selective processing of 2-dimensional visual-stimuli in young and old subjects—Electrophysiological analysis. *Psychophysiology*, *32*(2), 108–120.

Kennedy, K. M., & Raz, N. (2009). Aging white matter and cognition: Differential effects of regional variations in diffusion properties on memory, executive functions, and speed. *Neuropsychologia*, *47*(3), 916–927.

Lampit, A., Hallock, H., & Valenzuela, M. (2014). Computerized cognitive training in cognitively healthy older adults: A systematic review and meta-analysis of effect modifiers. *PLoS Medicine*, *11*(11), e1001756.

Langenecker, S. A., Nielson, K. A., & Rao, S. M. (2004). fMRI of healthy older adults during Stroop interference. *Neuroimage*, *21*(1), 192–200.

Lee, Y., Grady, C. L., Habak, C., Wilson, H. R., & Moscovitch, M. (2011). Face processing changes in normal aging revealed by fMRI adaptation. *Journal of Cognitive Neuroscience*, *23*(11), 3433–3447.

Li, L., Gratton, C., Fabiani, M., & Knight, R. T. (2013). Age-related frontoparietal changes during the control of bottom-up and top-down attention: An ERP study. *Neurobiology of Aging*, *34*(2), 477–488.

Lorenzo-Lopez, L., Amenedo, E., & Cadaveira, F. (2008). Feature processing during visual search in normal aging: Electrophysiological evidence. *Neurobiology of Aging*, *29*(7), 1101–1110.

Lorenzo-Lopez, L., Gutierrez, R., Moratti, S., Maestu, F., Cadaveira, F., & Amenedo, E. (2011). Age-related occipito-temporal hypoactivation during visual search: Relationships between mN2pc sources and performance. *Neuropsychologia*, *49*(5), 858–865.

Lustig, C., Shah, P., Seidler, R., & Reuter-Lorenz, P. A. (2009). Aging, training, and the brain: A review and future directions. *Neuropsychology Review*, *19*(4), 504–522.

Macpherson, H. N., White, D. J., Ellis, K. A., Stough, C., Camfield, D., Silberstein, R., & Pipingas, A. (2014). Age-related changes to the neural correlates of working memory which emerge after midlife. *Frontiers in Aging Neuroscience*, *6*, 70.

Madden, D. J., Costello, M. C., Dennis, N. A., Davis, S. W., Shepler, A. M., Spaniol, J., ... Cabeza, R. (2010). Adult age differences in functional connectivity during executive control. *Neuroimage*, *52*(2), 643–657.

Madden, D. J., & Hoffman, J. M. (1997). Application of positron emission tomography to age-related cognitive changes. In K. R. R. Krishmann & P. M. Doraiswamy (Eds.), *Brain Imaging in clinical psychiatry*. New York: M. Dekker.

Madden, D. J., Spaniol, J., Whiting, W. L., Bucur, B., Provenzale, J. M., Cabeza, R., ... Huettel, S. A. (2007). Adult age differences in the functional neuroanatomy of visual attention: A combined fMRI and DTI study. *Neurobiology of Aging*, *28*(3), 459–476.

Madden, D. J., Turkington, T. G., Provenzale, J. M., Hawk, T. C., & Hoffman, J. M. (1997). Selective and divided visual attention: Age-related changes in regional cerebral blood flow measured by (H_2O)-O-15 PET. *Human Brain Mapping*, *5*(6), 389–409.

Madden, D. J., & Whiting, W. L. (2004). Age-related changes in visual attention. In P. T. Costa & I. C. Siegler (Eds.), *Recent advances in psychology and aging* (pp. 41–88). Amsterdam: Elsevier.

McDowd, J. M., & Craik, F. I. (1988). Effects of aging and task difficulty on divided attention performance. *Journal of Experimental Psychology: Human Perception and Performance*, *14*(2), 267–280.

McDowd, J. M., & Shaw, R. J. (2000). Attention and aging: A functional perspective. In F. I. Craik & T. A. Salthouse (Eds.), *Handbook of aging and cognition* (2nd ed., pp. 221–292). Hillsdale, NJ: Erlbaum.

Mishra, J., de Villers-Sidani, E., Merzenich, M., & Gazzaley, A. (2014). Adaptive training diminishes distractibility in aging across species. *Neuron*, *84*(5), 1091–1103.

Muller-Oehring, E. M., Schulte, T., Rohlfing, T., Pfefferbaum, A., & Sullivan, E. V. (2013). Visual search and the aging brain: Discerning the effects of age-related brain volume shrinkage on alertness, feature binding, and attentional control. *Neuropsychology*, *27*(1), 48–59.

Nagel, I. E., Preuschhof, C., Li, S. C., Nyberg, L., Backman, L., Lindenberger, U., & Heekeren, H. R. (2009). Performance level modulates adult age differences in brain activation during spatial working memory. *Proceedings of the National Academy of Sciences of the United States of America*, *106*(52), 22552–22557.

Nielson, K. A., Langenecker, S. A., & Garavan, H. (2002). Differences in the functional neuroanatomy of inhibitory control across the adult life span. *Psychology and Aging*, *17*(1), 56–71.

Nielson, K. A., Langenecker, S. A., Ross, T. J., Garavan, H., Rao, S. M., & Stein, E. A. (2004). Comparability of functional MRI response in young and old during inhibition. *Neuroreport*, *15*(1), 129–133.

Nissen, M. J., & Corkin, S. (1985). Effectiveness of attentional cueing in older and younger adults. *Journals of Gerontology*, *40*(2), 185–191.

O'connell, R. G., Balsters, J. H., Kilcullen, S. M., Campbell, W., Bokde, A. W., Lai, R., ... Robertson, I. H. (2012). A simultaneous ERP/fMRI investigation of the P300 aging effect. *Neurobiology of Aging*, *33*(10), 2448–2461.

Park, D. C., & Reuter-Lorenz, P. (2009). The adaptive brain: Aging and neurocognitive scaffolding. *Annual Review of Psychology*, *60*, 173–196.

Passow, S., Westerhausen, R., Hugdahl, K., Wartenburger, I., Heekeren, H. R., Lindenberger, U., & Li, S. C. (2014). Electrophysiological correlates of adult age differences in attentional control of auditory processing. *Cerebral Cortex*, *24*(1), 249–260.

Piefke, M., Onur, O. A., & Fink, G. R. (2012). Aging-related changes of neural mechanisms underlying visual-spatial working memory. *Neurobiology of Aging*, *33*(7), 1284–1297.

Prakash, R. S., Erickson, K. I., Colcombe, S. J., Kim, J. S., Voss, M. W., & Kramer, A. F. (2009). Age-related differences in the involvement of the prefrontal cortex in attentional control. *Brain and Cognition, 71*(3), 328–335.

Prakash, R. S., Heo, S., Voss, M. W., Patterson, B., & Kramer, A. F. (2012). Age-related differences in cortical recruitment and suppression: Implications for cognitive performance. *Behavioural Brain Research, 230*(1), 192–200.

Quigley, C., Andersen, S. K., Schulze, L., Grunwald, M., & Muller, M. M. (2010). Feature-selective attention: Evidence for a decline in old age. *Neuroscience Letters, 474*(1), 5–8.

Quigley, C., & Muller, M. M. (2014). Feature-selective attention in healthy old age: A selective decline in selective attention? *Journal of Neuroscience, 34*(7), 2471–2476.

Rabbitt, P., Lunn, M., Pendleton, N., Horan, M., Scott, M., Thacker, N., … Jackson, A. (2007). White matter lesions account for all age-related declines in speed but not in intelligence. *Neuropsychology, 21*(3), 363–370.

Rabbitt, P., Mogapi, O., Scott, M., Thacker, N., Lowe, C., Horan, M., … Lunn, D. (2007). Effects of global atrophy, white matter lesions, and cerebral blood flow on age-related changes in speed, memory, intelligence, vocabulary, and frontal function. *Neuropsychology, 21*(6), 684–695.

Raz, N., Ghisletta, P., Rodrigue, K. M., Kennedy, K. M., & Lindenberger, U. (2010). Trajectories of brain aging in middle-aged and older adults: Regional and individual differences. *NeuroImage, 51*(2), 501–511.

Reuter-Lorenz, P. A., & Cappell, K. A. (2008). Neurocognitive aging and the compensation hypothesis. *Current Directions in Psychological Science, 17*(3), 177–182.

Reuter-Lorenz, P. A., Jonides, J., Smith, E. E., Hartley, A., Miller, A., Marshuetz, C., & Koeppe, R. A. (2000). Age differences in the frontal lateralization of verbal and spatial working memory revealed by PET. *Journal of Cognitive Neuroscience, 12*(1), 174–187.

Reuter-Lorenz, P., & Lustig, C. (2005). Brain aging: Reorganizing discoveries about the aging mind. *Current Opinion in Neurobiology, 15*, 245–251.

Rizio, A. A., & Dennis, N. A. (2014). The cognitive control of memory: Age differences in the neural correlates of successful remembering and intentional forgetting. *PLoS ONE, 9*(1), e87010.

Rypma, B., & D'Esposito, M. (2000). Isolating the neural mechanisms of age-related changes in human working memory. *Nature Neuroscience, 3*(5), 509–515.

Salthouse, T. A. (1996). The processing-speed theory of adult age differences in cognition. *Psychological Review, 103*(3), 403–428.

Salthouse, T. A., & Madden, D. J. (2007). Information processing speed and aging. In J. Deluca & J. Kalmar (Eds.), *Information processing speed in clinical populations* (pp. 221–241). New York: Psychology Press.

Sander, M. C., Werkle-Bergner, M., & Lindenberger, U. (2012). Amplitude modulations and inter-trial phase stability of alpha-oscillations differentially reflect working memory constraints across the lifespan. *Neuroimage, 59*(1), 646–654.

Schmitz, T. W., Cheng, F. H. T., & De Rosa, E. (2010). Failing to ignore: Paradoxical neural effects of perceptual load on early attentional selection in normal aging. *Journal of Neuroscience, 30*(44), 14750–14758.

Singh, G., Pichora-Fuller, M. K., & Schneider, B. A. (2008). The effect of age on auditory spatial attention in conditions of real and simulated spatial separation. *Journal of the Acoustical Society of America, 124*(2), 1294–1305.

Smith, E. E., Geva, A., Jonides, J., Miller, A., Reuter-Lorenz, P., & Koeppe, R. A. (2001). The neural basis of task-switching in working memory: Effects of performance and aging. *Proceedings of the National Academy of Sciences of the United States of America, 98*(4), 2095–2100.

Smith, P. J., Blumenthal, J. A., Hoffman, B. M., Cooper, H., Strauman, T. A., Welsh-Bohmer, K., … Sherwood, A. (2010). Aerobic exercise and neurocognitive performance: A meta-analytic review of randomized controlled trials. *Psychosomatic Medicine, 72*(3), 239–252.

Stern, Y. (2002). What is cognitive reserve? Theory and research application of the reserve concept. *Journal of the International Neuropsychological Society, 8*(3), 448–460.

Strouse, A., Ashmead, D. H., Ohde, R. N., & Grantham, D. W. (1998). Temporal processing in the aging auditory system. *Journal of the Acoustical Society of America, 104*(4), 2385–2399.

Sullivan, E. V., Adalsteinsson, E., & Pfefferbaum, A. (2006). Selective age-related degradation of anterior callosal fiber bundles quantified in vivo with fiber tracking. *Cerebral Cortex*, *16*(7), 1030–1039.

Toepper, M., Gebhardt, H., Bauer, E., Haberkamp, A., Beblo, T., Gallhofer, B., ... Sammer, G. (2014). The impact of age on load-related dorsolateral prefrontal cortex activation. *Frontiers in Aging Neuroscience*, *6*, 9.

Turken, A. U., Whitfield-Gabrieli, S., Bammer, R., Baldo, J. V., Dronkers, N. F., & Gabrieli, J. D. E. (2008). Cognitive processing speed and the structure of white matter pathways: Convergent evidence from normal variation and lesion studies. *Neuroimage*, *42*(2), 1032–1044.

Verhaeghen, P., Marcoen, A., & Goossens, L. (1992). Improving memory performance in the aged through mnemonic training: A meta-analytic study. *Psychology and Aging*, *7*(2), 242–251.

Verhaeghen, P., Steitz, D. W., Sliwinski, M. J., & Cerella, J. (2003). Aging and dual-task performance: A meta-analysis. *Psychology and Aging*, *18*(3), 443–460.

Wais, P. E., & Gazzaley, A. (2014a). Distractibility during retrieval of long-term memory: Domain-general interference, neural networks and increased susceptibility in normal aging. *Frontiers in Psychology*, *5*, 280.

Wais, P. E., & Gazzaley, A. (2014b). External distraction impairs categorization performance in older adults. *Psychology and Aging*, *29*(3), 666–671.

Wais, P. E., Martin, G. M., & Gazzaley, A. (2012). The impact of visual distraction on episodic retrieval in older adults. *Brain Research*, *1430*, 78–85.

Wang, Y., Fu, S. M., Greenwood, P., Luo, Y. J., & Parasuraman, R. (2012). Perceptual load, voluntary attention, and aging: An event-related potential study. *International Journal of Psychophysiology*, *84*(1), 17–25.

West, R. L. (1996). An application of prefrontal cortex function theory to cognitive aging. *Psychological Bulletin*, *120*(2), 272–292.

Wild-Wall, N., Hahn, M., & Falkenstein, M. (2011). Preparatory processes and compensatory effort in older and younger participants in a driving-like dual task. *Human Factors*, *53*(2), 91–102.

Wolf, D., Zschutschke, L., Scheurich, A., Schmitz, F., Lieb, K., Tuscher, O., & Fellgiebel, A. (2014). Age-related increases in Stroop interference: Delineation of general slowing based on behavioral and white matter analyses. *Human Brain Mapping*, *35*(5), 2448–2458.

Zanto, T. P., Chadick, J. Z., Satris, G., & Gazzaley, A. (2013). Rapid functional reorganization in human cortex following neural perturbation. *Journal of Neuroscience*, *33*(41), 16268–16274.

Zanto, T. P., Hennigan, K., Ostberg, M., Clapp, W. C., & Gazzaley, A. (2010). Predictive knowledge of stimulus relevance does not influence top-down suppression of irrelevant information in older adults. *Cortex*, *46*, 561–574.

Zanto, T. P., Pan, P., Liu, H., Bollinger, J., Nobre, A. C., & Gazzaley, A. (2011). Age-related changes in orienting attention in time. *The Journal of Neuroscience: The Official Journal of the Society for Neuroscience*, *31*(35), 12461–12470.

Zanto, T. P., Rubens, M. T., Thangavel, A., & Gazzaley, A. (2011). Causal role of the prefrontal cortex in top-down modulation of visual processing and working memory. *Nature Neuroscience*, *14*(5), 656–661.

Zanto, T. P., Toy, B., & Gazzaley, A. (2010). Delays in neural processing during working memory encoding in normal aging. *Neuropsychologia*, *48*(1), 13–25.

Zhuravleva, T. Y., Alperin, B. R., Haring, A. E., Rentz, D. M., Holcomb, P. J., & Daffner, K. R. (2014). Age-related decline in bottom-up processing and selective attention in the very old. *Journal of Clinical Neurophysiology*, *31*(3), 261–271.

28

Working Memory Training

Megan Spencer-Smith and Torkel Klingberg

How Can We Understand the Current Literature on Working Memory Training?

The benefits of a working memory training programme are being evaluated by an increasing number of published trials. So far, these trials have typically involved small numbers of participants and are important for providing insight into the potential benefits of a programme as well as its feasibility and acceptability. Small sample trials, however, lack the statistical power required to determine treatment effects. For this reason, care should be taken when focusing on the significance or non-significance of the effect of a training programme on cognitive, academic, or behavioural outcomes. One approach that has been used to overcome such an important methodological limitation is a meta-analysis, which provides the advantage of increasing power to examine effects of an intervention by pooling together the results of several trials.

So far, six groups have reported on systematic reviews and meta-analyses examining the effects of a working memory training programme on aspects of cognitive, behavioural, and academic functions in trials involving recruitment of children. Although there are some similarities across these reviews, there are important differences in the inclusion criteria for the trials, including the types of working memory programme employed, participants, trial design, control group type, and outcome measures employed. Here we provide a brief overview of the meta-analyses.

In one of the initial meta-analyses, Melby-Lervåg and Hulme (2013) evaluated the effectiveness of working memory training programmes on aspects of cognitive and academic functioning. They identified 23 trials comparing a working memory training programme and a control group. The meta-analysis included trials evaluating different computer-based working memory training programmes, including methods commercially available or designed for research purposes varying in duration and intensity. Most of these trials examined the effects of Cogmed, a 5-week programme of five training sessions per week each involving 8 of 12 verbal and visuospatial working memory tasks. Other trials included in the analysis evaluated programmes including Jungle Memory, an 8-week programme of four training sessions per week focusing on three tasks; Cognifit, a 6-week programme with 24 sessions focusing on auditory, visual, and cross-modal working memory tasks, and different n-back training programmes, for example, a spatial n-back programme performed over 4 weeks with five sessions per week. The experience of the control group differed across the included

The Wiley Handbook of Cognitive Control, First Edition. Edited by Tobias Egner.

trials, and included no-treatment conditions and treatment conditions, such as computerised language training, self-paced reading, and non-adaptive working memory training. In some of the included trials, participants were randomly assigned to the intervention or control group, and others were non-randomised. The included trials were described as involving young children, children, young adults, and older adults who were unselected or from a group with learning disorders.

Schwaighofer, Fischer, and Bühner (2015) provided an extension and update of the earlier meta-analysis reported by Melby-Lervåg and Hulme (2013), evaluating the effects of working memory training programmes on aspects of cognitive and academic outcomes for children. They identified 47 trials comparing an adaptive working memory training programme of at least 2 weeks duration (including Cogmed n = 14, Jungle Memory n = 2, N-Back training n = 12, and Cognifit n = 2) and a control group (active and/or passive) using either a randomised or a non-randomised design. Participants could be of any clinical status and below 75 years of age.

In another early meta-analysis, Rapport, Orban, Kofler, and Friedman (2013) focused on trials evaluating cognitive training programmes compared with a control group (none, non-adaptive, and adaptive) for children with ADHD using either a randomised or a non-randomised design. Of the 25 identified trials included in the meta-analysis, 9 were categorised as targeting working memory (Cogmed n = 8).

Sonuga-Barke and colleagues (2013) also reported a meta-analysis at an early stage, which has been recently updated by Cortese and colleagues (2015). This group focused on randomised controlled trials evaluating the effects of cognitive training programmes for children and adolescents with ADHD on aspects of cognitive and academic functions as well as ADHD symptoms. Of the 16 identified trials, 7 evaluated an adaptive working memory training programme (Cogmed RM) compared with a control group (non-adaptive working memory training, treatment as usual, and adaptive math training).

Peijnenborgh, Hurks, Aldenkamp, Vles, and Hendriksen (2015) performed a meta-analysis evaluating the effectiveness of an adaptive working memory training programme for children and adolescents with learning disabilities for aspects of cognitive and academic functions. They identified 13 trials with an intervention group (Cogmed n = 10, Jungle Memory n = 2, Braingame and Brian n = 1) and a control group (no training, non-adaptive working memory training, and non-working memory training). In 10 of these trials, children and adolescents were randomly assigned to the intervention or control group, and 3 were non-randomised. The included trials were of children and adolescents with ADHD only (n = 10), ADHD and learning disability not specified (n = 1), and learning difficulties not specified (n = 2).

We have previously reported a meta-analysis focusing on trials evaluating the effects of the Cogmed working memory training programme (Spencer-Smith & Klingberg, 2015). Given the recent surge in published trials, in the next section of this chapter we provide an update of our meta-analysis. Here we identified 14 trials that randomised individuals to Cogmed (JM, RM, and QM versions) or a control group (non-adaptive working memory training, treatment as usual, and wait list). The identified trials were of children, adolescents, and adults, and included both typically developing as well as clinical samples of individuals with ADHD and low working memory.

In the following sections of this chapter, we summarise the results of these six meta-analyses evaluating the effects of a working memory training programme, focusing on outcomes examined in the literature that are particularly relevant for ADHD: working memory, inhibitory control, academic performance, and behavioural symptoms of ADHD. We then go on to describe the results of the meta-analyses examining individual differences, focusing on outcomes for participants of different age groups and clinical status.

Benefits of a Working Memory Training Programme for Inattention in Daily Life: An Update

In this section, we examine whether a working memory training programme can improve inattention in daily life by updating our previous systematic review and meta-analysis of the literature (Spencer-Smith & Klingberg, 2015). We restrict our review to trials evaluating the Cogmed method (Klingberg et al., 2005) to reduce the influence of the different working memory training programmes on the findings (see the above outline of the meta-analysis by Melby-Lervåg & Hulme, 2013). The Cogmed programme involves computerised adaptive training of working memory in 20 or more sessions, performed 5 days a week over a 5-week period. Each session is 30 to 45 min and involves training on 8 of 12 verbal and visuospatial working memory tasks. The level of task difficulty is adapted according to a built-in algorithm on a trial-by-trial basis so that the individual is training on each task at a level that is close to their current capacity. There are three age-specific versions of the Cogmed programme: JM for preschoolers (4 to 7 years), which requires training for 10 to 15 min each session, and RM and QM for children, adolescents, and adults (7 years and older), which require training for 30 to 45 min each session. The three versions are similar, with only the user interface differing across the versions; the JM version does not include the verbal training tasks that are included in the RM and QM versions. Reinforcement is built into the programme, such as small weekly rewards for completing the training sessions. A training aide (typically a teacher or parent) supervises the user to ensure task adherence, and breaks are taken. A certified training coach monitors the training by tracking the user's progress online. The coach has weekly meetings with the user and training aide to review training progress, solve any difficulties with task adherence, and ensure compliance, which is an important part of the overall effectiveness of cognitive training.

Methods

We used PRISMA (Preferred Reporting Items for Systematic Reviews and Meta-Analyses) guidelines to identify trials to include in the meta-analysis. We followed the same protocol as outlined in our earlier systematic review and meta-analysis, and provide a brief summary here.

Study Search We previously included studies from February 1, 2005 (when the Cogmed method was first published) to December 31, 2013. Here, our final search date was May 31, 2015. The trials were identified by searching electronic databases Medline and PsycInfo using keyword searches '*working memory training*' or '*working memory and training*', searching relevant journals, and contacting researchers in the field to request unpublished data.

Study Selection The inclusion criteria were as follows: (1) The trial used the full Cogmed programme (JM, RM, or QM) and training schedule (minimum 20 sessions). (2) The trial included an intervention and control group using either a randomised or non-randomised design. We only included trials that used a passive control group (the group continued with treatment as usual, or no new treatment was provided), waitlist, or active and non-adaptive control group, where the aim of the trial was to evaluate the effectiveness of Cogmed compared with no treatment. (3) The trial reported estimate of inattention in daily life before and after training for intervention and control groups. We did not include a trial where the follow-up testing was performed 20 weeks after training for the intervention group and 4 weeks after training for the control group (Ludqvist, Grundström, Samuelsson, & Rönnberg, 2010). (4) The trial reported means, standard deviations, and sample sizes for outcomes so that effect sizes could be calculated. When this information was missing, we contacted the authors to obtain additional information. (5) The trial data had not been previously reported, to avoid duplication. (6) We considered trials of participants of any age (children, adolescents, and adults) and clinical status (healthy, ADHD diagnosis, and impaired working memory). We did not include a trial that recruited a group of adolescents born extremely low birth weight for the intervention group and a group of adolescents born normal birth weight for the control group (Løhaugen et al., 2011).

Coding Measures of Inattention in Daily Life When a study reported more than one estimate, we included the measure most commonly reported in other included studies. For example, we preferentially included parent ratings as the primary measure of interest for studies of children and adolescents. Only when parent ratings were not available were teacher ratings were included (Roughan & Hadwin, 2011). We included teacher ratings on the Attentional Control scale from the DuPaul ADHD rating scale rather than teacher ratings on the total score from the Strengths and Difficulties Questionnaire in the analyses (Roughan & Hadwin, 2011).

Coding Measures of Working Memory Performance We performed an analysis to examine if training effects for inattention in daily life (far-transfer) were accompanied by training effects for visuospatial and verbal working memory performance (near-transfer). When a study reported more than one estimate of visuospatial working memory or verbal working memory, we included the measure most commonly reported in other included trials.

Data Extraction and Statistical Analysis Trial information was entered into Review Manager 5.3. The primary outcome measure of effect size was the pooled standardised difference in means (SMD, which allows comparison of different measurement scales) for inattention in daily life between the intervention and control groups after working memory training. Review Manager uses Hedge's adjusted *g* to calculate the SMD, which includes an adjustment for small sample bias. Post-training scores were analysed in line with the Cochrane handbook (The Cochrane Collaboration, 2011). Subgroups with at least four effect sizes were included in the analyses. A random-effects model was used to calculate the pooled as well as individual trial effect sizes and their 95% confidence intervals. An effect size of –0.2 was considered small, –0.5 moderate, and –0.8 large (Cohen, 1992). Heterogeneity (variability in training effects) between studies was evaluated using the I^2 statistic, which represents the percentage of the variability in the effect size explained by heterogeneity between studies rather than sampling error, where 25% is mild heterogeneity, 50% moderate, and 75% high (Higgins, Thompson, Deeks, & Altman, 2003). Funnel plots are typically used to examine risk of publication bias when more than 10 group comparisons are included in a meta-analysis (The Cochrane Collaboration, 2011), with the estimated effect size (SMD) plotted against a measure of sample size (standard error) for individual studies. In the current meta-analysis, this approach is problematic because the SMD is calculated using sample size, and therefore the SMD and the standard error are correlated. Therefore, the funnel plot for the primary analysis of interest should be interpreted cautiously. Subgroup analyses were also performed to determine whether the intervention effect varied in relation to participant characteristics: (1) Age: Participants were coded as children and adolescents or adults. (2) Clinical status: Participants in the study were coded as healthy, diagnosed with ADHD, or at risk for working memory impairment (i.e., previous research indicates the study sample is at risk for working memory impairment).

We performed analyses to confirm effects of the training on visuospatial and verbal working memory measures, indicating that a training effect on inattention in daily life in the primary analysis was accompanied by near-transfer effects. Here a positive effect size indicated improved working memory performance.

Results

A total of 1,034 articles were identified and reviewed for inclusion: 1,034 articles were identified by systematically searching electronic databases (261 duplicates were removed), 4 articles were identified by searching relevant journals, and 1 was article identified from contacting researchers for unpublished data. This updated dataset included 15 trials that met inclusion criteria with 16 group comparisons of the Cogmed programme, an additional 3 trials compared with our previous systematic review and meta-analysis. All trials were randomised. See Table 28.1 for a summary of the methodological and participant characteristics of the identified trials. One trial was not included in analyses because inattention in daily life ratings was not reported in the paper owing to missing data (Björkdahl Akerlund, Svensson, & Esbjornsson, 2013). The final dataset used in the analyses included 14 studies and 15 group comparisons of the Cogmed programme.

Inattention in Daily Life After Training Figure 28.1 shows the 14 trials with 15 group comparison effect sizes and the pooled effect size comparing estimates of inattention in daily life for the intervention and control groups after completing Cogmed working memory training (up to 4 weeks post-training). The effect sizes for the 15 group comparisons ranged from 0.38 to –2.93. The meta-analysis showed that the pooled effect size was small to moderate and significant (SMD = –0.40, 95% CI –0.68, –0.12, $p < .005$), with heterogeneity between studies moderate to high (I^2 61%) and significant ($\chi^2 = 35.98$, df 14, $p = .001$).

A funnel plot, presented in Figure 28.2, shows that the trials with the smallest samples show the largest effect sizes, and these are scattered towards the bottom of the left of the plot (Hardy, Willard, & Bonner, 2011; Mawjee et al., 2014; Roughan & Hadwin, 2011; Westerberg et al., 2007). Three of these trials were included in analyses of working memory outcomes, and did not always show the largest effect sizes (see Figures 28.4 and 28.5). Asymmetry in a funnel plot might reflect: (a) selection biases, (b) true heterogeneity, (c) data irregularities, (d) artefacts, and (e) chance (The Cochrane Collaboration, 2011). In an analysis of interest, when the four trials with the largest effect sizes were excluded from the meta-analysis, the training effect remained significant and the effect size small.

We examined long-term effects of the working memory training. In our updated dataset, seven trials reported inattention in daily life following a delay after the WM training programme (ranging from 2 to 4 months), which did not include any additional trials from our previous review. One study was not included in this analysis because the M and SD were not obtained. The forest plot in Figure 28.3 shows the seven trials with eight group comparison effect sizes and the pooled effect size comparing estimates of inattention in daily life for the intervention and control groups. The effect sizes ranged from –0.13 to –0.82. The meta-analysis showed that the pooled effect size was small to moderate and significant (SMD = –0.33, 95% CI –0.57, –0.09, $p = .006$), with heterogeneity between studies low (I^2 0%) and non-significant ($\chi^2 = 1.16$, df 6, $p = .95$).

In summary, the meta-analysis indicated a small to moderate and significant effect of the working memory training programme on inattention in daily life (SMD –0.40, $p = .005$) and initial evidence for persisting benefits (SMD –0.33, $p = .006$).

Working Memory Performance After the Training for Included Studies In our updated dataset, 10 trials with 11 group comparison effect sizes included in the primary analysis of interest reported visuospatial working memory performance for the intervention and control groups after the training. Figure 28.4 shows the pooled effect size comparing estimates of visuospatial working memory performance for intervention and control groups after completing

Table 28.1 Participant and Methodological Characteristics of Identified Studies.

	Participants			*Risk of bias*			*Training*		*Testing*		*Outcome measure*		
	Age in years	*Status*	*Recruitment source*	*Design*	*Blinded a) Rater b) Tester*	*Control group*	*Version*	*Site*	*Immediate follow-up*	*Delayed follow-up*	*Inattentive behaviour (rater)*	*Visuospatial WM*	*Verbal WM*
Beck, Hanson, Puffenberger, Benninger, & Benninger, 2010*	7–17	ADHD[b]	Private school for students with ADHD and/or learning disability (LD)	RCT	a) baseline, b) NR	Waitlist	RM	Home	1 month	4 months	DSM-IV Inattention (parent)	None	None
Björkdahl et al., 2013[+]	22–63	WM impairment (brain injury)	Outpatient rehabilitation clinic	RCT	a) NR, b) NR	Passive	QM	NR	1 week	3 months	Working memory questionnaire (self)	None	WAIS-III Digit span backward
Brehmer et al., 2012	a) 20–30, b) 60–70	Healthy	Newspaper advertisement	RCT	a) yes, b) yes	Active and non-adaptive	QM	Home	NR	3 months	CFQ (self)	WAIS-R Span board backward	WAIS-R Digit span backward
Chacko et al., 2013	7–11	ADHD	Newspaper advertisements	RCT	a) yes, b) yes	Active and non-adaptive	NR	Home	3 weeks	None	DBD Inattention (parent)	AWMA Spatial recall	AWMA Listening recall
Egeland et al., 2013	10–12	ADHD[b]	Outpatient clinics at two hospitals	RCT	a) NR, b) yes	Waitlist	NR	School	NR	8 months	DuPaul Inattention (parent)	None	None
Green et al., 2012	7–14	ADHD[b, c]	Advertising, psychologists, psychiatrists, institute tracking	RCT	a) yes, b) yes	Active and non-adaptive	RM	Home	NR	None	Connors ADHD (parent)	None	None
Gropper, Gotlieb, Kronitz, & Tannock, 2014	19–52	ADHD[b] (ADHD/LD)	Student disability services at 3 post-secondary institutions	RCT	a) NR, b) NR	Waitlist	QM	Home	3 weeks	2 months	CFQ (self)	CANTAB Spatial WM (errors)	WISC-IV Digit span
Grunewaldt, Løhaugen, Austeng, Brubakk, & Skranes, 2013	5–6	WM impairment (VLBW)	Admission records of a hospital NICU	RCT[e]	a) NR, b) yes	Waitlist	JM	Home	1 month	None	DuPaul Inattention (parent)	Spatial span backward[a]	Digit span backward[a]

Hardy, Willard, Allen, & Bonner, 2013	8–16	WM impairment (Cancer)	Patients of the hospital Division of Pediatric Hematology-Oncology	RCT	a) yes[d], b) yes	Active and non-adaptive	RM	Home	NR	3 months	Connors Inattention (parent)	WRAML Symbolic WM	WRAML Verbal WM
Klingberg, 2005	7–12	ADHD	Referrals from paediatricians, psychiatrists, special school teachers	RCT	a) yes, b) yes	Active and non-adaptive	RM	Home or school	NR	3 months	DSM-IV Inattention (parent)	WAIS-R Span board	WAIS-R Digit span
Roughen, 2011	15–17	WM impairment (SEB)	Schools	RCT	a) no, b) NR	Passive	RM	School	3 weeks	3 months	DuPaul Attentional Control (teacher)	None	None
Mawjee et al., 2014	18–35 years	ADHD[b]	E-mails to students with ADHD registered with Student Disability Services, brochures	RCT	a) baseline, b) NR	Waitlist	RM	Home[g]	3 weeks	None	CFQ (self)	Spatial span backward	Digit span backward
Steeger, Gondoli, Gibson, & Morrissey, 2016	11–15	ADHD[b]	Local schools	RCT	a) yes, b) yes	Active and non-adaptive[f]	RM	Home	2 weeks	None	DuPaul Inattention (mother)	Spatial span backward	WISC-IV Digit span backward
van Dongen-Boomsma, Vollebregt, Buitelaar, & Slaats-Willemse, 2014,	5–7	ADHD[h]	Child and adolescent psychiatry university centre, advertisements on websites, magazines, newspapers.	RCT	a) yes, b) yes	Active and non-adaptive	JM	Home	2 weeks	None	ADHD Rating Scale (teacher)	Knox cubes Leiden Diagnostic Test (LDT) backward	WISC-III Digit span backward
Westerberg et al., 2007	34–65	WM impairment (stroke)	Stroke Rehabilitation Unit	RCT	a) baseline, b) baseline	Passive	RM	Home	NR	None	CFQ (self)	WAIS-R Span board	WAIS-R Digit span

Note. ADHD, Attention deficit hyperactivity disorder; CFQ, Cognitive Failures Questionnaire; DBD, Disruptive Behaviours Disorders Rating Scale, ELBW, extremely low birth weight; LD, learning disorder; NR, not reported; RCT, randomised controlled trial; SEB, social and emotional behavioural difficulties; VLBW, very low birth weight; WM, working memory

[+] not included in analyses because authors did not report inattention ratings due to missing data.

* not included in the analysis of delayed effects of training because scores required for the analysis were not reported in the paper and at the time of submission authors had not provided information.

[a] described by authors as a standard neuropsychological test.

[b] some participants prescribed stimulant medication.

[c] participants assessed for LD, and one met criteria.

[d] families informed pre-training there were two versions of the programme reflecting different levels of difficulty.

[e] trial is a RCT at post-training testing, and at long-term follow-up is a stepped wedge design.

[f] control group reported in this meta-analysis involved non-adaptive Cogmed and parent lectures on adolescent development, and the intervention group involved adaptive Cogmed and parent lectures on adolescent development. The study included two additional groups not reported here: adaptive Cogmed and behavioural parent intervention, and non-adaptive Cogmed and behavioural parent intervention.

[g] at place of residence, 64% living at home.

[h] medication free.

Study or subgroup	Cogmed Mean	Cogmed SD	Cogmed Total	Control Mean	Control SD	Control Total	Weight	Std. Mean difference IV, Random, 95% CI
Beck 2010	5.72	2.92	27	7.5	1.79	24	8.0%	−0.71 [−1.28, −0.15]
Brehmer 2012a	25.66	12.37	29	31.69	12.39	26	8.3%	−0.48 [−1.02, 0.06]
Brehmer 2012b	31.85	9.77	26	38.79	19.11	19	7.7%	−0.47 [−1.07, 0.13]
Chacko 2013	16.51	5.84	38	14.24	6.06	36	9.0%	0.38 [−0.08, 0.84]
Egeland 2013	15	5.6	33	16.2	6.2	34	8.8%	−0.20 [−0.68, 0.28]
Green 2012	67	12.2	12	70.9	8.3	14	6.2%	−0.37 [−1.15, 0.41]
Gropper unpublished	48.75	17.72	39	55.65	15.11	23	8.4%	−0.41 [−0.93, 0.12]
Grunewaldt 2013	6.9	5.3	9	5.1	4.5	11	5.4%	0.35 [−0.54, 1.24]
Hardy 2013	60.8	15.42	11	74	8.08	6	4.5%	−0.93 [−1.99, 0.12]
Klingberg 2005	13	6.5	17	15.5	7.2	19	7.2%	−0.36 [−1.02, 0.30]
Mawjee 2014	44.67	6.38	8	69	8.98	9	2.8%	−2.93 [−4.41, −1.46]
Roughan 2011	12.57	11.12	7	28.88	11.69	8	4.0%	−1.34 [−2.50, −0.18]
Steeger et al 2015	13.9	5.1	23	12.78	3.7	26	8.0%	0.25 [−0.31, 0.81]
van Dongen–Boomsma 2014	12.6	6.5	18	14	4.7	14	6.8%	−0.24 [−0.94, 0.47]
Westerberg 2007	29.2	12.1	9	43	13.8	9	4.8%	−1.01 [−2.01, −0.02]
Total (95% CI)			**306**			**278**	**100.0%**	**−0.40 [−0.68, −0.12]**

Heterogeneity: Tau² = 0.18; chi² = 35.98, df = 14 (P = 0,001); I² = 61%
Test for overall effect: Z = 2.79 (P = 0.005)

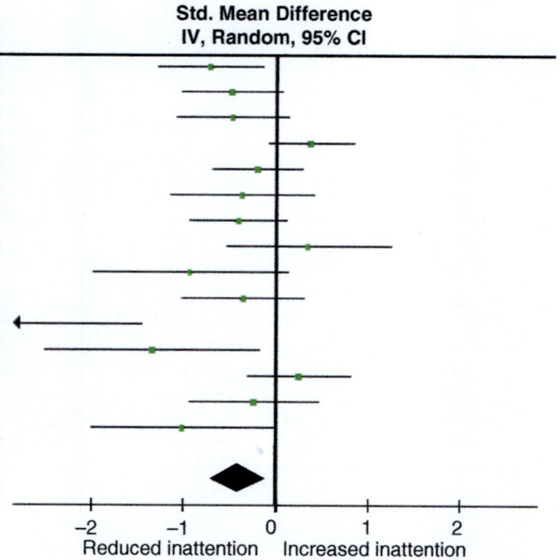

Figure 28.1 Forest plot for inattention in daily life after training. The overall pooled effect size (standardised mean difference, displayed as a diamond) as well as individual study effect sizes (displayed as rectangles) and their 95% confidence intervals (represented by horizontal lines) are shown.

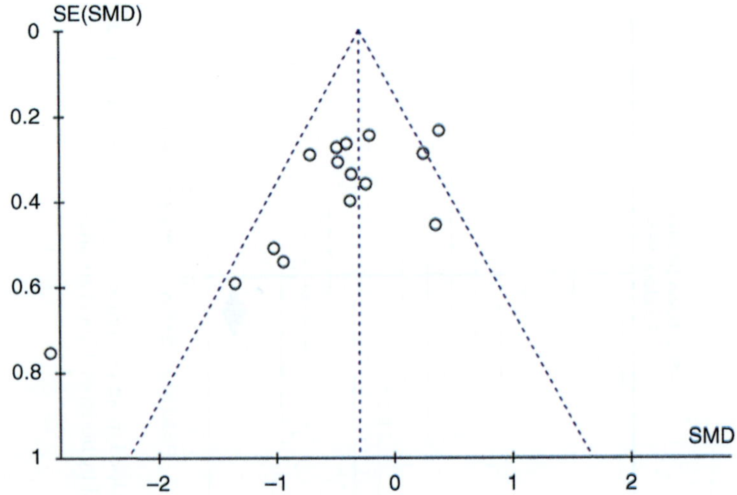

Figure 28.2 Funnel plot for inattention in daily life immediately after training. The overall pooled effect size (standardised mean difference, represented by the vertical dashed line) and its 95% confidence intervals (represented by diagonal dashed lines) are presented, showing the expected distribution of studies in the absence of heterogeneity or of selection biases.

training was moderate to large and significant (SMD = 0.59, 95% CI 0.31, 0.87, p = .0001), with heterogeneity between studies moderate (I^2 49%) and non-significant ($\chi^2 = 19.74$, df 10, p = .03).

The same 10 trials with 11 group comparisons reported verbal working memory performance for the intervention and control groups after the training. Figure 28.5 shows the pooled effect size comparing estimates of verbal working memory performance for intervention and control groups after completing training was moderate and significant (SMD = 0.48, 95% CI 0.33, 0.66, p = .00001), with heterogeneity between studies low (I^2 0%) and non-significant ($\chi^2 = 10.73$, df 11, p = .47).

In summary, the updated meta-analysis shows that a generalising effect of the working memory training programme for inattention in daily life was observed in the context of a significant effect of the training on visuospatial and verbal working memory performance.

Conclusions

The major finding of this updated meta-analysis is that benefits of a working memory training programme generalise to improvements in everyday functioning. We showed that inattention in daily life was significantly lower after completing the Cogmed programme compared with a control programme (SMD −0.40). We also showed that effects of the training for inattention in daily life persisted and remained significant at 2 to 8 months after training (SMD −0.33). The training effect on inattentive behaviour found in this meta-analysis is not inconsistent with findings of non-significant effects reported in some of the small sample studies included in the meta-analysis (e.g., Egeland, Aarlien, & Saunes, 2013). The low statistical power in most of the training studies published to date highlights the need for large sample studies, meta-analyses in the future, and to acknowledge and estimate the risk of type-II errors when reporting small sample trials.

Study or subgroup	Cogmed Mean	Cogmed SD	Cogmed Total	Control Mean	Control SD	Control Total	Weight	Std. Mean difference IV, Random, 95% CI
Brehmer 2012a	25.55	12.94	28	30.27	11.78	26	19.5%	−0.38 [−0.91, 0.16]
Brehmer 2012b	32.23	8.79	26	37.79	16.2	19	15.8%	−0.44 [−1.04, 0.16]
Egeland 2013	15.3	5.3	33	16.5	5.6	34	24.6%	−0.22 [−0.70, 0.26]
Gropper unpublished	46.17	23.69	24	53.52	18.35	21	16.3%	−0.34 [−0.93, 0.25]
Hardy 2013	57.8	13.59	11	68.3	8.35	6	5.2%	−0.82 [−1.87, 0.22]
Klingberg 2005	13.6	7.9	17	14.5	5.6	20	13.5%	−0.13 [−0.78, 0.52]
Roughan 2011	17.71	12	7	23.71	15.81	7	5.0%	−0.40 [−1.46, 0.66]
Total (95% CI)			**146**			**133**	**100.0%**	**−0.33 [−0.57, −0.09]**

Heterogeneity: $Tau^2 = 0.00$; $chi^2 = 1.61$, $df = 6$ ($P = 0.95$); $I^2 = 0\%$

Test for overall effect: $Z = 2.73$ ($P = 0.006$)

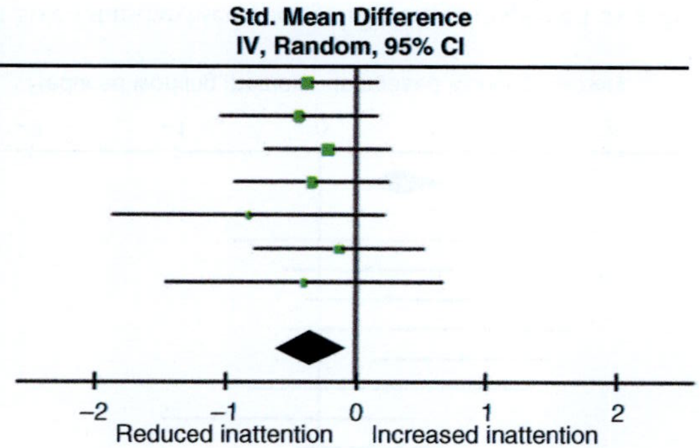

Figure 28.3 Forest plot for inattention in daily life following a delay after training. The overall pooled effect size (standardised mean difference, displayed as a diamond) as well as individual study effect sizes (displayed as rectangles) and their 95% confidence intervals (represented by horizontal lines) are shown.

Study or subgroup	Cogmed Mean	Cogmed SD	Cogmed Total	Control Mean	Control SD	Control Total	Weight	Std. Mean Difference IV, Random, 95% CI
Brehmer 2012a	7.14	0.97	29	5.83	0.42	26	10.0%	1.70 [1.07, 2.32]
Brehmer 2012b	5.69	0.65	26	5.26	0.67	19	10.2%	0.64 [0.03, 1.25]
Chacko 2013	107	19.21	37	99.97	17.15	36	12.7%	0.38 [−0.08, 0.84]
Gropper unpublished	0.185	0.96	39	−0.205	1.18	23	11.7%	0.37 [−0.15, 0.89]
Grunewaldt 2013	2.3	2	9	2.7	1.6	11	6.7%	−0.21 [−1.10, 0.67]
Hardy 2013	12.4	4.57	10	10.4	2.3	7	5.8%	0.50 [−0.49, 1.48]
Klingberg 2005	5.22	1.01	20	4.28	0.81	24	9.8%	1.02 [0.38, 1.65]
Mawjee 2014	7.75	1.28	8	6.56	1.67	9	5.7%	0.75 [−0.24, 1.75]
Steeger et al 2015	8.09	2	23	7.04	2.3	26	10.8%	0.48 [−0.09, 1.05]
van Dongen-Boomsma 2014	4.9	2.2	26	4.1	2.3	19	10.4%	0.35 [−0.25, 0.95]
Westerberg 2007	6.2	1	9	5.7	1.8	9	6.2%	0.33 [−0.60, 1.26]
Total (95% CI)			**236**			**209**	**100.0%**	**0.59 [0.31, 0.87]**

Heterogeneity: $Tau^2 = 0.11$; $Chi^2 = 19.74$, $df = 10$ ($P = 0.03$); $I^2 = 49\%$

Test for overall effect: $Z = 4.10$ ($P < 0.0001$)

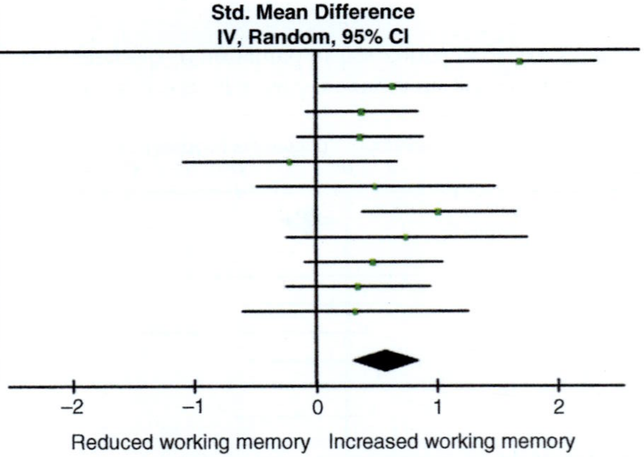

Figure 28.4 Forest plot for visuospatial working memory after training. The overall pooled effect size (standardised mean difference, displayed as a diamond) as well as individual study effect sizes (displayed as rectangles) and their 95% confidence intervals (represented by horizontal lines) are shown.

Study or subgroup	Cogmed Mean	Cogmed SD	Cogmed Total	Control Mean	Control SD	Control Total	Weight	Std. Mean Difference IV, Random, 95% CI
Brehmer 2012a	6.76	1.66	29	5.67	1.41	26	11.1%	0.69 [0.15, 1.24]
Brehmer 2012b	5.94	1.58	26	5.5	1.43	19	9.3%	0.28 [−0.31, 0.88]
Chacko 2013	104.37	11.88	37	103.43	14.56	36	15.7%	0.07 [−0.39, 0.53]
Gray et al	7.43	2.97	32	5.96	2.39	20	10.2%	0.52 [−0.04, 1.09]
Gropper unpublished	10.24	3.06	39	7.77	2.79	23	11.5%	0.82 [0.29, 1.36]
Grunewaldt 2013	1.9	1.2	9	1.1	1.3	11	4.0%	0.61 [−0.30, 1.52]
Hardy 2013	9.9	3.75	10	9.8	2.56	7	3.5%	0.03 [−0.94, 0.99]
Klingberg 2005	4.08	0.89	20	3.73	0.73	24	9.2%	0.43 [−0.17, 1.03]
Mawjee 2014	12.75	2.49	8	8.78	2.49	9	2.7%	1.51 [0.40, 2.63]
Steeger et al 2015	9.09	1.9	23	7.87	2.1	26	10.0%	0.60 [0.02, 1.17]
van Dongen–Boomsma 2014	2.9	0.9	22	2.4	1	21	8.9%	0.52 [−0.09, 1.13]
Westerberg 2007	5.8	1	9	5.7	0.9	9	3.9%	0.10 [−0.82, 1.02]
Total (95% CI)			**264**			**231**	**100.0%**	**0.48 [0.30, 0.66]**

Heterogeneity: $Tau^2 = 0.00$; $Chi^2 = 10.73$, df = 11 (P = 0.47); $I^2 = 0\%$
Test for overall effect: Z = 5.14 (P < 0.00001)

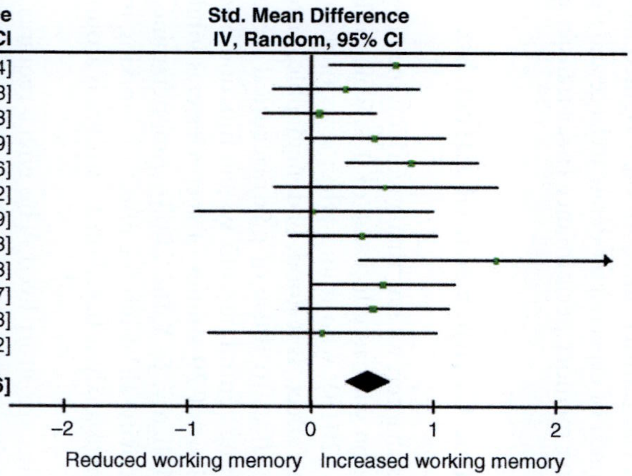

Figure 28.5 Forest plot for verbal working memory after training. The overall pooled effect size (standardised mean difference, displayed as a diamond) as well as individual study effect sizes (displayed as rectangles) and their 95% confidence intervals (represented by horizontal lines) are shown.

Importantly, the generalising (far transfer) effect of the Cogmed method to inattention in daily life was observed in the context of expected moderate to large and significant training effect in visuopatial and verbal working memory performance (near-transfer effects).

A Relative Comparison of Effect Sizes

In the preceding section of this chapter, we saw that a working memory training programme improves symptoms of inattention with an effect size of 0.4, which is in agreement with the meta-analysis of Cortese et al. (2015), who found effect sizes of the order of 0.3–0.5. But how should we evaluate such an effect: Is it good, bad, negligible, or too small?

We could compare such an effect to those of pharmaceutical treatments. It is generally acknowledged that drugs affecting central nervous system function have rather limited effect size. For example, a meta-analysis of 29 studies of acethylesteras inhibitors found an effect size of 0.1 (Trinh, Hoblyn, Mohanty, & Yaffe, 2003). Amphetamine and methylphenidate have a strong effect on hyperactivity, although the effect on working memory is weaker, and according to the most recent meta-analysis based on 13 studies is around 0.26 (Coghill et al., 2014); see Figure 28.6.

However, the effects of drugs are short-lived. The effect requires constant medication, and most parents and children do not continue with medication for long periods of time. A more relevant comparison to cognitive training might be the long-term effects of physical training. Physical training has received a lot of attention in research and media and been suggested as one of the few non-pharmacological treatments to improve cognitive function. An early meta-analysis (Colcombe & Kramer, 2003) found large effects for several cognitive functions. Later meta-analyses have been largely negative. We found five meta-analyses from the last 5 years that included measures of long-term effects on working memory. These found generally small effect sizes, ranging from 0.05 to 0.3, with an average effect size of 0.13 (Table 28.2).

Sonuga-Barke et al. (2013) performed an overview of non-pharmacological treatments for ADHD and found generally small or non-significant effects. Although it is not feasible

Study or Subgroup	Std. Mean Difference IV, Random 95% CI
Bedard et al, 2004 (36)	−0.22 [−0.76, 0.33]
Bedard et al, 2004 (36)	−0.41 [−0.96, 0.14]
Bedard et al, 2007 (67)	−0.40 [−0.79, 0.00]
Bedard et al, 2007 (67)	0.03 [−0.36, 0.42]
Bedard et al, 2007 (67)	−0.53 [−0.93, −0.13]
Bedard et al, 2007 (67)	−0.22 [−0.61, 0.17]
Coghill et al, 2007 (27)	−0.20 [−0.54, 0.14]
Douglas et al, 1988 (65)	−0.02 [−0.69, 0.65]
Douglas et al, 1988 (65)	−0.50 [−1.19, 0.18]
Mehta et al, 2004 (66)	−0.31 [−1.06, 0.43]
Rhodes et al, 2006 (35)	−0.38 [−0.95, 0.19]
Zeiner et al, 1999 (50)	−0.30 [−0.77, 0.16]
Zeiner et al, 1999 (50)	−0.05 [−0.51, 0.41]
Total (95% CI)	**−0.26 [−0.39, −0 .13]**

Heterogeneity: $Tau^2 = 0.00$; $chi^2 = 6.83$, df = 12 (P = 0.87); $I^2 = 0\%$

Test for overall effect: Z = 3.91 (P<0.0001)

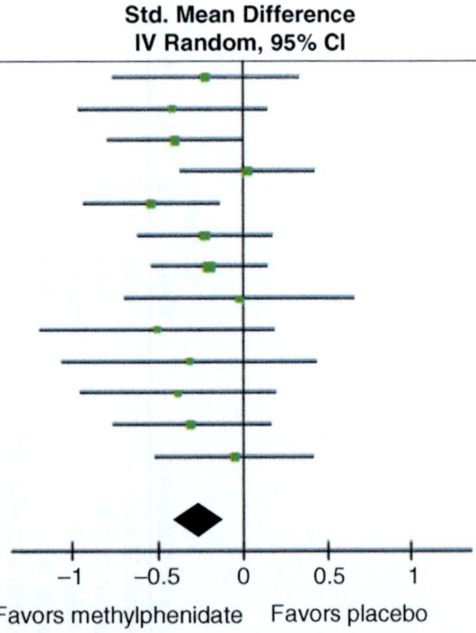

Figure 28.6 The effect of methylphenidate on working memory. Source: Coghill 2014. Reproduced with permission of Elsevier.

Table 28.2 Effect of Physical Training on Working Memory Performance.

Article	*Year*	*Subjects*	*Domain*	*Effect size*
Young, Angevaren, Rusted, & Tabet	2015	>55 years	WM	0.30
Verburgh, Königs, Scherder, & Oosterlaan Verbugth et al.	2015	Children, young adults	WM	0.05
Kelly et al.	2014	Old, healthy	WM	0.06
Roig, Nordbrandt, Geertsen, & Nielsen	2013	All ages	WM	0.15
Angevaren, Aufdemkampe, Verhaar, Aleman, & Vanhees	2008	Old, healthy	WM	0.10

to make an exhaustive list of all possible types of suggested treatments or interventions, we can conclude that changing cognitive aspects in a permanent way is difficult, and the systematic and sustained effects that are found are generally small.

Statistical Considerations

The previous section illustrated that effect sizes for treatments or interventions affecting central nervous system functioning are often small. We should similarly not expect effect sizes from cognitive training to be greater than small. This has consequences for statistical power. Although this is common sense and a well-known statistical property, it deserves repetition and illustration.

In the following examples, we performed a simulation using two populations that differed in the dependent variable by 0.3 standard deviations. From these two samples, we drew 200 samples of different sample sizes. From each sample, we calculated the effect size and the significance of the difference. If we expect effect sizes from a working memory training programme to have effects of the order of 0.3, these simulations illustrate the kind of variability we would expect.

As can be seen in Figure 28.7, the outcome of these simulated 'studies' vary widely, both in terms of the estimated effect size and significance levels. Even with sample sizes of around 100 subjects in each group, around 50% of studies are not significant, and the estimated effect sizes range from 0 to 0.8. In line with these outcomes, any 'review of the literature' of such studies will find contradictory evidence. This is to be expected if the literature consists of underpowered studies. Simulations tell us that in order to have a power of 80% with an effect size of 0.3, we would need around 175 subjects in each group. Negative studies with fewer subjects are simply inconclusive.

Benefits of a Working Memory Training Programme: Evidence From Meta-Analyses

Here we summarise results of six meta-analyses reported by independent groups examining effects of a working memory training programme (Cortese et al., 2015; Melby-Lervåg & Hulme, 2013; Peijnenborgh et al., 2015; Rapport et al., 2013; Schwaighofer et al., 2015; this chapter by Spencer-Smith & Klingberg). We describe results for the immediate effects and, when available, results for the long-term effects of training focusing on outcomes for working memory (often described as a near-transfer domain, given this is the targeted skill for training), inhibition, academic performance, and behavioural symptoms of ADHD (often described as far-transfer domains, given these skills and behaviours were not the target of the training).

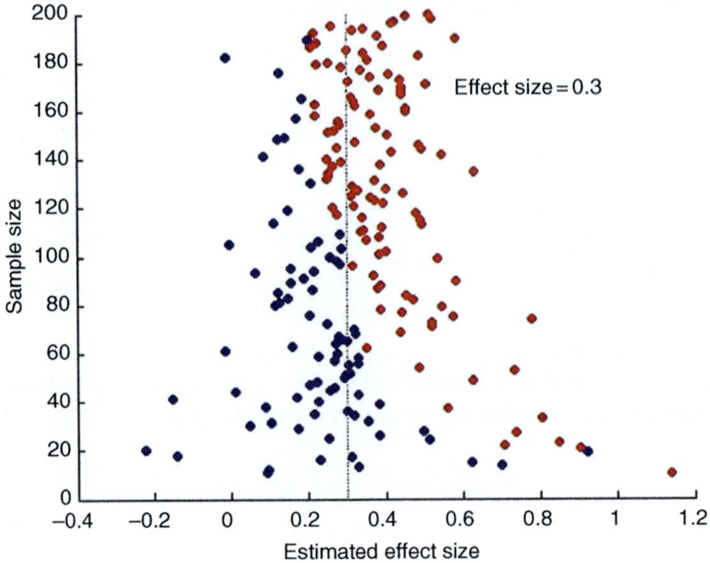

Figure 28.7 Simulation with random samples of difference sizes from two populations known to differ by 0.3 standard deviations. Each dot represents that estimated effect size of the samples. Red dots = significant difference; blue dots = non-significant between sample t-test.

Working Memory

Trials evaluating a working memory training programme have most commonly reported on outcomes for working memory performance, which provide important information about the benefits of the training for the targeted skill. Meta-analyses of these trials have consistently identified a moderate effect on both visuospatial and verbal working memory performance (Cortese et al., 2015; Melby-Lervåg & Hulme, 2013; Peijnenborgh et al., 2015; Rapport et al., 2013; Schwaighofer et al., 2015; this chapter by Spencer-Smith & Klingberg).

Melby-Lervåg and Hulme (2013) reported a moderate effect on visuospatial working memory (*d* 0.55, 95% CI 0.32, 0.72, $p < .001$; I^2 58.9%; $p < .001$, $n = 18$) and a large effect on verbal working memory (*d* 0.79, 95% CI 0.50, 1.09, $p < .001$; I^2 83.1%; $p < .001$; $n = 21$) in children, adolescents, and adults unselected or from a group with learning disabilities. Here, significant heterogeneity between the included trials was noted, which might in part reflect the variability in methodological characteristics of the included trials. One important source of variability in this analysis is the schedule of the working memory training programme; for example, the analysis included trials using a range of programmes such as a programme involving daily training for 12 min over 2 weeks (Loosli, Buschkuehl, Perrig, & Jaeggi, 2012) and a programme involving training three times per week for 30 min over 8 weeks (Alloway & Alloway, 2009). The researchers acknowledged this methodological limitation, and in a series of sub-analyses showed variability in effect sizes for different training programmes. For visuospatial working memory, the effect of the different training programmes was significant.

Focusing on children, Schwaighofer et al. (2015) in their analyses distinguished between training effects on short-term and working memory. For short-term memory, a small to moderate effect was reported for verbal short-term memory and a large effect for visuospatial short-term memory, all of which were significant (*g* 0.37, 95% CI 0.19, 0.56, $p<.001$; I^2 66.82; n = 32 effect sizes; *g* 0.72, 95% CI 0.56, 0.92, $p<.001$; I^2 64.8%, n = 25 effect sizes). For working memory, a moderate effect was identified for verbal working memory (*g* 0.55, 95% CI 0.33, 0.78, $p<.001$; I^2 83.73; n = 42 effect sizes) and a large effect for visuospatial working memory (*g* 0.63, 95% CI 0.35, 0.90, $p<.001$; I^2 79.5%; n = 19 effect sizes), and again these were significant effects.

Rapport and colleagues (2013) in their analysis indicated a moderate to large effect on working memory processes in children with ADHD (*d* 0.63, 95% CO 0.46, 0.80; n = 8 effect sizes). Cortese and colleagues (2015) reported a moderate effect on visuospatial (SMD 0.47, 95% CI 23, 70, $p<.01$; I^2 69%, $p<.001$; n = 5 effect sizes) and verbal working memory (SMD 0.57, 95% CI 0.29, 0.82, $p<.001$; I^2 32%, $p=.19$; n = 6 effect sizes) in children and adolescents with ADHD. Peijnenborgh and colleagues (2015) reported a moderate and significant effect on visuospatial (*g* 0.63, 95% CI 0.40, 0.85, $p<.01$; I^2 27.6%, $p=.20$; n = 9 effect sizes) and verbal working memory (*g* 0.64, 95% CI 0.38, 0.90, $p<.01$; I^2 42.5%, $p=.20$; n = 9 effect sizes) in children and adolescents with learning disabilities, most of whom had a diagnosis of ADHD.

In our meta-analysis (updated in this chapter), we chose to focus our review on a specific training programme to eliminate the variability introduced into the analysis from different schedules of a working memory training programme, highlighted by analyses conducted by Melby-Lervåg and Hulme described above. Our meta-analysis identified a moderate effect on both visuospatial (SMD 0.59, 95% CI 0.31, 0.87, $p=.0001$; I^2 49%, $p=.03$; n = 11 effect sizes) and verbal working memory (SMD 0.48, 95% CI 0.33, 0.66, $p=.00001$; I^2 0%, $p=.47$; n = 12 effect sizes).

Four meta-analyses have performed analyses on trials that report long-term follow-up assessments to examine if the immediate effects persist (Melby-Lervåg & Hulme, 2013; Peijnenborgh et al., 2015; Rapport et al., 2013; Schwaighofer et al., 2015). The results of these analyses provide some inconsistent findings. Melby-Lervåg and Hulme (2013) reported a small to moderate and significant long-term effect for visuospatial working memory (*d* 0.41, 95% CI 0.13, 0.69, $p<.001$; I^2 0%, $p=.48$; n = 4), but a small and non-significant long-term effect for verbal working memory (*d* 0.31, 95% CI −0.19, 0.80, $p=.22$; I^2 71.9%, $p<.01$; n = 6) for unselected participants and participants from groups with learning disabilities. Schwaighofer et al. (2015) observed significant long-term effects that were small for verbal short-term memory and working memory (*g* 0.22, 95% CI 0.02, 0.42, $p<.05$; n = 9 effect sizes; *g* 0.35, 95% CI 0.17, 0.68, $p<.05$; I^2 67; n = 11 effect sizes), large for visuospatial short-term memory and small to moderate for visuospatial working memory (*g* 0.78, 95% CI 0.33, 1.23, $p<.001$; n = 7 effect sizes; *g* 0.41, 95% CI 0.00, 0.81, $p<.05$; I^2 61.9%; n = 6 effect sizes) for children. In children with ADHD, Rapport and colleagues report a large long-term effects for working memory (*d* 0.71, 95% CI 0.45, 0.97; n = 3 effect sizes). Peijnenborgh and colleagues (2015) identified moderate and significant long-term effects for verbal working memory (*g* 0.54, 95% CI 0.16, 0.91, $p=.01$; I^2 60.8%, $p=.04$; n = 5 effect sizes) and moderate long-term effects for visuospatial working memory that approached significance (*g* 0.39, 95% CI 0.00, 0.78, $p=.05$; I^2 63.4%, $p=.03$; n = 5 effect sizes) for children with learning disabilities.

In summary, there is consistent evidence that working memory training programmes improve performance in verbal and visuospatial working memory. This result is important for confirming that a working memory training programme successfully improves performance in the target function. There are encouraging results from the few published trials with long-term follow-up that suggest these immediate benefits can persist, although with a reduced effect.

Inhibitory Control

Inhibitory control, an executive function that requires inhibiting a prepotent response (Chapter 6 by Verbruggen & Logan in this volume), has been described as a core cognitive impairment in ADHD (Sjöwall, Roth, Lindqvist, & Thorell, 2013; Wåhlstedt, Thorell & Bohlin, 2009; see also Chapter 33 by de Zeeuw & Durston in this volume). Two meta-analyses have examined the effects of a working memory training programme on inhibitory control and both highlight that few trials examine the effects of training for this cognitive function (Melby-Lervåg & Hulme, 2013; Peijnenborgh et al., 2015). The results of the meta-analyses provide some inconsistent findings.

In an analysis including trials of very young children through to adults from groups with learning disabilities and unselected, Melby-Lervåg and Hulme (2013) reported a small and significant immediate effect on inhibitory control (*d* 0.32, 95% CI 0.11, 0.53, $p<.01$; I^2 0%, $p=.51$, $n=10$ effect sizes), and a small long-term effect that was non-significant (*d* 0.09, 95% CI –0.19, 0.37; $n=4$ effect sizes). Although Schwaighofer et al. (2015) largely conducted their meta-analysis in line with the approach by Melby-Lervåg and Hulme, they chose not to include an analysis of inhibitory control outcomes. This decision was based on theoretical grounds, for example, they explain that inhibition is a multi-component construct. In an analysis of children with learning disabilities, Peijnenborgh and colleagues (2015) reported a small immediate effect that was non-significant (*g* 0.19, 95% CI -0.13, 0.51, $p=.26$; I^2 0%, $p=.95$, $n=4$ effect sizes), and a large long-term effect that did not reach statistical significance (*g* 0.74, 95% CI –0.09, 1.57, $p=.08$; I^2 74.9%, $p=.05$; $n=2$ effect sizes).

In summary, there is some initial evidence, based on the small number of published trials evaluating benefits for inhibitory control, that a working memory training programme can improve inhibitory control.

Academic Performance

There is an increasing interest in the effects of a working memory training programme for children's academic performance and achievement; however, few trials have examined these outcomes. This is highlighted by the small number of trials included in the three meta-analyses that report on training effects for aspects of academic performance (Melby-Lervåg & Hulme, 2013; Peijnenborgh et al., 2015; Schwaighofer et al., 2015).

Melby-Lervåg and Hulme (2013) reported small effects that were non-significant for both mathematical performance and word decoding immediately after training (mathematics: *d* 0.07, 95% CI –0.13, 0.27, I^2 0%, $p=.83$; $n=7$ effect sizes; word decoding: *d* 0.13, 95% CI –.07, 0.35, I^2 0%, $p=.85$; $n=7$ effect sizes) and at long-term follow-up (mathematics: *d* 0.18, 95% CI –0.11, 0.47, $n=3$ effect sizes; word decoding: *d* 0.13, 95% CI –0.17, 0.42, $n=3$ effect sizes). Similarly, Schwaighofer et al. (2015) reported a pattern of small and non-significant effects for mathematical performance and word decoding immediately after training (mathematics: *d* 0.09, 95% CI –0.09 to 0.27, I^2 49.7%; $n=15$ effect sizes; word decoding: *d* 0.08, 95% CI –0.06, –0.22; $n=14$ effect sizes) and at long-term follow-up (mathematics: *d* 0.08, 95% CI –0.12 to 0.28, I^2 0%, $p=.83$; $n=8$ effect sizes; word decoding: *d* 0.21, 95% CI –0.21, 0.45; $n=5$ effect sizes). In the analysis of children with ADHD, Peijnenborgh and colleagues (2015) observed for word decoding an immediate small and significant effect (*g* 0.36, 95% CI 0.04, 0.68, $p=.03$; I^2 0%, $p=.68$; $n=4$ effect sizes) and a long-term moderate and significant effect (*g* 0.48, 95% CI 0.18, 0.12, $p=.01$; I^2 19%, $p=.29$; $n=3$ effect sizes). For arithmetic, Peijnenborgh and colleagues identified a pattern of small effects that were non-significant both immediately (*g* 0.25, 95% CI –0.06, 0.56, $p=.11$; I^2

0%, $p = .82$; $n = 3$ effect sizes) and at long-term follow-up (g 0.29, 95% CI −0.06, 0.64, $p = .11$; I^2 0%, $p = .68$; $n = 2$ effect sizes).

In summary, the pooled effect on mathematics reported in meta-analyses has ranged from 0.07 to 0.25, and in none of the analyses has this effect been significant. For word decoding, the evidence is less consistent, and there is some suggestion that training can improve word decoding and this benefit can persist.

Behavioural Symptoms of ADHD

Given that several disorders across the lifespan, including ADHD, are characterised by inattentive behaviour, the benefits of a working memory training programme for symptoms such as these are of particular clinical interest. Two meta-analyses have evaluated the effects of training for aspects of behavioural symptoms of ADHD (Cortese et al., 2015; this chapter by Spencer-Smith & Klingberg). Both of these meta-analyses use ratings of behavioural symptoms from parents and teachers. Cortese and colleagues (2015) reported two different analyses, one with the main outcome used in the study, and one with a rater who was not involved in the training and thus considered to be blinded. For inattentive measures, the outcome of the first analysis was 0.47 CI −0.09 – 0.66, and for the second 0.32 (CI −0.01 – 0.66). In our own meta-analysis, we identified a small to medium effect of Cogmed working memory training on inattentive behaviour (SMD −0.40, 95% CI −0.68, −0.12, $p < .005$; I^2 61%, $p = .001$; $n = 15$ effect sizes), and this immediate effect persisted at long-term follow-up (SMD −0.33, 95% CI −0.57, −0.09, $p = .006$, I^2 0%, $p = .95$; $n = 7$ effect sizes).

In summary, the pooled effect for inattentive symptoms reported in meta-analyses has ranged from 0.32 to 0.47.

Conclusions and Future Directions for Working Memory Training Research

Studies of the effect of working memory training serve two aims. One aim is scientific use as a tool to study the principles of how cognitive functions are improved by training, and the underlying neural basis of this change. This might be informative for a general understanding of how experience and environment affect cognition. It has been suggested that the principles of cognitive training are one important factor behind the cognitive and brain development seen during childhood and adolescence (Klingberg, 2014).

Another outcome of the working memory training studies is its possible usefulness as a clinical or educational intervention. The first working memory training studies (Klingberg, Forssberg, & Westerberg, 2002; Klingberg et al., 2005) focused on inattentiveness in ADHD as a possible application area. Following the claims and publicity surrounding the study by Jaeggi, Buschkuehl, Jonides, and Perrig (2008), much attention was focused on the possible changes of non-verbal IQ, and whether working memory training ‘makes you smarter’. A third application area is the possible improvement of academic abilities such as reading comprehension and mathematical performance.

Starting with the latter discussion of academic abilities, the idea of enhancement is not far-fetched given the high correlation between working memory capacity and academic performance, as well as the brain regions underlying these functions. Indeed, there are several promising studies, such as a study including more than 170 children, suggesting a small but reliable improvement of arithmetic with an effect size of around 0.2 (Bergman-Nutley & Klingberg, 2014). However, the meta-analyses described in this chapter do not suggest any systematic improvement of mathematical performance after working memory training. This might reflect the fact that mathematics is a broad concept, encompassing many different abilities, and even if working memory training focusing on visuospatial tasks would improve one specific aspect of mathematics, such as arithmetic, this is not evident when a large number of studies using different tests of mathematics are combined. It might also be that enhancing working memory capacity is not sufficient for improved mathematical abilities, unless the cognitive training is combined with content-specific training.

The question of enhancement of IQ has not been reviewed in this chapter. As is the case with mathematics, non-verbal IQ is estimated by a range of different measures. If training has an effect on one specific aspect of reasoning, as suggested by a review by Au et al. (2015), it might take time to accumulate systematic evidence. The recent review suggested that working memory training using the n-back task has a systematic effect with an effect size of 0.24. However, for studies using active control groups, this effect decreased to 0.06.

The current overview suggests to us that one promising practical use of working memory training is to decrease inattentive symptoms. Meta-analyses described in this chapter have reported a decrease in symptoms ranging from 0.32 to 0.47 standard deviations, which is well in line with previous estimates of 0.3 to 0.5 by Cortese et al. (2015). Future analyses could suggest different responses for the different subgroups of individuals with distinct neuropsychological profiles, various comorbidities, and individuals prescribed different pharmaceutical treatments. So far, data suggest that training might be more difficult in children with oppositional defiant disorder (Chaco et al., 2013), and that children with learning difficulties and low IQ have a smaller response (Söderqvist, Nutley, Ottersen, Grill, & Klingberg, 2012), whereas both young and old healthy adults benefit (Brehmer, Westerberg, & Bäckman, 2012).

A challenge for this area is the development of ecologically relevant measures of inattentive symptoms. So far, most studies rely on questionnaires from teachers and parents, which are subjective and generally have a low inter-rater reliability. Furthermore, they might be less sensitive to cognitive aspects of behaviour, where we might expect effects from cognitive training. This was illustrated in the study by Green et al. (2012), where objective, quantified measures of inattentive behaviour picked up improvements from cognitive training, whereas ratings from teachers did not. When noisy measures are used, the signal-to-noise ratio, and thus the estimated effect size, is lower. One of the goals for future research should be to develop ecologically valid and objective measures of attention in everyday life.

Future studies and more data could reveal subgroups that are more or less responsive to training, and the less responsive groups might need longer training periods than those currently used. Optimisation of the tasks and training regimes used is a work that has barely started. Further research is thus warranted, and cognitive training remains a promising non-pharmacological treatment for inattentive symptoms.

References

Alloway, T. P., & Alloway, R. G. (2009). The efficacy of working memory training in improving crystallized intelligence. *Nature Precedings*. Retrieved from http://precedings.nature.com/documents/3697/version/1/files/npre20093697-1.pdf

Angevaren, M., Aufdemkampe, G., Verhaar, H. J., Aleman, A., & Vanhees, L. (2008). Physical activity and enhanced fitness to improve cognitive function in older people without known cognitive impairment. *Cochrane Database of Systematic Reviews*, (*3*), CD005381.

Au, J., Sheehan, E., Tsai, N., Duncan, G., Buschkuehl, M., & Jaeggi, S.M. (2015) Improving fluid intelligence with training on working memory: A meta-analysis. *Psychonomic Bulletin & Review, 22*, 366–377.

Beck, S., Hanson, C., Puffenberger, S., Benninger, K., & Benninger, W. (2010). A controlled trial of working memory training for children and adolescents with ADHD. *Journal of Clinical Child and Adolescent Psychology, 39*(6), 825–836.

Bergman-Nutley, S., & Klingberg, T. (2014) Effect of working memory training on working memory, arithmetic and following instructions. *Psychological Research, 78*, 869–877.

Björkdahl, A., Akerlund, E., Svensson, S., & Esbjörnsson, E. (2013). A randomized study of computerized working memory training and effects on functioning in everyday life for patients with brain injury. *Brain Injury, 27*(13–14), 1658–1665.

Brehmer, Y., Westerberg, H., & Bäckman, L. (2012). Working-memory training in younger and older adults: Training gains, transfer, and maintenance. *Frontiers in Human Neuroscience, 6*, 63.

Chacko, A., Bedard, A.-C., Marks, D., Feirsen, N., Uderman, J., Chimiklis, A.,...Ramon, M. (2013). A randomized clinical trial of Cogmed Working Memory Training in school-age children with ADHD: A replication in a diverse sample using a control condition. *Journal of Child Psychology and Psychiatry, 55*(3), 247–255.

Coghill, D. R., Seth, S., Pedroso, S., Usala, T., Currie, J., & Gagliano, A. (2014). Effects of methylphenidate on cognitive functions in children and adolescents with attention-deficit/hyperactivity disorder: Evidence from a systematic review and a meta-analysis. *Biological Psychiatry, 76*(8), 603–615.

Cohen, J. (1992). A power primer. *Psychological Bulletin, 112*(1), 1155–1159.

Colcombe, S., & Kramer, A. F. (2003). Fitness effects on the cognitive function of older adults: A meta-analytic study. *Psychological Science, 14*(2), 125–130.

Collaboration, T. C. (2011). *Cochrane handbook for systematic reviews of interventions version 5.1.0 Updated March 2011*: Available from http://www.cochrane-handbook.org.

Cortese, S., Ferrin, M., Brandeis, D., Buitelaar, J., Daley, D., Dittmann, R. W.,...Sonuga-Barke, E. J. S. (2015). Cognitive training for Attention-Deficit/Hyperactivity Disorder: Meta-analysis of clinical and neuropsychological outcomes from randomized controlled trials. *Journal of the American Academy of Child and Adolescent Psychiatry, 54*(3), 164–174.

Egeland, J., Aarlien, A., & Saunes, B.-K. (2013). Few effects of far transfer of working memory training in ADHD: A randomized controlled trial. *PLoS ONE, 8*(10), e75660.

Green, C., Long, D., Green, D., Iosif, A.-M., Dixon, J., Miller, M.,...Schweitzer, J. (2012). Will working memory training generalize to improve off-task behavior in children with attention-deficit/hyperactivity disorder? *Neurotherapeutics, 9*(3), 639–648.

Gropper, R. J., Gotlieb, H., Kronitz, R., & Tannock, R. (2014). Working memory training in college students with ADHD or LD. *Journal of Attention Disorders, 18*(4), 331–345.

Grunewaldt, K., Løhaugen, G., Austeng, D., Brubakk, A.-M., & Skranes, J. (2013). Working memory training improves cognitive function in VLBW preschoolers. *Pediatrics, 131*(3), e747–754.

Hardy, K. K., Willard, V. W., Allen, T. M., & Bonner, M. J. (2013). Working memory training in survivors of pediatric cancer: A randomized pilot study. *Psychooncology, 22*(8), 1856–1865.

Hardy, K., Willard, V., & Bonner, M. (2011). Computerized cognitive training in survivors of childhood cancer: A pilot study. *Journal of Pediatric Oncology Nursing, 28*(1), 27–33.

Higgins, J., Thompson, S., Deeks, J., & Altman, D. (2003). Measuring inconsistency in meta-analyses. *British Medical Journal, 327*(7417), 557–560.

Jaeggi, S. M., Buschkuehl, M., Jonides, J., & Perrig, W. J. (2008) Improving fluid intelligence with training on working memory. *Proceedings of the National Academy of Sciences of the United States of America, 105*, 6829–6833.

Kelly, M. E., Loughrey, D., Lawlor, B. A., Robertson, I. H., Walsh, C., & Brennan, S. (2014). The impact of exercise on the cognitive functioning of healthy older adults: A systematic review and meta-analysis. *Ageing Research Reviews, 16*, 12–31.

Klingberg, T., Fernell, E., Olesen, P., Johnson, M., Gustafsson, P., Dahlström, K.,...Westerberg, H. (2005). Computerized training of working memory in children with ADHD: A randomized, controlled trial. *Journal of the American Academy of Child & Adolescent Psychiatry, 44*(2), 177–186.

Klingberg, T., Forssberg, H., & Westerberg, H. (2002). Training of working memory in children with ADHD. *Journal of Clinical and Experimental Neuropsychology, 24*(6), 781–791.

Klingberg, T. (2014) Childhood cognitive development as a skill. *Trends in Cognitive Sciences, 18*, 573–579.

Loosli, S., Buschkeuhl, M., Perrig, W., & Jaeggi, S. (2012). Working memory training involves reading processes in typically developing children. *Child Neuropsychology, 18*(1), 62–78.

Løhaugen, G., Antonsen, I., Haberg, A., Gramstad, A., Vik, T., Brubakk, A., & Skranes, J. (2011). Computerized working memory training improves function in adolescents born at extremely low birth weight. *Journal of Pediatrics, 158*(4), 555–561.e554.

Ludqvist, A., Grundström, K., Samuelsson, K., & Rönnberg, J. (2010). Computerized working memory in a group of patients suffering from acquired brain injury. *Brain Injury, 24*(10), 1173–1183.

Mawjee, K., Woltering, S., Lai, N., Gotlieb, H., Kronitz, R., & Tannock, R. (2014). Working memory training in ADHD: Controlling for engagement, motivation, and expectancy of improvement (pilot study). *Journal of Attention Disorders, pii*: 1087054714557356.

Melby-Lervåg, M., & Hulme, C. (2013). Is working memory training effective? A meta-analytic review. *Developmental Psychology, 49*(2), 270–291.

Peijnenborgh, J. C. A., Hurks, P. M., Aldenkamp, A. P., Vles, J. S. H., & Hendriksen, J. G. M. (2015). Efficacy of working memory training in children and adolescents with learning disabilities: A review study and meta-analysis. *Neuropsychological Rehabilitation, 26*(5–6), 645–672.

Rapport, M. D., Orban, S. A., Kofler, M. J., & Friedman, L. M. (2013). Do programs designed to train working memory, other executive functions, and attention benefit children with ADHD? A meta-analytic review of cognitive, academic, and behavioural outcomes. *Clinical Psychology Review, 33*(8), 1237–1252.

Roig, M., Nordbrandt, S., Geertsen, S. S., & Nielsen, J. B. (2013). The effects of cardiovascular exercise on human memory: A review with meta-analysis. *Neuroscience & Biobehavioral Reviews, 37*(8), 1645–1666.

Roughan, L., & Hadwin, J. (2011). The impact of working memory training in young people with social, emotional and behavioural difficulties. *Learning and Individual Differences, 21*(6), 759–764.

Schwaighofer, M., Fischer, F., & Bühner, M. (2015). Does working memory training transfer? A meta-analysis including training conditions as moderators. *Educational Psychologist, 50*(2), 138–166.

Sjöwall, D., Roth, L., Lindqvist, S, & Thorell, L. (2013). Multiple deficits in ADHD: Executive dysfunction, delay aversion, reaction time variability, and emotional deficits. *Journal of Child Psychology and Psychiatry, 54*(6), 619–627

Sonuga-Barke, E. J., Bradeis, D., Cortese, S., Daley, D., Ferrin, M., Holtmann, M.,...European ADHD Guidelines Group. (2013). Nonpharmacological interventions for ADHD: Systematic review and meta-analyses of randomized controlled trials of dietary and psychological treatments. *American Journal of Psychiatry, 170*(3), 275–289.

Söderqvist, S., Nutley, S. B., Ottersen, J., Grill, K. M., & Klingberg, T. (2012). Computerized training of non-verbal reasoning and working memory in children with intellectual disability. *Frontiers in Human Neuroscience, 6*, 271.

Spencer-Smith, M., & Klingberg, T. (2015). Benefits of a working memory training program for inattention in daily life: A systematic review and meta-analysis. *PLoS ONE, 10*(3), e0119522.

Steeger, C. M., Gondoli, D. M., Gibson, B. S., & Morrissey, R. A. (2016). Combined cognitive and parent training interventions for adolescents with ADHD and their mothers: A randomized controlled trial. *Child Neuropsychology, 22*(4), 394–419.

Trinh, N. H., Hoblyn, J., Mohanty, S., & Yaffe, K. (2003). Efficacy of cholinesterase inhibitors in the treatment of neuropsychiatric symptoms and functional impairment in Alzheimer disease: A meta-analysis. *JAMA, 289*(2), 210–216.

van Dongen-Boomsma, M., Vollebregt, M. A., Buitelaar, J. K., & Slaats-Willemse, D. (2014). Working memory training in young children with ADHD: A randomized placebo-controlled trial. *Journal of Child Psychology and Psychiatry, 55*(8), 886–896.

Verburgh, L., Königs, M., Scherder, E. J., & Oosterlaan, J. (2014). Physical exercise and executive functions in preadolescent children, adolescents and young adults: A meta-analysis. *British Journal of Sports Medicine, 48*(12), 973–979.

Wåhlstedt, C., Thorell, L., & Bohlin, G. (2009). Heterogeneity in ADHD: Neuropsychological pathways, comorbidity and symptom domains. *Journal of Abnormal Child Psychology, 37*(4), 551–564.

Westerberg, H., Jacobaeus, H., Hirvikoski, T., Clevberger, P., Ostensson, M. L., Bartfai, A., & Klingberg, T. (2007). Computerized working memory training after stroke—A pilot study. *Brain Injury, 21*(1), 21–29.

Young, J., Angevaren, M., Rusted, J., & Tabet, N. (2015). Aerobic exercise to improve cognitive function in older people without known cognitive impairment. *Cochrane Database of Systematic Reviews, 4*, CD005381.

29

Cognitive Control in the Injured Brain

Lesley K. Fellows

Cognitive control refers to the collection of processes that allow behaviour to escape from the control of the immediate environment. Broadly speaking, these processes support the ability to maintain goals distant in time or place, and to reach those goals by anticipating the required sub-goals, taking actions different from prepotent or habitual responses, monitoring progress, and making adjustments in the face of changing internal or external conditions. Executive function is another term applied to more or less the same construct. Simply put, the capacity of cognitive (or executive) control is needed to carry out any behaviour that is not evidently prompted by the environment, right here, right now. Although either of these terms is sometimes used to refer to more specific constructs, we will take this broad conceptualisation as our starting point.

History

The notion that some brain regions might contribute to behaviour not by sensing or responding but rather by preventing 'automatic' responses was expressed as early as the nineteenth century. One of the first neurophysiologists, David Ferrier, speculated that the frontal lobe might carry out such a function, on the basis of the finding that, although stimulation of the frontal cortex of anaesthetised dogs produced no observable behaviour, ablations in this area altered behaviour. He wrote: 'But besides the power to act in response to feelings or desires there is also the power to inhibit or restrain action, notwithstanding the tendency of feelings and desires to manifest themselves in active motor outbursts' (Ferrier, 1876, p. 282). It is likely that Ferrier's speculation on a frontal role in higher-order behavioural control owed as much to the clinical knowledge that brain injury could lead to disordered, disinhibited behaviour as it did to his experimental findings. He acknowledged the influence of the clinical observations of the neurologist Hughlings Jackson to his work in general, and was surely aware of clinical cases of frontal lobe injury. For example, the famous case of Phineas Gage, a construction foreman who had an iron rod blown through his brain, had been reported a few years earlier. Harlow, the physician who published that report, captured the existing clinical gestalt, describing Gage post-injury as '... fitful, irreverent, indulging at times in the grossest profanity (which was not previously his custom), manifesting but little deference for his fellows, impatient of restraint or advice when it conflicts with his desires ...' (Harlow, 1868). The details of Gage's clinical presentation, supported by modern reconstructions of his injury (Damasio, Grabowski, Frank, Galaburda, & Damasio, 1994), argue that he suffered extensive

The Wiley Handbook of Cognitive Control, First Edition. Edited by Tobias Egner.

prefrontal damage. Although at the time his case was most notable for the fact that he survived, his was one of many cases that contributed to the idea that higher-order behaviours could be disrupted despite intact basic neurological functions, and that the frontal lobes had an important role in this behavioural control.

More systematic observations of case series, along with efforts to conceptualise cognitive functions more specifically, and measure these with tasks drawn from the rapidly developing field of experimental psychology, contributed a great deal to our understanding of structure–function relationships in the human brain in general, and of the underpinnings of cognitive control in particular. These efforts were accelerated by the large-scale armed conflicts of the twentieth century, which sadly produced many people with focal brain injuries, and perhaps more happily brought them into contact with some brilliant and insightful physicians and psychologists. Systematic observations or experimental studies in other clinical populations, including patients who had undergone frontal lobotomies, added further to this body of knowledge.

Modern conceptualisations of cognitive control can be directly traced to scientists influenced by clinical observations. In particular, Luria articulated the idea that the prefrontal cortex contributed importantly to a 'supervisory' attentional function, programming, monitoring, and adjusting behaviour, on the basis of his studies of dysfunction in a variety of clinical populations, notably head-injured veterans (Luria, 1973). Building on this tradition, and again on the basis of research in prefrontal-damaged patients as well as insights from cognitive psychology, Norman and Shallice proposed a more detailed model of a 'supervisory attentional system' related to the frontal lobes (Norman & Shallice, 1986). Stuss & Benson focused on the frontally mediated components of higher-order control within a similar 'executive' framework, proposing drive, sequencing, and control as key concepts (Stuss & Benson, 1984).

Although the frontal lobes received the most attention in this body of work, to the point where the functions under study were often referred to as 'frontal-executive', none of these scientists espoused what Luria termed a 'narrow localizationist' view (Luria, 1973). That is, that higher-order cognitive processes could be related exclusively to some specific brain region, frontal or otherwise. Rather, the working model has been that specific brain regions make specialised contributions to particular cognitive functions, with several regions contributing to a given higher-order process. (For a recent more detailed summary tracing the development of this idea, see Stuss, 2011).

The Injured Brain: Source of Basic Knowledge and Translational Target

People with injured brains have taught us a great deal about the biological basis of behaviour in general and of cognitive control in particular. They also stand to gain from this fundamental knowledge, to the extent that it can be applied to improve diagnosis or suggest avenues for treatment. Cognitive neuroscience has primarily drawn on clinical evidence from patients with focal brain lesions, studied experimentally, where structure–function relationships are more straightforward. However, many of the most common brain disorders arise from diffuse or multifocal dysfunction, as in neurodegenerative diseases or multiple sclerosis, and, at least presumptively, mental illnesses such as schizophrenia and depression. Impairments in cognitive control have been reported in all of these conditions, and many others besides. This work may or may not directly inform our fundamental understanding of the neurobiology of cognitive control, but it is of high translational importance, providing brain-based explanatory models, diagnostic tools, and starting points for rational treatment development for these common and often debilitating conditions.

This chapter will consider cognitive control in the injured brain from both fundamental and translational-applied perspectives. We will begin with a brief overview of the clinical phenomenology of impaired cognitive control, then review major cognitive neuroscience models of cognitive control, asking to what extent these models can be related to the clinical phenomena conceptually, and whether they have empirical support from studies of the effects of focal brain injury. Finally, we will discuss how current ideas about cognitive control are being applied in diagnosis and treatment of brain disorders.

Cognitive Control at the Bedside

'Frontal' Behavioural Syndrome

Clinicians have long recognised a constellation of symptoms and signs that tend to occur together in disorders affecting either the frontal lobes themselves, or subcortical regions connected to the frontal lobes, such as the mediodorsal thalamus and the basal ganglia (Barbas, 2015). Such patients generally show behaviour that is suboptimally tuned to their environmental context. Generally they follow instructions and well-learned routines quite well, but have difficulty in open-ended situations that require flexible behavioural adjustments in response to changes in the environment or context. Social interactions require a high degree of context-sensitive behavioural adjustment, and are often the most revealing of these behavioural difficulties. These patients may be perceived as socially inappropriate, overfamiliar, lacking in insight and empathy, and showing poor judgement. These broad characterisations of aberrant behaviour do not lend themselves well to analysis at the level of brain mechanisms, and, as we shall see, may or may not correspond closely with current ideas about cognitive control. However, there have been efforts to identify more specific elements of impaired behaviour, still at the clinical level, that might be more tractable for mechanistic analysis.

Disinhibition

An 'in your face' feature of the frontal syndrome, disinhibition is closest to Ferrier's and his contemporaries' idea about what the frontal lobes contribute to behaviour, and has been a powerful, albeit controversial, trope in this field ever since (Bari & Robbins, 2013). As the name implies, in clinical usage, this simply refers to release of behaviours that, while they might be suited to some situations, are inappropriate for the current setting. Common examples are overfamiliarity in formal situations like the doctor's office or in interactions with strangers, or swearing or making sexual remarks to an extent that was not the person's habit pre-injury, a pattern noted as far back as the case of Phineas Gage. Milder forms may occur, such as making insensitive or impolite comments to close family that the person might have self-censored more heavily prior to the injury. Even a casual analysis of this description suggests that such behaviours might emerge from more than one deficit: that is, difficulty inhibiting an inappropriate response, a failure to appreciate that the response is not, in fact, appropriate, or a deficit in generating more appropriate alternative responses. As we shall see, this multiplicity of potential mechanisms for clinically observable behavioural difficulties is a recurring theme.

Impulsivity

Itself a complex construct (Bickel, Jarmolowicz, Mueller, Gatchalian, & McClure, 2012; Evenden, 1999), impulsivity is often defined as 'acting without thinking', and has a connotation of rapidity: that is, that the 'act' has occurred too quickly to be prevented by the 'thinking'. This has obvious descriptive correspondence with cognitive control, and conceptual overlap with disinhibition (Bari & Robbins, 2013; Bickel et al., 2012). Actions that might be considered impulsive can be seen in those with frontal damage, although inappropriately rapid responses are not a particularly striking feature of the frontal syndrome, overall. Impulsivity in the speeded sense is perhaps more common in certain neuropsychiatric disorders, such as attention-deficit hyperactivity disorder or drug addiction (e.g., Chapter 32 by Chaarani et al. in this volume). Impaired cognitive control may be central to both these disorders, although this claim is debated, as is whether it relates to frontal dysfunction specifically in either case (Barkley, 2010; Bickel et al., 2012; Tang, Posner, Rothbart, & Volkow, 2015).

Perseveration

A classic behavioural finding after frontal lobe damage is the tendency to repeat a specific response or a particular behavioural repertoire despite receiving negative feedback, or to have difficulty departing from an overlearned sequence, as in alternation tasks such as the Trails B Test, which requires shifting between well-learned letter and number sequences. This is a hallmark of behavioural inflexibility, also famously captured by the Wisconsin Card Sorting Task (discussed below), and often reported by caregivers as an element of the frontal behavioural syndrome. As we shall see below, this much-studied phenomenon can arise from disruption of several more specific component abilities.

Abulia

Literally 'lack of will', this refers to a reduction in observable responses despite intact motor or language abilities required to produce them. It may be considered a contrast to impulsivity, where actions occur without much (or any) consideration of the consequences. Patients, if questioned, may report a sort of goal-lessness, a state that does not seem to be distressing to them, although it may be very distressing to those around them. The most extreme form of abulia, termed akinetic mutism, is associated with a complete absence of responses, although the patient is awake and capable of moving and speaking. This severe avolitional state is classically related to acute bilateral anterior cerebral artery stroke, a mercifully rare event that typically damages a large swathe of bilateral medial frontal lobes, and may affect the genu of the corpus callosum, frontal poles, and medial orbitofrontal cortex as well in some cases. It is also seen, at least transiently, after more focal injury to supplementary and pre-supplementary motor areas within the dmPFC (Potgieser, de Jong, Wagemakers, Hoving, & Groen, 2014).

Apathy

The term apathy is typically applied to capture a deficit in the emotional or motivational aspect of goal-directedness. In theory, such a patient could identify a goal, but lacks the drive or 'activation' to work towards it. A general lack of emotional responsiveness is often also present. To the observer, abulia and apathy will look much the same, and some consider them as describing the same behavioural change, or at least behaviours that overlap or lie on the same continuum (Marin, 1996).

Utilisation Behaviour

In its complete form, this is a striking example of cognitive control gone awry. Also termed 'environmentally bound' behaviour, patients act on objects (typically within their reach) in their environment according to the obvious affordances of those objects, and without consideration of the context. Subtle examples would be grasping the handle of a coffee mug on the desk in front of them, and drinking from it, apparently oblivious of the contextual detail that the mug belongs to the doctor. Similarly, they might answer the phone ringing on the desk, although it is not their phone. If objects are placed within reach without further instruction, they typically do whatever the object suggests: that is, pick up a pen and begin drawing or writing, in the absence of an instruction to do so, and without any evidence of a broader goal. They may also copy bizarre gestures or poses made by the examiner without being instructed to do so, and without remark or curiosity (Lhermitte, 1983; T Shallice, Burgess, Schon, & Baxter, 1989).

Alien Limb

Alien limb is another striking symptom, traditionally associated with damage to the dorsomedial frontal lobe, perhaps due to associated anterior corpus callosum disruption, but also a feature of parietal injury. In part, it can be understood as utilisation behaviour carried out by one limb (typically the contra-lesional arm). In this case, the actions taken by the limb (again, actions that are suggested by the environment, such as flicking on and off a light switch) are not congruent with the patient's goals, giving them the experience that the limb 'has a mind of its own'. Indeed, in extreme cases the unaffected arm, under the person's goal-directed control, may be used to try to stop the affected arm from taking these undesired actions (for a recent review, see Graff-Radford et al., 2013).

Anosognosia

This term refers to a lack of awareness of a patient's own neurological deficits (Gasquoine, 2015). It is not a specifically frontal finding: for example, it often accompanies neglect following parietal lobe injury, but it is a feature of the frontal lobe syndrome. Unlike the anosognosia seen with parietal lesions, where the patient may vigorously deny striking physical evidence of a deficit (paralysis of the arm, for example), the deficits of those with frontal lobe damage may be harder to recognise or put into words even for those observing the patient, because they are typically events occurring in specific contexts, rather than constantly present. Interestingly, the same patients who do not recognise their own behaviour as inappropriate while it is ongoing can identify such behaviour as inappropriate 'in principle' in a hypothetical scenario, or even when reviewing a video of themselves (Beer, John, Scabini, & Knight, 2006). These observations are important in thinking about the brain basis of self-monitoring, considered a key aspect of cognitive control.

Confabulation

Confabulation is an even more extreme behaviour that may reflect a monitoring failure, this time in the context of memory. Here, the patient does more than deny that their behaviour is inappropriate: They invent novel, sometimes bizarre, explanations for what they experience (or imagine), believe fervently in their own version of reality, and resist efforts by others to set the record straight. This can be associated with ventromedial frontal damage, although is not a constant feature of such damage (Gilboa et al., 2006).

Cognitive Control: From Bedside to Bench

The behavioural phenomena that can follow frontal lobe damage provide a rich starting point for investigating brain structure–function relationships, on the one hand suggesting candidate brain structures critical for cognitive control, and on the other, providing preliminary insights into the more basic behavioural elements that are drawn upon to produce flexible, goal-directed behaviour. However, clearly even the more specific of these clinical descriptions remain too general to be analysed at a formal or mechanistic level.

As chapters elsewhere in this volume amply illustrate, in the past decades, cognitive neuroscience has drawn upon many methods in working towards a more complete, brain-based understanding of cognitive control. Broadly speaking, there are two general directions in this work. One has conceptualised cognitive control as a single, domain-general process that can be flexibly applied to a variety of inputs (Duncan, 2001; Miller & Cohen, 2001; see also Chapter 13 by Stokes, Buschman, & Miller in this volume). The brain basis of this ability was at first localised to a few nodes, particularly within the prefrontal cortex (Duncan, 2010), and more recently is being re-envisaged in relation to more elaborate network models of complex behaviour (Hampshire & Sharp, 2015). The other has focused on splitting cognitive control into still more narrowly defined processes, establishing that these processes are in fact distinct, and seeking to relate them to specific brain regions, circuits, and mechanisms at the level of neuronal population codes. Although these two broad approaches are sometimes pitted against each other, there is good reason to believe that they are not mutually exclusive, and that both can provide useful frameworks for studying cognitive control, and for applying the findings in the clinic.

The field has moved far from bedside observations in patients, drawing on many new sources of evidence, most notably functional neuroimaging of the healthy brain, and borrowing ideas and formal models not only from psychology, but also computer science, ecology, and economics, among other fields. These efforts have provided fresh insights, new conceptual models, and analytic frameworks, and more specific behavioural measures. If this knowledge is to contribute to improving human health and well-being, in the end we must succeed in closing the loop that began at the bedside. It is therefore pertinent for both fundamental and translational reasons to ask whether current models of cognitive control are supported by evidence from patients with brain injury.

Lesion Studies in Humans

This question cannot be satisfactorily answered simply by matching clinical phenomenology as set out at the start of this chapter with cognitive neuroscience models, although researchers sometimes have trouble resisting this temptation. As already briefly discussed, these general clinical syndromes can, in principle, have many underlying explanations. A more rigorous approach is to test cognitive neuroscience models directly with lesion methods. As with any method, lesion approaches have strengths and limitations (for a detailed review, see Fellows, 2012). We do not claim here that these methods are some sort of gold standard, but they do have unique inferential strengths (Chatterjee, 2005; Fellows et al., 2005), and a robust model of cognitive control needs converging support from a variety of methods. Loss-of-function experiments, such as lesion studies, allow two important claims to be tested: first, that a brain region or circuit is necessary for a given process, and second, that putatively different behavioural processes are, in fact, dissociable. Finally, lesion evidence provides a direct bridge to clinical application of these models, for example, by allowing the study of how disruption of a particular component process is expressed in real-world function. There are several forms of lesion evidence, from single case studies to large-scale group designs allowing mapping of dysfunction at the level of individual voxels. For reasons discussed at length elsewhere (Fellows, 2012), group studies generally provide the most robust evidence, and will be emphasised here.

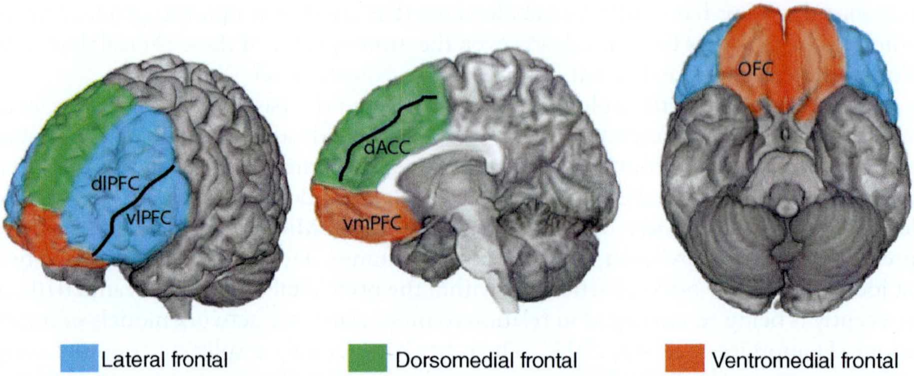

Figure 29.1 Schematic illustration of frontal lobe anatomical regions discussed in this chapter. Broad subdivisions are indicated by colours. More specific sub-regions are indicated, again at a schematic level, with labels superimposed on the brain: dorsolateral prefrontal cortex (dlPFC), ventromedial PFC (vmPFC), dorsal anterior cingulate cortex (dACC), and orbitofrontal cortex (OFC). Source: Adapted from Tsuchida, 2012. Reproduced with permission of Ami Tsuchida.

Such studies have largely focused on the frontal lobes, in part because of the rich clinical literature relating this area to cognitive control, although as mentioned earlier, there is no doubt that non-frontal regions also contribute to these processes. Over the decades, as methods improved and brain-based models were refined, this work has aimed for finer anatomical resolution. However, there are practical limits to the resolution that can be attained in human lesion studies. For the purposes of providing an overview of this literature, I will use a very coarse division of the frontal lobes that roughly matches those inherent limitations, illustrated schematically in Figure 29.1. Note that the terms as used here encompass the underlying white matter, because lesions generally disrupt both grey and white matter, and do not necessarily match the typically more specific usage of the same labels in the human neuroimaging literature.

Group-Based Study Designs

Human lesion studies are, in part, 'experiments of nature', in that the lesions themselves are not, for obvious reasons, under direct experimental control. As such, lesion location can be treated as either the dependent or independent variable. In behaviour-driven approaches, patients are recruited systematically from some predefined clinical population, and then classified as showing or not showing a given deficit (say, the frontal behavioural syndrome, or impaired performance on an experimental task), and the pattern of lesion location is then compared in the two groups. This is a challenging design to carry out effectively, because it is prone to several sources of bias. In some cases, lesion location is not considered at all: Instead, the experiment tests whether two or more processes can be dissociated in patients with brain injury or dysfunction (for example, in neurodegenerative or psychiatric disorders), or even healthy variation, an approach that can be very useful in establishing whether putative component processes are, in fact, distinct.

Lesion location-driven designs instead recruit patients with damage to a pre-specified region of interest (ROI), and then ask whether performance on a given task differs compared to performance of one or more control groups. Experiments that have a lesioned control group (i.e., patients with damage sparing the ROI in question) allow the strongest inferences, because effects cannot be ascribed to non-specific factors that might accompany any brain injury. This design also allows testing for dissociations. Indeed, one particularly robust approach tests two ROIs and two component processes, asking whether the structure–function relationships can be doubly dissociated. This not only addresses questions about whether component processes are distinct, but also sheds light on the brain basis of each, in one experiment, and also protects against certain methodological limitations of simpler ROI-based designs.

ROI approaches are practically limited in their anatomical resolution, because the boundaries of the region must be defined a priori. Although a tiny ROI could be selected in theory, in practice, this is generally not feasible: Patients suited to this kind of research do not exist in large numbers. ROIs are usually selected to balance theory and pragmatic factors, so that an adequate sample size will be achieved over a reasonable time frame.

Modern statistical methods, in part borrowed from functional MRI analytic approaches, have recently been applied to improve the anatomical resolution of group studies. Voxel-based lesion-symptom mapping is the most common approach: Here, a group of patients is characterised on a given task, and then a statistical map is created that tests whether those with damage to a given voxel (or cluster of voxels) differ significantly from those with damage sparing that voxel (Bates et al., 2003; Rorden, Karnath, & Bonilha, 2007). This method suffers from certain problems, in part because damage tends to co-vary across specific voxels owing to common lesion causes (e.g., strokes follow a vascular territory, so damage on one area typically co-varies with damage to another). Methods to mitigate this are available (Mah, Husain, Rees, & Nachev, 2014). Perhaps the major problem, though, is that these statistical approaches require large sample sizes that may be practically difficult to achieve. Nonetheless, these methods have been applied in several studies of cognitive control, providing interesting evidence that will discussed below.

Cognitive Control as a Domain-General Process

This approach builds on integrative views of complex behaviour arising from psychology: that there is a common factor ('g') underlying human intelligence (Hunt, 2011; Spearman, 1904), as captured by IQ tests, and, perhaps relatedly (Barbey et al., 2012), a common, multipurpose 'executive' system that serves to hold and manipulate a range of inputs (J. Duncan, Emslie, Williams, Johnson, & Freer, 1996; John Duncan & Owen, 2000; Miller & Cohen, 2001). The focus is on what process or processes diverse control-requiring tasks have in common, rather than in how they may differ. There is good evidence for a strong relationship between performance of various tests of cognitive control in healthy individuals, including across the age span (McCabe, Roediger, McDaniel, Balota, & Hambrick, 2010), but also evidence for distinctions (Miyake & Friedman, 2012). Functional neuroimaging studies have argued for a common set of interacting regions supporting this putatively common cognitive control process. Variously termed the 'cognitive control', 'multiple-demand network', or 'executive attention' network, it includes, at least, the ventrolateral PFC, dorsomedial frontal lobe (DMF), and parietal cortex (John Duncan, 2010; Hampshire & Sharp, 2015; Niendam, Laird, Ray, Dean, & Carter, 2013; see also Chapter 18 by Cole in this volume).

The strong prediction of this general model is that damage to any of these brain areas would disrupt performance of a range of tasks to the extent that they have a common requirement for cognitive control. There have been relatively few direct tests of this claim in patients with focal brain damage. One recent large-scale study of Vietnam veterans with penetrating head injuries compared a group with damage that included left lateral PFC damage to those with prefrontal damage that spared that region on general intelligence measured by the Wechsler Adult Intelligence Scale and a range of executive tasks drawn from a clinical battery. Although the sample size was large by the standards of this sort of study, there were no significant differences between dlPFC- and non-dlPFC-injured groups on the raw task measures, although the dlPFC group did differ from healthy controls on latent variables related to 'g' and a common 'executive' latent variable derived from five executive function tasks. These findings do not provide unambiguous evidence in favour of a unitary model of cognitive control: The design leaves the question open, given the use of executive tasks that are not uniquely tapping specific cognitive control processes, and that the lesioned control group presumably included those with damage to the DMF and white matter tracts that are also part of the putative cognitive control network. Indeed, another, presumably overlapping sample (n = 182) drawn from the same source was analysed using a voxel-based lesion-symptom mapping (VLSM) approach, reported in a different paper: Here, impairment on individual executive function tasks was related to damage to broad areas of the frontal lobe (extending well beyond the dlPFC and DMF), and the temporal-parietal lobe, as well as the underlying white matter between these regions, principally in the left hemisphere (Barbey et al., 2012). Individual tasks had somewhat different patterns of associated brain injury, and one (the tower test) did not show any significant structure–function associations.

Another large sample study used VLSM to test for lesion–function relationships across clinical neuropsychological tests putatively requiring cognitive control: the Trails test, which requires switching between well-learned patterns; the Stroop task; the Wisconsin Card Sorting Test (WCST); and the Controlled Oral Word Association test, which involves generating words according to a particular category, also called verbal fluency. That study found specific, largely distinct structure–function relationships for all four tasks. They defined a common variable across the tasks, which they termed the cognitive control factor, and found a significant relationship between worse performance as captured by that factor and a region of the left DMF, among other regions, including a restricted area in the OFC, motor cortex, and subcortical regions. The study was certainly powered to detect effects in the 'cognitive control' network defined above; it is interesting that only the left DMF is in common across that network and the regions related to a common 'control' factor in this study. That said, the tasks used, although well-established, clinically useful tests, vary in how closely they cleave to current views of cognitive control, and by and large, are not those used to identify the 'cognitive control' network in the functional neuroimaging research that has mainly driven this view.

In a recent study, we used three tasks drawn from the experimental literature, meant to tap the three sub-components of cognitive control identified through latent variable analysis of healthy individual variability: inhibition/interference control (Stroop task), attention shifting (letter-number naming), and updating in working memory (spatial search; Friedman & Miyake, 2004). We administered these to a group of patients with frontal damage (N = 45), and demographically matched healthy participants, asking whether damage to a putative cognitive control network would affect performance of all three tasks. We had good lesion coverage of key nodes, including the lateral frontal lobe (LF) and DMF. We found that DMF damage was not significantly associated with poor performance on any of these tasks. Left LF damage was selectively associated with poor performance in both the Stroop and attention-shifting tasks, but not updating in working memory (as captured by the spatial search task). The latter task relied (weakly) on the intactness of other frontal regions (Tsuchida & Fellows, 2012). This result would seem to argue against the strongest unitary view of cognitive control. Furthermore, in a separate region-of-interest study that focused on the right and left vlPFC, we measured interference control with two widely used tasks: the Stroop and Eriksen flanker tasks. We found a double dissociation in performance, with left vlPFC damage affecting Stroop interference, but sparing flanker interference, and vice versa (Geddes, Tsuchida, Ashley, Swick, & Fellows, 2014). This argues that regions thought to comprise a putatively 'general' cognitive control network are either not critical at all for widely used measures of cognitive control, or are selectively engaged depending on the details of the task, for example, whether the material over which control is required is verbal or non-verbal.

The lesion evidence for a common cognitive control network is still preliminary, but so far rather weak, with some studies presenting clear challenges to the strongest form of this model. The question deserves more study: Lesion approaches have a lot to offer in further establishing the details of which tasks (and which measures drawn from those tasks) have a shared reliance on specific sub-regions. Work to date suggests that there is important nuance in aspects of task design, nature of the stimuli, and probably other factors such as timing, response requirements, and extent of practice that can affect whether performance is disrupted by damage to a given brain area. Sorting these details out will allow a much more specific model of regional contributions to cognitive control, and indeed will help the field reach a clearer understanding of the nature of the control process (or control processes).

Component Processes of Cognitive Control

If lesion studies to date provide, at best, rather mixed evidence for a multipurpose cognitive control network, do they provide stronger evidence for regional contributions to task-specific control? A larger literature has addressed this question over several decades of work, aiming to isolate specific components of cognitive control, and test their dissociability both behaviourally and in relation to their brain localisation (Stuss, 2011). The list of candidate component processes studied in this way is long, but a shorter list has received the most attention, and is most pertinent to current views of cognitive control. These include classic 'cognitive' abilities such as maintenance, manipulation and updating in working memory, task switching, response inhibition, top-down attentional (interference) control, and detecting and adjusting to response conflict or errors. Recent work has expanded this list to include processes linked to 'hot' aspects of control, such as valuation, motivation, and drive, (i.e., the 'why' rather than the 'how' of goal-directed behaviour), with less consensus on how these should be conceptualised and distinguished (Bickel et al., 2012; Fellows, 2007; Glascher et al., 2012). Further complicating this list, as discussed above, there is evidence for material specificity for at least some of these processes, with different brain regions critical depending on the material.

Determining whether these putative processes are distinct requires being able to successfully isolate them behaviourally. This has been a challenge for the field. Although the most widely used tasks derive from clinical neuropsychology, and were developed in part to assess localisable neural processes, most are at least somewhat 'dirty', in that they call upon more than one process (McCabe et al., 2010). Many have been around for a long time, bringing the advantages of established psychometrics and robust norms, but the disadvantages of being only loosely related to current views of cognitive control. Even narrowly defined, theoretically grounded behavioural measures, including many related to specific patterns of regional BOLD activation, may not be pure measures of the constructs they target. For example, the much-studied trial-by-trial conflict adaptation effect in the Stroop or flanker tasks (the so-called Gratton effect [Gratton, Coles, & Donchin, 1992]) has been claimed to be partly or entirely due to automatic priming effects related to stimulus or response repetition (Mayr & Awh, 2009; Mayr, Awh, & Laurey, 2003), or more general expectancy effects (Egner, 2007; for an in-depth discussion, see Chapter 4 by Egner in this volume). Similarly, the stop-signal reaction time measure from the stop-signal task, widely interpreted as indexing response inhibition (Chapter 6 by Verbruggen & Logan in this volume), has been alternatively proposed to reflect a more general capacity to change (rather than inhibit) behaviour (Banich & Depue, 2015; although in the latter case, both constructs are considered examples of cognitive control). Details of task design may be critical in determining the extent to which cognitive control is engaged: For example, the extent of practice or whether conditions are manipulated across blocks or trials (e.g., Floden, Vallesi, & Stuss, 2011). With these provisos in mind, we next provide a critical overview of lesion evidence that prefrontal sub-regions are necessary for performance of specific component processes relevant to cognitive control.

Working Memory: Maintaining, Manipulating, Updating

Holding information in mind when it is not present in the environment, that is, working memory, seems to be a fundamental basis for cognitive control. Both the goal itself, and the information needed to reach the goal, may need to be held in working memory. Greater demands are placed on control as the amount of information (load) increases, and as that information must be protected from distraction, manipulated (for example, re-ordered), or updated. The content of working memory—conventionally, an important distinction has been made between verbal and non-verbal information—may also be important.

Simply holding information across a delay, as in the clinically widely used digit span or spatial span tests, is not consistently affected by frontal lobe damage in humans (D'Esposito, Cooney, Gazzaley, Gibbs, & Postle, 2006). As demands increase, however, intact frontal lobes become necessary for task performance (Baier et al., 2010; Milner, Petrides, & Smith, 1985; Thompson-Schill et al., 2002). Tasks with substantial lesion evidence include the self-ordered pointing task (Ross, Hanouskova, Giarla, Calhoun, & Tucker, 2007), variations on the n-back task also widely used in functional imaging studies, and clinical neuropsychological tests, including subtests of the Wechsler Adult Intelligence Scale.

Updating in working memory, as tested by a letter 2-back task, has been shown in a relatively large sample lesion study to rely critically on only two regions with the frontal lobes: the left LF, and DMF (including the dACC). The pattern of behavioural impairment differed by lesion location: Left lateral damage was associated with an increase in 'misses', whereas DMF damage was associated with an increase in false alarms (Tsuchida & Fellows, 2009; Figure 29.2). At the 'big picture' level, one could conclude from these data that the DMF and left LF regions are critical nodes in a network required for letter 2-back performance. However, clearly these two regions do not contribute identically to performance, implying distinct mechanistic contributions: Other evidence suggests that the left lateral PFC is likely involved in maintaining (or selecting) the currently relevant (verbal) information (Glascher et al., 2009; Thompson-Schill et al., 2002), whereas the DMF may be contributing to setting an optimal response threshold in the face of uncertainty (Sheth et al., 2012). Thus, even this narrowly defined task engages at least two definable processes, which in turn rely on distinct brain regions.

Furthermore, other tasks that also require updating in working memory clearly do not rely critically on the left LF. For example, a widely used spatial working memory task (Owen, Morris, Sahakian, Polkey, & Robbins, 1996) instead seems to rely weakly on other frontal regions. Other work shows a mixed picture across several working memory-requiring tasks (reviewed in Tsuchida & Fellows, 2012). Overall, specific frontal regions are clearly critical for performance of some working memory tasks, and those same regions are equally clearly *not* required for all working memory tasks, when directly tested in the same patient sample.

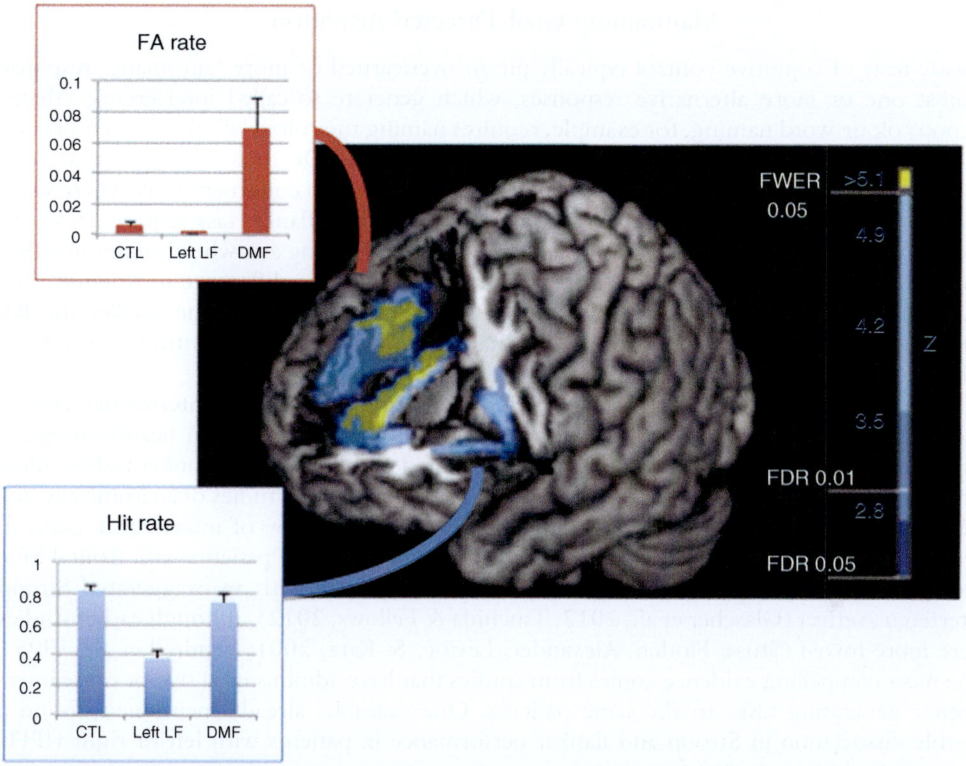

Figure 29.2 Voxel-based lesion-symptom statistical map (Brunner-Munzel tests, thresholded at a false discovery rate > 0.05) from a study of 29 patients with focal frontal lobe damage who completed a letter 2-back task, overlaid on a rendering of the Montreal Neurological Institute (MNI) brain. Damage to voxels within two regions, the left LF and DMF, was associated with significantly worse overall 2-back performance (d'L) compared to those with damage sparing those regions. The insets show that specific behavioural impairments differed between those with damage to the DMF (who had significantly higher false alarm [FA] rates) and the left LF (who had significantly lower hit rates), highlighting the distinct contributions of these two regions to task performance. Source: Adapted from Tsuchida and Fellows, 2009.

This supports a material specific account of control in working memory, and identifies the brain areas that are necessary for these more specific processes.

Controlled Attention: Maintaining and Shifting

A second major component of cognitive control is generally framed as attentional, broadly involving two related processes: strategically filtering sensory information to match current goals, and shifting that selective attention to newly goal-relevant information when the context changes. These processes share at least a superficial similarity to those outlined for working memory, above. The difference here is that the sensory stimuli are present in the environment, rather than held in working memory. It may be that at least of some of the brain processes involved are shared, however (Frank & Badre, 2015).

Maintaining Goal-Directed Attention

Classic tests of cognitive control typically pit an overlearned or more 'automatic' response against one or more alternative responses, which generate so-called interference effects. Stroop colour-word naming, for example, requires naming the colour of ink a word is printed in, while (on incongruent trials) *not* reading the word itself. On these trials, where the ink colour and word are different, responses are slower compared to congruent trials where word and ink colour invoke the same response. Similarly, the Eriksen flanker task requires choosing a left or right response based on a central arrow. Multiple flanking arrowheads point in either the same (congruent) or opposite (incongruent) direction: The difference in reaction time (RT) between these two trials types captures the interference effect. The smaller the RT difference, the more effectively the individual is controlling their attention to the task-relevant aspect of the stimulus.

Is there evidence that specific brain regions are critical for minimising interference effects, and if so, is this task- or content-specific, or a general-purpose ability? In healthy subjects, individual differences in interference resolution (tested with Stroop and flanker tasks) reflect a common latent variable (Friedman & Miyake, 2004). However, studies of anatomically discrete lesions have, by and large, supported a more modular view of interference control. There is strong evidence from recent, large-scale lesion studies of patients with frontal lobe damage that left lateral PFC damage is specifically associated with an exaggerated Stroop interference effect (Glascher et al., 2012; Tsuchida & Fellows, 2012), although earlier studies were more mixed (Stuss, Floden, Alexander, Levine, & Katz, 2001; Vendrell et al., 1995). The most compelling evidence comes from studies that have administered two or more interference generating tasks to the same patients. One example, already mentioned, found a double dissociation in Stroop and flanker performance in patients with left or right vlPFC damage (Geddes et al., 2014).

The DMF, including the dACC, has also been implicated in interference control. The lesion evidence for a role for this region is mixed, and only a few studies have tested whether any deficits generalise across tasks. Highly selective cingulotomy targeting the dACC (lesions for treatment of neuropsychiatric disorders) did not disrupt overall interference effects in a task with demands similar to both Stroop and flanker tasks (Sheth et al., 2012). Larger lesions involving the DMF had no significant impact on the Stroop effect, nor on the modulation of this effect by a manipulation of the level of response conflict (Fellows & Farah, 2005b; Tsuchida & Fellows, 2012), although another study of a similar population did find exaggerated interference in the Simon task (di Pellegrino, Ciaramelli, & Làdavas, 2007). However, Sheth et al. did demonstrate an acute change post-cingulotomy in trial-by-trial effects of conflict, again highlighting the level of nuance required in operationalising behavioural measures (and the theoretical accounts that guide them), an area of little consensus in the field as a whole.

Shifting Attention as Goals Change

A hallmark of cognitive control is its flexibility. As the context changes, it is often necessary to attend to different features of the environment. This ability is captured by attention-shifting tasks. These vary substantially in the details of their requirements, but at their core involve responding to different stimulus features depending on the trial. Most often, the goal is provided to the participant through instruction. The RT or error rate increases on trials that require a shift from one attentional 'set' to another, compared to trials where the same set is repeated, typically referred to as a shift cost (for a review, see Shallice, Stuss, Picton, Alexander, & Gillingham, 2009; Chapter 2 by Monsell in this volume). A classic example is the Wisconsin

Card Sorting Task, which adds the requirement of identifying the relevant stimulus feature (attentional set) through trial and error Another widely used clinical task, the Trails B Task, also (presumptively) taps the ability to shift between sets, as the subject connects an alternating series of numbers and letters. The failure to shift attentional set yields a pattern of perseveration, where the individual continues to select on the basis of the previously successful or overlearned set. Performance on these clinical tests is sensitive to the frontal lobe—particularly the LF—injury, although not with great specificity, likely reflecting the relative complexity of these tasks (Demakis, 2003; Kopp et al., 2015).

Studies using experimental tasks aimed at isolating attentional shifting provide stronger support for a specific, critical role of the LF, although the data are somewhat mixed. In part this may be because the concept of attention shifting turns out to be far from simple in its operationalisation, with different 'shifts', and different neural circuits, engaged across tasks (Shallice et al., 2009; Wager, Jonides, Smith, & Nichols, 2005). One study we carried out showed, across 45 patients with focal frontal damage, a specific association between left LF damage and exaggerated shift cost in a simple shifting task requiring a verbal response (letter or number naming; Tsuchida & Fellows, 2012), a finding consistent with earlier studies that found that left frontal damage (and not right frontal damage) resulted in higher switch costs across several tasks (Mayr, Diedrichsen, Ivry, & Keele, 2006; Rogers et al., 1998; Shallice et al., 2009). Another study found effects of both right and left LF damage on a more complex shifting task, proposing different contributions by the two regions (Aron, Monsell, Sahakian, & Robbins, 2004).

Response Selection

Goal Adjustment

A different kind of behavioural flexibility is called for when the association between a goal and a feature of the environment changes. In contrast to attentional shifting, where feedback (or instruction) signals the need to change the focus to other features, in this case, the same feature remains relevant, but an adjustment is needed in interpreting its behavioural meaning. The classic test of this form of behavioural flexibility, sometimes termed affective shifting (Dias, Robbins, & Roberts, 1996), is called reversal learning. In its simplest form, widely studied in animal models, two cues (e.g., different colours) are associated with reward or non-reward, and the organism must choose between them. Once learning is demonstrated by the consistent choice of the rewarded stimulus, the feedback associations are reversed, so that the previously rewarding colour is no longer rewarded, and vice versa. The animal must continue to focus on colour as relevant, but update the goal-meaning/reward associations of the colours. Early lesion work in animal models established that the ventromedial frontal (VMF) region was critical for reversal learning when feedback associations were fully deterministic (Butters, Butter, Rosen, & Stein, 1973), a specific effect confirmed in humans with VMF damage compared to those with LF damage (Fellows & Farah, 2003; Hornak et al., 2004). Again, these patients show a kind of perseveration, choosing stimuli based on a once-useful association that they seem to be unable to quickly update in the face of unexpected negative feedback. Subsequent work has used more subtle tasks in which the stimulus–feature associations are probabilistic. This is more demanding, requiring the participant to interpret the meaning of a given feedback event within the broader context of reward expectations in the task as a whole. Large frontal lobe lesion studies in humans have consistently found that such tasks, including the widely used Iowa Gambling Task (Bechara, Damasio, Tranel, & Damasio, 1997; Fellows & Farah, 2005a), rely critically on the VMF region, with the medial OFC perhaps being the most consistently implicated sub-region within the VMF cortex, particularly for tasks that focus on reversal learning more narrowly (Fellows & Farah, 2003; Glascher et al., 2012; Tsuchida, Doll, & Fellows, 2010; note there is uncertainty about this localisation, with very selective lesion studies in macaques raising the possibility that disruption of underlying white matter may drive the behavioural effect [Rudebeck, Saunders, Prescott, Chau, & Murray, 2013]). In these probabilistic tasks, patients with VMF lesions no longer show perseveration, instead maladaptively shifting their responses too often. This highlights how disruption of the same fundamental process—here, optimally interpreting feedback to develop stimulus–reward associations suited to the task context—can show itself in apparently opposite patterns of behaviour even in controlled experimental tasks, that is, perseveration on the one hand and impersistence on the other.

An interesting wrinkle in relation to affective shifting is that one can apply a nearly identical paradigm to test the ability to associate outcomes to specific actions, rather than specific stimulus features, and that this seemingly similar form of behavioural flexibility no longer requires intact OFC, instead being disrupted by DMF (in monkeys, specifically the dACC) damage (Camille, Tsuchida, & Fellows, 2011; Rudebeck et al., 2008). Once again, this emphasises the task-specific nature of prefrontal contributions, even to more narrowly defined aspects of cognitive control.

Response Inhibition

Inhibition is a central concept in cognitive control (Chapter 6 by Verbruggen & Logan in this volume), of particular interest in relation to neuropsychiatric disorders and Parkinson's disease (Dagher & Robbins, 2009; Chapter 32 by Chaarani et al. in this volume). The necessary neural substrates of response inhibition, defined as preventing a motor (usually manual) response, have been studied in considerable detail (Chapter 16 by Forstmann & Alkemade in this volume), although consensus has been surprisingly elusive. Common tasks in this literature include go-no go tasks and the stop-signal task. Echoing a theme seen for other aspects of cognitive control, above, there has been some work on whether the neural substrates of response inhibition are 'response-general', that is, a common circuit engaged to stop all manner of responses, or response specific: That is, different circuits are responsible for stopping manual, oculomotor, emotional, and even memory recall (as in the 'think-no-think' paradigm [Munakata et al., 2011]).

Focal lesion studies of response inhibition have, between them, emphasised two prefrontal regions as important: the DMF and the right vlPFC. There has also been considerable work on basal ganglia contributions to manual response inhibition, highlighting how this area of study must integrate basic aspects of action selection with higher-order, response-related processes (Frank & Badre, 2015).

Broad surveys of prefrontal contributions to response inhibition provide a fairly coherent account: One study of 43 patients with focal lesions involving the PFC found that the FA rate on a manual go-no go task, a behavioural measure that would seem the most obvious index of inhibition, was associated with damage to the DMF—specifically, the premotor cortex (area 6; Picton et al., 2007). This is also consistent with our finding of increased FAs in patients with dmPFC damage in an n-back task with an (implicit) go-no go response requirement, discussed above (Tsuchida & Fellows, 2009). The Picton et al. study also examined go-no go performance in terms of RT variability, finding this was related to damage to a more extensive region of the dmPFC as well as the right vlPFC. Whether this reflects inhibition, or a more general requirement to maintain 'task set' or simply vigilance, is unclear. A study of right and left lateral frontal-damaged patients found that the right vlPFC was selectively associated with prolonged SSRT in the stop-signal task (Aron, Fletcher, Bullmore, Sahakian, &

Robbins, 2003), leading to a strong claim that this region is uniquely critical for response inhibition (Aron, Robbins, & Poldrack, 2004). However, this claim is not without controversy, in part because of focal lesion evidence implicating other PFC regions in inhibition in seemingly similar tasks, and because of a lack of definitive replication of the initial finding in right vlPFC patients (Krämer et al., 2013; Swick & Chatham, 2014). More generally, the field still struggles with how to conceptualise inhibition (Aron & Verbruggen, 2008; Braver, 2012), and how domain-general conceptions of inhibitory control need to be framed to explain the data (Banich & Depue, 2015).

Response Adjustment

Two broad research streams, both very active areas, are pertinent to considering cognitive control in relation to responses: conflict monitoring and error processing. In contrast to the largely clinical inspiration for the cognitive processes studied so far, the concept of conflict monitoring arose from cognitive psychology and computational modelling as a solution to the problem that cognitive control seemed to require a 'homunculus' to signal the need to engage that control (Botvinick, Braver, Barch, Carter, & Cohen, 2001). The conflict monitoring theory proposes that competing (incompatible) responses signal conflict, which engages control mechanisms, in turn reducing conflict (e.g., Chapter 5 by Bugg in this volume). This is proposed to occur on a rapid time scale, explaining trial-by-trial adjustments in performance as in the Gratton effect (see Chapter 4 by Egner in this volume). The initial cognitive neuroscience work on this theory implicated the dACC specifically in conflict monitoring (M. Botvinick, Nystrom, Fissell, Carter, & Cohen, 1999). A great deal of research has been conducted around this model, defying quick summary (Mansouri, Tanaka, & Buckley, 2009). The lesion work on the topic has tested whether damage to the DMF disrupts the behavioural evidence of conflict monitoring, that is, measures of the degree to which cognitive control is being adjusted as task requirements change. Initial studies did not observe effects of dACC lesions on performance adjustments induced by block-wise manipulations of conflict level (Fellows & Farah, 2005b; Diane Swick & Turken, 2002), arguing against a critical role for this region in conflict monitoring, at least construed as a general process, though lesion effects on trial-by-trial performance adjustments were observed in a more recent study pre- and post-surgical cingulotomy in humans (Sheth et al., 2012). That argument rests on the behavioural measures in fact being reliable indicators of control adjustment, and, as discussed above, there has been considerable debate about these behavioural measures. (For an extensive review on ACC function and models, see Chapter 15 by Brown in this volume).

There is also an extensive literature on error processing that is likely pertinent to cognitive control (see Chapter 17 by Ullsperger and Chapter 14 by M.X. Cohen in this volume). EEG and fMRI studies have repeatedly implicated the DMF in error processing, leading to the conceptualisation that errors are simply an extreme case of (failed) response conflict monitoring, explaining why the dmPFC was engaged in both contexts. Lesion studies have examined whether the DMF plays a critical role in error processing, and attempted to better define that critical contribution: We showed that patients with such damage are aware of their errors and show slowing after errors, suggesting that trial-by-trial 'control' (as indexed by post-error slowing) is preserved. However, they show clear deficits in rapid error correction, within trial. They were able to make these corrections, but were significantly slower than healthy controls (Modirrousta & Fellows, 2008). We recently studied the kinematics of error responses in a similar sample of patients, adding further evidence that the impairment after dmPFC damage selectively affects the speed of the corrective response, within trial rather than the initiation or arrest of the erroneous response (Hochman, Wang, Milner, & Fellows, 2015). Other work showed that patients with left ventrolateral PFC damage had diminished post-error slowing, and reductions in evoked potential measures of error processing (Wessel, Klein, Ott, & Ullsperger, 2013).

These experiments use choice RT tasks, where the participant can establish that an error is occurring while it is under way, presumably by further stimulus processing after the erroneous response has been initiated, but before it is completed. The evidence reviewed above (and other work beyond the scope of this chapter) suggests that response adjustment related to correction can occur very quickly, and relies on the intact DMF, and perhaps the (left) vlPFC. Responses can also be judged to be errors in the context of feedback, for example, in tasks where the participant learns about response correctness by experiencing a subsequent reward or punishment. There has been recent interest in connecting these two views (Shenhav, Botvinick, & Cohen, 2013; Silvetti, Alexander, Verguts, & Brown, 2013; Chapter 10 by Kool, Shenhav, & Botvinick in this volume). As mentioned above, DMF lesions in humans do disrupt feedback-driven learning when the task is entirely in an action framework, but not when the feedback is linked to the stimulus (Camille et al., 2011), providing at least a preliminary point of contact between these conceptualisations of DMF contributions to response adjustment.

Synthesis

This chapter has provided an overview of the brain basis of cognitive control as studied with focal lesion methods in humans. We considered work that addresses cognitive control as a common, multiple-demand system engaged across tasks, and a larger body of evidence focused on testing the necessary contributions of specific (prefrontal) regions to component processes of cognitive control. Although lesion studies can provide important tests of either of these perspectives, they have not been brought to bear in any exhaustive way to test the multiple-demand view. However, at least some of the work that has been done cannot be easily reconciled with the strongest form of that view. It is clearly the case that performance across tasks that would all seem to engage cognitive control can be dissociated in patients with focal lesions, and that damage to specific regions implicated in the 'multiple-demand' network need not disrupt classic measures of control such as RT indices of interference in the Stroop and flanker tasks.

Taken together, the lesion evidence provides a fairly compelling account for a necessary role of some specific prefrontal sub-regions in some potential component processes of cognitive control, at least for some measures and some tasks. This chapter aimed to highlight where there is more consensus and replication in the lesion literature, and where there is controversy or outright failure to support component process models of cognitive control primarily arising from functional imaging data.

Methodological Considerations

Where current models are not supported by lesion studies, we need to consider two potential methodological explanations: First, the behavioural measures may be inadequate. If the tasks do not index the component process of interest sufficiently well, null findings in lesioned samples can be difficult to interpret. Recent efforts to use similar or identical experimental tasks to those used with other cognitive neuroscience methods would seem useful in this regard, providing more direct tests of current models than tasks drawn from clinical neuropsychology that often engage several control processes (for a nice example, see Volle et al., 2012). However, the psychometric properties of newer, more focused experimental measures may be suboptimal, or at least poorly characterised. The lesion literature reviewed here thus emphasises the importance of continuing to work not only towards well-specified, detailed models of cognitive control, but also of assuring rigorous operationalisation of candidate component processes. Even tasks carrying the same label, but varying in the nuances of their design, may be more or less sensitive to effects of specific prefrontal damage. This behavioural measurement problem is obviously important for the field as a whole, and, as we shall see below, it is crucial for effectively translating this work to clinical application.

A second methodological consideration is specific to lesion studies: Null effects could, in principle, be due to some 're-organisation' or recovery process. Most of the work reviewed here was carried out in adults, at least several months post-injury. Clinically, we know that patients tend to recover function to varying degrees over the days and weeks after such injuries. However, except perhaps after tiny lesions, it is unlikely that this functional recovery is based on full recovery of the underlying brain processes. If the brain is finding 'workarounds' to continue to support the activities the person wishes to undertake, these are typically revealed in slower RTs, higher error rates, or other evidence of the use of alternative strategies. Indeed, the enterprise of experimental neuropsychology relies on this observation. Real-life function may recover, either through engagement of alternative brain processes or environmental adaptations, but specific cognitive processes, carefully measured, can be shown to be impaired. Could cognitive control processes be exceptions to this general rule? Could their flexibility make them more resistant to the effects of focal damage? Clinically, the answer may very well be yes: There are potentially many ways to solve a higher-order behavioural challenge, and this could facilitate recovery. If one control module is not functioning well, another could be engaged. For example, if rapid error correction is impaired owing to DMF injury, the person could slow down and use explicit feedback to make behavioural adjustments. That this takes another 100–200 ms may have no detectable effect in many real-life settings (although presumably would continue to be disabling in, say, an Olympic table tennis player). However, it is surely detectable by a well-designed task in the laboratory. Indeed, lesion studies may be particularly useful in identifying those tasks and behavioural measures that have tight and specific structure–function relationships.

Conceptual Considerations

The rather weak support for domain-general views of cognitive control from lesion studies does not mean that some multi-demand type model might not be found that can encompass findings from studies of the effects of focal injury. Alternative ways of conceptualising how cognitive control might be organised have been proposed, variations on the idea that prefrontal representations are organised according to different levels of abstraction or in terms of acting on different kinds of inputs with a hierarchical or cascade organisation, rather than distinct cognitive processes (Badre & D'Esposito, 2009; Koechlin, Ody, & Kouneiher, 2003; O'Reilly, Noelle, Braver, & Cohen, 2002). These models are to some extent a middle ground between highly localised, domain-specific views of control, on one hand, and very general, multi-demand network views on the other. There has been some lesion work within these frameworks, providing preliminary support for at least two of these models (Azuar et al., 2014; Badre, Hoffman, Cooney, & D'Esposito, 2009; Koechlin et al., 2003). More research is needed, but these possibilities, particularly because they have been accompanied by efforts to give these models computational implementations, seem potentially fruitful.

Another 'middle ground' perspective proposes that both domain-specific and domain-general control processes may coexist (Wilson, Gaffan, Browning, & Baxter, 2010), although this is not easy to reconcile with the lesion evidence that argues against domain-general views, at least if specific and general control mechanisms are closely anatomically related (Fedorenko, Duncan, & Kanwisher, 2013).

A detailed mechanistic understanding of cognitive control must address the processes contributed by specific brain regions, and how those regions interact. Lesion studies are suited to contribute to this undertaking, although do have inherent limitations in their anatomical resolution. However, cognitive control processes in healthy subjects and in psychopathologies or non-focal neurological disorders might well vary for reasons other than regional dysfunction. It is clear that neuromodulators, particularly the monoamines, play a role in moderating neural responses in prefrontal and subcortical systems engaged in cognitive control (Arnsten, Wang, & Paspalas, 2015; Bowirrat et al., 2012; Cools & Robbins, 2004; Dagher & Robbins, 2009; Roberts, 2011; Chapter 19 by Robbins & Kehagia in this volume). To the extent that cognitive control requires interaction between distant brain regions, and requires that interaction to be precisely timed, it will be susceptible to failure due to variation or injury that affects that connectivity. The patterns of impairment in those settings may be very different from those seen after focal brain damage. Both of these sources of variation may be better conceptualised and studied in terms of more general, network-level models. Behavioural measures optimal for assessing these aspects of cognitive control may differ from those suited to detecting regionally specific effects.

Back to the Bedside?

The literature just reviewed, and the wider literature addressed in this book as a whole, have contributed to an understanding of the neural circuits of goal-directed behaviour that has been very influential in models of the many brain disorders, both neurologic and psychiatric, that have maladaptive behaviours as a central feature. To varying degrees, the tasks and explanatory frameworks have been directly applied in these disorders, (and, as discussed at the outset of this chapter, were often initially inspired by clinical observations). These direct links between fundamental and clinical-applied research are exciting, holding promise for improved the diagnosis, understanding, and treatment of brain disorders. However, it is worth briefly discussing the limitations and potential pitfalls of these translational efforts as well.

Perhaps the main challenge is that there is still little consensus about how best to model cognitive control or its several putative component processes, with considerable variation in the extent and quality of converging evidence for various existing models. In considering a model for clinical application, it is important to choose one with good evidence relevant to that application. Thus, if the aim is to characterise cognitive control in a particular disorder within an anatomical framework, tasks that have been shown to be sensitive to anatomically based dysfunction, whether lesions or disruption by transcranial magnetic stimulation, would support the strongest inferences. As mentioned above, the patterns of dysfunction (and the tasks most suited to assess these) may vary if the pathogenesis relates to non-localised or multifocal dysfunction, as might be seen in multiple sclerosis or Parkinson's disease, for example. Practically, it is also important to choose behavioural measures that are psychometrically robust, and to have a clear hypothesis about the patterns expected across measures to avoid multiple comparison problems.

These considerations apply to studies aiming to test the explanatory or diagnostic utility of models of cognitive control in clinical populations. They are even more pressing when applied to assessment of the impact of treatments. Both pharmacological and non-pharmacological approaches may improve cognitive control, and if the domain-general account holds, it may be possible to train control in one context and improve it in others (e.g., Tang et al., 2015; Chapter 28 by Spencer-Smith & Klingberg in this volume). Here, we face the additional challenge of understanding the clinical relevance of cognitive control measures drawn from fundamental cognitive neuroscience. If we successfully intervene to improve, for example, Stroop interference control, what might this mean for the person's real-life function? At the population level, there is evidence that cognitive control measured by behavioural tasks of this kind relates to important health outcomes: As just two examples, variation in executive task performance in the normal range, seemingly across all components of cognitive control, can predict higher body mass index (Vainik, Dagher, Dubé, & Fellows, 2013), and people with major depressive disorder perform worse on executive tasks than healthy controls (Snyder, 2014).However, there is much less information about how predictive such measures might be at the individual level, and less still about whether longitudinal change in performance on these tasks relates in a detectable way to function. Rigorously addressing these questions is an important emerging challenge for the field.

Relating task performance to symptoms detected by patients and caregivers would represent an intermediate step in establishing the ecological validity of this cognitive neuroscience-based approach to higher-order behaviour. A reasonable start would be to go back to the clinical phenomenology reviewed at the outset of this chapter, and ask whether it relates to either general or component measures of cognitive control in clinical populations. Here we encounter another problem: the need for valid instruments to assess clinical behaviours, whether through self- or informant-report. Many questionnaires that ask a person to rate the frequency and severity of various 'frontal' behaviours are available, but their utility remains a matter of debate. There is relatively weak evidence that these can isolate specific components of behaviour, such as impulsivity, disorganisation, or apathy (Malloy & Grace, 2005). This problem may be unavoidable, in that clinical symptoms may blur across component constructs: As mentioned above, even in experimental tasks, disruption of a single cognitive process may express itself in different behaviours depending on the context. In any case, there is some evidence that everyday difficulties do relate to cognitive control as measured by more specific tasks, although this has been studied primarily in relation to clinical neuropsychology tasks rather than the more specific constructs that have been the primary focus of this chapter (Burgess, Alderman, Evans, Emslie, & Wilson, 1998; Chaytor, Schmitter-Edgecombe, & Burr, 2006).

Conclusion

The study of cognitive control in the injured brain is an enterprise that stretches back to the very origins of human neuroscience. Modern cognitive neuroscience has built on this foundation, and now provides a range of sophisticated perspectives on cognitive control. These models vary in how well they explain experimental data from human participants with focal brain injury. The latter literature contributes considerable evidence for domain- and process-specific accounts of cognitive control, although there is still controversy around many of the details of those claims. Overall, lesion evidence provides only weak support for an exclusively domain-general perspective, with the existence of double dissociations across cognitive control-requiring tasks posing a particular challenge to those accounts. As cognitive neuroscience models are further refined, it will be important to validate their applicability in well-defined neurological and psychopharmacological contexts, to in turn apply them to better understand and treat the many psychiatric disorders that are marked by impaired cognitive control. The field as a whole has major potential for clinical application, but much remains to be done to fully realise that potential.

Acknowledgements

I thank Ami Tsuchida and Maiya Geddes for their contributions to some of the ideas about the effects of focal damage on cognitive control that are expressed here. The work from my laboratory was supported by the Canadian Institutes of Health Research (MOP-119291, MOP-97821) and salary support from the Fonds de Recherche du Quebec-Santé.

References

Arnsten, A., Wang, M., & Paspalas, C. (2015). Dopamine's actions in primate prefrontal cortex: Challenges for treating cognitive disorders. *Pharmacological Reviews*, *67*, 681–96.

Aron, A. R., Fletcher, P. C., Bullmore, E. T., Sahakian, B. J., & Robbins, T. W. (2003). Stop-signal inhibition disrupted by damage to right inferior frontal gyrus in humans. *Nature Neuroscience*, *6*(2), 115–116.

Aron, A. R., Monsell, S., Sahakian, B. J., & Robbins, T. W. (2004). A componential analysis of task-switching deficits associated with lesions of left and right frontal cortex. *Brain*, *127*(7), 1561–1573.

Aron, A. R., Robbins, T. W., & Poldrack, R. A. (2004). Inhibition and the right inferior frontal cortex. *Trends in Cognitive Sciences*, *8*(4), 170–177.

Aron, A. R., & Verbruggen, F. (2008). Stop the presses: Dissociating a selective from a global mechanism for stopping: Research article. *Psychological Science*, *19*(11), 1146–1153.

Azuar, C., Reyes, P., Slachevsky, A., Volle, E., Kinkingnehun, S., Kouneiher, F., ... Levy, R. (2014). Testing the model of caudo-rostral organization of cognitive control in the human with frontal lesions. *NeuroImage*, *84*, 1053–1060.

Badre, D., & D'Esposito, M. (2009). Is the rostro-caudal axis of the frontal lobe hierarchical? *Nature Reviews. Neuroscience*, *10*(9), 659–669.

Badre, D., Hoffman, J., Cooney, J. W., & D'Esposito, M. (2009). Hierarchical cognitive control deficits following damage to the human frontal lobe. *Nature Neuroscience*, *12*(4), 515–522.

Baier, B., Karnath, H.-O., Dieterich, M., Birklein, F., Heinze, C., & Müller, N. G. (2010). Keeping memory clear and stable—The contribution of human basal ganglia and prefrontal cortex to working memory. *The Journal of Neuroscience: The Official Journal of the Society for Neuroscience*, *30*(29), 9788–9792.

Banich, M. T., & Depue, B. E. (2015). Recent advances in understanding neural systems that support inhibitory control. *Current Opinion in Behavioral Sciences*, *1*, 17–22.

Barbas, H. (2015). General cortical and special prefrontal connections: Principles from structure to function. *Annual Review of Neuroscience*, *38*, 269–289.

Barbey, A. K., Colom, R., Solomon, J., Krueger, F., Forbes, C., & Grafman, J. (2012). An integrative architecture for general intelligence and executive function revealed by lesion mapping. *Brain*, *135*(4), 1154–1164.

Bari, A., & Robbins, T. W. (2013). Inhibition and impulsivity: Behavioral and neural basis of response control. *Progress in Neurobiology*, *108*, 44–79.

Barkley, R. A. (2010). Differential diagnosis of adults with ADHD: The role of executive function and self-regulation. *The Journal of Clinical Psychiatry*, *71*(7), e17.

Bates, E., Wilson, S. M., Saygin, A. P., Dick, F., Sereno, M. I., Knight, R. T., & Dronkers, N. F. (2003). Voxel-based lesion-symptom mapping. *Nature Neuroscience*, *6*(5), 448–450.

Bechara, A., Damasio, H., Tranel, D., & Damasio, A. R. (1997). Deciding advantageously before knowing the advantageous strategy. *Science (New York, N.Y.)*, *275*, 1293–1295.

Beer, J. S., John, O. P., Scabini, D., & Knight, R. T. (2006). Orbitofrontal cortex and social behavior: Integrating self-monitoring and emotion-cognition interactions. *Journal of Cognitive Neuroscience, 18*(6), 871–879.

Bickel, W. K., Jarmolowicz, D. P., Mueller, E. T., Gatchalian, K. M., & McClure, S. M. (2012). Are executive function and impulsivity antipodes? A conceptual reconstruction with special reference to addiction. *Psychopharmacology, 221*(3), 361–387.

Botvinick, M. M., Braver, T. S., Barch, D. M., Carter, C. S., & Cohen, J. D. (2001). Conflict monitoring and cognitive control. *Psychological Review, 108*(3), 624–652.

Botvinick, M., Nystrom, L. E., Fissell, K., Carter, C. S., & Cohen, J. D. (1999). Conflict monitoring versus selection-for-action in anterior cingulate cortex. *Nature, 402*(6758), 179–181.

Bowirrat, A., Chen, T. J. H., Oscar-Berman, M., Madigan, M., Chen, A. L. H., Bailey, J. A., … Blum, K. (2012). Neuropsychopharmacology and neurogenetic aspects of executive functioning: Should reward gene polymorphisms constitute a diagnostic tool to identify individuals at risk for impaired judgment? *Molecular Neurobiology, 45*(2), 298–313.

Braver, T. S. (2012). The variable nature of cognitive control: A dual mechanisms framework. *Trends in Cognitive Sciences, 16*(2), 106–113.

Burgess, P. W., Alderman, N., Evans, J., Emslie, H., & Wilson, B. A. (1998). The ecological validity of tests of executive function. *Journal of the International Neuropsychological Society: JINS, 4*(6), 547–558.

Butters, N., Butter, C., Rosen, J., & Stein, D. (1973). Behavioral effects of sequential and one-stage ablations of orbital prefrontal cortex in the monkey. *Experimental Neurology, 39*, 204–214.

Camille, N., Tsuchida, A., & Fellows, L. K. (2011). Double dissociation of stimulus-value and action-value learning in humans with orbitofrontal or anterior cingulate cortex damage. *Journal of Neuroscience, 31*(42), 15048–15052.

Chatterjee, A. (2005). A madness to the methods in cognitive neuroscience? *Journal of Cognitive Neuroscience, 17*(6), 847–9.

Chaytor, N., Schmitter-Edgecombe, M., & Burr, R. (2006). Improving the ecological validity of executive functioning assessment. *Archives of Clinical Neuropsychology, 21*(3), 217–227.

Cools, R., & Robbins, T. W. (2004). Chemistry of the adaptive mind. *Philosophical Transactions. Series A, Mathematical, Physical, and Engineering Sciences, 362*(1825), 2871–2888.

D'Esposito, M., Cooney, J. W., Gazzaley, A., Gibbs, S. E. B., & Postle, B. R. (2006). Is the prefrontal cortex necessary for delay task performance? Evidence from lesion and FMRI data. *Journal of the International Neuropsychological Society: JINS, 12*(2), 248–260.

Dagher, A., & Robbins, T. W. (2009). Personality, addiction, dopamine: Insights from Parkinson's disease. *Neuron, 61*(4), 502–510.

Damasio, H., Grabowski, T., Frank, R., Galaburda, A. M., & Damasio, A. R. (1994). The return of Phineas Gage: Clues about the brain from the skull of a famous patient. *Science (New York, N.Y.), 264*(5162), 1102–1105.

Demakis, G. J. (2003). A meta-analytic review of the sensitivity of the Wisconsin Card Sorting Test to frontal and lateralized frontal brain damage. *Neuropsychology, 17*(2), 255–264.

Di Pellegrino, G., Ciaramelli, E., & Làdavas, E. (2007). The regulation of cognitive control following rostral anterior cingulate cortex lesion in humans. *Journal of Cognitive Neuroscience, 19*(2), 275–286.

Dias, R., Robbins, T. W., & Roberts, A. C. (1996). Dissociation in prefrontal cortex of affective and attentional shifts. *Nature, 380*(6569), 69–72.

Duncan, J. (2001). An adaptive coding model of neural function in prefrontal cortex. *Nature Reviews Neuroscience, 2*(11), 820–829.

Duncan, J. (2010). The multiple-demand (MD) system of the primate brain: Mental programs for intelligent behaviour. *Trends in Cognitive Sciences, 14*(4), 172–179.

Duncan, J., Emslie, H., Williams, P., Johnson, R., & Freer, C. (1996). Intelligence and the frontal lobe: the organization of goal-directed behavior. *Cognitive Psychology, 30*(3), 257–303.

Duncan, J., & Owen, A. M. (2000). Common regions of the human frontal lobe recruited by diverse cognitive demands. *Trends in Neurosciences, 23*(10), 475–483.

Egner, T. (2007). Congruency sequence effects and cognitive control. *Cognitive, Affective & Behavioral Neuroscience, 7*(4), 380–390.

Evenden, J. L. (1999). Varieties of impulsivity. *Psychopharmacology, 146*(4), 348–361.

Fedorenko, E., Duncan, J., & Kanwisher, N. (2013). Broad domain generality in focal regions of frontal and parietal cortex. *Proceedings of the National Academy of Sciences of the United States of America, 110*(41), 16616–16621.

Fellows, L. K. (2007). Advances in understanding ventromedial prefrontal function: The accountant joins the executive. *Neurology, 68*(13), 991–995.

Fellows, L. K. (2012). Group Studies in Experimental Neuropsychology. In H. Cooper, P. M. Camic, D. L. Long, A. T. Panter, D. Rindskopf, & K. J. Sher (Eds.), *American Psychological Association handbook of research methods in psychology, Vol. 2: Research Designs: Quantitative, qualitative, neuropsychological and biological* (Vol. 2, pp. 647–659). Washington, DC.

Fellows, L. K., & Farah, M. J. (2003). Ventromedial frontal cortex mediates affective shifting in humans: Evidence from a reversal learning paradigm. *Brain, 126*, 1830–1837.

Fellows, L. K., & Farah, M. J. (2005a). Different underlying impairments in decision-making following ventromedial and dorsolateral frontal lobe damage in humans. *Cerebral Cortex (New York, N.Y.: 1991), 15*, 58–63.

Fellows, L. K., & Farah, M. J. (2005b). Is anterior cingulate cortex necessary for cognitive control? *Brain, 128*, 788–796.

Fellows, L. K., Heberlein, A. S., Morales, D. A., Shivde, G., Waller, S., & Wu, D. H. (2005). Method matters: An empirical study of impact in cognitive neuroscience. *Journal of Cognitive Neuroscience, 17*, 850–858.

Ferrier, D. (1876). *The functions of the brain*. New York: G.P. Putnam's Sons.

Floden, D., Vallesi, A., & Stuss, D. T. (2011). Task context and frontal lobe activation in the Stroop task. *Journal of Cognitive Neuroscience, 23*(4), 867–879.

Frank, M., & Badre, D. (2015). How cognitive theory guides neuroscience. *Cognition, 135*, 14–20.

Friedman, N. P., & Miyake, A. (2004). The relations among inhibition and interference control functions: A latent-variable analysis. *Journal of Experimental Psychology. General, 133*(1), 101–135.

Gasquoine, P. (2015). Blissfully unaware: Anosognosia and anosodiaphoria after acquired brain injury. *Neuropsychological Rehabilitation, 26*(2), 261–285.

Geddes, M. R., Tsuchida, A., Ashley, V., Swick, D., & Fellows, L. K. (2014). Material-specific interference control is dissociable and lateralized in human prefrontal cortex. *Neuropsychologia, 64*, 310–319.

Gilboa, A., Alain, C., Stuss, D. T., Melo, B., Miller, S., & Moscovitch, M. (2006). Mechanisms of spontaneous confabulations: A strategic retrieval account. *Brain, 129*(6), 1399–1414.

Glascher, J., Adolphs, R., Damasio, H., Bechara, A., Rudrauf, D., Calamia, M., … Tranel, D. (2012). Lesion mapping of cognitive control and value-based decision making in the prefrontal cortex. *Proceedings of the National Academy of Sciences of the United States of America, 109*(36), 14681–14686.

Gläscher, J., Tranel, D., Paul, L. K., Rudrauf, D., Rorden, C., Hornaday, A., … Adolphs, R. (2009). Lesion mapping of cognitive abilities linked to intelligence. *Neuron, 61*(5), 681–691.

Graff-Radford, J., Rubin, M. N., Jones, D. T., Aksamit, A. J., Ahlskog, J. E., Knopman, D. S., … Josephs, K. A. (2013). The alien limb phenomenon. *Journal of Neurology, 260*(7), 1880–1888.

Gratton, G., Coles, M. G., & Donchin, E. (1992). Optimizing the use of information: Strategic control of activation of responses. *Journal of Experimental Psychology. General, 121*(4), 480–506.

Hampshire, A., & Sharp, D. J. (2015). Contrasting network and modular perspectives on inhibitory control. *Trends in Cognitive Sciences, 19*(8), 445–452.

Harlow, J. (1868). Recovery from the passage of an iron bar through the head. *Publications of the Massachusetts Medical Society, 2*, 327–347.

Hochman, E., Wang, S., Milner, T., & Fellows, L. (2015). Double dissociation of error inhibition and correction deficits after basal ganglia or dorsomedial frontal damage in humans. *Neuropsychologia, 69*, 130–139.

Hornak, J., O'Doherty, J., Bramham, J., Rolls, E. T., Morris, R. G., Bullock, P. R., & Polkey, C. E. (2004). Reward-related reversal learning after surgical excisions in orbito-frontal or dorsolateral prefrontal cortex in humans. *Journal of Cognitive Neuroscience, 16*, 463–478.

Hunt, E. (2011). *Human intelligence*. Cambridge: Cambridge University Press.

Koechlin, E., Ody, C., & Kouneiher, F. (2003). The architecture of cognitive control in the human prefrontal cortex. *Science, 302*(5648), 1181–1185.

Kopp, B., Rosser, N., Tabeling, S., Sturenburg, H., de Haan, B., Karnath, H.-O., & Wessel, K. (2015). Errors on the trail making test are associated with right hemispheric frontal lobe damage in stroke patients. *Behavioral Neurology, 309235.*

Krämer, U. M., Solbakk, A.-K., Funderud, I., Løvstad, M., Endestad, T., & Knight, R. T. (2013). The role of the lateral prefrontal cortex in inhibitory motor control. *Cortex; A Journal Devoted to the Study of the Nervous System and Behavior, 49*(3), 837–849.

Lhermitte, F. (1983). 'Utilization behaviour' and its relation to lesions of the frontal lobes. *Brain: A Journal of Neurology, 106*(Pt. 2), 237–255.

Luria, A. R. (1973). *The working brain: An introduction to neuropsychology.* New York: Basic Books.

Mah, Y.-H., Husain, M., Rees, G., & Nachev, P. (2014). Human brain lesion-deficit inference remapped. *Brain: A Journal of Neurology, 137*(Pt. 9), 2522–2531.

Malloy, P., & Grace, J. (2005). A review of rating scales for measuring behavior change due to frontal systems damage. *Cognitive and Behavioral Neurology, 18*(1), 18–27.

Mansouri, F. A., Tanaka, K., & Buckley, M. J. (2009). Conflict-induced behavioural adjustment: A clue to the executive functions of the prefrontal cortex. *Nature Reviews. Neuroscience, 10*(2), 141–152.

Marin, R. S. (1996). Apathy: Concept, syndrome, neural mechanisms, and treatment. *Seminars in Clinical Neuropsychiatry, 1*(4), 304–314.

Mayr, U., & Awh, E. (2009). The elusive link between conflict and conflict adaptation. *Psychological Research, 73*(6), 794–802.

Mayr, U., Awh, E., & Laurey, P. (2003). Conflict adaptation effects in the absence of executive control. *Nature Neuroscience, 6*(5), 450–452.

Mayr, U., Diedrichsen, J., Ivry, R., & Keele, S. W. (2006). Dissociating task-set selection from task-set inhibition in the prefrontal cortex. *Journal of Cognitive Neuroscience, 18*(1), 14–21.

McCabe, D. P., Roediger, H. L., McDaniel, M. A., Balota, D. A., & Hambrick, D. Z. (2010). The relationship between working memory capacity and executive functioning: Evidence for a common executive attention construct. *Neuropsychology, 24*(2), 222–243.

Miller, E. K., & Cohen, J. D. (2001). An integrative theory of prefrontal cortex function. *Annual Review of Neuroscience, 24,* 167–202.

Milner, B., Petrides, M., & Smith, M. L. (1985). Frontal lobes and the temporal organization of memory. *Human Neurobiology, 4*(3), 137–142.

Miyake, A., & Friedman, N. P. (2012). The nature and organization of individual differences in executive functions: Four general conclusions. *Current Directions in Psychological Science, 21*(1), 8–14.

Modirrousta, M., & Fellows, L. K. (2008). Dorsal medial prefrontal cortex plays a necessary role in rapid error prediction in humans. *The Journal of Neuroscience: The Official Journal of the Society for Neuroscience, 28*(51), 14000–14005.

Munakata, Y., Herd, S., Chatham, C., Depue, B. E., Banich, M. T., & O'Reilly, R. C. (2011). A unified framework for inhibitory control. *Trends in Cognitive Sciences, 15*(10), 453–459.

Niendam, T. A., Laird, A. R., Ray, K. L., Dean, Y. M., & Carter, C. S. (2013). *Meta-analytic evidence for a superordinate cognitive control network subserving diverse executive functions, 12*(2), 241–268.

Norman, W., & Shallice, T. (1986). Attention to action. In R. Davidson, G. Schwartz, & D. Shapiro (Eds.), *Consciousness and self regulation: Advances in research and theory* (pp. 1–18). New York: Plenum.

O'Reilly, R. C., Noelle, D. C., Braver, T. S., & Cohen, J. D. (2002). Prefrontal cortex and dynamic categorization tasks: representational organization and neuromodulatory control. *Cerebral Cortex (New York, N.Y.: 1991), 12*(3), 246–57. Retrieved from http://www.ncbi.nlm.nih.gov/pubmed/11839599

Owen, A. M., Morris, R. G., Sahakian, B. J., Polkey, C. E., & Robbins, T. W. (1996). Double dissociations of memory and executive functions in working memory tasks following frontal lobe excisions, temporal lobe excisions or amygdalo-hippocampectomy in man. *Brain, 119*(5), 1597–1615.

Picton, T. W., Stuss, D. T., Alexander, M. P., Shallice, T., Binns, M. A., & Gillingham, S. (2007). Effects of focal frontal lesions on response inhibition. *Cerebral Cortex, 17*(4), 826–838.

Potgieser, A. R. E., de Jong, B. M., Wagemakers, M., Hoving, E. W., & Groen, R. J. M. (2014). Insights from the supplementary motor area syndrome in balancing movement initiation and inhibition. *Frontiers in Human Neuroscience, 8,* 960.

Roberts, A. C. (2011). The importance of serotonin for orbitofrontal function. *Biological Psychiatry, 69*(12), 1185–1191.

Rogers, R. D., Sahakian, B. J., Hodges, J. R., Polkey, C. E., Kennard, C., & Robbins, T. W. (1998). Dissociating executive mechanisms of task control following frontal lobe damage and Parkinson's disease. *Brain, 121*(5), 815–842.

Rorden, C., Karnath, H.-O., & Bonilha, L. (2007). Improving lesion-symptom mapping. *Journal of Cognitive Neuroscience, 19*(7), 1081–1088.

Ross, T. P., Hanouskova, E., Giarla, K., Calhoun, E., & Tucker, M. (2007). The reliability and validity of the self-ordered pointing task. *Archives of Clinical Neuropsychology, 22*(4), 449–458.

Rudebeck, P. H., Behrens, T. E., Kennerley, S. W., Baxter, M. G., Buckley, M. J., Walton, M. E., & Rushworth, M. F. S. (2008). Frontal cortex subregions play distinct roles in choices between actions and stimuli. *The Journal of Neuroscience: The Official Journal of the Society for Neuroscience, 28*, 13775–13785.

Rudebeck, P. H., Saunders, R. C., Prescott, A. T., Chau, L. S., & Murray, E. A. (2013). Prefrontal mechanisms of behavioral flexibility, emotion regulation and value updating. *Nature Neuroscience, 16*, 1140–1145.

Shallice, T., Burgess, P. W., Schon, F., & Baxter, D. M. (1989). The origins of utilization behaviour. *Brain: A Journal of Neurology, 112*(Pt. 6), 1587–1598.

Shallice, T., Stuss, D. T., Picton, T. W., Alexander, M. P., & Gillingham, S. (2009). Mapping task switching in frontal cortex through neuropsychological group studies. *Frontiers in Neuroscience, 2*(JUL), 79–85.

Shenhav, A., Botvinick, M. M., & Cohen, J. D. (2013). The expected value of control: An integrative theory of anterior cingulate cortex function. *Neuron, 79*(2), 217–240.

Sheth, S. A., Mian, M. K., Patel, S. R., Asaad, W. F., Williams, Z. M., Dougherty, D. D., … Eskandar, E. N. (2012). Human dorsal anterior cingulate cortex neurons mediate ongoing behavioural adaptation. *Nature, 488*(7410), 218–221.

Silvetti, M., Alexander, W., Verguts, T., & Brown, J. W. (2013). From conflict management to reward-based decision making: Actors and critics in primate medial frontal cortex. *Neuroscience and Biobehavioral Reviews, 46*(Pt. 1), 44–57.

Snyder, H. R. (2014). Major depressive disorder is associated with broad impairments on neuropsychological measures of executive function: A meta-analysis and review. *Psychological Bulletin, 139*(1), 81–132.

Spearman, C. (1904). General intelligence, objectively determined and measured. *American Journal of Psychology, 15*, 201–93.

Stuss, D. T. (2011). Functions of the frontal lobes: relation to executive functions. *Journal of the International Neuropsychological Society: JINS, 17*(5), 759–765.

Stuss, D. T., & Benson, D. F. (1984). Neuropsychological studies of the frontal lobes. *Psychological Bulletin, 95*(1), 3–28.

Stuss, D. T., Floden, D., Alexander, M. P., Levine, B., & Katz, D. (2001). Stroop performance in focal lesion patients: Dissociation of processes and frontal lobe lesion location. *Neuropsychologia, 39*(8), 771–786.

Swick, D., & Chatham, C. (2014). Ten years of inhibition revisited. *Frontiers in Human Neuroscience, 8*, 329.

Swick, D., & Turken, A. U. (2002). Dissociation between conflict detection and error monitoring in the human anterior cingulate cortex. *Proceedings of the National Academy of Sciences of the United States of America, 99*, 16354–16359.

Tang, Y.-Y., Posner, M. I., Rothbart, M. K., & Volkow, N. D. (2015). Circuitry of self-control and its role in reducing addiction. *Trends in Cognitive Sciences, 19*(8), 439–444.

Thompson-Schill, S. L., Jonides, J., Marshuetz, C., Smith, E. E., D'Esposito, M., Kan, I. P., … Swick, D. (2002). Effects of frontal lobe damage on interference effects in working memory. *Cognitive, Affective & Behavioral Neuroscience, 2*, 109–120.

Tsuchida, A. (2012). *Fractionating frontal lobe function*. Montreal, Canada: McGill University.

Tsuchida, A., Doll, B. B., & Fellows, L. K. (2010). Beyond reversal: A critical role for human orbitofrontal cortex in flexible learning from probabilistic feedback. *The Journal of Neuroscience, 30*, 16868–16875.

Tsuchida, A., & Fellows, L. K. (2009). Lesion evidence that two distinct regions within prefrontal cortex are critical for n-back performance in humans. *Journal of Cognitive Neuroscience, 21*, 2263–2275.

Tsuchida, A., & Fellows, L. K. (2012). Are core component processes of executive function dissociable within the frontal lobes? Evidence from humans with focal prefrontal damage. *Cortex; A Journal Devoted to the Study of the Nervous System and Behavior, 49*, 1790–1800.

Vainik, U., Dagher, A., Dubé, L., & Fellows, L. K. (2013). Neurobehavioural correlates of body mass index and eating behaviours in adults: A systematic review. *Neuroscience and Biobehavioral Reviews, 37*(3), 279–299.

Vendrell, P., Junqué, C., Pujol, J., Jurado, M. A., Molet, J., & Grafman, J. (1995). The role of prefrontal regions in the Stroop task. *Neuropsychologia, 33*(3), 341–352.

Volle, E., De Lacy Costello, A., Coates, L. M., McGuire, C., Towgood, K., Gilbert, S., … Burgess, P. W. (2012). Dissociation between verbal response initiation and suppression after prefrontal lesions. *Cerebral Cortex, 22*(10), 2428–2440.

Wager, T. D., Jonides, J., Smith, E. E., & Nichols, T. E. (2005). Toward a taxonomy of attention shifting: individual differences in fMRI during multiple shift types. *Cognitive, Affective & Behavioral Neuroscience, 5*(2), 127–143.

Wessel, J. R., Klein, T. A., Ott, D. V. M., & Ullsperger, M. (2013). Lesions to the prefrontal performance-monitoring network disrupt neural processing and adaptive behaviors after both errors and novelty. *Cortex, 50*, 45–54.

Wilson, C. R. E., Gaffan, D., Browning, P. G. F., & Baxter, M. G. (2010). Functional localization within the prefrontal cortex: Missing the forest for the trees? *Trends in Neurosciences, 33*(12), 533–540.

30

Transdiagnostic Impairment in Cognitive Control Neurocircuits

Behaviour, Structure, and Function[1]

Lisa M. McTeague, Madeleine S. Goodkind, and Amit Etkin

Cognitive control or executive functions refer to those processes integral to the effortful deployment of cognitive resources for flexible, adaptive responding to shifting contingencies. As such, cognitive control undergirds the self-regulation imperative for successful, dynamic accommodation to the demands of daily life (cf. Diamond, 2013, for a review). Even among healthy individuals, cognitive control capacity predicts endeavours and success in educational performance and attainment (Duncan et al., 2007), occupational stability and advancement (Foxall, 2014), health promotion (McClernon, Addicott, & Sweitzer, 2015; Riggs, Spruijt-Metz, Sakuma, Chou, & Pentz, 2010), as well as broader measures of overall quality of life (Davis, Marra, Najafzadeh, & Lui-Ambrose, 2010). In light of the influence on functional status in the context of well-being, cognitive control/executive functions are likely integral to the development, resistance to, maintenance, and remediation of psychopathology. That is, as phasic (or prolonged) distress manifests in the context of mental health or illness, cognitive control neurocircuits are likely recruited in the service of symptom regulation. In fact, meta-analyses of functional neuroimaging studies of adaptive emotion regulation demonstrate the recruitment of networks characteristic of cognitive control—a frontoparietal network containing the dorsolateral prefrontal and posterior parietal cortices, and a cingulo-opercular network containing the dorsal anterior cingulate cortex (dACC), anterior insula, and the anterior prefrontal cortex (Kohn et al., 2014).

The foundation of intact cognitive control has characteristic patterns in both behaviour and functional neurocircuitry. As such, cognitive control perturbations amid psychopathology are potentially most pronounced within indices of these processes common to adaptive functioning and moreover, evident across disorders (i.e., transdiagnostically). In this chapter, we discuss behavioural evidence of cognitive dyscontrol transdiagnostically, focusing on psychotic, bipolar, and unipolar depressive, anxiety, and substance use disorders. To contextualise behavioural correlates of cognitive control anomalies in mental illness relative to underlying structure, we discuss our recent findings from a meta-analysis (Goodkind et al., 2015) demonstrating transdiagnostic grey matter reductions in a dorsal cingulate-anterior insula-based network. We also highlight the demonstrated functional relevance of this network in terms of impaired cognitive control performance. Next, we focus on the neurocircuitry implicated across psychotic, bipolar, and unipolar depressive, anxiety, and substance use disorders in functional neuroimaging studies of cognitive dyscontrol. Finally, we briefly

The Wiley Handbook of Cognitive Control, First Edition. Edited by Tobias Egner.

discuss the potential to leverage cognitive control and its neurocircuitry for innovating powerful and broadly applicable interventions to ameliorate distress and functional interference.

Clues to Core Cognitive Control Dysfunction Common Across Psychiatric Disorders

A Common Underlying Cognitive Control Factor: Behaviour

Latent variable analysis of performance on a wide array of neuropsychological tasks has shown that intact cognition has a characteristic pattern of interrelated executive functions throughout the lifespan from childhood (Lehto, Juujarvi, Kooistra, & Pulkkinen, 2003) through middle (Miyake et al., 2000) and older adulthood (Adrover-Roig, Sesé, Barceló, & Palmer, 2012). That is, superordinate cognitive control can largely be explained by interrelated subprocesses. For example, Miyake and colleagues (2000, Miyake & Friedman, 2012) have demonstrated that updating (i.e., monitoring and refreshing working memory store), inhibition (resisting prepotent responses), and shifting (switching between mental sets) largely explain cognitive processes. Alternatively, Alvarez and Emory (2006) have highlighted the synergy of working memory, inhibition, and selective attention. The fact that diverse functions show coherence suggests that these heterogeneous processes may work in tandem to promote cognitive well-being, but may also share a vulnerability to dysfunction.

In fact, latent variable analysis of task batteries spanning multiple domains of cognitive control suggest that hallmark executive functions are explained not only as subprocesses such as updating, inhibition, and shifting, but also an underlying common factor reflecting general cognitive control capacity (Miyake & Friedman, 2012). The extent of impairment in this common factor has yet to be examined in most psychiatric disorders. Historically, hallmark symptoms of individual disorders have prompted hypotheses about domain-specific impairments in neuropsychological profiles per disorder (e.g., poor resistance to interference in posttraumatic stress disorder (PTSD) due to intrusions). As such, investigations on respective disorders have typically focused on one or two exemplar tasks of a given domain, precluding latent variable analysis to discern contributions of a common cognitive control factor. However, in sum, the extant literature on neuropsychological performance shows broad (i.e., domain nonspecific) rather than distinct performance impairments in studies of individual psychiatric disorders.

Evidence of broad impairments for respective disorders and classes of related disorders, albeit with variations in severity, was clearly demonstrated by Snyder, Miyake, and Hankin (2015). The authors summarised the effect sizes of meta-analyses of cognitive control/executive functions by individual disorder. For consistency with prevailing latent variable models, results for individual tasks were aggregated into domains of shifting, inhibition, updating, and working memory manipulation and maintenance. Measures of planning and verbal fluency, which typically recruit multiple executive functions and thus do not clearly load onto one of the other factors, were also summarised across studies as they have been examined frequently in clinical samples. Schizophrenia showed the most pronounced and consistently cross-domain deficits, consistent with the severe functional impairment characteristic of the disorder (Harvey & Strassnig, 2012). Specifically, aggregating across eight meta-analyses, schizophrenia demonstrated impairments with large effect sizes on all measures except the relatively less demanding process of working memory maintenance, for which a medium-effect-sized impairment emerged. Slightly less severe, but nonetheless consistent cross-domain deficits were observed in the average of ten meta-analyses of bipolar disorder.

Notably, eight of the ten meta-analyses examined euthymic bipolar disorder, suggesting that cognitive control deficits are likely more pronounced during acute depression or mania. Unipolar depression showed the same pattern but was marked by medium-effect-sized deficits across all domains, with the exception of a lesser impairment on working memory maintenance. Aggregating across three meta-analyses of obsessive-compulsive disorder (OCD) and one of PTSD showed medium to small effect sizes across all domains. Neurocognitive functioning in other (DSM-IV-defined) anxiety disorders (i.e., specific phobia, social anxiety, generalised anxiety, and panic disorders) have been the subject of few investigations, possibly reflecting the file drawer problem or an important presumption that cognitive control is uninterrupted in these presentations and thus not examined. Finally, multiple substance use disorders demonstrated deficits foremost in inhibition, but also shifting and working memory. Notably, Stavro, Pelletier, & Potvin (2013) revealed that deficits were typically moderate up to 1 year of abstinence, and lessened for users abstinent for at least 1 year. Despite the paucity of work directly comparing different disorders, on the whole, these findings have prompted a growing appreciation of the likelihood of shared deficits in cognitive control capacity across a wide range of Axis I disorders (e.g., Cáceda, Nemeroff, & Harvey, 2014; Cole, Repovs, & Anticevic, 2014; Etkin, Gyurak, & O'Hara, 2013; Goschke, 2014; Snyder et al., 2015).

A Common Underlying Psychopathology Factor: Symptoms

A general liability for cognitive dyscontrol, which cuts across diagnostic boundaries, may be related to transdiagnostic vulnerability to mental illness. The historic neo-Kraeplinian approach of the Diagnostic and Statistical Manual of Mental Disorders (e.g., DSM-III, APA, 1980; DSM-5, APA, 2013) and the corresponding International Classification of Diseases of the World Health Organization (ICD-9, WHO, 1977 ICD-10, WHO, 2007) conceptualises disorders as discontinuous entities with presumably distinct aetiologies and biomarkers. However, burgeoning evidence from large-scale phenotypic studies has robustly demonstrated common liabilities across disorders (e.g., Krueger, 1999). For example, examination of disorder patterns in epidemiological samples has revealed a latent internalising dimension, typically expressed as anxiety or unipolar depressive disorders. A latent externalising dimension is manifest in substance use and conduct disorders, whereas psychotic disorders load on a third independent thought disorder/psychosis dimension (Kotov et al., 2011; Markon, 2010; Wright et al., 2013). Bipolar disorders and symptoms, likely owing to the relative prominence of psychotic, depressive, and mania features in different samples, have at times loaded on the internalising dimension (Wright et al., 2013) and at others on the psychosis dimension (Kotov et al., 2011). Importantly, recent extensions of this work have shown that these dimensions are, in turn, strongly related to an underlying general psychopathology dimension (Carragher et al., 2015; Lahey et al., 2012). Twin studies have revealed a likely genetic basis to these higher- and lower-order dimensions (Kendler, Prescott, Myers, & Neale, 2003; Kendler et al., 2011). Notably, the general psychopathology factor robustly accounts for lifespan functional impairment, family history of mental illness, and prospective psychopathology (Caspi et al., 2014; Kim & Eaton, 2015) above and beyond current symptom-based predictions. It is also a more robust predictor of disorder-nonspecific factors such as childhood maltreatment, which likely catalyse disorder expression.

Of particular relevance to a potential transdiagnostic liability to cognitive dyscontrol, early childhood and adult neurocognitive functioning is far more strongly related to the underlying general psychopathology factor than the internalising, externalising, or thought disorder dimensions (Caspi et al., 2014). More specifically, higher loadings on the general psychopathology factor predict worse performance on neuropsychological tasks of recall, recognition, and working memory, sustained attention and vigilance, and motor planning. General psychopathology also predicts limited early academic achievement as well as lifespan IQ. Finally, at the extreme of this general psychopathology factor, neurocognitive impairments are concurrent with microvascular brain abnormalities, neurological soft signs, receptive language impairment, and IQ deficits (Caspi et al., 2014).

Notably, Snyder and colleagues (2015) underscore that the common cognitive control liability is distinct from rather than subsumed by the transdiagnostic vulnerability to psychopathology. Accordingly, neurocognitive functioning has meaningfully predicted longitudinal socio-occupational functional outcomes across disorders above and beyond symptom-based predictions (Gyurak et al., 2015; Miller et al., 2015; Tabarés-Seisdedos et al., 2008). In cases of severely impairing symptomatology owing to acute episodes of late onset schizophrenia, neurocognitive function is preserved in relation to early onset schizophrenia (Rajji, Ismail, & Mulsant, 2009). Further, impairments are evident even prior to the prodromal phase of psychosis (Bora & Murray, 2014). Relatedly, poor cognitive control is not only a consequence of substance abuse, but also an antecedent (e.g., Peeters, Vollebergh, Wiers, & Field, 2014). In essence, acute symptom expression does not correspond clearly to manifested neurocognitive deficits. Instead, cognitive control deficits may be an intermediate phenotype or endophenotype of latent psychopathology vulnerability (e.g., Buckholtz & Meyer-Lindenberg, 2012; Kozak & Cuthbert, 2015; Nolen-Hoeksema & Watkins, 2011). Towards this end, individual differences in general cognitive control capability are almost completely (i.e., 99%) heritable, rendering this factor among the most heritable of psychological traits, exceeding even IQ (Friedman et al., 2008).

A Common Underlying Structure: Grey Matter Volume

The preponderance of evidence for common, largely heritable liabilities to experiencing general psychopathology as well as cognitive dyscontrol prompts the question of whether there are accompanying structural anomalies seated within the neurocircuitry subserving cognitive control. To explore the possibility of such a common substrate, we recently completed a meta-analysis of volumetric differences across Axis I patient and matched control groups (Goodkind et al., 2015). Structural neuroimaging meta-analyses have been reported for a number of psychiatric disorders (e.g., Hamilton et al., 2012; Rotge et al., 2010; Selvaraj et al., 2012). However, these have either focused on single disorders or compared two phenotypically related disorders such as unipolar and bipolar depressive disorders (e.g., Kempton et al., 2011). As such, interpretation of findings has often reflected disorder-specific neural circuit models. Likewise, other meta-analyses have not provided spatially unbiased information across the brain (e.g., manual volumetric tracing studies). In meta-analytically summarising a more complete spectrum of psychopathology across the entire brain, we hypothesised that transdiagnostic commonalities might emerge.

In brief, studies were selected if they (a) used voxel-based morphometry (VBM) to analyse grey matter in patients with a psychiatric diagnosis, (b) included a comparison between these patients and matched healthy control participants, (c) performed a whole-brain analysis, and (d) reported coordinates in a defined stereotaxic space (e.g., Talaraich space or Montreal Neurological Institute space). A psychotic disorders category comprised studies of schizophrenia, schizoaffective, schizophreniform, and delusional disorders. A non-psychotic disorders category comprised studies of bipolar depressive disorders, unipolar depressive disorders (major depression, dysthymia), anxiety disorders, and substance use disorders (mixed substance abuse and/or dependence disorders). Studies were selected to capture lifespan patterns and thus included participants ranging from childhood through older adulthood. Included in the meta-analysis were peak voxel coordinates from published studies that compared a psychiatric group to healthy participants, which thus represented indicators of regional grey matter volume change associated with that diagnosis. The final sample included nearly 200 peer-reviewed papers and 16,000 patients and matched healthy controls.

The revised activation likelihood estimation (ALE) algorithm was implemented to identify consistent patterns of grey matter differences between patients and controls across studies (Eickhoff et al., 2009; Eickhoff, Bzdok, Laird, Kurth, & Fox, 2012). This algorithm aims to identify areas showing a convergence of reported coordinates across experiments higher than expected under a random spatial association.

Across all studies, the clear majority (85%) of peak voxels represented decreased grey matter in patients compared with controls. Consistent grey matter decreases in patients were found in the bilateral anterior insula, dorsal anterior cingulate (dACC) and dorsomedial prefrontal cortex (dmPFC), ventromedial PFC (vmPFC), thalamus, amygdala, hippocampus, superior temporal gyrus, and parietal operculum. By contrast, grey matter increases in patients were found exclusively in the striatum. To hone in on those regions common to both psychotic and non-psychotic disorders, we performed a conjunction that revealed a more targeted shared pattern of grey matter loss in the bilateral anterior insula and dACC (Figure 30.1). Furthermore, presence of grey matter loss in one region predicted grey matter loss in the

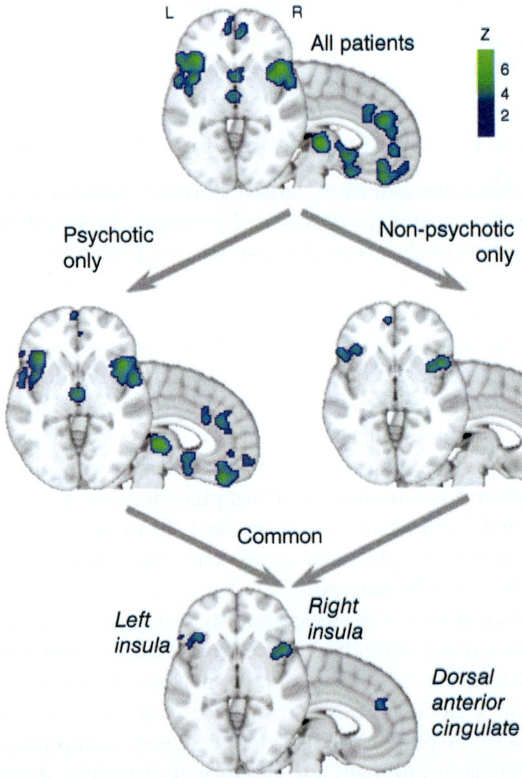

Figure 30.1 Shared patterns of decreased grey matter from the voxel-based morphometry (VBM) meta-analysis. Patient versus healthy participant comparisons (top) for studies pooled across all diagnoses, (middle) separately by psychotic or non-psychotic diagnosis studies, and (bottom) from a conjunction across the psychotic and non-psychotic diagnostic groups. Common grey matter loss is evident across diagnoses in the anterior insula and dorsal anterior cingulate (dACC). Source: Goodkind 2015. Reproduced with permission of American Medical Association.

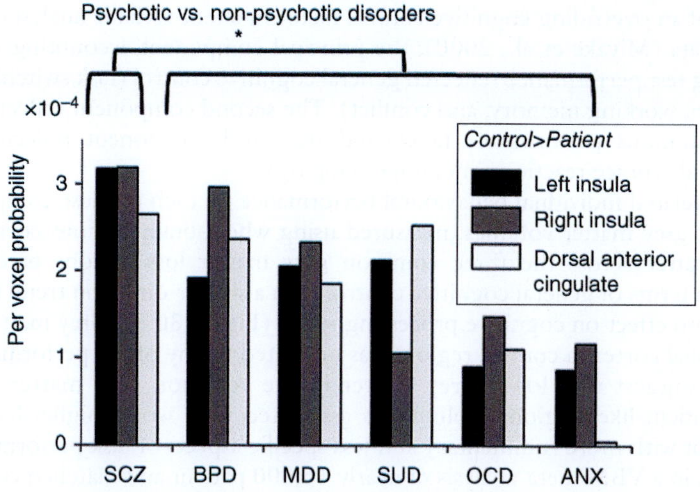

Figure 30.2 Extracted per voxel probabilities of decreased grey matter in the VBM meta-analysis, separated by diagnosis and common grey matter loss region (left and right anterior insula, dACC). Values represent the probability of identifying a grey matter abnormality for an average voxel within the region of interest, derived from the modelled activation maps. SCZ = schizophrenia, BPD = bipolar disorder, MDD = major depressive disorder/dysthymia, SUD = substance use disorder, OCD = obsessive-compulsive disorder, ANX = anxiety disorders. Source: Goodkind 2015. Reproduced with permission of American Medical Association.

other regions (Robinson, Laird, Glahn, Lovallo, & Fox, 2010), suggesting that this grey matter loss might occur in a coordinated fashion across a structural network inclusive of these regions. By contrast, the grey matter increases in the striatum of patients were evident only in the psychotic disorders group.

As the psychotic versus non-psychotic distinction does not account for more refined phenotypic differences, a follow-up analysis of the extracted per voxel probability of observing grey matter decreases in each of the three common regions (right and left insula, dACC) revealed similar magnitude effects across all non-psychotic diagnoses, as shown in Figure 30.2. Schizophrenia spectrum disorders showed significantly greater reductions of grey matter volume in these regions relative to the non-psychotic disorders as a whole. Notable given the dramatic differences in psychotropic use patterns between psychotic and non-psychotic disorders, these effects were not accounted for by current medication usage. In summary, these results suggest that anterior insula and dACC grey matter loss represent a transdiagnostic neural abnormality, most pronounced in disorders prone to psychosis.

To assess whether cognitive control capacity may be determined, in part, by the structural integrity of this anterior insula/dorsal anterior cingulate–based network, we examined relationships between grey matter volume in these nodes in relation to neuropsychological performance. We utilised a dataset of 163 healthy adults from the BRAINnet database who had completed a computerised neurocognitive assessment battery that covered a broad range of basic and higher-level cognitive functions. As such, this served as a conservative test of whether reduced grey matter volume would predict behavioural decrements even in the context of healthy cognitive functioning and mental well-being.

To reduce the array of tasks representing overlapping cognitive domains, we conducted principal components analysis, which yielded three principal components. Consistent with identification of an overriding cognitive control factor in latent variable analyses of cognitive performance data (Miyake et al., 2000), the principal component accounting for the most variance among test performance reflected general cognitive control (task switching, interference, inhibition, working memory, and conflict). The second component reflected sustained attention (continuous performance task), and the third component reflected cognitive processing speed (choice reaction time, finger tapping).

We then correlated individual behavioural performance on each of these components with subject-specific grey matter volumes, measured using whole-brain volume–corrected VBM. Lower grey matter across the three common grey matter loss regions predicted worse performance in terms of general cognitive control with a similar-direction trend for sustained attention, but no effect on cognitive processing speed (Figure 30.3). Grey matter volume in the primary visual cortex, a control region, was unrelated to any of the performance indices. These results suggest that lower grey matter in the common grey matter loss regions (i.e., more 'patient-like' regional volume) is associated with worse higher-level cognitive control, but not with more rudimentary and less specific aspects of task performance.

In summary, in a VBM meta-analysis of nearly 16,000 patient and matched control participants, we identified a transdiagnostic pattern of grey matter loss in the bilateral anterior insula and dACC. Further, lower grey matter volume in this network was systematically related to behavioural decrements on neuropsychological tasks—foremost for higher-order cognitive control. Taken together, these findings suggest a coordinated structural perturbation of this anterior insula/dorsal anterior cingulate–based network across disorders, likely contributing to transdiagnostic cognitive control deficits.

A Common Underlying Neurocircuit Disruption: Task Activation

Our VBM meta-analysis results do not indicate whether these nodes of common grey matter loss show parallel functional disruptions. A corresponding meta-analysis of cognitive control tasks is clearly warranted. In the interim, we examine the extant disorder-specific literature for evidence of abnormal activation in patients during cognitive control tasks. In short, we consider whether findings augur a common functional substrate of cognitive dyscontrol, which overlaps with the disrupted structural integrity of the anterior insula/dorsal anterior cingulate–based network.

Activity in the dACC and anterior insula is thought, in part, to tag salient events and coordinate dynamic interaction of large-scale neural networks in response to contextual demands (Jiang, Beck, Heller, & Egner, 2015; Menon & Uddin, 2010; Power, Schlaggar, Lessov-Schlaggar, & Petersen, 2013; Sridharan, Levitin, & Menon, 2008). Consistent with the essential role of switchboard or hub, weakening of the structural integrity would indeed affect a multitude of processes, possibly contributing to the wide array of potential symptom presentations in mental illness.

Importantly, the dACC (and the posteriorly adjacent medial frontal cortex) as well as the bilateral anterior insula have not only been conceptualised as a 'salience network' in the human brain (Seeley et al., 2007) but also as key nodes in the superordinate or 'multiple demand' cognitive control/processing network. Additional cortical regions reliably recruited as part of the multiple demand network include the dorsolateral prefrontal cortex (dlPFC)/middle frontal gyrus, inferior frontal junction, inferior frontal gyrus/ventrolateral prefrontal cortex (vlPFC), inferior parietal/intraparietal sulcus, and mid-cingulate cortex extending into the pre-supplementary motor area (pre-SMA; Duncan, 2010; Duncan & Owen, 2000; Müller, Langner, Cieslik, Rottschy, & Eickhoff, 2015). Similar to the latent or common cognitive control factor observed in behavioural measures of cognitive processing, the activity of this network suggests a 'common core' recruited across diverse cognitive challenges and

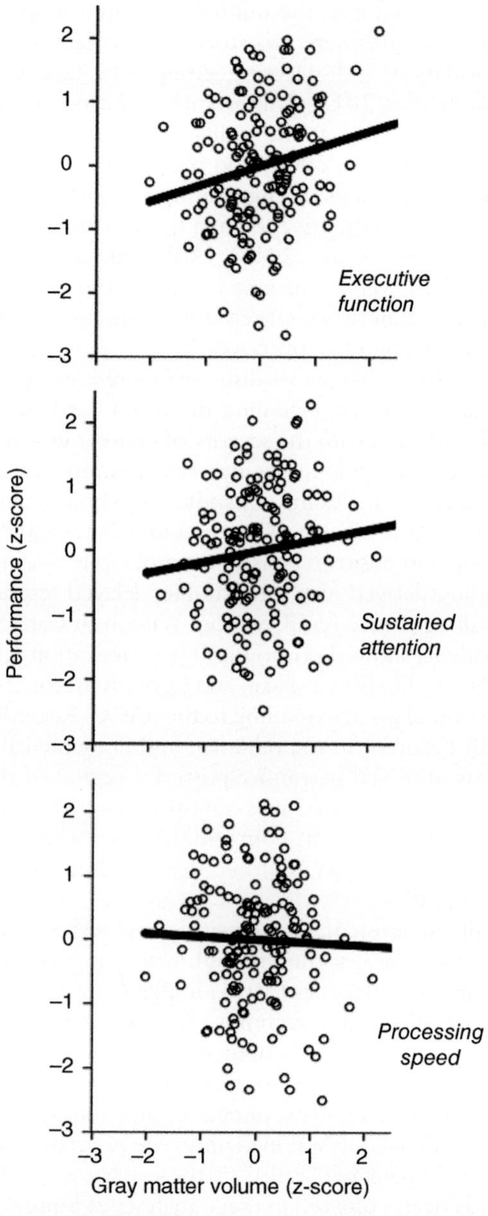

Figure 30.3 Among 163 healthy participants from the BRAINnet Foundation Database database who had completed a computerized neurocognitive assessment battery covering a broad range of basic and higher-level cognitive functions, lower grey matter across the three common grey matter loss regions (left and right anterior insula, dACC) predicted worse performance in terms of general executive function, with a similar trend for sustained attention, but no effect on general cognitive and performance speed. Even among healthy individuals, reduced grey matter in the key regions was associated with reduced cognitive control. Source: Goodkind 2015. Reproduced with permission of American Medical Association.

domains (Duncan, 2013). For example, the multiple demand network pattern of activation emerges in tasks ranging from arithmetic, to storing information in working memory, to inhibiting irrelevant information and selecting competing task-relevant responses (e.g., Fedorenko, Duncan, & Kanwisher, 2013). Importantly, the canonical pattern is observed not only at the single task level, but also at the individual level, suggesting that the patterns observed at group (Duncan & Owen, 2000) and meta-analytic levels (Müller et al., 2015; Niendam et al., 2012) are not artefacts of aggregating findings. Furthermore, the multiple demand network is sensitive to cognitive demand, showing activation proportional to increasing difficulty (Fedorenko et al., 2013). Although individual nodes show some differentiation in terms of phasic versus sustained responses under different task parameters (Cole & Schneider, 2007), the network largely functions in concert, thus providing the foundation for a broad array of cognitive processes.

Next we consider the possibility that across disorders there is evidence of functional impairment in the multiple demand network, including the dACC and anterior insula. Relative to other Axis I disorders and consistent with the severity of neurocognitive and functional impairment, schizophrenia has received the most intensive examination of cognitive control/executive function neurocircuitry, including a handful of whole-brain meta-analyses. In the most recent, Minzenberg, Laird, Thelen, Carter, and Glahn (2009) analysed functional magnetic resonance imaging and positron emission tomography findings from 41 studies of a broad array of tasks including delayed match-to-sample, delayed response, go/no-go, mental arithmetic, N-back, oddball, sequence recall, Stroop, Wisconsin Card Sort, and word generation tasks. Across tasks, patients showed pronounced hypoactivation relative to control participants in the bilateral dlPFC, right vlPFC, visual, and right premotor cortices as well as inferior parietal lobule and medial frontal gyrus extending to the dACC. Regarding subcortical regions, patient hypoactivation also occurred in the putamen and dorsomedial thalamus. In contrast, patient hyperactivation was observed in a more posterior region of the dACC extending to the supplementary motor area. Increased mid-cingulate/pre-SMA among patients relative to control participants may reflect a compensatory process for maintaining intact performance amid deficiencies in other multiple demand network nodes (i.e., proactive/reactive control; Jiang et al., 2015).

Very similar patterns of impairment seated within and extending beyond the multiple demand network emerge in bipolar disorder, though with less pervasive disruption compared to schizophrenia. Wegbreit and colleagues (2014) implemented a whole-brain meta-analysis on 44 cognitive processing studies, again, spanning numerous domains and tasks (e.g., Stroop, N-back, go/no-go, paced motor task, sentence completion, delayed match-to-sample, and stop signal). Patient hypoactivation emerged in the bilateral inferior frontal gyrus/vlPFC, superior frontal gyrus, visual cortex, putamen, precuneus, lingual gyrus, and inferior parietal lobule. Patient hyperactivation was evident in the pregenual ACC and mid-cingulate cortex. Underscoring the domain generality of these patterns, similar multiple demand network hypoactivation has been observed in meta-analyses of bipolar disorder circumscribed to episodic memory (Ragland et al., 2009) and response inhibition (Hajek, Alda, Hajek, & Ivanoff, 2013).

Major depressive disorder also shows aberrant recruitment of multiple demand network nodes, though even less extensive than bipolar disorder and consistent with the moderate-effect-sized neuropsychological deficits. In a whole-brain meta-analysis of 19 cognitive control tasks similar in domain and task breadth to the schizophrenia and bipolar studies reviewed previously, patients showed hypoactivation in the right inferior frontal gyrus and left thalamus. Patient hyperactivation emerged in the medial and bilateral middle frontal gyrus, putamen, and left thalamus (Palmer, Crewther, Carey, & START Project Team, 2015). Regarding the lateral and medial prefrontal patient hyperactivation, a separate meta-analysis of working memory tasks in major depression demonstrated that hyperactivation of these regions was attributable to patient samples showing worse performance and thus possibly reflects compensatory efforts (Wang et al., 2015).

Both anxiety disorders and substance use disorders have less often been the subject of functional neuroimaging investigation of cognitive control, and meta-analytical summaries are as yet unavailable. In the case of anxiety disorders, the majority of whole-brain functional imaging studies of cognitive control tasks have focused on DSM-IV defined OCD and PTSD, paralleling the neuropsychological publication biases. A search in PubMed for whole-brain neuroimaging investigations of cognitive control/executive function tasks of cognitive demand in anxiety disorders revealed 26 studies with 32 tasks. Nineteen of these studies were of OCD, 6 of PTSD, and only 1 of generalised social anxiety disorder. All in all, these anxiety disorders tended to show reduced lateral and medial prefrontal activation in tasks reflecting the canonical dimensions of shifting, inhibition, and updating (e.g., Chen, Li, Xu, & Liu, 2009; Falconer et al., 2008; Gu et al., 2008; Kang et al., 2013; Roth et al., 2007; Remijnse et al., 2013). In contrast, dACC/anterior insula involvement was alternately marked by hyper- (Fitzgerald et al., 2005; Maltby, Tolin, Worhunsky, O'Keefe, & Kiehl; 2005; Marsh et al., 2014; Roth et al., 2007) and hypoactivation (Chen et al., 2009; Gu et al., 2008; Kang et al., 2013; Remijnse et al., 2013) in different samples. The variability observed between these neuroimaging studies strongly underscores the need for additional investigations of cognitive control circuits in anxiety and related disorders, especially as PTSD, OCD, and traditionally defined anxiety disorders may be more heterogeneous than originally thought. In latent variable analysis of underlying symptom dimensions, no consensus has emerged regarding whether OCD and PTSD are better construed as predominantly fear or anxious-misery disorders (Cox, Clara, & Enns, 2002; Forbes et al., 2010). Additionally, PTSD and OCD were reclassified from anxiety disorders into separate spectra in DSM-5.

Similar to anxiety disorders, the literature on substance use disorders and functional cognitive control circuits suffers from a limited corpus and pronounced disorder heterogeneity. For example, neuropsychological deficits, central nervous system effects, and neurotoxicity vary dramatically across substances of abuse. Specifically, a search in PubMed for whole-brain neuroimaging investigations of cognitive control/executive function tasks of cognitive demand in substance disorders revealed 35 studies of stimulant (n = 13), alcohol (n = 9), cannabis (n = 6), polysubstance (n = 4), and opiate (n = 3) abuse/dependence also varying in terms of active or abstinent users. Although generalities about substance use are necessarily limited by this disorder diversity, these patients relative to other disorders tended to show more uniform attenuations of the multiple demand network and related regions (e.g., Akine et al., 2007; Bach et al., 2012; Bolla et al., 2003; Bolla, Eldreth, Matochik, & Cadet, 2005).

The discussion so far has focused exclusively on transdiagnostic impairments in cognitive control in the absence of concurrent affective or motivational demands. Not surprisingly, the frontal-cingulate-parietal-insular network proposed here to contribute to transdiagnostic deficits in cognitive control overlaps with regions identified as aberrant during emotional processing tasks in many individual disorders (anxiety disorders; Etkin & Wager, 2007; bipolar disorders; Satterthwaite et al., 2015). The insula and anterior cingulate, in particular, feature prominently in intact (Hayes, Duncan, Xu, & Northoff, 2014) as well as disordered emotional responding (Etkin & Wager, 2007). Increasingly, however, anterior cingulate and insular cortices are recognised as more broadly deployed—assigning salience for the coordination of goal-directed behaviour amid complex cognitive/affective experiences (Medford & Critchley, 2010; Menon & Uddin, 2010; Sridharan et al., 2008). Recent work on causal network interactions across multiple cognitive tasks (Cai et al., 2015; Chen et al., 2015; Jiang et al., 2015) suggests that the anterior insula amplifies salience detection in the anterior and mid-cingulate in a manner proportional to both cognitive demand and individual capacity, and, in turn prompts activation of the broader frontoparietal network, specifically the dlPFC and posterior parietal cortex. Further support for the deep functional integration of regions traditionally dichotomised into 'emotion' versus 'cognition' nodes, Cieslik and colleagues (2013) observed—across multiple cognitive paradigms—that the middle fontal gyrus (representing the relevant aspect of the dlPFC) has two functional subregions, with the anterior region preferentially connected to the salience network-based anterior cingulate and the posterior region with other regions identified here—the executive control network-based intraparietal sulci. In light of the heavy interconnectedness of the neurocircuitry undergirding emotional and cognitive processing, it is no surprise that across disorders patient reports of real-world impairment typically reflect this intersection.

Summary

Cumulative findings have revealed that psychiatric disorders share a common factor or vulnerability to dysfunction. Similarly, historic conceptualisations of distinct cognitive domains have been superseded by models that include a higher-order common cognitive control/executive function factor. Separate studies suggest that this common cognitive control factor may be related to deficits in a frontal-cingulate-parietal-insular network recruited for a wide diversity of cognitive demands (see also Chapter 13 by Stokes et al. and Chapter 18 by Cole in this volume). In addition, the nodes in this multiple demand network (the dACC and insula), which function as a switchboard for communication between large-scale networks, share a common vulnerability to grey matter reduction, especially at extreme levels of general psychopathology. Although burgeoning evidence of common factors calls into question historic parcellations of cognition and disease, it is important to note that shared variance also reflects the inherent impurity in real-world psychiatric distress and cognitive processing. For example, a bias for domain generality over specificity in brain function and behaviour may in part reflect that study tasks are grouped into domains based on conceptualisations of the predominant underlying executive function. Naturally, however, performance of most tasks involves a synergy of cognitive processes that spans domains such as working memory, set shifting, and performance monitoring. In a related vein, robust transdiagnostic effects in part reflect that polythetic diagnostic schemes, which yield tremendous phenotypic heterogeneity, in combination with high comorbidity hamper detection of profiles specific to putatively 'pure' disorder manifestations. Nonetheless, findings of common factors across modalities have been robust and, most importantly, have shown prospective prediction of distress and functional impairment that often exceed traditional 'distinct' factors.

From the above, large-scale investigations that enable dimensional factor analysis of common and specific symptom domains in conjunction with multivariate analysis of brain activation/deactivation patterns are warranted. At the forefront of this work and an exemplar for future investigations, Shanmugan and colleagues (2016) performed such an approach on data from a community sample of 1,129 individuals who completed a working memory task during fMRI acquisition. Severity on a general psychopathology factor reliably predicted attenuated recruitment of the multiple demand network. Symptom factors reflecting lower-order dimensions were also considered. That is, psychosis spectrum symptoms were associated with hypoactivation of the dlPFC, externalising symptoms with hypoactivation of the frontoparietal cortex and cerebellum, and anxious-misery symptoms with widespread hyperactivation of the executive network. Essentially, common and specific factors demonstrated dissociable patterns, which would have been obscured in a conventional nosological framework.

Conclusions

Consistent with principles of the Research Domain Criteria Project (Kozak & Cuthbert, 2016), the current review implicates disruption of a cognitive control network across disorders. Importantly, this network parallels the multiple demand network intrinsic to adaptive, flexible cognition. Given our prior findings of transdiagnostic grey matter loss in overlapping regions of this network, a parallel is suggested across structural and functional measures of brain dysfunction. Also highlighted is the particular vulnerability of the anterior cingulate and insular cortices to perturbations, especially in the context of highly burdensome general psychopathology. Most importantly, these findings signify promise for psychosocial, pharmacological, or neuromodulatory interventions that target the foundation of intact, dynamic cognition seated in the frontal-cingulate-parietal-insular network. Such targeted approaches could be powerful for ameliorating not only symptoms, but also the broad functional impairments and diminished quality of life prevalent across psychiatric disorders.

Acknowledgements

LM was funded by National Institute of Mental Health K23 MH104849. AE was funded by the Sierra-Pacific Mental Illness Research, Education and Clinical Center (MIRECC) at the Palo Alto VA.

Note

1 Disclosures: The authors declare no competing financial interests relevant to this research. Dr. Etkin reports consulting funds for unrelated work from Otsuka, Acadia, and Takaeda as well as a research grant support from Brain Resource.

References

Adrover-Roig, D., Sesé, A., Barceló, F., & Palmer, A. (2012). A latent variable approach to executive control in healthy ageing. *Brain and Cognition, 78*, 284–299.

Akine, Y., Kato, M., Muramatsu, T., Umeda, S., Mimura, M., Asai, Y.,...Suhara, T. (2007). Altered brain activation by a false recognition task in young abstinent patients with alcohol dependence. *Alcohol: Clinical & Experimental Research, 31*, 1589–1597.

Alvarez, J. A., & Emory, E. (2006). Executive function and the frontal lobes: A meta-analytic review. *Neuropsychology Review, 16*, 17–42.

American Psychiatric Association. (1980). *Diagnostic and statistical manual of mental disorders* (3rd ed.). Washington, DC: American Psychiatric Association.

American Psychiatric Association. (2013). *Diagnostic and statistical manual of mental disorders* (5th ed.). Arlington, VA: American Psychiatric Association.

Bach, P., Vollstädt-Klein, S., Frischknecht, U., Hoerst, M., Kiefer, F., Mann, K.,...Hermann, D. (2012). Diminished brain functional magnetic resonance imaging activation in patients on opiate maintenance despite normal spatial working memory task performance. *Clinical Neuropharmacology, 35*, 153–160.

Bolla, K. I., Eldreth, D. A., London, E. D., Kiehl, K. A., Mouratidis, M., Contoreggi, C.,...Ernst, M. (2003). Orbitofrontal cortex dysfunction in abstinent cocaine abusers performing a decision-making task. *Neuroimage, 19*, 1085–1094.

Bolla, K. I., Eldreth, D. A., Matochik, J. A., & Cadet, J. L. (2005). Neural substrates of faulty decision-making in abstinent marijuana users. *Neuroimage, 26*, 480–492.

Bora, E., & Murray, R. M. (2014). Meta-analysis of cognitive deficits in ultra-high risk to psychosis and first-episode psychosis: Do the cognitive deficits progress over, or after, the onset of psychosis? *Schizophrenia Bulletin, 40,* 744–755.

Buckholtz, J. W., & Meyer-Lindenberg, A. (2012) Psychopathology and the human connectome: Toward a transdiagnostic model of risk for mental illness. *Neuron, 74,* 990–1004.

Cáceda, R., Nemeroff, C. B., & Harvey, P. D. (2014). Toward an understanding of decision making in severe mental illness. *Journal of Neuropsychiatry & Clinical Neuroscience, 26,* 196–213.

Cai, W., Chen, T., Ryali, S., Kochalka, J., Li, C. S., & Menon, V. (2015). Causal interactions within a frontal-cingulate-parietal network during cognitive control: Convergent evidence from a multisite-multitask investigation. *Cerebral Cortex,* in press.

Carragher, N., Teesson, M., Sunderland, M., Newton, N. C., Krueger, R. F., Conrod, P. J.,...Slade, T. (2015). The structure of adolescent psychopathology: A symptom-level analysis. *Psychological Medicine,* in press.

Caspi, A., Houts, R. M., Belsky, D. W., Goldman-Mellor, S. J., Harrington, H., Israel, S.,...Moffitt, T. E. (2014). The p Factor: One general psychopathology factor in the structure of psychiatric disorders? *Clinical Psychological Science, 2,* 119–137.

Chen, S., Li, L., Xu, B., & Liu, J. (2009). Insular cortex involvement in declarative memory deficits in patients with post-traumatic stress disorder. *BMC Psychiatry, 9,* 39.

Chen, T., Michels, L., Supekar, K., Kochalka, J., Ryali, S., & Menon, V. (2015). Role of the anterior insular cortex in integrative causal signaling during multisensory auditory-visual attention. *European Journal of Neuroscience, 41,* 264–274.

Cieslik, E. C., Zilles, K., Caspers, S., Roski, C., Kellermann, T. S., Jakobs, O.,...Eickhoff, S. B. (2013). Is there 'one' DLPFC in cognitive action control? Evidence for heterogeneity from co-activation-based parcellation. *Cerebral Cortex, 23,* 2677–2689.

Cole, M. W., Repovš, G., & Anticevic, A. (2014). The frontoparietal control system: A central role in mental health. *Neuroscientist, 20*(6), 652–664.

Cole, M. W., & Schneider, W. (2007). The cognitive control network: Integrated cortical regions with dissociable functions. *Neuroimage, 37*(1), 343–360.

Cox, B. J., Clara, I. P., & Enns, M. W. (2002). Posttraumatic stress disorder and the structure of common mental disorders. *Depression & Anxiety, 15,* 168–171.

Davis, J. C., Marra, C. A., Najafzadeh, M., & Lui-Ambrose, T. (2010). The independent contribution of executive functions to health related quality of life in older women. *BMC Geriatrics, 10,* 16–23.

Diamond, A. (2013). Executive functions. *Annual Review of Psychology, 64,* 135–168.

Duncan, G. J., Dowsett, C. J., Claessens, A., Magnuson, K., Huston, A. C., Klebanov, P.,...Japel, C. (2007). School readiness and later achievement. *Developmental Psychology, 43,* 1428–1446.

Duncan, J. (2010). The multiple-demand (MD) system of the primate brain: Mental programs for intelligent behaviour. *Trends in Cognitive Science, 14,* 172–179.

Duncan, J. (2013). The structure of cognition: Attentional episodes in mind and brain. *Neuron, 80*(1), 35–50.

Duncan, J., & Owen, A. M. (2000). Common regions of the human frontal lobe recruited by diverse cognitive demands. *Trends in Neuroscience, 23,* 475–483.

Eickhoff, S. B., Bzdok, D., Laird, A. R., Kurth, F., & Fox, P. T. (2012). Activation likelihood estimation meta-analysis revisited. *Neuroimage, 59,* 2349–2361.

Eickhoff, S. B., Laird, A. R., Grefkes, C., Wang, L. E., Zilles, K., & Fox, P. T. (2009). Coordinate-based activation likelihood estimation meta-analysis of neuroimaging data: A random-effects approach based on empirical estimates of spatial uncertainty. *Human Brain Mapping, 30,* 2907–2926.

Etkin, A., Gyurak, A, & O'Hara, R. (2013). A neurobiological approach to the cognitive deficits of psychiatric disorders. *Dialogues in Clinical Neuroscience, 15,* 419–429.

Etkin, A., & Wager, T. D. (2007). Functional neuroimaging of anxiety: A meta-analysis of emotional processing in PTSD, social anxiety disorder, and specific phobia. *American Journal of Psychiatry, 164,* 1476–1488.

Falconer, E., Bryant, R., Felmingham, K. L., Kemp, A. H., Gordon, E., Peduto, A., & Williams, L. M. (2008). The neural networks of inhibitory control in posttraumatic stress disorder. *Journal of Psychiatry & Neuroscience, 33,* 413–422.

Fedorenko, E., Duncan, J., & Kanwisher, N. (2013). Broad domain generality in focal regions of frontal and parietal cortex. *Proceedings of the National Academy of Sciences of the United States of America, 110*, 16616–16621.

Fitzgerald, K. D., Welsh, R. C., Gehring, W. J., Abelson, J. L., Himle, J. A., Liberzon, I., & Taylor. S. F. (2005). Error-related hyperactivity of the anterior cingulate cortex in obsessive-compulsive disorder. *Biological Psychiatry, 57*, 287–294.

Foxall, G. R. (2014) Cognitive requirements of competing neuro-behavioral decision systems: Some implications of temporal horizon for managerial behavior in organizations. *Frontiers in Human Neuroscience, 8*, 184.

Forbes, D., Parslow, R., Creamer, M., O'Donnell, M., Bryant, R., McFarlane, A.,...Shalev, A. (2010). A longitudinal analysis of posttraumatic stress disorder symptoms and their relationship with fear and anxious-misery disorders: implications for DSM-V. *Journal of Affective Disorders, 127*, 147–152.

Friedman, N. P., Miyake, A., Young, S. E., Defries, J. C., Corley, R. P., & Hewitt, J. K. (2008). Individual differences in executive functions are almost entirely genetic in origin. *Journal of Experimental Psychology: General, 137*, 201–225.

Goodkind, M., Eickhoff, S. B., Oathes, D. J., Jiang, Y., Chang, A., Jones-Hagata, L. B.,...Etkin, A. (2015). Identification of a common neurobiological substrate for mental illness. *JAMA Psychiatry, 72*, 305–315.

Goschke, T. (2014). Dysfunctions of decision-making and cognitive control as transdiagnostic mechanisms of mental disorders: Advances, gaps, and needs in current research. *International Journal of Methods in Psychiatric Research. 23*, 41–57.

Gu, B. M., Park, J. Y., Kang, D. H., Lee, S. J., Yoo, S. Y., Jo, H. J.,... Kwon, J. S. (2008). Neural correlates of cognitive inflexibility during task-switching in obsessive-compulsive disorder. *Brain, 131*, 155–164.

Gyurak, A., Patenaude, B., Korgaonkar, M. S., Grieve, S. M., Williams, L. M., & Etkin, A. (2015). Frontoparietal activation during response inhibition predicts remission to antidepressants in patients with major depression. *Biological Psychiatry*, in press.

Hamilton, J. P., Etkin, A., Furman, D. J., Lemus, M. G., Johnson, R. F., & Gotlib, I. H. (2012). Functional neuroimaging of major depressive disorder: A meta-analysis and new integration of base line activation and neural response data. *American Journal of Psychiatry, 169*, 693–703.

Hajek, T., Alda, M., Hajek, E., & Ivanoff, J. (2013). Functional neuroanatomy of responseinhibition in bipolar disorders--combined voxel based and cognitive performance meta-analysis. *Journal of Psychiatric Research, 47*, 1955–1966.

Harvey, P. D., & Strassnig, M. (2012). Predicting the severity of everyday functional disability in people with schizophrenia: Cognitive deficits, functional capacity, symptoms, and health status. *World Psychiatry, 11*, 73–79.

Hayes, D. J., Duncan, N. W., Xu, J., & Northoff, G. (2014). A comparison of neural responses to appetitive and aversive stimuli in humans and other mammals. *Neuroscience & Biobehavioral Reviews, 45*, 350–368.

Jiang, J., Beck, J., Heller, K., & Egner, T. (2015). An insula-frontostriatal network mediates flexible cognitive control by adaptively predicting changing control demands. *Nature Communications, 6*, 8165.

Kang, D. H., Jang, J. H., Han, J. Y., Kim, J. H., Jung, W. H., Choi, J. S.,...Kwon, J. S. (2013). Neural correlates of altered response inhibition and dysfunctional connectivity at rest in obsessive-compulsive disorder. *Progress in Neuro-Psychopharmacology & Biological Psychiatry, 40*, 340–346.

Kempton, M. J., Salvador, Z., Munafò, M. R., Geddes, J. R., Simmons, A., Frangou, S., & Williams, S. C. (2011). Structural neuroimaging studies in major depressive disorder. Meta-analysis and comparison with bipolar disorder. *Archives of General Psychiatry, 68*, 675–690.

Kendler, K. S., Prescott, C. A., Myers, J., & Neale, M. C. (2003). The structure of genetic and environmental risk factors for common psychiatric and substance use disorders in men and women. *Archives of General Psychiatry, 60*, 929–937.

Kendler, K. S., Aggen, S. H., Knudsen, G. P., Røysamb, E., Neale, M. C., & Reichborn-Kjennerud, T. (2011). The structure of genetic and environmental risk factors for syndromal and subsyndromal common DSM-IV axis I and all axis II disorders. *American Journal of Psychiatry, 168*, 29–39.

Kim, H., & Eaton, N. R. (2015). The hierarchical structure of common mental disorders: Connecting multiple levels of comorbidity, bifactor models, and predictive validity. *Journal of Abnormal Psychology, 124,* 1064–1078.

Kohn, N., Eickhoff, S. B., Scheller, M., Laird, A. R., Fox, P. T., & Habel, U. (2014). Neural network of cognitive emotion regulation—An ALE meta-analysis and MACM analysis. *Neuroimage, 87,* 345–355.

Kotov, R., Chang, S. W., Fochtmann, L. J., Mojtabai, R., Carlson, G. A., Sedler, M. J., & Bromet, E. J. (2011). Schizophrenia in the internalizing-externalizing framework: a third dimension? *Schizophrenia Bulletin, 37,* 1168–1178.

Kozak, M. J., & Cuthbert, B. N. (2015). The NIMH research domain criteria initiative: Background, issues, and pragmatics. *Psychophysiology,* in press.

Kozak, M. J., & Cuthbert, B. N. (2016). The NIMH research domain criteria initiative: Background, issues, and pragmatics. *Psychophysiology, 53*(3), 286–297.

Krueger, R. F. (1999). The structure of common mental disorders. *Archives of General Psychiatry, 56,* 921–926.

Lahey, B. B., Applegate, B., Hakes, J. K., Zald, D. H., Hariri, A. R., & Rathouz, P. J. (2012). Is there a general factor of prevalent psychopathology during adulthood? *Journal of Abnormal Psychology, 121,* 971–977.

Lehto, J. E., Juujarvi, P., Kooistra, L., & Pulkkinen, L. (2003). Dimensions of executive functioning: Evidence from children, *British Journal of Developmental Psychology, 21,* 59–80.

Maltby, N., Tolin, D. F., Worhunsky, P., O'Keefe, T. M., & Kiehl, K. A. (2005). Dysfunctional action monitoring hyperactivates frontal-striatal circuits in obsessive-compulsive disorder: An event-related fMRI study. *Neuroimage, 24,* 495–503.

Markon, K. E. (2010). Modeling psychopathology structure: A symptom level analysis of Axis I and II disorders. *Psychological Medicine, 40,* 273–288.

Marsh, R., Horga, G., Parashar, N., Wang, Z., Peterson, B. S., & Simpson, H. B. (2014). Altered activation in fronto-striatal circuits during sequential processing of conflict in unmedicated adults with obsessive-compulsive disorder. *Biological Psychiatry, 75,* 615–622.

McClernon, F. J., Addicott, M. A., & Sweitzer, M. M. (2015). Smoking abstinence and neurocognition: Implications for cessation and relapse. *Current Topics in Behavioral Neuroscience, 23,* 193–227.

Medford, N., & Critchley, H. D. (2010). Conjoint activity of anterior insular and anterior cingulate cortex: Awareness and response. *Brain Structure & Function, 214,* 535–549.

Menon, V., & Uddin, L. Q. (2010). Saliency, switching, attention and control: A network model of insula function. *Brain Structure & Function, 214,* 655–667.

Miller, S., McTeague, L. M., Gyurak, A., Patenaude, B., Williams, L. M., Grieve, S. M.,...Etkin, A. (2015). Cognition-childhood maltreatment interactions in the prediction of antidepressant ouctomes in major depressive disorder patients: Results from the ISPOT-D trial. *Depression & Anxiety. 32,* 594–604.

Minzenberg, M. J., Laird, A. R., Thelen, S., Carter, C. S., & Glahn, D. C. (2009). Meta-analysis of 41 functional neuroimaging studies of executive function in schizophrenia. *Archives of General Psychiatry, 66,* 811–822.

Miyake, A., & Friedman, N. P. (2012). The nature and organization of individual differences in executive functions: Four general conclusions. *Current Directions in Psychological Science, 21,* 8–14.

Miyake, A., Friedman, N. P., Emerson, M. J., Witzki, A. H., Howerter, A., & Wager, T. D. (2000). The unity and diversity of executive functions and their contributions to complex 'Frontal Lobe' tasks: A latent variable analysis. *Cognitive Psychology, 41,* 49–100.

Müller, V. I., Langner, R., Cieslik, E. C., Rottschy, C., & Eickhoff, S. B. (2015). Interindividual differences in cognitive flexibility: Influence of gray matter volume, functional connectivity and trait impulsivity. *Brain Structure & Function, 220*(4), 2401–2414.

Niendam, T. A., Laird, A. R., Ray, K. L., Dean, Y. M., Glahn, D. C., & Carter, C. S. (2012). Meta-analytic evidence for a superordinate cognitive control network subserving diverse executive functions. *Cognitive Affective & Behavioral Neuroscience, 12,* 241–268.

Nolen-Hoeksema, S., & Watkins, E. R. (2011). A heuristic for developing transdiagnostic models of psychopathology: Explaining multifinality and divergent trajectories. *Perspectives in Psychological Science, 6*, 589–609.

Palmer, S. M., Crewther, S. G., Carey, L. M., & START Project Team. (2015). A meta-analysis of changes in brain activity in clinical depression. *Frontiers in Human Neuroscience, 8*, 1045.

Peeters, M., Vollebergh, W. A., Wiers, R. W., & Field, M. (2014). Psychological changes and cognitive impairments in adolescent heavy drinkers. *Alcohol and Alcoholism, 49*, 182–186.

Power, J. D., Schlaggar, B. L., Lessov-Schlaggar, C. N., & Petersen, S. E. (2013). Evidence for hubs in human functional brain networks. *Neuron, 79*, 798–813.

Ragland, J. D., Laird, A. R., Ranganath, C., Blumenfeld, R. S., Gonzales, S. M., & Glahn, D. C. (2009). Prefrontal activation deficits during episodic memory in schizophrenia. *American Journal of Psychiatry. 166*, 863–874.

Rajji, T. K., Ismail, Z., & Mulsant, B. H. (2009). Age at onset and cognition in schizophrenia: Meta-analysis. *British Journal of Psychiatry, 195*, 286–293.

Remijnse, P. L., van den Heuvel, O. A., Nielen, M. M., Vriend, C., Hendriks, G. J., Hoogendijk, W. J.,… Veltman, D. J. (2013). Cognitive inflexibility in obsessive-compulsive disorder and major depression is associated with distinct neural correlates. *PLoS ONE, 8*, e59600.

Riggs, N. R., Spruijt-Metz, D., Sakuma, K. L., Chou, C. P., & Pentz, M. A. (2010). Executive cognitive function and food intake in children. *Journal of Nutrition Education Behavior, 42*, 398–403.

Robinson, J. L., Laird, A. R., Glahn, D. C., Lovallo, W. R., & Fox, P. T. (2010). Metaanalytic connectivity modeling: Delineating the functional connectivity of the human amygdala. *Human Brain Mapping, 31*, 173–184.

Rotge, J. Y., Langbour, N., Guehl, D., Bioulac, B., Jaafari, N., Allard, M.,…Burbaud, P. (2010). Gray matter alterations in obsessive-compulsive disorder: An anatomic likelihood estimation meta-analysis. *Neuropsychopharmacology, 35*, 686–691.

Roth, R. M., Saykin, A. J., Flashman, L. A., Pixley, H. S., West, J. D., & Mamourian, A. C. (2007). Event-related functional magnetic resonance imaging of response inhibition in obsessive-compulsive disorder. *Biological Psychiatry, 62*, 901–909.

Satterthwaite, T. D., Kable, J. W., Vandekar, L., Katchmar, N., Bassett, D. S., Baldassano, C. F.,…Wolf, D. H. (2015). Common and dissociable dysfunction of the reward system in bipolar and unipolar depression. *Neuropsychopharmacology, 40*, 2258–2268.

Seeley, W. W., Menon, V., Schatzberg, A. F., Keller, J., Glover, G. H., Kenna, H.,…Greicius, M. D. (2007). Dissociable intrinsic connectivity networks for salience processing and executive control. *Journal of Neuroscience, 27*, 2349–2356.

Selvaraj, S., Arnone, D., Job, D., Stanfield, A., Farrow, T. F., Nugent, A. C.,…McIntosh, A. M. (2012). Grey matter differences in bipolar disorder: A meta-analysis of voxel-based morphometry studies. *Bipolar Disorders, 14*, 135–145.

Shanmugan, S., Wolf, D. H., Calkins, M. E., Moore, T. M., Ruparel, K., Hopson, R. D.,…Satterthwaite, T. D. (2016). Common and dissociable mechanisms of executive system dysfunction across psychiatric disorders in youth. *American Journal of Psychiatry*, in press.

Snyder, H. R., Miyake, A., & Hankin, B. L. (2015). Advancing understanding of executive function impairments and psychopathology: Bridging the gap between clinical and cognitive approaches. *Frontiers in Psychology, 6*, 328.

Sridharan, D., Levitin, D. J., & Menon, V. (2008). A critical role for the right fronto-insular cortex in switching between central-executive and default-mode networks. *Proceedings of the National Academy of Sciences of the United States of America, 105*, 12569–12574.

Stavro, K., Pelletier, J., & Potvin, S. (2013). Widespread and sustained cognitive deficits in alcoholism: A meta-analysis. *Addiction Biology, 18*, 203–213.

Tabarés-Seisdedos, R., Balanzá-Martínez, V., Sánchez-Moreno, J., Martinez-Aran, A., Salazar-Fraile, J., Selva-Vera, G.,…Vieta, E. (2008). Neurocognitive and clinical predictors of functional outcome in patients with schizophrenia and bipolar I disorder at one-year follow-up. *Journal of Affective Disorders, 109*, 286–299.

Wang, X. L., Du, M. Y., Chen, T. L., Chen, Z. Q., Huang, X. Q., Luo, Y.,…Gong, Q. Y. (2015). Neural correlates during working memory processing in major depressive disorder. *Progress in Neuro-Psychopharmacology & Biological Psychiatry, 56*, 101–108.

Wegbreit, E., Cushman, G. K., Puzia, M. E., Weissman, A. B., Kim, K. L., Laird, A. R., & Dickstein, D. P. (2014). Developmental meta-analyses of the functional neural correlates of bipolar disorder. *JAMA Psychiatry, 71*, 926–935.

World Health Organization. (1977). *International classification of diseases v.9.* World Health Organization.

World Health Organization. (2007). *International classification of diseases, v. 10.* World Health Organization. (http://apps.who.int/classifications/apps/icd/icd10online).

Wright, A. G., Krueger, R. F., Hobbs, M. J., Markon, K. E., Eaton, N. R., & Slade, T. (2013). The structure of psychopathology: Toward an expanded quantitative empirical model. *Journal of Abnormal Psychology, 122*, 281–294.

31

Cognitive Control in Schizophrenia

Psychological and Neural Mechanisms

Deanna M. Barch and Julia M. Sheffield

Cognitive Control in Schizophrenia: Core Psychological and Neural Mechanisms

When we think of the core symptoms of psychotic illnesses such as schizophrenia, we think of people who hear voices, see visions and have false beliefs about reality (i.e., delusions). However, psychiatrists and psychologists have long recognised that abnormalities in cognitive function are a critical aspect of schizophrenia, one of the most debilitating psychiatric disorders. In particular, clinicians have identified one of the crucial aspects of cognitive impairment in schizophrenia as a difficulty controlling thoughts and actions, which immediately brings to mind concepts related to cognitive control and executive function. Further, a strong emphasis on understanding the nature and mechanisms of cognitive function in schizophrenia is in part due to the now robust body of research suggesting that cognitive impairment in schizophrenia is one of the most critical determinants of quality of life and function in schizophrenia, potentially more so than the severity of other aspects/symptoms of schizophrenia such as hallucinations and delusions (Lepage, Bodnar, & Bowie, 2014; Nuechterlein et al., 2011).

Anyone who has ever conducted a study on cognitive function in schizophrenia (or even psychosis more broadly) immediately recognises one of the most challenging aspects of this domain of research: At least on the surface, individuals with this illness appear to have deficits in a wide array of domains, such as working memory, language function, executive function, episodic memory, processing speed, attention, inhibition, and sensory processing (Forbes, Carrick, McIntosh, & Lawrie, 2009; Mesholam-Gately, Giuliano, Goff, Faraone, & Seidman, 2009; Sheffield et al., 2014), with such deficits clearly present even in unmedicated individuals (Fatouros-Bergman, Cervenka, Flyckt, Edman, & Farde, 2014). The reality is that there are likely a number of different mechanisms that contribute to cognitive impairment in schizophrenia, and no one single mechanism, at either the psychological or neural level, is going to fully explain the diverse array of impairment that can be found in this illness. However, it is equally unlikely that every domain of impairment in schizophrenia reflects a dissociable deficit, each with its own psychological and neural pathophysiology. Instead, it may be the case that there are several core mechanisms that each contribute to impairments in a number of cognitive deficits in psychosis, and identifying one or more of these domains will help us to conceptualise the nature of cognitive dysfunction in this illness and other

The Wiley Handbook of Cognitive Control, First Edition. Edited by Tobias Egner.

psychotic disorders, identify developmental trajectories to impairment, and point to pathways towards more effective intervention for cognitive dysfunction in psychosis.

Proactive Control, Goal Representation, and Context Processing in Schizophrenia

In the spirit of identifying core mechanisms of cognitive dysfunction in schizophrenia, we and others have argued that one such mechanism is a deficit in the ability to actively represent goal information needed to guide behaviour in working memory, and that this deficit reflects impairments in the function of the dorsolateral prefrontal cortex (DLPFC), its interactions with other brain regions such as the parietal cortex, the thalamus, and the striatum, and the influence of neurotransmitter systems such as dopamine, GABA, and glutamate (Barch, Braver, Carter, Poldrack, & Robbins, 2009; Edwards, Barch, & Braver, 2010; Lesh, Niendam, Minzenberg, & Carter, 2011; see also Chapter 19 by Robbins & Kehagia in this volume). As described in more detail below, we refer to this as a deficit in '*proactive control*', a construct that we think succinctly captures the impairments that individuals with schizophrenia experience in a wide array of cognitive and life function domains that require individuals to represent, maintain, and use goal-relevant information to guide thoughts and actions (see also Chapter 9 by Chiew & Braver in this volume). In the sections below, we provide a discussion of the evidence for such impairment in schizophrenia and its relationship to specific neural substrates. We then discuss how such deficits in proactive control might contribute to, and manifest in, a number of putatively different cognitive domains, such as executive control, working memory, episodic memory, and conflict monitoring/error processing/action monitoring.

The Roots of Proactive Control: Context Processing

In previous work based in part upon computational modelling, Cohen and colleagues put forth the hypothesis that intact function of dopamine in the DLPFC was responsible for the processing of context, and that a disturbance in this mechanism was responsible for a range of cognitive deficits in schizophrenia (e.g., Barch et al., 2001; Braver, Barch, & Cohen, 1999; J. D. Cohen, Barch, Carter, & Servan-Schreiber, 1999; see also Chapter 9 by Chiew & Braver in this volume). In this framework, context refers to prior task-relevant information, including task instructions, the results of processing prior stimuli, and goals that are represented and maintained in working memory in a form that can bias the selection of an appropriate behavioural response. In many ways, this hypothesis about context-processing deficits in schizophrenia was very similar to earlier suggestions about the nature of cognitive deficits in this disorder put forth by researchers such as Shakow (Shakow, 1962). For example, Shakow suggested that 'we see particularly the various difficulties created by context … It is as if, in the scanning process which takes place before the response to a stimulus is made, the schizophrenic is unable to select out the material relevant for optimal response' (Shakow, 1962, p. 4).

One critical insight that emerged from this work was that a deficit in one aspect of executive control could contribute to deficits in cognitive domains that are often treated as independent. Particularly, we have argued that deficits in working memory, attention, inhibition, and language processing in schizophrenia can all be understood in terms of a deficit in context processing, as each of these domains requires the active representation of such context information for effective function (see for full discussion., Braver et al., 1999; J. D. Cohen et al., 1999). As described in more detail below, numerous prior studies have provided support for these hypotheses concerning context-processing deficits in schizophrenia (for a review, see Barch & Braver, 2007).

Proactive Control

In more recent years, the key role of context processing in cognition and in schizophrenia has been reconceptualised somewhat more broadly as the function of proactive cognitive control (Braver, Gray, & Burgess, 2007; Braver, Paxton, Locke, & Barch, 2009; Edwards et al., 2010; Haddon & Killcross, 2007). This conceptualisation builds upon earlier ideas of context processing to argue for flexible mechanisms of cognitive control that allow humans to deal with the diversity of challenges that we face in everyday life. In this theory, termed dual mechanisms of control (DMC; Braver et al., 2007; Braver et al., 2009; Edwards et al., 2010), a distinction is made between proactive and reactive modes of cognitive control. The proactive control mode can be thought of as a form of 'early selection', in which goal-relevant information is actively maintained in a sustained or anticipatory manner, before the occurrence of cognitively demanding events. This allows for optimal biasing of attention, perception, and action systems in a goal-driven manner. By goal information, we mean information about what one needs to accomplish in this particular task situation or the intended outcome of a series of actions or mental operations. In real-life settings, such goals may include the main points you wish to communicate in a conversation, or the need to organise a shopping trip so that you can make sure to get everything you need. In contrast, in the reactive mode, attentional control is recruited as a 'late correction' mechanism that is mobilised only when needed, such as after a high-interference event is detected (e.g., you encounter unexpected distracting stimuli and need to retrieve the topic of your conversation; see Chapter 4 by Egner and Chapter 17 by Ullsperger in this volume). Thus, proactive control relies on the *anticipation* and prevention of interference before it occurs, whereas reactive control relies on the detection and resolution of interference after its onset.

This theory postulates that proactive control depends on actively representing information in the lateral prefrontal cortex (Braver, 2012), and that the updating and maintenance of such information depends on interactions with other brain regions (e.g., the dorsal parietal cortex, dorsal anterior cingulate cortex, and subcortical regions; Cole et al., 2013), as well as precise inputs from neurotransmitter systems such as dopamine into the prefrontal cortex (see Figure 31.1). As such, proactive control may be particularly vulnerable to disruption, because it is resource demanding, dependent upon coordinated interactions with other regions, and upon precise dopamine–prefrontal interactions. Thus, we have suggested that populations characterised by disordered prefrontal and dopamine function (such as schizophrenia) may rely more heavily on reactive control, as it may be more robust in the face of such dysfunction (Edwards et al., 2010).

Evidence for Impaired Proactive Control in Schizophrenia

There are a number of different methods that one could use to assess proactive control in schizophrenia, each with varying strengths and weaknesses. One task that has been used in numerous studies is a version of the classic continuous performance test (Rosvold, Mirsky, Sarason, Bransome, & Beck, 1956) known as the AX-CPT (Cohen et al., 1999; Servan-Schreiber, Cohen, & Steingard, 1996), which has been specifically designed as a measure of context representation and maintenance (see also Chapter 9 by Chiew & Braver in this

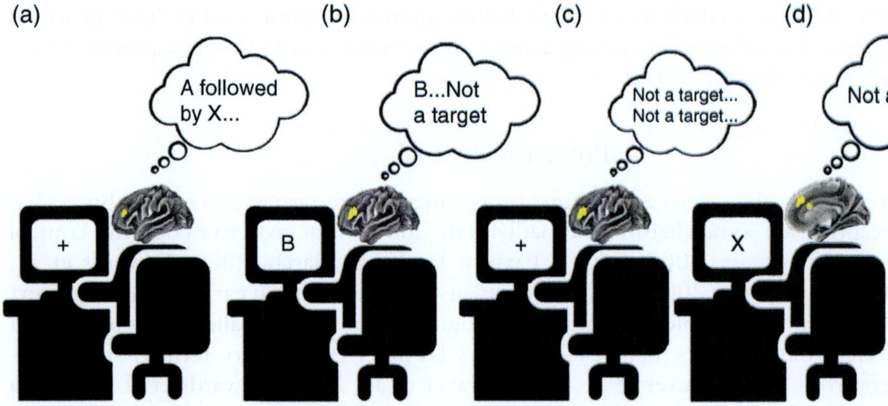

Figure 31.1 Brain regions relevant for understanding proactive control. Illustration of neural activity underlying the use of proactive control, during the AX-CPT task. Panel a: Continual maintenance of task rules activates the DLPFC. Panel b: When the subject observes an invalid cue (B cue), DLPFC activity increases, in order to prepare for future response and anticipate the need to override a prepotent response. Panel c: Proactive control requires the maintenance of cue information during the delay, leading to continually increased activity of the DLPFC. Panel d: When a target probe (X probe) is observed, the subject must override the prepotent response, leading to activity in the anterior cingulate cortex as well as the DLPFC, in order to facilitate successful conflict monitoring to select the correct response (non-target).

volume). In recent years, MacDonald and colleagues also developed a 'non-verbal' version of this task called the dot probe expectancy task (DPX) that uses modified Braille-type stimuli, but with the same structure and conceptual underpinnings (Henderson et al., 2012; A. W. MacDonald, 3rd et al., 2005). As shown in Figure 31.2a, in this task, participants are presented with cue–probe pairs and told to make a target response to an X (probe) but only when it follows an A (cue), and a non-target response otherwise. A correct response to X depends upon maintaining the 'context' or goal information provided by the cue (A or not-A). One change to the standard AX-CPT was to increase the frequency of target (AX) trials so that they occur with a high frequency (70%; see Figure 31.2b), with the remaining 30% of trials distributed across three types of non-target trials (BX, AY, and BY, where B refers to any non-A cue and Y refers to any non-X probe). This creates two types of biases that can be used to probe the integrity of proactive control. The first bias, or prepotent response, is that participants expect to make a target response when they see an X probe, because this is the correct response on most of the trials (87.5% of trials in which an X is presented). On BX trials, participants have to use the context provided by the B cue to inhibit this bias to respond 'target' to an X (which would lead to a false alarm). Thus, impaired proactive control will lead to poor performance on BX trials, because the context provided by the B cue would not be available to override the tendency to want to respond to the target X. The second bias is that participants expect to make a target response after they see an A cue, because most of the time an X follows the A cue. However, on trials in which the A is not followed by an X, this predictive aspect of proactive control actually *creates* the tendency to false alarm. Thus, intact goal representations will hurt performance on AY trials, because context induces an invalid expectancy, leading to worse AY than BX performance. In contrast, individuals with impaired proactive control should show worse BX than AY performance. In other words,

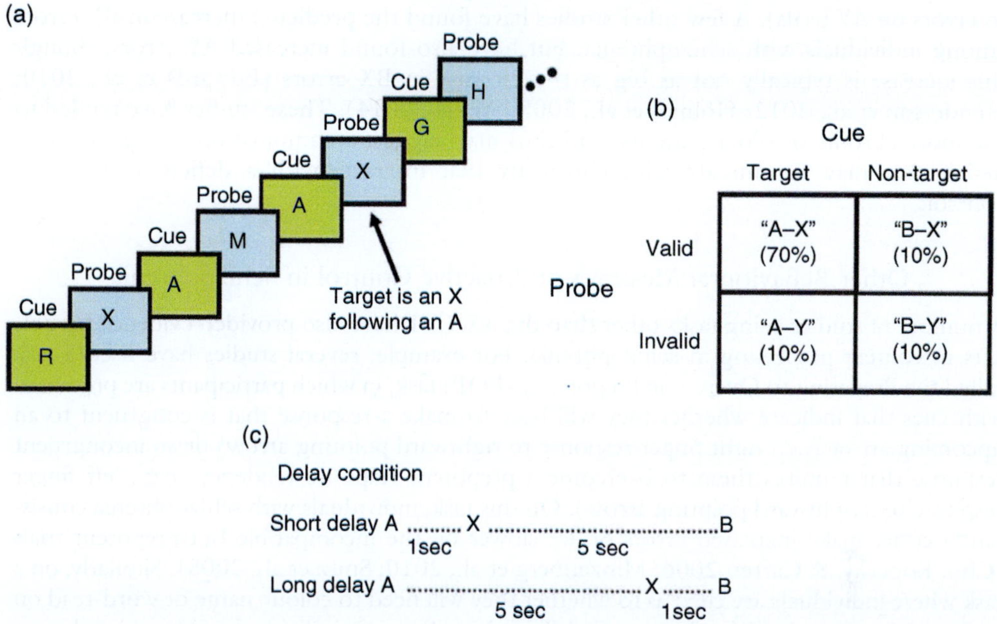

Figure 31.2 Illustration of the AX Continuous Performance Test. Panel a: Schematic of stimuli presentation across time in the AX-CPT. Panel b: Illustration of relative frequencies of different trial types in the AX-CPT. Panel c: Illustration of the timing of the stimuli in task conditions of the AX-CPT with either a short or a long delay between the cue and the probe. Source: Adapted from Barch & Braver, 2007a. Reproduced with permission of Cambridge University Press.

individuals with impaired context representations (such as individuals with schizophrenia) should show worse BX performance, but *better* AY performance, than individuals with intact proactive control.

Numerous previous studies have found results with this AX-CPT (or the DPX) task that provide evidence consistent with an impairment in proactive control in schizophrenia. More specifically, a number of studies have demonstrated that as predicted, individuals with schizophrenia (both medicated and mediation naive) made significantly more BX errors than healthy controls, but either no differences (J. D. Cohen et al., 1999; Fornito, Yoon, Zalesky, Bullmore, & Carter, 2011; Javitt, Rabinowicz, Silipo, & Dias, 2007; Javitt, Shelley, Silipo, & Lieberman, 2000; Jones, Sponheim, & MacDonald, 2010; Lesh et al., 2013; A. MacDonald et al., 2005; A. W. MacDonald, 3rd et al., 2005; Servan-Schreiber et al., 1996; Stratta, Daneluzzo, Bustini, Casacchia, & Rossi, 1998; Stratta, Daneluzzo, Bustini, Prosperini, & Rossi, 2000; Poppe, Carter, Minzenberg, & MacDonald, 2015; Yoon et al., 2008; Yoon et al., 2012; Yoon et al., 2014) or even significantly *fewer* AY errors (Barch et al., 2001; Barch, Carter, & Cohen, 2003), particularly at the longer delay. This pattern of both impaired and intact or even improved performance suggests that goal representation provided by the cues (A or B) are less available or less able to influence processing in individuals with schizophrenia, leading to an inability to override the incorrect tendency to respond target to the X on BX trials. In addition, such impairments in the proactive use of goal representations leads to less predictive use of goals on AY trials (which normally leads to errors on AY trials). A few other studies have found the predicted increase in BX errors among individuals with schizophrenia, but have also found increased AY errors, though this increase is typically not as big as the increase in BX errors (Edwards et al., 2010; Henderson et al., 2012; Holmes et al., 2005; Strauss, 2014). These studies have tended to use more chronic or older patients, and thus may suggest a combination of a generalised deficit that may accumulate with chronicity that interacts with a deficit in proactive control.

Other Behavioural Measures of Proactive Control in Schizophrenia

A number of studies using tasks other than the AX-CPT have also provided evidence for deficits in context processing in schizophrenia. For example, several studies have used a task called the Preparing to Overcome Prepotency (POP) task, in which participants are presented with cues that indicate whether they will have to make a response that is congruent to an upcoming arrow (i.e., right finger response to rightward pointing arrow) or an incongruent response that requires them to overcome a prepotent response tendency (e.g., left finger response to a rightward pointing arrow). On this task, individuals with schizophrenia consistently either make increased errors or are slower on the incompatible but prepotent trials (Cho, Konecky, & Carter, 2006; Minzenberg et al., 2010; Snitz et al., 2005). Similarly, on a task where individuals are cued as to whether they will need to colour-name or word-read on an upcoming Stroop trial—the so called 'Switching Stroop', individuals with schizophrenia make more colour naming errors on incongruent trials (Barch et al., 2003; J. D. Cohen et al., 1999), especially when they have to update task representations on a trial-by-trial basis (Elvevåg, Duncan, & McKenna, 2000). There is also evidence that individuals with schizophrenia show impairments in the use of episodic and contextual control cues in a complex cognitive control task developed by Koechlin and colleagues (Barbalat et al., 2011; Barbalat, Chambon, Franck, Koechlin, & Farrer, 2009).

Work from the language domain also suggests a deficit in the ability to use previously presented language to guide semantic interpretations among individuals with schizophrenia. Such results do not specifically point to a problem with proactive control per se, as these could result from a more passive semantic processing failure. Nonetheless, such findings are consistent with evidence for proactive control deficits. Specifically, a number of studies have examined an ERP component called the N400 that is thought to index the degree of consistency or relatedness between a stimulus (e.g., word) and the preceding context (which could be a single word, a sentence or an entire discourse; C. Brown & Hagoort, 1993; Kutas & Hillyard, 1980). The N400 is a negative-going component of an ERP waveform that occurs approximately 400 ms after the onset of the stimulus of interest. In cognitively intact individuals, words that are not consistent with the preceding context elicit a larger N400 than words that are consistent with the preceding context. A number of studies examining the N400 in schizophrenia have found evidence of reduced N400 priming (Condray, Steinhauer, Cohen, van Kammen, & Kasparek, 1999; Grillon, Rezvan, & Glazer, 1991; Salisbury, O'Donnell, McCarley, Nestor, & Shenton, 2000; Sitnikova, Salisbury, Kuperberg, & Holcomb, 2002; for related studies, see Adams et al., 1993; Niznikiewicz et al., 1997; Ohta, Uchiyama, Matsushima, & Toru, 1999; Olichney, Iragui, Kutas, Nowacki, & Jeste, 1997; Strandburgh et al., 1997; Titone, Holzman, & Levy, 2002; Titone, Levy, & Holzman, 2000). These N400 impairments may be more apparent with long times between prior context and integration, potentially reflecting an ability to maintain information (Mohammad & DeLisi, 2013). Further, there is also some evidence that such impairments may extend to the P600, which may index more rule-like language structure (Sitnikova, Perrone, Goff, & Kuperberg, 2010).

Individuals with schizophrenia also consistently show evidence of impairments on anti-saccade tasks, where they have to use cues to prepare to either make a prosaccade in the direction of flashed dot, or to overcome that prepotent response tendency in order to saccade in the opposite direction (Barton et al., 2002; Franke, Reuter, Schulz, & Kathmann, 2007; Greenzang, Manoach, Goff, & Barton, 2007; Manoach et al., 2002; Manoach et al., 2013). However, this antisaccade literature has also pointed to a set of results that are not so clearly consistent with a deficit in proactive control in schizophrenia. Specifically, many of the antisaccade studies have used task-switching paradigms where the prosaccade and antisaccade trials are interleaved. One might expect that having to switch between rules on a trial-by-trial basis would systematically impair performance among individuals with schizophrenia. However, only a handful of studies using task-switching paradigms have shown evidence for increased task- or response-switching costs (Elvevåg et al., 2000; Franke et al., 2007; Meiran, Levine, Meiran, & Henik, 2000), whereas many others have not shown any evidence of increased task-switching costs (Barton et al., 2002; Greenzang et al., 2007; Jamadar, Michie, & Karayanidis, 2010; Karayanidis et al., 2006; Kieffaber et al., 2006), at least not when overall longer RTs are taken into account. Further, the individuals with schizophrenia seem to show as much benefit as controls in there being a longer period between the task switch cue and the trial, which has traditionally been thought of as reflecting the ability to engage in preparatory task set reconfiguration (Manoach et al., 2002; Manoach et al., 2013; Meiran, 2000). The fact that individuals with schizophrenia seem able to use this cueing time to prepare in advance is not consistent with a deficit in proactive control. Interestingly, Ravizza and colleagues have shown task-switching impairments in schizophrenia when the task involved more complex rule switching, but not in a more perceptually based task switch (Ravizza, Moua, Long, & Carter, 2010; Wylie, Clark, Butler, & Javitt, 2010). They argue that task switch deficits will be more apparent the more complex the rule that needs to be updated. It may also be the case that there needs to be stronger prepotency to be overcome in order to clearly reveal evidence for difficulties using advance preparation to prepare for a new task challenge.

Neural Mechanisms of Impaired Proactive Control in Schizophrenia

As noted above, the DMC theory (Braver, 2012) and a number of other theories of cognitive control (Koechlin, Ody, & Kouneiher, 2003; Badre & D'Esposito, 2009) postulate that proactive control depends on actively representing information in the lateral prefrontal cortex, and that the updating and maintenance of such information depends on precise inputs from neurotransmitter systems such as dopamine into the prefrontal cortex. Thus, a logical hypothesis is that impairments in proactive control among individuals with schizophrenia reflect impairments in the function of the prefrontal cortex and its interactions with other brain systems relevant for cognitive control. In support of this hypothesis, there is ample evidence for an association between impairments in DLPFC activity and deficits of proactive control in schizophrenia (Barbalat et al., 2009; Edwards et al., 2010; Lesh et al., 2013; Minzenberg, Laird, Thelen, Carter, & Glahn, 2009; Poppe et al., 2015; Yoon et al., 2008; Yoon et al., 2012), for both medicated (Holmes et al., 2005) and unmedicated patients (Barch et al., 2001; Lesh et al., 2013; A. MacDonald et al., 2005; Snitz et al., 2005). There is also evidence for reduced gamma synchrony in the frontal cortex during performance of cognitive control tasks in schizophrenia (Cho et al., 2006; Minzenberg et al., 2010).

Further support for a role for impairments in the DLPFC in cognitive control in schizophrenia comes from an elegant meta-analysis conducted by Minzenberg and colleagues (Minzenberg et al., 2009) of imaging studies of executive control and proactive control.

The results of this meta-analysis provided robust evidence for reduced activity in the DLPFC in schizophrenia across a wide variety of specific tasks, again suggesting that there may be some common mechanisms at play across a variety of individual tasks. Importantly, however, the DLPFC is clearly not the only region relevant for proactive control, and the interactions of the DLPFC with other brain regions are also critical (Cole et al., 2013). Consistent with this more circuit-based perspective, there is growing evidence for a critical role for impaired connectivity between the DLPFC and other cognitive-control-related brain regions (Cole, Anticevic, Repovs, & Barch, 2011; Fornito et al., 2011; Repovs, Csernansky, & Barch, 2011; Yoon et al., 2008). There is also evidence for a role for altered activity and connectivity with subcortical regions rich with dopaminergic innervation, and evidence for the positive impacts of dopamine enhancement on cognitive control in psychosis (Barch & Carter, 2005; McClure et al., 2010). These pieces of evidence are consistent with the hypothesis that dopaminergic inputs into the DLPFC may also be critical for intact proactive control. At the same time, however, there is also intriguing evidence for GABA-ergicly mediated (Lewis et al., 2008) impairment in neural oscillations that may support representations in the DLPFC (Cho et al., 2006; Minzenberg et al., 2010).

Proactive Control and Risk for Psychosis

Other research has found that individuals at risk for schizophrenia, and who presumably share liability for this disorder, also display deficits in context processing on the AX-CPT task. For example, research has shown that non-psychotic siblings of patients with schizophrenia show a similar performance pattern on the AX-CPT to that found in schizophrenia patients, demonstrating increased BX errors and RTs, as well as either no difference, or even a decrease, in the number of AY errors (Delawalla, Csernansky, & Barch, 2008; A. W. MacDonald, 3rd et al., 2005; A. W. MacDonald, Pogue-Geile, Johnson, & Carter, 2003; Poppe et al., 2015) as well as impairments in other cognitive control tasks (A. W. MacDonald, 3rd, Becker, & Carter, 2006). Other research has shown that individuals at high clinical risk for psychosis also show a similar profile (Niendam et al., 2014). In our own work, we have examined context-processing deficits in individuals with schizotypal personality disorder, a disorder thought to share genetic liability with schizophrenia. We have also found a pattern of deficits on the AX-CPT among individuals with schizotypal personality disorder that is indicative of a disturbance in proactive control processing. Like individuals with schizophrenia, the individuals with schizotypal personality disorder displayed increased BX errors, but not AY errors, combined with increased BX RTs, but not AY RTs (Barch et al., 2004; McClure, Barch, Flory, Harvey, & Siever, 2008; McClure et al., 2007). However, individuals with other personality disorders do not show such deficits (McClure et al., 2008). Importantly, individuals at risk for psychosis also show impairments in the function of the DLPFC during performance of cognitive control tasks, though the pattern is sometimes one of hyperactivity versus hypoactivity (Delawalla et al., 2008; A. W. MacDonald et al., 2006; Niendam et al., 2014; Poppe et al., 2015). Taken together, these behavioural and imaging data suggest that deficits in proactive control may be part of the composite of factors associated with genetic risk for psychosis, and not just present in manifest illness.

Working Memory in Schizophrenia

As described above, we have suggested that there may be common mechanisms contributing to deficits across a variety of cognitive domains in schizophrenia. In particular, the types of deficits in DLPFC function and connectivity that we postulate to be central to deficits in proactive control are also thought to be critical in cognitive domains such as working memory and episodic memory, and thus may be part of the mechanisms leading to impairments in these domains in schizophrenia. Although there is a huge amount of work on executive function and proactive cognitive control in schizophrenia, there is even more data on working memory deficits in this illness (Forbes et al., 2009; Lee & Park, 2005). However, simply saying that individuals with schizophrenia have deficits on working memory tasks is not particularly informative, as working memory is a complex construct with different subprocesses (see also Miyake, this volume). To understand what specific aspects of working memory might be impaired in schizophrenia and how this might relate to the proactive control impairments described above, it is helpful to refer to a model of working memory function, such as Baddeley's model (Baddeley, 2000), which still remains one of the most commonly used. This model proposes four major components of working memory: (a) the visuospatial sketchpad, a short-term storage buffer for visual information; (b) the phonological loop, a short-term storage buffer for verbal information; (c) a central executive that supports the manipulation and transformation of information held within the storage buffers; and (d) an episodic buffer, in which complex, multi-modal events are integrated and stored on-line (Baddeley, 2000). The central executive has been associated with the function of the DLPFC in numerous studies, whereas the storage buffers have been associated with both inferior frontal and posterior parietal function (Wager & Smith, 2003).

Behavioural Evidence

There is robust evidence that individuals with schizophrenia have difficulty with processes attributed to the central executive component of working memory (Barch, 2005). A critical argument supporting the presence of central executive dysfunction in schizophrenia is the fact that patients with this illness have deficits on working memory tasks with all different material types, with relatively little evidence for selective deficits with one material type over another (Forbes et al., 2009; Lee & Park, 2005). Another argument is that individuals with schizophrenia consistently show deficits on tasks designed to measure a range of functions ascribed to the central executive, including manipulation (Horan et al., 2008; J. Kim, Glahn, Nuechterlein, & Cannon, 2004), interference control and/or dual-task coordination (e.g., Smith, Eich, Cebenoyan, & Malapani, 2011), and information updating and temporal indexing (e.g., Galletly, McFarlane, & Clark, 2007).

However, similar to the literature on proactive control in schizophrenia, there is some conflicting evidence. In particular, one would predict that individuals with schizophrenia should have challenges using predictive cues to help them select task-relevant versus task-irrelevant information for encoding into working memory. Nevertheless, some studies found evidence that individuals with schizophrenia had an intact ability to use predictive cues to guide management of the contents of working memory (Gold et al., 2006; Smith et al., 2011). In contrast, other studies found strong evidence for impaired working memory capacity (Gold et al., 2010; Gold, Wilk, McMahon, Buchanan, & Luck, 2003), and robust evidence that individuals with schizophrenia have difficulties inhibiting the impact of salient distractors in working memory (Hahn et al., 2010). As with the proactive control literature, it is not quite clear why there is not more evidence for impaired predictive control over working memory in schizophrenia, though one could again speculate as to whether introducing a prepotency manipulation might reveal susceptibilities in these mechanisms. Consistent with this hypothesis, we have seen that individuals with schizophrenia have particular difficulty on 'lure' trials during working memory, where there is a sense of familiarity from prior exposure, but the stimuli are no longer task relevant (Brahmbatt, Haut, Csernansky, & Barch, 2004; Ceaser & Barch, 2015). In addition, Smith and colleagues found similar evidence for difficulties ignoring or removing previously relevant information from working memory among individuals with schizophrenia (Eich, Nee, Insel, Malapani, & Smith, 2014; Smith et al., 2011). Thus, it is possible that future studies combining proactive cue manipulations with saliency or prepotency may reveal greater evidence for difficulties in the proactive control of the contents of working memory in schizophrenia.

Neural Evidence

Similar to the literature on cognitive control in schizophrenia, there is very robust evidence for altered DLPFC function among individuals with schizophrenia and their relatives during performance of working memory tasks that require central executive type functions, though the exact nature of the DLPFC impairment can vary as a function of both sample and task characteristics (Glahn et al., 2005; Van Snellenberg, Torres, & Thornton, 2006). For example, work comparing working memory performance during a task requiring updating and maintenance of information to control conditions with similar verbal and spatial processing requirements suggested reduced activity in the bilateral DLPFC and dorsal parietal cortex in schizophrenia (Barch & Csernansky, 2007). However, this study found relatively intact activity of ventrolateral prefrontal and parietal regions typically associated with the function of domain-specific storage buffers (e.g., verbal rehearsal, visual spatial scratchpad; Barch & Csernansky, 2007). Further, there is even evidence for reduced DLPFC activity during encoding or maintenance of information during working memory tasks that do not necessarily place heavy demands on central executive function (e.g., spatial delayed response tasks; Anticevic, Repovs, & Barch, 2013; Driesen et al., 2008), suggesting the possibility that central executive or proactive control mechanisms could be contributing to impaired performance in task contexts that might not normally be associated with strong proactive control demands in healthy individuals.

Importantly, as with proactive control, there is evidence suggesting a key role for impaired connectivity between the DLPFC and other working memory related regions (e.g., parietal cortex, thalamus, and striatum) in understanding working memory in schizophrenia (Barch & Csernansky, 2007; Hanlon et al., 2012; Henseler, Falkai, & Gruber, 2010; Karlsgodt et al., 2008; Kyriakopoulos et al., 2012), as well as evidence for altered gamma and theta oscillatory activity in prefrontal regions (e.g., Barr et al., 2010; Basar-Eroglu et al., 2007; Senkowski & Gallinat, 2015).

As alluded to above, numerous studies have also documented reduced working memory capacity in schizophrenia, including studies using change detection measures (e.g., Fuller et al., 2009; Gold, Fuller, Robinson, Braun, & Luck, 2007; Gold et al., 2006; Gold et al., 2010; Hahn et al., 2010) that are thought to be a relatively 'pure' measure of storage capacity (Cowan, 2010). Such capacity deficits cannot be solely accounted for by impaired control of attention (Gold et al., 2006) or by impaired precision of visual representations (Gold et al., 2010). Recent work in healthy individuals has shown that when the contributions of rehearsal, grouping, and chunking are reduced, working memory capacity is in the range of 3–5 items (Cowan, 2010). Several different lines of research point to the parietal cortex as a key node in the working memory network that drives these capacity limitations (Pisella, Berberovic, & Mattingley, 2004; Todd & Marois, 2004, 2005; Vogel & Machizawa, 2004). Work on the coding or signal properties of the parietal cortex that are responsible for this limitation suggests several potential mechanisms. One proposal is that items are maintained in working memory as unique patterns of coordinated firing across many neurons (Lisman, 2010). In such models, the number of high-frequency oscillations (i.e., the gamma range) that can be embedded in a lower frequency oscillation (e.g., theta) determines the number of distinct items that can be maintained in memory (Lisman, 2010). Alternatively, another hypothesis has been that the lateral inhibition mechanisms involved in neuronal representation limit the number of items that can be simultaneously maintained (Edin et al., 2009). Importantly, there is also data to suggest that DLPFC mechanisms can help increase capacity by virtue of enhanced excitatory input that competes with lateral inhibition mechanisms (Edin et al., 2009). As such, it is plausible that one mechanism contributing to impaired working memory capacity in schizophrenia is a reduction in DLPFC input that can normally serve to boost working memory capacity, a hypothesis that remains to be tested.

Episodic Memory in Schizophrenia

When one discusses episodic memory functions, a typical focus is on encoding, binding, relational, and retrieval mechanisms that are frequently associated with the function of medial temporal regions, including the hippocampus (Eichenbaum, Yonelinas, & Ranganath, 2007; Konkel & Cohen, 2009). In schizophrenia, there is meta-analytic evidence for greater impairments in relational than in item memory (Achim & Lepage, 2003). More recently, clinical researchers have begun to use tasks derived from the animal literature on hippocampal function, such as the transitive inference test, which measures the ability to learn the relationships among hierarchically arranged stimulus pairs. Individuals with schizophrenia are impaired on critical conditions of this task requiring relational processing, but not on conditions that require simpler associative reinforcement mappings (Coleman et al., 2010; Titone, Ditman, Holzman, Eichenbaum, & Levy, 2004). Other work has used eye-movement measures of relational memory shown to be impaired in patients with hippocampal lesions (Hannula & Ranganath, 2008, 2009), to identify relational memory impairments in schizophrenia (Hannula et al., 2010; Williams et al., 2010).

Many researchers have taken this body of work to suggest that episodic memory impairments in schizophrenia reflect medial temporal lobe deficits, with a specific focus on the hippocampus (e.g., Heckers & Konradi, 2010). However, it has become increasing clear that prefrontal structures also make important contributions to episodic memory (see also Chapter 20 by Dudukovic & Kuhl in this volume). Damage to the prefrontal cortex can lead to episodic memory deficits, although episodic memory is typically not the only cognitive function impaired in these individuals (e.g., Janowsky, Shimamura, Kritchevsky, & Squire, 1989). Such findings have contributed to the hypothesis that prefrontal cortex damage alters episodic memory by impairing strategic contributions to memory formation and retrieval. For example, studies have shown activation of ventral prefrontal regions such as Brodmann's areas 45 and 47 when participants are asked to process verbal information using semantic elaboration strategies (e.g., Wagner et al., 1998) that promote subsequent memory. In addition, one of the most compelling findings supporting a key role for such prefrontal structures in episodic memory are results showing that the increased activation during encoding in frontal regions such as BA 45 and 47 are very strongly predictive of subsequent memory performance (e.g., Wagner et al., 1998). Further, there is recent work suggesting the DLPFC may contribute specifically to successful relational memory formation and retrieval (e.g., Blumenfeld, Parks, Yonelinas, & Ranganath, 2011; Murray & Ranganath, 2007), though other theories suggest alternative DLPFC contributions to episodic memory (Balconi, 2013).

Such findings in healthy individuals raise the possibility that at least some episodic memory impairments among individuals with schizophrenia also reflect deficits in prefrontally mediated cognitive functions that contribute to successful memory encoding and retrieval, including strategic and proactive control mechanisms that may facilitate memory formation.

Consistent with this hypothesis, a number of studies suggest that individuals with schizophrenia are impaired in their ability to generate effective mnemonic strategies (for review, see Barch, 2005). However, when provided with strategies that promote successful episodic encoding, individuals with schizophrenia are typically able to benefit as much as controls from these strategies (e.g., Bonner-Jackson & Barch, 2011; Bonner-Jackson, Yodkovik, Csernansky, & Barch, 2008; Ragland et al., 2005), and can even show intact item recognition when provided with support for effective encoding (Mathews & Barch, 2004). Further, a meta-analysis of brain activity alterations during episodic memory performance in schizophrenia showed consistent evidence for reduced activation in both the ventrolateral PFC and DLPFC, but did not find consistent evidence for altered hippocampal activity (Ragland et al., 2009). Recent work on relational memory encoding and retrieval has shown evidence for impaired DLPFC function associated with impaired relational memory function in schizophrenia (Ragland et al., 2012; Ragland et al., 2015). Although these findings by no means exclude medial temporal lobe contributions to impaired episodic memory in schizophrenia, they do suggest a need to also take into account a role for prefrontally mediated cognitive functions. These findings also do not directly point to a specific contribution of proactive control in episodic memory impairments, but the findings on strategic changes and the role of the DLPFC suggests that this may be a fruitful avenue for future research on episodic memory in schizophrenia.

Conflict Detection/Error Monitoring as a Mechanism of Cognitive Control Impairments in Psychosis

An alternative (or potentially complimentary) theory about the nature of cognitive control deficits in schizophrenia suggests that disturbances in the ability to detect conflict or errors in ongoing processing, due at least in part to the function of the anterior cingulate cortex (ACC; see also Chapter 15 by Brown and Chapter 10 by Kool et al. in this volume), may lead to deficits in the ability to regulate and control a range of other components of executive control in schizophrenia. Further, some theorists have argued that monitoring deficits of this type might specifically contribute to symptoms such as delusions of alien control (Frith & Done, 1989). Such hypotheses regarding the potential importance of conflict detection and/or error monitoring in schizophrenia again stem in part from the basic cognitive neuroscience literature on the functional role of the ACC. One such hypothesis about the role of the ACC is based primarily on studies of the error-related negativity (ERN), an evoked response potential component that occurs with errors in a variety of experimental paradigms that involve speeded responses (Falkenstein, Hohnsbein, & Hoorman, 1995; Gehring, Coles, Meyer, & Donchin, 1990). The ERN has been consistently localised to the ACC (Dahaene, Posner, & Tucker, 1994). In earlier work, Coles and colleagues suggested that the ERN may index the function of a comparator process that compares the actual output with the 'correct' or intended output and then signals a mismatch when an error occurs. An update on this idea explicitly incorporates information about the dopaminergic inputs to the ACC to hypothesise that

> when human participants commit errors in reaction time tasks, the mesencephalic dopamine system conveys a negative reinforcement learning signal to the frontal cortex, where it generates the ERN by disinhibiting the apical dendrites of motor neurons in the anterior cingulate cortex. Furthermore, we suggest that the error signals are used to train the anterior cingulate cortex, ensuring that control over the motor system will be released to a motor controller that is best suited for the task at hand. (p. 679, Holroyd & Coles, 2002)

An alternative theory regarding the functional role of the ACC has been driven more by findings from functional neuroimaging studies, and posits that the ACC plays a crucial role in the monitoring and detection of conflict in ongoing processing (Botvinick, Braver, Barch, Carter, & Cohen, 2001), as well as potentially indexing the likelihood of errors (J. W. Brown & Braver, 2005, 2008) or the opportunity cost or expected value of cognitive control (Holroyd, 2013; Shenhav, Botvinick, & Cohen, 2013; see also Chapter 15 by Brown in this volume). A crucial aspect of these hypotheses is that the output of the ACC's response to conflict or cost computations serves to indicate the need for additional cognitive control function. In this theory, the actual implementation of control adjustments following conflict or error detection has been linked to the function of the DLPFC, such that the degree of DLPFC activity predicts the degree of conflict adaptation (Egner & Hirsch, 2005; Kerns, 2006; Kerns et al., 2004; Marco-Pallares, Camara, Munte, & Rodriguez-Fornells, 2008), with some evidence that the degree of ACC activity can also predict conflict adaptation in some conditions (C. Kim, Johnson, & Gold, 2014). Further, the degree of ACC activity on previous high-conflict trials predicts the degree of DLPFC activity on the subsequent trial (Hendrick, Ide, Luo, & Li, 2010; Kerns, 2006; Kerns et al., 2004; C. Kim et al., 2014). These various theories differ in the specific functions that they attribute to the ACC. Nonetheless, they all would predict that ACC disturbances that contribute to failures to detect conflict and/or errors would have a detrimental effect on a range of cognitive functions, in that detecting and responding to errors/conflict may be a critical means by which control is engaged and regulated.

At the behavioural level, there is very mixed evidence in schizophrenia regarding impairments on indices thought to reflect the ability to detect or respond to conflict or errors. For example, some studies have found that individuals with schizophrenia detect and/or correct errors as frequently as healthy controls (Kopp & Rist, 1994), whereas other studies have found reduced rates of error correction among individuals with schizophrenia (Turken, Vuilleumier, Mathalon, Swick, & Ford, 2003). In addition, some studies have found that individuals with schizophrenia show an intact 'Rabbit' effect, which is the slowing of responses on trials following errors (Alain, McNeely, Yu, Christensen, & West, 2002; Laurens, Ngan, Bates, Kiehl, & Liddle, 2003; Mathalon, Fedor, et al., 2002; Morris, Yee, & Nuechterlein, 2006; Polli et al., 2006), whereas other studies have found that this effect is reduced in patients with schizophrenia (Alain et al., 2002; Becerril, Repovs, & Barch, 2011; Carter, MacDonald, III, Ross, & Stenger, 2001; Kerns et al., 2005). In contrast, ERP, fMRI, and PET studies have provided more consistent evidence for altered conflict/error processing. For example, a number of studies have shown reduced ERN amplitudes or ACC activity in patients with schizophrenia on error trials (Alain et al., 2002; Bates, Kiehl, Laurens, & Liddle, 2002; Bates, Liddle, Kiehl, & Ngan, 2004; Becerril et al., 2011; Carter et al., 2001; Horan, Foti, Hajcak, Wynn, & Green, 2012; Kerns et al., 2005; Kopp & Rist, 1994, 1999; Krawitz, Braver, Barch, & Brown, 2011; Laurens et al., 2010; Laurens et al., 2003; Mathalon, Dedor, et al., 2002; Mathalon, Jorgensen, Roach, & Ford, 2009; Minzenberg, Gomes, Yoon, Swaab, & Carter, 2014; Morris, Heerey, Gold, & Holroyd, 2008; Morris, Holroyd, Mann-Wrobel, & Gold, 2011; Morris et al., 2006; Perez et al., 2012; Polli et al., 2008; Simmonite et al., 2012), even when these same individuals show intact behavioural indices or error detection (Kopp & Rist, 1994, 1999; Laurens et al., 2003; Mathalon, Fedor, et al., 2002; Morris et al., 2006).

Interestingly however, individuals with schizophrenia show intact feedback related negativity (Horan et al., 2012; Morris et al., 2008; Morris et al., 2011) in the context of impaired ERN. This pattern of findings has been interpreted as suggesting that individuals with schizophrenia can process error-related information when it is provided externally, but not when it is dependent on internal representations of tasks or outcomes. Our own recent work on ACC function is consistent with this interpretation, as we found intact ACC function and post-error adjustments in a task that provided information about the required response on a trial-to-trial basis (Becerril & Barch, 2013), thus providing external support for representations about correct versus incorrect outcomes.

At the same time, it is important to note that many studies have shown robust activation of the ACC among individuals with schizophrenia, and even in some cases increased ACC activation. For example, two major meta-analyses of cognitive task-related activation in schizophrenia—one focused on executive function and one focused on episodic memory—both showed consistent evidence for *increased* ACC activity in individuals with schizophrenia as compared to controls. Such findings raise interesting questions about the relationship between deficits in ACC and DLPFC function in schizophrenia, in that theories about primary deficits in either area would also predict deficits in the function of the other region as a consequence. For example, as described above, if ACC activity is important for the recruitment of control processes supported by the DLPFC, one would predict abnormal DLPFC activity in individuals with schizophrenia who have ACC deficits. In contrast, impaired DLPFC function and proactive control could contribute to increased conflict and errors, which in turn would predict increased ACC activity in individuals with schizophrenia according to the conflict monitoring or even the expected cost of control theory (if impaired control makes it more 'costly').

However, impaired DLPFC function and proactive control might predict reduced ACC activity according to the error-detection theory, if reduced DLPFC activity reflected degraded representations of the predictive information needed to drive an error correcting dopamine signals that in turn elicits ACC activity. Of course, individuals with schizophrenia may experience deficits in both ACC and DLPFC activity that are of equal relevance to understanding cognition, and potentially reflect the common importance of dopaminergic or other neurotransmitter inputs to both the DLPFC and ACC. Clearly, more research is needed that focuses on the relationship between conflict/error detection and the engagement of control processing in individuals with schizophrenia in order to help provide a better understanding of the dynamic processes that give rise to deficits in cognitive control, working memory and episodic memory in schizophrenia.

Conclusion

As this selective review highlights, individuals with schizophrenia show significant deficits on a wide variety of tasks that index various aspects of cognitive control, executive control, and working memory. We have suggested that there is a common mechanism contributing to these deficits: an impairment in proactive control that can influence performance in a wide variety of cognitive domains. Further, we have suggested that at the neural level, a common denominator of such deficits may be impaired function of the DLPFC, its connectivity with other brain regions, and the neurotransmitter systems that support precise updating and maintenance of goal representations that enable proactive control. However, as we have described in prior work, there are at least two ways to conceptualise this hypothesis about the role of proactive control in the cognitive impairment suffered by individuals with schizophrenia. One way to think about these relationships is that the unifying factor might be the *psychological* mechanism of goal representation/proactive control, as illustrated in Figure 31.3a. In other words, this hypothesis would suggest that the influence of DLPFC function on the range of cognitive domains we have highlighted in this review is *mediated* through the creation of proactive control deficits, which in turn contributes to impairments in a variety of other cognitive domains. However, alternatively, as shown in Figure 31.3b, it

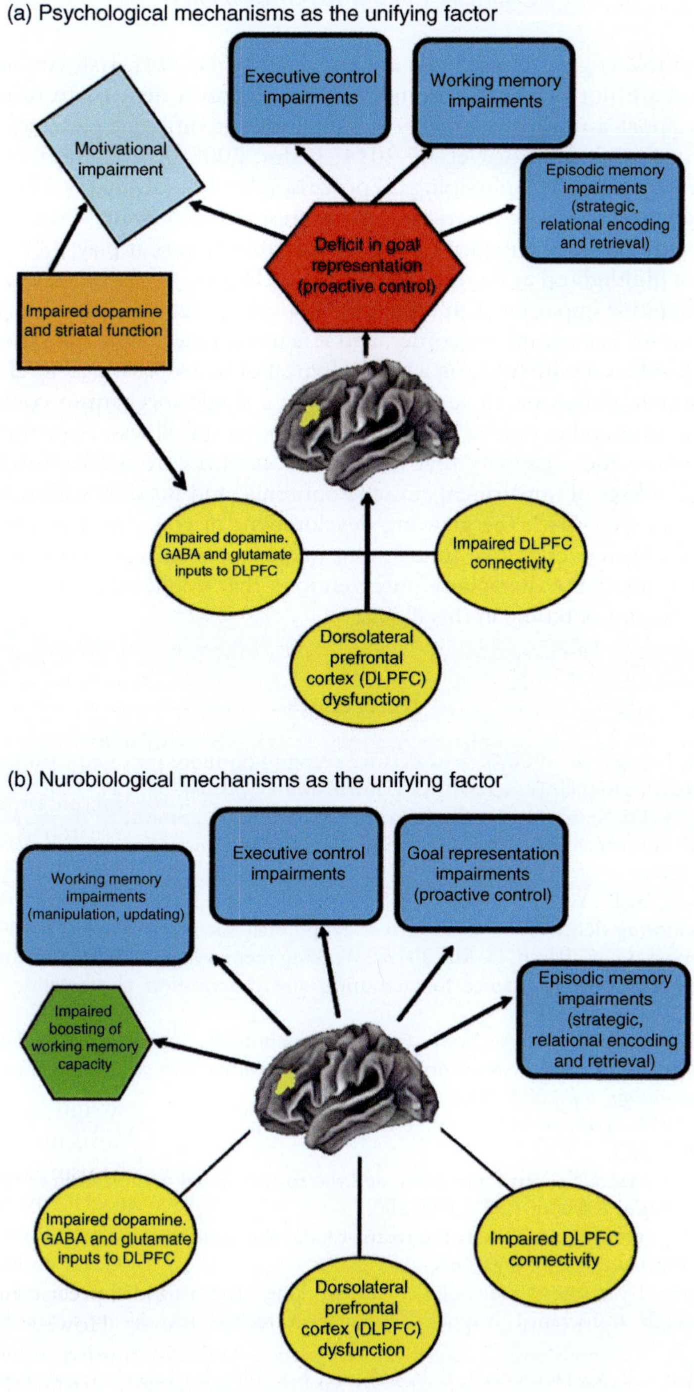

Figure 31.3 Potential common mechanisms of cognitive dysfunction in schizophrenia. Illustration of two potential mechanisms of cognitive dysfunction in schizophrenia, linking deficits in goal maintenance/proactive control, DLPFC function, and other cognitive impairments. (a) In this model, impairments in proactive control mediate the association between DLPFC dysfunction and deficits in other cognitive domains in schizophrenia. (b) Alternatively, DLPFC dysfunction may directly influence cognitive function in many domains in schizophrenia (including proactive control); however, deficits in other cognitive domains are not mediated by impairments in proactive control. Source: Barch 2012. Reproduced with permission of Elsevier.

may be that multiple cognitive functions are impacted by DLPFC dysfunction directly, but that these effects are not mediated through proactive control deficits. In other words, the DLPFC may support a number of different functions (or different parts of the PFC may support different functions; Azuar et al., 2014; Badre, 2008; Koechlin, 2014; Koechlin & Summerfield, 2007), and pathophysiological processes that broadly impair PFC function may thus impact a range of cognitive functions. Further research will be needed to compare these options and to understand the mediating and moderating factors at play.

Further, as we highlighted at the beginning of this chapter, we do not mean to assert that all aspects of cognitive impairment in schizophrenia can be fully explained by these mechanisms. Schizophrenia is a complex disorder, and it is not yet clear how the various symptoms that comprise this disorder interact, on a psychological or neurobiological level. Therefore, it would be an oversimplification to suggest that only a single mechanism could explain the diversity of symptoms and cognitive impairments found in this illness. Nonetheless, we think it important to raise the possibility that there is a common core mechanism that can help explain at least a subset of impairments in schizophrenia, and possibly within related mental health disorders as well. With the growing development of cognitive training and medications to directly improve cognitive functioning in schizophrenia, we hope that our model outlines a useful target for therapeutic interventions that could serve to broadly enhance cognitive function and outcome in this illness.

References

Achim, A. M., & Lepage, M. (2003). Is associative recognition more impaired than item recognition memory in schizophrenia? A meta-analysis. *Brain and Cognition*, *53*, 121–124.

Adams, J., Fauz, S. F., Nestor, P. G., Shenton, M. E., Marcy, B., Smith, S. R., & McCarley, R. W. (1993). ERP abnormalities during semantic processing in schizophrenia. *Schizophrenia Research*, *10*(3), 247–257.

Alain, C., McNeely, H. E., Yu, H., Christensen, B. K., & West, R. (2002). Neurophysiological evidence of error monitoring deficits in patients with schizophrenia. *Cerebral Cortex*, *12*, 840–846.

Anticevic, A., Repovs, G., & Barch, D. M. (2013). Working memory encoding and maintenance deficits in schizophrenia: Neural evidence for activation and deactivation abnormalities. *Schizophrenia Bulletin*, *39*(1), 168–178.

Azuar, C., Reyes, P., Slachevsky, A., Volle, E., Kinkingnehun, S., Kouneiher, F.,...Levy, R. (2014). Testing the model of caudo-rostral organization of cognitive control in the human with frontal lesions. *Neuroimage*, *84*, 1053–1060.

Baddeley, A. D. (2000). The episodic buffer: A new component of working memory? *Trends in Cognitive Science*, *4*, 417–423.

Badre, D. (2008). Cognitive control, hierarchy, and the rostro-caudal organization of the frontal lobes. *Trends in Cognitive Sciences*, *12*(5), 193–200.

Badre, D., & D'Esposito, M. (2009). Is the rostro-caudal axis of the frontal lobe hierarchical? *Nature Reviews Neuroscience*, *10*(9), 659–669.

Balconi, M. (2013). Dorsolateral prefrontal cortex, working memory and episodic memory processes: Insight through transcranial magnetic stimulation techniques. *Neuroscience Bulletin*, *29*(3), 381–389.

Barbalat, G., Chambon, V., Domenech, P. J., Ody, C., Koechlin, E., Franck, N., & Farrer, C. (2011). Impaired hierarchical control within the lateral prefrontal cortex in schizophrenia. *Biological Psychiatry*, *70*(1), 73–80.

Barbalat, G., Chambon, V., Franck, N., Koechlin, E., & Farrer, C. (2009). Organization of cognitive control within the lateral prefrontal cortex in schizophrenia. *Archives of general psychiatry*, *66*(4), 377–386.

Barch, D. M. (2005). The cognitive neuroscience of schizophrenia. In T. Cannon & S. Mineka (Eds.), *Annual Review of Clinical Psychology* (Vol. *1*, pp. 321–353). Washington, DC: American Psychological Association.

Barch, D. M., & Braver, T. S. (2007). Cognitive control in schizophrenia: Psychological and neural mechanisms. In R. W. Engle, G. Sedek, U. von Hecker, & A. M. McIntosh (Eds.), *Cognitive limitations in aging and psychopathology* (pp. 122–159). Cambridge: Cambridge University Press.

Barch, D. M., Braver, T. S., Carter, C. S., Poldrack, R. A., & Robbins, T. W. (2009). CNTRICS final task selection: Executive control. *Schizophrenia Bulletin*, *35*(1), 115–135.

Barch, D. M., & Carter, C. S. (2005). Amphetamine improves cognitive function in medicated individuals with schizophrenia and in healthy volunteers. *Schizophrenia Research*, *77*(1), 43–58.

Barch, D. M., Carter, C. S., Braver, T. S., McDonald, A., Sabb, F. W., Noll, D. C., & Cohen, J. D. (2001). Selective deficits in prefrontal cortex regions in medication naive schizophrenia patients. *Archives of General Psychiatry*, *50*, 280–288.

Barch, D. M., Carter, C. S., & Cohen, J. D. (2003). Context processing deficit in schizophrenia: Diagnostic specificity, 4-week course, and relationships to clinical symptoms. *Journal of Abnormal Psychology*, *112*(1), 132–143.

Barch, D. M., & Csernansky, J. G. (2007). Abnormal parietal cortex activation during working memory in schizophrenia: Verbal phonological coding disturbances versus domain-general executive dysfunction. *American Journal of Psychiatry*, *164*(7), 1090–1098.

Barch, D. M., Mitropoulou, V., Harvey, P. D., New, A. S., Silverman, J. M., & Siever, L. J. (2004). Context-processing deficits in schizotypal personality disorder. *Journal of Abnormal Psychology*, *113*(4), 556–568.

Barr, M. S., Farzan, F., Tran, L. C., Chen, R., Fitzgerald, P. B., & Daskalakis, Z. J. (2010). Evidence for excessive frontal evoked gamma oscillatory activity in schizophrenia during working memory. *Schizophrenia Research*, *121*(1–3), 146–152.

Barton, J. J., Cherkasova, M. V., Lindgren, K., Goff, D. C., Intriligator, J. M., & Manoach, D. S. (2002). Antisaccades and task switching: Studies of control processes in saccadic function in normal subjects and schizophrenic patients. *Annals of the New York Academy of Sciences*, *956*, 250–263.

Basar-Eroglu, C., Brand, A., Hildebrandt, H., Karolina Kedzior, K., Mathes, B., & Schmiedt, C. (2007). Working memory related gamma oscillations in schizophrenia patients. *International Journal of Psychophysiology*, *64*(1), 39–45.

Bates, A. T., Kiehl, K. A., Laurens, K. R., & Liddle, P. F. (2002). Error-related negativity and correct response negativity in schizophrenia. *Clinical Neurophysiology*, *113*(9), 1454–1463.

Bates, A. T., Liddle, P. F., Kiehl, K. A., & Ngan, E. T. (2004). State dependent changes in error monitoring in schizophrenia. *Journal of Psychiatric Research*, *38*(3), 347–356.

Becerril, K. E., & Barch, D. M. (2013). Conflict and error processing in an extended cingulo-opercular and cerebellar network in schizophrenia. *NeuroImage: Clinical*, *3*, 470–480.

Becerril, K. E., Repovs, G., & Barch, D. M. (2011). Error processing network dynamics in schizophrenia. *Neuroimage*, *54*(2), 1495–1505.

Blumenfeld, R. S., Parks, C. M., Yonelinas, A. P., & Ranganath, C. (2011). Putting the pieces together: The role of dorsolateral prefrontal cortex in relational memory encoding. *Journal of Cognitive Neuroscience*, *23*(1), 257–265.

Bonner-Jackson, A., & Barch, D. M. (2011). Strategic manipulations for associative memory and the role of verbal processing abilities in schizophrenia. *Journal of the International Neuropsychological Society: JINS*, *17*(5), 796–806.

Bonner-Jackson, A., Yodkovik, N., Csernansky, J. G., & Barch, D. M. (2008). Episodic memory in schizophrenia: The influence of strategy use on behavior and brain activation. *Psychiatry Research*, *164*(1), 1–15.

Botvinick, M. M., Braver, T. S., Barch, D. M., Carter, C. S., & Cohen, J. C. (2001). Conflict monitoring and cognitive control. *Psychological Review*, *108*, 624–652.

Brahmbatt, S., Haut, K. M., Csernansky, J. G., & Barch, D. M. (2004). *Working memory deficits in individuals with schizophrenia and their high risk siblings.* Paper presented at the Society for Neuroscience, San Diego.

Braver, T. S. (2012). The variable nature of cognitive control: A dual mechanisms framework. *Trends in Cognitive Sciences*, *16*(2), 106–113.

Braver, T. S., Barch, D. M., & Cohen, J. D. (1999). Cognition and control in schizophrenia: A computational model of dopamine and prefrontal function. *Biological Psychiatry*, *46*, 312–328.

Braver, T. S., Gray, J. R., & Burgess, G. C. (2007). Explaining the many varieties of working memory variation: Dual mechanisms of cognitive control. In A. R. Conway, C. Jarrold, M. J. Kane, A. Miyake & J. Towse (Eds.), *Variation in working memory*. Oxford: Oxford University Press.

Braver, T. S., Paxton, J. L., Locke, H. S., & Barch, D. M. (2009). Flexible neural mechanisms of cognitive control within human prefrontal cortex. *Proceedings of the National Academy of Sciences of the United States of America, 106*(18), 7351–7356.

Brown, C., & Hagoort, P. (1993). The processing nature of the N400: Evidence from masked priming. *Journal of Cognitive Neuroscience, 5*, 34–44.

Brown, J. W., & Braver, T. S. (2005). Learned predictions of error likelihood in the anterior cingulate cortex. *Science, 307*(5712), 1118–1121.

Brown, J. W., & Braver, T. S. (2008). A computational model of risk, conflict, and individual difference effects in the anterior cingulate cortex. *Brain Research, 1202*, 99–108.

Carter, C. S., MacDonald, A. W., III, Ross, L. L., & Stenger, V. A. (2001). Anterior cingulate cortex activity and impaired self-monitoring of performance in patients with schizophrenia: An event-related fMRI study. *American Journal of Psychiatry, 158*, 1423–1428.

Ceaser, A. E., & Barch, D. M. (2015). Striatal activity is associated with deficits of cognitive control and aberrant salience for patients with schizophrenia. *Frontiers in Human Neuroscience, 9*, 687.

Cho, R. Y., Konecky, R. O., & Carter, C. S. (2006). Impairments in frontal cortical gamma synchrony and cognitive control in schizophrenia. *Proceedings of the National Academy of Sciences of the United States of America, 103*(52), 19878–19883.

Cohen, J. D., Barch, D. M., Carter, C. S., & Servan-Schreiber, D. (1999). Context-processing deficits in schizophrenia: Converging evidence from three theoretically motivated cognitive tasks. *Journal of Abnormal Psychology, 108*(1), 120–133.

Cole, M. W., Anticevic, A., Repovs, G., & Barch, D. (2011). Variable global dysconnectivity and individual differences in schizophrenia. *Biological psychiatry, 70*(1), 43–50.

Cole, M. W., Reynolds, J. R., Power, J. D., Repovs, G., Anticevic, A., & Braver, T. S. (2013). Multi-task connectivity reveals flexible hubs for adaptive task control. *Nature Neuroscience, 16*(9), 1348–1355.

Coleman, M. J., Titone, D., Krastoshevsky, O., Krause, V., Huang, Z., Mendell, N. R.,...Levy, D. L. (2010). Reinforcement ambiguity and novelty do not account for transitive inference deficits in schizophrenia. *Schizophrenia Bulletin, 36*(6), 1187–1200.

Condray, R., Steinhauer, S. R., Cohen, J. D., van Kammen, D. P., & Kasparek, A. (1999). Modulation of language processing in schizophrenia: Effects of context and haloperidol on the event-related potential. *Biological Psychiatry, 45*(10), 1336–1355.

Cowan, N. (2010). The magical mystery four: How is working memory capacity limited, and why? *Current Directions in Psychological Science: A Journal of the American Psychological Society, 19*(1), 51–57.

Dahaene, S., Posner, M. I., & Tucker, D. M. (1994). Localization of a neural system for error detection and compensation. *Psychological Science, 5*, 303–305.

Delawalla, Z., Csernansky, J. G., & Barch, D. M. (2008). Prefrontal cortex function in nonpsychotic siblings of individuals with schizophrenia. *Biological Psychiatry, 63*(5), 490–497.

Driesen, N. R., Leung, H. C., Calhoun, V. D., Constable, R. T., Gueorguieva, R., Hoffman, R.,... Krystal, J. H. (2008). Impairment of working memory maintenance and response in schizophrenia: Functional magnetic resonance imaging evidence. *Biological Psychiatry, 64*(12), 1026–1034.

Edin, F., Klingberg, T., Johansson, P., McNab, F., Tegner, J., & Compte, A. (2009). Mechanism for top-down control of working memory capacity. *Proceedings of the National Academy of Sciences of the United States of America, 106*(16), 6802–6807.

Edwards, B. G., Barch, D. M., & Braver, T. S. (2010). Improving prefrontal cortex function in schizophrenia through focused training of cognitive control. *Frontiers in Human Neuroscience, 4*, 32.

Egner, T., & Hirsch, J. (2005). The neural correlates and functional integration of cognitive control in a Stroop task. *Neuroimage, 24*(2), 539–547.

Eich, T. S., Nee, D. E., Insel, C., Malapani, C., & Smith, E. E. (2014). Neural correlates of impaired cognitive control over working memory in schizophrenia. *Biological Psychiatry, 76*, 146–153.

Eichenbaum, H., Yonelinas, A. R., & Ranganath, C. (2007). The medial temporal lobe and recognition memory. *Annual Review of Neuroscience, 30*, 123–152.

Elvevåg, B., Duncan, J., & McKenna, P. J. (2000). The use of cognitive context in schizophrenia: An investigation. *Psychological Medicine, 30*, 885–897.

Falkenstein, M., Hohnsbein, J., & Hoorman, J. (1995). Event related potential correlates of errors in reaction tasks. In G. Karmos, M. Molnar, V. Csepe, I. Czigler, & J. E. Desmedt (Eds.), *Perspectives of event-related potentials research* (pp. 287–296). Amsterdam: Elsevier Science B. V.

Fatouros-Bergman, H., Cervenka, S., Flyckt, L., Edman, G., & Farde, L. (2014). Meta-analysis of cognitive performance in drug-naive patients with schizophrenia. *Schizophrenia Research, 158*(1–3), 156–162.

Forbes, N. F., Carrick, L. A., McIntosh, A. M., & Lawrie, S. M. (2009). Working memory in schizophrenia: A meta-analysis. *Psychological Medicine, 39*(6), 889–905.

Fornito, A., Yoon, J., Zalesky, A., Bullmore, E. T., & Carter, C. S. (2011). General and specific functional connectivity disturbances in first-episode schizophrenia during cognitive control performance. *Biological Psychiatry, 70*(1), 64–72.

Franke, C., Reuter, B., Schulz, L., & Kathmann, N. (2007). Schizophrenia patients show impaired response switching in saccade tasks. *Biological Psychology, 76*(1–2), 91–99.

Frith, C. D., & Done, J. (1989). Experiences of alien control in schizophrenia reflect a disorder in the central monitoring of action. *Psychological Medicine, 19*, 359–363.

Fuller, R. L., Luck, S. J., Braun, E. L., Robinson, B. M., McMahon, R. P., & Gold, J. M. (2009). Impaired visual working memory consolidation in schizophrenia. *Neuropsychology, 23*(1), 71–80.

Galletly, C. A., McFarlane, A. C., & Clark, C. R. (2007). Impaired updating of working memory in schizophrenia. *International Journal of Psychophysiology: Official Journal of the International Organization of Psychophysiology, 63*(3), 265–274.

Gehring, W. J., Coles, M. G. H., Meyer, D. E., & Donchin, E. (1990). The error-related negativity: An event-related potential accompanying errors. *Psychophysiology, 27*, S34.

Glahn, D. C., Ragland, J. D., Abramoff, A., Barrett, J., Laird, A. R., Bearden, C. E., & Velligan, D. I. (2005). Beyond hypofrontality: A quantitative meta-analysis of functional neuroimaging studies of working memory in schizophrenia. *Human Brain Mapping, 25*(1), 60–69.

Gold, J. M., Fuller, R. L., Robinson, B. M., Braun, E. L., & Luck, S. J. (2007). Impaired top-down control of visual search in schizophrenia. *Schizophrenia Research, 94*(1–3), 148–155.

Gold, J. M., Fuller, R. L., Robinson, B. M., McMahon, R. P., Braun, E. L., & Luck, S. J. (2006). Intact attentional control of working memory encoding in schizophrenia. *Journal of Abnormal Psychology, 115*(4), 658–673.

Gold, J. M., Hahn, B., Zhang, W. W., Robinson, B. M., Kappenman, E. S., Beck, V. M., & Luck, S. J. (2010). Reduced capacity but spared precision and maintenance of working memory representations in schizophrenia. *Archives of General Psychiatry, 67*(6), 570–577.

Gold, J. M., Wilk, C. M., McMahon, R. P., Buchanan, R. W., & Luck, S. J. (2003). Working memory for visual features and conjunctions in schizophrenia. *Journal of Abnormal Psychology, 112*(1), 61–71.

Greenzang, C., Manoach, D. S., Goff, D. C., & Barton, J. J. (2007). Task-switching in schizophrenia: Active switching costs and passive carry-over effects in an antisaccade paradigm. *Experimental Brain Research, 181*(3), 493–502.

Grillon, C., Rezvan, A., & Glazer, W. M. (1991). N400 and semantic categorization in schizophrenia. *Biological Psychiatry, 29*, 467–480.

Haddon, J. E., & Killcross, S. (2007). Contextual control of choice performance: Behavioral, neurobiological, and neurochemical influences. *Annals of the New York Academy of Sciences, 1104*, 250–269.

Hahn, B., Robinson, B. M., Kaiser, S. T., Harvey, A. N., Beck, V. M., Leonard, C. J.,...Gold, J. M. (2010). Failure of schizophrenia patients to overcome salient distractors during working memory encoding. *Biological Psychiatry, 68*(7), 603–609.

Hanlon, F. M., Houck, J. M., Klimaj, S. D., Caprihan, A., Mayer, A. R., Weisend, M. P.,...Tesche, C. D. (2012). Frontotemporal anatomical connectivity and working-relational memory performance predict everyday functioning in schizophrenia. *Psychophysiology, 49*(10), 1340–1352.

Hannula, D. E., & Ranganath, C. (2008). Medial temporal lobe activity predicts successful relational memory binding. *Journal of Neuroscience, 28*(1), 116–124.

Hannula, D. E., & Ranganath, C. (2009). The eyes have it: Hippocampal activity predicts expression of memory in eye movements. *Neuron, 63*(5), 592–599.
Hannula, D. E., Ranganath, C., Ramsay, I. S., Solomon, M., Yoon, J., Niendam, T. A.,…Ragland, J. D. (2010). Use of eye movement monitoring to examine item and relational memory in schizophrenia. *Biological Psychiatry, 68*(7), 610–616.
Heckers, S., & Konradi, C. (2010). Hippocampal pathology in schizophrenia. *Current Topics in Behavioral Neurosciences, 4*, 529–553.
Henderson, D., Poppe, A. B., Barch, D. M., Carter, C. S., Gold, J. M., Ragland, J. D.,…MacDonald, A. W., 3rd. (2012). Optimization of a goal maintenance task for use in clinical applications. *Schizophrenia Bulletin, 38*(1), 104–113.
Hendrick, O. M., Ide, J. S., Luo, X., & Li, C. S. (2010). Dissociable processes of cognitive control during error and non-error conflicts: A study of the stop signal task. *PLoS ONE, 5*(10), e13155.
Henseler, I., Falkai, P., & Gruber, O. (2010). Disturbed functional connectivity within brain networks subserving domain-specific subcomponents of working memory in schizophrenia: Relation to performance and clinical symptoms. *Journal of Psychiatric Research, 44*(6), 364–372.
Holmes, A. J., MacDonald, A., 3rd, Carter, C. S., Barch, D. M., Andrew Stenger, V., & Cohen, J. D. (2005). Prefrontal functioning during context processing in schizophrenia and major depression: An event-related fMRI study. *Schizophrenia Research, 76*(2–3), 199–206.
Holroyd, C. B. (2013). Theories of anterior cingulate cortex function: Opportunity cost. *Behavioral and Brain Sciences, 36*(6), 693–694; discussion 707–626.
Holroyd, C. B., & Coles, M. G. (2002). The neural basis of human error processing: Reinforcement learning, dopamine, and the error-related negativity. *Psychological Review, 109*(4), 679–709.
Horan, W. P., Braff, D. L., Nuechterlein, K. H., Sugar, C. A., Cadenhead, K. S., Calkins, M. E.,… Green, M. F. (2008). Verbal working memory impairments in individuals with schizophrenia and their first-degree relatives: Findings from the Consortium on the Genetics of Schizophrenia. *Schizophrenia Research, 103*(1–3), 218–228.
Horan, W. P., Foti, D., Hajcak, G., Wynn, J. K., & Green, M. F. (2012). Impaired neural response to internal but not external feedback in schizophrenia. *Psychological Medicine, 42*(8), 1637–1647.
Jamadar, S., Michie, P., & Karayanidis, F. (2010). Compensatory mechanisms underlie intact task-switching performance in schizophrenia. *Neuropsychologia, 48*(5), 1305–1323.
Janowsky, J. S., Shimamura, A. P., Kritchevsky, M., & Squire, L. R. (1989). Cognitive impairment following frontal lobe damage and its relevance to human amnesia. *Behavioral Neuroscience, 103*(3), 548–560.
Javitt, D. C., Rabinowicz, E., Silipo, G., & Dias, E. C. (2007). Encoding vs. retention: Differential effects of cue manipulation on working memory performance in schizophrenia. *Schizophrenia Research, 91*(1–3), 159–168.
Javitt, D. C., Shelley, A., Silipo, G., & Lieberman, J. A. (2000). Deficits in auditory and visual context-dependent processing in schizophrenia. *Archives of General Psychiatry, 57*, 1131–1137.
Jones, J. A., Sponheim, S. R., & MacDonald, A. W., 3rd. (2010). The dot pattern expectancy task: Reliability and replication of deficits in schizophrenia. *Psychological Assessment, 22*(1), 131–141.
Karayanidis, F., Nicholson, R., Schall, U., Meem, L., Fulham, R., & Michie, P. T. (2006). Switching between univalent task-sets in schizophrenia: ERP evidence of an anticipatory task-set reconfiguration deficit. *Clinical Neurophysiology, 117*(10), 2172–2190.
Karlsgodt, K. H., van Erp, T. G., Poldrack, R. A., Bearden, C. E., Nuechterlein, K. H., & Cannon, T. D. (2008). Diffusion tensor imaging of the superior longitudinal fasciculus and working memory in recent-onset schizophrenia. *Biological Psychiatry, 63*(5), 512–518.
Kerns, J. G. (2006). Anterior cingulate and prefrontal cortex activity in an FMRI study of trial-to-trial adjustments on the Simon task. *Neuroimage, 33*(1), 399–405.
Kerns, J. G., Cohen, J. D., MacDonald, A. W., 3rd, Cho, R. Y., Stenger, V. A., & Carter, C. S. (2004). Anterior cingulate conflict monitoring and adjustments in control. *Science, 303*(5660), 1023–1026.
Kerns, J. G., Cohen, J. D., MacDonald, A. W., 3rd, Johnson, M. K., Stenger, V. A., Aizenstein, H., & Carter, C. S. (2005). Decreased conflict- and error-related activity in the anterior cingulate cortex in subjects with schizophrenia. *American Journal of Psychiatry, 162*(10), 1833–1839.

Kieffaber, P. D., Kappenman, E. S., Bodkins, M., Shekhar, A., O'Donnell, B. F., & Hetrick, W. P. (2006). Switch and maintenance of task set in schizophrenia. *Schizophrenia Research, 84*(2–3), 345–358.

Kim, C., Johnson, N. F., & Gold, B. T. (2014). Conflict adaptation in prefrontal cortex: Now you see it, now you don't. *Cortex, 50*, 76–85.

Kim, J., Glahn, D. C., Nuechterlein, K. H., & Cannon, T. D. (2004). Maintenance and manipulation of information in schizophrenia: Further evidence for impairment in the central executive component of working memory. *Schizophrenia Research, 68*(2–3), 173–187.

Koechlin, E. (2014). An evolutionary computational theory of prefrontal executive function in decision-making. *Philosophical Transactions of the Royal Society of London. Series B: Biological Sciences, 369*(1655).

Koechlin, E., & Summerfield, C. (2007). An information theoretical approach to prefrontal executive function. *Trends in Cognitive Sciences, 11*(6), 229–235.

Koechlin, E., Ody, C., & Kouneiher, F. (2003). The architecture of cognitive control in the human prefrontal cortex. *Science, 302*(5648), 1181–1185.

Konkel, A., & Cohen, N. J. (2009). Relational memory and the hippocampus: Representations and methods. *Frontiers in Neuroscience, 3*(2), 166–174.

Kopp, B., & Rist, F. (1994). Error-correcting behavior in schizophrenic patients. *Schizophrenia Research, 13*(1), 11–22.

Kopp, B., & Rist, F. (1999). An event-related brain potential substrate of disturbed response monitoring in paranoid schizophrenic patients. *Journal of Abnormal Psychology, 108*(2), 337–346.

Krawitz, A., Braver, T. S., Barch, D. M., & Brown, J. W. (2011). Impaired error-likelihood prediction in medial prefrontal cortex in schizophrenia. *NeuroImage, 54*(2), 1506–1517.

Kutas, M., & Hillyard, S. A. (1980). Reading senseless sentences: Brain potentials reflect semantic incongruity. *Science, 207*(11), 203–205.

Kyriakopoulos, M., Dima, D., Roiser, J. P., Corrigall, R., Barker, G. J., & Frangou, S. (2012). Abnormal functional activation and connectivity in the working memory network in early-onset schizophrenia. *Journal of the American Academy of Child & Adolescent Psychiatry, 51*(9), 911–920 e912.

Laurens, K. R., Hodgins, S., Mould, G. L., West, S. A., Schoenberg, P. L., Murray, R. M., & Taylor, E. A. (2010). Error-related processing dysfunction in children aged 9 to 12 years presenting putative antecedents of schizophrenia. *Biological Psychiatry, 67*(3), 238–245.

Laurens, K. R., Ngan, E. T., Bates, A. T., Kiehl, K. A., & Liddle, P. F. (2003). Rostral anterior cingulate cortex dysfunction during error processing in schizophrenia. *Brain, 126*, 610–622.

Lee, J., & Park, S. (2005). Working memory impairments in schizophrenia: A meta-analysis. *Journal of Abnormal Psychology, 114*(4), 599–611.

Lepage, M., Bodnar, M., & Bowie, C. R. (2014). Neurocognition: Clinical and functional outcomes in schizophrenia. *Canadian Journal of Psychiatry, 59*(1), 5–12.

Lesh, T. A., Niendam, T. A., Minzenberg, M. J., & Carter, C. S. (2011). Cognitive control deficits in schizophrenia: Mechanisms and meaning. *Neuropsychopharmacology: Official Publication of the American College of Neuropsychopharmacology, 36*(1), 316–338.

Lesh, T. A., Westphal, A. J., Niendam, T. A., Yoon, J. H., Minzenberg, M. J., Ragland, J. D., Solomon, M., & Carter, C. S. (2013). Proactive and reactive cognitive control and dorsolateral prefrontal cortex dysfunction in first episode schizophrenia. *NeuroImage: Clinical, 2*, 590–599.

Lewis, D. A., Cho, R. Y., Carter, C. S., Eklund, K., Forster, S., Kelly, M. A., & Montrose, D. (2008). Subunit-selective modulation of GABA type A receptor neurotransmission and cognition in schizophrenia. *American Journal of Psychiatry, 165*(12), 1585–1593.

Lisman, J. (2010). Working memory: The importance of theta and gamma oscillations. *Current Biology, 20*(11), R490–492.

MacDonald, A., Carter, C. S., Kerns, J. G., Ursu, S., Barch, D. M., Holmes, A. J.,...Cohen, J. D. (2005). Specificity of prefrontal dysfunction and context processing deficits to schizophrenia in a never medicated first-episode psychotic sample. *American Journal of Psychiatry, 162*, 475–484.

MacDonald, A. W., 3rd, Becker, T. M., & Carter, C. S. (2006). Functional magnetic resonance imaging study of cognitive control in the healthy relatives of schizophrenia patients. *Biological Psychiatry, 60*(11), 1241–1249.

MacDonald, A. W., 3rd, Goghari, V. M., Hicks, B. M., Flory, J. D., Carter, C. S., & Manuck, S. B. (2005). A convergent-divergent approach to context processing, general intellectual functioning, and the genetic liability to schizophrenia. *Neuropsychology, 19*(6), 814–821.

MacDonald, A. W., Pogue-Geile, M. F., Johnson, M. K., & Carter, C. S. (2003). A specific deficit in context processing in the unaffected siblings of patients with schizophrenia. *Archives of General Psychiatry, 60*, 57–65.

Manoach, D. S., Lee, A. K., Hamalainen, M. S., Dyckman, K. A., Friedman, J. S., Vangel, M.,...Barton, J. J. (2013). Anomalous use of context during task preparation in schizophrenia: A magnetoencephalography study. *Biological Psychiatry, 73*(10), 967–975.

Manoach, D. S., Lindgren, K. A., Cherkasova, M. V., Goff, D. C., Halpern, E. F., Intriligator, J., & Barton, J. J. (2002). Schizophrenic subjects show deficient inhibition but intact task switching on saccadic tasks. *Biological Psychiatry, 51*(10), 816–826.

Marco-Pallares, J., Camara, E., Munte, T. F., & Rodriguez-Fornells, A. (2008). Neural mechanisms underlying adaptive actions after slips. *Journal of Cognitive Neuroscience, 20*(9), 1595–1610.

Mathalon, D. H., Dedor, M., Faustman, W. O., Gray, M., Askari, N., & Ford, J. M. (2002). Response-monitoring dysfunction in schizophrenia: An event-related brain potential study. *Journal of Abnormal Psychology, 111*, 22–41.

Mathalon, D. H., Fedor, M., Faustman, W. O., Gray, M., Askari, N., & Ford, J. M. (2002). Response-monitoring dysfunction in schizophrenia: An event-related brain potential study. *Journal of Abnormal Psychology, 111*(1), 22–41.

Mathalon, D. H., Jorgensen, K. W., Roach, B. J., & Ford, J. M. (2009). Error detection failures in schizophrenia: ERPs and FMRI. *International Journal of Psychophysiology, 73*(2), 109–117.

Mathews, J. R., & Barch, D. M. (2004). Episodic memory for emotional and nonemotional words in schizophrenia. *Cognition and Emotion, 18*(6), 721–740.

McClure, M. M., Barch, D. M., Flory, J. D., Harvey, P. D., & Siever, L. J. (2008). Context processing in schizotypal personality disorder: Evidence of specificity of impairment to the schizophrenia spectrum. *Journal of Abnormal Psychology, 117*(2), 342–354.

McClure, M. M., Barch, D. M., Romero, M. J., Minzenberg, M. J., Triebwasser, J., Harvey, P. D., & Siever, L. J. (2007). The effects of guanfacine on context processing abnormalities in schizotypal personality disorder. *Biological Psychiatry, 61*(10), 1157–1160.

McClure, M. M., Harvey, P. D., Goodman, M., Triebwasser, J., New, A., Koenigsberg, H. W.,...Siever, L. J. (2010). Pergolide treatment of cognitive deficits associated with schizotypal personality disorder: Continued evidence of the importance of the dopamine system in the schizophrenia spectrum. *Neuropsychopharmacology, 35*(6), 1356–1362.

Meiran, N. (2000). Modeling cognitive control in task-switching. *Psychological Research, 63*(3–4), 234–249.

Meiran, N., Levine, J., Meiran, N., & Henik, A. (2000). Task set switching in schizophrenia. *Neuropsychology, 14*(3), 471–482.

Mesholam-Gately, R. I., Giuliano, A. J., Goff, K. P., Faraone, S. V., & Seidman, L. J. (2009). Neurocognition in first-episode schizophrenia: A meta-analytic review. *Neuropsychology, 23*(3), 315–336.

Minzenberg, M. J., Firl, A. J., Yoon, J. H., Gomes, G. C., Reinking, C., & Carter, C. S. (2010). Gamma oscillatory power is impaired during cognitive control independent of medication status in first-episode schizophrenia. *Neuropsychopharmacology, 35*(13), 2590–2599.

Minzenberg, M. J., Gomes, G. C., Yoon, J. H., Swaab, T. Y., & Carter, C. S. (2014). Disrupted action monitoring in recent-onset psychosis patients with schizophrenia and bipolar disorder. *Psychiatry Research, 221*(1), 114–121.

Minzenberg, M. J., Laird, A. R., Thelen, S., Carter, C. S., & Glahn, D. C. (2009). Meta-analysis of 41 functional neuroimaging studies of executive function in schizophrenia. *Archives of General Psychiatry, 66*(8), 811–822.

Mohammad, O. M., & DeLisi, L. E. (2013). N400 in schizophrenia patients. *Current Opinion in Psychiatry, 26*(2), 196–207.

Morris, S. E., Heerey, E. A., Gold, J. M., & Holroyd, C. B. (2008). Learning-related changes in brain activity following errors and performance feedback in schizophrenia. *Schizophrenia Research, 99*(1–3), 274–285.

Morris, S. E., Holroyd, C. B., Mann-Wrobel, M. C., & Gold, J. M. (2011). Dissociation of response and feedback negativity in schizophrenia: Electrophysiological and computational evidence for a deficit in the representation of value. *Frontiers in Human Neuroscience*, *5*, 123.

Morris, S. E., Yee, C. M., & Nuechterlein, K. H. (2006). Electrophysiological analysis of error monitoring in schizophrenia. *Journal of Abnormal Psychology*, *115*(2), 239–250.

Murray, L. J., & Ranganath, C. (2007). The dorsolateral prefrontal cortex contributes to successful relational memory encoding. *Journal of Neuroscience*, *27*(20), 5515–5522.

Niendam, T. A., Lesh, T. A., Yoon, J., Westphal, A. J., Hutchison, N., Daniel Ragland, J.,…Carter, C. S. (2014). Impaired context processing as a potential marker of psychosis risk state. *Psychiatry Research*, *221*(1), 13–20.

Niznikiewicz, M. A., O'Donnell, B. F., Nestor, P. G., Smith, L., Law, S., Karapelou, M.,…McCarley, R. W. (1997). ERP assessment of visual and auditory language processing in schizophrenia. *Journal of Abnormal Psychology*, *106*(1), 85–94.

Nuechterlein, K. H., Subotnik, K. L., Green, M. F., Ventura, J., Asarnow, R. F., Gitlin, M. J.,…Mintz, J. (2011). Neurocognitive predictors of work outcome in recent-onset schizophrenia. *Schizophrenia Bulletin*, *37*(Suppl. 2), S33–40.

Ohta, K., Uchiyama, M., Matsushima, E., & Toru, M. (1999). An event-related potential study in schizophrenia using Japanese sentences. *Schizophrenia Research*, *40*, 159–170.

Olichney, J. M., Iragui, V. J., Kutas, M., Nowacki, R., & Jeste, D. V. (1997). N400 abnormalities in late life schizophrenia and related psychoses. *Biological Psychiatry*, *42*, 13–23.

Perez, V. B., Ford, J. M., Roach, B. J., Woods, S. W., McGlashan, T. H., Srihari, V. H.,…Mathalon, D. H. (2012). Error monitoring dysfunction across the illness course of schizophrenia. *Journal of Abnormal Psychology*, *121*(2), 372–387.

Pisella, L., Berberovic, N., & Mattingley, J. B. (2004). Impaired working memory for location but not for colour or shape in visual neglect: A comparison of parietal and non-parietal lesions. *Cortex*, *40*(2), 379–390.

Polli, F. E., Barton, J. J., Thakkar, K. N., Greve, D. N., Goff, D. C., Rauch, S. L., & Manoach, D. S. (2008). Reduced error-related activation in two anterior cingulate circuits is related to impaired performance in schizophrenia. *Brain*, *131*(Pt. 4), 971–986.

Polli, F. E., Barton, J. J., Vangel, M., Goff, D. C., Iguchi, L., & Manoach, D. S. (2006). Schizophrenia patients show intact immediate error-related performance adjustments on an antisaccade task. *Schizophrenia Research*, *82*(2–3), 191–201.

Poppe, A. B., Carter, C. S., Minzenberg, M. J., & MacDonald, A. W., 3rd. (2015). Task-based functional connectivity as an indicator of genetic liability to schizophrenia. *Schizophrenia Research*, *162*(1–3), 118–123.

Ragland, J. D., Blumenfeld, R. S., Ramsay, I. S., Yonelinas, A., Yoon, J., Solomon, M.,…Ranganath, C. (2012). Neural correlates of relational and item-specific encoding during working and long-term memory in schizophrenia. *NeuroImage*, *59*(2), 1719–1726.

Ragland, J. D., Gur, R. C., Valdez, J. N., Loughead, J., Elliott, M., Kohler, C.,…Gur, R. E. (2005). Levels-of-processing effect on frontotemporal function in schizophrenia during word encoding and recognition. *American Journal of Psychiatry*, *162*(10), 1840–1848.

Ragland, J. D., Laird, A. R., Ranganath, C., Blumenfeld, R. S., Gonzales, S. M., & Glahn, D. C. (2009). Prefrontal activation deficits during episodic memory in schizophrenia. *American Journal of Psychiatry*, *166*(8), 863–874.

Ragland, J. D., Ranganath, C., Harms, M. P., Barch, D. M., Gold, J. M., Layher, E., Lesh, T. A., MacDonald, A. W., 3rd, Niendam, T. A., Phillips, J., Silverstein, S. M., Yonelinas, A. P., & Carter, C. S. (2015). Functional and neuroanatomic specificity of episodic memory dysfunction in schizophrenia: A functional magnetic resonance imaging study of the relational and item-specific encoding task. *JAMA Psychiatry*, *72*(9), 909–916.

Ravizza, S. M., Moua, K. C., Long, D., & Carter, C. S. (2010). The impact of context processing deficits on task-switching performance in schizophrenia. *Schizophrenia Research*, *116*(2–3), 274–279.

Repovs, G., Csernansky, J. G., & Barch, D. M. (2011). Brain network connectivity in individuals with schizophrenia and their siblings. *Biological Psychiatry*, *69*(10), 967–973.

Rosvold, H. E., Mirsky, A. F., Sarason, I., Bransome, E. D. Jr., & Beck, L. H. (1956). A continuous performance test of brain damage. *Journal of Consulting Psychology*, *20*(5), 343–350.

Salisbury, D., O'Donnell, B. F., McCarley, R. W., Nestor, P. G., & Shenton, M. E. (2000). Event-related potentials elicited during a context-free homograph task in normal versus schizophrenic subjects. *Psychophysiology*, *37*, 456–463.

Senkowski, D., & Gallinat, J. (2015). Dysfunctional prefrontal gamma-band oscillations reflect working memory and other cognitive deficits in schizophrenia. *Biological Psychiatry*, *77*(12), 1010–1019.

Servan-Schreiber, D., Cohen, J. D., & Steingard, S. (1996). Schizophrenic deficits in the processing of context: A test of a theoretical model. *Archives of General Psychiatry*, *53*, 1105–1112.

Shakow, D. (1962). Segmental set: A theory of the formal psychological deficit in schizophrenia. *Archives of General Psychiatry*, *6*, 1–17.

Sheffield, J. M., Gold, J. M., Strauss, M. E., Carter, C. S., MacDonald, A. W., 3rd, Ragland, J. D.,... Barch, D. M. (2014). Common and specific cognitive deficits in schizophrenia: Relationships to function. *Cognitive, Affective & Behavioral Neuroscience*, *14*(1), 161–174.

Shenhav, A., Botvinick, M. M., & Cohen, J. D. (2013). The expected value of control: An integrative theory of anterior cingulate cortex function. *Neuron*, *79*(2), 217–240.

Simmonite, M., Bates, A. T., Groom, M. J., Jackson, G. M., Hollis, C., & Liddle, P. F. (2012). Error processing-associated event-related potentials in schizophrenia and unaffected siblings. *International Journal of Psychophysiology*, *84*(1), 74–79.

Sitnikova, T., Perrone, C., Goff, D., & Kuperberg, G. R. (2010). Neurocognitive mechanisms of conceptual processing in healthy adults and patients with schizophrenia. *International Journal of Psychophysiology*, *75*(2), 86–99.

Sitnikova, T., Salisbury, D. F., Kuperberg, G., & Holcomb, P. J. (2002). Electrophysiological insights into language processing in schizophrenia. *Psychophysiology*, *39*, 851–860.

Smith, E. E., Eich, T. S., Cebenoyan, D., & Malapani, C. (2011). Intact and impaired cognitive-control processes in schizophrenia. *Schizophrenia Research*, *126*(1–3), 132–137.

Snitz, B. E., MacDonald, A., 3rd, Cohen, J. D., Cho, R. Y., Becker, T., & Carter, C. S. (2005). Lateral and medial hypofrontality in first-episode schizophrenia: Functional activity in a medication-naive state and effects of short-term atypical antipsychotic treatment. *American Journal of Psychiatry*, *162*(12), 2322–2329.

Strandburgh, R. J., Marsh, J. T., Brown, W. S., Asarnow, R. F., Guthrie, D., Harper, R.,...Nuechterlein, K. H. (1997). Event-related potential correlates of linguistic information processing in schizophrenics. *Biological Psychiatry*, *42*, 596–608.

Stratta, P., Daneluzzo, E., Bustini, M., Casacchia, M., & Rossi, A. (1998). Schizophrenic deficits in the processing of context. *Archives of General Psychiatry*, *55*, 186–187.

Stratta, P., Daneluzzo, E., Bustini, M., Prosperini, P., & Rossi, A. (2000). Processing of context information in schizophrenia: Relation to clinical symptoms and WCST performance. *Schizophrenia Research*, *44*, 57–67.

Titone, D., Ditman, T., Holzman, P. S., Eichenbaum, H., & Levy, D. L. (2004). Transitive inference in schizophrenia: Impairments in relational memory organization. *Schizophrenia Research*, *68*(2–3), 235–247.

Titone, D., Holzman, P. S., & Levy, D. L. (2002). Idiom processing in schizophrenia: Literal implausibility saves the day for idiom priming. *Journal of Abnormal Psychology*, *11*(2), 313–320.

Titone, D., Levy, D. L., & Holzman, P. S. (2000). Contextual insensitivity in schizophrenic language processing: Evidence from lexical ambiguity. *Journal of Abnormal Psychology*, *109*(4), 761–767.

Todd, J. J., & Marois, R. (2004). Capacity limit of visual short-term memory in human posterior parietal cortex. *Nature*, *428*(6984), 751–754.

Todd, J. J., & Marois, R. (2005). Posterior parietal cortex activity predicts individual differences in visual short-term memory capacity. *Cognitive, Affective & Behavioral Neuroscience*, *5*(2), 144–155.

Turken, A. U., Vuilleumier, P., Mathalon, D. H., Swick, D., & Ford, J. M. (2003). Are impairments of action monitoring and executive control true dissociative dysfunctions in patients with schizophrenia? *American Journal of Psychiatry*, *160*(10), 1881–1883.

Van Snellenberg, J. X., Torres, I. J., & Thornton, A. E. (2006). Functional neuroimaging of working memory in schizophrenia: Task performance as a moderating variable. *Neuropsychology*, *20*(5), 497–510.

Vogel, E. K., & Machizawa, M. G. (2004). Neural activity predicts individual differences in visual working memory capacity. *Nature*, *428*(6984), 748–751.

Wager, T. D., & Smith, E. E. (2003). Neuroimaging studies of working memory: A meta-analysis. *Cognitive Affective and Behavioral Neuroscience*, *3*(4), 255–274.

Wagner, A. D., Schacter, D., Rotte, M., Koutstaal, W., Maril, A., Dale, A. M.,...Buckner, R. L. (1998). Building memories: Remembering and forgetting of verbal experiences as predicted by brain activity. *Science*, *281*(5380), 1188–1191.

Williams, L. E., Must, A., Avery, S., Woolard, A., Woodward, N. D., Cohen, N. J., & Heckers, S. (2010). Eye-movement behavior reveals relational memory impairment in schizophrenia. *Biological Psychiatry*, *68*(7), 617–624.

Wylie, G. R., Clark, E. A., Butler, P. D., & Javitt, D. C. (2010). Schizophrenia patients show task switching deficits consistent with N-methyl-d-aspartate system dysfunction but not global executive deficits: Implications for pathophysiology of executive dysfunction in schizophrenia. *Schizophrenia Bulletin*, *36*(3), 585–594.

Yoon, J. H., Minzenberg, M. J., Ursu, S., Ryan Walter, B. S., Wendelken, C., Ragland, J. D., & Carter, C. S. (2008). Association of dorsolateral prefrontal cortex dysfunction with disrupted coordinated brain activity in schizophrenia: Relationship with impaired cognition, behavioral disorganization, and global function. *American Journal of Psychiatry*, *165*(8), 1006–1014.

Yoon, J. H., Nguyen, D. V., McVay, L. M., Deramo, P., Minzenberg, M. J., Ragland, J. D.,...Carter, C. S. (2012). Automated classification of fMRI during cognitive control identifies more severely disorganized subjects with schizophrenia. *Schizophrenia Research*, *135*(1–3), 28–33.

Yoon, J. H., Westphal, A. J., Minzenberg, M. J., Niendam, T., Ragland, J. D., Lesh, T.,...Carter, C. S. (2014). Task-evoked substantia nigra hyperactivity associated with prefrontal hypofunction, prefrontonigral disconnectivity and nigrostriatal connectivity predicting psychosis severity in medication naive first episode schizophrenia. *Schizophrenia Research*, *159*(2–3), 521–526.

32

The Neural Basis of Response Inhibition and Substance Abuse

Bader Chaarani, Philip A. Spechler, Kelsey E. Hudson, John J. Foxe, Alexandra S. Potter, and Hugh Garavan

Introduction

Cognitive control, also commonly referred to as 'executive functions', is an umbrella term comprising a wide range of cognitive processes and behavioural competencies that include verbal reasoning, problem solving, planning, sequencing, the ability to sustain attention, resistance to interference, utilisation of feedback, multitasking, cognitive flexibility, and the ability to deal with novelty (Chan, Shum, Toulopoulou, & Chen, 2008). Cognitive control takes on many forms, including, but not limited to, inhibitory control, which broadly refers to counteracting behaviours that are typically elicited from action-related cues (see Chapter 6 by Verbruggen & Logan in this volume).

Inhibitory control has been repeatedly implicated in drug addiction and may be especially relevant for treatment success. Through a combination of pre-existing or drug-induced impairments, the addicted individual may lack the cognitive ability to reduce or cease drug use. Suppressing powerful drug urges and the behaviours that lead to relapse may be one circumstance in which reliance on a functioning inhibitory control system is most pressing. For example, a substance-using individual may fail to inhibit their behaviour and impulsively purchase a pack of cigarettes at a convenience store, in addition to their intended items. In this regard, cognitive functions may also be of significant importance in understanding the transition from recreational to dependent use, in addition to the relapse so typical of those attempting abstinence. Furthermore, the impact of drugs of abuse on cognitive abilities may provide insight into the drug's mechanisms of action, and inform the transition towards maladaptive processing and dependence. As cognitive deficits may exacerbate dependence, targeting these systems in particular may help guide therapeutic interventions.

In this chapter, we will focus on inhibitory control, largely operationalised as response inhibition, and its contribution to substance abuse, highlighting results from two classic tasks of response inhibition: the stop-signal task, and the go/no-go task. First, we will briefly offer an overview of cognitive control and the specifics of response inhibition. After laying the groundwork, we will investigate the structural brain changes and the genetic contribution to response inhibition, as described by structural neuroimaging and genetic markers based techniques. Finally, we will lead into an in-depth discussion on the relationship between response inhibition and substance abuse disorders as studied with functional neuroimaging and event-related potentials, where response inhibition will be considered as a means to characterise the cognitive control deficits of substance-dependent individuals.

The Wiley Handbook of Cognitive Control, First Edition. Edited by Tobias Egner.

Response Inhibition Tasks

Inhibitory control is broadly conceptualised as the ability to suppress or countermand a thought, action, or feeling. Many investigators study inhibitory control using carefully designed tasks like the stop-signal task (SST), or the go/no-go task, that measure an individual's ability to suppress a prepotent motor response. During the SST, subjects perform a primary task such as identifying with button press responses if a visually presented arrow (the target stimulus) points to the left or the right. On a minority of trials, often one quarter of trials, a unique auditory or visual stimulus (the stop signal) follows the target and instructs the subject to countermand their response. Task difficulty is manipulated by varying the delay between the target and stop stimulus, such that the longer the delay the more difficult it is to inhibit the response. By calculating how fast subjects respond on trials without a stop signal and the average stop-signal delay on trials in which they successful inhibit 50% of the time, one can estimate the speed of the response inhibition process, known as the stop-signal reaction time (SSRT; Logan & Cowan, 1984; see also Chapter 6 by Verbruggen & Logan in this volume)). During go/no-go tasks, subjects are presented with a continuous stream of stimuli, the majority of which require a button press execution (go trial), and a minority requiring no response (no-go trial) with the inhibitory demand being induced through the prepotency to respond even on no-go trials. Although both tasks are arguably very rudimentary examples of inhibitory control, there is evidence, outlined below, that the neural circuitry subserving response inhibition is also involved in other types of cognitive and emotional inhibition, thereby indicating that it may serve as a reasonable probe for more complex inhibitory demands, including those related to resisting drugs. As the neural circuitry of response inhibition is relatively well understood (see Chapter 16 by Forstmann & Alkemade in this volume) and yields reliable and sensitive behavioural measures of inhibitory ability, it has generated a significant number of studies focused on the role of response inhibition in addiction (Smith, Mattick, Jamadar, & Iredale, 2014).

The Neurobiology of Control

Neuroimaging research has identified the dorsolateral prefrontal cortex (dlPFC) as a brain region critical for cognitive control. Evidence suggests that the dlPFC is implicated during dual-task coordination (D'Esposito et al., 1995; Mansouri, Tanaka, & Buckley, 2009), task switching (Badre & Wagner, 2006; Dove, Pollmann, Schubert, Wiggins, & Yves von Cramon, 2000; Sohn, Ursu, Anderson, Stenger, & Carter, 2000), memory updating (Edin et al., 2009; Salmon et al., 1996), and response sequencing, monitoring, and manipulation (Kim, Chung, & Kim, 2013; Owen, Evans, & Petrides, 1996). This is consistent with the human lesion literature implicating the frontal lobes in organising, regulating, and producing coherent behaviour (Bryden & Roesch, 2015; Luria & Pribram, 1973; Stuss & Benson, 1987; see also Chapter 29 by Fellows in this volume). Frontal lobe damaged patients appear to lose important aspects of autonomous cognitive control as evidenced by the loss of behavioural control to environmental contingencies (e.g., capture errors and utilisation behaviours; Lhermitte, 1986). Although the focus of this chapter will be on prefrontal systems mediating inhibitory control, these systems operate in conjunction with extensive parietal, premotor, cingulate, subcortical, and cerebellar networks. Further, despite the evidence implicating the frontal lobes in cognitive control, the assignment of specific frontal loci to specific functions is far from resolved due, perhaps, to one of the defining characteristics of the frontal lobes being their ability to flexibly adapt to task demands (see Chapter 13 by Stokes, Buschman, & Miller in this volume). Dosenbach and colleagues suggest that different brain networks are involved in distinct aspects of control, with the frontoparietal cortex implicated in initiating and adapting behaviour, whereas sustained stable task performance is associated with the anterior cingulate cortex (ACC), anterior insula, frontal operculum, and anterior prefrontal cortex (Dosenbach et al., 2007).

With a specific focus on inhibitory control, a body of research (lesion, transcranial magnetic stimulation [TMS], and fMRI methodologies) implicates the right inferior frontal cortex (rIFC) in motor response inhibition (A. R. Aron, Fletcher, Bullmore, Sahakian, & Robbins, 2003; Chambers et al., 2006; Garavan, Hester, Murphy, Fassbender, & Kelly, 2006; Garavan, Ross, & Stein, 1999). More broadly, the rIFC is one node of a motor inhibition network that also includes the pre-supplementary motor area (pre-SMA), and subthalamic nucleus (STN; A. R. Aron, Robbins, & Poldrack, 2014). The exact causal pathways of these regions is unclear (Duann, Ide, Luo, & Li, 2009a; Neubert, Mars, Buch, Olivier, & Rushworth, 2010; Swann et al., 2012), but research suggests that the STN receives input from both the rIFC and pre-SMA, and the STN inhibits motor activity at the basal ganglia (A. R. Aron & Poldrack, 2006; Schmidt, Leventhal, Mallet, Chen, & Berke, 2013; see also Chapter 16 by Forstmann & Alkemade in this volume). Human lesion studies provide converging evidence that lesions in the right pre-SMA (Floden & Stuss, 2006; Nachev, Wydell, O'Neill, Husain, & Kennard, 2007) and the right inferior frontal gryus (IFG) sub-region pars opercularis impair response inhibition (A. R. Aron et al., 2003; A. R. Aron, Robbins, & Poldrack, 2004; A. R. Aron et al., 2014). A study using TMS found that temporary deactivation of the right IFG pars opercularis selectively impaired the ability to stop an already initiated action, whereas the deactivation of the same region did not affect physiological arousal or the ability to execute responses, confirming the important role of the IFG in the regulation of response inhibition (Chambers et al., 2006). In addition, Cai and colleagues showed that stimulation of the right pre-SMA slowed the implementation of stopping (measured via SSRT) but had no influence on modulation of response tendencies and suggested that this region impairs stopping behaviour through a specific disruption of response inhibition (Cai, George, Verbruggen, Chambers, & Aron, 2012). These studies are supported by the temporal and spatial precision afforded by electrocorticography studies, which have found that the rIFC responds prior to successful inhibition (Swann et al., 2009, 2012). Recent studies suggest that this may reflect a broader role for this region in detecting attentionally salient events (Hampshire, Chamberlain, Monti, Duncan, & Owen, 2010), although it may be the case that in order to evoke right IFG activity, the salience of these events must be relevant to response control (Dodds, Morein-Zamir, & Robbins, 2011).

Although typically not activated in imaging studies of motor response inhibition, there is considerable evidence of a role for the orbitofrontal cortex (OFC) in impulse control. For example, OFC damage in a rodent model increases SSRT (Eagle et al., 2008), whereas patients with lesion damage to the OFC show increased self-report and cognitive measures of impulsivity and altered time perception relative to healthy controls and non-OFC lesioned patients (Berlin, Rolls, & Kischka, 2004). That said, many behaviours that appear impulsive might not be driven by a deficit in impulsivity per se. For example, Torregrossa and colleagues argue that the most robust deficit in OFC-damaged animals is in reversal learning. Seemingly impulsive behaviours, such as perseverative responding, and failure to alter responding when rewards for a learned behaviour are devalued, may in fact reflect impairment in the ability to update the value of an outcome, especially under changing circumstances (Torregrossa, Quinn, & Taylor, 2008).

There is evidence that regions implicated in motor inhibition and, in particular, the right frontal cortex, are involved in aspects of inhibitory control beyond response inhibition. This includes the suppression of drug cravings elicited by a cocaine video: Brain activation in the rIFC was increased when inhibiting a craving response and was negatively coupled with activation levels in the right nucleus accumbens (Volkow et al., 2010). In a think/no-think paradigm, in which paired associates are actively suppressed, activation in the rIFC was associated with suppression of the sensory components of memories (Depue, Curran, & Banich, 2007). De Fockert and colleagues showed that increasing working memory load increased activity levels in the bilateral inferior and middle frontal gyri while simultaneously increasing the distraction caused by (and sensory processing of) irrelevant faces (de Fockert, Rees, Frith, & Lavie, 2001). Hester and colleagues modified this paradigm to show that irrelevant drug stimuli produced heightened activity in the visual cortex in cocaine users relative to drug-naïve controls (Hester & Garavan, 2009). Critically, those users with the greatest levels of activity in the right prefrontal cortex showed the smallest behavioural interference caused by the distracting drug stimuli. In a similar manner, a study of the ability to ignore ecstasy-related stimuli produced greater occipital activation, but reduced right prefrontal activation in ecstasy users relative to controls (Roberts & Garavan, 2013). Tabibnia and colleagues identified the rIFC in a number of inhibitory deficits of methamphetamine-dependent subjects (Tabibnia et al., 2011a). Results indicated lower grey matter in the rIFC in dependent subjects relative to controls, where grey matter was correlated with drug craving, response inhibition performance, and a test of affect regulation. Finally, Behan and colleagues have recently shown that the rIFC is more active when subjects suppress reward anticipation (Behan, Stone, & Garavan, 2015). Here, a novel task required subjects to prepare for either a target to which they must respond as fast as possible to receive a reward, or a stop signal indicating they should make no response. A psychophysiological interaction analysis suggested the possibility of having to inhibit, rather than respond quickly, produced activity increases in the rIFC, which were correlated with activity decreases in the ventral striatum. Further, the rIFC activity was adjacent to a distinct rIFC region associated with motor inhibition. In sum, this brief review suggests that the rIFC may have a broad role in inhibitory processes that extend beyond motor inhibition. That said, there remains a lack of a comprehensive theory regarding the similarities and differences between the various types of inhibitory control with their neurobiological and psychological overlap. Further research probing the multiple types of inhibitory control in the same sample may be a valuable advance.

Structural Brain Correlates of Response Inhibition

Functional imaging studies have shown a primarily right-lateralised network implicated in response inhibition in healthy subjects, involving three key regions: the right IFG, the pre-SMA, and the STN (A. Aron & Poldrack, 2005; A. R. Aron, Behrens, Smith, Frank, & Poldrack, 2007; A. R. Aron et al., 2014; Chevrier, Noseworthy, & Schachar, 2007). In the following, we briefly describe the microstructural and structural brain matter characteristics associated with response inhibition on the basis of evidence from diffusion-weighted imaging (DWI) and voxel-wise anatomical studies. For a more in-depth review of the neuroanatomy and neurochemistry of response inhibition, see Chapter 16 by Forstmann & Alkemade in this volume.

DWI has been used to assess the white matter microstructure of humans and animal models for a better understanding of the white matter microstructure in the pathways connecting the three key regions implicated in response inhibition. Evidence from tractography studies in humans (A. R. Aron et al., 2007; Bari & Robbins, 2013; Johansen-Berg et al., 2004) and in monkeys (Inase, Tokuno, Nambu, Akazawa, & Takada, 1999) suggests

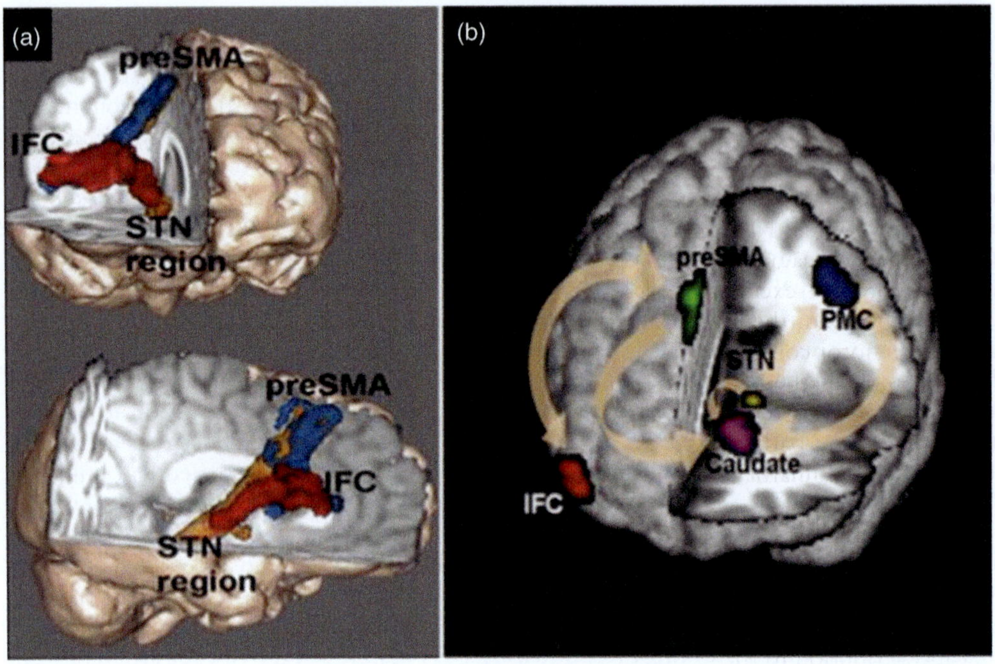

Figure 32.1 (a) 3D rendering of a diffusion-weighted tractography analysis of the network involved in response inhibition as measured by the SST showing white matter tracts between the IFC, the STN, and the pre-SMA (A. R. Aron et al., 2014). (b) Granger causality analysis results showing functional connectivity between cortical and subcortical areas during response inhibition (Duann, Ide, Luo, & Li, 2009b). Abbreviations: pre-SMA, pre-supplementary motor area; IFC, inferior frontal cortex; STN, subthalamic nucleus; PMC, primary motor cortex. Source: Bri 2013. Reproduced with permission of Elsevier.

the presence of direct fibre connections between the IFG, the STN, and the pre-SMA (Figure 32.1). More importantly, studies in humans using DWI and fMRI confirmed that both the IFG and the STN are connected with the pre-SMA (Figure 32.1) which corresponds with the loci of brain activity assessed with the SST (A. R. Aron et al., 2007; Bari & Robbins, 2013), suggesting that this three-part network participates in slowing or completely inhibiting responses.

Other DWI studies assessed associations between response inhibition performance and regional white matter microstructure in typically developing children. Results indicated that white matter microstructural changes, indexed by the radial diffusivity in the fronto-striatal network, have been correlated with faster reaction times in the go/no-go task (Liston et al., 2006). Moreover, a faster response inhibition as measured with the SST was significantly associated with a higher fractional anisotropy (FA) and a lower perpendicular diffusivity in both the right IFG and the right pre-SMA (Madsen et al., 2010), possibly reflecting a faster neural conduction within more densely packed or better myelinated fibre tracts. Paradoxically, higher FA was associated with worse inhibitory performance in adults and adolescents within several clusters in posterior brain regions and one in the brainstem, but with better cognitive flexibility, suggesting a dichotomous relationship between FA and these two aspects of cognitive control (Treit, Chen, Rasmussen, & Beaulieu, 2014). Some studies suggest that individual differences in behavioural indices of response inhibition can predict the activity in the right IFC as well as the density of white matter tracts necessary to suppress the initial activation and execution of the incorrect response (Forstmann et al., 2008).

More recently, studies using DWI (Cummins et al., 2012; Forstmann et al., 2008) in adults and in adolescents (Seghete, Herting, & Nagel, 2013) showed that an individual's ability to stop an action was associated with greater integrity of the white matter microstructure and predicted by structural connectivity between cortical and subcortical regions of the inhibitory control network. Using the SST, DWI, and analysis of effective connectivity, Rae and colleagues (Rae, Hughes, Anderson, & Rowe, 2015) demonstrated that the IFG modulates an excitatory influence on the pre-SMA-to-STN link during successful response inhibition, thereby amplifying the downstream polysynaptic inhibition from the STN to the motor cortex. Further, the mean diffusivity in tracts between pre-SMA and the STN, and between the IFG and STN, predicted individual differences in stopping efficiency. Finally, the white matter tract between pre-SMA and STN correlated with effective connectivity of the same pathway, providing important cross-modal validation of the network dynamics and modulatory role of the prefrontal cortex that underpin individual differences in inhibitory control.

Brain voxel-wise anatomical data have also been used to assess the relationship between response inhibition and grey matter volume (GMV) in healthy and psychologically impaired populations. Here, we will briefly discuss structural magnetic resonance imaging (MRI) findings in relation to psychopathology as a means to laying the groundwork related to the anatomical contribution to the response inhibition literature.

As noted above, in healthy subjects, the GMV of the IFG pars opercularis assessed using voxel-based morphometry was correlated with SST performance (Tabibnia et al., 2011b). Voxel-wise MRI correlations of SSRT and GMV demonstrated that ADHD patients, in addition to having longer SSRTs than controls, had SSRTs that were highly correlated with the anterior cingulate, striatal, and medial temporal volumes (McAlonan et al., 2009). Similarly, in another study of response inhibition operationalised by go/no-go and SST, patients with ADHD had significant inhibitory deficits and showed altered functional activation and reduced GMV in the right IFG (A. Aron & Poldrack, 2005). In bipolar disorder patients, better inhibitory control measured with the Stroop task correlated positively with GMV in the right parietal cortical regions (Haldane, Cunningham, Androutsos, & Frangou, 2008). In comorbid posttraumatic stress disorder (PTSD) and mild traumatic brain injury (mTBI), voxel-based and surface-based morphometry indicated volumetric reductions in the bilateral anterior amygdala predicting poorer inhibitory control as measured by performance on the go/no-go task (B. E. Depue et al., 2014). In animal studies involving both anatomical T1 and DWI, slower performance on motor tasks in monkeys was significantly correlated with lower volumetric measures in cortical, subcortical, and cerebellar areas as well as decreased FA in several major fibre bundles (Sridharan et al., 2012).

Drug-induced morphometric changes in regions related to alterations in inhibitory control have been investigated in methamphetamine and cocaine users. In methamphetamine-dependent subjects, the GMV in the rIFG was correlated with methamphetamine craving where, compared to healthy subjects, lower rIFG volumes and worse motor inhibitory control were observed (Tabibnia et al., 2011b). In monkeys, methamphetamine exposure, compared to saline, increased GMV within the right putamen and was positively correlated with changes in inhibitory control performance (Groman, Morales, Lee, London, & Jentsch, 2013).

These studies indicate that a better response inhibition performance seems to be positively correlated to the integrity of the white matter microstructure and GMV, mainly in the IFG. They also provide additional structural evidence highlighting the role of a functional-anatomical network in the right hemisphere for response control, and largely supporting the traditional findings stating that the IFG, the STN, and the pre-SMA are the main inhibitory control mediators.

Response Inhibition and Genetic Markers

Few studies have assessed the associations between genetic markers and response inhibition. The small extant literature has mainly targeted impaired serotonin (5-HT) function (Landrø et al., 2015), as well as contributions from the noradrenaline and dopamine systems (Cummins et al., 2012). Impaired serotonin (5-HT) function has been shown to reliably contribute to the neurobiology of impaired executive control processes (Robbins, 2005) and impulsive behaviours (Carver, Johnson, & Joormann, 2009). Stoltenberg and colleagues found in 199 young adults who donated buccal cells for genetic analyses that the longest SSRT was associated with the T/T genotype at the intron-8 polymorphism (rs1386483) of the tryptophan hydroxylase-2 gene (the rate-limiting enzyme for serotonin biosynthesis), suggesting that individuals homozygous for the T variant may have reduced tryptophan hydroxylase-2 function and correspondingly lower central serotonin levels.

Other studies have directly studied the potential role of the 5-HT-transporter-linked polymorphic region (5-HTTLPR) in the promoter region of the human 5-HT transporter gene. The 5-HTTLPR has short and long variants (Heils et al., 1996). A recent study on 141 healthy adults involving 5-HTTLPR and an additional single nucleotide polymorphism (rs25531) revealed that 5-HTTLPR polymorphisms predict response inhibition in the SST performance. Subjects carrying the high expressive allele exhibited significantly faster response inhibition as compared to subjects carrying low expressive alleles (Landrø et al., 2015).

Conflicting results have however also been reported for the association between 5-HTTLPR polymorphism and inhibitory control in healthy subjects under laboratory conditions. Whereas some studies using variants of continuous performance or response inhibition tasks reported that the short allele of the 5-HTTLPR may mediate impairments in impulse control (Paaver et al., 2007; Walderhaug, Herman, Magnusson, Morgan, & Landrø, 2010), others from rodent and humans found no association between 5-HTTLPR variants and measures of inhibitory control and impulsivity (Clark et al., 2005; Lage et al., 2011), leading to the suggestion that serotonin is not a critical neuromodulator for stop-signal inhibition (Bari, Eagle, Mar, Robinson, & Robbins, 2009; Chamberlain et al., 2006; Nandam et al., 2011).

There is also pharmacological evidence in rodents and humans highlighting important roles for noradrenaline and dopamine in modulating response inhibition (Cummins et al., 2012), by targeting both prefrontal and striatal nodes of the response inhibition network (Nandam et al., 2011). In a study by Cummins and colleagues (Cummins et al., 2012) on a total of 405 healthy adult participants who performed the SST, allelic variation in polymorphisms of the dopamine transporter gene (SLC6A3: rs37020; rs460000) predicted individual differences in SSRT. Furthermore, activity in frontal regions (anterior frontal, superior frontal, and superior medial gyri) and caudate varied additively with the T-allele of polymorphism rs37020. Moreover, Whelan and colleagues observed a relationship between SSRT and activity in a right frontal network in a very large sample of adolescents (Whelan et al., 2012). Activity in that network was also shown to be related to polymorphisms in *rs36024*, an SNP located in intron 4 of the *SLC6A2* gene, which encodes the norepinephrine transporter (NET).

In sum, these behavioural-genetic studies suggest a potential role of serotonin, noradrenaline, and dopamine transporter polymorphisms in regulating response inhibition and impulsive behavioural traits in healthy people. However, further investigations of the reported associations on larger samples providing sufficient statistical power for the genetic analyses are still required.

Response Inhibition and Drugs of Abuse

1-Functional Neuroimaging

Response inhibition tasks have been especially important in elucidating the functional neurobiology of addiction-related deficits in cognitive control. The neuroimaging literature suggests impaired response inhibition among individuals dependent on many classes of drugs, including nicotine, alcohol, cannabis, cocaine, and MDMA/ecstasy.

Nicotine Although findings in the literature are mixed, a recent meta-analysis found a small but significant effect relating cigarette smoking to response inhibition deficits (Smith et al., 2014). Results from neuroimaging investigations have generally found alterations in the neural circuitry associated with response inhibition in smokers compared to non-smoking controls (de Ruiter, Oosterlaan, Veltman, van den Brink, & Goudriaan, 2012; Galván, Poldrack, Baker, McGlennen, & London, 2011; Luijten, O'Connor, Rossiter, Franken, & Hester, 2013; Nestor, McCabe, Jones, Clancy, & Garavan, 2011). For example, Nestor and colleagues (Nestor et al., 2011) found that smokers showed reduced activation compared to non-smokers during response inhibition in a go/no-go task in a widely distributed network including the ACC, left IFG, bilateral inferior parietal lobules, and bilateral insula. This is similar to the findings of de Ruiter and colleagues (de Ruiter et al., 2012), who found reduced activation of the rostral ACC during inhibition in smokers.

One interesting line of research has examined the relationship between neural activity during successful response inhibition and craving for cigarettes. Berkman, Falk, and Lieberman (2011) demonstrated that subjects with greater task-related neural activity in nodes of the response inhibition network (bilateral inferior frontal gyrus, SMA, putamen, and left caudate) smoked less in response to subsequent, naturally occurring occasions of cigarette craving. These results suggest that functioning in the circuitry underlying motor inhibition translated to greater behavioural control in response to craving. Further, these investigators found an inverse relationship between amygdala activation during response inhibition and behaviour, such that subjects with greater amygdala activation had a stronger positive relationship between craving and smoking behaviour. These findings link altered patterns of neural activation with behavioural constructs known to be critical in addiction. Further, as studies have reported hypo-activation in the neural circuitry for response inhibition without differences in task performance, this study underscores the potential utility of neuroimaging as a sensitive measure of neurobiological alterations related to impulsive behaviour.

Alcohol Alcohol abusers have increased commission error rates compared to non-drinkers or social drinkers on go/no-go tasks (Kamarajan et al., 2005; Murphy & Garavan, 2011), and longer SSRTs on the SST compared to controls (e.g., Goudriaan, Grekin, & Sher, 2011; Lawrence, Luty, Bogdan, Sahakian, & Clark, 2009; Rubio et al., 2007). However, mixed results have been reported with a number of studies showing no difference in response inhibition related to alcohol consumption (Li, Luo, Yan, Bergquist, & Sinha, 2009; Papachristou et al., 2013; Schmaal et al., 2013). It has been suggested (Smith & Mattick, 2013) that this may relate to sex differences, based on evidence that heavy drinking may be preferentially associated with impaired response inhibition in females (Nederkoorn, Baltus, Guerrieri, & Wiers, 2009; Smith & Mattick, 2013; Townshend & Duka, 2005). That said, few studies have been sufficiently powered to specifically examine sex differences in response inhibition related to alcohol consumption. Nonetheless, Smith and colleagues reported an overall impairment in response inhibition in their meta-analysis, and suggested that a dose response relationship may exist between impaired response inhibition and drinking patterns (Smith et al., 2014). However, there have been no systematic studies addressing this possibility.

Studies using functional neuroimaging to examine response inhibition in problem drinkers are limited. Li and colleagues found no performance differences on SSRT, but lower activation in the left dlPFC in alcohol-dependent patients (Li et al., 2009). However, these subjects were all successfully abstinent in alcohol treatment at the time of scanning, making it difficult to determine if activation patterns were related to alcohol withdrawal or early recovery from alcohol dependence. Recent findings have shown that alcohol use disorders are associated with lower activation in the inferior frontal gyrus, insula, inferior parietal lobule, and ACC compared to controls (Claus, Feldstein Ewing, Filbey, & Hutchison, 2013). When comparing heavy to light alcohol consumption in college drinkers, the heavy drinkers showed impaired performance and altered patterns of neural activity during response inhibition in areas including the ACC, portions of the frontal lobe, hippocampus, and thalamus (Ahmadi et al., 2013). Structural neuroimaging experiments have suggested that chronic alcohol abuse is associated with global volume reduction, cortical and subcortical grey matter reductions, and enlargement of the ventricles. The volume losses in the frontal, cerebellar, and subcortical regions are believed to play a critical role in individual differences related to task performance (Chanraud et al., 2007; Scheurich, 2005; Sullivan, 2003). Therefore, as the neural architecture supporting response inhibition deteriorates, behavioural inhibition capacity is likely to suffer.

Cannabis Studies in both adolescent and adult cannabis users have found little evidence for disrupted cognitive performance (Grant, Chamberlain, Schreiber, & Odlaug, 2012; Jager, Block, Luijten, & Ramsey, 2010; Schweinsburg et al., 2010; Tapert et al., 2007); however, see Moreno et al., 2012. Interestingly, several studies have demonstrated that although there are inconsistent effects of cannabis use on inhibitory performance, there are neural differences that can be detected via fMRI (Behan et al., 2014; Hester, Nestor, & Garavan, 2009; G. M. P. Roberts & Garavan, 2010; Schweinsburg et al., 2008; Tapert et al., 2007). For example, Roberts and Garavan investigated neural activity using fMRI during response inhibition in adolescent cannabis users and non-drug using controls. Although users had equal performance to control subjects, the users had increased activation in frontal and parietal regions during successful inhibitions. This pattern of activation was interpreted to indicate increased neural resources required of the users to achieve performance levels comparable to controls (G. M. P. Roberts & Garavan, 2010). Similar results were found in a study of college students (cannabis users compared to non-drug users) where there was equal task performance but increased activation in the right inferior parietal lobule, the right putamen, and the supplementary motor area in the users (Hester et al., 2009).

Some evidence suggests that heavier use, earlier onset, and greater cumulative cannabis consumption is associated with smaller increases in activation in frontal and parietal regions compared to lighter users or those who begin using later (Schweinsburg et al., 2008, 2010). Such findings indicate that there may be an interaction of brain development and cannabis exposure on brain function and may additionally suggest a compensatory mechanism in heavy cannabis users (Jacobus, Bava, Cohen-Zion, Mahmood, & Tapert, 2009).

Cocaine There is strong evidence that cocaine users have poorer response inhibition than non-users. This is observed in studies using the SST (Colzato, van den Wildenberg, & Hommel, 2007; Fillmore & Rush, 2002; Li, Milivojevic, Kemp, Hong, & Sinha, 2006; Morie, De Sanctis, Garavan, & Foxe, 2014; Vonmoos et al., 2013) and in go/no-go tasks (Fernández-Serrano, Perales, Moreno-López, Pérez-García, & Verdejo-García, 2011; Hester & Garavan, 2004; Hester, Simões-Franklin, & Garavan, 2007; Kaufman, Ross, Stein, & Garavan, 2003; Lane, Moeller, Steinberg, Buzby, & Kosten, 2007). A review by Spronk and colleagues calculated pooled effect sizes for both SSRT on the SST and errors of commission on go/no-go tasks and found a moderate pooled effect size (0.50) of cocaine user status on the SST and a moderate to large (0.64) pooled effect size for errors of commission on the go/no-go task (Spronk, van Wel, Ramaekers, & Verkes, 2013). fMRI studies have generally shown reduced neural activity in the PFC including rostral ACC and SMA (Hester & Garavan, 2004; Kaufman et al., 2003; Li et al., 2007).

Using independent component analysis on a SST, Elton and colleagues discriminated cocaine users from non-users on the basis of activity patterns decomposed into eleven components. Two of these components were specifically related to stop-signal success, and cocaine users exhibited decreased activation in these networks compared to controls. One network comprised the bilateral IFG, angular gyri, middle temporal, and posterior parietal gyri, and the other network comprised the dlPFC, ventrolateral PFC, dorsomedial PFC, anterior insula, and middle temporal gyrus (Elton et al., 2014).

MDMA/Ecstasy A meta-analysis found that overall there is a small effect size on inhibitory errors in heavy MDMA users compared to controls (Smith et al., 2014). Among individual studies, there are several that reported no behavioural performance differences (Geusau, Stalenhoef, Huizinga, Snel, & Ridderinkhof, 2004; Quednow et al., 2006; C. A. Roberts, Fairclough, Fisk, Tames, & Montgomery, 2013; G. M. P. Roberts & Garavan, 2010). However, two of these studies used neuroimaging and found altered neural processing in MDMA users. For example, Roberts and colleagues found that ecstasy/polydrug users showed altered EEG patterns suggestive of attentional or inhibitory deficits (C. A. Roberts et al., 2013). Similarly, Roberts and Garavan (G. M. P. Roberts & Garavan, 2010) found intact performance but increased activation in the response inhibition network (the right DLPFC, inferior frontal gyrus, and parietal lobule) in recreational ecstasy users. Other studies of current MDMA users have reported impaired behavioural performance in response inhibition (Hoshi et al., 2007). Taken together, the available literature suggests a small impairment in response inhibition associated with MDMA use and altered neural processing in users with intact behavioural performance.

2-Event-Related Potentials

As discussed above, individuals with pathological substance use demonstrate fMRI-related differences in response inhibition compared with controls. However, fMRI is unable to adequately capture the temporal dynamics associated with response inhibition. Therefore, cognitive neuroscientists have been keen to examine response inhibition in addicted individuals using the high temporal precision afforded by event-related potentials (ERPs). Changes in brain activity related to inhibition have been consistently associated with two ERP components: the N2 and the P3 (Kok, Ramautar, De Ruiter, Band, & Ridderinkhof, 2004). The N2 component is characterised by a negative wave occurring 200–300 ms after a stimulus is presented, and is thought to be generated by the ACC (Huster, Westerhausen, Pantev, & Konrad, 2010; Kok et al., 2004) and the IFG (Lavric, Pizzagalli, & Forstmeier, 2004). The N2 is believed to relate to behavioural outcomes of inhibitory control, and is interpreted as an index of the early cognitive processes necessary for recruiting and implementing inhibitory control. The P3 component is characterised by a positive wave occurring 300–500 ms after a stimulus is presented, and is believed to originate near motor cortices (Huster et al., 2010; Kok et al., 2004). Consequently, the P3 component appears to reflect the later stages of inhibition that are more closely related to inhibiting the motor system in the premotor cortex (Band & van Boxtel, 1999). Consequently, less marked N2 or P3 amplitudes in addicted individuals, compared with healthy controls, may be interpreted as an indication of deficits in inhibitory control. Further, it is argued that amplitude differences in the absence of behavioural impairment may provide evidence for the utility of ERP components as sensitive measures of cognitive control (Buzzell, Fedota, Roberts, & McDonald, 2014; Gamma, Brandeis, Brandeis, & Vollenweider, 2005). ERP studies that have utilised classic response inhibition paradigms have demonstrated modest evidence for abnormalities in brain function in individuals with various forms of addiction.

Nicotine ERP studies with individuals who are addicted to nicotine show inconsistent associations with deficits in response inhibition. Luijten and colleagues (Luijten et al., 2014; Luijten, Littel, & Franken, 2011) demonstrated tentative evidence that N2 amplitudes may be lower in individuals with nicotine dependence than controls; however, results for P3 amplitudes were contradictory. Results from one study found that reduced N2 amplitudes during a go/no-go task in smokers, compared to controls, were accompanied by decreased task performance, and no differences between groups were found in P3 amplitudes. These results were interpreted as representing an impairment of inhibition in smokers. Results from another study found that P3 amplitudes during no-go trials were lower in individuals with nicotine dependence (Evans, Park, Maxfield, & Drobes, 2009), and no performance differences were found between the two groups. Results obtained by Buzzell and colleagues (Buzzell et al., 2014) also found a decreased N2 amplitude during no-go trials in smokers as compared with non-smokers. No group differences were found for P3 amplitudes during no-go trials. Task performance, as indicated by reaction time and accuracy, did not differ in smokers versus non-smokers, which the authors interpreted to reflect the low levels of nicotine dependence in the sample. Further, the authors noted that the group differences in N2 amplitudes, accompanied by the absence of behavioural impairments, provided evidence that the N2 was a sensitive measure of cognitive control impairment control in smokers, even in individuals who had low levels of nicotine dependence. To date, there is evidence that N2 amplitudes are lower in individuals with nicotine dependence, whereas P3 results remain contradictory. Evidence also indicates that initial detection of errors may be compromised in nicotine-dependent individuals when there are limited cognitive resources for monitoring errors such as during an exposure to smoking cues (e.g., Luijten et al., 2011).

Alcohol Evidence for ERP differences associated with response inhibition is largely mixed for individuals who are alcohol dependent, and many of the ERP studies that have been conducted examined currently-abstaining individuals who participated during their time in alcohol treatment programmes. Owing to the paucity of research specifically addressing N2 amplitudes in individuals with alcohol dependence, no strong conclusions can be drawn regarding early inhibitory processes. Evidence is mixed for deficits on P3 amplitudes and for performance and accuracy differences during response inhibition tasks in individuals with alcohol dependence; thus, the results appear to reflect similar contradictory findings in behavioural research on alcohol dependence (Luijten et al., 2014).

Alcohol-dependent individuals have been shown to perform with decreased accuracy on go/no-go tasks compared with healthy controls (Kamarajan et al., 2005); however, contrasting results have also been reported, with some research finding no differences in response inhibition between the groups on go/no-go tasks and flanker tasks (Fallgatter, Wiesbeck, Weijers, Boening, & Strik, 1998; Karch et al., 2007). In several studies, P3 amplitudes were smaller during no-go trials in alcohol-dependent individuals as compared with controls (Cohen, Porjesz, Begleiter, & Wang, 1997; Colrain et al., 2011; Kamarajan et al., 2005); however, in some studies, P3 amplitudes were also smaller for the go trials in alcoholic individuals (Cohen et al., 1997; Kamarajan et al., 2005; Pfefferbaum, Rosenbloom, & Ford, 1987), indicating that alcoholic individuals may present with overall neural deficits rather than specific deficits in response inhibition. Indeed, there is some evidence to suggest that alcohol-dependent individuals may in fact activate altogether inappropriate brain circuitry during cognitive tasks (e.g., Kamarajan et al., 2005).

Cannabis Few ERP studies have focused on response inhibition differences in individuals with cannabis dependence. In a study investigating whether ERPs differed in heavy ($M = 340$ times per year, average 2.3 joints per occasion) versus occasional ($M = 55$ times per year, average 1.2 joints per occasion) cannabis users after THC administration, Theunissen and colleagues (Theunissen et al., 2012) demonstrated increased stop reaction time during the SST in both heavy and occasional users. In regards to neural differences between the groups, N2 amplitudes during the SST were not affected by cannabis use history; however, the authors noted that this finding may have resulted because the SST used in the study was not well suited to eliciting a clear and reliable N2. In contrast, Battisti and colleagues (Battisti et al., 2010) found that cannabis users (16.4 years of use) demonstrated increased errors on colour-incongruent words on a Stroop task compared with healthy controls, but no performance differences from colour-congruent or neutral trials. The authors further investigated predictors of performance, and found that poorer incongruent trial performance was predicted by an earlier age of onset of cannabis use. Also, their ERP findings suggested that cannabis users, in comparison to healthy controls, displayed an altered sustained potential related to conflict resolution, providing evidence for disruption to normal neural functioning. Taken together, these studies indicate that chronic use of cannabis may impair the brain's ability to respond optimally in the presence of events that require conflict resolution; however, methodological differences and small sample sizes require further investigation in this domain.

Cocaine ERPs have been under-studied in relation to response inhibition in cocaine-dependent individuals. Evidence suggests that neural deficits may be present in both the early and late stages of the inhibition process; however, it remains unclear whether this may also result in behavioural deficits. One study measured accuracy as well as N2 and P3 amplitudes during a flanker task that incorporated no-go trials in current cocaine-dependent users compared with controls (Sokhadze, Stewart, Hollifield, & Tasman, 2008). Results indicated that there were no differences between groups in accuracy during the task; however, reaction times in both the congruent and incongruent go trials were slower in the cocaine-dependent group. The authors also found that no-go N2 and P3 amplitudes, relative to go amplitudes, were less pronounced in cocaine-dependent individuals than in controls.

MDMA/Ecstasy Within the last two decades, the likelihood of cognitive dysfunction in chronic ecstasy users has become a focus of neuroimaging and neuropsychological research (Gamma et al., 2005); however, little emphasis has been placed on electrophysiological measures in this research, and even fewer studies have focused on response inhibition within this domain. Gamma and colleagues (Gamma et al., 2005) examined response inhibition in regular ecstasy/polydrug users compared with controls by measuring the P3 in a go/no-go paradigm. Ecstasy users demonstrated lower P3 amplitudes and a less anterior location of no-go P3 peaks; however, these effects became weaker and were no longer statistically significant after controlling for age, educational level, and lifetime cannabis use. Further, task performance did not differ between the groups, and there were no significant correlations between ERP and task performance variables with lifetime drug use. The study also posed several limitations; notably, participants were polydrug users, and the two groups were not well matched on a number of key variables (e.g., cannabis use), which may have limited the specificity of the results. More research is needed within this domain to assess electrophysiological differences in ecstasy users.

Conclusions and Future Directions

Researchers have been able to reliably detect differences in the structural and functional neurobiology of inhibitory control between healthy and substance-abusing populations and often do so when inhibitory control task performance fails to delineate users from non-users. Structural MRI suggests that better response inhibition performance is proportional to the integrity of the white matter microstructure and the grey matter macrostructure, mainly within the rIFG. Moreover, these grey and white matter findings within the response inhibition network, rIFC–pre-SMA–STN, have been observed in ADHD and PTSD. Aside from methamphetamine, there is a lack of structural MRI research informing the impact of other drugs of abuse on the brain anatomy subserving response inhibition. Furthermore, few studies have explored the genetic biomarkers related to response inhibition, such as SNPs related to impulsivity (McGue et al., 2013). To date, limited data has found genetic variation in serotonin, noradrenaline, and dopamine transporter genes related to the development of frontostriatal inhibition networks. These variations may represent a key risk mechanism for disorders of behavioural inhibition.

The structural and functional neuroimaging and genetic studies coupled with response inhibition performance provide enough evidence that inhibitory control abnormalities detected in human and animal models are due, at least in part, to the underlying effects of white and grey matter atrophy, as well as genetic polymorphisms. Still, larger studies on structural and genetic data are needed to replicate these findings. All things considered, a combined right IFG-STN-pre-SMA structural and functional response inhibition phenotype is a particularly good candidate for cognitive and inhibitory control prediction.

Across many classes of drugs, fMRI and ERP studies of response inhibition suggest that neural deficits in inhibitory control characterise substance dependence. N2 amplitudes are mostly decreased in individuals with addiction compared with controls, whereas P3 amplitudes have generally shown mixed results. Broadly, results support the claim that individuals with addictions demonstrate deficits related to inhibition.

The preceding review suggests that deficits in inhibitory control characterise substance dependence. There are, however, drug-specific effects that require further elaboration (e.g., the mixed findings in cannabis users). The integration of functional activation, functional connectivity, and brain structural data is important, but so too is a much richer phenotypic characterisation of the users including their drug use histories (age of onset, polydrug use), mental health co-morbidities, family and environmental influences, and so on. In reviewing the literature, there persists a lack of a comprehensive understanding on how the various types of inhibitory control relate to one another, psychologically and neurobiologically. More assessments of drug use and other types of inhibitory control (e.g., delaying gratification) or inhibitory control in reward-related contexts may yield new insights. It is a conundrum that although different aspects of inhibitory control appear to be uncorrelated with one another (e.g., self-report personality measures, impulsive choice and impulsive responding; Reynolds,

Ortengren, Richards, & de Wit, 2006), drug users score highly impulsive on all. Combining this with the evidence that inhibitory control is related to reward processes, such as drug-induced euphoria and drug self-administration (Cervantes, Laughlin, & Jentsch, 2013; Weafer & de Wit, 2013), we can conclude that more conceptual work is required to integrate these constructs. Finally, as noted above, relating laboratory-based measures of inhibitory control to drug urges and craving in the natural environment is an important extension of the existing research.

References

Ahmadi, A., Pearlson, G. D., Meda, S. A., Dager, A., Potenza, M. N., Rosen, R., … Stevens, M. C. (2013). Influence of alcohol use on neural response to Go/No-Go task in college drinkers. *Neuropsychopharmacology, 38*(11), 2197–2208.

Aron, A., & Poldrack, R. (2005). The cognitive neuroscience of response inhibition: Relevance for genetic research in attention-deficit/hyperactivity disorder. *Biological Psychiatry, 57.*

Aron, A. R., Behrens, T. E., Smith, S., Frank, M. J., & Poldrack, R. A. (2007). Triangulating a cognitive control network using diffusion-weighted magnetic resonance imaging (MRI) and functional MRI. *The Journal of Neuroscience, 27*(14), 3743–3752.

Aron, A. R., Fletcher, P. C., Bullmore, E. T., Sahakian, B. J., & Robbins, T. W. (2003). Stop-signal inhibition disrupted by damage to right inferior frontal gyrus in humans. *Nature Neuroscience, 6*(2), 115–116.

Aron, A. R., & Poldrack, R. A. (2006). Cortical and subcortical contributions to stop signal response inhibition: Role of the subthalamic nucleus. *The Journal of Neuroscience, 26*(9), 2424–2433.

Aron, A. R., Robbins, T. W., & Poldrack, R. A. (2004). Inhibition and the right inferior frontal cortex. *Trends in Cognitive Sciences, 8*(4), 170–177.

Aron, A. R., Robbins, T. W., & Poldrack, R. A. (2014). Inhibition and the right inferior frontal cortex: One decade on. *Trends in Cognitive Sciences, 18*(4), 177–185.

Badre, D., & Wagner, A. D. (2006). Computational and neurobiological mechanisms underlying cognitive flexibility. *Proceedings of the National Academy of Sciences of the United States of America, 103*(18), 7186–7191.

Band, G. P. H., & van Boxtel, G. J. M. (1999). Inhibitory motor control in stop paradigms: Review and reinterpretation of neural mechanisms. *Acta Psychologica, 101*(2–3), 179–211.

Bari, A., Eagle, D. M., Mar, A. C., Robinson, E. S. J., & Robbins, T. W. (2009). Dissociable effects of noradrenaline, dopamine, and serotonin uptake blockade on stop task performance in rats. *Psychopharmacology, 205*(2), 273–283.

Bari, A., & Robbins, T. W. (2013). Inhibition and impulsivity: Behavioral and neural basis of response control. *Progress in Neurobiology, 108*, 44–79.

Battisti, R. A., Roodenrys, S., Johnstone, S. J., Pesa, N., Hermens, D. F., & Solowij, N. (2010). Chronic cannabis users show altered neurophysiological functioning on Stroop task conflict resolution. *Psychopharmacology, 212*(4), 613–624.

Behan, B., Connolly, C. G., Datwani, S., Doucet, M., Ivanovic, J., Morioka, R., … Garavan, H. (2014). Response inhibition and elevated parietal-cerebellar correlations in chronic adolescent cannabis users. *Neuropharmacology, 84*, 131–137.

Behan, B., Stone, A., & Garavan, H. (2015). Right prefrontal and ventral striatum interactions underlying impulsive choice and impulsive responding. *Human Brain Mapping, 36*(1), 187–198.

Berkman, E. T., Falk, E. B., & Lieberman, M. D. (2011). In the trenches of real-world self-control neural correlates of breaking the link between craving and smoking. *Psychological Science, 22*(4), 498–506.

Berlin, H. A., Rolls, E. T., & Kischka, U. (2004). Impulsivity, time perception, emotion and reinforcement sensitivity in patients with orbitofrontal cortex lesions. *Brain, 127*(5), 1108–1126.

Bryden, D. W., & Roesch, M. R. (2015). Executive control signals in orbitofrontal cortex during response inhibition. *The Journal of Neuroscience, 35*(9), 3903–3914.

Buzzell, G. A., Fedota, J. R., Roberts, D. M., & McDonald, C. G. (2014). The N2 ERP component as an index of impaired cognitive control in smokers. *Neuroscience Letters*, *563*, 61–65.

Cai, W., George, J. S., Verbruggen, F., Chambers, C. D., & Aron, A. R. (2012). The role of the right presupplementary motor area in stopping action: Two studies with event-related transcranial magnetic stimulation. *Journal of Neurophysiology*, *108*(2), 380–389.

Carver, C. S., Johnson, S. L., & Joormann, J. (2009). Two-mode models of self-regulation as a tool for conceptualizing effects of the serotonin system in normal behavior and diverse disorders. *Current Directions in Psychological Science*, *18*(4), 195–199.

Cervantes, M. C., Laughlin, R. E., & Jentsch, J. D. (2013). Cocaine self-administration behavior in inbred mouse lines segregating different capacities for inhibitory control. *Psychopharmacology*, *229*(3), 515–525.

Chamberlain, S. R., Müller, U., Blackwell, A. D., Clark, L., Robbins, T. W., & Sahakian, B. J. (2006). Neurochemical modulation of response inhibition and probabilistic learning in humans. *Science (New York, N.Y.)*, *311*(5762), 861–863.

Chambers, C. D., Bellgrove, M. A., Stokes, M. G., Henderson, T. R., Garavan, H., Robertson, I. H., … Mattingley, J. B. (2006). Executive 'brake failure' following deactivation of human frontal lobe. *Journal of Cognitive Neuroscience*, *18*(3), 444–455.

Chanraud, S., Martelli, C., Delain, F., Kostogianni, N., Douaud, G., Aubin, H.-J., … Martinot, J.-L. (2007). Brain morphometry and cognitive performance in detoxified alcohol-dependents with preserved psychosocial functioning. *Neuropsychopharmacology: Official Publication of the American College of Neuropsychopharmacology*, *32*(2), 429–438.

Chan, R. C. K., Shum, D., Toulopoulou, T., & Chen, E. Y. H. (2008). Assessment of executive functions: Review of instruments and identification of critical issues. *Archives of Clinical Neuropsychology: The Official Journal of the National Academy of Neuropsychologists*, *23*(2), 201–216.

Chevrier, A. D., Noseworthy, M. D., & Schachar, R. (2007). Dissociation of response inhibition and performance monitoring in the stop signal task using event-related fMRI. *Human Brain Mapping*, *28*(12), 1347–1358.

Clark, L., Roiser, J. P., Cools, R., Rubinsztein, D. C., Sahakian, B. J., & Robbins, T. W. (2005). Stop signal response inhibition is not modulated by tryptophan depletion or the serotonin transporter polymorphism in healthy volunteers: Implications for the 5-HT theory of impulsivity. *Psychopharmacology*, *182*(4), 570–578.

Claus, E. D., Feldstein Ewing, S. W., Filbey, F. M., & Hutchison, K. E. (2013). Behavioral control in alcohol use disorders: Relationships with severity. *Journal of Studies on Alcohol and Drugs*, *74*(1), 141–151.

Cohen, H. L., Porjesz, B., Begleiter, H., & Wang, W. (1997). Neurophysiological correlates of response production and inhibition in alcoholics. *Alcoholism, Clinical and Experimental Research*, *21*(8), 1398–1406.

Colrain, I. M., Sullivan, E. V., Ford, J. M., Mathalon, D. H., McPherson, S.-L., Roach, B. J., … Pfefferbaum, A. (2011). Frontally mediated inhibitory processing and white matter microstructure: Age and alcoholism effects. *Psychopharmacology*, *213*(4), 669–679.

Colzato, L. S., van den Wildenberg, W. P. M., & Hommel, B. (2007). Impaired inhibitory control in recreational cocaine users. *PLoS ONE*, *2*(11), e1143.

Cummins, T. D. R., Hawi, Z., Hocking, J., Strudwick, M., Hester, R., Garavan, H., … Bellgrove, M. A. (2012). Dopamine transporter genotype predicts behavioural and neural measures of response inhibition. *Molecular Psychiatry*, *17*(11), 1086–1092.

de Fockert, J. W., Rees, G., Frith, C. D., & Lavie, N. (2001). The role of working memory in visual selective attention. *Science*, *291*(5509), 1803–1806.

Depue, B. E., Curran, T., & Banich, M. T. (2007). Prefrontal regions orchestrate suppression of emotional memories via a two-phase process. *Science*, *317*(5835), 215–219.

Depue, B. E., Olson-Madden, J. H., Smolker, H. R., Rajamani, M., Brenner, L. A., & Banich, M. T. (2014). Reduced amygdala volume is associated with deficits in inhibitory control: A voxel- and surface-based morphometric analysis of comorbid PTSD/mild TBI. *BioMed Research International*, *2014*, 691505.

de Ruiter, M. B., Oosterlaan, J., Veltman, D. J., van den Brink, W., & Goudriaan, A. E. (2012). Similar hyporesponsiveness of the dorsomedial prefrontal cortex in problem gamblers and heavy smokers during an inhibitory control task. *Drug and Alcohol Dependence*, *121*(1–2), 81–89.

D'Esposito, M., Detre, J. A., Alsop, D. C., Shin, R. K., Atlas, S., & Grossman, M. (1995). The neural basis of the central executive system of working memory. *Nature*, *378*(6554), 279–281.

Dodds, C. M., Morein-Zamir, S., & Robbins, T. W. (2011). Dissociating inhibition, attention, and response control in the frontoparietal network using functional magnetic resonance imaging. *Cerebral Cortex*, *21*(5), 1155–1165.

Dosenbach, N. U. F., Fair, D. A., Miezin, F. M., Cohen, A. L., Wenger, K. K., Dosenbach, R. A. T., … Petersen, S. E. (2007). Distinct brain networks for adaptive and stable task control in humans. *Proceedings of the National Academy of Sciences of the United States of America*, *104*(26), 11073–11078.

Dove, A., Pollmann, S., Schubert, T., Wiggins, C. J., & Yves von Cramon, D. (2000). Prefrontal cortex activation in task switching: An event-related fMRI study. *Cognitive Brain Research*, *9*(1), 103–109.

Duann, J.-R., Ide, J. S., Luo, X., & Li, C. R. (2009a). Functional connectivity delineates distinct roles of the inferior frontal cortex and presupplementary motor area in stop signal inhibition. *The Journal of Neuroscience*, *29*(32), 10171–10179.

Duann, J.-R., Ide, J. S., Luo, X., & Li, C. R. (2009b). Functional connectivity delineates distinct roles of the inferior frontal cortex and presupplementary motor area in stop signal inhibition. *The Journal of Neuroscience*, *29*(32), 10171–10179.

Eagle, D. M., Baunez, C., Hutcheson, D. M., Lehmann, O., Shah, A. P., & Robbins, T. W. (2008). Stop-signal reaction-time task performance: Role of prefrontal cortex and subthalamic nucleus. *Cerebral Cortex*, *18*(1), 178–188.

Edin, F., Klingberg, T., Johansson, P., McNab, F., Tegnér, J., & Compte, A. (2009). Mechanism for top-down control of working memory capacity. *Proceedings of the National Academy of Sciences of the United States of America*, *106*(16), 6802–6807.

Elton, A., Young, J., Smitherman, S., Gross, R. E., Mletzko, T., & Kilts, C. D. (2014). Neural network activation during a stop-signal task discriminates cocaine-dependent from non-drug-abusing men. *Addiction Biology*, *19*(3), 427–438.

Evans, D. E., Park, J. Y., Maxfield, N., & Drobes, D. J. (2009). Neurocognitive variation in smoking behavior and withdrawal: Genetic and affective moderators. *Genes, Brain, and Behavior*, *8*(1), 86–96.

Fallgatter, A. J., Wiesbeck, G. A., Weijers, H. G., Boening, J., & Strik, W. K. (1998). Event-related correlates of response suppression as indicators of novelty seeking in alcoholics. *Alcohol and Alcoholism (Oxford, Oxfordshire)*, *33*(5), 475–481.

Fernández-Serrano, M. J., Perales, J. C., Moreno-López, L., Pérez-García, M., & Verdejo-García, A. (2011). Neuropsychological profiling of impulsivity and compulsivity in cocaine dependent individuals. *Psychopharmacology*, *219*(2), 673–683.

Fillmore, M. T., & Rush, C. R. (2002). Impaired inhibitory control of behavior in chronic cocaine users. *Drug and Alcohol Dependence*, *66*(3), 265–273.

Floden, D., & Stuss, D. T. (2006). Inhibitory control is slowed in patients with right superior medial frontal damage. *Journal of Cognitive Neuroscience*, *18*(11), 1843–1849.

Forstmann, B. U., Jahfari, S., Scholte, H. S., Wolfensteller, U., van den Wildenberg, W. P. M., & Ridderinkhof, K. R. (2008). Function and structure of the right inferior frontal cortex predict individual differences in response inhibition: A model-based approach. *Journal of Neuroscience*, *28*(39), 9790–9796.

Galván, A., Poldrack, R. A., Baker, C. M., McGlennen, K. M., & London, E. D. (2011). Neural correlates of response inhibition and cigarette smoking in late adolescence. *Neuropsychopharmacology*, *36*(5), 970–978.

Gamma, A., Brandeis, D., Brandeis, R., & Vollenweider, F. X. (2005). The P3 in 'ecstasy' polydrug users during response inhibition and execution. *Journal of Psychopharmacology (Oxford, England)*, *19*(5), 504–512.

Garavan, H., Hester, R., Murphy, K., Fassbender, C., & Kelly, C. (2006). Individual differences in the functional neuroanatomy of inhibitory control. *Brain Research*, *1105*(1), 130–142.

Garavan, H., Ross, T. J., & Stein, E. A. (1999). Right hemispheric dominance of inhibitory control: An event-related functional MRI study. *Proceedings of the National Academy of Sciences of the United States of America*, *96*(14), 8301–8306.

Geusau, N. A. von, Stalenhoef, P., Huizinga, M., Snel, J., & Ridderinkhof, K. R. (2004). Impaired executive function in male MDMA ('ecstasy') users. *Psychopharmacology*, *175*(3), 331–341.

Goudriaan, A. E., Grekin, E. R., & Sher, K. J. (2011). Decision making and response inhibition as predictors of heavy alcohol use: A prospective study. *Alcoholism: Clinical and Experimental Research*, *35*(6), 1050–1057.

Grant, J. E., Chamberlain, S. R., Schreiber, L., & Odlaug, B. L. (2012). Neuropsychological deficits associated with cannabis use in young adults. *Drug and Alcohol Dependence*, *121*(1–2), 159–162.

Groman, S. M., Morales, A. M., Lee, B., London, E. D., & Jentsch, J. D. (2013). Methamphetamine-induced increases in putamen gray matter associate with inhibitory control. *Psychopharmacology*, *229*(3), 527–538.

Haldane, M., Cunningham, G., Androutsos, C., & Frangou, S. (2008). Structural brain correlates of response inhibition in Bipolar Disorder I. *Journal of Psychopharmacology (Oxford, England)*, *22*(2), 138–143.

Hampshire, A., Chamberlain, S. R., Monti, M. M., Duncan, J., & Owen, A. M. (2010). The role of the right inferior frontal gyrus: Inhibition and attentional control. *NeuroImage*, *50*(3), 1313–1319.

Heils, A., Teufel, A., Petri, S., Stöber, G., Riederer, P., Bengel, D., & Lesch, K. P. (1996). Allelic variation of human serotonin transporter gene expression. *Journal of Neurochemistry*, *66*(6), 2621–2624.

Hester, R., & Garavan, H. (2004). Executive dysfunction in cocaine addiction: Evidence for discordant frontal, cingulate, and cerebellar activity. *The Journal of Neuroscience*, *24*(49), 11017–11022.

Hester, R., & Garavan, H. (2009). Neural mechanisms underlying drug-related cue distraction in active cocaine users. *Pharmacology Biochemistry and Behavior*, *93*(3), 270–277.

Hester, R., Nestor, L., & Garavan, H. (2009). Impaired error awareness and anterior cingulate cortex hypoactivity in chronic cannabis users. *Neuropsychopharmacology*, *34*(11), 2450–2458.

Hester, R., Simões-Franklin, C., & Garavan, H. (2007). Post-error behavior in active cocaine users: Poor awareness of errors in the presence of intact performance adjustments. *Neuropsychopharmacology*, *32*(9), 1974–1984.

Hoshi, R., Mullins, K., Boundy, C., Brignell, C., Piccini, P., & Curran, H. V. (2007). Neurocognitive function in current and ex-users of ecstasy in comparison to both matched polydrug-using controls and drug-naïve controls. *Psychopharmacology*, *194*(3), 371–379.

Huster, R. J., Westerhausen, R., Pantev, C., & Konrad, C. (2010). The role of the cingulate cortex as neural generator of the N200 and P300 in a tactile response inhibition task. *Human Brain Mapping*, *31*(8), 1260–1271.

Inase, M., Tokuno, H., Nambu, A., Akazawa, T., & Takada, M. (1999). Corticostriatal and corticosubthalamic input zones from the presupplementary motor area in the macaque monkey: Comparison with the input zones from the supplementary motor area. *Brain Research*, *833*(2), 191–201.

Jacobus, J., Bava, S., Cohen-Zion, M., Mahmood, O., & Tapert, S. F. (2009). Functional consequences of marijuana use in adolescents. *Pharmacology Biochemistry and Behavior*, *92*(4), 559–565.

Jager, G., Block, R. I., Luijten, M., & Ramsey, N. F. (2010). Cannabis use and memory brain function in adolescent boys: A cross-sectional multicenter functional magnetic resonance imaging study. *Journal of the American Academy of Child & Adolescent Psychiatry*, *49*(6), 561–572.e3.

Johansen-Berg, H., Behrens, T. E. J., Robson, M. D., Drobnjak, I., Rushworth, M. F. S., Brady, J. M., … Matthews, P. M. (2004). Changes in connectivity profiles define functionally distinct regions in human medial frontal cortex. *Proceedings of the National Academy of Sciences of the United States of America*, *101*(36), 13335–13340.

Kamarajan, C., Porjesz, B., Jones, K. A., Choi, K., Chorlian, D. B., Padmanabhapillai, A., … Begleiter, H. (2005). Alcoholism is a disinhibitory disorder: Neurophysiological evidence from a Go/No-Go task. *Biological Psychology*, *69*(3), 353–373.

Karch, S., Graz, C., Jager, L., Karamatskos, E., Flatz, A. S. W., Lutz, J., … Mulert, C. (2007). Influence of anxiety on electrophysiological correlates of response inhibition capacities in alcoholism. *Clinical EEG and Neuroscience*, *38*(2), 89–95.

Kaufman, J. N., Ross, T. J., Stein, E. A., & Garavan, H. (2003). Cingulate hypoactivity in cocaine users during a GO-NOGO task as revealed by event-related functional magnetic resonance imaging. *The Journal of Neuroscience*, *23*(21), 7839–7843.

Kim, C., Chung, C., & Kim, J. (2013). Task-dependent response conflict monitoring and cognitive control in anterior cingulate and dorsolateral prefrontal cortices. *Brain Research, 1537*, 216–223.

Kok, A., Ramautar, J. R., De Ruiter, M. B., Band, G. P. H., & Ridderinkhof, K. R. (2004). ERP components associated with successful and unsuccessful stopping in a stop-signal task. *Psychophysiology, 41*(1), 9–20.

Lage, G. M., Malloy-Diniz, L. F., Matos, L. O., Bastos, M. A. R., Abrantes, S. S. C., & Corrêa, H. (2011). Impulsivity and the 5-HTTLPR polymorphism in a non-clinical sample. *PLoS ONE, 6*(2), e16927.

Landrø, N. I., Jonassen, R., Clark, L., Haug, K. B. F., Aker, M., Bø, R., ... Stiles, T. C. (2015). Serotonin transporter polymorphisms predict response inhibition in healthy volunteers. *Neuroscience Letters, 584*, 109–112.

Lane, S. D., Moeller, F. G., Steinberg, J. L., Buzby, M., & Kosten, T. R. (2007). Performance of cocaine dependent individuals and controls on a response inhibition task with varying levels of difficulty. *The American Journal of Drug and Alcohol Abuse, 33*(5), 717–726.

Lavric, A., Pizzagalli, D. A., & Forstmeier, S. (2004). When 'go' and 'No-Go' are equally frequent: ERP components and cortical tomography. *The European Journal of Neuroscience, 20*(9), 2483–2488.

Lawrence, A. J., Luty, J., Bogdan, N. A., Sahakian, B. J., & Clark, L. (2009). Impulsivity and response inhibition in alcohol dependence and problem gambling. *Psychopharmacology, 207*(1), 163–172.

Lhermitte, F. (1986). Human autonomy and the frontal lobes. Part II: Patient behavior in complex and social situations: The 'environmental dependency syndrome'. *Annals of Neurology, 19*(4), 335–343.

Li, C. R., Huang, C., Yan, P., Bhagwagar, Z., Milivojevic, V., & Sinha, R. (2007). Neural correlates of impulse control during stop signal inhibition in cocaine-dependent men. *Neuropsychopharmacology, 33*(8), 1798–1806.

Li, C. R., Luo, X., Yan, P., Bergquist, K., & Sinha, R. (2009). Altered impulse control in alcohol dependence: neural measures of stop signal performance. *Alcoholism: Clinical and Experimental Research, 33*(4), 740–750.

Li, C. R., Milivojevic, V., Kemp, K., Hong, K., & Sinha, R. (2006). Performance monitoring and stop signal inhibition in abstinent patients with cocaine dependence. *Drug and Alcohol Dependence, 85*(3), 205–212.

Liston, C., Watts, R., Tottenham, N., Davidson, M. C., Niogi, S., Ulug, A. M., & Casey, B. J. (2006). Frontostriatal microstructure modulates efficient recruitment of cognitive control. *Cerebral Cortex (New York, N.Y.: 1991), 16*(4), 553–560.

Logan, G. D., & Cowan, W. B. (1984). On the ability to inhibit thought and action: A theory of an act of control. *Psychological Review, 91*(3), 295–327.

Luijten, M., Littel, M., & Franken, I. H. A. (2011). Deficits in inhibitory control in smokers during a Go/No-Go task: An investigation using event-related brain potentials. *PLoS ONE, 6*(4), e18898.

Luijten, M., Machielsen, M. W. J., Veltman, D. J., Hester, R., de Haan, L., & Franken, I. H. A. (2014). Systematic review of ERP and fMRI studies investigating inhibitory control and error processing in people with substance dependence and behavioural addictions. *Journal of Psychiatry & Neuroscience: JPN, 39*(3), 149–169.

Luijten, M., O'Connor, D. A., Rossiter, S., Franken, I. H. A., & Hester, R. (2013). Effects of reward and punishment on brain activations associated with inhibitory control in cigarette smokers. *Addiction, 108*(11), 1969–1978.

Luria, A. R., & Pribram, K. H. (1973). The frontal lobes and the regulation of behavior. In K. H. Pribram & A. R. Luria (Eds.), *Psychophysiology of the frontal lobes* (pp. 3–26). New York: Academic Press.

Madsen, K. S., Baaré, W. F. C., Vestergaard, M., Skimminge, A., Ejersbo, L. R., Ramsøy, T. Z., ... Jernigan, T. L. (2010). Response inhibition is associated with white matter microstructure in children. *Neuropsychologia, 48*(4), 854–862.

Mansouri, F. A., Tanaka, K., & Buckley, M. J. (2009). Conflict-induced behavioural adjustment: A clue to the executive functions of the prefrontal cortex. *Nature Reviews Neuroscience, 10*(2), 141–152.

McAlonan, G. M., Cheung, V., Chua, S. E., Oosterlaan, J., Hung, S., Tang, C., ... Leung, P. W. L. (2009). Age-related GMV correlates of response inhibition and shifting in attention-deficit hyperactivity disorder. *The British Journal of Psychiatry: The Journal of Mental Science, 194*(2), 123–129.

McGue, M., Zhang, Y., Miller, M. B., Basu, S., Vrieze, S., Hicks, B., … Iacono, W. G. (2013). A genome-wide association study of behavioral disinhibition. *Behavior Genetics, 43*(5), 363–373.

Moreno, M., Estevez, A. F., Zaldivar, F., Montes, J. M. G., Gutiérrez-Ferre, V. E., Esteban, L., … Flores, P. (2012). Impulsivity differences in recreational cannabis users and binge drinkers in a university population. *Drug and Alcohol Dependence, 124*(3), 355–362.

Morie, K. P., De Sanctis, P., Garavan, H., & Foxe, J. J. (2014). Executive dysfunction and reward dysregulation: A high-density electrical mapping study in cocaine abusers. *Neuropharmacology, 85*, 397–407.

Murphy, P., & Garavan, H. (2011). Cognitive predictors of problem drinking and AUDIT scores among college students. *Drug and Alcohol Dependence, 115*(1–2), 94–100.

Nachev, P., Wydell, H., O'Neill, K., Husain, M., & Kennard, C. (2007). The role of the pre-supplementary motor area in the control of action. *NeuroImage, 36*, T155–T163.

Nandam, L. S., Hester, R., Wagner, J., Cummins, T. D. R., Garner, K., Dean, A. J., … Bellgrove, M. A. (2011). Methylphenidate but not atomoxetine or citalopram modulates inhibitory control and response time variability. *Biological Psychiatry, 69*(9), 902–904.

Nederkoorn, C., Baltus, M., Guerrieri, R., & Wiers, R. W. (2009). Heavy drinking is associated with deficient response inhibition in women but not in men. *Pharmacology Biochemistry and Behavior, 93*(3), 331–336.

Nestor, L., McCabe, E., Jones, J., Clancy, L., & Garavan, H. (2011). Differences in "bottom-up" and "top-down" neural activity in current and former cigarette smokers: Evidence for neural substrates which may promote nicotine abstinence through increased cognitive control. *NeuroImage, 56*(4), 2258–2275.

Neubert, F.-X., Mars, R. B., Buch, E. R., Olivier, E., & Rushworth, M. F. S. (2010). Cortical and subcortical interactions during action reprogramming and their related white matter pathways. *Proceedings of the National Academy of Sciences of the United States of America, 107*(30), 13240–13245.

Owen, A. M., Evans, A. C., & Petrides, M. (1996). Evidence for a two-stage model of spatial working memory processing within the lateral frontal cortex: A positron emission tomography study. *Cerebral Cortex, 6*(1), 31–38.

Paaver, M., Nordquist, N., Parik, J., Harro, M., Oreland, L., & Harro, J. (2007). Platelet MAO activity and the 5-HTT gene promoter polymorphism are associated with impulsivity and cognitive style in visual information processing. *Psychopharmacology, 194*(4), 545–554.

Papachristou, H., Nederkoorn, C., Havermans, R., Bongers, P., Beunen, S., & Jansen, A. (2013). Higher levels of trait impulsiveness and a less effective response inhibition are linked to more intense cue-elicited craving for alcohol in alcohol-dependent patients. *Psychopharmacology, 228*(4), 641–649.

Pfefferbaum, A., Rosenbloom, M., & Ford, J. M. (1987). Late event-related potential changes in alcoholics. *Alcohol (Fayetteville, N.Y.), 4*(4), 275–281.

Quednow, B. B., Kühn, K.-U., Hoppe, C., Westheide, J., Maier, W., Daum, I., & Wagner, M. (2006). Elevated impulsivity and impaired decision-making cognition in heavy users of MDMA ("Ecstasy"). *Psychopharmacology, 189*(4), 517–530.

Rae, C. L., Hughes, L. E., Anderson, M. C., & Rowe, J. B. (2015). The prefrontal cortex achieves inhibitory control by facilitating subcortical motor pathway connectivity. *Journal of Neuroscience, 35*(2), 786–794.

Reynolds, B., Ortengren, A., Richards, J. B., & de Wit, H. (2006). Dimensions of impulsive behavior: Personality and behavioral measures. *Personality and Individual Differences, 40*(2), 305–315.

Robbins, T. W. (2005). Controlling stress: How the brain protects itself from depression. *Nature Neuroscience, 8*(3), 261–262.

Roberts, C. A., Fairclough, S., Fisk, J. E., Tames, F. T., & Montgomery, C. (2013). Electrophysiological indices of response inhibition in human polydrug users. *Journal of Psychopharmacology (Oxford, England), 27*(9), 779–789.

Roberts, G. M. P., & Garavan, H. (2010). Evidence of increased activation underlying cognitive control in ecstasy and cannabis users. *NeuroImage, 52*(2), 429–435.

Roberts, G. M. P., & Garavan, H. (2013). Neural mechanisms underlying ecstasy-related attentional bias. *Psychiatry Research: Neuroimaging, 213*(2), 122–132.

Rubio, G., Jiménez, M., Rodríguez-Jiménez, R., Martínez, I., Iribarren, M. M., Jiménez-Arriero, M. A., ... Avila, C. (2007). Varieties of impulsivity in males with alcohol dependence: The role of Cluster-B personality disorder. *Alcoholism, Clinical and Experimental Research*, *31*(11), 1826–1832.

Salmon, E., Linden, M. V. der, Collette, F., Delfiore, G., Maquet, P., Degueldre, C., ... Franck, G. (1996). Regional brain activity during working memory tasks. *Brain*, *119*(5), 1617–1625.

Scheurich, A. (2005). Neuropsychological functioning and alcohol dependence. *Current Opinion in Psychiatry*, *18*(3), 319–323.

Schmaal, L., Joos, L., Koeleman, M., Veltman, D. J., van den Brink, W., & Goudriaan, A. E. (2013). Effects of modafinil on neural correlates of response inhibition in alcohol-dependent patients. *Biological Psychiatry*, *73*(3), 211–218.

Schmidt, R., Leventhal, D. K., Mallet, N., Chen, F., & Berke, J. D. (2013). Canceling actions involves a race between basal ganglia pathways. *Nature Neuroscience*, *16*(8), 1118–1124.

Schweinsburg, A. D., Nagel, B. J., Schweinsburg, B. C., Park, A., Theilmann, R. J., & Tapert, S. F. (2008). Abstinent adolescent marijuana users show altered fMRI response during spatial working memory. *Psychiatry Research: Neuroimaging*, *163*(1), 40–51.

Schweinsburg, A. D., Schweinsburg, B. C., Medina, K. L., McQueeny, T., Brown, S. A., & Tapert, S. F. (2010). The Influence of recency of use on fMRI response during spatial working memory in adolescent marijuana users. *Journal of Psychoactive Drugs*, *42*(3), 401–412.

Seghete, K. L. M., Herting, M. M., & Nagel, B. J. (2013). White matter microstructure correlates of inhibition and task-switching in adolescents. *Brain Research*, *1527*, 15–28.

Smith, J. L., & Mattick, R. P. (2013). Evidence of deficits in behavioural inhibition and performance monitoring in young female heavy drinkers. *Drug and Alcohol Dependence*, *133*(2), 398–404.

Smith, J. L., Mattick, R. P., Jamadar, S. D., & Iredale, J. M. (2014). Deficits in behavioural inhibition in substance abuse and addiction: A meta-analysis. *Drug and Alcohol Dependence*, *145*, 1–33.

Sohn, M.-H., Ursu, S., Anderson, J. R., Stenger, V. A., & Carter, C. S. (2000). The role of prefrontal cortex and posterior parietal cortex in task switching. *Proceedings of the National Academy of Sciences of the United States of America*, *97*(24), 13448–13453.

Sokhadze, E., Stewart, C., Hollifield, M., & Tasman, A. (2008). Event-related potential study of executive dysfunctions in a speeded reaction task in cocaine addiction. *Journal of Neurotherapy*, *12*(4), 185–204.

Spronk, D. B., van Wel, J. H. P., Ramaekers, J. G., & Verkes, R. J. (2013). Characterizing the cognitive effects of cocaine: A comprehensive review. *Neuroscience & Biobehavioral Reviews*, *37*(8), 1838–1859.

Sridharan, A., Willette, A. A., Bendlin, B. B., Alexander, A. L., Coe, C. L., Voytko, M. L., ... Johnson, S. C. (2012). Brain volumetric and microstructural correlates of executive and motor performance in aged rhesus monkeys. *Frontiers in Aging Neuroscience*, *4*, 31.

Stuss, D. T., & Benson, D. F. (1987). The frontal lobes and control of cognition and memory. In E. Perecman (Ed.), *The frontal lobes revisited* (pp. 141–158). New York: Erlbaum.

Sullivan, E. V. (2003). Compromised pontocerebellar and cerebellothalamocortical systems: Speculations on their contributions to cognitive and motor impairment in nonamnesic alcoholism. *Alcoholism: Clinical and Experimental Research*, *27*(9), 1409–1419.

Swann, N. C., Cai, W., Conner, C. R., Pieters, T. A., Claffey, M. P., George, J. S., ... Tandon, N. (2012). Roles for the pre-supplementary motor area and the right inferior frontal gyrus in stopping action: Electrophysiological responses and functional and structural connectivity. *NeuroImage*, *59*(3), 2860–2870.

Swann, N. C., Tandon, N., Canolty, R., Ellmore, T. M., McEvoy, L. K., Dreyer, S., ... Aron, A. R. (2009). Intracranial EEG reveals a time- and frequency-specific role for the right inferior frontal gyrus and primary motor cortex in stopping initiated responses. *The Journal of Neuroscience*, *29*(40), 12675–12685.

Tabibnia, G., Monterosso, J. R., Baicy, K., Aron, A. R., Poldrack, R. A., Chakrapani, S., ... London, E. D. (2011a). Different forms of self-control share a neurocognitive substrate. *The Journal of Neuroscience*, *31*(13), 4805–4810.

Tabibnia, G., Monterosso, J. R., Baicy, K., Aron, A. R., Poldrack, R. A., Chakrapani, S., ... London, E. D. (2011b). Different forms of self-control share a neurocognitive substrate. *The Journal of Neuroscience: The Official Journal of the Society for Neuroscience*, *31*(13), 4805–4810.

Tapert, S. F., Schweinsburg, A. D., Drummond, S. P. A., Paulus, M. P., Brown, S. A., Yang, T. T., & Frank, L. R. (2007). Functional MRI of inhibitory processing in abstinent adolescent marijuana users. *Psychopharmacology*, *194*(2), 173–183.

Theunissen, E. L., Kauert, G. F., Toennes, S. W., Moeller, M. R., Sambeth, A., Blanchard, M. M., & Ramaekers, J. G. (2012). Neurophysiological functioning of occasional and heavy cannabis users during THC intoxication. *Psychopharmacology*, *220*(2), 341–350.

Torregrossa, M. M., Quinn, J. J., & Taylor, J. R. (2008). Impulsivity, compulsivity, and habit: The role of orbitofrontal cortex revisited. *Biological Psychiatry*, *63*(3), 253–255.

Townshend, J. M., & Duka, T. (2005). Binge drinking, cognitive performance and mood in a population of young social drinkers. *Alcoholism: Clinical and Experimental Research*, *29*(3), 317–325.

Treit, S., Chen, Z., Rasmussen, C., & Beaulieu, C. (2014). White matter correlates of cognitive inhibition during development: A diffusion tensor imaging study. *Neuroscience*, *276*, 87–97.

Volkow, N. D., Fowler, J. S., Wang, G.-J., Telang, F., Logan, J., Jayne, M., ... Swanson, J. M. (2010). Cognitive control of drug craving inhibits brain reward regions in cocaine abusers. *NeuroImage*, *49*(3), 2536–2543.

Vonmoos, M., Hulka, L. M., Preller, K. H., Jenni, D., Schulz, C., Baumgartner, M. R., & Quednow, B. B. (2013). Differences in self-reported and behavioral measures of impulsivity in recreational and dependent cocaine users. *Drug and Alcohol Dependence*, *133*(1), 61–70.

Walderhaug, E., Herman, A. I., Magnusson, A., Morgan, M. J., & Landrø, N. I. (2010). The short (S) allele of the serotonin transporter polymorphism and acute tryptophan depletion both increase impulsivity in men. *Neuroscience Letters*, *473*(3), 208–211.

Weafer, J., & de Wit, H. (2013). Inattention, impulsive action, and subjective response to D-amphetamine. *Drug and Alcohol Dependence*, *133*(1), 127–133.

Whelan, R., Conrod, P. J., Poline, J.-B., Lourdusamy, A., Banaschewski, T., Barker, G. J., ... IMAGEN Consortium (2012). Adolescent impulsivity phenotypes characterized by distinct brain networks. *Nature Neuroscience*, *15*(6), 920–925.

33

Cognitive Control in Attention Deficit Hyperactivity Disorder

Patrick de Zeeuw and Sarah Durston

Introduction

It may seem redundant to emphasise the importance of cognitive control in understanding attention deficit hyperactivity disorder (ADHD). After all, deficits in self-control are a defining feature of the disorder that is described by two clusters of behavioural symptoms: problems in directing and sustaining attention and elevated distractibility on the one hand, and motor hyperactivity and impulsive behaviour on the other hand (American Psychiatric Association, 2013). For a thorough description of the clinical presentation and developmental course of the disorder, the reader is referred to recent overview papers (De Zeeuw & Durston, in press; Spencer, Biederman, & Mick, 2007). Neurobiological and neuropsychological theories of ADHD date back to the early 1900s and have been described in other reviews (Lange, Reichl, Lange, Tucha, and Tucha, 2010). Arguably, deficits in cognitive control are the—or at least one of the—longest-standing neurobiological theories of ADHD. However, the casual reader of the cognitive literature on ADHD is likely to be overwhelmed by the interrelated terminology used to describe the cognitive and neural constituents of behavioural problems in self-control. This underscores that the relationship between behavioural impulsiveness and cognitive and brain processes is not clear-cut. The purpose of this chapter is not to provide an exhaustive review, but rather to introduce the major theoretical perspectives and accompanying research that aim to explain ADHD symptoms from deficits in cognitive control, and to provide an overview of current topics of debate.

Theoretical Considerations and Models of ADHD

Terminology

Impulsiveness, impulse control, behavioural inhibition, executive functioning, and cognitive control are all terms that are often used near-interchangeably in the clinical, neuropsychological, and neuroscience literature on ADHD. This is problematic as it confuses behavioural symptoms and the underlying cognitive processes, and thereby confounds cause and consequence. Behaviourally, impulsivity refers to (a) acting before thinking, (b) acting in a way that provides immediate reinforcement, even in the face of long-term adverse effects, or (c) the inability to withhold (i.e., stop) prepotent or ongoing responses (Barkley, 1997;

The Wiley Handbook of Cognitive Control, First Edition. Edited by Tobias Egner.

Schachar, Tannock, & Logan, 1993). Here, ongoing responses are behaviours that are already being executed and require interruption, and prepotent responses are actions that have previously been useful or reinforced, but that are not useful in the current situation owing to changes in the context (Barkley, 1997). This description illustrates that deficient cognitive control cannot simply be equated to impulsivity: Deficits in reward processing or decision making may also underlie impulsivity as defined under (a) and (b) (also see the section on multiple-pathway theories of ADHD below). However, the majority of research on cognitive control in ADHD utilises the last definition of impulsivity (c): stopping prepotent or ongoing responses. (For a thorough review of the cognitive psychology literature on response inhibition, see Chapter 6 by Verbruggen & Logan in this volume). The terms behavioural inhibition and inhibitory control are often used to refer to the construct underlying this definition (Aron, 2007; Barkley, 1997; Schachar et al., 1993). In contrast, cognitive control is typically taken to refer to a broader set of processes, including error monitoring (reviewed in Chapter 17 by Ullsperger in this volume), conflict monitoring and interference control (see Chapter 5 by Bugg, Chapter 4 by Egner, and Chapter 10 by Kool, Shenhav, and Botvinick in this volume), and is tightly tied to the construct of executive functioning (Aron, 2007). In ADHD, deficits in this broader defined construct have commonly been found (Willcutt, Doyle, Nigg, Faraone, & Pennington, 2005). However, cognitive research in ADHD has often operationalised cognitive control as behavioural inhibition (definition [c] above). Therefore, this chapter mainly discusses studies that have focused on that aspect of cognitive control.

Cognitive control may appear to be particularly relevant to those patients with ADHD who have hyperactive/impulsive behaviour (the hyperactive/impulsive and combined presentations of ADHD), in contrast to those with a more inattentive presentation. It is important to note that the vast majority of cognitive studies in ADHD have focused on the hyperactive/impulsive and/or combined presentations.

Theoretical Models of Cognitive Control in ADHD

Theoretical accounts of cognitive control in ADHD vary widely, with some aiming to explain changes in brain processes related to cognitive control, whereas others attempt to show how deficits in cognitive control can explain the behavioural expression. Cognitive control was perhaps the first cognitive deficit to be suggested to underlie ADHD: George Still, who is often credited with the first modern description of the syndrome, wrote that a deficiency in volitional inhibition of behaviour and 'moral control' may underlie the disorder (Pennington & Ozonoff, 1996; Still, 1902). Empirical work on inhibitory control in ADHD (definition c) originated in the 1970s, when neuropsychological studies showed that attention deficits were not the only cognitive deficits in ADHD, and that they did not provide a clear explanation of its behavioural expression (Barkley, 1997; Sergeant, Oosterlaan, & van der Meere, 1999). Seminal work by Douglas (1972) and Quay (1997) drew on observations of individuals with ADHD performing tasks taxing 'frontal lobe function' to first propose that the inhibition of inappropriate responses may be a core deficit. In 1997, Barkley described a unifying theory of ADHD, where behavioural inhibition was placed at the centre of cognitive dysfunction in ADHD. This has arguably been a major impetus for cognitive and neuroimaging studies of ADHD.

One particularly appealing aspect of Barkley's model is that it integrates the neuropsychological and behavioural levels. The model places behavioural inhibition at the top of a hierarchy of self-regulatory behaviour. Its purpose is to inhibit or stop prepotent or ongoing behavioural output from interfering with normal functioning, thereby permitting appropriate development of working memory, internalisation of speech and (ultimately) self-regulation of motivation, arousal, and motor control (Barkley, 1997; for a review of the construct of self-regulation vis-à-vis cognitive control, see Chapter 25 by Davisson & Hoyle in this volume). As such, the model predicts that behaviour in ADHD is context dependent and strongly guided by external events or reinforcers, rather than by internal representations of situational context, rules, or behavioural repertoire. Support for Barkley's model comes from research that shows children with ADHD have trouble adapting their behaviour in response to feedback or after they have made an error (e.g., post-error slowing, adaptation of speed–accuracy balance; Mulder et al., 2010; Sergeant, 2000; van Meel, Oosterlaan, Heslenfeld, & Sergeant, 2005).

These theoretical contributions led to a multitude of studies on the role of cognitive control in ADHD in the 1990s and 2000s. However, this empirical research did not support the hypothesis that problems with cognitive control (or executive function more broadly) were the central deficit in ADHD (Nigg, 2001; Willcutt et al., 2005). First, it was found that deficits were not specific to ADHD, as similar deficits were also found in autism and learning disorders (Geurts, Verte, Oosterlaan, Roeyers, & Sergeant, 2004; Rommelse, Geurts, Franke, Buitelaar, & Hartman, 2011). Second, meta-analyses concluded that deficits in cognitive control were not universal in ADHD, as deficits in executive functioning (a broader category) did not affect more than half of children with ADHD (Nigg, Willcutt, Doyle, & Sonuga-Barke, 2005) and could therefore not be central to the disorder.

Nevertheless, this large body of research did show that, although deficits in cognitive control are not universal in ADHD, they are perhaps the most robust finding at a group level. Indeed, later theories of ADHD have incorporated these deficits within broader neuropsychological frameworks (Sergeant, 2005; Sonuga-Barke, 2002) and neural-system-based accounts (Durston, Van Belle, & De Zeeuw, 2011; Makris, Biederman, Monuteaux, & Seidman, 2009; Nigg & Casey, 2005; Rubia, 2011). The unifying characteristic in most of these multiple-pathway accounts is that deficits in multiple, partially separable brain systems may lead to symptoms of ADHD. As such, they denounce the concept of a core deficit or final common pathway to the disorder. Most of these accounts include networks supporting temporal and reinforcement processing and/or salience/attention networks in addition to cognitive control networks.

Neuropsychological Studies of Cognitive Control in ADHD

Neuropsychological studies of ADHD initially focused on attention. When they found that attention deficits were not a core deficit, their focus shifted to 'frontal' or more executive functions (Douglas, 1972; Pennington & Ozonoff, 1996; Sergeant et al., 1999), including cognitive control (Pennington & Ozonoff, 1996; Sergeant, Geurts, & Oosterlaan, 2002; Willcutt et al., 2005). Cognitive control has often been operationalised as behavioural inhibition and has been studied using tasks such as go/no-go (GNG) paradigms, and the stop-signal task (SST; Alderson, Rapport, & Kofler, 2007; Huang-Pollock, Karalunas, Tam, & Moore, 2012; Willcutt et al., 2005). A GNG task is a variation on a continuous performance task (Huang-Pollock et al., 2012), where a subject responds to a continuous stream of go-stimuli (go trials) with a button press but has to refrain from responding when a no-go target appears (no-go trials). In order to induce prepotency of the go-response, go trials typically outnumber no-go trials. In the SST, a subject is asked to respond as quickly as possible to a stimulus but to refrain from responding when a stop-signal (visual or auditory) follows the target. As such, the SST measures inhibition of an ongoing response. By varying the delay between the onset of the cue and the stop-signal, the duration of the stop process can be estimated, a measure that is known as the stop-signal reaction time (SSRT; see Chapter 6 by Verbruggen & Logan in this volume).

Many studies have reported worse performance on these tasks for both children and adults with ADHD compared to controls, with more commission errors (i.e., responding to a no-go trial) on GNG tasks, and longer SSRTs (Frazier, Demaree, & Youngstrom, 2004; Hervey, Epstein, & Curry, 2004; Willcutt et al., 2005). However, it is not only inhibition measures that are affected, as other performance measures are often also worse, particularly for child samples, with greater numbers of omission errors (i.e., not responding to targets), longer response times, and greater variability in response times. In fact, it has been suggested that this latter aspect may be particularly useful in differentiating between subjects with and without ADHD (Castellanos, Sonuga-Barke, Milham, & Tannock, 2006; de Zeeuw et al., 2008; Klein, Wendling, Huettner, Ruder, & Peper, 2006).

Measures of executive functions and cognitive control, such as behavioural inhibition, can be assessed from an early age and improve greatly during (early) development. Most aspects reach adult, or near-adult performance, by mid-adolescence (Davidson, Amso, Anderson, & Diamond, 2006). This suggests that deficits in cognitive control in ADHD may reflect a developmental delay, analogous to similar suggestions for brain development (Shaw et al., 2007). If this were the case, fewer deficits in cognitive control would be expected in adult ADHD samples. However, systematic reviews report that deficits in cognitive control are also found in adulthood, both for behavioural inhibition and more complex forms, such as interference control (Boonstra, Oosterlaan, Sergeant, & Buitelaar, 2005; van Lieshout, Luman, Buitelaar, Rommelse, & Oosterlaan, 2013). However, research also shows there may be differential developmental profiles between these functions, with many improving with age, and even normalising for some individuals, but with deficits in others continuing into adulthood (Balint et al., 2009; Bramham et al., 2012; van Lieshout et al., 2013).

Neuroimaging of Cognitive Control in ADHD

Cognitive control, behavioural inhibition in particular, is supported by prefrontal-striatal-thalamic brain circuitry, mainly involving connections between the inferior frontal (IFC) and dorsolateral prefrontal cortex (DLPFC), and pre/supplementary motor cortices (PM/SMA) with striatum (STR, dorsal caudate), via the external globus pallidus (GPe), with connections via the subthalamic nucleus (STN), internal globus pallidus (GPi) and thalamus (TH), and back to frontal cortex (Arnsten & Rubia, 2012; Aron, 2007, 2011; Chambers, Garavan, & Bellgrove, 2009; Chapter 16 by Forstmann & Alkemade in this volume). Neuroimaging studies have reported changes, typically reductions, in activation in this network in ADHD (Cortese et al., 2012; Hart, Radua, Nakao, Mataix-Cols, & Rubia, 2013; McCarthy, Skokauskas, & Frodl, 2014). The role of the anterior cingulate cortex (ACC) is also relevant to ADHD and lies in performance monitoring, by detecting errors and incongruencies between outcome and expectations (MacDonald, Cohen, Stenger, & Carter, 2000; Shenhav, Botvinick, & Cohen, 2013; Chapter 10 by Kool, Shenhav, and Botvinick in this volume). As such, the ACC is particularly relevant to tasks with more complex rules and changing contextual cues, such as varying GNG ratios or interference conditions (for a thorough review of models of ACC function, see Chapter 15 by Brown in this volume).

A number of recent meta-analyses have documented widespread differences in brain function in ADHD, across a range of tasks, as well as in rest (Cortese et al., 2012; Hart et al., 2013; McCarthy et al., 2014). Functional neuroimaging studies of cognitive control have typically employed GNG or SST paradigms. Hart et al. (2013) performed a meta-analysis on studies using whole-brain analyses of activation related to response inhibition and found decreased brain activation in ADHD throughout the cognitive control network, including the right IFG, basal ganglia (left caudate mainly), right SMA and ACC, and right thalamus

(Hart et al., 2013). They also reported on interference inhibition (from studies using Stroop, Simon, and Erikson flanker tasks), where they reported largely overlapping results, as well as decreases in activation in the parietal lobe, related to ADHD. A second meta-analysis applied slightly different inclusion criteria and analysed stop and GNG paradigms separately (McCarthy et al., 2014). They reported reduced activity in the bilateral IFG, and right middle and superior frontal gyrus (including the DLPFC) for the SST and in the left medial frontal gyrus and right caudate for the GNG task (McCarthy et al., 2014).

In all, these meta-analyses clearly point to reduced activation of the cognitive control network in ADHD, particularly in prefrontal and striatal regions. Interestingly, these are also regions where structural neuroimaging studies have found smaller brain volumes relatively consistently (Durston, de Zeeuw, & Staal, 2009; Frodl & Skokauskas, 2012; Nakao, Radua, Rubia, & Mataix-Cols, 2011; Valera, Faraone, Murray, & Seidman, 2007). Although a detailed discussion of these anatomical findings is beyond the scope of this chapter, we do note that coinciding decreases in volume and activation raise the possibility of a (causal) interaction. However, studies combining structural and functional imaging that might address this are still scarce.

Studies investigating structural and functional connectivity of the cognitive control network have often found differences between ADHD and control groups. However, this literature is not yet conclusive, given methodological changes over the past years and differences between studies in approach. Diffusion tensor imaging studies have often reported decreased structural prefrontal–striatal connectivity (Ashtari et al., 2005; de Zeeuw, Mandl, Hulshoff Pol, van Engeland, & Durston, 2012; Konrad et al., 2010; Onnink et al., 2015; Pavuluri et al., 2009), although others have reported increases (Chaim et al., 2014; Davenport, Karatekin, White, & Lim, 2010; Lawrence et al., 2013; Silk, Vance, Rinehart, Bradshaw, & Cunnington, 2009; Tamm, Barnea-Goraly, & Reiss, 2012), or sometimes both increases and decreases in these regions (Lei et al., 2014). A meta-analysis concluded that there were decreases in structural connectivity in the cortico-striatal network (van Ewijk, Heslenfeld, Zwiers, Buitelaar, & Oosterlaan, 2012), but the field is still relatively young and the level of consistency is not yet high.

Studies investigating functional connectivity have used both task and resting data. Resting state studies are not yet conclusive, although those that are methodologically strongest have reported reduced connectivity in the cognitive control network in ADHD (Posner, Park, & Wang, 2014). Studies investigating functional connectivity during the execution of a relevant task have generally reported decreased functional connectivity between key regions of the network. However, it should be noted that other studies have suggested that decreases in functional connectivity may be widespread in ADHD (Posner et al., 2014), calling into question the specificity of the networks supporting cognitive control in this regard.

The Role of Development

The vast majority of studies have investigated children, with less literature available on adolescents and adults with ADHD. Imaging studies on cognitive control in adults with ADHD have typically also reported decreased activation in the cognitive control network (Banich et al., 2009; Bush et al., 1999; Burgess et al., 2010; Morein-Zamir et al., 2014) and changes (mostly decreases) in structural and functional connectivity (Casey et al., 2007; Kessler, Angstadt, Welsh, & Sripada, 2014; McCarthy et al., 2013; Schulz et al., 2014). The meta-analyses on functional imaging of cognitive control mentioned earlier also reported on age effects. Hart et al. (2013) split their sample of combined GNG and SST studies into child and adult studies. They reported that decreases in activation of the SMA, caudate, and putamen were only found for children and that adults showed a greater decrease in activation of the right IFG and right thalamus. Although it is tempting to interpret these findings in the context of developmental changes, they may also reflect the differentiation and specialisation characteristic of typical network development: Developmental studies have shown that cognitive functions tend to be supported by diffuse networks in childhood that specialise and become more integrated during development (Durston & Casey, 2006; Fair et al., 2007). As such, qualitative differences between children and adults with ADHD may reflect more typical developmental effects rather than anything specific to maturation in ADHD.

The meta-analysis of McCarthy and colleagues (2014) analysed stop-signal and GNG studies separately and found different developmental results: For stop-signal studies, they found that older children with ADHD showed more right IFG activation than younger children. Although the authors did not include adult studies in this analysis and warn against overinterpretation, their result appears to contrast with those of Hart and colleagues (2013). For GNG studies, they reported that—in children—increasing age was related to less pronounced reductions of activity in the left inferior temporal gyrus and left IFG, whereas in adults, older age was associated with increased activity in the bilateral lingual gyrus and left fusiform gyrus, and reduced activation in the left thalamus (McCarthy et al., 2014). It is not entirely clear what led to the differences between these two meta-analyses, considering that there is significant overlap in the studies included. There are, of course, methodological differences between the two: Hart and colleagues (2013) collapsed across cognitive control tasks, whereas McCarthy and colleagues (2014), analysed stop-signal and GNG tasks separately. The developmental analyses were also conducted differently and were based entirely on cross-sectional data. In all, the best conclusion is perhaps that the developmental course of brain activation related to cognitive control in ADHD is not yet clear, and does not show a pattern that is robust in the face of methodological differences between studies.

Clinical Presentation

One final issue that has often been overlooked in the literature is the role of the clinical presentation of ADHD. McCarthy and colleagues attempted to address this in their meta-analysis (2014). Their results suggest that there may be differences between studies based on the composition of the sample (McCarthy et al., 2014). However, the authors rightly concluded that the number of participants with inattentive and hyperactive-impulsive presentation was too low across studies to lend much credence to these results yet.

Endophenotypes: Is There a Role for Cognitive Control in Explaining Heritability in ADHD?

Cognitive control and associated differences in brain activation form the endophenotype that has perhaps been most often put forward for ADHD. An endophenotype is a (biological) phenotype intermediate between phenomenology (behavioural symptoms) and genetic variation associated with the disorder (Durston et al., 2009). These phenotypes are theorised to bear a closer relationship to genetic variance that is associated with the disorder than behavioural symptoms.

One of the major requirements for an endophenotype is that it shows familial clustering with the disorder and co-segregates with it (Durston et al., 2009). In neuropsychological studies of cognitive control, such familial segregation has indeed been shown with unaffected siblings of individuals with ADHD showing similar performance to probands across cognitive control tasks (Kuntsi et al., 2010; Pironti et al., 2014; Rommelse, Altink, et al., 2008; Slaats-Willemse,

Swaab-Barneveld, de Sonneville, van der Meulen, & Buitelaar, 2003). For example, Rommelse and colleagues (Rommelse, Altink, et al., 2008) investigated a large sample of children with ADHD, their affected and unaffected siblings and controls on a range of tasks, including the SST. They showed a gradient of performance with the best performance in controls, followed by unaffected siblings, affected siblings, with the worst performance in probands (Rommelse, Altink, et al., 2008).

Neuroimaging studies have investigated brain activation and functional connectivity during cognitive control tasks using similar sibling designs (Durston, Mulder, Casey, Ziermans, & van Engeland, 2006; Mulder et al., 2008; Mulder, van Belle, van Engeland, & Durston, 2011; van Rooij et al., 2015). In general, these studies have found decreased activation of the cognitive control network in children with ADHD and (to varying extent) in their unaffected siblings. An innovative study compared parent–child dyads with and without ADHD with functional imaging and functional and structural connectivity in frontostriatal networks (Casey et al., 2007). This study found a tight association between frontostriatal structural connectivity within parent–child dyads with ADHD, frontostriatal activity during a GNG task and performance on the task. This suggests that both dysfunction of the frontostriatal system and associated cognitive control deficits may be heritable (Casey et al., 2007).

Other studies have investigated whether these findings can be extended to molecular genetics. Some have investigated whether (often dopaminergic) candidate genes could explain neuropsychological performance, or related brain activity, in ADHD, but results have been mixed (Durston, 2010; Kebir & Joober, 2011). Other studies have tried to use neuropsychological test scores to identify new genetic susceptibility loci for ADHD (Rommelse, Arias-Vasquez, et al., 2008). Overall, this literature is not yet consistent, and this has led some authors to question the validity of the endophenotype model (Coghill, Hayward, Rhodes, Grimmer, & Matthews, 2014). There have been other criticisms of this model, including the observation that cognitive endophenotypes in ADHD may not be stable over development (Nikolas & Nigg, 2015) and the fact that stronger genetic control of endophenotypes has proved hard to show (Flint & Munafo, 2007). Some authors have even suggested that cognitive deficits and their endophenotypic properties may be epiphenomena, with their causal role in ADHD being uncertain (Coghill, Hayward, et al., 2014; van Lieshout et al., 2013). In addition, some authors have questioned whether cognitive control is the best endophenotype for ADHD, as deficits in response speed and accuracy have also been shown to occur across tasks in ADHD (Rommelse et al., 2007). Moreover, nearly all cognitive control tasks are serial (two) choice reaction time tasks that require efficient information processing and a balance between speed and accuracy. Some recent studies using different task types have suggested that such basic processes may be affected in ADHD (Huang-Pollock et al., 2012; Metin et al., 2013; Mulder et al., 2010), bringing into question the central role and/or specificity of cognitive control deficits in ADHD.

Other Issues in Understanding Deficits in Cognitive Control in ADHD

Effects of Medication on Cognitive Control in ADHD

Psychostimulant medication (methylphenidate, amphetamine) is a common treatment option for moderate to severe cases of ADHD and is often effective (Swanson, Baler, & Volkow, 2011; Taylor et al., 2004). The therapeutic efficacy of stimulants is most likely due to their properties as dopamine agonists in the striatum and parts of the PFC, where they increase availability of dopamine in the synapse by blocking the dopamine reuptake transporter

(Swanson et al., 2011). This seems to imply that ADHD is somehow related to a hypodopaminergic state. However, it should be noted that the mechanism leading to this state is not fully understood, nor has it been universally demonstrated (Swanson et al., 2011). Nonetheless, dopaminergic neurotransmission is ubiquitous in frontostriatal brain circuits (Chapter 16 by Forstmann & Alkemade and Chapter 19 by Robbins & Kehagia in this volume), which are central to cognitive control and ADHD. Therefore, it is relevant whether stimulant medication has an ameliorating effect on cognitive control deficits in ADHD.

Recently, Coghill and colleagues (2014) published a meta-analysis of studies of the acute effects of methylphenidate on a variety of cognitive tasks, including behavioural response inhibition (GNG, SST) and more complex forms of cognitive control, such as interference control (i.e., flanker and Stroop tasks). They found clear superiority of active methylphenidate over placebo across a wide range of measures (reaction time, reaction time variability, and accuracy on no-go/stop trials; Coghill, Seth, et al., 2014). These effects were not limited to tasks tapping cognitive control, but rather spanned multiple cognitive domains.

Meta-analyses of functional imaging studies of cognitive control in ADHD have found that pharmacological treatment history affects their results. Hart and colleagues (2013) reported a correlation between the percentage of medicated participants per study and greater normalisation of activity in the right caudate nucleus relative to typically developing controls. McCarthy and colleagues (2014) reported that, for the stop-task, activity in the right superior frontal gyrus correlated with the percentage of previously medicated participants across studies. Furthermore, activity in the cerebellum was greater for individuals with ADHD who had previously received more medication. They found a similar pattern of normalisation in the right caudate nucleus and superior frontal gyrus for GNG studies (McCarthy et al., 2014).

However, normalisation effects calculated using the percentage of medicated participants should be interpreted with some caution: McCarthy and colleagues (2014) warned that these effects could be partly accounted for by insufficiently short washout periods prior to the fMRI experiment. The acute effects of stimulant medication could then lead to normalisation of activity, rather than treatment history (McCarthy et al., 2014). Indeed, a subsequent meta-analysis of the acute effects of stimulant medication during cognitive control tasks found normalisation of activity in the bilateral IFG, right anterior cerebellum, and fusiform gyrus (Rubia et al., 2014).

In sum, stimulant medications appear to have an acute effect on cognition and associated neural activity (Coghill, Seth, et al., 2014; Hart et al., 2013; Schweren, de Zeeuw, & Durston, 2013). It has been suggested that long(er)-term treatment with stimulants may normalise deficits in the cognitive control networks and thereby contribute to an age-dependent decline in (particularly hyperactive/impulsive) symptoms (Faraone, Biederman, & Mick, 2006). This is largely consistent with the literature on medication effects on brain function and brain structure in ADHD more generally (Nakao et al., 2011; Schweren et al., 2013). However, a number of issues render such an interpretation premature: First, it has not been shown that normalisation of cognitive control mediates changes in clinical symptoms. Second, meta-analyses indicate that some of the effects of long-term medication that have been reported in fMRI studies of cognitive control may actually be due to acute effects. Third, virtually no studies have adopted a developmental framework, which will be important for understanding the contribution of treatment to the normalisation of brain circuits and improvements in symptoms with age. Finally, there are large methodological differences between studies, and concerns have been raised that publication bias may be keeping negative results out of the literature (Schweren et al., 2013). To conclude, there are credible mechanisms for the effects of stimulant medication on cognitive control in ADHD. Yet, it remains an open question whether their clinical effect is mediated specifically through cognitive control systems.

Alternative Explanations for Cognitive Control Deficits in ADHD

We have already hinted that it is being questioned whether deficits in cognitive control are a core deficit in ADHD, or whether they may be secondary to more basic information processing deficits (Huang-Pollock et al., 2012; Metin et al., 2013; Mulder et al., 2010; Rommelse et al., 2007). In some sense, this returns to theoretical views that were first proposed some decades ago, such as the cognitive energetic model (CEM) (Sergeant, 2000). The suggestion from this literature is that problems in the regulation of basic information processing may explain higher-order deficits in ADHD. For example, the studies of Metin et al. (2013) and Huang-Pollock et al. (2012) suggest that other processes, unrelated to decision making or response selection, can have a strong influence on performance in ADHD, such as perceptual processing or motor preparation, but perhaps also state regulation of arousal, and effort/motivation, as was proposed by the CEM (Sergeant, 2000).

A role for the regulation of arousal and attention in cognitive control deficits in ADHD ties in with studies that show how specific properties of a task affect performance. A meta-analysis of the effect of event rate on GNG performance showed that the speed of trial presentation affected group differences between children with ADHD and controls (Metin, Roeyers, Wiersema, van der Meere, & Sonuga-Barke, 2012). Specifically, the effect size for differences in mean reaction time and no-go accuracy (but not response variability) were larger in slow GNG tasks (with long inter-stimulus intervals, ISIs) than in faster tasks. These results underscore that cognitive control deficits in ADHD are context dependent, consistent with the clinical phenomenology. Similarly, we showed that children with ADHD were less able than typically developing controls to adapt their task strategy (in this case, the regulation of the speed–accuracy trade-off) in response to contextual demands, such as task instructions (Mulder et al., 2010). The hypothesis of state regulation deficits as underlying cognitive deficits also ties in with research showing that task performance in ADHD may be characterised by periodic lapses of attention, perhaps particularly during tasks with low event rates (Sonuga-Barke & Castellanos, 2007). There is some evidence to suggest that this may be related to neural 'default mode network' intrusions (Helps et al., 2010; Hoekzema et al., 2014; Spinelli et al., 2011). The default mode network is a network that is typically deactivated during cognitive tasks and whose activity is associated with mind-wandering and self-referential processing and, as such, may interfere with appropriate task performance.

Proactive Versus Reactive Inhibition

As we have pointed out, studies of cognitive control in ADHD have often operationalised it as behavioural inhibition (or inhibitory control): the withholding of a prepotent or ongoing response. This is a form of *reactive control*, where direct and complete shutdown of an action is required in response to an environmental cue. However, this is arguably a rather narrow definition of cognitive control and may therefore form a limited model for more complex forms of self-control required in everyday life (Aron, 2007, 2011). Recently, more sophisticated models of cognitive control have differentiated between this type of reactive control and versus *proactive control*: the preparation of a reactive cognitive control network when it seems likely that reactive control may be required (Aron, 2011; Kenemans, 2015; see also Chapter 9 by Chiew & Braver in this volume). Traditional GNG and SSTs tap into reactive control only. As such, we do not yet know whether cognitive control deficits in ADHD are due to a reduced ability to actively prepare and recruit the system as a whole (proactive control) or rather reduced ability to 'hit the brakes' (reactive control). This question is particularly pertinent, as the Diagnostic and Statistical Manual of Mental Disorders (DSM) definition of ADHD describes situations where proactive control is required rather than reactive control.

The first studies addressing this have suggested that both reactive and proactive control may be impaired (Bhaijiwala, Chevrier, & Schachar, 2014; Pani et al., 2013), or alternatively that the control deficit (proactive vs. reactive) may differ between subtypes of ADHD (Kenemans, 2015). Interestingly, reduced activation of the DLPFC in ADHD has been reported in studies using more complex demands on cognitive control, which are more likely to require proactive control (Hart et al., 2013). This is in keeping with the view that frontostriatal circuitry may be particularly important for the maintenance of complex response sets across trials, rather than for trial-by-trial reactivity.

The proactive/reactive discussion also ties into state regulation theories of ADHD discussed previously: one of the mechanisms for proactive control has been suggested to be to regulate downstream attention and perceptual systems to increase task focus (perhaps by decreasing default mode network intrusion, or increasing the efficiency of perceptual processing for relevant stimuli; see Kenemans, 2015). The distinction between reactive and proactive control and its relation to state regulation in ADHD seem to hold promise for explaining why cognitive control deficits in ADHD often appear to be context dependent and will likely form a focus for future research. (For a parallel interest in this construct in the context of schizophrenia, see Chapter 31 by Barch & Sheffield in this volume).

The Relationship Between Cognitive Control and Clinical Outcome

The neuropsychological and imaging studies described above show that deficits in cognitive control are associated with ADHD. The suggestion that they may form an endophenotype in ADHD requires them to have a place in the causal chain: Changes in genes lead to changes in brain development and through cognitive deficits to changes in behaviour. We have discussed how this model has come under fire, with investigators suggesting that cognitive deficits in ADHD may be an epiphenomenon rather than causally related to the disorder. One important reason for this suggestion has been the unclear relationship between cognitive control and clinical progression in ADHD: On the one hand, cognitive control as assessed early in childhood does seem to predict later diagnoses or symptom levels of ADHD to some degree (van Lieshout et al., 2013). In addition, deficits in executive functioning (including cognitive control) have been found to predict and even mediate the relationship between ADHD symptoms and functional impairment, such as academic achievement (Sjowall & Thorell, 2014). On the other hand, the developmental progression into adulthood is far more dynamic and varies greatly between individuals. The full diagnosis of ADHD persists in roughly a third of cases (Faraone et al., 2006), and the role of cognitive control is unclear. For cognitive control to have a mediatory role in the causal chain, cognitive control would be expected to normalise in line with the remittance of ADHD symptoms (Halperin & Schulz, 2006). However, a recent review of neurocognitive predictors for ADHD showed that cognitive control does not predict clinical persistence or remittance of ADHD (van Lieshout et al., 2013). This suggests that deficits in cognitive control may best be conceptualised as an epiphenomenon in ADHD, rather than as a causal factor. However, cognitive deficits have been shown to mediate functional problems cross-sectionally and may therefore also play a role in mediating functional impairment over time, even if they not constitute core symptoms of ADHD (Miller & Hinshaw, 2010). In sum, the jury is still out on the role of cognitive control in clinical outcome in ADHD. What is clear is that is likely to be an important avenue for future studies.

Neurobiological Heterogeneity

In this chapter, we have discussed cognitive control as a focus of cognitive research in ADHD. We have discussed that the central role of deficits in cognitive control in ADHD has been questioned, particularly by studies showing that such deficits do not occur in all affected individuals. Furthermore, impulsiveness as a symptom of ADHD does not equate directly to deficits in cognitive control at the neural or cognitive level. There are two other important cognitive deficits that have been reported in ADHD and may equally relate to behavioural impulsivity. First, deficits in temporal processing, where the neural representation of time is disrupted, have been frequently reported in ADHD (Durston et al., 2011; Noreika, Falter, & Rubia, 2013) and may be related to impulsive behaviour. Second, the seminal works of Sagvolden (Sagvolden, Johansen, Aase, & Russell, 2005) and Tripp and Wickens (2007) have shown how deficits in reinforcement processing might lead to hyperactive and impulsive behaviours. This has led numerous authors to suggest that ADHD may not relate to a single core cognitive deficit and to propose multiple-pathway theoretical accounts suggesting that dysfunction in several distinct brain systems may independently lead to the behavioural phenotype (Durston et al., 2011). Simultaneously, it has been broadly acknowledged that deficits in brain and cognitive development do not segregate between psychiatric diagnoses. Rather, deficits in cognitive control (as well as other cognitive deficits) are found across many disorders (see Chapter 31 by Barch & Sheffield, Chapter 32 by Chaarani et al., and Chapter 30 by McTeague et al. in this volume).

Conclusion

Deficits in cognitive control have long been put forward as the cognitive core of ADHD. In this chapter, we aimed to provide a balanced overview of this history and to discuss the current state of the art. Neuropsychological and neuroimaging studies have confirmed that cognitive control is impaired in ADHD at the group level. However, it has become clear that it is not universally affected in ADHD and that other cognitive functions are also affected, including some that may be more basic. In addition, the relationship between cognitive control and behavioural phenomenology and clinical outcome are not clear. In short, the theory on the role of cognitive control in ADHD is in a transition phase, with investigators looking towards more sophisticated models of its involvement, including models involving proactive as well as reactive inhibition, state regulation, and multiple-pathway models for ADHD. Although it is now safe to conclude that deficits in cognitive control are not the final common pathway to ADHD, it is clear that they will continue to play an important role in our thinking about this disorder.

References

Alderson, R. M., Rapport, M. D., & Kofler, M. J. (2007). Attention-deficit/hyperactivity disorder and behavioral inhibition: A meta-analytic review of the stop-signal paradigm. *Journal of Abnormal Child Psychology, 35*(5), 745–758.

American Psychiatric Association. (2013). *Diagnostic and statistical manual of mental disorders* (5th ed.). Arlington: American Psychiatric Publishing.

Arnsten, A. F., & Rubia, K. (2012). Neurobiological circuits regulating attention, cognitive control, motivation, and emotion: Disruptions in neurodevelopmental psychiatric disorders. *Journal of the American Academy of Child and Adolescent Psychiatry, 51*(4), 356–367.

Aron, A. R. (2007). The neural basis of inhibition in cognitive control. *Neuroscientist, 13*(3), 214–228.

Aron, A. R. (2011). From reactive to proactive and selective control: Developing a richer model for stopping inappropriate responses. *Biological Psychiatry, 69*(12), e55–68.

Ashtari, M., Kumra, S., Bhaskar, S. L., Clarke, T., Thaden, E., Cervellione, K. L., ... Ardekani, B. A. (2005). Attention-deficit/hyperactivity disorder: A preliminary diffusion tensor imaging study. *Biological Psychiatry, 57*(5), 448–455.

Balint, S., Czobor, P., Komlosi, S., Meszaros, A., Simon, V., & Bitter, I. (2009). Attention deficit hyperactivity disorder (ADHD): Gender- and age-related differences in neurocognition. *Psychological Medicine, 39*(8), 1337–1345.

Banich, M. T., Burgess, G. C., Depue, B. E., Ruzic, L., Bidwell, L. C., Hitt-Laustsen, S., ... Willcutt, E. G. (2009). The neural basis of sustained and transient attentional control in young adults with ADHD. *Neuropsychologia, 47*(14), 3095–3104.

Barkley, R. A. (1997). Behavioral inhibition, sustained attention, and executive functions: Constructing a unifying theory of ADHD. *Psychological Bulletin, 121*(1), 65–94.

Bhaijiwala, M., Chevrier, A., & Schachar, R. (2014). Withholding and canceling a response in ADHD adolescents. *Brain and Behavior, 4*(5), 602–614.

Boonstra, A. M., Oosterlaan, J., Sergeant, J. A., & Buitelaar, J. K. (2005). Executive functioning in adult ADHD: A meta-analytic review. *Psychological Medicine, 35*(8), 1097–1108.

Bramham, J., Murphy, D. G., Xenitidis, K., Asherson, P., Hopkin, G., & Young, S. (2012). Adults with attention deficit hyperactivity disorder: An investigation of age-related differences in behavioural symptoms, neuropsychological function and co-morbidity. *Psychological Medicine, 42*(10), 2225–2234.

Burgess, G. C., Depue, B. E., Ruzic, L., Willcutt, E. G., Du, Y. P., & Banich, M. T. (2010). Attentional control activation relates to working memory in attention-deficit/hyperactivity disorder. *Biological Psychiatry, 67*(7), 632–640.

Bush, G., Frazier, J. A., Rauch, S. L., Seidman, L. J., Whalen, P. J., Jenike, M. A., ... Biederman, J. (1999). Anterior cingulate cortex dysfunction in attention-deficit/hyperactivity disorder revealed by fMRI and the counting Stroop. *Biological Psychiatry, 45*(12), 1542–1552.

Casey, B. J., Epstein, J. N., Buhle, J., Liston, C., Davidson, M. C., Tonev, S. T., ... Glover, G. (2007). Frontostriatal connectivity and its role in cognitive control in parent-child dyads with ADHD. *American Journal of Psychiatry, 164*(11), 1729–1736.

Castellanos, F. X., Sonuga-Barke, E. J., Milham, M. P., & Tannock, R. (2006). Characterizing cognition in ADHD: Beyond executive dysfunction. *Trends in Cognitive Sciences, 10*(3), 117–123.

Chaim, T. M., Zhang, T., Zanetti, M. V., da Silva, M. A., Louza, M. R., Doshi, J., ... Busatto, G. F. (2014). Multimodal magnetic resonance imaging study of treatment-naive adults with attention-deficit/hyperactivity disorder. *PLoS ONE, 9*(10), e110199.

Chambers, C. D., Garavan, H., & Bellgrove, M. A. (2009). Insights into the neural basis of response inhibition from cognitive and clinical neuroscience. *Neuroscience and Biobehavioral Reviews, 33*(5), 631–646.

Coghill, D. R., Hayward, D., Rhodes, S. M., Grimmer, C., & Matthews, K. (2014). A longitudinal examination of neuropsychological and clinical functioning in boys with attention deficit hyperactivity disorder (ADHD): Improvements in executive functioning do not explain clinical improvement. *Psychological Medicine, 44*(5), 1087–1099.

Coghill, D. R., Seth, S., Pedroso, S., Usala, T., Currie, J., & Gagliano, A. (2014). Effects of methylphenidate on cognitive functions in children and adolescents with attention-deficit/hyperactivity disorder: Evidence from a systematic review and a meta-analysis. *Biological Psychiatry, 76*(8), 603–615.

Cortese, S., Kelly, C., Chabernaud, C., Proal, E., Di Martino, A., Milham, M. P., & Castellanos, F. X. (2012). Toward systems neuroscience of ADHD: A meta-analysis of 55 fMRI studies. *American Journal of Psychiatry, 169*(10), 1038–1055.

Davenport, N. D., Karatekin, C., White, T., & Lim, K. O. (2010). Differential fractional anisotropy abnormalities in adolescents with ADHD or schizophrenia. *Psychiatry Research, 181*(3), 193–198.

Davidson, M. C., Amso, D., Anderson, L. C., & Diamond, A. (2006). Development of cognitive control and executive functions from 4 to 13 years: Evidence from manipulations of memory, inhibition, and task switching. *Neuropsychologia, 44*(11), 2037–2078.

De Zeeuw, P., Aarnoudse-Moens, C., Bijlhout, J., Konig, C., Post Uiterweer, A., Papanikolau, A., ... Oosterlaan, J. (2008). Inhibitory performance, response speed, intraindividual variability, and response accuracy in ADHD. *Journal of the American Academy of Child and Adolescent Psychiatry, 47*(7), 808–816.

De Zeeuw, P., & Durston, S. (in press). Attention deficit hyperactivity disorder. In B. Hopkins (Ed.), *Cambridge encyclopedia of child development.* New York: Cambridge University Press.

De Zeeuw, P., Mandl, R. C., Hulshoff Pol, H. E., van Engeland, H., & Durston, S. (2012). Decreased frontostriatal microstructural organization in attention deficit/hyperactivity disorder. *Human Brain Mapping, 33*(8), 1941–1951.

Douglas, V. I. (1972). Stop, look and listen: The problem of sustained attention and impulse control in hyperactive and normal children. *Canadian Journal of Behavioural Science, 4*(4), 259–282.

Durston, S. (2010). Imaging genetics in ADHD. *Neuroimage, 53*(3), 832–838.

Durston, S., & Casey, B. J. (2006). A shift from diffuse to focal cortical activity with development. *Developmental Science, 9*(1), 18–20.

Durston, S., de Zeeuw, P., & Staal, W. G. (2009). Imaging genetics in ADHD: A focus on cognitive control. *Neuroscience and Biobehavioral Reviews, 33*(5), 674–689.

Durston, S., Mulder, M., Casey, B. J., Ziermans, T., & van Engeland, H. (2006). Activation in ventral prefrontal cortex is sensitive to genetic vulnerability for attention-deficit hyperactivity disorder. *Biological Psychiatry, 60*(10), 1062–1070.

Durston, S., Van Belle, J., & De Zeeuw, P. (2011). Differentiating fronto-striatal and fronto-cerebellar circuits in ADHD. *Biological Psychiatry, 69*(12), 1178–1184.

Fair, D. A., Dosenbach, N. U., Church, J. A., Cohen, A. L., Brahmbhatt, S., Miezin, F. M., ... Schlaggar, B. L. (2007). Development of distinct control networks through segregation and integration. *Proceedings of the National Academy of Sciences of the United States of America, 104*(33), 13507–13512.

Faraone, S. V., Biederman, J., & Mick, E. (2006). The age-dependent decline of attention deficit hyperactivity disorder: A meta-analysis of follow-up studies. *Psychological Medicine, 36*(2), 159–165.

Flint, J., & Munafo, M. R. (2007). The endophenotype concept in psychiatric genetics. *Psychological Medicine, 37*(2), 163–180.

Frazier, T. W., Demaree, H. A., & Youngstrom, E. A. (2004). Meta-analysis of intellectual and neuropsychological test performance in attention-deficit/hyperactivity disorder. *Neuropsychology, 18*(3), 543–555.

Frodl, T., & Skokauskas, N. (2012). Meta-analysis of structural MRI studies in children and adults with attention deficit hyperactivity disorder indicates treatment effects. *Acta Psychiatrica Scandinavica, 125*(2), 114–126.

Geurts, H. M., Verte, S., Oosterlaan, J., Roeyers, H., & Sergeant, J. A. (2004). How specific are executive functioning deficits in attention deficit hyperactivity disorder and autism? *Journal of Child Psychology and Psychiatry, 45*(4), 836–854.

Halperin, J. M., & Schulz, K. P. (2006). Revisiting the role of the prefrontal cortex in the pathophysiology of attention-deficit/hyperactivity disorder. *Psychological Bulletin, 132*(4), 560–581.

Hart, H., Radua, J., Nakao, T., Mataix-Cols, D., & Rubia, K. (2013). Meta-analysis of functional magnetic resonance imaging studies of inhibition and attention in attention-deficit/hyperactivity disorder: Exploring task-specific, stimulant medication, and age effects. *JAMA Psychiatry, 70*(2), 185–198.

Helps, S. K., Broyd, S. J., James, C. J., Karl, A., Chen, W., & Sonuga-Barke, E. J. (2010). Altered spontaneous low frequency brain activity in attention deficit/hyperactivity disorder. *Brain Research, 1322*, 134–143.

Hervey, A. S., Epstein, J. N., & Curry, J. F. (2004). Neuropsychology of adults with attention-deficit/hyperactivity disorder: A meta-analytic review. *Neuropsychology, 18*(3), 485–503.

Hoekzema, E., Carmona, S., Ramos-Quiroga, J. A., Richarte Fernandez, V., Bosch, R., Soliva, J. C., ... Vilarroya, O. (2014). An independent components and functional connectivity analysis of resting state fMRI data points to neural network dysregulation in adult ADHD. *Human Brain Mapping, 35*(4), 1261–1272.

Huang-Pollock, C. L., Karalunas, S. L., Tam, H., & Moore, A. N. (2012). Evaluating vigilance deficits in ADHD: A meta-analysis of CPT performance. *Journal of Abnormal Psychology, 121*(2), 360–371.

Kebir, O., & Joober, R. (2011). Neuropsychological endophenotypes in attention-deficit/hyperactivity disorder: A review of genetic association studies. *European Archives of Psychiatry and Clinical Neuroscience, 261*(8), 583–594.

Kenemans, J. L. (2015). Specific proactive and generic reactive inhibition. *Neuroscience and Biobehavioral Reviews, 56*, 115–126.

Kessler, D., Angstadt, M., Welsh, R. C., & Sripada, C. (2014). Modality-spanning deficits in attention-deficit/hyperactivity disorder in functional networks, gray matter, and white matter. *Journal of Neuroscience, 34*(50), 16555–16566.

Klein, C., Wendling, K., Huettner, P., Ruder, H., & Peper, M. (2006). Intra-subject variability in attention-deficit hyperactivity disorder. *Biological Psychiatry, 60*(10), 1088–1097.

Konrad, A., Dielentheis, T. F., El Masri, D., Bayerl, M., Fehr, C., Gesierich, T., ... Winterer, G. (2010). Disturbed structural connectivity is related to inattention and impulsivity in adult attention deficit hyperactivity disorder. *European Journal of Neuroscience, 31*(5), 912–919.

Kuntsi, J., Wood, A. C., Rijsdijk, F., Johnson, K. A., Andreou, P., Albrecht, B., ... Asherson, P. (2010). Separation of cognitive impairments in attention-deficit/hyperactivity disorder into 2 familial factors. *Archives of General Psychiatry, 67*(11), 1159–1167.

Lange, K. W., Reichl, S., Lange, K. M., Tucha, L., & Tucha, O. (2010). The history of attention deficit hyperactivity disorder. *ADHD Attention Deficit and Hyperactivity Disorders, 2*(4), 241–255.

Lawrence, K. E., Levitt, J. G., Loo, S. K., Ly, R., Yee, V., O'Neill, J., ... Narr, K. L. (2013). White matter microstructure in subjects with attention-deficit/hyperactivity disorder and their siblings. *Journal of the American Academy of Child and Adolescent Psychiatry, 52*(4), 431–440 e434.

Lei, D., Ma, J., Du, X., Shen, G., Jin, X., & Gong, Q. (2014). Microstructural abnormalities in the combined and inattentive subtypes of attention deficit hyperactivity disorder: A diffusion tensor imaging study. *Scientific Reports, 4*, 6875.

MacDonald, A. W., 3rd, Cohen, J. D., Stenger, V. A., & Carter, C. S. (2000). Dissociating the role of the dorsolateral prefrontal and anterior cingulate cortex in cognitive control. *Science, 288*(5472), 1835–1838.

Makris, N., Biederman, J., Monuteaux, M. C., & Seidman, L. J. (2009). Towards conceptualizing a neural systems-based anatomy of attention-deficit/hyperactivity disorder. *Developmental Neuroscience, 31*(1–2), 36–49.

McCarthy, H., Skokauskas, N., & Frodl, T. (2014). Identifying a consistent pattern of neural function in attention deficit hyperactivity disorder: A meta-analysis. *Psychological Medicine, 44*(4), 869–880.

McCarthy, H., Skokauskas, N., Mulligan, A., Donohoe, G., Mullins, D., Kelly, J., ... Frodl, T. (2013). Attention network hypoconnectivity with default and affective network hyperconnectivity in adults diagnosed with attention-deficit/hyperactivity disorder in childhood. *JAMA Psychiatry, 70*(12), 1329–1337.

Metin, B., Roeyers, H., Wiersema, J. R., van der Meere, J., & Sonuga-Barke, E. (2012). A meta-analytic study of event rate effects on Go/No-Go performance in attention-deficit/hyperactivity disorder. *Biological Psychiatry, 72*(12), 990–996.

Metin, B., Roeyers, H., Wiersema, J. R., van der Meere, J. J., Thompson, M., & Sonuga-Barke, E. (2013). ADHD performance reflects inefficient but not impulsive information processing: A diffusion model analysis. *Neuropsychology, 27*(2), 193–200.

Miller, M., & Hinshaw, S. P. (2010). Does childhood executive function predict adolescent functional outcomes in girls with ADHD? *Journal of Abnormal Child Psychology, 38*(3), 315–326.

Morein-Zamir, S., Dodds, C., van Hartevelt, T. J., Schwarzkopf, W., Sahakian, B., Muller, U., & Robbins, T. (2014). Hypoactivation in right inferior frontal cortex is specifically associated with motor response inhibition in adult ADHD. *Human Brain Mapping, 35*(10), 5141–5152.

Mulder, M. J., Baeyens, D., Davidson, M. C., Casey, B. J., van den Ban, E., van Engeland, H., & Durston, S. (2008). Familial vulnerability to ADHD affects activity in the cerebellum in addition to the prefrontal systems. *Journal of the American Academy of Child and Adolescent Psychiatry, 47*(1), 68–75.

Mulder, M. J., Bos, D., Weusten, J. M., Belle, J. V., van Dijk, S. C., Simen, P., … Durston, S. (2010). Basic impairments in regulating the speed-accuracy tradeoff predict symptoms of attention-deficit/hyperactivity disorder. *Biological Psychiatry*, *68*(12), 1114–1119.

Mulder, M. J., van Belle, J., van Engeland, H., & Durston, S. (2011). Functional connectivity between cognitive control regions is sensitive to familial risk for ADHD. *Human Brain Mapping*, *32*(9), 1511–1518.

Nakao, T., Radua, J., Rubia, K., & Mataix-Cols, D. (2011). Gray matter volume abnormalities in ADHD: Voxel-based meta-analysis exploring the effects of age and stimulant medication. *American Journal of Psychiatry*, *168*(11), 1154–1163.

Nigg, J. T. (2001). Is ADHD a disinhibitory disorder? *Psychological Bulletin*, *127*(5), 571–598.

Nigg, J. T., & Casey, B. J. (2005). An integrative theory of attention-deficit/hyperactivity disorder based on the cognitive and affective neurosciences. *Development and Psychopathology*, *17*(3), 785–806.

Nigg, J. T., Willcutt, E. G., Doyle, A. E., & Sonuga-Barke, E. J. (2005). Causal heterogeneity in attention-deficit/hyperactivity disorder: Do we need neuropsychologically impaired subtypes? *Biological Psychiatry*, *57*(11), 1224–1230.

Nikolas, M. A., & Nigg, J. T. (2015). Moderators of neuropsychological mechanism in attention-deficit hyperactivity disorder. *Journal of Abnormal Child Psychology*, *43*(2), 271–281.

Noreika, V., Falter, C. M., & Rubia, K. (2013). Timing deficits in attention-deficit/hyperactivity disorder (ADHD): Evidence from neurocognitive and neuroimaging studies. *Neuropsychologia*, *51*(2), 235–266.

Onnink, A. M., Zwiers, M. P., Hoogman, M., Mostert, J. C., Dammers, J., Kan, C. C., … Franke, B. (2015). Deviant white matter structure in adults with attention-deficit/hyperactivity disorder points to aberrant myelination and affects neuropsychological performance. *Progress in Neuro-Psychopharmacology and Biological Psychiatry*, *63*, 14–22.

Pani, P., Menghini, D., Napolitano, C., Calcagni, M., Armando, M., Sergeant, J. A., & Vicari, S. (2013). Proactive and reactive control of movement are differently affected in attention deficit hyperactivity disorder children. *Research in Developmental Disabilities*, *34*(10), 3104–3111.

Pavuluri, M. N., Yang, S., Kamineni, K., Passarotti, A. M., Srinivasan, G., Harral, E. M., … Zhou, X. J. (2009). Diffusion tensor imaging study of white matter fiber tracts in pediatric bipolar disorder and attention-deficit/hyperactivity disorder. *Biological Psychiatry*, *65*(7), 586–593.

Pennington, B. F., & Ozonoff, S. (1996). Executive functions and developmental psychopathology. *Journal of Child Psychology and Psychiatry*, *37*(1), 51–87.

Pironti, V. A., Lai, M. C., Müller, U., Dodds, C. M., Suckling, J., Bullmore, E. T., & Sahakian, B. J. (2014). Neuroanatomical abnormalities and cognitive impairments are shared by adults with attention-deficit/hyperactivity disorder and their unaffected first-degree relatives. *Biological Psychiatry*, *76*(8), 639–647.

Posner, J., Park, C., & Wang, Z. (2014). Connecting the dots: A review of resting connectivity MRI studies in attention-deficit/hyperactivity disorder. *Neuropsychology Review*, *24*(1), 3–15.

Quay, H. C. (1997). Inhibition and attention deficit hyperactivity disorder. *Journal of Abnormal Child Psychology*, *25*(1), 7–13.

Rommelse, N. N., Altink, M. E., de Sonneville, L. M., Buschgens, C. J., Buitelaar, J., Oosterlaan, J., & Sergeant, J. A. (2007). Are motor inhibition and cognitive flexibility dead ends in ADHD? *Journal of Abnormal Child Psychology*, *35*(6), 957–967.

Rommelse, N. N., Altink, M. E., Oosterlaan, J., Buschgens, C. J., Buitelaar, J., & Sergeant, J. A. (2008). Support for an independent familial segregation of executive and intelligence endophenotypes in ADHD families. *Psychological Medicine*, *38*(11), 1595–1606.

Rommelse, N. N., Arias-Vasquez, A., Altink, M. E., Buschgens, C. J., Fliers, E., Asherson, P., … Franke, B. (2008). Neuropsychological endophenotype approach to genome-wide linkage analysis identifies susceptibility loci for ADHD on 2q21.1 and 13q12.11. *American Journal of Human Genetics*, *83*(1), 99–105.

Rommelse, N. N., Geurts, H. M., Franke, B., Buitelaar, J. K., & Hartman, C. A. (2011). A review on cognitive and brain endophenotypes that may be common in autism spectrum disorder and attention-deficit/hyperactivity disorder and facilitate the search for pleiotropic genes. *Neuroscience and Biobehavioral Reviews*, *35*(6), 1363–1396.

Rubia, K. (2011). 'Cool' inferior frontostriatal dysfunction in attention-deficit/hyperactivity disorder versus 'hot' ventromedial orbitofrontal-limbic dysfunction in conduct disorder: A review. *Biological Psychiatry*, *69*(12), e69–e87.

Rubia, K., Alegria, A. A., Cubillo, A. I., Smith, A. B., Brammer, M. J., & Radua, J. (2014). Effects of stimulants on brain function in attention-deficit/hyperactivity disorder: A systematic review and meta-analysis. *Biological Psychiatry*, *76*(8), 616–628.

Sagvolden, T., Johansen, E. B., Aase, H., & Russell, V. A. (2005). A dynamic developmental theory of attention-deficit/hyperactivity disorder (ADHD) predominantly hyperactive/impulsive and combined subtypes. *Behavioral and Brain Sciences*, *28*(3), 397–419; discussion 419-368.

Schachar, R., Tannock, R., & Logan, G. (1993). Inhibitory control, impulsiveness, and attention deficit hyperactivity disorder. *Clinical Psychology Review*, *13*, 721–739.

Schulz, K. P., Bédard, A. C., Fan, J., Clerkin, S. M., Dima, D., Newcorn, J. H., & Halperin, J. M. (2014). Emotional bias of cognitive control in adults with childhood attention-deficit/hyperactivity disorder. *Neuroimage Clinical*, *5*, 1–9.

Schweren, L. J., de Zeeuw, P., & Durston, S. (2013). MR imaging of the effects of methylphenidate on brain structure and function in attention-deficit/hyperactivity disorder. *European Neuropsychopharmacology*, *23*(10), 1151–1164.

Sergeant, J. (2000). The cognitive-energetic model: An empirical approach to attention-deficit hyperactivity disorder. *Neuroscience and Biobehavioral Reviews*, *24*(1), 7–12.

Sergeant, J. A. (2005). Modeling attention-deficit/hyperactivity disorder: A critical appraisal of the cognitive-energetic model. *Biological Psychiatry*, *57*(11), 1248–1255.

Sergeant, J. A., Geurts, H., & Oosterlaan, J. (2002). How specific is a deficit of executive functioning for attention-deficit/hyperactivity disorder? *Behavioural Brain Research*, *130*(1–2), 3–28.

Sergeant, J., Oosterlaan, J., & van der Meere, J. (1999). Information processing and energetic factors in attention-deficit/hyperactivity disorder. In H. C. Quay & A. E. Hogan (Eds.), *Handbook of disruptive behavior disorders* (pp. 75–104). New York: Kluwer Academic/Plenum Publishers.

Shaw, P., Eckstrand, K., Sharp, W., Blumenthal, J., Lerch, J. P., Greenstein, D., ... Rapoport, J. L. (2007). Attention-deficit/hyperactivity disorder is characterized by a delay in cortical maturation. *Proceedings of the National Academy of Sciences of the United States of America*, *104*(49), 19649–19654.

Shenhav, A., Botvinick, M. M., & Cohen, J. D. (2013). The expected value of control: An integrative theory of anterior cingulate cortex function. *Neuron*, *79*(2), 217–240.

Silk, T. J., Vance, A., Rinehart, N., Bradshaw, J. L., & Cunnington, R. (2009). White-matter abnormalities in attention deficit hyperactivity disorder: A diffusion tensor imaging study. *Human Brain Mapping*, *30*(9), 2757–2765.

Sjowall, D., & Thorell, L. B. (2014). Functional impairments in attention deficit hyperactivity disorder: The mediating role of neuropsychological functioning. *Developmental Neuropsychology*, *39*(3), 187–204.

Slaats-Willemse, D., Swaab-Barneveld, H., de Sonneville, L., van der Meulen, E., & Buitelaar, J. (2003). Deficient response inhibition as a cognitive endophenotype of ADHD. *Journal of the American Academy of Child and Adolescent Psychiatry*, *42*(10), 1242–1248.

Sonuga-Barke, E. J. (2002). Psychological heterogeneity in AD/HD—A dual pathway model of behaviour and cognition. *Behavioural Brain Research*, *130*(1–2), 29–36.

Sonuga-Barke, E. J., & Castellanos, F. X. (2007). Spontaneous attentional fluctuations in impaired states and pathological conditions: A neurobiological hypothesis. *Neuroscience and Biobehavioral Reviews*, *31*(7), 977–986.

Spencer, T. J., Biederman, J., & Mick, E. (2007). Attention-deficit/hyperactivity disorder: Diagnosis, lifespan, comorbidities, and neurobiology. *Ambulatory Pediatrics*, *7*(1 Suppl.), 73–81.

Spinelli, S., Joel, S., Nelson, T. E., Vasa, R. A., Pekar, J. J., & Mostofsky, S. H. (2011). Different neural patterns are associated with trials preceding inhibitory errors in children with and without attention-deficit/hyperactivity disorder. *Journal of the American Academy of Child and Adolescent Psychiatry*, *50*(7), 705–715 e703.

Still, G. (1902). The Goulstonian lectures on some abnormal psychical conditions in children. *Lecture I. Lancet*, *159*(4102), 1008–1012.

Swanson, J., Baler, R. D., & Volkow, N. D. (2011). Understanding the effects of stimulant medications on cognition in individuals with attention-deficit hyperactivity disorder: A decade of progress. *Neuropsychopharmacology, 36*(1), 207–226.

Tamm, L., Barnea-Goraly, N., & Reiss, A. L. (2012). Diffusion tensor imaging reveals white matter abnormalities in attention-deficit/hyperactivity disorder. *Psychiatry Research, 202*(2), 150–154.

Taylor, E., Dopfner, M., Sergeant, J., Asherson, P., Banaschewski, T., Buitelaar, J., ... Zuddas, A. (2004). European clinical guidelines for hyperkinetic disorder—First upgrade. *European Child and Adolescent Psychiatry, 13*(Suppl. 1), I7–30.

Tripp, G., & Wickens, J. R. (2007). Dopamine transfer deficit: A neurobiological theory of altered reinforcement mechanisms in ADHD. *Journal of Child Psychology and Psychiatry, 49*(7), 691–704.

Valera, E. M., Faraone, S. V., Murray, K. E., & Seidman, L. J. (2007). Meta-analysis of structural imaging findings in attention-deficit/hyperactivity disorder. *Biological Psychiatry, 61*(12), 1361–1369.

Van Ewijk, H., Heslenfeld, D. J., Zwiers, M. P., Buitelaar, J. K., & Oosterlaan, J. (2012). Diffusion tensor imaging in attention deficit/hyperactivity disorder: A systematic review and meta-analysis. *Neuroscience and Biobehavioral Reviews, 36*(4), 1093–1106.

Van Lieshout, M., Luman, M., Buitelaar, J., Rommelse, N. N., & Oosterlaan, J. (2013). Does neurocognitive functioning predict future or persistence of ADHD? A systematic review. *Clinical Psychology Review, 33*(4), 539–560.

Van Meel, C. S., Oosterlaan, J., Heslenfeld, D. J., & Sergeant, J. A. (2005). Telling good from bad news: ADHD differentially affects processing of positive and negative feedback during guessing. *Neuropsychologia, 43*(13), 1946–1954.

Van Rooij, D., Hoekstra, P. J., Mennes, M., von Rhein, D., Thissen, A. J., Heslenfeld, D., ... Hartman, C. A. (2015). Distinguishing adolescents with ADHD from their unaffected siblings and healthy comparison subjects by neural activation patterns during response inhibition. *American Journal of Psychiatry, 172*(7), 674–683.

Willcutt, E. G., Doyle, A. E., Nigg, J. T., Faraone, S. V., & Pennington, B. F. (2005). Validity of the executive function theory of attention-deficit/hyperactivity disorder: A meta-analytic review. *Biological Psychiatry, 57*(11), 1336–1346.

Index

Page references to Figures or Tables will be in *italics*, followed by the letters 'f' or 't' as appropriate, while references to Notes will be followed by the letter 'n' and Note number.

The Wiley Handbook of Cognitive Control, First Edition. Edited by Tobias Egner.